LIVING IN A
microbial world

Garland Science

LIVING IN A
microbial world

Bruce V. Hofkin

Garland Science
Vice President: Denise Schanck
Senior Editor: Michael Morales
Production Editing and Layout: EJ Publishing Services
Project Editor: Sigrid Masson
Assistant Editor & Photo Research: Monica Toledo
Copyeditors: Richard K. Mickey, Heather Whirlow Cammarn
Text Editors: Emma Jeffcock, Martha Cushmann, Eleanor Lawrence
Illustrator & Cover Design: Matthew McClements, Blink Studio
Indexer: Liza Furnival

Bruce V. Hofkin received his Ph.D. from the University of New Mexico where he is currently a faculty member in the Department of Biology. His primary research interest is the epidemiology of vector-borne diseases.

Library of Congress Cataloging-in-Publication Data

Hofkin, Bruce V.
    Living in a microbial world / Bruce V. Hofkin.
        p. cm.
    ISBN 978-0-8153-4175-8
    1. Microorganisms--Textbooks. 2. Microbiology--Textbooks. I. Title.
    QR41.2.H638 2009
    616.9'041--dc22

                          2010002066

Published by Garland Science, Taylor and Francis Group, LLC, an informa business
270 Madison Avenue, New York, NY 10016, USA, and
2 Park Square, Milton Park, Abingdon, OX14 4RN, UK.

Visit our website at http://www.garlandscience.com

Taylor & Francis Group, an informa business

Mixed Sources
Product group from well-managed forests, controlled sources and recycled wood or fiber
www.fsc.org   Cert no. SW-COC-002985
©1996 Forest Stewardship Council

Printed in the United States of America
15 14 13 12 11 10 9 8 7 6 5 4 3 2 1

Preface

In the era of the 24-hour news cycle, we are bombarded routinely with important stories about microorganisms. AIDS, SARS, "bird flu," *E. coli* outbreaks, meningitis, vaccines, antibiotic resistance, and most recently "swine flu," make alarming copy that scare the public, and in turn fuel greater media coverage from established news sources, columnists, and the blogosphere. With conflicting information swirling through the media, it has become increasingly important for people to understand the basic principles of microbiology, in order to separate rumor and conjecture from good science, and make rational decisions about potential microbial threats and their health.

Living in a Microbial World is a textbook written for students taking a general microbiology or microbiology-themed course for non-science majors. It teaches the essential concepts of microbiology through practical examples and a conversational writing style intended to make the material accessible to a wide audience. The book will enable non-scientists to understand important issues about microorganisms and disease that they will encounter throughout their lives, and I hope it will help them make informed decisions about health issues for themselves, their families, and communities. Learning the principles of microbiology pays other dividends as well. The core topics—such as cell structure and function, microbial genetics, metabolism, and microbial evolution—have broader application to the life sciences in general, and can help students understand the living world, as well as the basic science, in a more profound and rewarding manner. The reader will learn the role that microorganisms play not only in our health but also in ecosystem processes, our diet, industrial production, and human history. Topics that we hear about every day, from global warming to energy independence to bioterrorism, all have a microbial angle. This text is designed to provide the reader with the background needed to understand and discuss such topics with a genuine understanding rooted in science.

In order to make the science relevant to everyday life and the practical interests of a non-science audience, each chapter of the book contains a series of cases intended to motivate learning the microbiology concepts. The cases present microbiology in the news, in history, in literature, and in scenarios of everyday life. Each case ends with several questions intended to pique student interest, and the questions are then answered as the student reads the next section of the chapter. For example, to introduce the topic of pathogen transmission, the introductory case recalls the true story of how the Norwalk Virus was spread among players of opposing teams during a college football game. Similarly, microbial freshwater ecology is introduced with a description of a recent outbreak of avian botulism in Lake Erie. And drug resistance is discussed in the context of a mother who does not follow the doctor's directions, and stops giving her son antibiotics immediately after he feels better.

In addition to the cases, there are other features that should make the text inviting to the non-scientist. To emphasize the human context, there is a separate chapter on "Microbiology in History and the History of Microbiology," and epidemiology has its own chapter as well. Due to its foundational role in biology as a whole, and microbiology in particular, there is a chapter devoted entirely to microbial evolution, and evolutionary concepts are emphasized throughout the book. To streamline the text and make it more accessible, several traditional microbiology topics, which often receive their own chapters, have been integrated throughout other parts of the book. For example, in lieu of having a chapter on microbiology techniques, topics such as the Gram stain and the acid-fast stain are introduced in the discussion of cell wall structure. And throughout the book, examples of specific diseases are integrated into discussions of appropriate topics.

In these ways I have attempted to provide the reader with a meaningful, relevant, and contemporary text, through which they can explore the many wonders of the microbial world. It is my hope that after reading the book, students will share my fascination with this remarkable and diverse assembly of living things.

Like all textbooks, *Living in a Microbial World* has been a collaborative effort. Many people helped bring it to fruition, and these individuals are recognized in the acknowledgments. Any remaining errors, however, are solely the responsibility of the author. Please help us with these errors by contacting science@garland.com so that corrections can be made in the next printing.

Student And Instructor Resources

The following supplements are available for students and instructors. They can be accessed at http://www.garlandscience.com/LMW

The Art of *Living in a Microbial World*

The images and tables from the book are available in two convenient formats: PowerPoint® and JPEG. The PowerPoint slides can be customized for lectures.

Microbiology Movies

Short movies have been developed to complement material in a select number of chapters, with a special emphasis on molecular genetics, virology, and immunology. Each movie has a voice-over narration, and the text of the narration is available for download.

Student Quizzes

Short online quizzes are available for each chapter to test basic reading comprehension.

Flashcards

Online flashcards are available to test mastery of the key terms listed at the end of each chapter.

Glossary

The glossary at the end of the book is available online for quick searching and browsing.

Acknowledgments

Multiple drafts of every chapter were reviewed by professors around the country, and I would like to thank them for their detailed comments and helpful suggestions. In a particular I would like to thank:

Suzanne Anglehart, University of Wisconsin; Jason Arnold, Hopkinsville Community College; Linda Bruslind, Oregon State University; Alyssa Bumbaugh, Penn State Altoona; Jean Cardinale, Alfred University; Edward Cluett, Ithaca College; Eileen Gregory, Rollins College; Juanita Leonhard, Illinois College; Sylvia Franke McDevitt, Skidmore College; Amy Medlock, University of Georgia; Roderick Morgan, Grand Valley State University; Carolyn Peters, Spoon River College; Mark Schneegurt, Wichita State University; Von Sigler, University of Toledo; Jeanne Weidner, San Diego State University; and Jamie Welling, South Suburban College.

In addition, many colleagues and friends reviewed different sections of the book, and I am indebted to them for their help: Lee Couch, University of New Mexico; Richard Cripps, University of New Mexico; Christina Fridrick, University of New Mexico; Charlotte Kent, Centers for Disease Control and Prevention, Atlanta; Eric Loker, University of New Mexico; Colleen MacNamara, Central New Mexico Community College; Robert Miller, University of New Mexico; and Robert Sinsabaugh, University of New Mexico.

I would also like to thank the many individuals at Garland Science who helped to make *Living in a Microbial World* a reality. First and foremost, I wish to thank Michael Morales, my editor, who kept me on track and coordinated the overall effort. In addition to organizing the review process for each chapter, Mike produced the accompanying Web material, and through our regular conversations and collaborations, helped to guide both the style and the content of the text. Sigrid Masson, as the project editor, managed the flow of chapters through the production process and proofread the entire text. Monica Toledo provided editorial assistance and supplied enormous help with photo research, providing innumerable valuable insights and suggestions. All artwork as well as the book's cover was rendered by Matt McClements of Blink Studio. His artist's perspective and eye for detail greatly improved many of the original ideas for illustrations. Emma Jeffcock, Eleanor Lawrence, and Martha Cushman served as developmental editors for different portions of the book. Emma also did the layout for the entire text and incorporated the required corrections. Richard K. Mickey served as the final copy editor, and helped fine-tune the style. Adam Sendroff and

Lucy Brodie were in charge of both marketing and promotion. And Denise Schanck, Vice President of Garland Science, supported this project from the start to the end. I offer my deep and sincere gratitude to all of these remarkable and dedicated people.

Lastly, thank you to Leslie for being there.

Contents

Chapter 1   Living in a Microbial World   1

Chapter 2   The Chemistry of Life   17

Chapter 3   The Cell: Where Life Begins   45

Chapter 4   A Field Guide to the Microorganisms   71

Chapter 5   The Microbiology of History and the History of
Microbiology   107

Chapter 6   Microbial Genetics   133

Chapter 7   Metabolism and Growth   169

Chapter 8   Microbial Evolution: The Origin and Diversity of Life   203

Chapter 9   An Ecologist's Guide to Microbiology   229

Chapter 10 The Nature of Disease: A Pathogen's Perspective   259

Chapter 11 Host Defense   285

Chapter 12 Control of Microbial Growth   315

Chapter 13 Epidemiology: Who, What, When, Where, and Why?   343

Chapter 14 The Future Is Here: Microorganisms and
Biotechnology   367

Chapter 15 Guess Who's Coming to Dinner: Microorganisms
and Food   395

Chapter 16 Better Living With Microoorganisms: Industrial
and Applied Microbiology   417

Glossary   G:1

Figure Acknowledgments   F:1

Index   I:1

Detailed Contents

Chapter 1 Living in a Microbial World   1

The Science of Microbiology   3

All living things are composed of one or more cells   3

All living things display other observable characteristics   4

Microbiology involves the study of several distinct groups of living things   4

Viruses strain our notion of what it means to be "alive"   7

Microbiology is closely intertwined with the study of nonmicroorganisms   7

Microbiology is composed of many specialized subdisciplines   7

CASE: ART COMES ALIVE?   7

The Scientific Method   10

A proper scientific experiment involves a series of well-defined steps   10

CASE: FLEMING REVISITED   10

If a hypothesis cannot be disproved, it may eventually become a theory   13

Coming Up Next...   13

Key Terms   14

Concept Questions   14

Chapter 2 The Chemistry of Life   17

Atoms: the Basic Building Blocks of Matter   17

Atoms are made up of smaller components called subatomic particles   18

As an atom's stability increases, its energy decreases   19

An ionic bond is formed when electrons are transferred from one atom to another   20

CASE: BACTERIA, SALT, AND CYSTIC FIBROSIS   20

Covalent bonds form when atoms share electrons in their outermost shells   22

Covalent bonds can be classified as either polar or nonpolar   22

The Importance of Water   25

Acids and bases are compounds that increase or decrease the concentration of protons in water   25

CASE: FISHLESS IN PENNSYLVANIA   25

Organic Molecules: the Building Blocks of Life   29

Carbohydrates function as energy storage and structural molecules   30

CASE: BACTERIAL HORSE HELPERS   30

Lipids are relatively hydrophobic molecules, also used for energy storage and structure   33

Proteins participate in a variety of crucial biological processes   35

CASE: BAD BAMBOO   35

Biological reactions require enzymes functioning as catalysts   38

CASE: CALL IN THE CLOT BUSTERS   38

Nucleic acids direct the assembly of proteins   40

Coming Up Next...   41

Key Terms   42

Concept Questions   42

Chapter 3 The Cell: Where Life Begins   45

Basic Concepts in Cell Biology   46

As cells get larger, their efficiency decreases   48

CASE: SIZE DOES MATTER!   48

The Prokaryotic Cell   49

Many prokaryotes have extracellular structures,
  extending beyond the cell wall   50

CASE: PLAQUE ATTACK   50

Most prokaryotes are protected from the exterior
  environment by a rigid cell wall   52

CASE: AN OCCUPATIONAL HAZARD   53

Bacteria with different cell wall types can be
  distinguished by specific staining techniques   56

Some prokaryotes lack a cell wall   57

Each cell is surrounded by a plasma membrane   58

The liquid interior of the cell forms the cell's
  cytoplasm   59

CASE: THE ANTHRAX SCARE   60

The Eukaryotic Cell   62

The number of cells in a eukaryotic organism
  is variable   62

While plant and fungal cells have cell walls,
  animal cells do not   63

Like prokaryotes, all eukaryotic cells are surrounded
  by a membrane   63

The cytoplasm of eukaryotic cells contains a
  variety of membrane-bound organelles   64

CASE: PARROT FEVER   66

Coming Up Next...   68

Key Terms   68

Concept Questions   69

**Chapter 4 A Field Guide to the
Microorganisms**   **71**

Taxonomy: How Life Is Classified   71

Taxonomy is based on a system of hierarchical
  groupings   72

Modern classification reflects evolutionary
  relatedness   72

DNA analysis provides evidence for relatedness   73

CASE: STREAMS, SNAILS, AND SCHISTOSOMES   74

Kingdoms are organized among three domains   75

Organizing Microorganisms   76

Domain Bacteria   76

The earliest bacteria were adapted to life on the
  primitive Earth   77

CASE: THE HEAT WAS ON!   77

The tree spreads out   78

The most recently evolved lineages are found at the
  farthest branches of the tree   79

Domain Archaea   79

Eukaryotic Microorganisms   80

Protozoa are single-celled organisms within the
  domain eukarya   81

CASE: THE CAT'S OUT OF THE BAG ON A
PROTOZOAN PARASITE   81

Many fungi are involved in disease or ecological
  processes or are useful for industrial purposes   84

CASE: FIRST YOU SHAKE, THEN YOU ACHE   84

Fungi have one of two body plans and can be
  unicellular or multicellular   86

Kingdom Fungi is composed of four phyla   87

The Viruses   89

CASE: "DEATH" IN THE RUE MORGUE   89

Most viruses have one of a few basic structures   90

Specific viruses usually infect only certain hosts
  and certain cells within those hosts   91

Replication of animal viruses proceeds through
  a series of defined steps   92

Replication is different for DNA and RNA viruses   95

Not all viral infections cause symptoms or kill
  host cells   97

Viruses can damage host cells in several ways   98

Like animals, plants are susceptible to many viral
  infections   99

Bacteriophages are viruses that infect bacteria   99

Phages can influence the numbers of bacteria
  and even the diseases that they cause   101

Prions   102

CASE: LAST LAUGH FOR THE "LAUGHING
DEATH"   102

Coming Up Next...   104

Key Terms   104

Concept Questions   105

**Chapter 5 The Microbiology of History
and the History of Microbiology**   **107**

If Microbiologists Wrote History   107

CASE: THE SPANISH CONQUEST OF MEXICO   107

Disease influenced important events in ancient
  Greece and Rome   108

The Black Death altered European society forever   111

European explorers imported many diseases into
  the new world   112

Several important events of the 19th century owe
  their outcomes to microorganisms   115

Certain episodes in both World Wars I and II were influenced by microorganisms   117

Some infectious disease is caused by nonmicroorganisms   120

The Science of Microbiology: A Brief History   120

CASE: THE ASSASSINATION OF PRESIDENT GARFIELD   120

Microorganisms were first discovered in the 17th century   121

The science of microbiology was born in the second half of the 19th century   121

The "germ theory of disease" was advanced by the understanding of the need for hospital sanitation   124

The development of pure culture technique allowed rapid advances in microbiology   126

Koch was the first to link a specific microorganism to a particular disease   126

The germ theory led to major advances in disease control   127

With the development of the first vaccines, the field of immunology was born   128

Antimicrobial drugs are a 20th century innovation   129

Into the modern era   130

Coming Up Next...   130

Key Terms   130

Concept Questions   131

Chapter 6 Microbial Genetics   133

DNA: Structure and Organization   134

DNA is an information molecule   134

The structure of DNA is the key to how it functions   135

DNA is found on chromosomes   137

DNA Function: Replication   138

Each DNA molecule can be accurately copied into two new DNA molecules   138

CASE: DNA: FORM SUGGESTS FUNCTION   138

DNA replication begins at a site called the origin   140

DNA Function: Genes to Proteins   141

A cell's characteristics are largely determined by the proteins that the cell produces   141

In transcription, a gene's DNA is used to produce complementary RNA   143

CASE: CHRISTMAS TREES AND TRANSCRIPTION   143

Amino acids are assembled into protein during translation   147

CASE: PROTEIN SYNTHESIS: DEAD IN ITS TRACKS   147

Gene activity is often carefully regulated   151

The environment influences the nature of genetically determined characteristics   153

Sources of Genetic Variation   154

Both asexual and sexual organisms are able to generate new genetic combinations   154

CASE: GRIFFITH'S TRANSFORMING FACTOR   155

Genetic recombination in prokaryotes has great significance for humans   159

Mutations are the original source of genetic variation   160

CASE: END OF THE LINE FOR A LAST-LINE DEFENSE?   160

Mutations are caused by many factors   162

Many mutations are repaired before they can affect phenotype   163

Mutations are the raw material of evolution   165

Coming Up Next...   165

Key Terms   165

Concept Questions   166

Chapter 7 Metabolism and Growth   169

Metabolism is similar in all living things   170

Basic Concepts   170

Energy released from food molecules is used for other processes that require energy   170

Cells convert the energy in biological molecules into ATP   171

Energy is transferred from one molecule to another via oxidation and reduction   172

Cell Respiration   175

Carbohydrates are the primary source of energy used in cell respiration   175

Cell respiration occurs in three stages   176

In glycolysis, glucose is partially oxidized, forming two smaller molecules called pyruvate   177

In the Krebs cycle, pyruvate from glycolysis is completely oxidized and the energy released is transferred as electrons to NAD^+ and FAD   178

NADH and $FADH_2$ are oxidized in electron transport, providing the energy for ATP synthesis   179

CASE: SOMETHING'S FISHY IN THE FRIDGE   180

Protons that are pumped across a membrane in electron transport provide an energy source for ATP synthesis   182

Incomplete glucose oxidation results in fermentation   183

Molecules other than glucose can be used to generate ATP   185

CASE: NAME THAT BACTERIUM 185

Autotrophs: the Self-feeders 186

In photosynthesis, energy in sunlight is used to produce biological molecules 186

Chemoautotrophs use chemical energy the way that photoautotrophs use solar energy 188

Microbial Metabolism and Growth 188

Microbial growth refers to population growth 189

Oxygen, temperature, nutrient levels, and other environmental factors all influence microbial growth rate 191

CASE: CANINE FIRST AID 192

Microbial populations pass through a sequence of phases called the growth curve 196

As the environment changes, microbial metabolism and therefore the growth curve changes in response 197

CASE: MAKING YOGURT 197

Coming Up Next... 199

Key Terms 200

Concept Questions 200

Chapter 8 Microbial Evolution: The Origin and Diversity of Life 203

How Life Began 203

Conditions on the early Earth were very different than they are today 203

The first biological molecules were formed from nonbiological precursors 204

Genetic information may have originally been encoded in RNA instead of DNA 206

The first cells required a membrane and genetic material 207

The first prokaryotes are thought to have arisen approximately 3.5 billion years ago 208

Certain important metabolic pathways evolved in a defined sequence 208

Eukaryotes evolved from prokaryotic ancestors 210

CASE: CELLS WITHIN CELLS 210

Multicellular life arose from colonies of unicellular eukaryotes 212

The origin of viruses is obscure 213

Prions may have arisen from abnormal host proteins 214

Exobiology is the search for extraterrestrial life 214

Evolution: Explaining Life's Diversity 215

Natural selection is the driving force of evolution 216

Microorganisms are subject to the laws of natural selection 218

CASE: VANCOMYCIN RESISTANCE: THE SEQUEL 218

Natural selection can influence the virulence of disease-causing organisms 220

CASE: A TALE OF TWO COUNTRIES—AUSTRALIA, ENGLAND, RABBITS, AND THE MYXOMA VIRUS 220

Microorganisms often influence the evolution of their hosts 222

CASE: ALICE IN WONDERLAND 224

Coming Up Next... 225

Key Terms 226

Concept Questions 226

Chapter 9 An Ecologist's Guide to Microbiology 229

Basic Ecological Principles 230

Ecology is the study of how living things interact with each other and the environment 230

Energy and nutrients are passed between organisms in an ecosystem 230

Nutrients are recycled in an ecosystem, whereas energy is not 232

Microbial Ecology 232

Microorganisms live in microenvironments 233

Environmental conditions affect the growth rate of microorganisms 233

Many microorganisms live in biofilms 234

Microbial Habitats: Here, There, Everywhere 235

Soil often harbors rich microbial communities 235

Many microorganisms are adapted to life in freshwater 238

CASE: WATER BIRDS AND BOTULISM 238

Water pollution can lead to severe oxygen depletion 241

CASE: LAKE ERIE: A NEAR-DEATH EXPERIENCE 241

Although similar in many ways, the marine environment is distinct from freshwater 242

Cloud-dwelling bacteria may be important in promoting rainfall 245

Microorganisms and Biogeochemical Cycles 245

Carbon moves between living things and the environment 246

CASE: IRONING OUT GLOBAL WARMING? 246

Bacteria convert nitrogen into forms that plants can absorb 248

CASE: THE KEY TO A BOUNTIFUL HARVEST   248

Sulfur in organic material is returned to the environment by microorganisms   251

Like other cycles, the phosphorus cycle relies on microbial activity   251

Ecological Interactions Involving Microorganisms   253

Both organisms in a mutualism benefit from the relationship   253

In a commensal relationship one organism is benefited while the other is unaffected   254

Microorganisms in the same environment may compete for certain resources   254

CASE: A MICROSCOPIC DOG-EAT-DOG WORLD   254

Some microorganisms are predators   255

Coming Up Next…   256

Key Terms   257

Concept Questions   257

Chapter 10 The Nature of Disease: A Pathogen's Perspective   259

Basic Principles of Infectious Disease   260

Contact with microorganisms only rarely results in disease   260

Hosts are colonized with normal microbial flora that is usually harmless   260

CASE: SAYONARA SALMONELLA   260

The Process of Infectious Disease   263

A pathogen must achieve several objectives if it is to cause disease   263

Reservoirs provide a place for pathogens to persist before and after an infection   264

CASE: WHAT'S BUGGING KITTY?   264

The type of reservoir used by a pathogen has implications for disease control   266

Pathogens must reach a new host via one or more modes of transmission   268

CASE: REVENGE OF THE BLUE DEVILS   268

Pathogens gain access to the host through a portal of entry   270

Once they have entered, pathogens must adhere to the host   270

Most pathogens must increase in number before they cause disease   271

Successful pathogens must at least initially evade host defenses   272

Bacterial pathogens cause disease in several different ways   272

CASE: TROUBLE IN PARADISE   273

CASE: A RUPTURED APPENDIX   275

Viruses cause disease by interfering with the normal activities of the of the cells they infect   277

Different eukaryotic parasites affect their hosts in diverse ways   278

Pathogens leave the host through a portal of exit   279

Symptoms of disease often assist the pathogen in its transmission to a new host   280

CASE: WHO'S HURTING WHOM?   280

Coming Up Next…   281

Key Terms   282

Concept Questions   282

Chapter 11 Host Defense   285

When infection cannot be blocked, first innate immunity and then adaptive immunity is activated   286

CASE: INFLUENZA: EXPOSED!   286

Barriers to Entry   288

CASE: FRED   288

CASE: LAURA   290

The Innate Immune Response   290

If barriers to entry fail, the pathogen is confronted by elements of innate immunity   290

When innate mechanisms fail to eliminate an infection, the adaptive immune response is activated   293

Antigen-presenting cells activate those cells responsible for adaptive immunity   294

Antigen-presenting cells migrate to lymphatic organs to activate adaptive immunity   295

The Adaptive Immune Response   296

Antigen-presenting cells activate helper T cells to initiate an adaptive response   296

Some activated helper T cells activate cytotoxic T cells to initiate a cell-mediated response   298

Helper T cells also activate B cells to initiate a humoral immune response   298

In a humoral response, several different classes of antibodies may be produced   302

A successful adaptive response culminates in the elimination of the pathogen   303

An adaptive response is not always successful   304

CASE: BONNIE   304

Subsequent exposure to the same pathogen results in a stronger and faster adaptive response   304

Vaccines induce immunological memory without
    causing disease   306

HIV: A Problem of Immune System Destruction   307

CASE: OSCAR   307

Host Versus Pathogen: A Summary   311

Coming Up Next...   312

Key Terms   312

Concept Questions   313

Chapter 12 Control of Microbial Growth   315

Physical and Chemical Means of Control   316

Physical control of microorganisms involves
    manipulation of specific environmental factors
    such as temperature   316

CASE: COLD, HARD FACTS ABOUT COLD-
FILTERED BEER   318

Chemical methods can control microorganisms
    on living and nonliving material   320

A variety of chemicals have antiseptic and
    disinfectant properties   320

CASE: WASH YOUR HANDS FIRST!   321

Antimicrobial Chemotherapy   322

An ideal antimicrobial drug inhibits microorganisms
    without harming the host   323

A drug's mode and speed of action, the type of
    infection being treated, the potential for side
    effects, and the likelihood of drug resistance all
    influence drug selection   325

CASE: A MYSTERY ILLNESS   326

Antibiotics work by interfering with specific
    bacterial structures or enzymes   328

Selective toxicity, while possible, is harder to achieve
    against eukaryotic pathogens   331

Antiviral drugs must interfere with a particular step
    in the viral replicative cycle   332

The misuse of antibiotics has led to the problem
    of drug resistance   335

CASE: THE IMPORTANCE OF COMPLETING
PRESCRIPTIONS   335

New strategies provide options for circumventing
    antibiotic resistance   338

Coming Up Next...   339

Key Terms   340

Concept Questions   340

Chapter 13 Epidemiology: Who, What, When, Where, and Why?   343

The Birth of Epidemiology   344

The first modern epidemiological study identified
    cholera as a waterborne disease   344

Florence Nightingale found that improved hygiene
    reduced the likelihood of typhus   345

Epidemics   345

A single contaminated site can give rise to a
    common source epidemic   345

CASE: AN OUTBREAK OF HEPATITIS A   345

Host-to-host epidemics are spread from infected
    to noninfected individuals   347

Epidemics can occur for biological, environmental,
    and/or social reasons   347

CASE: CAUGHT RED-HANDED—AND WHITE-
BEAKED!   347

Epidemics become more likely when fewer people
    in a population are resistant   349

Epidemic outbreaks of influenza occur as the virus
    changes genetically   352

CASE: FLU SEASON   352

Investigating Disease Outbreaks   355

CASE: A NEW BUG ON THE BLOCK   356

A case definition helps health authorities determine
    if unusual cases are related   356

Time, place, and personal characteristics of a new
    disease provide clues to the disease's identity   357

Case–control studies can pinpoint a common risk
    factor among affected individuals   358

Emergent Diseases   359

Environmental, biological, behavioral, and social
    changes can result in emergent diseases   360

Emergent diseases can be categorized as one of
    four basic types   362

Bioterrorism   364

Coming Up Next...   364

Key Terms   365

Concept Questions   365

Chapter 14 The Future Is Here: Microorganisms and Biotechnology   367

The Analysis of DNA   368

Bacterial restriction enzymes have proven useful
    for cutting DNA at specific sites   368

DNA fragments can be separated by gel
 electrophoresis   369

Southern blotting can be used to identify a specific
 gene of interest   371

CASE: PINPOINTING A CANCER-CAUSING GENE   371

Sequencing techniques can be used to reveal the
 sequence of nucleotides in a DNA sample   373

CASE: TB—BC ("BEFORE COLUMBUS," THAT IS)   374

Large amounts of specific DNA sequences can be
 obtained with the polymerase chain reaction   374

The Sanger method can be used to sequence a
 specific sample of DNA   375

The entire genomes of many organisms have been
 sequenced   376

CASE: E. COLI'S DARK SIDE   376

Genetic Engineering: Whose Gene Is It Anyway?   379

CASE: BIOLOGICAL BLACKMAIL   379

Genes of interest can be cloned by inserting them
 into bacteria   379

DNA from any organism can be maintained in a
 DNA library   383

Insights and Applications   384

DNA technology has provided better understanding
 of genes and how they function   384

DNA technology has numerous medical, agricultural,
 and industrial applications   385

CASE: THE MONARCH BUTTERFLY: THE KING
IS DEAD OR LONG LIVE THE KING?   389

Ethics and Safety   392

Coming Up Next…   393

Key Terms   393

Concept Questions   393

Chapter 15 Guess Who's Coming to Dinner:
Microorganisms and Food   395

The Beginnings of a Beautiful Friendship   396

CASE: FLIGHT FROM EGYPT   396

Microbial activity can help to preserve the quality
 of some foods   396

Fermented dairy products and grains have been
 used for thousands of years   397

Microorganisms and Food Production   398

Some fungi and bacteria are consumed directly
 as food   398

CASE: WHAT'S FOR TUCKER, MATE?   398

In the absence of oxygen, some microorganisms
 undergo fermentation, releasing specific waste
 products   401

Many plant products can be fermented into various
 food items   401

CASE: HOMEMADE WINE   402

Fermented milk is the basis of making cheese and
 yogurt   407

Certain meat products, including salamis and cured
 hams, require fermentation   408

Production of Other Foods and Dietary
 Supplements   408

Coffee beans are readied for roasting through the
 use of bacteria   408

Many common dietary supplements are produced
 by microorganisms   408

CASE: MICROBIAL METABOLITES FOR A
FEATHERED FRIEND   408

Time to Eat: A Few Microbial Recipes   410

Sauerkraut   410

Pickles   410

Bagels   411

Yogurt   412

Greek feta cheese   412

Vegemite sandwich   413

Coming Up Next…   414

Key Terms   414

Concept Questions   414

Chapter 16 Better Living With
Microoorganisms: Industrial and
Applied Microbiology   417

Commercial Applications   418

CASE: UNEARTHING A NEW ANTIBIOTIC   418

Many microorganisms produce metabolites with
 commercial potential   419

Microorganisms producing promising metabolites
 must often be subjected to strain improvement   419

Potentially valuable microbes must also grow well
 in an industrial setting and must not pose undue
 risks to humans or the environment   421

A defined series of steps are followed to move
 production from the laboratory to the factory   422

Many industrially produced microbial metabolites
 have useful medical applications   423

Industrial microbial metabolites have a wide variety
 of other, nonmedical uses   425

Big Problems, Little Solutions   426

New strategies are required to combat
    environmental pollution   427

CASE: THE SWEET SMELL OF SUCCESS   427

Microorganisms are used to digest harmful
    chemicals through the process of
    bioremediation   428

Both the environmental context and the microbe
    being used determine how bioremediation is
    conducted   429

Bioremediation can prove valuable in many
    different settings   430

Microorganisms can help reduce solid waste
    and improve soil quality through composting   431

CASE: WASTE REDUCTION BEGINS AT HOME   431

Plastics may be replaced by biodegradable,
    microbially produced alternatives   432

Microorganisms may be used to help meet the
    demand for limited resources   433

Microorganisms may be able to stabilize soil,
    reducing earthquake damage   436

Coming Up Next...   436

Key Terms   436

Concept Questions   437

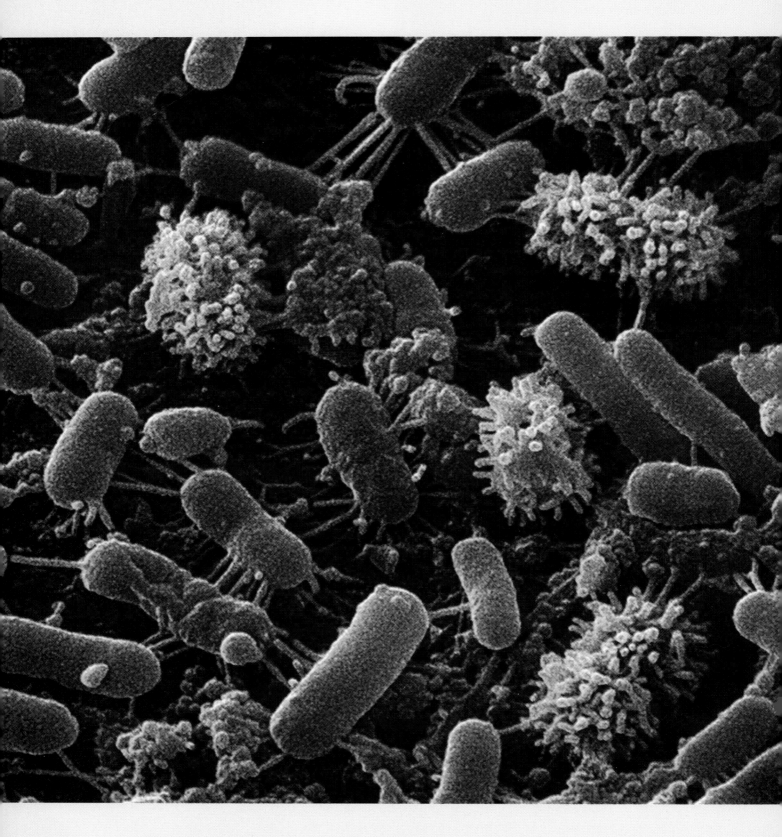

We live in a world teeming with microbial life, like these bacteria found on a kitchen sponge.

Chapter 1

Living in a Microbial World

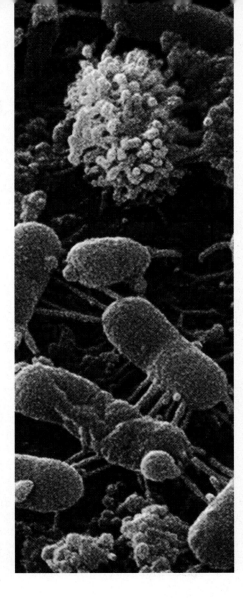

We live in a microbial world. Ask a friend to name a living thing and he or she will likely pick a dog, a daffodil, or some other familiar animal or plant. The vast majority of living things on this planet, however, are **microorganisms**—bacteria, viruses, fungi, and protozoa, too small to see with the unaided eye (**Figure 1.1a**). Just how far these organisms outnumber larger living things is staggering to think about. A single teaspoon of soil can hold millions or billions of bacteria and fungi. A drop of seawater teems with microscopic life. And larger animals and plants provide shelter for literally trillions of smaller living things. At this moment, more bacteria than all the humans that have ever lived are quietly going about their business in your intestine.

Microorganisms, also often referred to as **microbes**, live everywhere (see Figure 1.1b and c). They have been found frozen in ice and inside solid rock. They live in the deepest parts of the ocean and in boiling hot springs. They are found in habitats that are as acidic as battery acid or in briny water so salty that no fish could survive.

Many people lump all microorganisms together as "germs"—tiny organisms that are of little importance to us unless they make us sick. Nothing could be further from the truth. This book will not only demonstrate how diverse these fascinating organisms are but will also explore the many and often surprising ways they affect our lives. Indeed, it is difficult to open a newspaper or turn on the radio and not read or hear *something* about the way microorganisms impact humans (**Figure 1.2**).

It was recently announced, for example, that researchers have discovered that a common aquatic bacterium makes a sticky substance that might one day serve as a "biological superglue," which can work even in water. The bacterium, known as *Caulobacter crescentus*, uses the adhesive to attach to rocks or other surfaces. It has been estimated that a small surface—even a wet surface—about the size of two postage stamps that was coated with the substance could potentially hold about 70 tons. Another surprising recent report suggests that the number of calories that we obtain from food is to some degree determined by microorganisms living in our intestine. It has been understood for some time that such microbes assist in the digestion of certain food molecules. To examine this more closely, scientists developed a strain of laboratory animals that were entirely lacking these intestinal microorganisms. They next divided the animals into several groups and allowed different intestinal microorganisms or combinations of microorganisms to colonize each group of animals. All animals were fed an identical

Figure 1.1 The microbial world around us. (a) Bacteria growing on the head of a pin. (b) Microorganisms live almost everywhere. When surfaces in the environment (a doorknob, a body part, or a kitchen sink, for example) are swabbed with a moist sterile swab, microorganisms that adhere to the swab can be transferred to a special plate, known as a nutrient agar plate, that supports microbial growth. After approximately 24 hours at a favorable temperature, all of these samples of environmental material show heavy microbial growth. Each visible spot on the plate represents millions of microbial cells, known as a *colony* of a microorganism. (c) Even environments such as this hot spring, too extreme to support plant or animal life, are conducive to the survival of certain microorganisms.

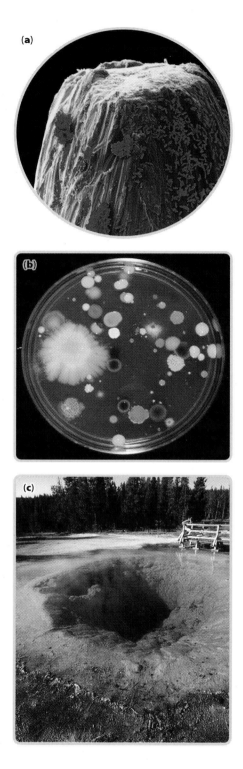

diet. Some groups of animals, however, gained weight faster than others because their food was more readily digested by certain microorganisms, making the food easier for the animals to absorb. The scientists concluded that to understand the effect that food has on weight we must consider an individual's "microbial ecology." They foresee the day when manipulating our intestinal microorganisms might even be used to treat obesity.

Certainly there is no shortage of news stories to remind us that some microorganisms cause disease. In the spring of 2009 a new strain of influenza dubbed swine flu, and more formally known as H1N1, reared its head and riveted the world's attention. In June of that year, the World Health Organization declared an influenza pandemic, the first such declaration in nearly half a century. We still hear regularly about the ongoing AIDS crisis. In the last few years we learned of a major breakthrough in the fight against cancer—an effective vaccine against cervical cancer, which kills about 250,000 women a year worldwide (see Figure 1.2). The great majority of cases of cervical cancer are caused by a virus called the human papillomavirus. Development of a vaccine that protects against the most common strains of this virus can largely prevent this cancer. But microorganisms are not necessarily the bad guys when it comes to our health. Consider the remarkable recent finding that the common soil bacterium *Mycobacterium vaccae* may be able to alleviate depression. Lung cancer patients, for instance, if injected with *M. vaccae*, reported that they felt better and had less nausea and pain. The bacteria apparently activate a set of chemical-releasing nerve cells in the brain—the same nerves targeted by the antidepression drug Prozac.

The study of any living thing starts with the fundamentals, and microorganisms are no exception. We will therefore start by investigating the basic chemistry of microorganisms, their structure, and how they grow. We will learn how they pass genetic information on to the next generation and how they acquire new characteristics over time. We will take a close look at how they impact, for better or worse, our environment, our health, our diet, and even our history. Along the way we will learn, just to provide a few examples,

- What the designation H1N1 in reference to swine flu actually signifies
- Why biological "dead zones," devoid of most sea life, are becoming an increasingly important problem in the world's oceans
- How microorganisms contributed to the fall of Rome, the conquest of the Americas, and the development of a middle class in England
- Why hydrogen peroxide is a good thing to keep in your medicine cabinet
- How disease-causing microbes develop resistance to antibiotics and how such resistance can be reversed
- What gives Camembert cheese its particular flavor
- Why "stone-washed" denim has nothing to do with stones

The answers to all these questions are to be found in the study of microbes. The reasons for studying microorganisms, however, go well beyond such examples, no matter how interesting they might be. When we learn about microorganisms, we are learning about life itself. As you proceed through this text, you will find that basic life processes are the same for all living things. Most of the major discoveries made by biologists about the nature of life and how it works were first made in microorganisms; biologists then found that the same discoveries apply to other forms of life. Biologists have been attracted to microorganisms because, in many respects, they are simpler, they are easier to work with, and they yield experimental results faster than animals or plants. Furthermore, many of the most important issues in the world today—issues that on the surface may seem to have little to do with biology, such as global warming, rising energy costs, and international terrorism—are tied in to a greater or lesser degree with microorganisms. Consequently, for the newcomer to biology, there is no better place to start than the world of microorganisms.

We will learn why life itself is dependent on microorganisms. Although we may not see them and although some of them cause more than their share of misery and disease, we cannot live without them. Microorganisms were here millions of years before us, and they will no doubt be here long after we are gone. Before we begin to tell their tale, however, let us consider what microbiology, the study of microorganisms, actually involves.

The Science of Microbiology

Microbiology is a field within the general discipline of biology, the study of life. Although the microbiologist is primarily interested in *micro*organisms—those living things too small to see with the unaided eye—in many respects microorganisms are not very different from more familiar organisms.

All living things are composed of one or more cells

One of the most fundamental principles in all of biology is the **cell theory**: the notion that all living things are composed of **cells**. Cells are highly organized structures that are always surrounded by a membrane. Cells are the basic unit of life, and all living things are composed of one or more cells (**Figure 1.3**). The interior of cells is a watery mixture of many molecules and

> **FDA NEWS RELEASE**
> **FOR IMMEDIATE RELEASE**
> June 8, 2006
>
> **FDA Licenses New Vaccine for Prevention of Cervical Cancer and Other Diseases in Females Caused by Human Papillomavirus**
>
> ***Rapid Approval Marks Major Advancement in Public Health***
>
> The Food and Drug Administration (FDA) today announced the approval of Gardasil, the first vaccine developed to prevent cervical cancer, precancerous genital lesions and genital warts due to human papillomavirus (HPV) types 6, 11, 16 and 18. The vaccine is approved for use in females 9-26 years of age. Gardasil was evaluated and approved in six months under FDA's priority review process—a process for products with potential to provide significant health benefits.

Figure 1.2 Microbiology in the news. In one of the most important science and health stories in recent years, a vaccine against cervical cancer was announced in 2006. Almost all human cervical cancer is caused by forms of the human papillomavirus. Vaccinating against the most common strains of the virus consequently can prevent some two-thirds of cervical cancers.

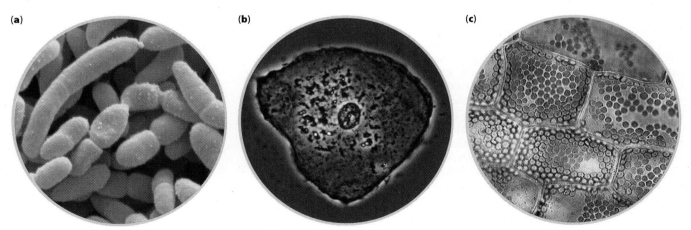

(a)   **(b)**   **(c)**

Figure 1.3 Cell types. All cells are surrounded by a cell membrane. (a) *Acetobacter,* an example of bacterial cells. These bacteria are commonly used to make vinegar. Because each individual cell constitutes an individual bacterium, most bacteria are considered to be *unicellular* organisms. Like almost all bacterial cells, these cells small and surrounded by a rigid cell wall. The genetic material of bacterial cells is not enclosed within a membrane. Cells that lack a membrane for this purpose are called *prokaryotic* cells. (b) An animal cell from the inside of the mouth. Note the densely stained nucleus in the cell's interior. The nucleus, which contains the cell's genetic material, is surrounded by a nuclear membrane. Cells with such a membrane are called *eukaryotic* cells. (c) Plant cells. In addition to a nucleus, plant cells also often contain chloroplasts (seen as green membrane-bound structures), the site of photosynthesis. Note the more regular shape of the plant cells, compared with animal cells. This is due to the presence of a cell wall, exterior to the cell membrane. Such a wall is absent in animal cells. Individual plants and animals are both composed of many cells and are consequently called *multicellular* organisms.

structures that carry out various activities. The microbes we will discuss in this text are most commonly **unicellular**, meaning they consist of only a single cell. Some microbes, on the other hand, as well as all animals and plants, are **multicellular**, meaning that they are composed of many cells.

Cells fall into two general types. Most microorganisms have simpler and usually smaller cells known as **prokaryotic cells** (see Figure 1.3). This name comes from the Greek for "before the nucleus," and it reminds us that in these cells the genetic material is not surrounded by a nuclear membrane. This is in contrast to the larger and more complex **eukaryotic cells** (from the Greek for "true nucleus"), in which a membrane surrounding the genetic material forms the nucleus. Eukaryotic cells also have a variety of other small structures called **organelles** ("little organs"), which carry out specific functions required by the cell. Eukaryotic cells are thus more compartmentalized than prokaryotic cells, with many specific activities taking place in discrete locations. In prokaryotic cells, although these same activities occur, the location within the cell *where* they occur is less defined. Eukaryotic cells are found in animals and plants, as well as some microorganisms.

All living things display other observable characteristics

What exactly is life? Perhaps surprisingly, such a simple question does not have an easy answer. One unifying characteristic of living things is that they are composed of cells. These cells, prokaryotic and eukaryotic alike, share a number of defining characteristics, and the presence of these characteristics usually allows us to separate the living world from the nonliving.

First of all, living cells typically have an internal environment that is different from their surroundings (**Figure 1.4**). It may be more or less salty or different in terms of acidity. Moreover, cells have mechanisms to maintain these differences, a process called **homeostasis**. Just to provide one familiar example of homeostasis, your body temperature is kept relatively constant, about 37°C, regardless of the environmental temperature. In humans, critical cellular processes function best at this temperature. Should temperature-regulating homeostasis fail, a cell's ability to carry out these processes will decline, and if homeostasis cannot be reestablished, the cell will ultimately die.

Furthermore, all living things have the ability to reproduce, which requires a blueprint, usually encoded in a molecule called **deoxyribonucleic acid**, frequently abbreviated as **DNA**. Living things can also respond to their environment: they can alter their behavior as environmental conditions change. Certain environmental cues may cause cells to move toward or away from a stimulus. Cells respond to chemical signals in their environment, and often these signals are released by other cells. This represents a form of cell-to-cell communication, and such communication is also a hallmark of living systems. Another important characteristic of living things is the ability to assimilate and use energy, a process known as **metabolism**. Finally, all life changes over time. Via the process of **evolution**, the characteristics of living things may change over many generations.

Microbiology involves the study of several distinct groups of living things

Microbiology is distinguished from other subdisciplines within biology in that it deals with primarily microscopic organisms. It actually covers quite a bit of ground because microorganisms are such a diverse lot—far more diverse than animals or plants.

Bacteria are small, most commonly single-celled prokaryotic organisms (**Table 1.1**). The other major group of prokaryotes is the **archaea**. Although

(a)

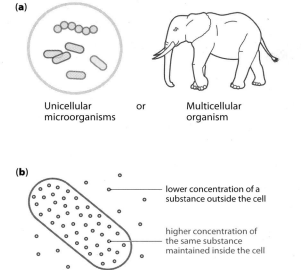

Unicellular   or   Multicellular
microorganisms   organism

Living things are composed of one or more cells.

(b)

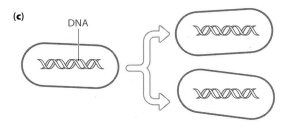

lower concentration of a substance outside the cell

higher concentration of the same substance maintained inside the cell

Cells and all living things are capable of homeostasis, or maintaining favorable internal conditions.

(c)

DNA

All living things are able to reproduce.

(d)

chemical signals to which cells may respond

Cells respond to chemical signals in the environment and those released by other cells.

(e)

Release of wastes

Use of energy

Energy uptake

Cells derive the energy they need for vital functions through metabolism.

(f)

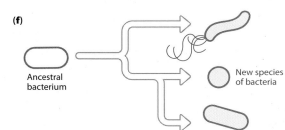

Ancestral bacterium

New species of bacteria

All life forms undergo evolution over time.

Figure 1.4 Some characteristics of life. (a) Living things are composed of one or more cells. (b) For some environmental factors, homeostasis allows cells to maintain different conditions within the cell than those of the surrounding environment. (c) All living things are able to reproduce. Genetic information is transferred between parent and offspring in the form of a molecule called DNA. (d) Cells respond to chemical signals that are released by other cells or that are present in the environment. (e) They also require energy to carry out various cellular functions. (f) They undergo evolutionary change over time.

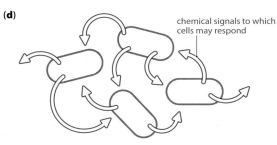

MICROORGANISM	CELL TYPE	CELL WALL	PHOTOSYNTHESIS
bacteria	prokaryotic	yes, almost all	some
archaea	prokaryotic	some	some
protozoa	eukaryotic	in some life cycle stages	no
fungi	eukaryotic	yes	no
algae	eukaryotic	yes	yes
viruses	acellular (not composed of cells)	no	no
prions	acellular (protein only)	no	no

Table 1.1 Some characteristics of major types of microorganisms. Bacteria and archaea have prokaryotic cells, while protozoa, fungi, and algae have eukaryotic cells. Neither viruses nor prions are composed of cells and both are therefore termed acellular.

archaea look much the same as bacteria, there are many structural differences between these two groups, and they are only distantly related.

Other microorganisms have eukaryotic cells (see Table 1.1). Eukaryotic microorganisms include many fungi and the protozoa (**Figure 1.5**). Algae likewise have eukaryotic cells, and the smaller microscopic algae are considered to be microorganisms as well. Animals and plants are also composed of more complex eukaryotic cells, meaning that, in spite of the size difference, the cells of microorganisms such as baker's yeast (a fungus) and malaria parasites (protozoa) have more in common with human or other animal cells than they do with prokaryotic bacteria or archaea.

The **viruses** are a particularly interesting and unusual group of microorganisms. Viruses are the ultimate parasite: they are utterly unable to reproduce unless they gain access to the interior of an appropriate cell. Once inside, however, they essentially hijack the cell and use the cell's machinery to carry out their own replication. The cell, forced to divert energy and other resources to viral replication, may eventually die. Once they have successfully replicated, new viral particles leave the cell and go on to infect other cells to repeat the replicative cycle. Because they rely so heavily on the cells that they parasitize, viruses are unusually simple. They consist of little more than genetic material surrounded by a protein coat, and they have few of the structures found in other microorganisms. Many of humanity's worst scourges are various viruses such as influenza and HIV (**Figure 1.6**). Finally, there are the **prions**, infectious proteins, which make even viruses look complex by comparison (**Figure 1.7**). Prions are responsible for a set of odd diseases such as mad cow disease.

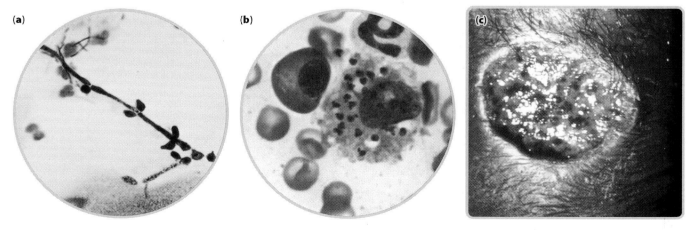

Figure 1.5 Eukaryotic microorganisms. (a) *Candida albicans*, the fungal pathogen responsible for vaginal yeast infections. (b) *Leishmania donovani* (small purple bodies) inside larger mammalian white blood cells. This protozoan parasite is transmitted through the bite of infected sand flies. (c) A skin lesion in a patient suffering from a leishmania infection.

Figure 1.6 Human immunodeficiency virus (HIV). Newly formed viral particles can be seen leaving the much larger infected human cell. Unlike other microorganisms, viruses are not composed of cells, and they are unable to replicate outside an appropriate cell of another organism. Consequently, they are called *obligate parasites* (organisms for which parasitism is an absolute requirement). Once inside an appropriate cell, viruses commandeer the host cell machinery to facilitate their own replication. Depending on the virus, such cellular hijacking may or may not kill the cell.

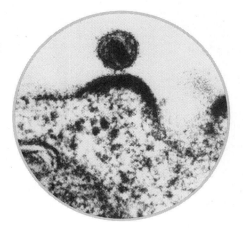

Viruses strain our notion of what it means to be "alive"

Not all living things display all the characteristics of life, and microorganisms especially test the boundaries separating the living from the nonliving world. Sometimes, for instance, the capacity for independent movement is considered a characteristic of life, but many microorganisms are incapable of such movement. Viruses are particularly problematic; they seem to straddle the border between life and nonlife. Viruses are not composed of cells. On their own, they are unable to reproduce, and they carry out little if any independent metabolism. Yet viruses are able to attach to and gain entry into the cells of other organisms. Once inside, they begin to replicate and certainly *act* like living things. So are they alive? Such a question is perhaps more philosophical than biological, but it highlights the fact that the division between life and nonlife is not as sharp as one might think.

Microbiology is closely intertwined with the study of nonmicroorganisms

The scope of microbiology does not end with the microorganisms themselves. In many areas of microbiology, we cannot neglect the biology of multicellular eukaryotes such as ourselves. When we discuss, for example, how certain microorganisms cause disease, it is important to understand not only what infectious microorganisms do to us but also how *we* defend against *them*. Consequently, we must also consider the immune system and how it helps us fend off a never-ending microbial onslaught.

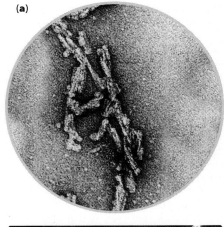

(a)

Because microbiology is such a large and complex field, most microbiologists specialize within one of many specific fields of expertise. Before we begin our study of the organisms themselves, let us briefly consider some of the diverse areas of study within the science of microbiology.

Microbiology is composed of many specialized subdisciplines

CASE: ART COMES ALIVE?

Paintings such as *Starry Night* by Vincent van Gogh and *The Wedding* by Marc Chagall are under attack. The yellow and orange paints used by these and other artists are slowly turning an ugly green or brown. Both van Gogh and Chagall used a mineral called crocoite to formulate their yellows and oranges. Some types of bacteria are able to degrade crocoite, and when they do, the discoloration occurs. Microbiologists are currently looking for ways to protect such masterpieces from further microbial assault without causing additional damage to the artwork.

1. **What type of microbiologist would concern himself or herself with this type of problem?**
2. **How might valuable artwork be protected from microbial damage?**
3. **Are other priceless artifacts aside from oil paintings at similar risk?**

Figure 1.7 Prions. (a) Prion proteins. Prions are unique infectious agents, consisting of protein only. (b) Prions cause neurological disorders. This sheep suffers from a prion disease known as scrapie. Infected sheep behave erratically, often scraping themselves against hard objects, causing the loss of their wool coat.

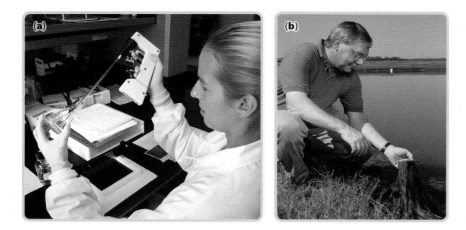

Figure 1.8 Microbiologists at work.
(a) Basic microbiology seeks to gain a more complete understanding of biological processes, focusing on microorganisms. Here, a microbiologist uses a device called a pipetter to transfer material from one tube to another. (b) Using knowledge gained through basic research, the applied microbiologist attempts to solve practical problems. Here, an environmental microbiologist investigates the effect of microorganisms on the water quality of a lake.

Microbiology, like biology in general, has both **basic** and **applied** components. A biologist studying topics in basic biology is interested in a more complete understanding of life processes. Learning exactly how specific cells utilize energy, or how different organisms are related to each other, are both examples of basic biological questions.

Much of our understanding of basic biological processes comes from the study of microorganisms (**Figure 1.8a**). With their simpler cellular structure, microorganisms provide the microbiologist with a useful starting point for teasing apart cellular processes. They are also easier to maintain in the laboratory than plants or animals, and they are relatively easy to manipulate experimentally. Because they reproduce so quickly, experimental results can be obtained far more rapidly with them than with slower growing and more slowly reproducing multicellular organisms. And finally, as we have already noted and as we will highlight throughout this text, many fundamental biological processes are very similar whether we are talking about a single-celled bacterium or a large animal composed of trillions of cells. Consequently, lessons learned from microorganisms are often useful in understanding the basic biology of animals, plants, and other organisms.

Certain bacteria and other living things that have been closely studied for these and similar reasons are referred to as **model organisms**. The bacterium *Escherichia coli*, for example, is sometimes referred to as the "microbial lab rat" (rats are an animal model organism), because of the intensity with which it has been studied. Many seminal biological discoveries, such as the discovery of DNA and how it functions, were first elucidated by use of this bacterial species. *E. coli* continues to reveal secrets with applications to living things in general. For instance, in Chapter 14 we will see how many breakthroughs of the biotechnology revolution came to us courtesy of this microorganism.

The basic biologist is not trying to solve a practical problem. That is the job of the applied biologist. Applied biologists use the knowledge gained in basic biology to solve real-world problems. For example, a basic researcher might determine the exact way that a certain virus enters human cells. An applied scientist may then use this information to design a new antiviral drug that would thwart the entry of the virus. The division between basic and applied science is, however, not a sharp one, and it is not uncommon for a scientist to work on both basic and applied aspects of a particular problem.

Within the broad area of basic microbiology, a scientist might wish to study any number of specific topics. A microbial geneticist, for example, may choose to study how the genetic material of a particular microorganism is turned on or shut off, in response to changing environmental conditions. A microbial ecologist might opt to investigate how a community of microorganisms living in the soil influences the availability of nutrients for plants. A microbiologist interested in adaptations to extreme environments may wish to know how certain microorganisms withstand levels of radiation that would kill an animal.

Applied microbiology likewise encompasses many diverse areas of interest. Medical microbiologists try to better understand the role of microorganisms in infectious disease, with an eye toward prevention or treatment. Public health microbiologists help educate the public about how to best avoid such disease, whereas epidemiologists track the spread of disease in the community and attempt to predict future outbreaks.

Food microbiologists, on the other hand, utilize microorganisms to produce a large portion of what we eat, from cheese to bread to soy sauce. Agricultural microbiologists often utilize knowledge gained through the study of microbial ecology to enhance crop growth or reduce the likelihood that valuable food plants will suffer from disease (see Figure 1.8b). Similarly, industrial microbiologists investigate the manner in which bacteria and fungi can be used to produce a wide variety of industrial and commercial products.

More and more, microbiologists, along with other biologists, are venturing into the new biological frontier of genetic engineering and recombinant DNA technology. Using a variety of techniques and molecular tools provided by microbes, these scientists have been able to introduce foreign DNA into new organisms, creating cells with novel capabilities. As we will discover, because of this powerful new technology, we now have bacteria that produce insulin for human diabetics and potential cures for genetic diseases, as well as a whole host of other potentially exciting, sometimes scary applications.

These are just a few of the many avenues of investigation open to the microbiologist. We will discover others as we proceed through the text. And what about the scientists who are investigating microbial damage to irreplaceable artwork (**Figure 1.9a**)? Those who work to protect paintings and other art objects from damage of any sort are generally called "conservation scientists," but this new field, in which microbial damage specifically is being confronted, as yet has no formal name. Perhaps it should be called "restoration microbiology" or even "artifact microbiology." Clearly, scientists in this field are at work on a problem in applied microbiology. Like other applied microbiologists, they are grounded in basic microbiology, in this instance the precise manner in which certain bacteria break down the crocoite in yellow and orange paints.

It is still not exactly clear how to prevent such damage. Chemicals that kill the bacteria can damage the delicate artwork as well. Furthermore, some antibacterial compounds kill only some types of bacteria, perhaps giving other, equally damaging organisms the chance to take over. One possible solution being considered is to seal artwork in oxygen-free, airtight cases, and then to pump in argon gas. Whether this will restrict the growth of all offending bacteria remains to be seen. And it is not just famous oil paintings that are threatened. It has only recently been recognized that famous historical sites, from the Mayan ruins in Mexico to the Angkor Wat temple in Cambodia, are slowly eroding and discoloring in response to bacteria and fungi growing on and even beneath their surfaces (Figure 1.9b). It appears that artifact microbiologists will have plenty to keep them busy for years to come.

Microbiology, like all fields of biology, is a **science**. By this we mean that microbiology is a process of learning about nature by observation and experiment. Scientists gain knowledge about the natural world in a precise manner known as the **scientific method**. Before we end this introductory chapter and begin *our* study of microorganisms in earnest, let us consider exactly how scientific knowledge is acquired.

Figure 1.9 Microbial damage to artworks and monuments. (a) *Starry Night*, painted by Vincent van Gogh in 1889. The faint areas of brownish and greenish discoloration are due to the breakdown of crocoite, a mineral used by van Gogh and other artists in the 19th century to formulate certain colors. It has recently become apparent that the breakdown of crocoite is caused by bacteria. (b) Microorganisms can cause the erosion and discoloration of some of our most cherished historical landmarks. The dark discoloration in these Mayan ruins in Yucatan, Mexico, is caused by microorganisms. Microbiologists are playing an increasing role in protecting or restoring such artworks and monuments.

The Scientific Method

A proper scientific experiment involves a series of well-defined steps

CASE: FLEMING REVISITED

The British scientist Alexander Fleming discovered the first antibiotic in 1929. He noticed that when certain molds contaminated his bacterial culture plates, the bacteria were unable to grow close to the mold. Fleming reasoned that the mold was secreting a substance that killed the bacteria. This proved to be the case, and the mold was identified as *Penicillium notatum*. Fleming named the secreted antibiotic penicillin (**Figure 1.10a**).

Two microbiology students, Rich and Shawn, learn of Fleming's work in their class, and later, when they must perform a small, independent research project as part of their laboratory grade, they decide to repeat Fleming's work with a different yet closely related mold, *Penicillium roqueforti*. This mold is commonly used in the production of blue cheese. The mold digests milk sugars, and the waste products that are released contribute to the taste of the cheese. The mold also gives blue cheese its characteristic blue streaks (Figure 1.10b). Does it also, like *P. notatum*, have antibacterial properties?

To find out, Rich and Shawn buy a block of blue cheese and transfer mold cells onto nutrient agar plates, which will support the growth of a wide range of bacteria as well as the *P. roqueforti* mold. They inoculate 12 plates with "lawns" of the bacterium *Staphylococcus epidermidis*. Lawns are prepared by swabbing a plate with a uniform coating of bacteria. The bacteria then grow over the entire surface of the plate (**Figure 1.11a**). Rich and Shawn, using an inoculating loop, next streak *P. roqueforti* in a zigzag pattern over six of their plates (Figure 1.11b). The other six plates are "streaked" with an uninoculated loop without the mold. All 12 plates are then incubated at 35°C for 24 hours. Rich and Shawn then examine their plates, comparing those with and without *P. roqueforti* for evidence of antimicrobial activity.

1. What hypothesis are Rich and Shawn testing?
2. Which of their nutrient agar plates are serving as controls? Which are the experimental plates? Why are the control plates critical?
3. What is the experimental variable in this experiment? What are the important control variables?
4. Why is it important to use six plates with and without the mold, rather than just one of each?

Figure 1.10 Penicillium molds.
(a) Penicillin secreted by the mold *Penicillium notatum* prevents the bacteria from growing near the mold colony. (b) *Penicillium roqueforti,* seen as blue streaks, is used to make blue cheese.

The scientific method and the process of scientific inquiry begin with an observation. A curator in a museum, for example, may have observed that the yellows in a famous van Gogh painting were beginning to turn brown. Fleming noticed that *Staphylococcus* was unable to grow close to a mold contaminant. Observations such as these lead to questions. *Why* was the color in the painting changing? *How* did the mold inhibit bacterial growth?

To answer such questions, the scientist formulates an appropriate **hypothesis**. A hypothesis is a tentative explanation for a specific question. To be valid, a hypothesis must be testable. In other words, it must be possible to collect evidence that either supports or refutes the hypothesis. To gather such evidence, one must be able to make predictions based on the hypothesis. Experiments or observations of future events are then used to determine whether the prediction was realized. If so, the hypothesis is supported. If the prediction proves to be inaccurate, the hypothesis is rejected as false. In the case of the damaged artwork, the initial hypothesis may have been that some air pollutant was affecting the yellow color. The prediction that results from this hypothesis is that if the air quality is carefully maintained, the degradation of the color will stop. Rich and Shawn are testing the

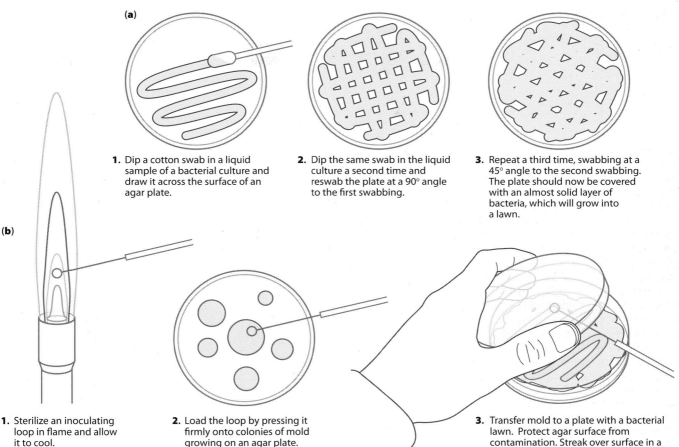

(a)

1. Dip a cotton swab in a liquid sample of a bacterial culture and draw it across the surface of an agar plate.

2. Dip the same swab in the liquid culture a second time and reswab the plate at a 90° angle to the first swabbing.

3. Repeat a third time, swabbing at a 45° angle to the second swabbing. The plate should now be covered with an almost solid layer of bacteria, which will grow into a lawn.

(b)

1. Sterilize an inoculating loop in flame and allow it to cool.

2. Load the loop by pressing it firmly onto colonies of mold growing on an agar plate.

3. Transfer mold to a plate with a bacterial lawn. Protect agar surface from contamination. Streak over surface in a zigzag motion.

hypothesis that *P. roqueforti* produces compounds that can inhibit bacterial growth. They therefore predict that *Staphylococcus* will be unable to grow close to the mold.

The next step is to test the hypothesis with a valid experiment. Such an experiment will compare an **experimental group** with a **control group** (**Figure 1.12**). The experimental group is the group in which a crucial factor will be manipulated. In the control group, that same factor will be left unchanged, as a means of comparison. If the manipulation of this factor results in a specific change as predicted by the hypothesis, and if that change does not occur in the unmanipulated control group, the hypothesis is supported.

For example, in Rich and Shawn's experiment, the experimental group consists of those bacterial lawns onto which the mold was streaked. No mold was placed on the control plates. Consequently, if their hypothesis that *P. roqueforti* is able to produce antibiotic compounds is correct, bacterial growth on the experimental and control plates should be different. If the hypothesis is incorrect, bacterial growth on the two types of plates will be similar, and the hypothesis can be rejected.

The experimental and control groups in any experiment should be identical for all factors except one. In Rich and Shawn's experiment, all plates contained the same nutrients and they were incubated at identical temperatures. These represent **controlled variables**. The only difference was the **experimental variable**, in this case, the presence or absence of mold. After 24 hours, the two students compared bacterial growth on their experimental plates streaked with mold with their control plates lacking it. Because they carefully controlled all variables except one, any differences in

Figure 1.11 Testing a mold for antimicrobial activity. (a) Prepare bacterial lawns on several agar plates. (b) Inoculate some of the plates with mold. After an incubation period, compare the inoculated and the uninoculated plates for evidence of inhibited bacterial growth.

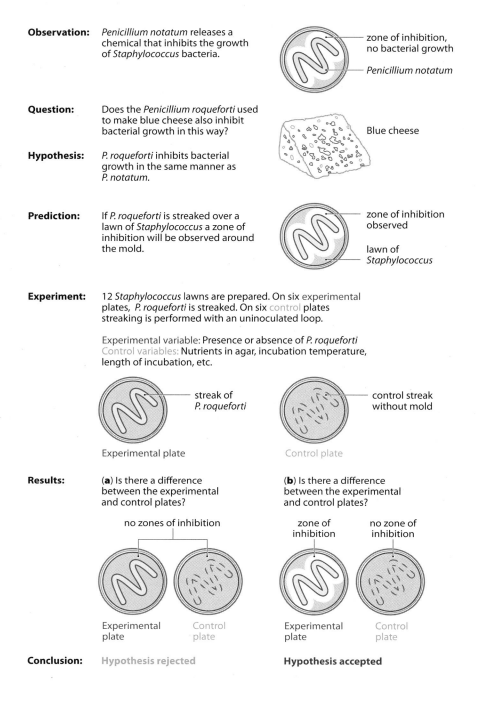

Observation: *Penicillium notatum* releases a chemical that inhibits the growth of *Staphylococcus* bacteria.

zone of inhibition, no bacterial growth

Penicillium notatum

Question: Does the *Penicillium roqueforti* used to make blue cheese also inhibit bacterial growth in this way?

Blue cheese

Hypothesis: *P. roqueforti* inhibits bacterial growth in the same manner as *P. notatum*.

Prediction: If *P. roqueforti* is streaked over a lawn of *Staphylococcus* a zone of inhibition will be observed around the mold.

zone of inhibition observed

lawn of *Staphylococcus*

Experiment: 12 *Staphylococcus* lawns are prepared. On six experimental plates, *P. roqueforti* is streaked. On six control plates streaking is performed with an uninoculated loop.

Experimental variable: Presence or absence of *P. roqueforti*
Control variables: Nutrients in agar, incubation temperature, length of incubation, etc.

streak of *P. roqueforti*

control streak without mold

Experimental plate

Control plate

Results: (**a**) Is there a difference between the experimental and control plates?

(**b**) Is there a difference between the experimental and control plates?

no zones of inhibition

zone of inhibition

no zone of inhibition

Experimental plate

Control plate

Experimental plate

Control plate

Conclusion: Hypothesis rejected

Hypothesis accepted

Figure 1.12 A controlled experiment to test a hypothesis. After making an initial observation that leads to a specific question, the scientist formulates a proper hypothesis. A proper hypothesis is a tentative answer to the question that has been posed, and it must predict future events. If these events do not occur, the hypothesis is rejected and a new hypothesis must be generated. If the events predicted by the hypothesis occur, the hypothesis is supported. In this example, the hypothesis that *P. roqueforti* can inhibit bacterial growth by secreting bacterial compounds is being tested experimentally. The experimenter prepares bacterial lawns and streaks *P. roqueforti* on half of these lawns. These represent the *experimental group*. On the remaining bacterial lawns, no *P. roqueforti* is added. These represent the *control group*, which will be compared with the experimental group to determine whether the added *P. roqueforti* has had any effect on bacterial growth. The only difference between the two groups of plates is the presence or absence of *P. roqueforti*. This difference is the *experimental variable*. All other factors (for example, temperature or provided nutrients) must be the same in the two groups. These are the *control variables*.

bacterial growth on the two groups of plates could be attributed to the mold rather than some other factor. Note that Rich and Shawn even "streaked" the control plates with an empty inoculating loop. This added controlled variable ensures that if they do see reduced bacterial growth on experimental plates, it must be due to the mold rather than the physical act of streaking.

Following completion of the experiment, the experimental and control plates are compared. Without the control group there would be no way to interpret the effect of the experimental variable. A significant inhibition of growth on the experimental plates indicates that the hypothesis may be tentatively accepted. No difference in bacterial growth on the two sets of plates indicates that the hypothesis should be rejected. When hypotheses are rejected in this way, a scientist has to develop a new hypothesis to explain the initial observation. A new experiment with a different experimental variable is then conducted. Keep in mind that in most situations there may be numerous plausible hypotheses. The task of the scientist, in this case, is to rule out as many of these competing hypotheses as possible. Ideally, only a single hypothesis remains to explain the initial observation.

For instance, if we were to test the hypothesis that air quality was responsible for the deterioration of the paintings, we might do so by monitoring color changes in two groups of paintings. The control paintings would be exposed to unpurified museum air as usual. Experimental paintings would be kept in a room of filtered air. The air quality would be the experimental variable. All other control variables, such as room temperature and amount of lighting, would be kept identical for both sets of paintings. If we actually carried out this experiment, we would probably find that filtering the museum air did not slow the rate of color change in the experimental paintings relative to the control paintings. At this point, our hypothesis would be rejected, and it would be necessary to consider an alternative explanation.

Returning to our case, recall that Rich and Shawn inoculated *groups* of experimental and control plates, rather than just one of each. It is almost always true that the larger the number of treatments, the more meaningful the results. Replication reduces the amount of variance due to unaccounted-for events. For example, Rich and Shawn might have accidentally prepared one of their bacterial lawns improperly. If they put too few organisms on an experimental plate, they may have incorrectly assumed that *P. roqueforti* was unable to inhibit the growth of *S. epidermidis*. In truth, these results might merely have reflected improper inoculation. By repeating the experiment, the likelihood of such erroneous interpretation can be reduced.

If a hypothesis cannot be disproved, it may eventually become a theory

If a hypothesis is rigorously and repeatedly tested and is never rejected, it may in time become part of a **theory**. A theory is an important principle supported by a large body of experimental evidence. You are no doubt familiar with many scientific theories, such as the theory of gravity. In this text we have already mentioned the cell theory, and we will encounter many biological theories such as the theory of biological evolution and the germ theory of disease. A theory can never be proven beyond all doubt, but a scientific theory is a powerful concept; a large body of evidence has accumulated to support it. Until proven otherwise, it remains the best explanation available to account for a given phenomenon. In other words, a theory has withstood the test of time. In everyday speech, however, a "theory" may refer to merely an idea or a guess that has not been tested or lacks any supporting evidence. Because the word has two very different meanings, it is often a source of great confusion among the general public. Indeed, it is common to hear nonscientists erroneously dismiss a fundamental scientific principle as "only a theory," when in fact it has the weight of scientific evidence behind it.

Coming Up Next...

Now that we have gained some insight into exactly what we will be studying, it is time to begin. In the next chapter we will learn some of the basic chemical principles that are crucial to a full understanding of microbiology. As we proceed through the fundamentals of biological chemistry, we will see that even when we are dealing with atoms, the basic units of all matter, the implications for microorganisms and the manner in which they interact with us are ever-present and often unexpected.

Key Terms

applied biology	hypothesis
archaeon	metabolism
bacterium	microbe
basic biology	microorganism
cell theory	model organism
cell	multicellular
control group	organelle
controlled variable	prion
deoxyribonucleic acid (DNA)	prokaryotic cell
eukaryotic cell	science
evolution	scientific method
experimental group	theory
experimental variable	unicellular
homeostasis	virus

Concept Questions

1. Exobiology is a new field of biology in which researchers try to determine what, if it exists, extraterrestrial life would be like. Exobiologists often study microorganisms in "extreme environments" (for example, environments that are extremely hot, acidic, salty), assuming that life on other planets might resemble these microorganisms. Frequently, such microorganisms are the only living thing in these environments. What is the rationale for studying these microorganisms if the goal is to understand what life on other planets might be like?

2. We frequently hear nonscientists refer to all microorganisms as "germs." Is it reasonable to lump all microorganisms together with this catch-all term? Why or why not?

3. Open a newspaper or magazine, or go to an online news site and find a story that in some way discusses bacteria, a virus, or another microorganism. Are the scientists whom the story describes engaged in basic or applied microbiology?

4. A drug manufacturer develops a new antiviral drug, which it hopes will be effective in preventing colds. First a group of 1000 volunteers is assembled. All are comparable in terms of age and general health. Of these volunteers, 500 (group A) are given the drug and told to take it as prescribed throughout the cold season. The remaining 500 individuals (group B) are given a placebo (a sugar pill that has no antiviral activity) to take. None of the volunteers know whether they are getting the drug or the placebo. After 5 months, all 1000 volunteers are asked whether or not they developed colds.
    a. What is the hypothesis being tested in this experiment?
    b. Which group is the experimental group? Which group is the control group?
    c. What is the experimental variable? What might some of the controlled variables be?
    d. What results might allow us to tentatively accept the hypothesis? What results would cause us to reject the hypothesis?

5. In a class you read an article that describes the prevailing theory as to why the dinosaurs went extinct. Data are provided to show that dinosaur eradication was caused by the impact of a gigantic meteorite that threw enormous quantities of dust into the atmosphere. This caused both a decline in the earth's temperature and a loss of photosynthetic vegetation, resulting in dinosaur extinction. Animals that were both better able to survive the cooler temperatures and less dependent on a huge diet of plant matter, such as primitive mammals, for example, survived. You tell a friend about this, and his response is "That's just a theory. I don't believe it." In what way is your friend misunderstanding you?

Enzymes from the heat-loving bacteria *Staphylothermus marinus* catalyze the chemical reactions needed for life.

Chapter 2

The Chemistry of Life

As we embark on our investigation of the microbial world, you may be asking yourself, "Why are we beginning with nonliving atoms and molecules?" In other words, what makes chemistry necessary if our goal is to understand microorganisms and the impact they have on us?

The answer is actually straightforward: all matter, living and nonliving, is composed of chemical compounds. The rules governing how these compounds form or interact are identical, whether we are discussing glaciers, global warming, geraniums, or giraffes. That means that if we really want to understand biological processes, some basic chemistry is essential. This may be particularly true in microbiology. Most of the organisms we will discuss in this text are too small to be seen with the unaided eye, and most of their impact is a result of chemical changes they cause in their environment. When we discuss the role microorganisms play in composting, how bacteria in your mouth can cause cavities, why certain intestinal bacteria are harmless while others can cause serious health problems, or the role specific microorganisms play in the production of sourdough bread or sauerkraut, we will really be considering aspects of microbial chemistry.

A comprehensive study of chemistry is well beyond the scope of this text. Rather, our goal in this chapter is to review those basic chemical principles necessary for us to fully appreciate how microorganisms interact with the world around them. We will begin by looking at atoms, the basic structural unit of matter, and then investigate how two or more atoms combine with each other to form molecules, the basic unit of compounds. One familiar molecule is water, and we will pay special attention to this remarkable substance, upon which all life is dependent. Finally, we will discuss the important biological molecules that are found in all living things, including microbes, and the roles they play in the "chemistry of life."

Atoms: the Basic Building Blocks of Matter

All matter in the universe is composed of one or more **elements**. An element is a substance that cannot be broken down into other substances by chemical reaction. There are approximately 92 naturally occurring elements known. Many, such as sulfur, neon, and calcium, are quite familiar to us. Others, such as lanthanum and vanadium, are more obscure. The most basic unit of an element is an **atom**. In other words, an atom is the smallest unit that can be identified as a specific element. We can speak, for instance, about an atom of magnesium, but if that atom is broken into its smaller component parts, it is no longer recognizable as magnesium.

Only about 25 elements are commonly found in microorganisms and other living things. The four most important are hydrogen, carbon, nitrogen, and oxygen, which make up approximately 96% of an organism's mass. The remaining 4% is composed of small amounts of other elements such as potassium and iron.

Atoms are made up of smaller components called subatomic particles

Atoms themselves are composed of still smaller subatomic particles. It is the number and kind of these particles in an atom that determine what kind of element it is and the types of chemical properties it possesses. Three of these subatomic particles, **electrons**, **neutrons**, and **protons**, are of particular importance. The smallest atoms, those of the element hydrogen, have only one proton, no neutrons, and one electron. An atom of uranium, on the other hand, is relatively large and is composed of 92 protons, 146 neutrons, and 92 electrons. Carbon typically has six protons, six neutrons, and six electrons. Nitrogen has seven each of these subatomic particles, while oxygen has eight.

The core of an atom, called the **nucleus**, is composed of protons and neutrons. All atoms can be identified by their characteristic **atomic number**, which corresponds to the number of protons in the nucleus. A proton carries a positive charge, while neutrons are electrically neutral. Electrons have negative charges and typically move around the nucleus at a specified distance in what is referred to as the electron's **shell** (**Figure 2.1**).

The protons and neutrons in the nucleus make up almost 100% of an atom's mass, and a given atom's **atomic mass** is simply the sum total of neutrons and protons. Thus, the atomic masses of hydrogen, carbon, nitrogen, and oxygen are typically 1, 12, 14, and 16, respectively (**Table 2.1**; also see Figure 2.1). Atoms on occasion can have different numbers of neutrons, which changes the atomic mass. For example, although carbon typically has six neutrons, sometimes it has seven or eight. These different forms of carbon have atomic masses of 12, 13, and 14, respectively. Atoms of the same element with different numbers of neutrons, and thus different atomic masses, are termed **isotopes** of the element. If, on the other hand, an atom gains or loses protons, it essentially becomes a different element, and it behaves in fundamentally different ways.

Recall that electrons have a negative charge. Consequently, the electrons are attracted to the positively charged protons in the nucleus. However, because electrons are in motion around the nucleus, and therefore have energy, they are not pulled into the nucleus. Electrons are found in characteristic "electron shells" at specific distances from the nucleus. Those electrons closest to the nucleus are held by the nucleus most tightly and have less energy than those electrons farther away from the nucleus. Each shell can hold only a specific number of electrons. In larger atoms with more

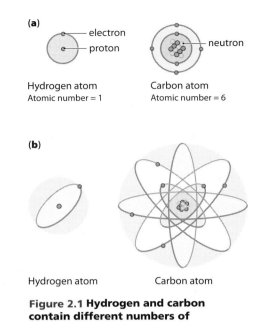

Figure 2.1 Hydrogen and carbon contain different numbers of subatomic particles. (a) An atom of hydrogen, with a single positively charged proton in its nucleus, has an atomic number of 1. Its atomic mass is also equal to 1. Carbon, with a nucleus consisting of six protons and six uncharged neutrons, has an atomic number of 6 and an atomic mass of 12. A complete atom consists of the tightly packed nucleus, composed of protons and neutrons, surrounded by orbiting, negatively charged electrons. The schematic representation is highly simplified.
(b) Electrons are actually located in specific *electron shells* surrounding the nucleus, and at any one time electrons may be in any of many possible locations.

ELEMENT	PROTONS	NEUTRONS	ELECTRONS	ATOMIC NO.	ATOMIC WT.
hydrogen (H)	1	0	1	1	1
carbon (C)	6	6	6	6	12
nitrogen (N)	7	7	7	7	14
oxygen (O)	8	8	8	8	16

Table 2.1 Atomic characteristics of elements common in living things.

electrons, additional shells are needed, and electrons in more distant shells are under less nuclear attraction. These more distant electrons consequently have more energy.

The first electron shell holds a maximum of two electrons. Hydrogen, with only one electron, has its single electron in this first shell (see Figure 2.1a). Helium, with two electrons, fills up this first shell, while lithium, having three electrons, has a full innermost shell and a single electron in the second shell. The second shell holds a maximum of eight electrons. Neon, with a total of 10 electrons, has a full first and second shell. The next larger atoms, sodium (11 electrons) and magnesium (12 electrons), start the process of filling up the third shell. The largest atoms yet known require up to eight shells to hold all their electrons.

As an atom's stability increases, its energy decreases

Energy always tends toward its lowest state. Consider the water in a river as it reaches a waterfall. As the water falls down, it loses energy; it cannot flow back up the waterfall on its own. It would have to be pumped back up through an input of mechanical energy. By moving down the waterfall, the water has moved from a higher to a lower energy state. Since it is easier for water to lose energy than it is to gain it, the lower energy state is less likely to change and is therefore considered to be more stable.

The same is true of atoms. They are more likely to interact with other atoms when they have more energy and are therefore less chemically stable. Atoms reach their lowest possible energy level, and thus their greatest chemical stability, when their outermost electron shell is full. When this shell is not full, an atom is **reactive**, meaning it will tend to gain or lose electrons (**Figure 2.2**). An atom tends to interact with other atoms and remain reactive until it either gains enough electrons to fill the outermost shell or loses enough outer electrons to lose its outer shell entirely. In either manner, an atom becomes stable and will no longer interact with other atoms.

Sodium, for instance, with 11 electrons, typically has a full inner shell (two electrons), a full second shell (eight electrons), and one electron in the third shell. In this uncombined state, sodium is electrically neutral because the 11 negative electrons are perfectly balanced by the 11 positively charged protons in the nucleus (**Figure 2.3**). Sodium is, however, reactive in this

Atomic number	Element	Energy level (electron shell)			
		I	II	III	IV
1	Hydrogen	O			
2	Helium	OO			
6	Carbon	OO	OOOO		
7	Nitrogen	OO	OOOOO		
8	Oxygen	OO	OOOOOO		
10	Neon	OO	OOOOOOOO		
11	Sodium	OO	OOOOOOOO	O	
12	Magnesium	OO	OOOOOOOO	OO	
15	Phosphorus	OO	OOOOOOOO	OOOOO	
16	Sulfur	OO	OOOOOOOO	OOOOOO	
17	Chlorine	OO	OOOOOOOO	OOOOOOO	
18	Argon	OO	OOOOOOOO	OOOOOOOO	
19	Potassium	OO	OOOOOOOO	OOOOOOOO	O
20	Calcium	OO	OOOOOOOO	OOOOOOOO	OO

Figure 2.2 Shell configurations of some atoms. The number of electrons in an atom's outermost shell determines its chemical reactivity. The red circles indicate electrons in unfilled outermost electron shells. Atoms with such electrons are considered chemically reactive. Electrons in filled shells are indicated by blue circles. Atoms with no reactive outer electrons, such as helium, neon, and argon, are considered nonreactive. Notice that the four elements most common in living organisms—hydrogen, carbon, nitrogen, and oxygen—are all chemically reactive.

form, and because of its lack of stability it tends to lose its outermost electron. If this occurs, sodium becomes stable because its 10 electrons fill both the first and second shells. But it is no longer electrically neutral. Because it has lost a negatively charged electron, there are now 11 protons but only 10 electrons, giving sodium a charge of +1. Charged yet stable atoms are called **ions**. Calcium, in its reactive state, has two electrons in its outermost shell. To achieve stability, a calcium atom loses both of these electrons. Because it now has two more protons than it has electrons, the charge on a stable calcium ion is +2. Fluorine, on the other hand, has seven outer electrons in a shell that can hold eight. Fluorine can become stable by gaining a single electron, and in the process it becomes an ion with a charge of –1. Note that some atoms such as helium and neon are stable without gaining or losing electrons, because their outermost shells are already filled (see Figures 2.2 and 2.3). Such atoms are already at maximum stability and will not form ions. Because they are stable, they do not interact with other atoms and are called the **inert gases**.

Perhaps the most important property of an atom is the manner in which it interacts or combines with other atoms. When two or more atoms combine, the linkage between them is called a **chemical bond**. An atom's electrons are of critical importance in determining the types of bonds an atom can form and with which other atoms it can form them. Not all bonds form in exactly the same way, and we will next inspect the principal types of bonds that atoms may form with each other.

An ionic bond is formed when electrons are transferred from one atom to another

CASE: BACTERIA, SALT, AND CYSTIC FIBROSIS

Cystic fibrosis is an inherited disease affecting one in 2400 Americans of European descent. In normal individuals, chlorine atoms, which exist as negatively charged chloride ions, can easily cross the membrane of lung cells because these cells have special chlorine-transporting proteins on their surfaces. Many different mutations can affect these transport proteins, and in many individuals with cystic fibrosis these transporters do not function properly. Chlorine is therefore unable to enter cells efficiently. Prevented from entering lung cells, the ions readily bind to sodium ions, which carry a positive charge. The combined chlorine and sodium form the salt sodium chloride. The high levels of salt are part of the reason that bacteria proliferate in the lungs of cystic fibrosis sufferers. Normally, cells lining the lung produce a natural protein antibiotic that protects the lungs from bacterial colonization. However, the high salt concentration in the lungs of cystic fibrosis patients inactivates the protein, allowing the bacteria to proliferate (Figure 2.4).

1. **Why do chlorine atoms form negatively charged ions, while sodium atoms form positively charged ions?**
2. **What exactly is a salt? Why do these charged atoms combine to form salt in the manner described?**
3. **How does the salt interfere with the antibacterial protein in the lungs of cystic fibrosis patients and thus increase the likelihood of bacterial infections?**

If atoms become stable simply by gaining or losing electrons, why don't they just go ahead and gain or lose them, and why should we ever encounter unstable, uncharged atoms in nature? An atom will not ordinarily lose electrons unless another atom that tends to gain electrons is nearby. The atom that requires an electron to achieve stability literally takes the electron from the atom that tends to lose electrons, forming both a positive and a negative ion in the process. In other words, atoms typically do not spontaneously gain or lose electrons. They do so because they are in the vicinity of other

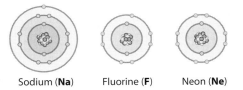

Sodium (**Na**)   Fluorine (**F**)   Neon (**Ne**)

Figure 2.3 Atoms of three elements. A sodium atom becomes stable by losing one electron from its outermost shell. Once this negatively charged electron is lost, the sodium atom becomes a positively charged ion. A fluorine atom becomes stable by gaining an electron to complete its outermost shell. By gaining an electron, the fluorine atom becomes a negatively charged ion. A neon atom with a complete outermost shell neither gains nor loses electrons; it is considered nonreactive and is an example of an inert gas.

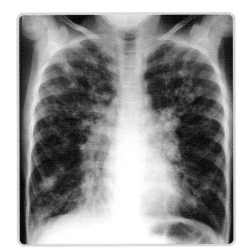

Figure 2.4 Chest X-ray of a cystic fibrosis patient. Light-colored areas in the lungs indicate areas of bacterial infection. Cystic fibrosis patients are often plagued by salty, thick mucus in the lining of the lungs and associated air passages that is difficult to cough up.

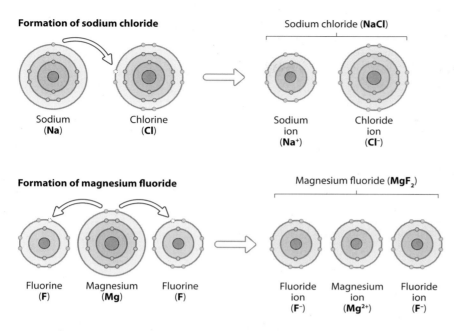

Formation of sodium chloride

Sodium chloride (**NaCl**)

Sodium
(**Na**)

Chlorine
(**Cl**)

Sodium
ion
(**Na⁺**)

Chloride
ion
(**Cl⁻**)

Formation of magnesium fluoride

Magnesium fluoride (**MgF₂**)

Fluorine
(**F**)

Magnesium
(**Mg**)

Fluorine
(**F**)

Fluoride
ion
(**F⁻**)

Magnesium
ion
(**Mg²⁺**)

Fluoride
ion
(**F⁻**)

Figure 2.5 Ionic bonding. Ionic bonds form between negative and positive ions, forming salts. The red arrows in this figure indicate the transfer of electrons from one atom to another. The atom losing one or more electrons becomes a positively charged ion, while the atom gaining one or more electrons becomes a negatively charged ion. Ions with opposing charges attract each other, forming an ionic bond. (a) Sodium chloride. (b) Magnesium achieves stability by losing its two outermost electrons, while each fluorine becomes stable by gaining one electron. Magnesium consequently transfers one electron to each of two fluorine atoms. Magnesium forms a +2 ion, while each fluorine forms a −1 ion, and the positively charged magnesium is attracted to the two negatively charged fluorines, forming the magnesium fluoride salt.

atoms that attract their electrons more strongly or weakly than they do. The degree to which an atom attracts electrons is described as that atom's electron **affinity**. Atoms that attract electrons strongly are said to have high electron affinity. Those that attract electrons weakly have low affinity.

When one or more electrons are transferred in this manner, the positive and negative ions attract each other by virtue of their opposite charges. Such an attraction, called an **ionic bond**, holds the two ions together, forming an ionic compound, also known as a **salt**. The salt in a cystic fibrosis patient's lungs is formed in just this way. Sodium (chemical symbol Na) has one outermost electron, which it can lose to chlorine (chemical symbol Cl) with seven outer electrons. Sodium thus forms a +1 ion (Na^+), while chlorine forms a −1 ion (Cl^-). The opposing charges hold the two atoms together. The ionic compound formed, sodium chloride (Na^+Cl^-), is shown in **Figure 2.5**. Note that even though each atom in the compound is charged, the negative and positive charges balance each other exactly, making the salt electrically neutral. In the case of cystic fibrosis, since the chlorine ions are unable to enter the cells, they combine with the sodium in this manner.

Sodium chloride is the chemical name for common table salt, one crystal of which contains many sodium and chlorine ions packed together. Sodium chloride is just one type of salt. Magnesium fluoride is a salt containing magnesium and fluorine atoms ionically bonded. Magnesium must lose two electrons from its outermost shell, while fluorine needs to gain one electron to complete its outermost shell in order to become stable. Therefore, in the presence of two fluorine atoms, magnesium forms ionic bonds with both negatively charged fluorines. Having lost two electrons, magnesium converts into a +2 ion (Mg^{2+}). Each fluorine forms a −1 ion, and the salt MgF_2 (one atom of magnesium bound to two fluorine atoms) is formed. Again, the +2 charge on the magnesium and the two −1 charges on the fluorine atoms balance exactly, meaning that the salt has an overall neutral charge (see Figure 2.5).

Now that we know why sodium chloride forms in the lungs of cystic fibrosis patients, we can explain why such patients often suffer from bacterial lung infections. Cells lining the lungs produce a protein that acts as a natural antibiotic, protecting the lungs from bacterial colonization. This compound, however, only works efficiently in relatively low-salt environments. In the lungs of cystic fibrosis patients, the salt tends to interfere with and inactivate the protein, opening the door to infection. Interestingly, the faulty chlorine transporters that result in salt formation in the lungs of cystic fibrosis patients are found in other parts of the body as well. This explains why the saliva and perspiration of people with this disease are also unusually salty.

Covalent bonds form when atoms share electrons in their outermost shells

Ionic bonds are not the only way atoms can interact. Consider two reactive atoms, both of which tend to achieve chemical stability by gaining rather than losing electrons. In this case, neither atom is likely to completely donate an electron to the other. The two atoms may, however, obtain a full outer shell, and thus become stable, by sharing outer electrons. As an example, consider hydrogen, the most abundant element on Earth. Hydrogen rarely exists as individual H atoms. A hydrogen atom can, however, bond to another hydrogen atom to form hydrogen gas (H_2).

Recall that each hydrogen atom has only a single electron in its first electron shell, which holds a maximum of two electrons. When two hydrogen atoms come in contact, since their ability to attract electrons is exactly the same (that is, they have the same affinity), neither can take an electron from the other. Rather, the two hydrogen atoms share their electrons, providing each atom with the two outer electrons required to achieve stability. When two atoms share electrons in their outer shells in this way, they have formed a **covalent bond** (**Figure 2.6**).

Hydrogen gas is actually rare, because hydrogen forms covalent bonds with other types of atoms more readily than it does with itself. An important example is water, the molecular formula of which is H_2O. A reactive oxygen atom has six outer electrons in a shell that holds a maximum of eight. Oxygen thus needs two additional electrons to become stable, and it can obtain them by sharing two electrons—one with each of two hydrogens (see Figure 2.6b). Each water molecule is composed of three atoms, while each molecule of hydrogen gas is composed of two atoms.

Hydrogen also commonly bonds to carbon, forming an important class of molecules called **hydrocarbons**. The simplest hydrocarbon is methane, which is composed of one carbon atom and four hydrogens (CH_4). Carbon, with four outer electrons, can form four covalent bonds by sharing each of these electrons with hydrogen (see Figure 2.6d). Methane is an important "greenhouse gas" that contributes to global warming. Because much of the methane in our atmosphere is a bacterial waste product, we will discuss this gas more fully in Chapter 9, where we investigate the role that microbes play in environmental processes.

Two atoms may form more than one covalent bond with each other in certain cases. For instance, atmospheric oxygen generally is in the form of O_2. Each oxygen atom has six outer electrons in its second shell. Both atoms require an additional two electrons to become stable. Consequently, they will share a total of four electrons, resulting in two covalent bonds. In carbon dioxide (CO_2), recall that each carbon atom can form four covalent bonds, while each oxygen can form two. The single carbon thus forms two bonds with each oxygen.

Covalent bonds can be classified as either polar or nonpolar

You are no doubt familiar with the word "polar." The Earth with its north and south poles and a magnet with a positively and negatively charged ends are common examples of polar objects because they are different at their two ends. Likewise, some covalent bonds are termed **polar covalent bonds**,

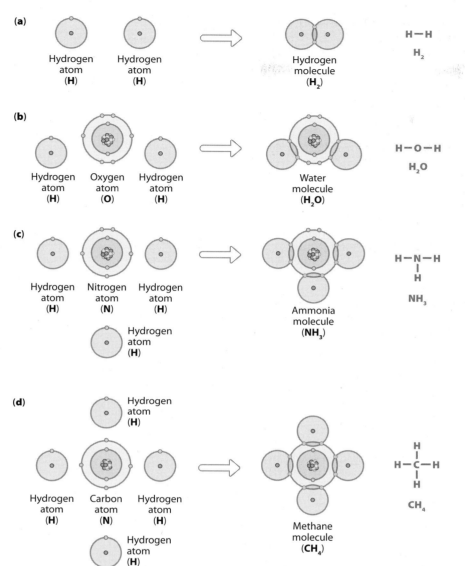

(a)

Hydrogen atom **(H)**   Hydrogen atom **(H)**   →   Hydrogen molecule **(H₂)**

H—H

H₂

(b)

Hydrogen atom **(H)**   Oxygen atom **(O)**   Hydrogen atom **(H)**   →   Water molecule **(H₂O)**

H—O—H

H₂O

(c)

Hydrogen atom **(H)**   Nitrogen atom **(N)**   Hydrogen atom **(H)**
Hydrogen atom **(H)**   →   Ammonia molecule **(NH₃)**

H—N—H
    |
    H

NH₃

(d)

Hydrogen atom **(H)**
Hydrogen atom **(H)**   Carbon atom **(N)**   Hydrogen atom **(H)**
Hydrogen atom **(H)**   →   Methane molecule **(CH₄)**

H
|
H—C—H
|
H

CH₄

Figure 2.6 Covalent bonding. Covalent bonds form between reactive atoms that share electrons. Note that in all examples the shared electrons (in blue) are in the outermost shells of the atoms involved. Unshared electrons are in gray. The nucleus is depicted in the center of each atom. In the simplified stick models next to each atomic model, each line represents a shared electron pair. (a) Each hydrogen atom needs one additional electron to achieve stability. To gain the necessary electron, the two atoms share their electrons to form a molecule of stable hydrogen gas (H₂). (b) Oxygen has a full inner shell, but in its second shell, which can hold eight electrons, there are only six. Oxygen can become stable by sharing two outer electrons with two atoms of hydrogen, forming a molecule of water (H₂O). (c) Nitrogen requires three outer electrons to fill its outer shell, and it may obtain these by sharing three electrons with three hydrogen atoms, forming ammonia (NH₃). (d) Carbon, with four outer electrons, can form four covalent bonds with hydrogen, forming methane (CH₄).

because the combination of participating atoms carries a slight positive charge at one end and a negative charge at the other. Other covalent bonds lack these opposing charges at the opposite ends. In other words, in terms of charge, the two ends are both neutral. Such bonds are called **nonpolar covalent bonds**.

Why do some covalently bound molecules end up with slight opposing charges at opposite ends while others do not? To explain, consider the examples of hydrogen gas (H₂), in which the covalent bond is nonpolar, and ammonia (NH₃), in which the single nitrogen atom is bound to three hydrogens by three polar covalent bonds.

When two hydrogen atoms covalently bond, forming a molecule of hydrogen gas, each atom has an identical affinity for the shared electrons. In other words, the electrons are attracted equally to the nucleus of each hydrogen atom, and they are shared equally. This is not the case, however when nitrogen binds hydrogen to form NH₃. Remember that protons in the nucleus attract the negatively charged electrons. Nitrogen has seven protons in its nucleus, while hydrogen has only one. Thus when nitrogen forms a covalent bond with hydrogen, the electrons are actually more attracted to the nitrogen than they are to the hydrogen. The difference in attraction is slight. Even though hydrogen has only a single proton to attract its shared electrons, those electrons are in hydrogen's first shell, and thus closer to the nucleus than they are in nitrogen. Thus nitrogen's edge in number of positively charged protons is to some degree offset by the closeness of hydrogen's nucleus to its electrons. Nevertheless, the difference in attraction is large enough to ensure that the shared electrons are *not* equally shared. The nitrogen tends to pull the electrons away from hydrogen and toward itself. Because the electrons are negatively charged and tend to cluster closer to nitrogen, the nitrogen atom gains a very slight, partial negative charge. At the same time, the three hydrogens to which nitrogen is bound obtain partial positive charges, because the negatively charged electrons are pulled away from them and toward the nitrogen.

This gives ammonia a *polarity*, meaning that there is a difference in electrical charge at each of its ends. The covalent bonds holding nitrogen to each hydrogen are therefore polar covalent bonds (**Figure 2.7**). The covalent bonds between hydrogen and oxygen in a molecule of water are similarly polar. When there is no charge difference at either end of a molecule, as is the case with H_2 gas, the result is a nonpolar covalent bond. Oxygen gas (O_2) and the carbon–hydrogen bonds found in hydrocarbons such as methane are also nonpolar. This may *sound* like a fairly trivial point, but in reality the consequences of this are enormous. As we will see, molecules that are composed of mostly polar covalent bonds and those that are composed of mostly nonpolar covalent bonds behave very differently.

Water is an important example of a polar molecule (see Figure 2.7). The partial positive charges on the hydrogen atoms of one molecule will be attracted to the partial negative charge on the oxygen of a different water molecule. This means that all water molecules are attracted to each other because of opposing partial charges. Such interactions between separate molecules are termed **hydrogen bonds** (**Figure 2.8**). Each hydrogen bond is very weak, but collectively, many hydrogen bonds can exert considerable attraction. For instance, when water is boiled, individual water molecules will not break off and leave solution as vapor until enough heat is applied to begin disruption of the hydrogen bonds holding the water molecules together.

It is not just water molecules that attract each other. Ammonia, as we have seen, has a partial negative charge on nitrogen and a partial positive charge on each hydrogen. Ammonia will form hydrogen bonds not only with water but also with other ammonia molecules as well. Nonpolar compounds, on the other hand, which lack partial charges, are unable to form hydrogen bonds with polar compounds such as water.

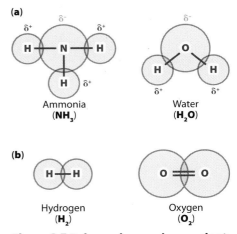

Figure 2.7 Polar and nonpolar covalent bonds. (a) Ammonia and water molecules are held together by polar covalent bonds. Because nitrogen in ammonia or oxygen in water attracts electrons more strongly than hydrogen, nitrogen and oxygen both obtain partial negative charges, represented by δ⁻. The hydrogens in these molecules, with less attraction (lower affinity) for the electrons, each obtain a partial positive (δ⁺) charge. (b) In hydrogen and oxygen molecules, electrons are shared equally. Because there are no partial charges, the covalent bonds holding the atoms together are nonpolar covalent bonds.

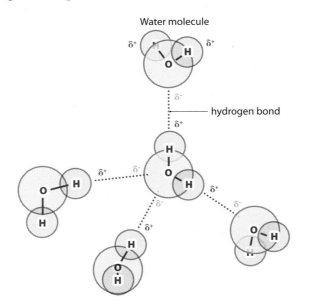

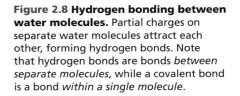

Figure 2.8 Hydrogen bonding between water molecules. Partial charges on separate water molecules attract each other, forming hydrogen bonds. Note that hydrogen bonds are bonds *between separate molecules*, while a covalent bond is a bond *within a single molecule*.

The Importance of Water

Water is fundamental to all life. As we will discuss in Chapter 8, life itself evolved in water, and all biochemical reactions take place in an aqueous or water environment. Environments that are totally lacking in water will be devoid of living things. If water is present, it is likely that life will be present as well. For instance, researchers have found microorganisms in pockets of liquid water over 10 meters deep in frozen Antarctic lakes, in boiling hot oceanic volcanic vents, and even in water vapor found in clouds.

The ability or inability of a compound to interact with water is a crucial characteristic of that compound. We have already seen that polar compounds can interact with water. Nonpolar molecules cannot because, lacking partial charges, they are unable to form hydrogen bonds with water. A simple example is ordinary "table sugar," also called sucrose ($C_{12}H_{22}O_{11}$). Sucrose is largely polar in nature. When you place a teaspoon of sugar in water, the individual sugar molecules form hydrogen bonds with the water and disperse through the solution; the sugar *dissolves*. When you drop a piece of fat into water, on the other hand, nothing happens. Large parts of fat molecules are pure hydrocarbons, which are nonpolar and thus unable to dissolve in water.

It is not only polar compounds that will dissolve in water; anything with electrical charges will be attracted to water's partial positive and negative charges. For instance, when you place an ionically bonded compound such as table salt (NaCl) into water, positively charged sodium ions are attracted to the partial negative charges on the oxygen in water molecules. Negatively charged chlorine ions, on the other hand, are attracted by the partial positive charges on the hydrogens. Ultimately these attractive forces pull apart the ionic compound, separating it into individual ions. Hence, the salt you place in water also dissolves. Consider the odd fact that even in "saltwater" there is little or no actual salt; it is largely dissolved into individual ions. It is only when water is removed by evaporation, for instance, that the positive and negative ions reassociate to form salt.

Polar covalent and ionic compounds that interact well with water in this fashion are termed **hydrophilic**, from the Greek term for "water-loving." Nonpolar covalent compounds that cannot dissolve in water are referred to as **hydrophobic** or "water-fearing" (**Figure 2.9**). While molecules such as salt or oil are purely hydrophilic or hydrophobic, respectively, many molecules have both hydrophilic and hydrophobic regions and will consequently interact with water to greater or lesser degrees.

Acids and bases are compounds that increase or decrease the concentration of protons in water

CASE: FISHLESS IN PENNSYLVANIA

Commercial mining operations can often wreak havoc on nearby streams and rivers. A case in point is the Shamokin Creek Basin in east-central Pennsylvania. Shamokin Creek drains an area of approximately 350 km^2 and at one time was considered to be prime fishing habitat. The area is also rich in anthracite coal, which was mined extensively in the area from about 1840 to 1950. In 1999, surveys were conducted to determine the ecological health of Shamokin Creek, as well as other streams in the area. Certain stretches of the creek were found to be devoid of all fish, with pH levels as low as 4.2 (**Figure 2.10**). Other indicators of ecological health such as populations of aquatic plants and invertebrates were also absent or greatly decreased.

1. **How do mines adversely affect the ecology of streams such as Shamokin Creek?**
2. **What exactly does a pH value of 4.2 mean?**
3. **What role do microorganisms play in the degradation of streams caused by mining?**

(a)

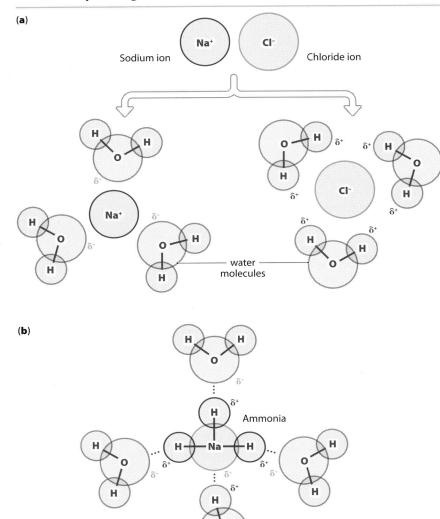

(b)

(c)

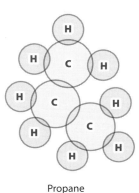 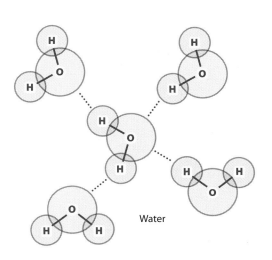

Figure 2.9 Solubility of molecules in water. Molecules with charges interact well with water and are therefore water-soluble. Such molecules are considered *hydrophilic*. Molecules without charges are unable to dissolve in water and are consequently water-insoluble and *hydrophobic*. (a) Ionic compounds such as sodium chloride are hydrophilic because the charges on each ion readily interact with the partial charges on water molecules. (b) Similarly, polar covalent molecules such as ammonia have charges that interact well with the charges found on water molecules. Polar covalent molecules are therefore also hydrophilic. (c) Alternatively, propane is an example of a nonpolar covalent molecule. Because it has no charges, it is unable to interact with water and is therefore hydrophobic.

Figure 2.10 Damage to a stream resulting from mining. This stream has been badly acidified through the activity of bacteria. Mining activity releases sulfur compounds into the environment, and certain bacteria thrive on these compounds. Acids are released as a waste product of bacterial metabolism, and as bacterial numbers explode, nearby streams become increasingly acidic.

As the example in our case illustrates, mines can be deathtraps as far as fish living in nearby streams are concerned. The culprits in this case are bacteria known as *Acidithiobacillus ferrooxidans*, which have profited from the manner in which mine waste was disposed of. Before we can fully understand, however, how mining and microorganisms conspire to harm freshwater ecosystems, we need to learn a little bit about acids and bases.

Recall that, in water, all atoms carry partial charges. Occasionally, the combined strength of attraction between oxygen and hydrogen on different water molecules actually pulls a water molecule apart. Oxygen atoms with partial negative charges can sometimes pull a hydrogen atom's proton off its molecule, leaving the shared electron pair behind with the remaining portion of the water molecule. This remnant of water, because it now has both of the previously shared electrons, picks up a negative charge and is termed a *hydroxide* or OH^- ion. By the same token, the proton that was pulled off into solution is a positive H^+ ion. Note that in pure water, every time an OH^- is formed, an H^+ is also generated. Thus, the concentration of these two ions stays exactly equal. The reaction is considered reversible, because an OH^- and an H^+ might easily attract each other and recombine to re-form H_2O. Protons are not pulled off water in this manner very often; it is estimated that in pure water, only one molecule in 554 million would be found in this **ionized state**.

Pure water, however, is exceedingly rare. Water usually exists as a **solution**. A solution is composed of a liquid (the **solvent**) and whatever is dissolved in it (the **solute**). Remember that for something to dissolve in water, it must be hydrophilic in nature. Many hydrophilic substances, when placed in water, form numerous hydrogen bonds with water molecules and might on occasion lose a proton into the solution. In this case, the concentration of H^+ is now higher than the concentration of OH^-.

Anything that raises the concentration of H^+ in this manner is termed an **acid**. As an example, consider hydrochloric acid (HCl). Hydrochloric acid is an ionic compound. Hydrogen loses an electron to chlorine, and the attraction between the positive hydrogen and negative chlorine forms the ionic bond. However, when placed in water, the positively charged hydrogen is readily pulled off into solution by the oxygens in water molecules, which each carry a partial negative charge. When the acid dissociates in this manner, negative chlorine ions are left in solution (**Figure 2.11a**).

In contrast to an acid, a **base** is any compound that *lowers* the concentration of H^+ ions in solution. For example, sodium hydroxide (NaOH) is a compound consisting of a positive sodium ion (Na^+) ionically bound to a negative hydroxide ion (OH^-). When placed into water, sodium hydroxide

(a)

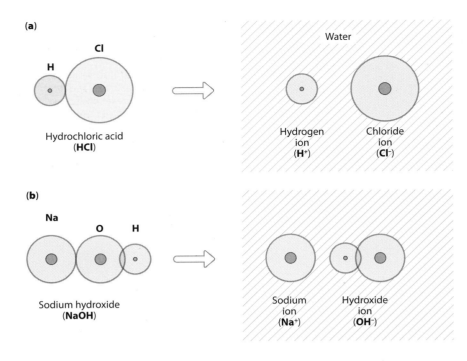

(b)

Figure 2.11 Concentration of H⁺ ions in acid and base solution. (a) If hydrochloric acid is added to water, it will dissociate into H⁺ and Cl⁻ ions, raising the concentration of H⁺ ions in solution. (b) When a base such as NaOH is placed in water, it will dissociate into Na⁺ and OH⁻ ions. Because OH⁻ will readily combine with free H⁺ ions to form water, the concentration of H⁺ in solution decreases.

dissociates into Na^+ and OH^-. The negatively charged hydroxide ions readily bind to any H^+ ions that may be present in solution and thus lower the acidity by essentially acting as proton sponges (see Figure 2.11b).

As you probably already know, acids and bases are corrosive and damaging to living tissue. Hydrogen ions are simply naked protons, and these positively charged particles are extremely reactive. They will essentially latch onto anything with a negative charge, and in doing so, they disrupt the structure of molecules to which they attach. Bases are also highly reactive (bleach, for example). Because of their high affinity for hydrogen ions, they tend to remove protons from various molecules, damaging their structure. Streams close to old mines are frequently devoid of life because of acids produced by bacteria. When ores are dug up, sulfur-containing compounds are often exposed to oxygen. This provides ideal conditions for bacteria such as *A. ferrooxidans*. Not only do they now have the oxygen they require, but this bacterial species uses sulfur compounds as an energy source. In the course of breaking down these sulfur-containing molecules, they release sulfuric acid as a waste. When sulfuric acid (H_2SO_4) enters the water, H^+ ions are pulled off into solution, leaving HSO_4^- behind. Over time, as the microbes continue to digest compounds in the mine tailings, the water becomes increasingly acidic until it can no longer support plant and animal life. As we will see, such acids are often waste products of bacterial metabolism, and their release into the environment explains processes as diverse as tooth decay and the production of yogurt.

To measure the acidity of any solution we rely on the **pH scale**. This scale provides a way to exactly quantify the concentration of H^+ ions. The scale runs from 1 to 14, with a pH of 7 indicating that the solution is neutral because the concentration of H^+ is equal to that of OH^-. As H^+ ions are released into solution by an acid, the value on the pH scale drops below 7, whereas the pH value is greater than 7 when the concentration of H^+ is less than that of OH^-. The pH scale is logarithmic, meaning that a solution of pH 6 is actually 10 times more acidic than a solution of pH 7. Healthy freshwater streams typically have a pH close to 7. The fact that parts of Shamokin Creek had pH readings of 4.2 highlights the severity of the problem. **Figure 2.12** provides the pH for some common substances.

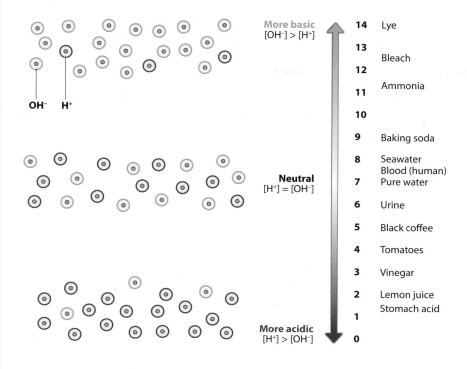

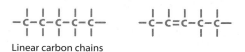

Figure 2.12 The pH scale, showing the pH of several common substances. A pH of 7 indicates that the concentrations of H$^+$ and OH$^-$ are equal. A pH greater than 7 indicates that the concentration of OH$^-$ exceeds that of H$^+$. A pH less than 7 means that the concentration of H$^+$ is greater than that of OH$^-$.

Organic Molecules: the Building Blocks of Life

Living things can be distinguished from inanimate matter by the presence of **organic** molecules. Such molecules are largely composed of carbon atoms bonded to one another and to atoms of other elements. Carbon, with four outer electrons, is able to form four stable covalent bonds. Carbon atoms therefore can link together to form complex chain and ring structures.

Carbon is unique in its ability to form such structures. The covalent bonds formed by carbon are very strong because the shared electron pairs are only in the second electron shell and therefore close to the positively charged nucleus. Because of the relative strength of these bonds, carbon compounds are quite stable in the environment and are thus ideal for serving as life's building blocks. As we will see in this section, all important biological molecules are based on modifications of hydrocarbon compounds. Several representative hydrocarbon structures are illustrated in **Figure 2.13**.

Silicon also has four outermost electrons and can also form four covalent bonds. You might consequently conclude that silicon compounds are an equally good choice as a backbone for biological molecules. Silicon, however, has its outermost electrons in the third electron shell. Because these electrons are further from the nucleus and therefore at higher energy, any covalent bonds formed by silicon will be inherently less stable than those formed by carbon. Because they are less stable, long chains of silicon cannot persist in the environment. In spite of this fact, science fiction writers and filmmakers sometimes describe their aliens as "silicon-based life forms," as opposed to the carbon-based life forms on our planet, but such beings could exist only in the realm of fantasy, because the chemistry just does not work.

Organic molecules differ in other ways besides their number of carbon and hydrogen atoms. In many cases, one or more hydrogen atoms of a hydrocarbon molecule are removed and replaced by a different atom or group of atoms. For example, the hydrocarbon ethane (C_2H_6) can be converted into ethyl alcohol (C_2H_5OH), also known as ethanol, by replacing one of the

Linear carbon chains

Branched carbon chain

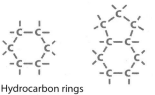

Hydrocarbon rings

Figure 2.13 Some hydrocarbon structures. Each carbon forms four covalent bonds, while each hydrogen forms one. Carbon has unique structural flexibility because of its ability to form long chains and other structures. Carbon can form single or double bonds with other carbons and may form linear chains, branched chains, or simple or complex ring structures. Each dash indicates a covalent bond between carbon and hydrogen.

hydrogens with an OH group (**Figure 2.14**). An OH group is known as a **hydroxyl** group. Ethane is nonpolar and hence hydrophobic, but when a hydrogen is replaced with a very polar hydroxyl group, the molecule becomes hydrophilic.

Four types of such "modified hydrocarbons" are of particular importance in microorganisms and other living things: **carbohydrates**, **lipids**, **proteins**, and **nucleic acids**. Due to the manner in which they are modified, these four classes of biological molecules have certain characteristic properties and are used by living things for specific purposes. Carbohydrates are important structural components of cells and are also the primary means by which cells store energy. Lipids also have both structural and energy storage functions. Proteins are used for a wide variety of tasks. They are important structural components of living things, and they are absolutely critical as biological catalysts called *enzymes* that are required to carry out specific biological reactions. Proteins also help transport materials into and out of cells, and they serve as important chemical messengers that cells use to communicate with each other. Proteins play a key role as antibodies in our defense against disease-causing microorganisms, and as we will shortly see, some microorganisms cause disease by releasing proteins that act as dangerous toxins. Nucleic acids determine which proteins a particular cell can make. Some nucleic acids contain information, analogous to a cookbook, containing "recipes" for various proteins. Other nucleic acids function as the "cook," using the recipes to construct the specified proteins. In the remainder of this chapter we will consider each of these four types of biological molecules.

Carbohydrates function as energy storage and structural molecules

CASE: BACTERIAL HORSE HELPERS

Cyrus is a 16 year old Arabian stallion who lives on a grass-filled pasture in Virginia. Much of his diet consists of the rich Kentucky bluegrass regularly available to him. When he suddenly loses his appetite, his owner becomes alarmed, especially after noticing that Cyrus' stools are looser and darker than normal. He calls the veterinarian, who suspects bacterial diarrhea. The veterinarian prescribes a course of ampicillin for Cyrus. He also provides the owner with a tube of Benebac® with instructions that Cyrus should be given a daily dose while he is on antibiotics. Although the bacteria that cause equine diarrhea are often resistant to ampicillin, Cyrus responds well to the treatment, and recovers fully without complications.

Benebac® is a veterinary preparation, commonly used in situations such as this. It consists of a paste containing live cultures of various bacterial species, considered to be "beneficial bacteria" for horses and other herbivores.
1. **What is meant by "beneficial bacteria"?**
2. **Why would the veterinarian recommend Benebac® for Cyrus, specifically while he is being treated for bacterial diarrhea?**

Sugars, starches, and cellulose are familiar examples of carbohydrates. This group of molecules is principally composed of carbon, hydrogen, and oxygen. Carbohydrates are important structural and energy-storage compounds in microorganisms and other living things. An important characteristic of sugars is their ability to dissolve in water. When you put table sugar into a cup of coffee, the sugar dissolves because the many polar regions on the sugar molecules form hydrogen bonds with the water in the coffee. This causes the individual sugar molecules to disperse throughout the solution.

The size of carbohydrate molecules spans the gamut from small to very large. The simplest carbohydrates are composed of a single sugar molecule

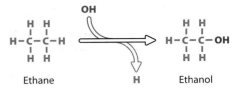

Figure 2.14 Substitution in hydrocarbon molecules. Ethane is composed entirely of nonpolar covalent bonds. It is therefore hydrophobic and unable to dissolve in water. If, however, a hydrogen in ethane is replaced by a hydroxyl (OH) group, the result is ethanol, which is not a hydrocarbon but one of a class of compounds called *alcohols*. Because the bond between O and H in the hydroxyl group is a polar covalent bond, each of these two atoms obtains a partial charge, allowing it to form hydrogen bonds with water. The molecule is now hydrophilic and can dissolve in water.

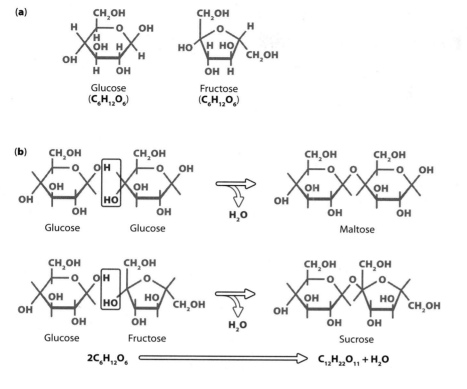

Figure 2.15 Monosaccharides and disaccharides. (a) Glucose and fructose both have the same chemical formula ($C_6H_{12}O_6$) but different structures. They are therefore *isomers* of each other.
(b) Two monosaccharides can be linked via condensation to form a disaccharide. In each condensation reaction an OH is removed from one sugar and an H is removed from the other sugar. The OH and H join together, forming a molecule of water.

and are termed **monosaccharides**. A given monosaccharide may contain from three to six carbons. Many important sugars, such as glucose and fructose, are six-carbon sugars. Both of these monosaccharides have the chemical formula $C_6H_{12}O_6$, but the manner in which these atoms are assembled is slightly different in each one. Molecules such as these, having identical chemical formulas but different structures, are known as **isomers** (**Figure 2.15a**).

Two monosaccharides can be covalently linked together to form a **disaccharide**. The reaction that forms the bond is called a **condensation reaction** (see Figure 2.15b). In such a reaction a molecule of water is removed as a bond is formed between the two sugars. Two glucose molecules can be joined in this way to form maltose (malt sugar), while a glucose and a fructose can form the disaccharide sucrose (ordinary table sugar). Lactose (milk sugar) is the disaccharide formed by joining glucose with a different six-carbon sugar called galactose.

Condensation reactions are reversible. In other words, a disaccharide can be split back into monosaccharides by reinserting a molecule of water. Such a reaction is termed a **hydrolysis reaction**.

To either remove or replace water to either form or break apart a disaccharide, a specific biological catalyst is required. Such catalysts are called **enzymes**. Enzymes are chiefly proteins (there are a few exceptions), and each enzyme is highly specific, participating in only one or a few reactions. The enzyme that acts on sucrose is known as *sucrase* and it participates only in this reaction. Bacteria capable of synthesizing lactose-digesting enzymes are able to break lactose into monosaccharides, which can then be used for energy. Other bacteria, lacking these enzymes, are not able to utilize this sugar. When dairy products (like milk) that contain lactose become spoiled, it is generally because lactose-digesting bacteria have broken down the disaccharide. The waste product of this process, lactic acid, gives the milk its sour taste.

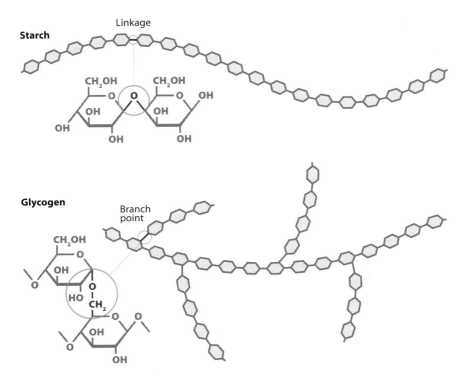

Figure 2.16 Representative polysaccharides. Both starch and glycogen are long polysaccharides composed entirely of glucose. The only difference is the extensive branching typical of glycogen.

Many carbohydrates consist of hundreds or thousands of monosaccharides, bonded together to form **polysaccharides**. Starch, for instance, is a polysaccharide consisting of many glucose molecules. Starch is the primary molecule used by plants for energy storage and is present in many foods that are derived from plants. When an animal eats a food containing starch, the carbohydrate is digested via hydrolysis. The fact that animals, including humans, can digest starch indicates that we have the specific enzymes necessary to do so. One such enzyme found in human saliva cleaves the large starch molecules into many maltose disaccharides. The maltose sugars can then be digested to glucose in the intestine by the enzyme maltase. These individual glucose sugars are used to meet immediate energy needs. Alternatively, an animal might store extra glucose in the form of glycogen as a reserve to meet future energy demands. Glycogen and starch are very similar in structure, both molecules consisting of covalently linked glucose sugars. Starch and glycogen differ only in the exact position of the bonds holding the individual sugars together (**Figure 2.16**).

Cellulose is another polysaccharide composed of glucose subunits. All plant cells are surrounded by a rigid cellulose wall, which provides the cell with structural support. Animals, therefore, encounter cellulose whenever they eat plants. Animals, however, lack the enzyme cellulase, so they are unable to digest the bonds between the glucose sugars in this polysaccharide. A certain amount of such indigestible material or "dietary fiber" in our diet has been equated with a healthy gastrointestinal tract.

Which brings us back to Cyrus. How can a horse or other herbivores, such as sheep and rabbits, survive on a diet of grass, which is mostly cellulose (**Figure 2.17**)? In the case of humans, even though cellulose is composed of glucose, and thus a potentially rich source of energy, we cannot use it, because we lack the necessary enzymes. Many bacteria, however, produce cellulase, allowing them use the glucose in cellulose. Herbivores such as Cyrus have rich bacterial flora in their intestine that do much of their digestion for them. When a horse eats grass, these beneficial intestinal bacteria break down the cellulose in the grass. There is far more glucose released

Figure 2.17 Herbivores feeding. Herbivores such as sheep are able to survive on a cellulose-rich diet, thanks to the billions of cellulose-digesting bacteria living in their intestines.

than the bacteria can use, and the excess is available to the horse. But in Cyrus' case, in addition to these normal bacterial flora, he was also infected with organisms causing his diarrhea. When the veterinarian gave him an antibiotic to destroy these pathogens, much of his normal flora was also killed. Consequently, he could temporarily become somewhat less able to digest grass. The bacteria found in the Benebac® paste, however, help to replace these organisms. Once Cyrus was removed from the antibiotics, additional intestinal bacteria were acquired from the environment as he grazed normally.

Lipids are relatively hydrophobic molecules, also used for energy storage and structure

Lipids are familiar to us as fats, oils, and cholesterol. Like carbohydrates, they are used by microorganisms and other living things for structure and energy storage. If you compare a lipid molecule and a carbohydrate molecule, however, you will note several important differences. First, although both molecules are primarily composed of hydrogen, carbon, and oxygen, the lipid contains less oxygen, and more carbon–hydrogen bonds, than does the carbohydrate. As we will see in Chapter 7, carbon–hydrogen bonds are the primary type of bonds that cells use as an energy source. The energy in the shared electrons in such bonds is harvested and ultimately utilized by cells for all energy-requiring processes, such as movement, transport of materials in and out of the cell, and construction of molecules. Because a lipid has more carbon–hydrogen bonds, it has more potential energy per unit weight than carbohydrates do. Calories are a measure of such potential energy. A slice of cheesecake has more calories than an equally sized portion of bread, reflecting the high lipid content of the cheesecake relative to the bread and the higher proportion of energy-yielding carbon–hydrogen bonds.

A second important difference is that lipids such as fats and oils will not dissolve in water (**Figure 2.18**). Because they lack hydrophilic regions, lipids are generally very hydrophobic. With few or no atoms capable of hydrogen-bonding with water, lipids will not disperse through solution but tend to aggregate together as far from water as they can.

An example of lipid structure is a typical molecule of fat (**Figure 2.19**). Each fat molecule is composed of one molecule called glycerol and three fatty

Figure 2.18 Oil slick on a water surface. Oil and water do not mix. In Oahu's Pearl Harbor, diesel fuel continues to leak from the USS *Arizona*, more than half a century after the ship sank. Diesel fuel, like other hydrocarbon fuels, is unable to interact with water owing to its hydrophobic nature. Consequently it forms an oil slick on the water's surface. New technologies using oil-digesting microbes are currently being considered as a way to combat oil spills. This topic will be more fully explored in Chapter 16.

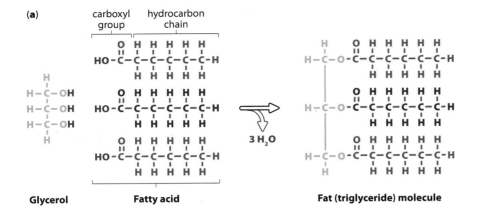

(a)

Glycerol   Fatty acid

Fat (triglyceride) molecule

Figure 2.19 Fats and fatty acids.
(a) Formation of a fat. Three fatty acids are linked to one molecule of glycerol via three condensation reactions. (b) Representative fatty acids. Note that although all three happen to have equal numbers of carbon atoms (18), they differ in the number of hydrogens. Saturated fatty acids, such as stearic acid, have no carbon–carbon double bonds and therefore are bound to a maximum number of hydrogens. In other words, they are "saturated" with hydrogen. Oleic acid, with one carbon–carbon double bond, is termed monounsaturated. Other fatty acids, such as linoleic acid, have more than one carbon–carbon double bond and are called polyunsaturated fatty acids. Each double bond reduces the potential number of carbon–hydrogen bonds. Note the twists in the structure of the unsaturated fatty acids, caused by the C=C double bonds.

(b) **Stearic acid** is a saturated fatty acid found in beef fat.

Oleic acid is a monosaturated fatty acid—it has one double bond—found in olive oil.

Linoleic acid is a polyunsaturated fatty acid—it has more than one double bond—found in sunflower seeds.

acid chains. Each fatty acid is linked to glycerol via a condensation reaction between the glycerol and the three fatty acids. As there are three reactions, three molecules of water are produced in fat synthesis. To digest a fat, instead of removing water, water is added via hydrolysis. The water introduced into these reactions reconstitutes the glycerol and fatty acids. Just as each reaction in the formation or breakdown of a carbohydrate requires a specific enzyme, so does lipid synthesis or digestion. Such enzymes are collectively called **lipases**.

The hydrocarbon chains in a fatty acid may be of variable length. For instance, stearic acid, found in beef fat, consists of 18 carbons. Butyric acid, found in butterfat, has only four carbons. Furthermore, the carbons in the fatty acid chain may be bound to other carbons via single or double covalent bonds. Because carbon can only form a total of four covalent bonds, if it is double-bonded to another carbon, it can form fewer bonds with hydrogen. In **saturated fats**, carbons are linked exclusively by single covalent bonds. Carbons are thus bound to the maximum number of hydrogens: they are "saturated" with as much hydrogen as they can possibly bind. If there are double bonds between the carbons in a fatty acid, the fat is termed **unsaturated** because it holds fewer than the maximum number of hydrogens (see Figure 2.19).

Saturated and/or longer fatty acids tend to make a fat more solid, because individual fat molecules can be packed together more tightly. Shorter hydrocarbon chains and/or double bonds make a fat more fluid and liquidlike due to the looser packing of the individual fat molecules. Indeed, the only difference between a fat and an oil is that the fat is solid at room temperature while the oil is liquid.

Consider the "natural style" peanut butter you might buy at the market. The unsaturated fatty acids in peanut oil give it a liquid consistency that must be stirred in with the non-lipid portion prior to use. Because some people might find this inconvenient, manufacturers often bubble hydrogen gas through the peanut oil. This converts the unsaturated hydrocarbons to "partially hydrogenated" and makes the product more solid at room temperature. Such hydrogenated oils are associated with increased risk of heart disease, but at least you don't have to stir your peanut butter before spreading it on an English muffin.

Proteins participate in a variety of crucial biological processes

CASE: BAD BAMBOO

On March 14, 2006, approximately 200 people attended a religious festival in rural Thailand. Within two days, large numbers of people began showing up at local area hospitals with symptoms including difficulty breathing, nausea, dry mouth, abdominal pain, and double vision. The patients, all of whom had attended the festival, were diagnosed with botulism. One hundred fourteen required hospitalization, and of these, 42 were placed on mechanical ventilators to allow them to breathe. Ten individuals were treated as outpatients and released. Subsequent investigation showed that all patients had consumed bamboo shoots, which had been home-canned locally by a women's group. Individuals who did not consume the bamboo shoots did not develop botulism. No deaths were reported, but approximately 25 individuals required hospitalization and mechanical ventilation for over a month.

1. What is the precise cause of botulism?
2. Is there anything specific about bamboo shoots that makes them a likely source of botulism? What, if any, is the link to home canning?
3. What other foods have similar risks and why?

Botulism is a rare disease, but when it strikes it can be spectacularly deadly. The culprit is the bacterium *Clostridium botulinum* (**Figure 2.20**), but the microorganism does not cause the disease directly. Rather, in appropriate environments, *C. botulinum* produces a protein that is highly toxic to the nerves that operate muscles. Since the ability to breathe is dependent on the action of the muscular diaphragm, exposure to the protein toxin generally results in the sorts of respiratory problems seen in some of the patients. Botulism toxin is perhaps the most lethal toxin known, but why the connection with certain foods? To answer that, we first need to understand something about a third class of biological molecules—the proteins.

Proteins are large molecules that, as we stated earlier, are involved in a variety of functions, ranging from structure to transport to host defense. They are essential as enzymes, and as we have seen in our bamboo case they can sometimes cause life-threatening medical problems.

The basic building blocks of proteins are **amino acids**. There are 20 different amino acids, and while they all have similar structure—all contain an NH_2 group and a COOH group—they differ in terms of their **R group** (**Figure 2.21**). It is the R group that gives each amino acid its unique properties. Those with polar R groups are hydrophilic, meaning they can form hydrogen bonds with each other or with other hydrophilic compounds. Those with positively or negatively charged R groups are likewise hydrophilic and will attract molecules or ions with opposite charges. Nonpolar R groups render an amino acid hydrophobic.

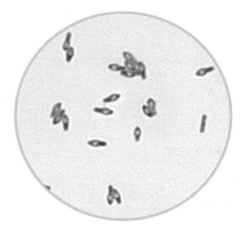

Figure 2.20 *Clostridium botulinum,* **the agent of botulism.** When exposed to oxygen, *C. botulinum* bacteria persist as metabolically inactive spores (small green structures). However, when such spores contaminate oxygen-free environments such as an improperly canned food item, they germinate into active bacterial cells (elongated red structures containing the spores). These cells then synthesize the protein toxin that causes botulism.

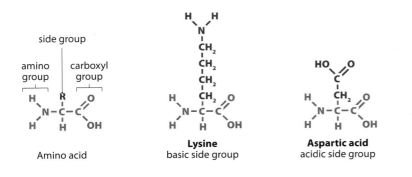

Amino acid

Lysine
basic side group

Aspartic acid
acidic side group

Valine
nonpolar,
hydrophobic side group

Serine
polar,
hydrophilic side group

The number of amino acids in a protein is highly variable. A small protein might have a few dozen, while a larger protein may have hundreds. Amino acids are linked together in a manner that should by now be familiar. A molecule of water is removed enzymatically, resulting in the formation of a covalent bond. An OH, removed from one amino acid, joins with an H, removed from an adjacent amino acid, to form the water molecule (**Figure 2.22**). (The loss of the water molecule is again known as *condensation*.) The new covalent bond is termed a **peptide bond**, and a short chain of amino acids (not yet a complete protein) is called a **peptide**.

A protein, however, is not merely a long, straight chain of amino acids. Rather, the amino acids fold up into a complex, and intricately folded, three-dimensional structure. The precise shape is crucial to a protein's function. Many proteins interact with other molecules much like pieces of a jigsaw puzzle. If the pieces don't fit, they cannot come together.

The precise sequence of amino acids in a protein is the starting point for determining the specific shape of that protein. As we have discussed, amino acids differ in their properties and some lend themselves to making bonds, such as hydrogen bonds, more than others. The order of amino acids in a protein, therefore, affects overall protein structure. Different types of bonds may be formed between different amino acids in the protein as they are brought into contact with each other (**Figure 2.23**). A protein that is in its precise, three-dimensional shape is said to be in its **native** conformation.

Anything that affects these numerous interactions may cause the protein to change shape, and any alteration of shape is likely to reduce protein function. If the three-dimensional shape of a protein is completely destroyed, the protein is rendered nonfunctional. A protein that has unfolded and lost its three-dimensional shape is said to be **denatured**. Many environmental factors, such as high temperature and extremes of pH, can disrupt weak

Figure 2.21 Amino acids, the subunits of proteins. Representative amino acids are shown. All 20 amino acids can be characterized as having basic, acidic, nonpolar, or polar side groups (also called R groups).

(a)

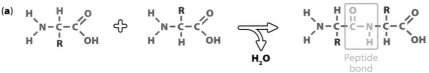

H₂O

Peptide bond

(b)

Phenylalanine
(Phe)

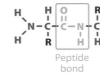

Threonine
(Thr)

Methionine
(Met)

Figure 2.22 Peptide bonding. (a) A peptide bond, highlighted in blue, forms by condensation between two amino acids. (b) A short peptide consisting of three amino acids, threonine (Thr), phenylalanine (Phe), and methionine (Met). Side groups of each of the three amino acids are highlighted in red.

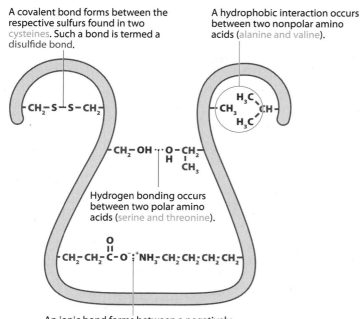

A covalent bond forms between the respective sulfurs found in two cysteines. Such a bond is termed a disulfide bond.

A hydrophobic interaction occurs between two nonpolar amino acids (alanine and valine).

Hydrogen bonding occurs between two polar amino acids (serine and threonine).

An ionic bond forms between a negatively charged, acidic amino acid (glutamic acid) and a positively charged, basic amino acid (lysine).

Figure 2.23 Protein three-dimensional structure. Many interactions between amino acids take place to give a protein molecule its intricately folded shape. This diagram shows several of them.

attractive forces and denature a protein (**Figure 2.24**). A familiar example is the process of milk spoilage. As we discussed earlier in this chapter, when lactose-digesting bacteria gain access to milk, they release an acid waste as they digest the milk sugar. Besides giving the milk a sour taste, the acid also denatures proteins in the milk, converting them from the liquid to the solid "curdled" state.

Like other proteins, the botulism toxin requires a precise three-dimensional shape to adversely affect anyone unlucky enough to consume it. Only when folded up properly can it latch on to nerve cells and prevent them from releasing the chemicals that cause muscle contraction. The bamboo shoots implicated in the outbreak in Thailand were tainted with protein toxin that was likewise in its native shape. Had the bamboo shoots been well cooked prior to consumption, it is likely that the toxin would have been denatured by the heat and consequently unable to cause disease.

The *C. botulinum* bacteria that produce the toxin are unable to grow in the presence of oxygen. Consequently, most cases of botulism are associated with improperly canned food such as the bamboo shoots. The bacteria are present in the environment as nongrowing spores (see Figure 2.20), but if they are able to get into a vacuum-packed can that was improperly sterilized, they begin to grow in the can and secrete toxin in the process. Spores on fresh food exposed to oxygen pose no risk because they are unable to grow. Interestingly, not all canned foods are equally risky, even without heating. You would have less to fear from highly acidic foods because low pH will also denature the toxin. Canned sardines packed in vinegar, for instance, would be less worrisome than sardines packed in oil. Canned tomatoes, because of their low pH, are an unlikely source of botulism contamination. But other canned foods such as mushrooms, which are at neutral pH and may be served cold, are occasionally implicated in botulism.

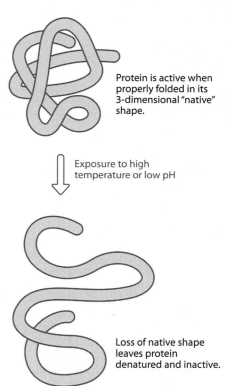

Protein is active when properly folded in its 3-dimensional "native" shape.

Exposure to high temperature or low pH

Loss of native shape leaves protein denatured and inactive.

Figure 2.24 Disruption of protein shape by environmental factors. Proteins can be denatured by conditions such as high temperature or low pH. Because protein function depends on shape, alterations in a protein's three-dimensional structure tend to reduce or eliminate protein activity.

Biological reactions require enzymes functioning as catalysts

CASE: CALL IN THE CLOT BUSTERS

Dr. Creek is a cardiologist in a large urban hospital. While on call late one night, he is summoned to the emergency room (ER) to treat an apparent heart attack victim. Upon arriving in the ER, he quickly confirms that the patient has indeed suffered a severe heart attack and he quickly administers supportive measures. Among the treatments he orders is the immediate administration of intravenous (IV) streptokinase over the next hour. Streptokinase is actually a bacterial enzyme that is often used to treat heart attack victims. Dr. Creek notes with satisfaction that blood flow to the heart tissue significantly increases, raising hopes for the patient's safe recovery.

1. What exactly is an enzyme?

2. How does streptokinase increase blood flow to the patient's heart tissue? Can any enzyme do this, or is there something special about streptokinase?

We have already referred to enzymes and the crucial role they play in biological reactions such as the condensation and hydrolysis reactions that form or digest carbohydrates and lipids. Almost all biological reactions require a specific enzyme if they are to take place, and almost all enzymes are proteins. Enzymes are so important that much of the difference that we observe between living things reflects the fact that different organisms have different enzymes and are therefore capable of different biological reactions.

We have recently discussed, for instance, the intestinal bacteria that are able to digest cellulose. Not all bacteria, however, can digest this sugar. Those that cannot lack the necessary enzymes. Fortunately for our heart attack victim and for cardiologists like Dr. Creek, certain species of *Streptococcus* bacteria produce the bacterial enzyme streptokinase. Streptokinase is a protein-digesting enzyme known as a **protease** (you will note that most enzymes end with "ase" and their names often provide clues as to what the enzyme acts upon). Most enzymes are very specific, and a protease will not necessarily digest any protein. In the case of streptokinase, the targeted protein is fibrin, an important component of blood clots. The enzyme interacts with fibrin, facilitating its digestion.

When *Streptococcus* bacteria invade host tissue, they use streptokinase to help free themselves when they are trapped inside a clot. These bacteria thus use streptokinase to help them spread through host tissues. Cardiologists take advantage of this bacterial enzyme to increase the survival rate of heart attack patients. Heart attacks are often caused when clots form in the arteries that supply the heart with blood. Without adequate blood supply, heart tissue quickly begins to die. But the immediate IV administration of streptokinase can dissolve such clots, restoring normal blood flow and preventing excessive damage to heart tissue.

When we say that enzymes facilitate specific biological reactions, this is not to say that a given reaction would *never* occur if the proper enzyme were missing. Enzymes, however, greatly speed up the rate at which reactions take place, sometimes up to a billion times. For all practical purposes, life itself is dependent on enzymes. Fibrin might eventually break down to its component amino acids without streptokinase, but neither *Streptococcus* bacteria nor heart attack patients can wait days or weeks until it does. To efficiently digest fibrin, the enzyme is required.

As we have stated, enzymes are highly specific. Each one participates in only one or a very small number of reactions. A cellulose-digesting enzyme

(a)

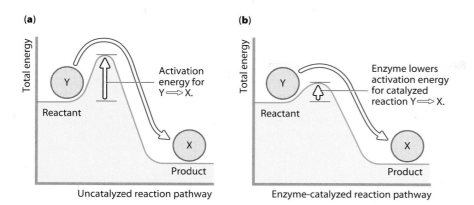

Uncatalyzed reaction pathway

(b)

Enzyme-catalyzed reaction pathway

Figure 2.25 Effect of enzymes on activation energy. (a) Compound Y (a substrate) is in a relatively stable state, and energy is required to convert it to compound X (a product), even though X is at a lower overall energy level than Y. This conversion will not take place unless compound Y can acquire enough activation energy from its surroundings to undergo the reaction that converts it into compound X. This energy may be provided by increasing the temperature. The precise amount of energy needed to allow the reaction to proceed is called the *activation energy*. The reverse reaction (X → Y) will require a much larger input of energy to occur and consequently will take place only rarely. (b) Energy barriers for specific reactions can be lowered by catalysts. Enzymes are particularly effective catalysts because they greatly reduce the activation energy for the specific reactions they catalyze.

does not digest other carbohydrates. Streptokinase digests fibrin but not other molecules. This means that cells need a large number of different enzymes to carry out all of their necessary biochemical reactions.

Enzymes are necessary because biological molecules, as a rule, are fairly stable. They have to be, or they would spontaneously decompose and life itself would become impossible. Charcoal (a hydrocarbon), for example, placed in your barbecue does not spontaneously burst into flames. Even though burning the charcoal *releases energy*, a small amount of energy must first be *added* to get charcoal burning. This energy input might be in the form of a match. Yet anyone who has ever lit a barbecue knows that one match rarely does the trick; you would have to use many matches to ordinarily light the coals. Or you could use a catalyst, like lighter fluid, to lower the number of matches you need to ignite the charcoal. With the catalyst, the amount of energy needed to get the reaction started, known as the **activation energy**, is greatly reduced. Enzymes also act as catalysts by lowering the activation energy needed for a reaction to occur (**Figure 2.25**).

Enzymes accomplish this by temporarily binding to the molecule to be altered (the **substrate**) at a special binding site on the enzyme called the **active site** (**Figure 2.26**). When the substrate is bound by the enzyme's active site, it is physically altered such that it begins to take on the shape of the final **product** of the reaction. In other words, the fit of the enzyme in the active site is not a perfect fit. When it is bound, the bonds in the substrate are stressed and can then be more readily altered into the product. In the case of streptokinase, these bonds are the peptide bonds in the fibrin protein. When these bonds are stressed, it becomes much easier for the required digestion to occur.

Living things employ a variety of mechanisms to ensure that enzymes are active only when particular reactions are needed. For example, in many situations, a non-protein molecule known as an **inhibitor** latches onto the enzyme when no additional product is needed. This essentially acts as an "off" switch, shutting the enzyme down until the product is used up. The inhibitor causes the enzyme to slightly and temporarily denature, by

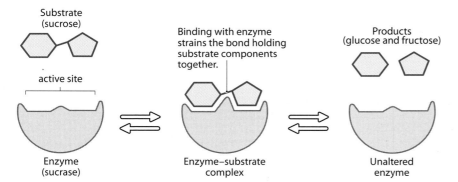

Figure 2.26 Enzyme action. In this example, the enzyme sucrase catalyzes the digestion of the disaccharide sucrose. The reaction is reversible: sucrase can also catalyze the synthesis of sucrose from the monosaccharides glucose and fructose. Enzyme specificity is a function of the active site, which can bind only specific substrates. Once the substrate is bound, an enzyme–substrate complex is formed. A change in the enzyme's shape strains the bonds in the substrate, reducing the energy necessary to break or rearrange them. The products are then released, leaving the enzyme unchanged and free to combine with new substrate molecules.

Figure 2.27 **Nucleotide structure.** (a) A single DNA nucleotide is composed of a molecule of deoxyribose, a phosphate, and a nitrogenous base. In RNA the sugar is ribose instead of deoxyribose. (b) Nitrogenous bases. Thymine is found only in DNA; in RNA, thymine is replaced by uracil.

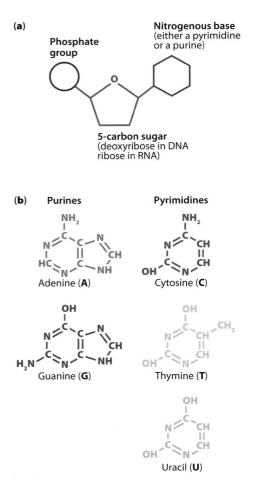

altering its three-dimensional, native shape. When more of the product is again needed, the inhibitor detaches and the enzyme regains its proper native shape. The enzyme is now once again "on" and able to catalyze its reaction.

Nucleic acids direct the assembly of proteins

A living thing produces hundreds or thousands of different proteins, and in a very real way, an organism's structure and function is a consequence of the proteins it makes. In a sense, proteins also determine what carbohydrates and lipids are made or digested by an organism, since the synthesis or degradation of these molecules requires specific enzymes. Different organisms, however, make different proteins. It is the task of the fourth and final group of biological molecules, the nucleic acids, to specify exactly which proteins an organism can produce.

Deoxyribonucleic acid (DNA) is the genetic material in microorganisms and other living things. DNA is composed of individual building blocks called **nucleotides**. Each nucleotide is composed of three principal parts: a 5-carbon sugar (deoxyribose), a phosphate (PO_4^-) group, and a nitrogen-containing molecule with basic properties, known as a nitrogenous base. A nucleotide many have any of four different nitrogenous bases. Cytosine and thymine are single-ring bases called **pyrimidines**, whereas the **purines**, adenine and guanine, have a double-ring structure (**Figure 2.27**).

A strand of DNA may contain thousands or millions of nucleotides, all covalently bound between the phosphate on one nucleotide and the sugar on the next. The entire DNA molecule consists of two such strands wound around each other like two spiral staircases in what is known as a **double helix (Figure 2.28)**. Note how the two helices orient toward each other, with backbones of sugar and phosphate on the outside and the nitrogenous bases projecting in toward the interior of the helix. The bases on opposite sides of the helix are able to hydrogen-bond with each other, holding the two helices together. This base pairing is highly specific, with a purine bonding only with a pyrimidine. Furthermore, adenine is able to properly bond only to thymine, while guanine pairs only with cytosine.

The sequence of nitrogenous bases on a strand of DNA specifies the exact amino acids that will be incorporated into a protein. The bases therefore represent a code, utilized for protein production.

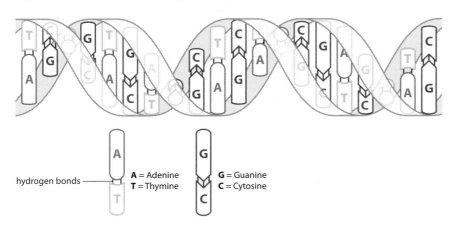

hydrogen bonds —

A = Adenine   G = Guanine
T = Thymine   C = Cytosine

Figure 2.28 **The double helix.** Note the specific hydrogen bonding of adenine with thymine and guanine with cytosine.

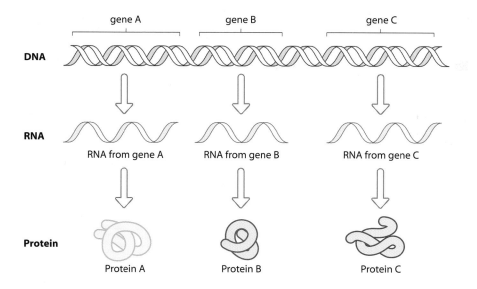

Much of an organism's DNA is organized into **genes**. A gene is a sequence of DNA nucleotides that codes for a single protein or part of a protein. A single bacterium may have several thousand genes, indicating that it can produce several thousand proteins.

Although DNA is composed of genes and contains the exact instructions for protein synthesis, DNA does not actually link the amino acids together to form the protein. Before a protein can actually be synthesized, the information encoded in a gene must be transferred to a second type of nucleic acid called **ribonucleic acid (RNA)**. RNA acts as a "messenger," carrying DNA's genetic message to the actual site where amino acids are assembled into proteins (**Figure 2.29**). This flow of genetic information, from DNA to RNA to protein, is a fundamental process common to almost all living things, microbial or otherwise. We will carefully consider this important concept in Chapter 6.

The structure of RNA is very similar to that of DNA, with three major differences. First, unlike DNA, RNA is usually a single-stranded molecule. Second, the sugar found in an RNA nucleotide is ribose rather than deoxyribose. Finally, RNA uses the nitrogenous base uracil in place of thymine (see Figure 2.27).

Coming Up Next...

In this chapter we have investigated some of the fundamental chemical principles that govern the behavior of all matter. In addition to gaining insight into basic atomic structure, we have also learned how atoms interact to form molecules. These include the biological molecules that are so crucial to the makeup of microorganisms and other living things.

Biological molecules, while they are life's fundamental building blocks, are not themselves alive. These molecules are assembled into cells—the basic units of life. In Chapter 3, we will take a close look at cell structure and function. We will investigate those properties shared by all cells, as well as the important differences found between the cells of different organisms. In doing so we will come to understand, among other things, why some cells divide faster than others, how an antibiotic such as penicillin will kill some but not all bacteria, and how some cells are able to persist for hundreds or even perhaps millions of years, as inert spores in ancient clay pottery or in the stomachs of amber-encased insects.

Key Terms

acid
activation energy
active site
amino acid
atom
atomic mass
atomic number
base
carbohydrate
chemical bond
condensation reaction
covalent bond
denatured
deoxyribonucleic acid (DNA)
disaccharide
double helix
electron
electron affinity
electron shell
element
enzyme
gene

hydrocarbon
hydrogen bond
hydrolysis reaction
hydrophilic
hydrophobic
hydroxyl
inert gas
inhibitor
ionic bond
ionized state
ion
isomer
isotope
lipase
lipid
monosaccharide
native conformation
neutron
nonpolar covalent bond
nucleic acid
nucleotide
nucleus

organic
peptide
peptide bond
pH scale
polar covalent bond
polysaccharide
product
protease
protein
proton
purine
pyrimidine
R group
reactive
ribonucleic acid (RNA)
salt
saturated fat
solute
solution
solvent
substrate
unsaturated fat

Concept Questions

1. Is an ionic bond more like a polar covalent bond or a nonpolar covalent bond?

2. Water striders are insects that are actually able to walk over the surface of still water. If such an insect landed on a pool of oil, would it be able to walk over the surface in a similar fashion? Why or why not?

3. If a large ship sinks in the ocean, given enough time, it will dissolve. Why is this? Would the same thing happen if the ship were submerged in a nonpolar solution?

4. When H_2SO_4 is placed into water, it converts into H^+ and HSO_4^-. Why does this happen only when the molecule is placed in water? On the basis of this information, what type of molecule is H_2SO_4?

5. Imagine a room with 100 seats in it. One hundred ten people enter the room. In situation A, 100 people race to occupy the seats, pushing out of the way the 10 others, who don't seem to care much about sitting down. These 10 people don't get a seat. In situation B, 110 people once again enter the room, but 10 people, for some reason, run screaming in fear out of the room, leaving the seats for the 100 other people. If you consider the seats as water molecules and the 10 people who don't get seats as hydrophobic molecules, which situation, A or B, is a more accurate representation of the way they would respond to water?

6. Hair is basically composed of the protein keratin. Imagine a woman gets her naturally very straight hair done in a perm. The hair stylist applies certain chemicals to the hair and then sets the hair in the newly desired shape. What do you think the function of the chemicals is? On her way home, the woman stops at a store to pick up some shampoo. Assuming she wants to maintain her perm for as long as possible, does it matter if she gets a pH-balanced shampoo or not?

7. Very few microorganisms are able to digest lignin, a very large and complex carbohydrate found in plant material. What do you think distinguishes those microorganisms that are able to digest lignin from those that cannot?

8. Like the cells of all living things, the cells of microorganisms are surrounded by a membrane, which is largely composed of lipid. Usually the lipids in these membranes contain saturated fatty acids. But in microorganisms that live in extremely cold water (where there might be a danger of freezing), the membranes contain a high proportion of unsaturated fatty acids. Why do you think this is the case?

9. You learn that a particular nucleic acid contains 30% guanine nucleotides and 20% adenine nucleotides. How much thymine and cytosine does it contain?

10. Lifeguards often keep a bottle of vinegar in their lifeguard station. If a swimmer is stung by a jellyfish, the lifeguard will pour vinegar on the skin to reduce the stinging. Knowing that jellyfish toxin is a protein, and that vinegar is acidic, explain why the vinegar is effective.

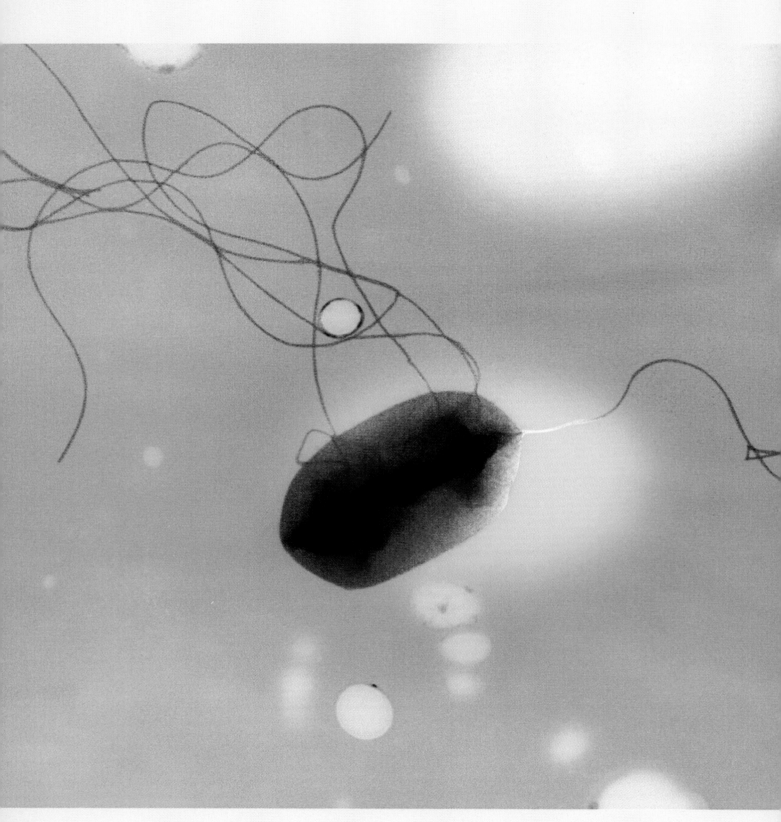

Escherichia coli 0157:H7 uses its flagella to move through the environment.

Chapter 3

The Cell: Where Life Begins

In 1665, the English scientist Robert Hooke peered through a crude microscope at a slice of cork from the bark of a tree. What he saw astounded him. The cork appeared to be divided up into small individual components, invisible to the naked eye. Because each of these units resembled a small room or chamber, he called them "cells."

Over time it became clear that what was true for trees was also true for all other living things. One of the foundations of modern biology, the *cell theory*, states that all living things, from the smallest bacterium to the largest redwood tree, are composed of one or more cells. As we will see in Chapter 4, viruses are acellular, and may therefore be an exception, depending on one's definition of life. However, even viruses must enter cells in order to reproduce, and for those organisms that clearly show all the features of living things, the cell theory has remained valid for well over a century.

The cell theory also states that the cell is the fundamental unit of life. What exactly does this mean? In Chapter 1 we reviewed some of the features that distinguish living things from inanimate objects. Living things respond to their environment and they utilize energy. They reproduce and pass genetic information on to progeny in the form of nucleic acid. All life has evolved from other living things, and a living organism is highly organized compared with a nonliving object. Cells possess all of these features. Biological molecules such as proteins or DNA, discussed in Chapter 2, do not; although such molecules and other subcellular components are organized into living cells, they themselves are not living.

If we really want to understand living things, microbial or otherwise, we must investigate those entities in which life begins: cells. Consequently, in this chapter we will take a close look at cell structure and how this structure relates to function. By understanding basic cell structure, we learn a great deal about how a particular organism grows, reproduces, and otherwise goes about the business of living. We will discuss the features that all cells share, as well as those that account for the enormous diversity of cell types seen in nature. As we will see, microorganisms with different cell structures present us with very different sets of problems when it comes to understanding and controlling the impact they have on us. Thus, in addition to a consideration of basic cell biology, we will learn why, when you have seen one cell, you most certainly have *not* "seen them all."

Basic Concepts in Cell Biology

All cells are one of two basic types. Most of the organisms with which you are most familiar, including animals, plants, and fungi, are composed of **eukaryotic cells**. On the other hand, many of the microorganisms we will consider in this text are composed of **prokaryotic cells** (**Figure 3.1**). When you visit the doctor because of "strep throat," or if you suffer through a case of "traveler's diarrhea" on an overseas trip, you are experiencing problems caused by prokaryotes. On the other hand, if you use vinegar on your salad, or enjoy a cup of yogurt for lunch, you can thank the prokaryotes that produced them. Much of the oxygen we breathe is produced by prokaryotes, and when animals and plants die, the nutrients in these organisms are returned to the environment, in large part due to prokaryotic activity. Bacterial cells are prokaryotic, as are the cells of another large group of microscopic, single-celled organisms, the archaea. Before we discuss the differences between prokaryotic and eukaryotic cells, however, let us consider some of the things they have in common.

Both prokaryotic and eukaryotic cells are composed of the same basic biological molecules—the carbohydrates, lipids, proteins, and nucleic acids described in Chapter 2. In both cell types, genetic information is encoded in DNA, and the production of proteins, while not identical, is very similar regardless of cell structure. Prokaryotes and eukaryotes both use the same sorts of nutrients to meet their nutritional and energy needs, and both harvest energy in similar ways.

All cells are surrounded by a membrane, composed largely of lipids and protein (see Figure 3.1). In other words, cells are membrane-bound. This **plasma membrane** (also commonly called the **cell membrane**) separates the inside of the cell from the outside world. The fluid-filled interior of the cell is called the **cytoplasm**. The cytoplasm contains a variety of small, subcellular structures that allow the cell to carry out its various functions. The nature of these structures varies considerably according to cell type. Frequently there are additional structural features exterior to the cell membrane. These may include a rigid cell wall or other structures involved in cell movement or adherence. The membrane also has special structures to allow substances to move from outside to inside the cell and vice versa.

There are also important differences that distinguish the two cell types. In addition to the cytoplasmic structures referred to above, eukaryotic cells generally have a variety of membrane-bound **organelles** ("little organs") located in the cytoplasm (see Chapter 1, p. 4). As we will see, each organelle is associated with specific activities. The most prominent of these structures is the **nucleus**, which surrounds the cell's genetic material. Membrane-bound organelles, including the nucleus, are absent in prokaryotic cells. The names "eukaryotic" and "prokaryotic" are derived from the Greek terms for "true nucleus" and "before the nucleus," respectively. Prokaryotic cells do, of course, have genetic material, but it is present in the cytoplasm rather than surrounded by its own membrane. They also contain a handful of other internal structures without membranes, such as the ribosomes, utilized to link amino acids into proteins. Figure 3.1 illustrates the principal prokaryotic and eukaryotic cell features that will be described in this chapter.

The first difference between prokaryotic and eukaryotic cells that you might notice is their size disparity. Most prokaryotic cells are only a few micrometers long. Eukaryotic cells are often more than 10 times larger (**Figure 3.2**). As we will shortly see, there are some fascinating exceptions to this rule.

Most cells, of course, are microscopic, and to see them requires the use of a microscope. Even with a microscope, cells would be difficult to see if they were not stained prior to observation. Staining the cell increases the contrast between the cell and its background, permitting easy viewing. We generally stain the specimen of cells we wish to view with a *basic* chemical such

(a) Typical bacterial cell (prokaryote)

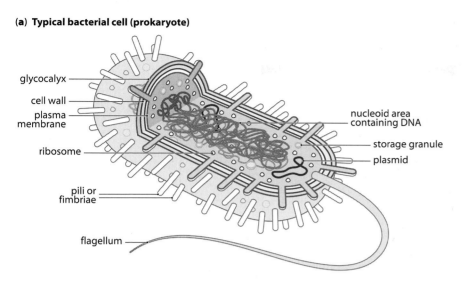

glycocalyx

cell wall

plasma
membrane

ribosome

pili or
fimbriae

flagellum

nucleoid area
containing DNA

storage granule

plasmid

Figure 3.1 Prokaryotic and eukaryotic cells. (a) A prokaryotic cell. Both bacteria and archaea have prokaryotic cells. The term "prokaryote" comes from the Greek for "before the nucleus," and prokaryotic cells are characterized by the lack of a nuclear membrane surrounding the DNA. Certain features of this cell, such as the pili and the flagella, are not present in all prokaryotes. (b) Animal and plant cells are examples of eukaryotic cells. They are different from prokaryotes in that their DNA is always surrounded by a nuclear membrane ("eukaryote" comes from the Greek for "true nucleus"). Eukaryotic cells are typically more complex than prokaryotic cells, with a number of subcellular membrane-bound structures called *organelles*, which carry out various cell functions. Not all eukaryotic cells have all organelles.

(b) Typical animal cell (eukaryote)

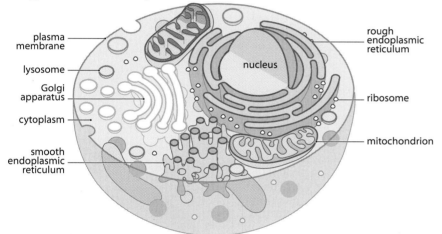

plasma
membrane

lysosome

Golgi
apparatus

cytoplasm

smooth
endoplasmic
reticulum

rough
endoplasmic
reticulum

nucleus

ribosome

mitochondrion

Typical plant cell (eukaryote)

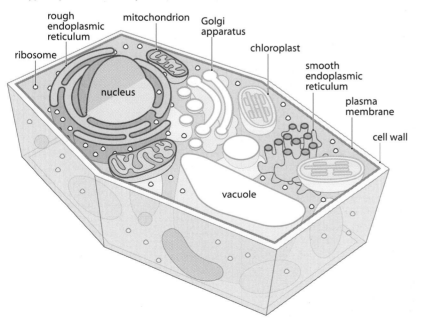

rough
endoplasmic
reticulum

mitochondrion

Golgi
apparatus

chloroplast

ribosome

nucleus

smooth
endoplasmic
reticulum

plasma
membrane

cell wall

vacuole

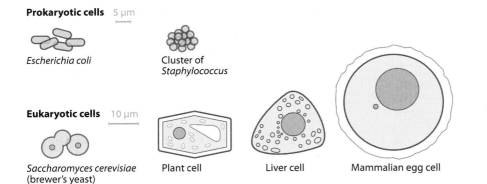

Prokaryotic cells 5 μm

Escherichia coli

Cluster of *Staphylococcus*

Eukaryotic cells 10 μm

Saccharomyces cerevisiae (brewer's yeast)

Plant cell

Liver cell

Mammalian egg cell

Figure 3.2 Sizes of various cells. Typical prokaryotic cells may be between 2 and 8 μm in length [1 micrometer (μm) = 1/1,000,000 or 1 × 10⁻⁶ meter]. The size of eukaryotic cells varies greatly but on average is about 10 times that of prokaryotic cells. To provide a sense of comparison, consider that your thumb is approximately 20,000 μm across.

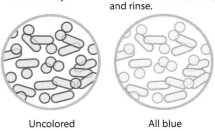

Unstained cells on a microscope slide usually appear too faint to study.

To view many kinds of cells, we apply a basic stain such as methylene blue for 1 minute, drain, and rinse.

Uncolored

All blue

Figure 3.3 The simple stain. Cells are stained prior to examination under the microscope to increase contrast with the illuminated background. The basic stains, carrying positive charges, adhere to the negatively charged plasma membrane of the cells.

as methylene blue. Recall from Chapter 2 that bases accept protons and therefore acquire positive charges when in solution. Plasma membranes, on the other hand, carry negative charges. The positively charged stain adheres to the negatively charged plasma membrane, providing the necessary color and contrast for the cell to stand out from its background. The simple stain procedure is diagrammed in **Figure 3.3**.

As cells get larger, their efficiency decreases

CASE: SIZE DOES MATTER!

In 1988, researchers discovered what were by far the largest known bacteria. *Epulopiscium fishelsoni* cells can be up to 0.57 mm long and 0.06 mm thick, making them easily visible to the naked eye (**Figure 3.4**). These behemoth cells have a volume more than a million times greater than that of more typical bacteria. The gigantic bacteria were found living in the intestines of surgeonfish on Australia's Great Barrier Reef. These unusual microorganisms digest the algae on which the fish feed. In this way they are similar to the normal gut bacteria found in herbivorous land animals such as horses or rabbits. Since 1988 *E. fishelsoni* has been found in other surgeonfish from several locations including Hawaii, Japan, and the Red Sea. Besides their gigantic proportions, these bacteria are unusual in other ways. For instance, their cytoplasm is packed with an intricately folded system of what appear to be membranes.

1. How would the metabolism of *E. fishelsoni* compare with that of ordinary-sized bacteria?
2. Would you expect these cells to reproduce more quickly or more slowly than other prokaryotes?
3. What, if any, is the relationship between the large size of *E. fishelsoni* and the membrane-like structures in the cytoplasm?

Newcomers to microbiology often make the mistake of assuming that since eukaryotic cells are usually so much larger and more complex than prokaryotic cells, they must somehow be better at carrying out their functions. They are not. In many ways prokaryotic cells, by virtue of their simplicity, are actually more efficient. Small size in particular gives prokaryotes an edge when it comes to metabolic efficiency—the ability to use energy effectively.

All cells must absorb nutrients from their environment and excrete waste products. As cell size increases, the need for nutrients and waste excretion

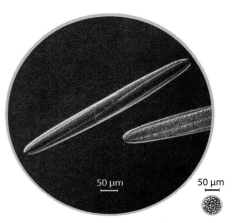

50 μm   50 μm

Figure 3.4 *Epulopiscium fishelsoni*. These gigantic bacteria, first discovered living in the digestive system of fish in Australia, have since been found in other locations as well. Other species of unusually large bacteria have been found living off the southwest coast of Africa. The large bacterial cell is approximately 600 μm (0.6 mm) long. More typically sized bacteria, seen in the small image to the lower right, are provided for comparison. Such cells, with a length of about 4 μm, are each less than 1% the length of the *E. fishelsoni* cell.

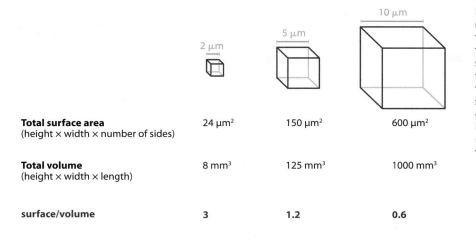

	2 μm	5 μm	10 μm
Total surface area (height × width × number of sides)	24 μm²	150 μm²	600 μm²
Total volume (height × width × length)	8 mm³	125 mm³	1000 mm³
surface/volume	3	1.2	0.6

Figure 3.5 Effect of surface-to-volume ratio on metabolic efficiency. Consider the cubes depicted to represent cells of three different sizes. Although the total surface area increases as size increases, the volume increases more rapidly, meaning the surface-to-volume ratio decreases as size increases. The result is that as cells get larger, their membrane area cannot keep up with the increased demand for membrane transport. Metabolic efficiency therefore declines.

goes up. All traffic into and out of the cell must cross the plasma membrane (see Figure 3.1). Having a large amount of membrane relative to cell volume ensures an efficient transport of materials into and out of the cell. But as cells increase in size, even though the total surface area goes up, the *ratio* of surface area to volume decreases (**Figure 3.5**). Small cells therefore have more membrane surface area per unit of volume and are thus more efficient at moving materials across the membrane. This superior transport into and out of the cell translates into a more efficient metabolism, which in turn means the cells can grow and reproduce more quickly. This is a principal reason why bacteria reproduce rapidly compared with eukaryotes. Typical bacteria such as *Escherichia coli* might divide every 30 minutes or so under ideal conditions. The time between divisions for a eukaryotic cell is more frequently measured in hours or days. *Epulopiscium fishelsoni*, the giant species discussed in our case, would, like eukaryotes, have a relatively low surface-to-volume ratio. This results in a relatively inefficient metabolism, and we would expect this species to reproduce very slowly compared with more typical bacteria. *Epulopiscium* is able to compensate for its large size to some degree by having a highly infolded plasma membrane that extends deep into the cytoplasm. The folds of the membrane increase the total area of the membrane, permitting the cells to circumvent some of the disadvantages associated with large size.

The Prokaryotic Cell

The shape of prokaryotic cells is distinctive, and while some have highly unusual morphologies, most cells fall into one of a few easily recognizable groups: rod-shaped **bacilli** (singular, bacillus), spherical **cocci** (singular, coccus), or spiral-shaped **spirilla** (singular, spirillum) (**Figure 3.6**).

When bacteria divide they often separate into individual cells, but in some species the newly formed cells remain joined together and form distinctive patterns. Cocci that divide and remain attached in long chains are called **streptococci**. Those that divide into grapelike clusters are called **staphylococci**. The prefixes "strepto-" and "staphylo-" come from the Greek for "chain" and "grape," respectively. Bacilli generally form individual rods, but some divide into long chains reminiscent of delicatessen sausage links. Such cells are called **streptobacilli**.

Most prokaryotic cells have structures exterior to the cell wall that function in protection, adherence, and movement. We will begin our tour of the prokaryotic cell by examining these extracellular structures. We will then move inward to the cell wall and the plasma membrane. Finally, we will explore those structures found in the cell's cytoplasm.

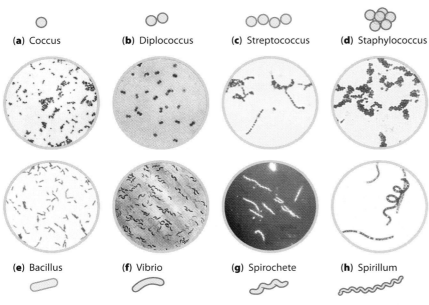

(a) Coccus   **(b)** Diplococcus   **(c)** Streptococcus   **(d)** Staphylococcus

(e) Bacillus   **(f)** Vibrio   **(g)** Spirochete   **(h)** Spirillum

Figure 3.6 Common bacterial shapes and cell arrangements. (A) Coccus or spherical bacterium. Each round sphere in the photograph is an individual bacterial cell. (B) Diplococcus. (C) Streptococcus. (D) Staphylococcus. (E) Bacillus. (F) Vibrio is a bacillus that has a characteristic "comma" shape. (G) Spirochete. (H) Spirillum.

Many prokaryotes have extracellular structures, extending beyond the cell wall

CASE: PLAQUE ATTACK

David and Jennifer are a happily married couple, both in their mid-forties. They have one son, Kyle, age 11. Like all married couples they have their differences. David is extremely conscientious (Jennifer would say obsessive) regarding dental care. He brushes his teeth without fail after each meal and regularly flosses, at least twice a day. Jennifer finds this behavior annoying, and when David is out of town on business, as he frequently is, she tends to be rather lax regarding her own and Kyle's dental hygiene. When David takes his son to the dentist for a routine checkup, he is shocked when the dentist reports that not only does Kyle have several cavities, but his teeth are also heavily covered with plaque (**Figure 3.7**). Two additional visits to the office will be required for cleaning and fillings. The dentist explains that cavities are caused by bacterial infections and that to reduce the likelihood of additional problems, Kyle must follow the traditional advice: brush regularly, and avoid sugary snacks. David is upset to hear about the cavities and asks the dentist the following questions:

1. What exactly is "dental plaque"?
2. What is the role of bacteria in plaque and cavity formation? What characteristics do these bacteria possess?
3. Why are sugar-rich foods associated with cavities?

Beyond the cell wall, many prokaryotes come equipped with a variety of structures or appendages, which are involved in tasks such as adherence to surfaces, protection, and movement. Here we will review some of the most important of these *extracellular* structures.

The Glycocalyx. Many bacterial cells are surrounded by a gel-like layer, composed of polysaccharides, called the **glycocalyx**. When the glycocalyx forms a thick, regular shell-like structure around the cell it is called a **capsule** (**Figure 3.8**). The capsule may help the cell to adhere to various surfaces, or it may provide protection against the host's immune system. For instance, *Streptococcus pneumoniae* can cause lower respiratory tract infections. This species also is responsible for many of the middle ear infections commonly seen in young children. Not all strains of *S. pneumoniae*, however, are able to cause disease. Those bacteria without capsules typically pose no threat to health; it is only encapsulated strains that are worrisome.

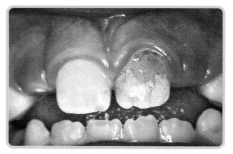

Figure 3.7 Dental plaque. Bacteria adhere to the enamel of teeth with secreted extracellular polysaccharides that make up the glycocalyx. The combination of polysaccharides and the bacteria that adhere to them are known as dental plaque.

To understand why, consider that as a first line of defense against invading microorganisms, white blood cells of the immune system try to engulf the invading organisms, quite literally swallowing them whole through a process called phagocytosis. Bacterial capsules, however, impede efficient phagocytosis, meaning encapsulated cells are better able to survive this initial immune onslaught and go on to reproduce and initiate disease. Cells lacking capsules are quickly engulfed and destroyed by phagocytic cells. The capsule is a good example of a **virulence factor**—a characteristic that increases the disease-causing capacity (or *virulence*) of the microorganism.

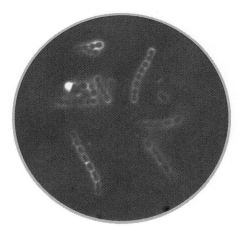

Figure 3.8 The bacterial capsule. The capsules produced by these streptococci are seen as the clear "halo" surrounding each cell.

Other bacterial cells secrete a glycocalyx that is irregular and diffuse. Such a structure is called a **slime layer**. Like a capsule, its primary function is to aid in adherence or to protect the bacteria. *Streptococcus mutans*, for instance, secretes a slime layer that allows the bacterial cells to adhere to teeth. Other bacteria can also adhere to this sticky, sugary material, and the combination of bacteria and glycocalyx on teeth is commonly referred to as "dental plaque." It is unlikely that David, the assiduous brusher and flosser in our case, has much plaque on his teeth because regular routine dental care removes this bacterial coating. Kyle, on the other hand, and perhaps Jennifer as well, by not flossing regularly, allow plaque buildup in tooth crevices. Sugar, or more specifically the sucrose found in sweet food products, adds to the problem, but perhaps not in the way you think. The sucrose itself does not directly cause plaque or cavities. *Streptococcus mutans* uses the sucrose in food to synthesize its glycocalyx. This problem is especially bad when the sucrose comes in the form of candy or other sticky substances. When a caramel or gum drop adheres to the teeth we essentially have the bacterial equivalent of room service. Cavities are formed when the metabolizing bacteria release an acid waste product that damages the tooth enamel. Perhaps Kyle and Jennifer should at the very least start chewing sugarless gum. Such gum may actually reduce the likelihood of cavities because the extra saliva generated while chewing helps to wash away bacterial acids and thus reduces the time during which such acids can damage tooth enamel.

Pili. Pili (singular, pilus) are short protein fibers that are arranged in a helical manner, forming a cylinder with a hollow core (**Figure 3.9**). These structures, also known as **fimbriae**, extend out beyond the cell wall and allow the cell to adhere to specific surfaces. Often they function as a crucial virulence factor, because if bacteria are unable to adhere in certain locations, the probability that they will cause disease is low. An excellent example is provided by *Escherichia coli*. Generally speaking, *E. coli* are harmless bacteria that occur by the billions in the human intestine. However, certain strains of *E. coli* have a variety of potent virulence factors, among which are especially efficient pili. These pili permit the cells to bind tightly to epithelial cells in the intestine. Other virulence factors may contribute to the watery diarrhea that many of us know as traveler's diarrhea.

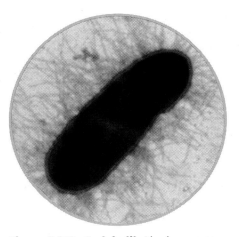

Figure 3.9 Bacterial pili. Also known as fimbriae, these structures, composed of protein, allow the cell to tightly adhere to surfaces. In some cases pili act as a virulence factor, because if disease-causing organisms cannot adhere to the surface of host tissues, they are far less likely to cause disease.

Sex pili, although made of the same protein as fimbriae, are longer and have an entirely different function. A bacterial cell known as the **donor** can grow a sex pilus, which connects it with another cell known as the **recipient**, in a process called **conjugation**. Once the cells are brought together by the sex pilus, small segments of DNA may be transferred from the donor to the recipient. The recipient is therefore genetically altered and may acquire characteristics present in the donor. Because this process is so important in bacterial genetics, and because the genes transferred via conjugation often code for important virulence factors, we will discuss conjugation more thoroughly in Chapter 6.

Flagella. **Flagella**, similar to pili, are thin protein tubes that extend out from the cell surface. They are, however, much longer than pili and they are used for locomotion rather than adherence. Prokaryotic flagella are able to spin, much like a boat's propeller, moving the cell through its environment. Each

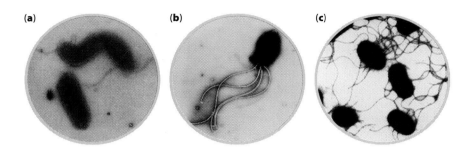

(a)   **(b)**   **(c)**

Figure 3.10 Bacterial flagella.
Flagellated bacteria may have a variable number of flagella, in any one of several arrangements. (a) Bacteria with a single flagellum are termed *monotrichous* ("single-haired"). (b) A *lophotrichous* ("having a tuft of hair") bacterium, with several flagella at one end. (c) These bacteria have flagella over their entire cell surface and are termed *peritrichous* ("hairy all around").

flagellum terminates in the cell's cytoplasm, where it is rotated by a "motor" protein in an energy-dependent manner. Depending on the species, a bacterium may have one, a few, or many flagella, and they may be present in any of several arrangements (**Figure 3.10**).

Not all bacteria have flagella, and those that do not are incapable of independent movement. Bacteria that can move independently are termed **motile**. Such bacteria use their flagella to move toward a favorable stimulus (exhibiting **positive taxis**) or away from unfavorable conditions (exhibiting **negative taxis**). For example, bacteria that are moving toward a favorable temperature are said to be exhibiting positive **thermotaxis**. A different bacterial species with different temperature requirements might move away from this same temperature, displaying negative thermotaxis. Movement toward or away from certain chemicals is likewise called positive or negative **chemotaxis**, respectively. Some bacteria require oxygen for survival, while for other species, oxygen is a toxin. These respective bacteria, if motile, might therefore demonstrate positive or negative **aerotaxis** (**Figure 3.11**).

Most prokaryotes are protected from the exterior environment by a rigid cell wall

Prokaryotic cells face many hazards in their environment. The plasma membrane in particular is susceptible to bursting if the cell absorbs excess water. Such bursting of the cell membrane is called **lysis**. Furthermore, many chemicals commonly found in a cell's environment might easily damage the plasma membrane. Bile, for instance, produced by the liver, aids in the breakdown of dietary fat in the intestine. Yet bacteria live in the intestine by the trillions, in spite of having a plasma membrane composed largely of lipid. Why are their membranes not digested as well? With very few exceptions, prokaryotes have a protective **cell wall** exterior to the plasma membrane. The cell wall thus serves as a barrier that protects the more vulnerable membrane beneath it from environmental perils.

The majority of bacteria have one of three basic cell wall types: Gram-positive, Gram-negative, or acid-fast. Although each type of cell wall allows bacteria to persist in the face of environmental hazards, they differ in important ways. To some extent cell wall type influences where the cells are found

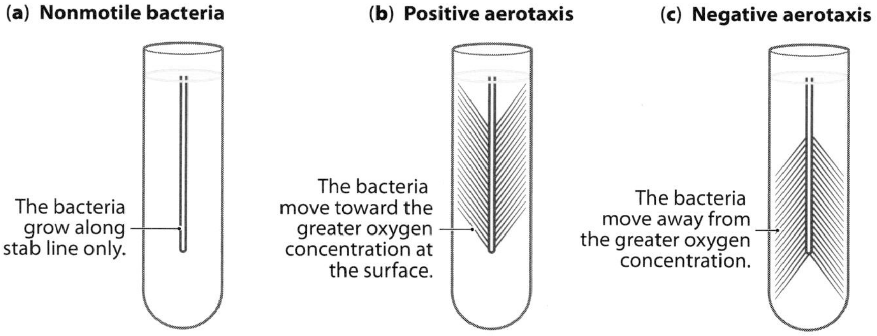

(a) Nonmotile bacteria   **(b) Positive aerotaxis**   **(c) Negative aerotaxis**

The bacteria grow along stab line only.

The bacteria move toward the greater oxygen concentration at the surface.

The bacteria move away from the greater oxygen concentration.

Figure 3.11 Experimental demonstration of bacterial motility. Motile bacteria have flagella while nonmotile bacteria do not. In this example, bacteria have been inoculated into a *semisolid deep*—a medium that permits bacterial growth and, because of its semisolid nature, permits motile bacteria to move. Bacteria are collected on an inoculating needle and stabbed directly into the medium. (a) Nonmotile bacteria can grow only along the stab line. Motile bacteria may show taxis. (b) A motile species that requires oxygen for growth will move *toward* the surface of the medium where oxygen is abundant. This is an example of positive aerotaxis. (c) A motile bacterial species for which oxygen is toxic will move *away* from the surface, displaying negative aerotaxis.

and the rate at which they grow. In certain cases, as we will see, the kinds of diseases caused by bacteria are a function of their cell walls, and treatment will often vary depending on cell wall type. A sound knowledge of cell wall structure and function is therefore crucial to our understanding of how these microorganisms live and how they affect us.

CASE: AN OCCUPATIONAL HAZARD

Dirk, a 24-year-old carpenter, accidentally cut his left hand while working with a circular saw. During the next week the injury site became red, swollen, and painful. A few days later, he began to experience chills and fever. He then sought medical attention from his doctor who, on the basis of the nature of the injury and Dirk's symptoms, suspected a *Staphylococcus* infection. The doctor sent a sample of pus from the wound to the laboratory for analysis. In the laboratory, the sample was Gram-stained and showed Gram-positive clusters of cocci, suggestive of *Staphylococcus*. The bacteria were grown in culture and confirmed as *Staphylococcus aureus*. A blood sample taken from Dirk was also positive for this species. The doctor prescribed a course of oxacillin, and within a few days Dirk had made a complete recovery.

1. **What about Dirk's case made the doctor suspect *Staphylococcus*?**
2. **What is a "Gram stain" and why was it used?**
3. **Why was oxacillin the appropriate drug in this case? Would it be recommended for an infection with Gram-negative bacteria as well?**

The Gram-positive cell wall. The major component of the **Gram-positive** cell wall is a complex molecule composed of sugars and amino acids called **peptidoglycan** (**Figure 3.12**). Multiple layers of this molecule surround the cell, protecting the cell from damage and helping it maintain its shape. While this molecule is found in the cell walls of both Gram-positive and Gram-negative bacteria, it is more abundant in Gram-positive cells.

Note in Figure 3.12b that the polysaccharide portion of peptidoglycan is composed of two alternating sugars, *N*-acetylglucosamine (NAG) and *N*-acetylmuramic acid (NAM). Attached to each NAM is a short chain of four amino acids. The sugar backbone of peptidoglycan is identical in all bacterial species but the amino acids in the short peptide chain may vary. The amino acid in the third position on one peptide is covalently linked to an amino acid in position four on the peptide on a neighboring layer of peptidoglycan. These linkages are via **pentaglycine bridges**, each composed of five glycine amino acids (Figure 3.12c). Pentaglycine bridges hold the individual layers of peptidoglycan together and give the cell wall its strength and its ability to both maintain cell shape and prevent cell rupture. Additionally, recall that in Chapter 2 we learned that sugars are hydrophilic in nature. Molecules like bile that might damage the hydrophobic lipids of the plasma membrane are themselves hydrophobic. Because bile is unable to interact with and cross the hydrophilic cell wall, it is prevented from reaching the vulnerable cell membrane.

The thick, hydrophilic Gram-positive cell wall may consist of up to 30 or 40 layers of cross-linked peptidoglycan. Those species with a greater number of layers are able to withstand increasingly harsh environmental conditions. For example, human skin is a fairly inhospitable environment for many bacteria. Not only is the skin too dry for most species, but salts deposited on the skin during perspiration can draw water out of bacterial cells. Those bacteria that do survive on the skin tend to be those Gram-positive species with especially thick cell walls. *Staphylococcus* in particular is an important part of the skin flora. This explains why Dirk's doctor suspected a staphylococcal ("staph") infection early on. Species such as *Staphylococcus aureus* may cause no particular problem on intact skin, but if the skin is broken these bacteria can gain access to subdermal tissue and the circulatory system, causing an infection as they did in Dirk.

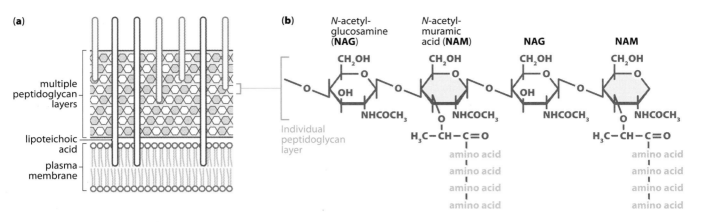

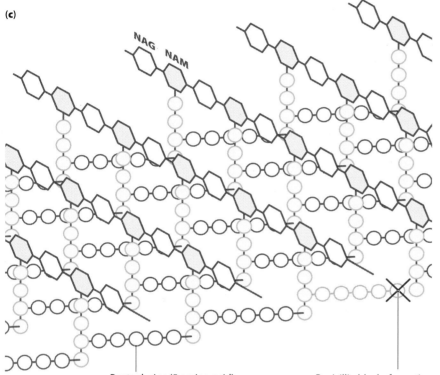

Pentaglycine (5-amino-acid) cross bridges hold peptidoglycan layers together.

Penicillin blocks formation of a pentaglycine bridge.

Figure 3.12 The Gram-positive cell wall. (a) The cell wall is exterior to the plasma membrane and consists of many layers of peptidoglycan. The number of peptidoglycan layers is variable. Molecules called lipoteichoic acid link the cell wall to the plasma membrane. (b) A single layer of peptidoglycan consists of a polysaccharide made of alternating *N*-acetylglucosamine (NAG) and *N*-acetylmuramic acid (NAM) sugars. A short chain of four amino acids is attached to each *N*-acetylmuramic acid. (c) Separate peptidoglycan molecules are linked by pentaglycine bridges, each composed of five glycine amino acids. Note that the pentaglycine bridges (open circles) form links between the short amino acid chains (in green) attached to each NAM. Penicillin prevents the formation of these pentaglycine bridges, weakening the cell wall of Gram-positive bacteria.

Peptidoglycan is a unique bacterial molecule; it is found in no other living system. As such, it makes an ideal target for antibacterial drugs, known as **antibiotics**. When antibiotics are developed to combat bacterial infections, it is crucial that the drug does not damage or kill human or other host cells in the process. Because peptidoglycan is lacking from all eukaryotic cells, the risk of unwanted side effects of drugs that target it is greatly reduced. The penicillins are a large and important group of antibiotics that inhibit peptidoglycan synthesis. Specifically, these drugs prevent the formation of pentaglycine bridges (see Figure 3.12c). Without these linkages between the peptidoglycan layers, the cell wall is much less rigid and the plasma membrane is prone to lysis. Oxacillin, an important member of the penicillins, works in this manner, which explains why it was so effective in Dirk's case. These antibiotics do not damage peptidoglycan that is already formed; they only inhibit the synthesis of new peptidoglycan. They are consequently most effective against actively dividing bacterial cells that are in the process of producing new peptidoglycan.

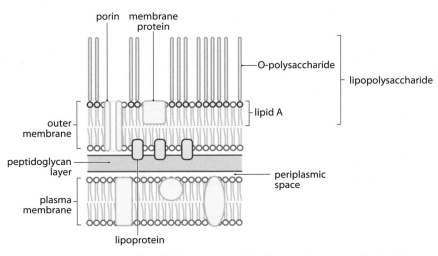

porin   membrane
protein

O-polysaccharide

lipopolysaccharide

outer
membrane

lipid A

peptidoglycan
layer

periplasmic
space

plasma
membrane

lipoprotein

Figure 3.13 The Gram-negative cell wall and outer membrane. In Gram-negative bacteria there is only a thin layer of peptidoglycan (in purple), sandwiched between the plasma membrane and an outer membrane. Like the plasma membrane, the outer membrane is composed principally of lipid and proteins and includes an important component called lipopolysaccharide (LPS). LPS consists of carbohydrates called O polysaccharides, which are anchored to the outer membrane by lipid A. Hydrophilic protein channels called porins permit the passage of small hydrophilic molecules and ions across the outer membrane. Lipoprotein attaches the outer membrane to the peptidoglycan layer.

The Gram-negative cell wall. Gram-negative bacteria protect their plasma membrane in a very different way. Like Gram-positive cells they have a peptidoglycan cell wall exterior to the plasma membrane, but in Gram-negative cells, this wall is a thin structure only one or two layers thick. Beyond this cell wall is a second **outer membrane (Figure 3.13)**. In some ways the structure of the outer membrane is similar to that of the plasma membrane, which we will describe shortly. However, the outer membrane contains a molecule unique to Gram-negative bacteria known as a **bacterial lipopolysaccharide** or **LPS**. As the name implies, this molecule is composed of both lipid and sugars. The lipid portion is known as **lipid A**. Sugars, called **O polysaccharides**, form the exteriormost portion of the outer membrane, and like all carbohydrates they are hydrophilic. Gram-negative cells are thus surrounded by a hydrophilic covering that prevents hydrophobic compounds such as bile from reaching and digesting either the outer or the plasma membrane.

Although most substances are unable to easily cross the outer membrane, proteins called **porins** permit the entry of certain materials (see Figure 3.13). Porins essentially form hydrophilic tunnels through the hydrophobic membrane, permitting small hydrophilic molecules and ions to enter. Amino acids, monosaccharides, and iron ions are examples of substances that can move through porins.

The LPS in the outer membrane is extremely important from a medical point of view. When Gram-negative bacteria die, their outer membranes break down and LPS is released into the environment. Lipid A in particular is recognized by the host immune system, and this recognition initiates a number of immunological events including fever and inflammation.

Such immune response is normal and beneficial when it occurs in a controlled manner, but when the host is suddenly exposed to a large amount of LPS, the overreaction of the immune system can result in shock or even death. This is why a ruptured appendix or a perforated bowel is such a medical emergency. The intestine is home to trillions of Gram-negative bacteria. Within the confines of the bowel, these organisms generally pose no medical risk and actually perform a number of beneficial functions. But when the integrity of the intestinal lining is breached, as it is when the appendix ruptures, bacteria flood into the normally sterile blood, causing a serious condition known as **bacterial sepsis**. As these bacteria die in large numbers, high levels of LPS are released into the bloodstream, increasing the likelihood of shock. Because of its toxic properties, LPS is also referred to as an **endotoxin**.

Treating a Gram-negative sepsis is tricky. Many antibiotics, such as the oxacillin used on Dirk, cannot cross the outer membrane and are thus useless in such a situation. Other antibiotics are able to cross this barrier, but even then, drugs must be selected with care. Many of our most powerful antibiotics, including the penicillins, act by killing bacteria and are therefore known as **bactericidal** agents. However, if bactericidal drugs were used to treat a Gram-negative sepsis, the cure might be worse than the disease, since endotoxin is released when Gram-negative bacteria die. High levels of endotoxin released by large numbers of dying bacteria could induce shock. **Bacteriostatic** drugs must therefore be used. These antibiotics prevent bacteria from reproducing; they do not directly kill bacterial cells. By keeping bacterial numbers in check, such drugs give the immune system an opportunity to contain the infection slowly. The steady, relatively low levels of endotoxin released in such instances, while associated with certain symptoms such as fever, are unlikely to pose a danger to the patient. Because treatment of Gram-positive and Gram-negative infections is often so different, clinicians must be able to quickly determine which type of infection they must treat.

The acid-fast cell wall. Although most bacteria have Gram-positive or Gram-negative cell walls, a few employ a third solution to the problem of plasma membrane protection. These **acid-fast** bacteria include in their ranks some of the most important **pathogens** (disease-causing organisms) afflicting humans and animals, including *Mycobacterium tuberculosis*, the causative agent of tuberculosis. The closely related *Mycobacterium leprae*, which causes leprosy, is also acid-fast. The acid-fast cell wall contains large amounts of waxes, interlaced with peptidoglycan. Waxes are essentially derivatives of very-long-chain fatty acids that give these cells a highly impervious wall. Acid-fast cells are thus very hardy and difficult to destroy. They are resistant to many toxic chemicals as well as the phagocytic activity of immune system cells.

All this protection, however, comes with a cost. Nutrients and oxygen also enter the cells very slowly, and thus the growth of acid-fast bacteria is slow. While typical Gram-positive or Gram-negative cells may divide in less than an hour under ideal conditions, *M. tuberculosis* might require 24 hours. Antibiotics used to treat tuberculosis are also absorbed slowly, which explains why the course of treatment is so long. A typical antibiotic prescription for Gram-positive or Gram-negative bacteria might be for 10 days. An infection with an acid-fast species might require treatment lasting six months or longer.

Bacteria with different cell wall types can be distinguished by specific staining techniques

When Dirk went to the doctor, a **Gram stain** was ordered as part of the laboratory analysis. This is the simplest way to quickly determine if bacteria are Gram-positive or Gram-negative. The simple stain discussed earlier in this chapter (see Figure 3.3) stains all cells in a similar manner. Although it can provide information about cell size and shape, it cannot distinguish Gram-positive from Gram-negative cells. The Gram stain, however, takes advantage of differences in cell wall structure, allowing us to distinguish between the two cell types. The first step is to stain all cells with the basic stain crystal violet. Consequently, at this stage, all cells will appear purple in color. When iodine is next applied, the crystal violet bound to the membrane forms large crystals that are unable to escape through the cell wall. The cells are subsequently exposed to an alcohol, which dissolves the outer membrane of Gram-negative cells; the Gram-positive cell wall is unaffected. The crystal violet can then be washed out of Gram-negative cells with distilled water, leaving them colorless; the Gram-positive cells will remain purple. Because colorless cells are difficult to see, a second basic stain, safranin, is applied that turns the Gram-negative cells pink. The Gram stain procedure is diagrammed in **Figure 3.14a**. Because most bacterial infections are caused by

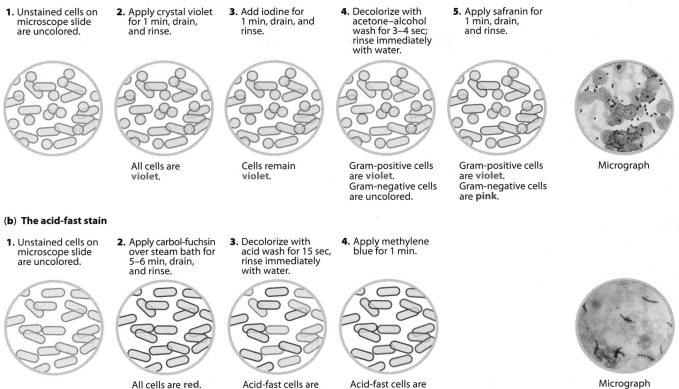

(a) The Gram stain

1. Unstained cells on microscope slide are uncolored.

2. Apply crystal violet for 1 min, drain, and rinse.

3. Add iodine for 1 min, drain, and rinse.

4. Decolorize with acetone–alcohol wash for 3–4 sec; rinse immediately with water.

5. Apply safranin for 1 min, drain, and rinse.

All cells are **violet**.

Cells remain **violet**.

Gram-positive cells are **violet**. Gram-negative cells are uncolored.

Gram-positive cells are **violet**. Gram-negative cells are **pink**.

Micrograph

(b) The acid-fast stain

1. Unstained cells on microscope slide are uncolored.

2. Apply carbol-fuchsin over steam bath for 5–6 min, drain, and rinse.

3. Decolorize with acid wash for 15 sec, rinse immediately with water.

4. Apply methylene blue for 1 min.

All cells are **red**.

Acid-fast cells are **red**. Non-acid-fast cells are uncolored.

Acid-fast cells are **red**. Non-acid-fast cells are **blue**.

Micrograph

either Gram-positive or Gram-negative species, and because these two types of infections are often treated differently, a Gram stain is often an appropriate way to quickly determine which of these two types of bacteria is causing a specific problem.

To identify acid-fast cells, a special **acid-fast stain** is used (see Figure 3.14b). Cells are initially stained with carbol-fuchsin, a reddish basic stain. Staining is performed over heat to allow the stain to penetrate the thick waxy acid-fast wall. Cells are then exposed to an acid wash. Gram-positive and Gram-negative cell walls are damaged by the acid, resulting in their decolorization. Acid-fast cells, on the other hand, are not damaged by the acid and they retain their red color. A blue counterstain, methylene blue, is used to visualize non-acid-fast cells. If a patient is displaying symptoms consistent with an infection caused by acid-fast bacteria, an acid-fast stain is a fast and efficient way to confirm the diagnosis.

Some prokaryotes lack a cell wall

A few bacteria lack a cell wall entirely. *Mycoplasma* species, for instance, have rigid lipid molecules called **sterols** in their plasma membranes to help stabilize their structure. Nevertheless, the lack of a sturdy cell wall means that these bacteria have an extremely variable shape. One species in particular, *M. pneumoniae*, causes a relatively mild form of lower respiratory tract infection called "atypical" or "walking" pneumonia. As you might guess, penicillin, which inhibits cell wall formation, has no effect on these organisms. Infections such as this must be treated with other antibiotics, which target structures other than the cell wall.

Figure 3.14 Common staining procedures to identify bacterial cell wall type. (a) The Gram stain takes advantage of the differences in Gram-positive and Gram-negative bacteria to distinguish between these two important groups. (b) The acid-fast stain differentiates between acid-fast and non-acid-fast bacteria. Because these staining procedures stain cells with different structures differently, they are both examples of *differential stains*.

Each cell is surrounded by a plasma membrane

The plasma membrane lies just below the cell wall. This thin, flexible structure separates the cell's cytoplasm from the external environment. Its primary responsibilities are to contain the cytoplasm and to regulate what enters and leaves the cell.

Plasma membrane structure is very similar in bacteria and eukaryotes. In both cell types, the plasma membrane contains a large proportion of lipids called **phospholipids** (**Figure 3.15a**). A phospholipid has a similar structure to a fat (see Chapter 2, p. 34) with one big exception: while a fat consists of three fatty acids covalently bound to a molecule of glycerol, a phospholipid has only two fatty acid chains. A **phosphate group** (PO_4^-) is bound to the glycerol in place of the third fatty acid. This gives a phospholipid a very important property: because the phosphate group carries a negative charge, it is hydrophilic. The fatty acid chains, on the other hand, are hydrophobic. As we learned in Chapter 2, the hydrophilic region will be attracted to water, while the hydrophobic regions will aggregate with other hydrophobic molecules, away from water. In cell membranes, this results in the formation of a lipid **bilayer**, consisting of two layers of phospholipids. The phosphate groups project out toward the extracellular environment and in toward the cytoplasm, where they can hydrogen-bond with water. The hydrophobic hydrocarbon chains of the fatty acids are sandwiched between the two layers of phosphates, away from water (Figure 3.15b).

The other principal components of the plasma membrane are proteins. Membrane proteins typically make up about half, but sometimes as much as 70%, of a membrane's mass. Many of these are **carrier proteins** that span the membrane from the **extracellular** side (the side of the membrane outside the cell) to the **cytoplasmic** side (the side on the cell's interior) and function in the transport of materials across the plasma membrane. Carrier proteins and other membrane proteins that extend across the membrane

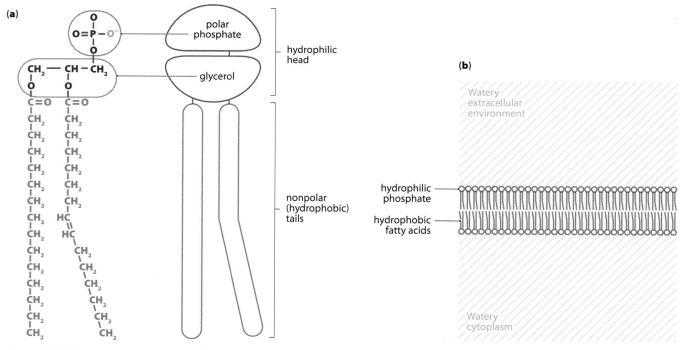

Figure 3.15 The plasma membrane. Phospholipids are the major lipid component of bacterial plasma membranes. (a) Diagram and chemical composition of a phospholipid. The phosphate group, because of its negative charge, is hydrophilic. The fatty acid chains, composed of carbon and hydrogen only, are nonpolar and thus hydrophobic in nature. (b) In a watery environment, phospholipids form a bilayer. Hydrophilic phosphate groups interact with water on both the exterior (extracellular) and interior (cytoplasmic) sides of the membrane. Hydrophobic fatty acids are found in the membrane's interior, away from water.

from the extracellular to the cytoplasmic side are collectively called **trans-membrane proteins**. Other proteins, called **peripheral proteins**, are bound to the membrane surface, where they may carry out diverse functions.

The phospholipids in the membrane are not bound to each other. The bilayer that they form merely results from the way in which the phosphates in phospholipids are attracted to water and the fatty acid tails are not. If it were not for water, the phospholipid bilayer would not form. The plasma membrane is therefore not a rigid structure. Rather, it has the consistency of oil. Because of its fluidity, it is in a constant state of flux, and the membrane phospholipids are continually moving around and changing position within the membrane. Neither are the proteins bound to the phospholipids. Their behavior in the phospholipid bilayer is akin to that of icebergs floating in a lipid sea. Because the membrane is dynamic and the constituent molecules are able to move, the membrane is said to conform to the **fluid mosaic model** of membrane structure (**Figure 3.16**).

Because of its vital role, one might think that the cell membrane would make a tempting target for antibiotics. Certainly many compounds can disrupt the plasma membrane and kill bacterial cells. Yet the plasma membrane is rarely the target of antibiotic activity. Since the basic structure of the plasma membrane is nearly identical in bacterial and eukaryotic cells, any antibiotic that disrupts bacterial membranes is likely to damage our membranes as well.

The liquid interior of the cell forms the cell's cytoplasm

The cytoplasm is the liquid portion inside the cell surrounded by the plasma membrane. It is composed primarily of water and is the site of most of the cell's metabolic reactions. Although the prokaryotic cytoplasm does not contain the membrane-bound organelles found in eukaryotic cells, a small number of important subcellular structures are present.

The bacterial chromosome. Bacteria typically have a single, circular loop of double-stranded DNA called the **bacterial chromosome** (**Figure 3.17**). Unlike the chromosomes of eukaryotic cells, which are surrounded by a nuclear membrane, bacterial chromosomes are free in the cytoplasm. The area in the cytoplasm containing the chromosome is called the **nucleoid area**. Many bacterial cells, in addition to their chromosome, have one or more much smaller, circular DNA molecules called **plasmids**. Although plasmids generally carry a very small number of genes, these genes often code for virulence factors, which can be passed to another bacterium via conjugation. This topic will be explored in Chapter 6.

Ribosomes. Ribosomes are composed of RNA and protein. Ribosomes are the site of protein synthesis, and it is on their surface that amino acids are linked together into proteins. The bacterial cytoplasm contains up to 20,000 ribosomes, which give the cell a grainy appearance when viewed with an electron microscope (see Figure 3.17).

Although ribosomes are found in both prokaryotic and eukaryotic cells, prokaryotic ribosomes are smaller and have a somewhat different structure than eukaryotic ribosomes. Many common antibiotics such as erythromycin bind to and interfere with the smaller bacterial ribosomes, arresting protein synthesis. Since these drugs do not bind to larger eukaryotic ribosomes, they generally can be used safely without risk to the host.

Storage granules and inclusion bodies. As the name implies, storage granules serve as repositories for various nutrients. Granules composed of carbohydrate or lipid, for instance, often serve as reservoirs for carbon, nitrogen, or phosphate. Inclusion bodies do not store nutrients but may have other specialized functions. Some aquatic bacteria, for example, have gas-

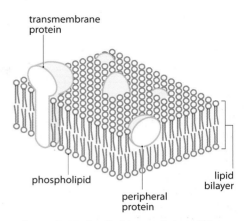

Figure 3.16 The fluid mosaic model of membrane structure. The plasma membrane, composed mainly of proteins embedded in a phospholipid bilayer, has the consistency of oil. Because of its fluid nature, membrane components are in a constant state of motion within the membrane. Proteins that extend all the way through the phospholipid bilayer are called *transmembrane proteins*. Proteins that attach to the membrane's surface are called *peripheral proteins*.

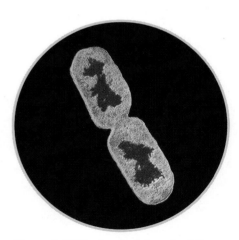

Figure 3.17 The bacterial nucleoid area. In this photo, taken with a transmission electron microscope and subsequently colored, the nucleoid area, containing the chromosome, appears red. Cell division of one original cell into two new daughter cells is almost complete. The grainy appearance of the cytoplasm is due to the presence of numerous ribosomes.

filled inclusion bodies in their cytoplasm. These protein-coated structures allow the cell to float near the surface, providing optimum conditions for their growth.

CASE: THE ANTHRAX SCARE

In late 2001, soon after the terrorist attacks of September 11th, the United States was in the grip of a new terror threat (**Figure 3.18**). Between October 4th and November 20th, twenty-two cases of anthrax were reported. Eleven of these cases were cutaneous anthrax, while eleven were pulmonary anthrax. Five of the patients who contracted pulmonary anthrax died. The source of the outbreak was traced to letters contaminated with *Bacillus anthracis* endospores. The letters were deliberately sent by the perpetrator to a variety of prominent individuals, including members of the media and the U.S. Senate. Twenty of the 22 patients (91%) were mail handlers or others who entered workplaces where contaminated mail was processed or received.

1.   Why is *Bacillus anthracis* such a potent bioterrorism agent?

Endospores. Many bacteria are capable of causing serious disease, but only a few are considered viable as biological weapons. Those that are, including *Bacillus anthracis*, frequently form structures called **endospores**. These unique bacterial structures form in response to nutrient depletion or other unfavorable environmental conditions. Under such circumstances, the bacterial cell first replicates its chromosome (**Figure 3.19**). The newly replicated chromosome is then walled off from the rest of the cytoplasm by a portion of the plasma membrane. This membrane barrier, which divides the cell into two unequal compartments, is called the **septum**. The smaller of these two compartments is next engulfed by the larger compartment, forming a double membrane around the new chromosome. Layers of peptidoglycan are produced between the two membrane layers, and the entire structure is next surrounded by a thick coat of protein. When endospore formation is complete the cell lyses, releasing the spore into the environment.

Endospores are metabolically inactive. Because of their thick coat they can remain dormant in the environment for years or decades, impervious to the effects of heat, desiccation, radiation, and chemical agents. If and when conditions for growth become favorable, an endospore can germinate. As conditions improve, the permeability of the coat increases, and as water enters, the endospore swells, eventually rupturing the coat. The developing

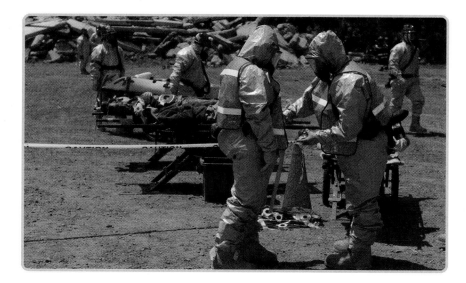

Figure 3.18 First responders to a possible bioterrorism event. Note the respirators and other protective clothing worn by emergency response personnel to prevent their exposure to possible infectious agents such as anthrax endospores.

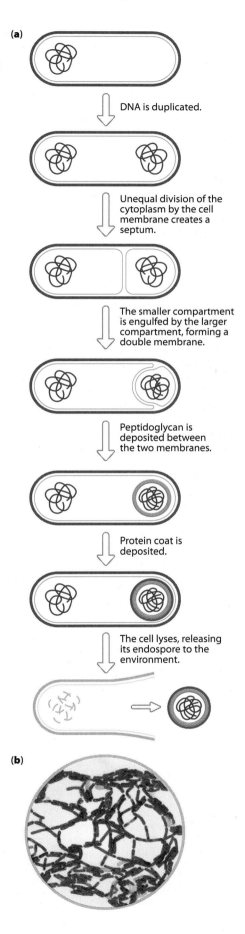

Figure 3.19 Bacterial endospores. (a) The process of endospore formation. The released endospore is protected by a double membrane, with peptidoglycan (in purple) between the two membranes, as well as an exterior protein coat (in blue). (b) Endospores (green) in the cytoplasm of bacilli (red).

(a)

DNA is duplicated.

Unequal division of the cytoplasm by the cell membrane creates a septum.

The smaller compartment is engulfed by the larger compartment, forming a double membrane.

Peptidoglycan is deposited between the two membranes.

Protein coat is deposited.

The cell lyses, releasing its endospore to the environment.

(b)

cell then emerges, resuming metabolic activity and reproduction. Endospores thus represent a unique adaptation, permitting cells to persist over long periods of severe conditions.

How long can endospores survive? This remains a difficult question to answer, but certainly most can survive at least a year, and many remain viable for much longer. In the mid-1970s endospores were recovered from compacted clay in Roman ruins dated from about 90 AD. When placed in a warm oxygen-rich environment, some of these endospores germinated and were identified as *Thermoactinomyces vulgaris*. Even more remarkable was the announcement in 1995 that viable endospores were recovered from the digestive system of a bee that had been encased in amber for 25 million years. Researchers allowed these endospores to germinate, and an analysis of their DNA indicated that they were a species of *Bacillus* that was no longer living today. In other words, the scientists had resurrected a formerly extinct species of bacteria!

Most common endospore formers are species of *Clostridium* and *Bacillus*. You will recall the case in Chapter 2, where a large number of individuals became dangerously ill after ingesting canned bamboo shoots contaminated with *Clostridium botulinum* endospores. Other important endospore formers include *C. tetani*, which causes tetanus, and *C. perfringens*, the causative agent of gas gangrene. Endospores of these organisms are found in the environment, but germination and growth can occur only in the absence of oxygen. When the endospores of *C. tetani*, for instance, enter the body through a puncture wound, the favorable, oxygen-free conditions permit germination. As the bacteria begin to grow they release a powerful toxin that results in tetanus or "lockjaw."

Bacillus anthracis, the causative agent of anthrax, is a Gram-positive rod that is also an endospore former. Its endospores can survive in the soil for decades. Anthrax is mainly a disease of animals, particularly livestock that might ingest the endospores while grazing. Humans occasionally become infected when viable endospores come in contact with broken skin (they may contract cutaneous anthrax) or are respired into the lungs (where they cause pulmonary anthrax). The *Bacillus* endospores germinate in these moist, oxygen-rich environments, and the emerging cells resume reproduction. Cutaneous anthrax, although it causes skin ulceration and other symptoms, is unlikely to be lethal. Pulmonary anthrax, on the other hand, results in a rapidly progressing, potentially fatal form of pneumonia, for which immediate treatment is required. Because it can be so deadly and because it forms endospores, *B. anthracis* is in some ways an ideal bioterrorism tool. Most non-endospore-forming bacteria, no matter how deadly, would be unable to persist in the environment long enough to infect many people. Consequently, many potential bioterrorism agents would not survive in an envelope long enough to be effectively spread through the mail. The viable *B. anthracis*, on the other hand, can be spread to unsuspecting individuals as it was in the 2001 mail attacks (**Figure 3.20**), because the endospores can easily survive their transit through the mail. This is not, however, as easy as it may sound. Creating an effective anthrax weapon requires that the endospore-containing material be produced in a "weaponized" form that is easily transmissible through the air. Fortunately, this requires a level of sophistication, knowledge, and technology that is not available to every would-be terrorist.

Figure 3.20 **The fall 2001 anthrax mail attacks**. This letter, which contained anthrax spores, was sent to Senator Tom Daschle of South Dakota. Other letters were mailed to several other prominent individuals.

The Eukaryotic Cell

Having considered prokaryotic cell structure and function, we will now turn our attention to eukaryotic cells. All eukaryotes have cells that are structurally similar. Typically they are larger than prokaryotic cells and are far more complex internally. Like prokaryotic cells, they are all bound by a plasma membrane. Their most distinctive feature is that they contain many organelles, each surrounded by a membrane, such as the nuclear membrane enveloping the genetic material. Eukaryotic cells may or may not have a cell wall or extracellular appendages (see Figure 3.1).

Different organelles perform different functions, and they allow the compartmentalization of eukaryotic cells; that is, they allow complex functions to take place in specific regions of the cell. For instance, powerful enzymes used to digest nutrients are restricted to specific organelles, ensuring that these enzymes do not damage the cell itself. Other organelles are involved with biomolecule synthesis, energy use, and intracellular transport.

The number of cells in a eukaryotic organism is variable

Most prokaryotes are **unicellular** (one-celled) organisms (see Chapter 1, p. 4). Some eukaryotes are also unicellular, but many others are **multicellular** (many-celled); some of the larger eukaryotic organisms consist of billions or trillions of cells. All eukaryotic cells, whether they are a unicellular organism or part of a larger multicellular plant or animal, share many important structural features. Eukaryotic cells are highly diverse, however, making it difficult to describe a "typical" example. Protozoa, for instance, are eukaryotes (**Figure 3.21**) and are generally unicellular; each individual cell must constitute a fully independent, self-contained entity, carrying out all critical life functions. It must, for instance, be able to find and ingest food and generate energy without assistance from other cells. In a multicellular animal or plant, on the other hand, all cells work together as a single organism. Consequently, different cells specialize in different functions. Fat cells, for example, store energy, while muscle cells allow movement and nerve cells conduct electrical impulses. In humans, there are several hundred cell types, each type specialized for specific functions. An individual liver or lung cell is incapable of independent living because of its dependence on other cells in the body. Each cell type has only those organelles and structures required to carry out its particular role.

In multicellular eukaryotes, groups of cells of one or more types often work together for particular functions. Such groupings are called **tissues**. Examples in animals include nervous, connective, muscle, and epithelial tissue. Nervous tissue consists not only of the nerve cells themselves (the neurons) but also a variety of other cell types including Schwann and glial cells that surround, protect, and nourish the neurons. Groups of tissues that work together to carry out specific tasks are called **organs**. Examples in animals include the heart, liver, brain, and skin.

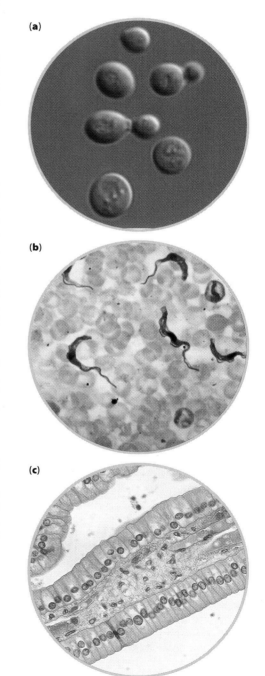

(a)

(b)

(c)

Figure 3.21 **Representative eukaryotic cells**. (a) *Saccharomyces cerevisiae*, referred to commonly as brewer's yeast, is a unicellular fungal microorganism. (b) *Trypanosoma brucei* (in dark purple), the causative agent of African sleeping sickness, is a unicellular protozoan parasite. Red blood cells are seen in pale red. (c) Epithelial tissue in a multicellular animal. Epithelial tissue is found lining body surfaces. Note the darkly stained nuclei.

Many organisms of interest to the microbiologist are eukaryotes. More children are killed each year by the protozoan *Plasmodium*, the causative agent of malaria, than any other single microorganism, prokaryotic or eukaryotic. Countless people are infected with other protozoans including amoebas and the trypanosomes that cause sleeping sickness (see Figure 3.21). Fungal microorganisms permit bread to rise, give flavor to cheese, and recycle nutrients in the environment. They also spoil our food and cause athlete's foot.

We will revisit these organisms in more detail in the next chapter. Here we will focus on important features of eukaryotic cells, especially those aspects that distinguish them from prokaryotic cells.

While plant and fungal cells have cell walls, animal cells do not

As previously discussed, the vast majority of prokaryotic cells are surrounded by a rigid cell wall. Among eukaryotes, cell walls are the rule in plants and fungi, where they help to maintain cell shape, as they do in prokaryotes. Plant cells are surrounded by a cell wall composed of cellulose, a polysaccharide consisting of a long chain of glucose molecules. Fungal cell walls may be composed of cellulose as well, but many are constructed out of chitin, a nitrogen-containing polysaccharide. The presence of a cell wall in large, multicellular eukaryotes is an adaptation to a stationary mode of life. Because neither plants nor fungi move, their cells profit from the protection and rigidity provided by the cell wall. Cell walls are lacking in both protozoa and animals. All animals are motile, at least during part of their lives. A rigid cell wall surrounding their cells would be incompatible with locomotion.

Like prokaryotes, all eukaryotic cells are surrounded by a membrane

The eukaryotic cell membrane is structurally similar to that described for prokaryotes (see Figure 3.16). Both are composed of phospholipid bilayers associated with proteins. As in prokaryotes, the plasma membrane separates the cell cytoplasm from the extracellular environment and regulates transport into and out of the cell. Many proteins in the eukaryotic cell membrane are covalently linked to polysaccharides that extend into the extracellular environment (**Figure 3.22**). These **glycoproteins** play important roles in cell-to-cell communication. Such communication is especially important in multicellular organisms, where activities in different cells must be coordinated.

Most eukaryotic plasma membranes contain lipids called sterols in addition to the phospholipids. Sterols are rarely found in prokaryotic plasma membranes. These molecules give the membrane added durability and flexibility. In animal cells, cholesterol serves this function, while ergosterol is found in the membranes of fungal cells. Because it is sufficiently different in structure from cholesterol, ergosterol is a frequent target for antifungal drugs.

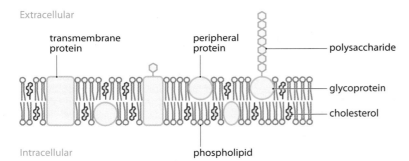

Figure 3.22 The animal cell membrane. The structure of the cell membrane in eukaryotes is very similar to that described for prokaryotes. In this diagram of the animal cell membrane, note the presence of membrane glycoproteins and cholesterol, which provide added flexibility and strength to the membrane. Cholesterol is characteristic of animal cell membranes only. It is absent from the cell membranes of other eukaryotes.

Many eukaryotic cells have either flagella or **cilia**, which permit cell movement or generate a water current over the cell surface (**Figure 3.23**). Cilia are shorter than flagella but otherwise have a similar structure; both are composed of numerous rod-shaped proteins called microtubules. They are able to beat back and forth very rapidly, as many as 40 times a second. Cilia may help move a cell through its environment, or in multicellular organisms they may move material over the cell surface. For instance, in the female mammal reproductive tract, the gentle sweeping of ciliated cells moves eggs from the ovary through the fallopian tube to the uterus.

The cytoplasm of eukaryotic cells contains a variety of membrane-bound organelles

The fluid-filled cytoplasm contains many organelles, as well as a variety of cytoplasmic proteins and other structures such as ribosomes. A system of protein filaments in the cytoplasm called the **cytoskeleton**, act as a sort of scaffolding for the cell (**Figure 3.24**). These long, slender protein filaments assemble into a dense mesh immediately beneath the plasma membrane. They also form an extensive network throughout the cytoplasm that anchors organelles at specified locations, and they function as a sort of microscopic rail system, along which intracellular transport occurs.

The cytoskeleton is often cited as a unique feature of eukaryotic cells. Recent findings, however, bring this assumption into question. Within the last few years, researchers have identified proteins in bacteria that are structurally very similar to the cytoskeletal proteins in eukaryotic cells. Furthermore, it has been found that, as in eukaryotes, these proteins form a framework beneath the plasma membrane. These results have since been confirmed in all other prokaryotic species that have been studied. Should this important finding gain acceptance across the scientific community, it appears that biology teachers will have to rewrite at least one lecture on cell structure!

We will conclude our tour of the eukaryotic cell by briefly describing the principal organelles found in the cytoplasm that distinguish eukaryotic from prokaryotic cells. Additional information about these important structures will be provided in subsequent chapters, when we discuss their specific roles in microbial processes or in microbe–host interaction. Remember that many eukaryotic cells have very specific functions, especially in multicellular organisms. Consequently cells differ in terms of both the types and numbers of organelles present.

The nucleus. The word "eukaryotic" comes from the Greek for "true nucleus." The nucleus is the largest organelle in the cell and it is bounded by a double-layered membrane called the **nuclear envelope**, which surrounds the DNA. The envelope has many pores in its surface to facilitate transport between the cytoplasm and the interior of the nucleus. Among its various functions, the nuclear membrane is thought to help the cell regulate the rate at which the information stored in DNA is used to make proteins.

The endomembrane system. We have learned that cells are usually busy places, with lots of transport of substances in and out. This transport occurs in several ways. Certain small, hydrophobic molecules can simply move across the plasma membrane. Specialized transport proteins in the membrane move other relatively small hydrophilic substances into and out of the cell as needed. Prokaryotes, however, are unable to either import or export large molecules. Consider the peptidoglycan found in bacterial cell walls. Bacteria are unable to secrete preassembled peptidoglycan. Rather, they secrete small subunits of this material, which then assemble outside the cell. Likewise, bacteria cannot import large molecules. Such molecules must be digested outside the cell. Smaller subunits can then be brought in via transport proteins. Eukaryotes, on the other hand, can move many large

Figure 3.23 Eukaryotic cilia. Cilia may provide motility in single-celled organisms. In multicellular organisms they help to move fluid over the surface of the ciliated cell.

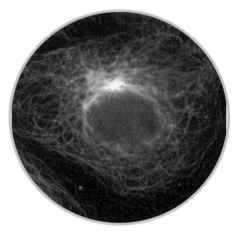

Figure 3.24 The cytoskeleton. The cytoskeleton appears as a dense meshwork of long protein fibers. The fluorescent-microscopic technique used to take this image reveals only the cytoskeleton and leaves other structures obscured. Ordinarily the cytoskeletal elements would extend throughout the eukaryotic cell cytoplasm.

Figure 3.25 **The endoplasmic reticulum**. (a) Electron micrograph of rough ER. The dark, grainy regions on the surface of the ER are ribosomes. (b) An antibody-secreting *plasma cell*. This immune system cell is specialized for the production of protein antibodies, which are secreted by the cell into the blood. The darkly stained, large nucleus is seen in the center of the cell. Note the cytoplasm filled with extensive rough ER. (Courtesy of Leilo Orci, University of Geneva, Switzerland.)

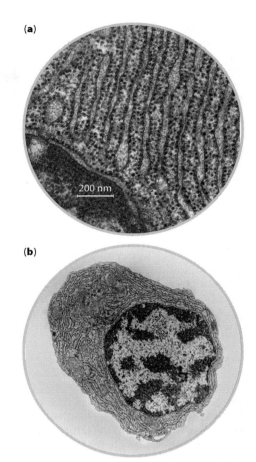

(a)

200 nm

(b)

molecules or particles directly into or out of the cell. To do this, they rely on a group of organelles collectively known as the **endomembrane system**. Cells that are actively involved in such import or export appear to have cytoplasm that is crammed with membrane, much of which is part of this cellular transport system.

To explain how this intricate system works, we will first consider how a large protein is exported out of the cell. We will then describe the opposite problem: how cells transport large substances in, and what then happens to them once they are inside the cell.

As we have discussed, amino acids are linked into proteins on structures called ribosomes. If a protein is destined to be exported from the cell, soon after protein synthesis begins, the ribosome attaches to the complex system of double membranes that extend throughout the cytoplasm called the **endoplasmic reticulum (ER)**. The presence of ribosomes on part of the ER surface gives it a "knobby" appearance, and it is called **rough ER** (**Figure 3.25a**). Other regions of the ER lacking ribosomes are called **smooth ER**. Both types of ER are seen in most eukaryotic cells. In multicellular organisms, cells specialized for the production and secretion of specific proteins can be recognized by their especially extensive network of rough ER. Immune system cells called plasma cells, specialized for the production and secretion of antibodies into the blood, are an example (Figure 3.25b).

As a protein is being produced on the surface of rough ER, the protein is threaded through the ER membrane into the interior space, or **lumen**, of the ER (**Figure 3.26**). Once in the lumen, the protein is transported through the ER, toward the membrane. Along the way, the protein assumes its three-dimensional shape and may be modified by the addition of sugars. When it reaches the furthest portion of the ER, the protein buds off in a small,

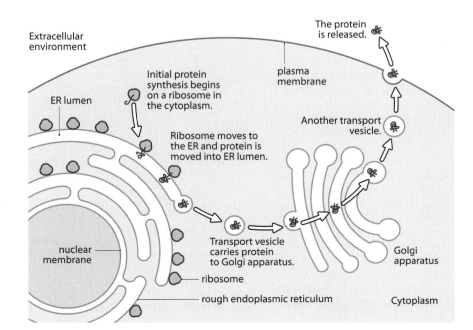

Extracellular environment

ER lumen

Initial protein synthesis begins on a ribosome in the cytoplasm.

plasma membrane

The protein is released.

Another transport vesicle.

Ribosome moves to the ER and protein is moved into ER lumen.

Transport vesicle carries protein to Golgi apparatus.

Golgi apparatus

nuclear membrane

ribosome

rough endoplasmic reticulum

Cytoplasm

Figure 3.26 **The path of exported proteins in eukaryotic cells**. Protein synthesis begins on a ribosome (in dark gray) that is free in the cytoplasm. If the protein (in red) is to be used within the cell, the ribosome remains free in the cytoplasm, but if the protein is to be exported, the ribosome attaches to the ER. As the protein is being made, it is threaded into the ER lumen. Once inside the lumen, it is transported through the ER. Along the way the protein folds up into its specific three-dimensional shape. Eventually the protein buds off the ER in a transport vesicle, which moves along the cytoskeleton to the Golgi apparatus. The transport vesicle fuses with the membranes of the Golgi apparatus and moves through the Golgi's series of membrane-bound sacs. The protein, which may be modified within the Golgi apparatus, buds off in another transport vesicle and moves to the plasma membrane. When the transport vesicle fuses with the plasma membrane, the protein is released. This process is known as *exocytosis*.

membrane-enclosed sphere called a **transport vesicle**. The vesicle moves along the cytoskeleton until it reaches a different organelle called the **Golgi apparatus**.

The primary function of the Golgi apparatus is to receive materials from the ER and direct them to their ultimate destination. This organelle is composed of a series of flattened membrane-bound sacs (see Figure 3.26). When a transport vesicle containing a protein buds off the ER, it fuses with the Golgi apparatus, releasing its protein into the series of membrane sacs. As the protein is transported through this stack of membranes, it is further modified and routed to its final destination. This destination may be the plasma membrane, where the protein will serve as a membrane protein, or it may be another organelle within the cell. Alternatively the protein may be transported to the cell surface in another membrane-bound vesicle that will fuse with the plasma membrane and release the protein to the extracellular environment. This process, known as **exocytosis**, is unique to eukaryotes; it does not occur in bacteria. All vesicle transport from the ER to the Golgi apparatus and from the Golgi apparatus to the final destination is along cytoskeletal elements. Other molecules are brought into the cell when they first bind to the cell surface and are next surrounded by portions of the plasma membrane. The membrane forms a vesicle around the cell, which is then brought into the cytoplasm by **endocytosis**.

CASE: PARROT FEVER

Thinking that their two sons should have a pet, Joe and Charlotte buy the boys a pair of cockatiels. The birds initially seem healthy but within a few weeks they appear inactive and listless. Concerned about the birds, Joe takes them to the veterinarian, who diagnoses a case of psittacosis, otherwise known as parrot fever. The veterinarian recommends giving the birds food that has been impregnated with the antibiotic tetracycline. He also tells Joe to exercise great care while cleaning the cage for the next few weeks. Curious about the disease, Joe reads up on psittacosis. He finds that it is caused by the bacterium *Chlamydia psittaci*, which is most common in parrots and related birds. Pet stores that sell birds are a regular source of *C. psittaci* infection. Joe also learns that these bacteria are unusual in that during an infection, they are engulfed but not easily destroyed by phagocytic immune system cells.

1. **How do the bacteria resist destruction by phagocytic cells?**
2. **Why did the veterinarian caution Joe about cleaning the cockatiels' cage?**

When a large molecule is brought into a cell by endocytosis, the transport vesicle containing the imported material is usually routed along the cytoskeleton to another organelle called the **lysosome** (**Figure 3.27**). The lysosomes contain powerful digestive enzymes. The transport vesicle and the lysosome fuse, releasing the vesicle's cargo into the lysosome interior. The lysosome's enzymes then digest whatever was carried in the transport vesicle. In many cases, the digested products are nutrients, which are then available to the cell.

Lysosomes play an especially important role in certain white blood cells of the immune system. When such cells contact invading microorganisms such as a bacterium or a virus, they first adhere to and then engulf them by extending a portion of the plasma membrane around the intruder. The membrane fuses around the microbe and seals it off in a transport vesicle. The vesicle then fuses with the lysosome and the microbe is digested.

This process, called **phagocytosis** (**Figure 3.28**), is a type of endocytosis, and it is crucial to a successful immune response. However, some microorganisms have devised insidious strategies to avoid destruction in this manner. For instance, the *C. psittaci* infecting the cockatiels in the case are

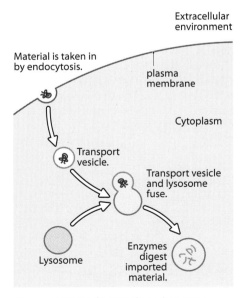

Figure 3.27 Endocytosis. Eukaryotes can bring large molecules into the cell via endocytosis. The adherence of the foreign material to the plasma membrane causes the membrane to engulf the material, eventually enclosing it in a transport vesicle. The vesicle, which moves along the cytoskeleton, may then fuse with a lysosome. Enzymes within the lysosome digest the imported material.

ingested by phagocytic cells called macrophages. Once ingested, these bacteria are at least temporarily able to prevent fusion of the transport vesicle and the lysosome. If they reproduce sufficiently in the vesicle, they may kill the macrophage, and the dying cell releases chemicals that can stimulate inflammation.

It is not clear how the bacteria are able to prevent lysosome fusion, but it is probably due to the secretion of a compound that alters the membrane of the transport vesicle in a way that blocks fusion. Tetracycline works well against *C. psittaci*, but the treatment can last as long as a month or more. During that time, infected birds may still be shedding bacteria in their feces. If a human breathes in dust from these droppings, he or she can develop pneumonia. Hence the veterinarian's warning about cleaning the birds' cage.

Chloroplasts and mitochondria. Chloroplasts and mitochondria are involved in the production and utilization of energy by the cell. Within chloroplasts, carbon dioxide and water are converted into sugar, using the energy found in sunlight. This process is known as **photosynthesis**. Chloroplasts are found in the photosynthetic cells of green plants and algae. Even in plants, chloroplasts are not found in all cells but only in leaves and stems, where photosynthesis occurs. Mitochondria, on the other hand, are found in all but a few eukaryotic cells. These organelles are involved in the release of energy from the chemical bonds in biological molecules. We will return to the topic of energy use and production in Chapter 7.

The structure of chloroplasts and mitochondria is similar (**Figure 3.29**). In both of these organelles a double membrane surrounds a space known as the **stroma** in chloroplasts and the **matrix** in mitochondria. Within the stroma are stacks of membranous vesicles called **thylakoids**. The various reactions of photosynthesis and energy release take place either on the surface of these membranes or in the stroma or matrix.

Some bacteria are also capable of photosynthesis, and many carry out processes associated with the mitochondrion as well. Because they do not have organelles, these processes take place in the cytoplasm or in association with the plasma membrane.

Chloroplasts and mitochondria are especially interesting because in many ways these organelles resemble bacterial cells themselves. They are approximately the same size and shape as some bacteria and they contain their own DNA. The DNA is organized in a loop as it is in prokaryotes. Additionally,

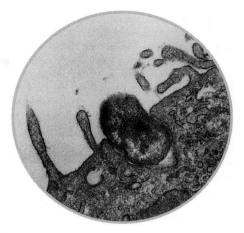

Figure 3.28 A macrophage phagocytosing a bacterial cell. Macrophages are white blood cells that play an important role in defense against microorganisms. Macrophages are able to engulf bacteria and other foreign material in a type of endocytosis called phagocytosis. The bacterium is the small, oval-shaped body in the center of the image. The macrophage is extending its membrane around the bacterium as it begins to ingest it. Normally, the ingested bacteria are killed by lysosomal enzymes. *Chlamydia psittaci* and certain other bacteria are able to at least temporarily survive inside macrophages because they can block the fusion of the transport vesicle to the lysosome.

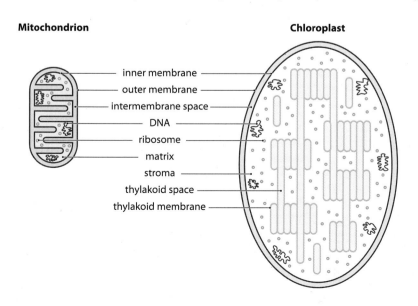

Mitochondrion   Chloroplast

- inner membrane
- outer membrane
- intermembrane space
- DNA
- ribosome
- matrix
- stroma
- thylakoid space
- thylakoid membrane

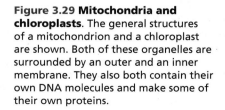

Figure 3.29 Mitochondria and chloroplasts. The general structures of a mitochondrion and a chloroplast are shown. Both of these organelles are surrounded by an outer and an inner membrane. They also both contain their own DNA molecules and make some of their own proteins.

chloroplasts and mitochondria produce some of their own proteins, and these proteins are made on ribosomes that match those of bacteria in terms of size and mass. These similarities have given rise to the proposition that originally chloroplasts and mitochondria *were* bacteria that entered into a symbiotic relationship with other cells over a billion years ago. Although this idea was first proposed in the nineteenth century, it was ridiculed as nonsense until the last few decades. Gradually, however, the evidence in favor of this concept, known as **endosymbiosis**, the union of a bacterial cell and larger cells into a mutually beneficial relationship, has built to such a degree that it is generally accepted by the scientific community. The precise manner in which endosymbiosis is thought to have occurred will be explored in Chapter 8.

Coming Up Next...

In this chapter we have seen that all living things can be classified according to cell type, either prokaryotic or eukaryotic. This division is in fact the most basic way to divide the living world. Both of these large groupings can be further divided into numerous subcategories, and in Chapter 4 we will investigate how scientists organize living things into a meaningful system of classification or **taxonomy**. After discussing the basic principles of classification, we will introduce each of the primary groups or **taxa** of microorganisms. In this microbial "field guide" we will focus on how they are related to one another as well as how they may interact with us.

Key Terms

acid-fast bacterium	exocytosis	phospholipid
acid-fast stain	extracellular	photosynthesis
aerotaxis	fimbria	pilus
antibiotic	flagellum	plasma membrane
bacillus	fluid mosaic model	plasmid
bacterial chromosome	glycocalyx	porin
bacterial lipopolysaccharide (LPS)	glycoprotein	positive taxis
bacterial sepsis	Golgi apparatus	prokaryotic cell
bactericidal	Gram stain	recipient
bacteriostatic	Gram-negative bacterium	ribosome
bilayer	Gram-positive bacterium	rough ER
capsule	lipid A	septum
carrier protein	lysis	sex pilus
cell membrane	lysosome	slime layer
cell wall	matrix	smooth ER
chemotaxis	motile	spirillum
cilium	multicellular	staphylococcus
coccus	negative taxis	sterol
conjugation	nuclear envelope	streptobacillus
cytoplasm	nucleoid area	streptococcus
cytoplasmic	nucleus	stroma
cytoskeleton	O polysaccharide	taxon
donor	organelle	taxonomy
endocytosis	organ	thermotaxis
endomembrane system	outer membrane	thylakoid
endoplasmic reticulum (ER)	pathogen	tissue
endospore	pentaglycine bridge	transmembrane protein
endosymbiosis	peptidoglycan	transport vesicle
endotoxin	peripheral protein	unicellular
ER lumen	phagocytosis	virulence factor
eukaryotic cell	phosphate group	

Concept Questions

1. If asked to give a succinct, one-sentence definition of a cell, what would you say?

2. You look at a cell under the microscope at high magnification. How might you tell if the cell were prokaryotic or eukaryotic?

3. Why are large organisms composed of billions or trillions of small cells? Why aren't they composed of a relatively few, much larger cells?

4. Sometimes, when doing a Gram stain, erroneous results are obtained because of errors in the Gram staining process. One mistake might be that the cells are exposed to alcohol for too short a time. If this mistake is made, would you more likely expect Gram-positive cells to be incorrectly identified as Gram-negative or Gram-negative cells to be incorrectly identified as Gram-positive?

5. Consider two new antibiotics. Antibiotic A disrupts cell membranes. Antibiotic B interferes with the synthesis of peptidoglycan. Which of these drugs do you think is safer to use to treat an infection in an animal and why?

6. A bacterium shows strong negative taxis to acidic conditions. It most likely
    a. is an acid-fast bacterium
    b. has pili
    c. has flagella
    d. is a coccus of some type

7. Which of the following doctors has made a mistake?
    a. Dr. Adams prescribes penicillin for a *Streptococcus* (Gram-positive) infection.
    b. Dr. Baker orders a drug for a Gram-negative infection of the blood. The drug causes rapid killing of the bacterial cells.
    c. Dr. Curtis uses a bactericidal drug to treat a *Staphylococcus* (Gram-positive) infection.
    d. Dr. Duran treats a tuberculosis patient with a drug that inhibits cell wall synthesis.

8. You visit a small island, which just days earlier experienced a volcanic explosion. Nothing appears to be alive on the island, but you collect soil samples and place them on culture media in the laboratory. Within 24 hours you find actively growing bacterial colonies on several of your plates. The bacteria that you find are most likely to
    a. have sex pili
    b. have a glycocalyx
    c. have capsules
    d. be Gram-negative
    e. have endospores

9. Enterotoxic *E. coli* can cause severe food poisoning, unlike other strains of *E. coli*, which are generally part of the normal bacterial flora. The enterotoxic strains have a number of virulence factors that allow them to cause disease. These include fimbrae and toxins that inhibit protein production in eukaryotic cells. If an anti-pilin vaccine could be developed against enterotoxic *E. coli*, how might it prevent infection by this organism?

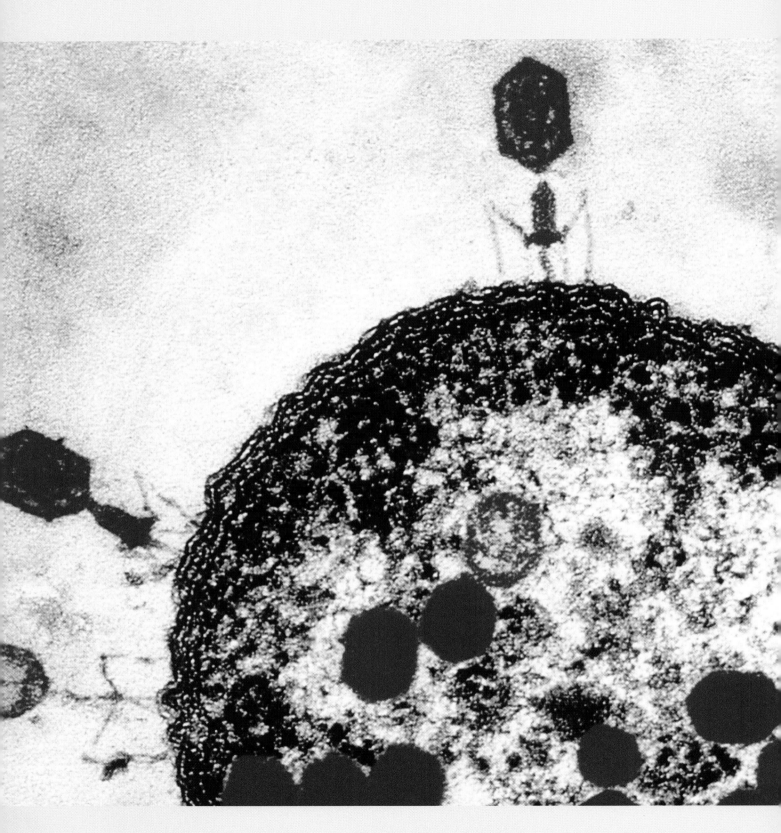

The T2 bacteriophage attaches to the bacterium and injects its DNA into the host.

Chapter 4

A Field Guide to the Microorganisms

By now you are probably beginning to comprehend how diverse microorganisms are. They are found in every conceivable environment, from frozen Antarctic lakes to superheated thermal vents on the ocean floor where the water is at or above the boiling point. They live in desert soils, on your skin, and under your toenails. There are microorganisms that thrive on toxic chemicals and others that digest petroleum. It is safe to say that microorganisms exploit every possible habitat and resource on the planet.

Microorganisms even strain our notion of what the word "life" means. Viruses, far smaller than bacteria, are little more than nucleic acid surrounded by protein. Viruses are not composed of cells, and they display only some of the characteristics that are attributed to living things. Even smaller subviral agents, such as the infectious proteins called *prions*, cause a bizarre set of diseases including the infamous "mad cow" disease.

In this chapter we will formally introduce the principal groups of microorganisms to be considered throughout this text. Before we begin, however, we need to get organized. Because all living things, including microbes, are classified in specific ways, we will start by investigating the principles of modern biological classification and how it reflects the relatedness of all living things.

Taxonomy: How Life Is Classified

The science of biological classification, known as **taxonomy**, attempts to group organisms together on the basis of similarity. Accordingly, an accurate taxonomy reflects the relationship between organisms. Taxonomy also gives all scientists a common language with which to communicate about the living world.

If you have ever spent time in another English-speaking country, you have no doubt occasionally been struck by the different names that are used for animals and plants in different places. Even within North America, crayfish might also be called "crawfish" or "crawdads" depending on where you are. These interchangeable terms may not be too confusing, but would you know what Australians were talking about if they mentioned "yabbies"? Of course such problems only get worse when you travel to non-English-speaking countries. Among biologists such problems can be more serious than minor misunderstandings. When two scientists are sharing information about an organism, there must be no doubt that they are discussing the same thing. In microbiology specifically, the consequences of such miscommunication can literally be life-threatening. For instance, antibiotics that work well against one species of *Staphylococcus* bacteria may have little or no effect against others. Two doctors discussing treatment of a *Staphylococcus* infection must be certain they are talking about the same microorganism or an inappropriate medication might be prescribed.

Taxonomy is based on a system of hierarchical groupings

Although attempts to classify animals and plants date back to Aristotle in the fourth century BC, the modern era of taxonomy began with the Swedish naturalist Carolus Linnaeus in the mid-1700s. Linnaeus developed a classification scheme based on a hierarchical sequence of groupings. In this system, the smallest, most fundamental group is the **species** (plural, species). All members of a particular species group are essentially the same type of organism. Groups of species that appeared to be similar were placed in the same **genus** (plural, genera). Thus, while the domestic dog is a unique species, *Canis familiaris*, it is placed in the genus *Canis* along with other canines such as the wolf (*Canis lupus*) and the coyote (*Canis latrans*). Note that each species is assigned two names in a manner called *binomial nomenclature*, first introduced by Linnaeus. The first name indicates the organism's genus, while the second identifies it as a particular species within that genus. Both the genus and the species names are Latinized and are either underlined (when written) or printed in italics. The genus name is always capitalized, while the species name is always written in lowercase. Although the genus name must be written out in full the first time it is used, it may subsequently be abbreviated.

Moving up the hierarchy, those genera considered to be similar are placed in a common **family**, and related families compose an **order**. For instance, the fox is considered to be different enough from dogs to merit its own genus, *Vulpes*. Both genera, *Canis* and *Vulpes*, are in the same family of dog-like animals known as Canidae, and all Canidae are in the order Carnivora along with other similar families such as Felidae (the cats) and Ursidae (the bears). Carnivores are just one order in the **class** Mammalia (the mammals). Others include the order Primates (monkeys, apes, and humans) and the order Cetacea (whales and dolphins). Mammals, along with the classes Aves (birds) and Osteichthyes (bony fish), as well as other animals having a backbone or a backbone-like structure at least some time in their lives, are grouped together in the **phylum** (plural, phyla) Chordata, and all animals including Chordata are in the **kingdom** Animalia (**Figure 4.1**). Animals are one of several kingdoms, to be discussed in this chapter.

Modern classification reflects evolutionary relatedness

Although the basic strategies of hierarchal classification and binomial nomenclature have remained the same since first proposed by Linnaeus, as

	Bacterium	**Corn**	**Housefly**	**Domestic cat**
Kingdom	None designated for bacteria	Plantae	Animalia	Animalia
Phylum	Proteobacteria	Anthophyta	Arthropoda	Chordata
Class	Zymobacteria	Monocotyledonae	Insecta	Mammalia
Order	Enterobacteriales	Commelinales	Diptera	Carnivora
Family	Enterobacteriaceae	Poaceae	Muscidae	Felidae
Genus	*Escherichia*	*Zea*	*Musca*	*Felis*
Species	*E. coli*	*Z. mays*	*M. domestica*	*F. silvestris*

| *E. coli* | *Z. mays* | *M. domestica* | *F. silvestris* |

Figure 4.1 Taxonomy's hierarchy. A hierarchical classification of a bacterium, a plant, and two animals. In such a hierarchy, similar species are placed in the same genus, and similar genera are placed in the same family. Similar families are placed in the same order, and so on. Both the housefly and the domestic cat, for instance, are different enough to be placed in separate phyla, but they share certain characteristics that place them both in the kingdom Animalia.

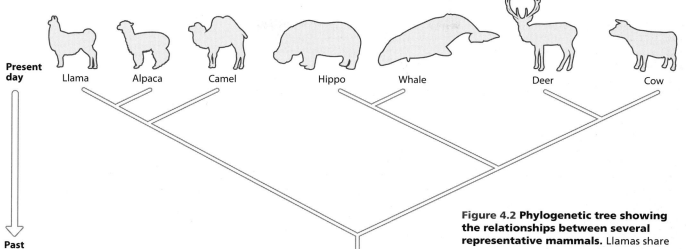

Figure 4.2 Phylogenetic tree showing the relationships between several representative mammals. Llamas share a more recent common ancestor with alpacas than they do with other animals in this tree. Thus llamas and alpacas are more closely related to each other than they are to other animals. Llamas and alpacas, along with the next closest group, the camels, form the family Camelidae. Others in this tree share a common ancestor more recently with each other than any of them do with members of Camelidae.

our understanding of living things has increased, the way we think about and use taxonomy has changed. Prior to the development of evolutionary theory, to be described in Chapter 8, organisms were grouped on the basis of similar visible features. No attempt to group organisms according to evolutionary relatedness was made, as these relationships were not recognized. However, most modern taxonomic schemes have tried to organize living things according to their evolutionary ancestry. Taxonomy based on a shared evolutionary history is known as **phylogenetics**. In such a taxonomy, species within a genus share a common evolutionary ancestor more recently than do species in separate genera. Two genera within the same family are more closely related than genera in different families, and so on, up the taxonomic hierarchy. Evolutionary relationships are often depicted as a tree (**Figure 4.2**). In such a "tree of life" the trunk represents the ancestral organisms of all species in the tree. Branches indicate where groups of organisms separated from each other in the course of their evolution. In a complete taxonomic tree, the small terminal branches would represent individual species. By examining a well-constructed phylogenetic tree, the evolutionary history of organisms can be traced back, and the relatedness of different organisms can be inferred.

DNA analysis provides evidence for relatedness

Originally, similar anatomical and biochemical features were taken as an indication of a similar evolutionary past. Since the 1970s, however, taxonomists have increasingly relied on genetic similarities to determine phylogenies. By comparing certain genes between organisms, scientists now have a far more accurate means of determining who is related to whom. Recall from our discussion of nucleic acids in Chapter 2 that a gene is a specific amount of DNA providing the code for the production of a particular protein. DNA is composed of nucleotides, and the sequence of the nucleotides in a specific gene dictates the order of amino acids in the protein to be produced. Even if two organisms produce the same protein, however, they might actually make it in slightly different forms. These differences are due to mistakes, known as **mutations**, that occur in reproducing the DNA. Mutations are changes in the sequence of nucleotides in a gene, and they occur with time in all organisms.

If two species arose from a common ancestor only recently, there will not have been much time for each to acquire mutations independently. Their genetic material will therefore still be very similar. But if two species are only distantly related, the number of differences in their DNA will be correspondingly greater. Because more time has passed since they shared a common ancestor, each species will have accumulated mutations on its own for a longer period. Taxonomists use the number of differences seen in a gene (mutations in the nucleotide sequence) as an indicator of relatedness.

CASE: STREAMS, SNAILS, AND SCHISTOSOMES

Schistosomiasis is a debilitating disease caused by parasitic flatworms in the genus *Schistosoma*. Over 200 million people are infected worldwide, primarily in the tropics. In Latin America, the causative agent is *Schistosoma mansoni*. These parasites develop as larvae in freshwater snails in the genus *Biomphalaria*. After a period of larval development and reproduction, the parasites leave the snail and enter the water. Humans become infected when they enter contaminated water bodies (Figure 4.3). The larval flatworm burrows through the skin and migrates to the veins surrounding the intestine. Following mating, large numbers of eggs are produced that pass into the intestine and are released with the feces. Upon hatching, the larval worm seeks out an appropriate snail, and the cycle is complete. The adult worms can survive for many years in the human host, producing millions of eggs. Many of the eggs fail to pass into the intestine and get swept by the circulatory system to the liver, where they become lodged. Inflammation caused by the eggs can eventually lead to severe liver disease. Interestingly, not all *Biomphalaria* snails are susceptible to infection. The most susceptible species in South America is *Biomphalaria glabrata*. Others, such as *Biomphalaria obstructa*, are resistant. It is unknown whether or not *Biomphalaria temascalensis* found in the area of Temascal, Oaxaca, in southern Mexico, is subject to infection.

Dr. Randall DeJong, a biologist at the National Institutes of Health, has examined the taxonomic relationship of *Biomphalaria* snails in Latin America. He proceeded by extracting DNA from the three species above and determining the nucleotide sequence for an rRNA gene from each type of snail. He then found that the sequences for *B. obstructa* and *B. temascalensis* are almost identical. The sequence for *B. glabrata* is significantly different from that of the other two species.

1. **What is an rRNA gene and why is it useful in this analysis? How does a comparison of genes between species help deduce their taxonomic relationship?**
2. **Which two of these three snail species share a more recent common ancestor?**
3. **In the area of Temascal, *B. temascalensis* is the only *Biomphalaria* species found. Do you think there is a likelihood of schistosomiasis in these areas?**
4. **How would the rRNA gene Dr. DeJong studied in this snail genus compare with the same gene in other taxonomic groups of animals?**

Cells typically possess structures called ribosomes (see Chapter 3, p. 59), which are used to assemble amino acids into proteins. Part of each ribosome is composed of RNA, and the genes that code for this RNA are called rRNA genes. These genes are very useful in taxonomic analysis for several reasons. First, they are found in all living things, so they can be used to compare very different organisms. Second, rRNA genes mutate slowly and at a more or less constant and predictable rate. Separate species accumulate unique gene mutations, and the longer ago they diverged, the more different their genes will be. Because the number of mutations is time-dependent, scientists like Dr. De Jong can actually deduce how long ago separate species were formed. DNA can thus serve as a "molecular clock," which can be used to construct phylogenetic trees.

So, is schistosomiasis likely in the Temascal, Oaxaca, area? Dr. DeJong's data indicate that *B. obstructa* and *B. temascalensis* diverged only recently, while the ancestor of *B. glabrata* branched off from the others relatively early in

Figure 4.3 Schistosomiasis. (a) Adult male and female schistosomes. The slender female worm (indicated by the lower arrow) lies in a groove along the surface of the male's body (indicated by the upper arrow). In infected humans, these paired schistosomes are found in the mesenteric veins surrounding the intestine. (b) Humans become infected when they come in contact with water contaminated with schistosome larvae. (c) *Biomphalaria glabrata*. These aquatic snails are the most important schistosome larval host in South America.

the evolution of this snail genus. Because the closely related *B. obstructa* is resistant to infection, it is likely that schistosomes cannot infect *B. temascalensis* either. Streams in this part of southern Mexico are therefore probably schistosome-free. Other *Biomphalaria* species extend into the southern United States, but they are also resistant, and an analysis of rRNA genes places them closer to *B. obstructa* and *B. temascalensis* than to *B. glabrata*.

If the rRNA gene of *Biomphalaria* snails were compared with that of other snails in different genera but the same family, there would no doubt be an increased number of sequence differences. Snails are members of the phylum Mollusca, which is divided into eight classes. Snails are in the class Gastropoda. Other familiar molluscan classes include Bivalvia (clams, oysters, etc.) and Cephalopoda (squid and octopuses). Any two snails would be more similar to each other in terms of rRNA gene sequence then they would be to a member of a different molluscan class. A snail and an octopus, however, would have greater similarity than either would to an animal in a different phylum such as a beetle (phylum Arthropoda).

Kingdoms are organized among three domains

Originally, all life was divided into two principal kingdoms: Animalia and Plantae (animals and plants). However many types of living things did not fit easily into either of these categories, and the recognition of major structural differences between eukaryotic and prokaryotic cells (see Chapter 3, p. 46) led to important changes in classification. Eventually, three more kingdoms were added—Protozoa (single-celled organisms but with eukaryotic cells similar to those of animals and plants), Fungi (eukaryotes such as mushrooms, molds, and yeasts), and Monera (the prokaryotes). This scheme became known as the five-kingdom classification system.

In the 1980s, researchers used the powerful new techniques of molecular biology to dramatically change the way we look at the relatedness of living things. By comparing rRNA genes from a very large number of organisms, these scientists found that living things could actually be divided into three groups. These very large divisions, encompassing all five kingdoms, are termed **domains**.

Originally, all prokaryotes were grouped together in the kingdom Monera. Molecular data, however, suggested that these organisms actually had two different evolutionary origins. The single kingdom was therefore reorganized into two very distantly related domains. Most of the former members of the kingdom Monera with which we are most familiar belong to the domain termed **Bacteria**. Other prokaryotes once considered to be bacteria are different enough to merit their own domain, **Archaea**. Additional study of archaea has revealed other important structural and biochemical differences from the bacteria. Animals and plants, along with protozoa and fungi, are grouped together in the domain **Eukarya** (**Figure 4.4**). All of these

Figure 4.4 The three domains of life.
A phylogenetic tree based on ribosomal RNA similarity, showing some of the principal groups in each of the three domains. Note that domain Archaea is somewhat closer to domain Eukarya than either is to domain Bacteria.

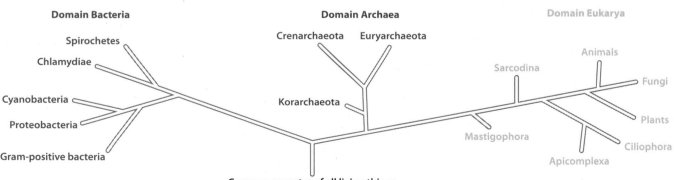

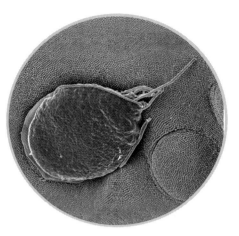

Figure 4.5 *Giardia intestinalis.* These flagellated protozoans use a ventral sucker to attach to the lining of the digestive tract, blocking proper nutrient absorption and causing intestinal distress. Note the circular mark on the intestinal lining left by a ventral sucker.

organisms have eukaryotic cells, and rRNA analysis suggests that they originally evolved from a common ancestor.

Having reviewed the fundamentals of classification, it is time to turn our attention to each of the principal groups of microorganisms. A complete review of all species is well beyond our scope. Rather, we will consider the important characteristics and the major divisions within each important lineage, providing representative examples where appropriate. Think of this section, then, as your microbial "who's who" or as a roster of our microbial players.

Organizing Microorganisms

How many species of microorganisms are there? No one knows, because so few have been described so far. There are currently about 6000 described species of bacteria. The real number of species no doubt numbers in the millions. A single ounce of seawater may contain thousands of species, most of them unknown to science. As for the archaea, scientists are only just beginning to understand their diversity. Until recently, it was believed that archaea were largely limited to "extreme" environments, such as very hot, salty, or acidic habitats (see Figure 1.1c). We now know that in many common environments, a large proportion of the microorganisms found turn out to be archaea. The sea, for example, teems with them.

The eukaryotic microorganisms are also a diverse lot. If, for example, while on a hike, you pause to drink from an apparently clear freshwater stream, you may become infected with a microorganism called *Giardia intestinalis* and develop a case of what is commonly called "hiker's disease" (**Figure 4.5**). The water becomes contaminated when beavers and other mammals pass *Giardia* cysts into the environment while defecating. If you enjoy drinking beer, on the other hand, you may be interested to know that different styles of beer are made by using different species of yeast in fermentation. To produce an English-style ale, the brewmaster adds *Saccharomyces cerevisiae* to a filtrate of the mixed malt and hops. A lager is the result if *Saccharomyces carlsbergensis* is used. Both *Giardia* and the two yeast species are examples of microorganisms in the domain Eukarya.

We will begin by reviewing the two prokaryotic domains before moving on to the eukaryotic microorganisms. We will conclude this chapter by considering the viruses and subviral infectious particles that straddle the gray area between the living and the nonliving world.

Domain Bacteria

Genetic analysis indicates that the domain Bacteria can be divided into more than a dozen major phylogenetic lineages. A phylogenetic tree showing the relationship of the major groups is provided in **Figure 4.6**. As discussed earlier, such trees are often constructed by comparing rRNA genes. The closer two groups are on the tree, the more similar are their rRNA sequences. Branches in the tree indicate where an ancestral lineage diverged into separate groups. To date, there is no firm consensus on which if any of the bacterial lineages should be grouped together as kingdoms. Depending on the source, each group has been referred to as a separate kingdom, a phylum, or even a class. Consequently, until this issue is resolved, it is probably preferable to refer to each of the large groups as simply distinct lineages, each one united by a common evolutionary history.

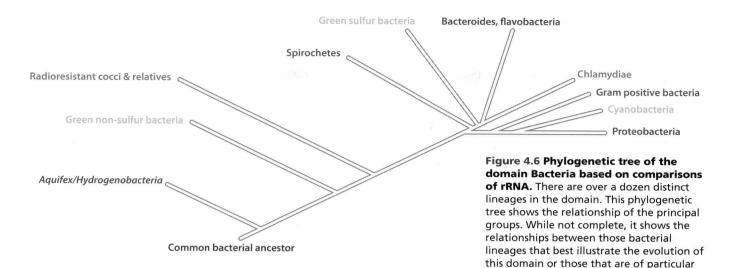

Figure 4.6 Phylogenetic tree of the domain Bacteria based on comparisons of rRNA. There are over a dozen distinct lineages in the domain. This phylogenetic tree shows the relationship of the principal groups. While not complete, it shows the relationships between those bacterial lineages that best illustrate the evolution of this domain or those that are of particular importance ecologically or medically.

The earliest bacteria were adapted to life on the primitive Earth

CASE: THE HEAT WAS ON!

It is estimated that the Earth was formed about 4.5 billion years ago. Yet it would be another 400 million years before there was even a solid piece of ground to stand on. Prior to that time the Earth's surface was an inferno of molten rock. About 4.1 billion years ago the Earth had cooled enough for the crust to solidify. Another 700 million years later, the first cells appeared (**Figure 4.7**). However, the Earth was still an inhospitable place. Temperatures were much higher than they are today, and without an ozone layer, the surface was bombarded by UV radiation.

1. What characteristics would have been typical of the earliest cells?
2. Are any of these characteristics present in the modern ancestors of these first prokaryotes?

Let us consider the bacteria living today that most resemble those ancient pioneers of 3.5 billion years ago. As seen in Figure 4.6, the ***Aquifex/Hydrogenobacter*** group were the first to branch off the main lineage within the domain Bacteria. As such they are the most primitive members of this domain, closest to the original ancestor of all bacteria. The term "primitive," when used in this context, does not imply that these organisms are somehow less complex or that they are poorly adapted to their environment. Rather, it means that this group has changed the least since the evolutionary origin of the domain.

These bacteria grow at very high temperatures (up to 95°C or 203°F) and are therefore known as **thermophiles** (from the Greek for "heat loving"). Their ability to tolerate high temperatures, along with their phylogenetic position near the base of the bacterial tree, makes sense in light of prevailing opinion that the early Earth was very hot. Continuing up the evolutionary tree, we next encounter the **green nonsulfur** bacteria, another group of thermophiles. Interestingly, although they grow well at elevated temperatures, they do not survive the extreme heat tolerated by members of the *Aquifex/Hydrogenobacter* lineage. This suggests that although the Earth was still

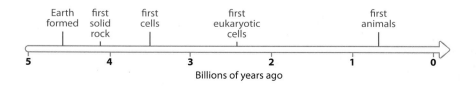

Figure 4.7 Origin of the Earth and appearance of the first cells. The Earth is thought to have formed approximately 4.5 billion years ago. The first cells appeared about 1 billion years later, about 3.5 billion years ago. Eukaryotes were first present approximately 2.5 billion years ago.

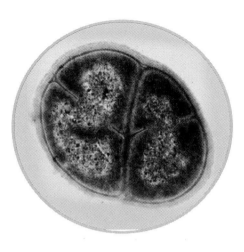

Figure 4.8 *Deinococcus radiodurans.* This species is highly resistant to radiation, in part because of its ability to repair radiation-induced damage to its DNA. Each cell is surrounded by a peptidoglycan cell wall typical of Gram-positive bacteria, as well as an outer membrane seen in Gram-negative species.

quite warm when this group evolved, it had already cooled to some degree. The two lineages discussed so far are composed entirely of thermophilic species. Thermophiles are found in other groups of bacteria, but only in these most primitive groups is the characteristic universal. Green nonsulfur bacteria are also photosynthetic, using sunlight as an energy source to produce sugars.

The **radioresistant cocci** and relatives are noteworthy because of their ability to tolerate radiation. One species, *Deinococcus radiodurans*, is even more radiation-resistant than the bacterial endospores discussed in Chapter 3. These microbes can survive up to 3 million rads of radiation, whereas for a human, exposure to fewer than 500 rads can be lethal. Not surprisingly, a number of unusual adaptations make this possible. Although unrelated to other Gram-positive bacteria, *D. radiodurans* has a multilayered Gram-positive cell wall. It also has an outer membrane, typically found only in Gram-negative bacteria (**Figure 4.8**). High doses of radiation are usually lethal primarily because they cause mutations in a cell's DNA, but *D. radiodurans* has a remarkable ability to repair its DNA, far exceeding our own. Such a characteristic would obviously have been useful when life first evolved because of the high levels of ambient radiation. The resistance seen in species such as *D. radiodurans* may be a legacy of those times.

The tree spreads out

At this point, our phylogenetic tree diverges into a number of distinct lineages. The **spirochetes** are a group of bacteria having an unusual helical and flexible shape (**Figure 4.9a**). They are found in many environments, and several species infect humans and animals, causing a handful of important diseases. Perhaps the most significant of these is syphilis, caused by *Treponema pallidum* (Figure 4.9b).

The **green sulfur bacteria**, like the green nonsulfur bacteria described above, are photosynthetic. As their name suggests, however, sulfur-containing compounds are used in the photosynthetic process. Note that the green sulfur and green nonsulfur bacteria are only distantly related (see Figure 4.6). This suggests that photosynthesis evolved more than once in taxonomically distinct groups. Indeed, it seems to have evolved several times in different lineages scattered across the bacterial domain.

If you have ever been told by your dentist that you are at risk for gum disease or gingivitis, it is likely that you can blame *Bacteroides gingivalis*, a member of the next lineage on our tree, **flavobacteria**. Because *B. gingivalis* is unable to grow in the presence of oxygen, it is called a **strict anaerobe**. Consequently, it flourishes in the oxygen-free environment under the gum line, where it causes the inflammation that results in disease. When you floss your teeth, you actually allow oxygen into this space, inhibiting the growth of the bacteria and reducing the likelihood of gingivitis.

Chlamydiae are interesting in several respects. First, they cannot reproduce unless they enter the cytoplasm of an appropriate host cell. Therefore they are examples of **obligate intracellular parasites**. Although they have a Gram-negative outer membrane, they are unrelated to other Gram-negative species. Unlike more familiar Gram-negative bacteria, they have no peptidoglycan layer between the plasma and outer membranes. *Chlamydia trachomatis* inhabits the human urogenital tract, and infections caused by this

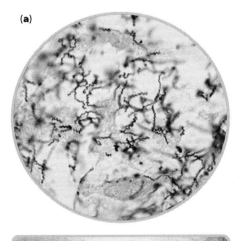

(a)

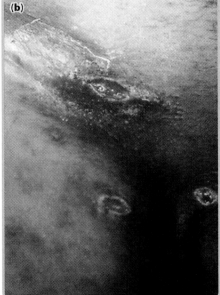

(b)

Figure 4.9 *Treponema pallidum.*
(a) *Treponema pallidum*, a representative spirochete, is the causative agent of syphilis. (b) Syphilitic lesions caused by *T. pallidum*.

species are perhaps the most common of all sexually transmitted diseases. This species can also infect the eye. The scarring that occurs on the cornea can result in trachoma, the leading cause of blindness in humans.

The most recently evolved lineages are found at the farthest branches of the tree

There are three remaining evolutionary lineages in our bacterial tree (see Figure 4.6). All three are more closely related to each other than they are to any of the lineages thus far discussed. The **Gram-positive bacteria** (see Figure 3.14a) contain many familiar genera such as *Staphylococcus* and *Streptococcus*. Because members of this lineage are so common and because they affect human health and welfare in so many ways, they will reappear often throughout this text.

We owe a large debt of gratitude to the **cyanobacteria** (**Figure 4.10**). Before their evolution, approximately 2.4 billion years ago, the atmosphere contained about 0.1% free oxygen. Today the air we breathe is about 20% oxygen, much of it due to the photosynthetic activity of cyanobacteria. These bacteria, therefore, helped change the Earth's environment in a way that made the later evolution of aerobic (oxygen-requiring) life possible. Today they continue to play a major role in ecological processes. Cyanobacteria release large amounts of oxygen, especially in aquatic and marine environments. Additionally, many species remove nitrogen from the atmosphere and convert it to a form that can be used by plants, a process called nitrogen fixation. The bluish-green coloration of cyanobacteria comes from their photosynthetic pigments.

The final branch on our tree, clustered with the Gram-positive bacteria and the cyanobacteria, is the **proteobacteria** (see Figure 4.6). This large and varied group contains most of the Gram-negative species with which we are most familiar (*Escherichia coli*, for example). Although they are all Gram-negative, proteobacteria span the gamut in terms of cell shape and metabolism. They may be cocci or rods, motile or nonmotile, aerobic or anaerobic. While some species are photosynthetic, the majority are not. The name "proteobacteria" refers to the Greek god Proteus, who could change his shape, alluding to the great diversity in this group. DNA analysis, however, confirms that these bacteria share a common origin. Like the Gram-positive bacteria, because so many proteobacteria impact human or animal health or play important roles in ecological processes, we will encounter many members of this group in subsequent chapters.

Domain Archaea

If you visited a very salty or "hypersaline" pool in Death Valley, or a hot spring in Yellowstone where the water is near boiling, you could be excused for thinking that nothing could survive in such a place. But you would be wrong. Such extreme environments are the habitat of choice for many species of archaea, often referred to as "extremophiles" (see Figure 1.1c). Their adaptations to harsh environments suggest that they also, like the thermophilic bacteria discussed earlier in our case, were among the Earth's earliest inhabitants. As previously discussed, it was not until recently that these organisms were considered to be anything other than unusual bacteria. Although archaea superficially resemble bacteria, phylogenetic analysis indicates that these two prokaryotic groups are only remotely related. Indeed, rRNA comparisons place domain Archaea somewhat closer to domain Eukarya than either group is to domain Bacteria.

The structure of archaea varies considerably. Cell walls may or may not be present, and although cell wall structure is variable, peptidoglycan is always

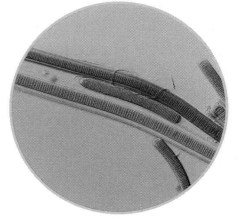

Figure 4.10 Cyanobacteria. Photosynthesis in the cyanobacteria evolved independently from photosynthesis in other bacterial lineages. Oxygen, released by cyanobacteria as a photosynthetic waste, helped to create an oxygen-rich atmosphere.

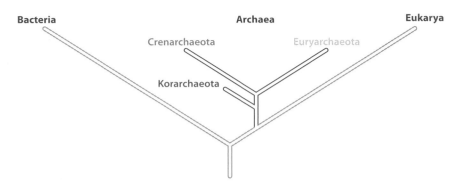

Figure 4.11 Phylogeny of the three principal lineages of Archaea. Korarchaeota are thought to be the most primitive of the three lineages, most similar to the original ancestor of all three domains. Crenarchaeota and Euryarchaeota are more closely related to each other than either is to Korarchaeota.

absent. Many are rods or cocci, but others have highly unusual shapes. They may or may not require oxygen, and some are photosynthetic.

Phylogenetically, domain Archaea is often divided into three main lineages (**Figure 4.11**), although this issue still remains to be settled definitively. We still know little about Korarchaeota, the most primitive lineage of Archaea, and those presumably most similar to the original ancestor of all prokaryotes and eukaryotes. Crenarchaeota are a lineage of extreme thermophiles and are thought to be the most heat-loving of all prokaryotes. Some members of Crenarchaeota are also extreme acidophiles, living in environments with pH below 1.0. That means that if it were hot enough, and with proper nutrients, they could thrive in battery acid! The third lineage of Archaea is Euryarchaeota. The extreme halophilic ("salt-loving") archaea are members of this group. *Halobacterium*, for instance, grows in salt deposits or other environments where salt concentration may reach 7 times that found in the ocean. The **methanogens**, which thrive in the oxygen-free environments found in sewage, swamps, and the intestines of many animals, are also Euryarchaeota. The name of this group refers to the fact that they release methane as a metabolic waste product. Methane gas, of course, is flammable; because it is also odorless, it presents a particular hazard. If you collect compost, you may know that it is advisable to keep your compost heap away from structures and closed spaces to prevent an explosion. Also, turning the compost heap from time to time allows oxygen to enter, thereby inhibiting methanogen growth. Methane is also a "greenhouse gas," helping to trap heat in the atmosphere. The methanogens in a single cow's intestines produce about 10 cubic feet of the gas each day. Billions of cows, all releasing tons of methane, are thought to be part of the global warming problem.

Eukaryotic Microorganisms

Note in **Figure 4.12** that initially life branched off in two directions. One lineage gave rise to domain Bacteria. The other lineage later branched into

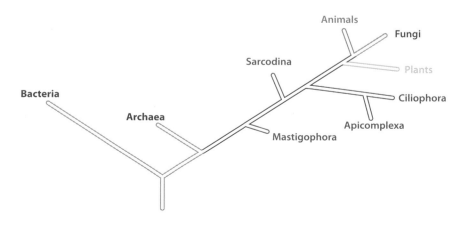

Figure 4.12 Domain Eukarya. A phylogeny of the Eukarya, based on rRNA similarity. Note that animals (in orange) and fungi (in black) are more closely related to each other than either group is to plants (in green). The four lineages in blue are all currently classified as members of the kingdom Protozoa. The exact relationship between these lineages is still under investigation.

Archaea and Eukarya. These last two domains are therefore more closely related to each other than either is to domain Bacteria. In the Eukarya, many of the more ancient lineages are placed in the kingdom Protozoa, single-celled eukaryotes, abundant as both parasites and free-living forms. Later in the tree, we see that modern animals, plants, fungi, and some protozoa were the result of a more recent burst of diversity.

Protozoa are single-celled organisms within the domain eukarya

CASE: THE CAT'S OUT OF THE BAG ON A PROTOZOAN PARASITE

Donna, a 32-year-old woman, has been married for three years. She has just learned that she is pregnant. Anxious to provide proper prenatal care, she reads extensively on the subject of how to best care for herself and her developing baby. In a parenting magazine, she is horrified to read that cats pose a risk to a developing fetus. Cats are often infected with the protozoan parasite *Toxoplasma gondii*, which can be transmitted to a pregnant woman, and across the placenta to the fetus. Birth defects due to *Toxoplasma* infection in a newborn can include deafness, vision problems, and other neurological abnormalities. Donna has always owned cats, and her current cat, Woody, has lived with her for years. The idea of parting with her feline friend is almost more than she can bear. She wisely, however, puts off making any important decisions about her cat until she consults with her doctor. While at the doctor's office she asks the following questions:

1. **Is it true that Woody may pose a risk to her baby?**
2. **If so, can Donna be tested for *Toxoplasma* infection, and if she is infected, can she be treated with antibiotics to protect both her and the baby?**
3. **Should she find a new home for Woody?**

The protozoa are microscopic, unicellular eukaryotes that lack cell walls. They are usually motile. Their lack of photosynthetic chloroplasts distinguishes protozoa from the unicellular algae, with whom they are often categorized in a "superkingdom" called Protista.

Protozoa are found in many environments including fresh and salt water and moist terrestrial habitats. They feed by ingesting other organisms or organic material. Protozoa often have complex life cycles. While many are free-living in the environment, others are parasites with great medical and veterinary significance. When conditions permit, they exist in an active feeding and reproducing stage as **trophozoites**. However, if environmental conditions deteriorate, many protozoa form metabolically inactive, protective structures called **cysts**. In parasitic species, cysts are most common among those that spend a portion of their life cycle outside the body of a host. Often these cysts are excreted in feces into the environment, where they might ultimately be ingested by a new host. Once the cyst is ingested, it will develop into the active trophozoite, completing the life cycle. Other protozoa reach new hosts by way of **vectors** such as mosquitoes, which ingest the parasite when they take a blood meal. The parasite is then transferred to the next host when the vector feeds a second time. Because protozoa that are carried by vectors are never exposed to the external environment, there is no cyst formation.

Phylogenetic analysis shows that the protozoa are not necessarily closely related to each other (see Figure 4.12). In this sense, the kingdom Protozoa can be compared to a grab bag where organisms that do not seem to fit elsewhere are placed. It is likely that further research may reorganize this kingdom, and what we now refer to collectively as "Protozoa" will one day be

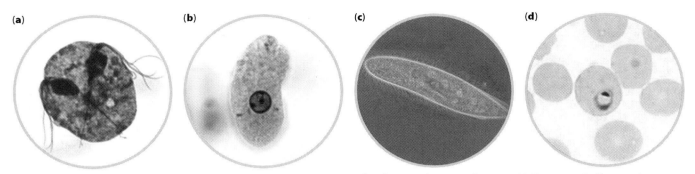

(a)   **(b)**   **(c)**   **(d)**

divided into several distinct lineages. By current convention, the four main groups within the protozoa are recognized on the basis of mode of locomotion.

Mastigophora. Mastigophora have one or more flagella, which they use for locomotion. They are commonly referred to as the flagellates (**Figure 4.13a**). Some are important pathogens, including *G. intestinalis*, the cause of hiker's diarrhea mentioned earlier (see Figure 4.5). Several species within the genus *Leishmania* (see Chapter 1, Figure 1.5b) can cause the disease leishmaniasis. These protozoa received a great deal of attention during the Persian Gulf War in 1991. The disease is found throughout the Middle East, and a number of service personnel became infected after being bitten by sand flies, which serve as vectors. The symptoms of leishmaniasis include severe ulceration and lesions on the skin (see Figure 1.5c). Because the *Leishmania* protozoa are found in the blood, Americans who served in the Gulf War were not permitted to donate blood upon their return.

Sarcodina. This group includes the amoebas, which move by extending portions of their cytoplasm called pseudopodia (see Figure 4.13b), changing shape as they move. Most amoebas are free-living. Others can cause severe dysentery. *Entamoeba histolytica*, for example, is thought to infect close to half a billion people and to cause up to 100,000 deaths annually. Amoebic dysentery is mainly a problem in parts of the world where sanitation is poor. Cysts are passed in the feces of infected individuals, where they may subsequently come in contact with food or water, to infect new hosts. Access to clean drinking water vastly reduces the risk of infection.

Ciliophora. These protozoa are recognized by the presence of cilia, which beat in a coordinated manner, directing locomotion. Perhaps your first view of the microbial world was a drop of pond water under the microscope. If so, you may have observed free-living *Paramecium* swimming furiously across the slide in search of prey (see Figure 4.13c). Most ciliophora are free-living and play no role in disease processes. One exception known to tropical fish hobbyists is *Ichthyophthirius multifiliis*. These parasitic protozoans penetrate the mucous coat on a fish's skin or gills and feed on epithelial and blood cells. The result is the fatal skin condition called "ich" or "white spot."

Apicomplexa. All protozoa in this fourth group are parasitic and most are intracellular parasites, spending at least part of their life cycle inside host cells. They are named for a complex of structures found at the apex of the cell. There are usually no structures related to locomotion, and most are nonmotile.

Some of the most lethal human pathogens are the apicomplexans in the genus *Plasmodium* that cause malaria (see Figure 4.13d). *Plasmodium* infects up to half a billion humans worldwide each year, killing as many as 1 million annually. The parasite is transmitted by the bite of specific vector mosquitoes. Many other animals besides humans are also affected. If you visit Hawaii, for instance, almost all of the beautiful birds you see near the coast are nonnative species. Many native birds can now thrive only at

Figure 4.13 Representative protozoans. (a) A mastigophoran, *Trichomonas vaginalis*, the most widespread agent of any protozoan disease, causes itching and inflammation of the female reproductive tract. It is transmitted via sexual contact. (b) *Entamoeba histolytica*, a sarcodinan, is a common cause of amoebic dysentery. It is transmitted through contaminated food and water and is a problem wherever adequate sanitation is not assured. (c) The common freshwater ciliophoran *Paramecium* is a free-living (nonparasitic) protozoan. (d) *Plasmodium*, the apicomplexan that causes malaria. The parasite in this photograph is the darkly stained ringlike structure. It is shown inside a human red blood cell.

elevations above 1500 meters where malaria-bearing mosquitoes are not common. Hawaii did not originally have the mosquitoes necessary to transmit malaria. They first arrived in about 1850, probably on a ship, and they brought *Plasmodium* with them. Indigenous birds, with no previous exposure, had no resistance to the disease and died in large numbers.

Toxoplasma gondii, introduced in our case, is also an apicomplexan. Like many parasites, *T. gondii* has a complex life cycle, involving more than one host (**Figure 4.14**). As trophozoites, *T. gondii* live in the intestine of cats, where they feed and reproduce sexually. Because sexual reproduction takes place in the cat, cats serve as the **definitive host** for this parasite. Reproduction results in the production of **oocysts**, resistant cystlike structures that pass out of the intestine with the cat's feces. Rodents or birds may accidentally consume these oocysts. Following ingestion, the oocyst ruptures, releasing the parasite into the intestine. The organisms penetrate the gut wall and spread throughout the body, eventually forming dormant tissue cysts in various organs. Infected rodents or birds, capable of reinfecting cats, are known as **intermediate hosts**. The cycle is completed when a cat kills and eats an infected animal, reestablishing the infection in the cat's intestine.

Humans can become infected if they inadvertently consume oocysts. Tissue cysts then form in the infected human. In an individual with a healthy

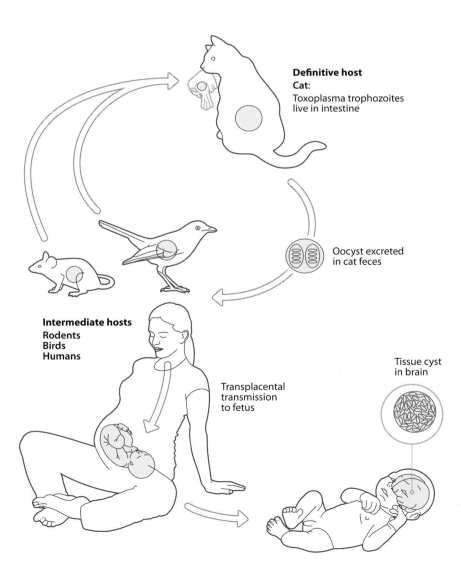

Definitive host
Cat:
Toxoplasma trophozoites live in intestine

Oocyst excreted in cat feces

Intermediate hosts
Rodents
Birds
Humans

Transplacental transmission to fetus

Tissue cyst in brain

Figure 4.14 Life cycle of *T. gondii*. Sexual reproduction of this parasite takes place in the cat, which therefore is called the *definitive host*. Cystlike oocysts are passed in the cat's feces, where they may be later consumed by a rodent or other small animal, which serves as the *intermediate host*. Here the oocysts form tissue cysts, which usually remain dormant until they reenter the definitive host, when a cat eats the intermediate host. Humans can also become infected if they consume oocysts. If a woman first becomes infected during pregnancy, there is a danger that the parasite may cross the placenta and infect a fetus before it is contained by the woman's immune system. Such infection of a developing fetus can result in serious birth defects.

immune system, these cysts usually remain dormant for life, without causing problems. Most long-time cat owners, including Donna, will probably test positive for *T. gondii* but will not have symptoms because the immune system usually keeps the parasite from causing illness. In an immunocompromised individual, such as an AIDS patient, *Toxoplasma* may cause severe inflammation, which can result in tissue damage and death.

Donna asked her doctor about antibiotics. Antibiotics would not cure a *Toxoplasma* infection because antibiotics are useful only to treat bacterial infections, and most function by attacking unique prokaryotic features such as the cell wall or bacterial ribosomes. They have no effect against eukaryotic pathogens. When a *Toxoplasma* infection must be treated, as it would in someone with AIDS, entirely different drugs capable of inhibiting eukaryotic cells must be used.

So, is it time for Woody to go? Since Donna has had cats for years, it is likely that she is already infected with *T. gondii*. In such a case her immune system will have contained the infection as dormant tissue cysts, preventing any spread to her fetus. She will also be resistant to new infections. There is no need to find a new home for her pet.

The real danger is if a woman becomes infected for the first time during pregnancy. In this case, the parasite may spread across the placenta before it is contained. Because a fetus does not have a fully functioning immune system, the infection can cause severe damage, resulting in birth defects. Consequently, a kitten is a poor choice for a baby shower gift, if the mother-to-be has never owned cats. A good precaution in any event is to change cat litter frequently during pregnancy. When oocysts pass out of an infected cat, they need to be exposed to oxygen for 24–48 hours before they can infect another animal. Fresh litter each day greatly reduces the risk of exposure to infective oocysts. As an added safety precaution, pregnant women might consider having another family member take over the task of changing the cat litter until the baby is born.

Many fungi are involved in disease or ecological processes or are useful for industrial purposes

CASE: FIRST YOU SHAKE, THEN YOU ACHE

On January 17, 1994, southern California was struck by a massive earthquake. With a magnitude of 6.7 on the Richter scale, the temblor that came to be known as the Northridge earthquake downed buildings and collapsed freeways. Fifty-eight people were killed and approximately 2000 were injured. As if to compound the misery, over the next eight weeks southern California was hit by an epidemic of "Valley fever." Two hundred three individuals were diagnosed with the respiratory infection that causes cough, chest pain, fever, night sweats, and anorexia. Three people died. Valley fever is caused by a soil-dwelling fungus called *Coccidioides immitis*. This organism is found in soils throughout the arid Southwestern United States (Figure 4.15). People become infected when they breathe in the fungal spores. There are typically only a few cases of Valley fever each year. The 1994 outbreak was attributed to the Northridge earthquake several weeks earlier.
1. **What is the link between the earthquake and the outbreak of Valley fever?**
2. **What, if anything, can be done to prevent similar future epidemics?**

Fungi are perhaps most familiar to us either as food or because of their use in baking and brewing. But the importance of fungi far exceeds their gastronomic value. If you have never seen a Costa Rican golden toad, for instance, you have probably missed your chance. This spectacularly colored

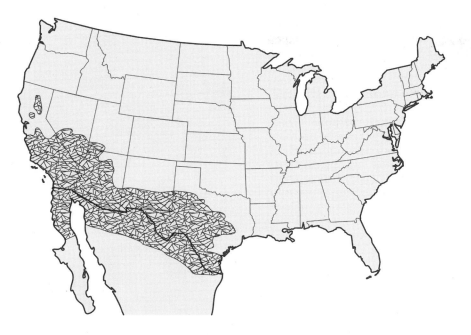

Figure 4.15 Distribution of Valley fever in North America. Valley fever is caused by the soil fungus *Coccidioides immitis*. The fungus thrives in the dry, desert soil of the American Southwest and northern Mexico.

amphibian is presumably extinct. If so, this would be the first known vertebrate extinction caused by a parasite. The culprit is thought to be an unusual fungus known as *Batrachochytrium dendrobatidis*. You might still find, on the other hand, an American chestnut tree to admire, but the vast stands of chestnuts that once blanketed the eastern United States are gone. Again, the reason is a fungal pathogen that was first introduced into New York in about 1900.

These examples illustrate the substantial impact that fungal parasites can have on the welfare of animals and plants. Yet the vast majority of fungi are not pathogenic, and the role played by many fungi in ecological processes can hardly be overstated. Fungi are ubiquitous in the soil environment, and many fungi extract nutrients from dead or decaying organic material in the soil. They are thus important as **decomposers**—organisms that break down and recycle nutrients in the environment (**Figure 4.16**). The fungi have no equal when it comes to converting large molecules such as proteins and complex carbohydrates into their constituent building blocks. These small precursor molecules are recycled back into the environment, where they become available to growing plants. Other fungi form important **mutualistic** relationships with specific living plants; the fungus and the plant live together in an association that benefits both partners. The fungi absorb food from the living plants, while in return they provide the plant with crucial nutrients. Forests would indeed be barren places were it not for the fungi unobtrusively recycling nutrients and augmenting plant nutrition.

Close to 100,000 species of fungi have been described, but the actual number of fungal species is no doubt many times greater than this. Hundreds of new species are discovered each year. Originally, fungi were believed to be primitive or degenerate plants that lacked chlorophyll and thus could not perform photosynthesis. This historical fact explains why phyla in the kingdom Fungi are sometimes still called "divisions" as they are in the kingdom Plantae. We now recognize that fungi are a separate lineage that differs from other eukaryotes in important ways. Ironically, a comparison of rRNA places fungi closer to the animals than to the plants (see Figure 4.12). These molecular data are supported by several morphological and biochemical features shared by fungi and animals but not by plants. For example, all fungi synthesize a resilient structural carbohydrate called chitin that is often found in fungal cell walls. Plants do not produce this molecule, but most animals do. The exoskeleton of insects and crustaceans, for example, is composed of

Figure 4.16 Fungi are important decomposers in the environment. This fungus (the irregular shaped growth in the middle of the photograph) is growing on a forest floor. Fungi secrete enzymes, which digest organic molecules in the plant material. The fungi then absorb the digested components to meet their own nutritional needs.

chitin. Additionally, plants store glucose in the form of starch to meet future energy demands (see Chapter 2, p. 32). Both animals and fungi assemble individual glucose molecules into a different polysaccharide called glycogen.

Photosynthetic organisms, like plants, are termed **autotrophs** ("self-feeders") because they can synthesize organic molecules from inorganic starting material. Fungi, alternatively, like animals, are **heterotrophs**, which must obtain organic compounds from other organisms. Unlike animals, however, which typically ingest their food, fungi absorb nutrients from the environment. Fungi secrete digestive enzymes into the environment, where they degrade organic matter into a form that the fungi can absorb. Those fungi that grow on dead material are termed **saprophytes**. Other parasitic fungi absorb nutrients directly from a living host organism. Mutualistic fungi live within and are supported by the living tissue of a host, and they provide nutrients or other digestive services in return. Although there are a few aquatic species, most fungi are found in terrestrial habitats.

Fungi have one of two body plans and can be unicellular or multicellular

The fungi have one of two basic body plans (**Figure 4.17**). While some are microorganisms, others can be quite large. The first group are single-celled fungi such as yeast. The second group are the **filamentous fungi**, which are made up of an extensively branched system of tubelike filaments known as **hyphae**. The entire network of hyphae making up a fungus is collectively called the **mycelium**. In the **molds**, the hyphae of the mycelium are loosely intertwined throughout the fungus. This gives molds their characteristic fuzzy appearance. **Fleshy fungi**, such as mushrooms, on the other hand, produce reproductive structures made of tightly packed hyphae. The remaining non-reproductive portion of the body may be a loosely organized mycelium, as seen in molds (**Figure 4.18**).

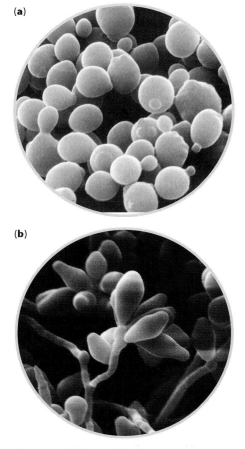

(a)

(b)

Figure 4.17 Fungal body plans.
(a) Single-celled yeast. (b) A filamentous fungus. The many tubelike hyphae collectively make up the fungal body, called the mycelium.

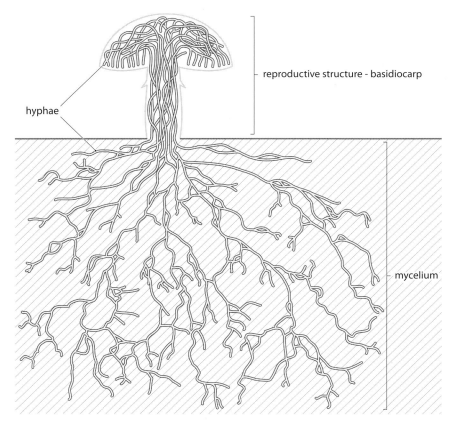

reproductive structure - basidiocarp

hyphae

mycelium

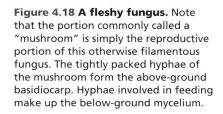

Figure 4.18 A fleshy fungus. Note that the portion commonly called a "mushroom" is simply the reproductive portion of this otherwise filamentous fungus. The tightly packed hyphae of the mushroom form the above-ground basidiocarp. Hyphae involved in feeding make up the below-ground mycelium.

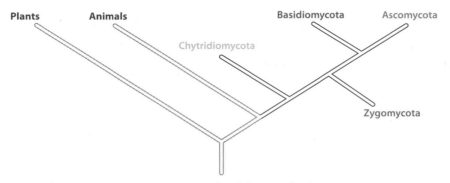

Figure 4.19 Phylogeny of the fungi. Kingdom Fungi is composed of four phyla, each indicated in a different color. Chytridiomycota is the most primitive phylum. Basidiomycota and Ascomycota are more related to each other than they are to the other two phyla.

Kingdom Fungi is composed of four phyla

The fungal kingdom is divided into four phyla. We will treat each of these phyla separately, concentrating on how they most affect us and our environment.

Chytridiomycota. Chytridiomycota, or "chytrids," are the most primitive of the four phyla; that is, they are believed to be the most similar to the original fungal ancestor (**Figure 4.19**). Chytridiomycota is the only phylum of fungi that is primarily aquatic. Most members of this phylum are saprophytic and function ecologically as decomposers, but a number of them are known parasites of invertebrates.

Although they may parasitize insects, until recently it was believed that the chytridiomycota did not infect vertebrates. Unfortunately, we now know this is not the case. *Batrachochytrium dendrobatidis*, mentioned earlier in this section, is a chytrid parasite of amphibians. The fungus infects skin cells of frogs and other amphibians (**Figure 4.20**), killing the animal by an unknown mechanism. As noted earlier, golden toads are now extinct in Central America, presumably due to a massive *B. dendrobatidis* infestation. It is still unclear why this problem has cropped up only recently, but a growing body of research has implicated global warming. Freshwater habitats in the Costa Rican highlands have become scarcer in recent years due to rising temperatures and reduced rainfall. Consequently frogs are forced to crowd into the few remaining water bodies. Such crowding increases the likelihood of parasite transmission from infected to uninfected individuals.

Zygomycota. Zygomycota are mostly saprophytic, but some species participate in mutualistic associations with the roots of plants. Others are parasitic in plants or animals. Perhaps the most familiar fungus in this phylum is the common black bread mold, *Rhizopus stolonifer* (**Figure 4.21**). When *R. stolonifer* spore lands on a piece of bread or other suitable substrate, the spore may germinate and produce specialized hyphae that spread across the bread, absorbing nutrients. This species is an important food spoiler. Tons of food must be thrown away each year due to damage caused by *R. stolonifer* and other food-decomposing fungi.

What the Zygomycota take away in food spoilage, they more than give back as plant mutualists. Many members of this phylum grow on plant roots and actually extend hyphae into root cells. In return for sugars provided by the plant, the fungus supplies the plant with phosphorus. Such associations between fungi and plant roots are called **mycorrhizal** associations, and they are found in up to 80% of all land plants. They are especially common in grasslands and tropical forests, where phosphorus is often scarce.

Basidiomycota. Basidiomycota include the familiar mushrooms and puffballs. "Mushroom" is actually the common name for the spore-producing structure called the **basidiocarp** (see Figure 4.18). This structure is composed of tightly packed hyphae and is produced solely for reproductive purposes. The much larger mycelium of the fungus remains below ground.

Figure 4.20 A golden toad. These amphibians are now believed to be extinct, the result of infection with the chytridiomycotan *Batrachochytrium dendrobatidis*. If true, this would be the first known case of a vertebrate extinction due to a parasite.

Figure 4.21 Zygomycota. The common bread mold, *Rhizopus stolonifer*. This saprophytic fungus is a common cause of food spoilage. Hyphae, extended into the bread, release digestive enzymes, which digest carbohydrates in the bread. The digested products are then absorbed by the fungal cells.

Like zygomycota, many species of basidiomycota form mycorrhizal associations with plants. In the basidiomycota, however, the fungal hyphae do not actually invade the root cells. Rather, a dense network of hyphae surrounds the roots. Once again, the fungus benefits by absorbing sugars from the plant. Mycorrhizal basidiomycota, however, typically provide the plant with nitrogen rather than phosphorus as seen in zygomycota. This type of mutualism is most common in colder northern and temperate forests. In such environments the growing season is short and decomposition of plant material is slow. Consequently, nitrogen tends to remain bound up inside plant material. Very little is available in the environment. Without the mutualistic fungi, plant growth would be severely limited by a lack of nitrogen. In the warmer climates where zygomycota provide plants with phosphorus, the longer growing season ensures more rapid organic breakdown and ample nitrogen availability.

Other members of this phylum are important plant parasites such as the smuts and rusts. Rust infections appear as numerous colored spots on the surface of a plant (**Figure 4.22**). The best known rusts cause *stem rust* on wheat and other grains. Smuts primarily attack the ovaries of grasses and eventually destroy the grain kernels. Many are important parasites of corn, cereals, and other crops. Wooden structures often fall prey to wood-rotting basidiomycota that can digest the cellulose or other complex carbohydrates found in wood.

Ascomycota. Most members of this phylum are saprophytic and serve as important decomposers. The majority of ascomycota produce hyphae, but some species are unicellular yeasts. These include the common brewer's yeast *S. cerevisiae*, used to make bread, beer, and wine.

A variety of important plant parasites are members of Ascomycota. An example is the fungus responsible for chestnut blight, mentioned earlier in this chapter (**Figure 4.23**). *Cryphonectria parasitica* reached New York on plants imported from Asia in about 1900. The spores of *C. parasitica* are spread by wind, rain, or birds. If they land in a crack or wound in the bark of a chestnut tree, the spores germinate, and as the hyphae grow, they form a lesion which may expand to the point of girdling the tree limb. If this occurs, everything above the lesion is killed. By 1940, over 3.5 billion American chestnut trees had died. Today, only a very few remain.

Fungal infections in animals are called **mycoses**, and several important mycoses in humans are caused by members of Ascomycota. Vaginal yeast infections are the result of an overgrowth by *Candida albicans*. In the throat, a *C. albicans* infection is called *thrush*. Interestingly, *C. albicans* is generally present in small numbers and is considered part of the normal human microbial flora. Its numbers are usually held in check by various bacteria. However, when the bacterial flora is disrupted, as it may be during antibiotic therapy, the yeast is able to proliferate.

Finally, the causative agent of Valley fever, *C. immitis*, belongs to the phylum Ascomycota. This soil fungus is endemic to the American Southwest and northern Mexico. Infections are acquired by inhalation of spore-laden dust. Most cases are asymptomatic, but respiratory symptoms resembling influenza or bronchitis are common. Occasionally the disease can develop into a far more serious progressive form, with the fungus spreading to other organs including bones, skin, viscera, and brain. Those most at risk include

Figure 4.22 Parasitic rust.
A representative basidiomycotan, orange rust, growing on a blackberry plant. Like other fungi, this rust secretes enzymes that digest organic molecules in the surrounding environment, which it then absorbs. In this case, those molecules are found in the tissue of its host plant.

farmers, highway workers, and others regularly exposed to large amounts of dust. During seismic events such as the Northridge earthquake, seemingly everyone is at elevated risk, due to the tons of soil that are catapulted into the atmosphere. Unfortunately, short of avoiding dust, little can be done to reduce the risk of infection. The fungus is ubiquitous in the soil, and eradicating it is out of the question. With earthquakes a part of life on the West Coast, Valley fever is no doubt here to stay.

The Viruses

CASE: "DEATH" IN THE RUE MORGUE

Edgar Allan Poe died October 7, 1849, at the age of 39. His death has always been attributed to the effects and complications of alcoholism and drug abuse. Michael Benitez at the University of Maryland, on the other hand, believes Poe died of rabies. Benitez came to this conclusion by studying records of Poe's symptoms at the time of his death. He was first delirious with tremors and hallucinations and then slipped into a coma. He emerged from his coma and was calm and lucid before he lapsed back into delirium. After 4 days, he died. These symptoms, say Benitez, are "tell-tale" for rabies but are not typical for alcoholism. Additionally, records show that Poe had abstained from drinking for 6 months prior to his death. We will never know for certain if rabies killed one of the greatest American authors (**Figure 4.24**). But can we continue to definitively blame Poe's death on the bottle? "Nevermore."

1. **What type of infection is rabies? How is the agent that causes rabies different than other infectious organisms we have already discussed?**
2. **Why are the symptoms associated with a rabies infection primarily neurological?**

Rabies, caused by rabies virus, has been recognized as a serious disease since ancient times. As far back as 4000 years ago the Mesopotamians had written laws detailing the responsibilities of dog owners, designed to reduce the risk of this dreaded illness. But the doctors who treated Edgar Allan Poe can be forgiven for not understanding what was affecting the author. Indeed, it would still be over 40 years before it was even suggested that something called a "virus" existed.

What exactly are viruses, and how do they differ from other infectious agents? Nobel Prize-winning immunologist Sir Peter Medawar has described them as "bad news wrapped in protein," which is a fairly good definition.

First of all, viruses are small (**Figure 4.25**). A typical bacterium might have a length of approximately 5 micrometers (μm; 1,000,000 μm = 1 meter). Viruses, on the other hand, range in size from about 20 to several thousand nanometers (nm) in length (1000 nm = 1 μm, so 1,000,000,000 nm = 1 meter). The largest animal virus, ironically the smallpox virus, is approximately 300 nm long and is barely visible with the strongest light microscope. The rabies virus is 170 nm long and about 70 nm wide.

Second, unlike prokaryotes and eukaryotes, viruses are *acellular*, which means they are not composed of cells. In its simplest form, a viral particle consists of only nucleic acid, enclosed in a protein coat, much as Medawar described them. As we will see shortly, many viruses are somewhat more complex than this, but viruses are substantially simpler than the cell-based microorganisms we have discussed so far.

Third, we have previously noted that all living things use DNA to code for the proteins they need to survive. Some viruses also encode their genetic information in DNA, but many rely on RNA as genetic material. Consequently, viruses are often broadly classified as either DNA viruses or RNA viruses.

Figure 4.23 Chestnut tree stem infected with chestnut blight. This fungal plant pathogen, responsible for the decline of chestnut trees in North America, is a member of the fungal phylum Ascomycota.

Figure 4.24 Edgar Allan Poe. Was his death due to rabies?

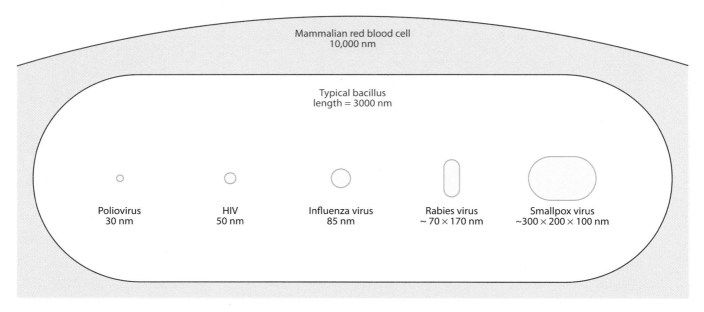

Fourth, all viruses are obligate intracellular parasites; they can reproduce and continue their existence only if they invade a host cell. A viral particle has only a small number of genes. This genetic material codes primarily for the structural proteins of the coat and other required proteins, including some enzymes needed for replication. Everything else is supplied by the host cell. Outside of a host cell, viruses are unable to replicate themselves, and they show few if any of the properties commonly attributed to living things. However, once inside an appropriate cell, viruses essentially commandeer the host cell's machinery and use it to replicate themselves. In this manner, an infected cell becomes a viral factory. The virus uses host enzymes and other resources such as amino acids, as well as the host's own structures like ribosomes, to make copies of itself. Newly assembled viral particles are released from the host cell in a process that often kills the cell or seriously compromises normal functioning.

Because they are acellular, and because they lack many fundamental characteristics of living things, one might ask if viruses are even "alive." The answer to such a question, however, is perhaps more philosophical than biological. Viruses certainly replicate and evolve as other living things do, and the way we treat viral disease is the same whether or not we consider them to be alive. Perhaps the most accurate way to view viruses is to think of them as straddling the border between the living and nonliving world.

Most viruses have one of a few basic structures

An individual, complete viral particle is called a **virion**. As previously stated, all viral particles are composed of genetic material surrounded by a protein coat. Depending on the type of virus, the genetic material may be DNA or RNA. The protein coat is called the **capsid**, and the individual subunits that make up the capsid are **capsomeres**. Each capsomere is composed of one or more proteins. Collectively, the nucleic acid and the capsid are called the **nucleocapsid**.

Most capsids have one of two basic shapes (**Figure 4.26**). Many viral coats are composed of 20 capsomeres, all in the shape of equilateral triangles. Such viruses have the appearance of geodesic spheres and are known as **icosahedral** (20-sided) viruses. **Helical** viruses have capsomeres that fit together to form a spiral around the enclosed nucleic acid. Apart from these two most common shapes, some viruses have a **complex** structure. Smallpox virus, for example, has a capsid that is composed of multiple layers of protein, neither icosahedral nor helical in shape.

Figure 4.25 Sizes of viruses. Viruses are significantly smaller than the typical bacterial and animal cells shown for comparison.

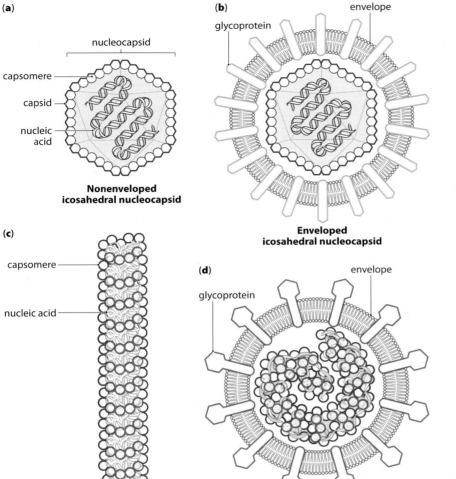

(a)

nucleocapsid

capsomere

capsid

nucleic acid

Nonenveloped icosahedral nucleocapsid

(b)

glycoprotein

envelope

Enveloped icosahedral nucleocapsid

(c)

capsomere

nucleic acid

Nonenveloped helical nucleocapsid

(d)

glycoprotein

envelope

Enveloped helical nucleocapsid

Figure 4.26 Viral structure. Most viruses have a capsid that can be described as either icosahedral or helical. Icosahedral capids consist of 20 identical subunits called capsomeres, all in the shape of equilateral triangles. In helical viruses the capsomere subunits form a helix or spiral around the nucleic acid. Viruses may also be either enveloped or nonenveloped. In enveloped viruses, the envelope is actually host membrane, acquired by the virus as it leaves the host cell. Viral proteins called *glycoproteins* (also called "spikes") are embedded within the envelope. Nonenveloped viruses leave host cells in a different way and do not acquire a layer of host membrane. (a) Nonenveloped icosahedral virus. (b) Enveloped icosahedral virus. (c) Nonenveloped helical virus. (d) Enveloped helical virus.

Many viruses are surrounded by an **envelope** that is composed of a phospholipid bilayer in which viral proteins called **glycoproteins** are embedded (see Figure 4.26). Other viruses are **nonenveloped** and lack this structure. As we will see, the presence or absence of an envelope and its associated glycoproteins tells us a great deal about how a particular virus enters and leaves a host cell. Many viral nucleocapsids contain a small number of enzymes. These are enzymes lacking in the host cell, which are required for successful viral replication.

Specific viruses usually infect only certain hosts and certain cells within those hosts

An important characteristic of a virus is its **host range**—the spectrum of hosts that it is able to infect. Sometimes, viruses are simply classified based on their host type, and are therefore categorized as animal viruses, plant viruses, etc. Even within a group such as animal viruses, however, a given virus cannot infect all animal species. For example, if you have a cat, you might justifiably be concerned about feline leukemia virus (FLV). While dangerous to cats, this virus is incapable of infecting humans. On the other hand, while humans are susceptible to influenza virus, cats are not.

Even within an appropriate host, only certain types of cells can be infected. For example, the rabies virus that may have killed Edgar Allan Poe can kill

(a)

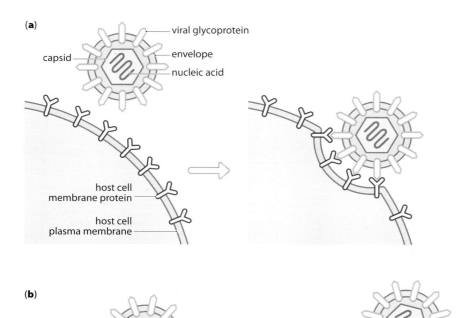

viral glycoprotein

capsid

envelope

nucleic acid

host cell
membrane protein

host cell
plasma membrane

(b)

Figure 4.27 Viral host-cell specificity. The ability of a specific virus to infect specific host cells is due in large part to the ability of the virus to bind to proteins found in the host-cell membrane. (a) The virus has glycoproteins that match the three-dimensional structure and chemical properties of the host-cell membrane protein. Chemical attraction between the virus and the host proteins binds the virus to the host cell. The virus is then able to enter the cell via one of several mechanisms. (b) The same virus is unable to bind a different cell type, because proteins found on this cell are not complementary to the viral glycoproteins. Consequently, this cell type is resistant to infection by this specific virus.

dozens of other mammalian species as well. However, rabies virus is able to infect only certain cell types within these hosts. The **specificity** of the virus describes the variety of cell types a particular virus can successfully infect. What determines specificity? Rabies virus is an enveloped virus. The glycoproteins that are embedded in the viral envelope are able to bind to proteins found in the plasma membrane of nerve cells (neurons). The protein in the plasma membrane of the neuron is similar to a "lock" while the glycoprotein is like a "key." Only cells that have the proper membrane protein can be infected by this virus (**Figure 4.27**). This explains why the symptoms accompanying rabies are largely neurological. Hepatitis B virus, on the other hand, lacks the necessary glycoprotein spike to bind to neurons and therefore cannot infect this cell type. Hepatitis B virus has a different envelope glycoprotein, one that allows it to bind to proteins on the surface of liver cells. While some viruses are very specific, others can infect many cell types; the molecule with which they attach to host cells is widespread and found on a variety of different cells. Nonenveloped viruses attach to target host cells via components of their capsid. In poliovirus, for instance, the point where three capsomeres come together forms a region called a canyon. Cells lining the intestine have proteins in their membranes that can fit snugly into these canyons, permitting infection by the virus.

Replication of animal viruses proceeds through a series of defined steps

The infection of a susceptible animal cell by a virus can conveniently be divided into a number of steps, which we will consider in sequence. Many of the differences that occur in the replicative cycle of different viruses depend on whether or not the virus has an envelope, and whether it has DNA or RNA in its nucleocapsid.

Attachment. The first step in the infection process is the attachment between a virion and its target host cell. A virion is not motile. It is merely transported passively until it happens to make contact with a cell bearing the proper receptor. For enveloped viruses, the glycoproteins in the envelope bind the proper receptor in the host-cell plasma membrane. In nonenveloped viruses, proteins forming part of the capsid often function as attachment sites. The receptor molecule to which a particular virus attaches is typically a membrane protein of some sort. Only cells bearing the correct receptor are vulnerable to a specific virus. For instance, human immunodeficiency virus (HIV) can infect only cells that bear proteins called CD4 on their surface.

Penetration. Following attachment, the virion must enter the host cell (**Figure 4.28**). For nonenveloped viruses, entry occurs via endocytosis (see Chapter 3, p. 66, for a discussion of endocytosis). The binding of the virion to the receptor causes the cell to transport the virus across the membrane and into the cytoplasm inside a membrane-bound vesicle. Enveloped viruses may also enter a cell via endocytosis, but many employ a different strategy called **fusion**. During fusion, viral attachment brings the plasma membrane and the viral envelope into close proximity. The two lipid bilayers actually fuse together, releasing the nucleocapsid into the cell cytoplasm. Many of the details of fusion remain unclear, but it appears that when viral glycoproteins bind to their receptors, the glycoproteins change their shape due to the interaction. This causes disruption of the envelope and plasma membrane, allowing them to fuse.

Figure 4.28 Entry of an animal virus into a host cell. (a) Entry via endocytosis. Following viral attachment to host-cell membrane receptors, the host cell transports the virus into its cytoplasm. Once endocytosis is complete, the viral particle is released into the cytoplasm. Both enveloped and nonenveloped viruses may enter a cell via endocytosis. (b) Entry via fusion. Following viral attachment, the host-cell plasma membrane and the viral envelope are brought into close proximity. Following fusion of the two lipid bilayers, the viral nucleocapsid is released into the host-cell cytoplasm. Only enveloped viruses enter host cells in this fashion.

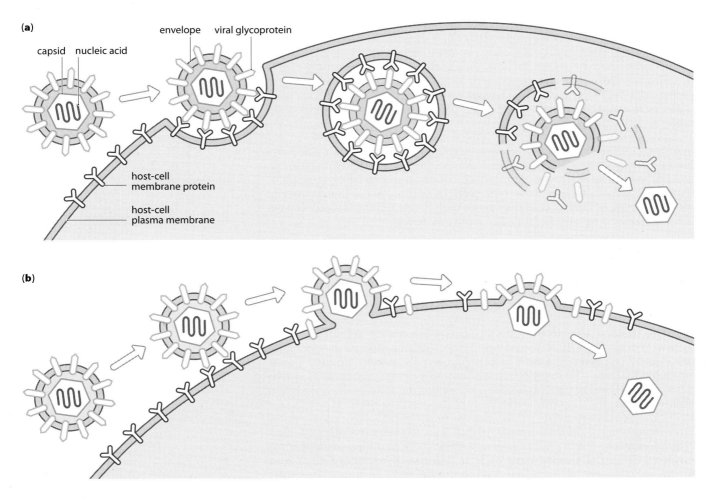

(a)

capsid   nucleic acid   envelope   viral glycoprotein

host-cell membrane protein

host-cell plasma membrane

(b)

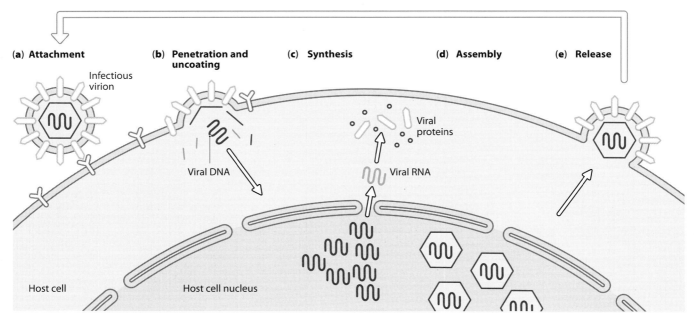

Figure 4.29 Replication of an enveloped DNA virus. (a) The virus attaches to an appropriate host cell via its envelope glycoproteins, which bind to host-cell membrane proteins. (b) In this example, following entry into the host cell by fusion, uncoating has released the viral DNA. (c) Synthesis of new viral DNA and viral proteins. Viral DNA generates viral RNA, which codes for the production of new viral proteins. (d) Newly synthesized viral capsid proteins move back to the nucleus, where they assemble around newly synthesized copies of viral DNA, creating new virions. Meanwhile, newly synthesized viral glycoproteins are transported to the cell plasma membrane. (e) New virions are released from the host cell by budding, during which they acquire their new envelope. These newly released virions may then infect other susceptible cells.

Uncoating. Prior to replication, the virus must shed its capsid (**Figure 4.29**). This process, called **uncoating**, may occur in endocytotic vesicles, where the pH is low. In other cases, host enzymes called proteases digest the viral protein coat. Once a virus has uncoated, we say that it has entered the *eclipse* phase. An intact viral particle no longer exists; the eclipse phase ends only when newly replicated virions are assembled at the end of the replicative cycle.

Synthesis. Following uncoating, two important tasks must be completed before new viral particles are assembled. First, the nucleic acid must be replicated. Second, new structural proteins must be produced (see Figure 4.29). Different viruses achieve these two goals in various ways that primarily reflect the type of nucleic acid they carry. We will return to this topic shortly, following our completion of the viral replicative cycle.

Assembly. Once new viral nucleic acid and structural proteins are made, new virions are formed in a process called **assembly** (see Figure 4.29). In some viruses the capsomeres first assemble to form the capsids, and the genetic material is then inserted. In others, the capsomeres latch on to the genetic material, ultimately producing the new viral particle.

Release. The final step in the viral replicative cycle is the **release** of newly formed virions from the infected cell. Nonenveloped viruses are generally released by cell **lysis**. The host cell literally explodes, releasing the newly assembled virions. The released virions may now contact other host cells, to begin a new round of viral replication. The cells that undergo lysis are killed in the process.

Enveloped viruses, on the other hand, are typically released from infected cells by **budding** (**Figure 4.30**). In this process the assembled nucleocapsid pushes through the plasma membrane. In doing so, it becomes coated with plasma membrane material that now forms the viral envelope. The virus

Figure 4.30 Viral release by budding. Stages in the release of an enveloped helical virion are shown. Enveloped viruses acquire their envelope as they bud through the plasma membrane of the infected host cell. The envelope is actually a portion of the host-cell plasma membrane. Viral glycoproteins, already produced during the synthesis stage (see Figure 4.29), become embedded in the plasma membrane and are acquired along with the lipid bilayer as newly assembled viral particles leave the cell.

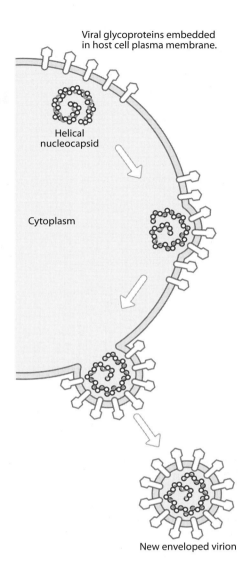

Viral glycoproteins embedded in host cell plasma membrane.

Helical nucleocapsid

Cytoplasm

New enveloped virion

has already coded for envelope glycoproteins; these will already be present in the plasma membrane, and the new virus will acquire them along with its envelope. Unlike lysis, budding does not necessarily result in cell death.

Replication is different for DNA and RNA viruses

The synthesis step described above is substantially different for DNA and RNA viruses. Here we consider some of the important details in the synthesis stage for these two viral groups.

DNA viruses. Once a DNA virus releases its DNA into the cell, some of the viral genes begin to code for protein. This process is similar to what is seen in both eukaryotic and prokaryotic cells and involves two crucial steps. First, the viral genes, composed of DNA, are used to produce a molecule of RNA that carries the gene's genetic instructions. Second, the RNA associates with host ribosomes, and the genetic message is converted into the specified viral protein. The details of this two-step process will be explored in Chapter 6. Some of the proteins that are made first are those needed to replicate the viral DNA. The viral DNA is then replicated, and hundreds or even thousands of new DNA molecules are produced. Finally, another round of protein synthesis takes place, primarily using host enzymes and resources. This time the proteins being produced are the structural proteins that will make up the completed virions. Examples of DNA viruses that replicate in this manner include the herpesviruses and the papillomaviruses that cause warts. Synthesis of a DNA virus is illustrated in **Figure 4.31**.

Figure 4.31 Synthesis in a DNA virus. To produce new progeny viral particles, the infecting virion must both replicate its DNA and synthesize necessary proteins. Initially, enzymes that the virus needs to replicate its DNA are produced. DNA replication then follows. Additional structural proteins such as capsomeres and glycoproteins are then produced, permitting the assembly of new virions.

1. Following penetration and uncoating, viral DNA enters the host-cell nucleus.

2. Viral enzymes are produced for replication.

3. The viral DNA replicates itself using viral enzymes.

4. Viral coat proteins are produced.

5. New viral particles assemble in the nucleus.

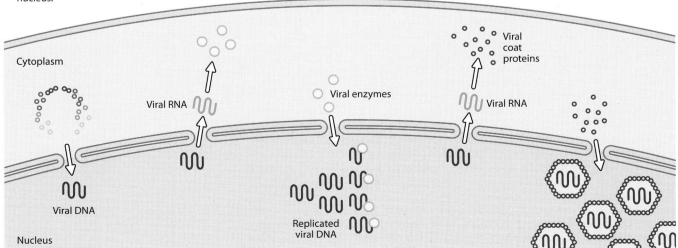

Cytoplasm

Viral RNA

Viral enzymes

Viral coat proteins

Viral RNA

Viral DNA

Replicated viral DNA

Nucleus

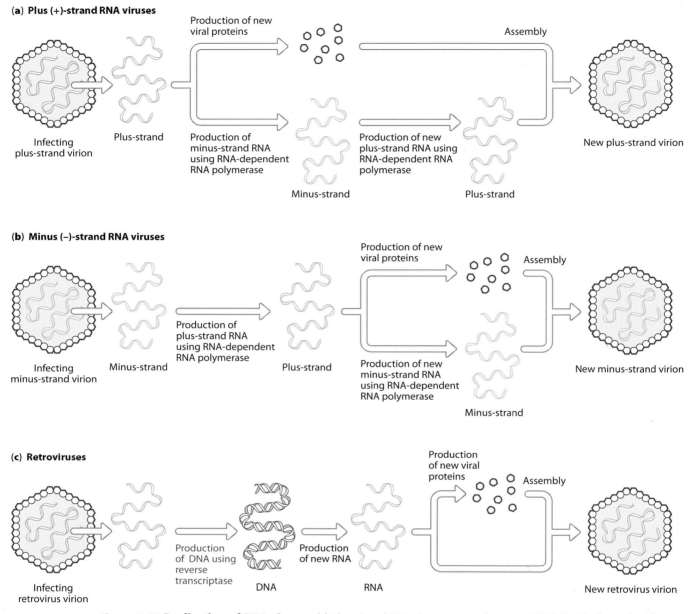

(a) Plus (+)-strand RNA viruses

Infecting plus-strand virion → Plus-strand

Production of new viral proteins → Assembly → New plus-strand virion

Production of minus-strand RNA using RNA-dependent RNA polymerase → Minus-strand → Production of new plus-strand RNA using RNA-dependent RNA polymerase → Plus-strand

(b) Minus (–)-strand RNA viruses

Infecting minus-strand virion → Minus-strand

Production of plus-strand RNA using RNA-dependent RNA polymerase → Plus-strand

Production of new viral proteins → Assembly → New minus-strand virion

Production of new minus-strand RNA using RNA-dependent RNA polymerase → Minus-strand

(c) Retroviruses

Infecting retrovirus virion

Production of DNA using reverse transcriptase → DNA → Production of new RNA → RNA

Production of new viral proteins → Assembly → New retrovirus virion

Figure 4.32 Replication of RNA viruses. (a) Plus-strand RNA viruses carry plus-strand RNA (in blue) that is able to code for proteins. This RNA is also used to make new minus-strand RNA (in green). The minus-strand RNA will be copied back into plus-strand RNA, to be incorporated into the newly formed virions during assembly. (b) Minus-strand RNA viruses carry minus-strand RNA (in green). Following infection, the minus-strand RNA is used as a template to produce plus-strand RNA (in blue), which is then used to make necessary proteins. The plus-strand RNA is also converted back into minus-strand RNA, which assembles with the proteins to form new progeny virions. (c) In retroviruses, plus-strand RNA (in blue) is converted into DNA (in red), which is then used to make many copies of new plus-strand RNA. The newly synthesized plus-strand RNA is used both for protein production and for incorporation into new progeny virions.

RNA viruses. Unlike any prokaryote or eukaryote, RNA viruses encode their genetic information in RNA. The RNA viruses can be subdivided into three categories (**Figure 4.32**). The **plus (+) strand RNA viruses** carry RNA that can be immediately used to make proteins. Once they uncoat, these viruses start producing viral proteins such as the capsomeres necessary for new virion assembly. The RNA is also used as a template to produce complementary **minus (–) strand RNA**. This minus-strand RNA is then used to produce many copies of plus-strand RNA. Once both new proteins and new genetic material are available, assembly of new virions can occur.

Only RNA viruses convert RNA to complementary RNA in this manner. To do so, they require an enzyme called **RNA-dependent RNA polymerase**; this type of enzyme is just about unique to RNA viruses, among all life forms. Note that DNA plays no role in the replication of these viruses. Some familiar plus-strand RNA viruses include poliovirus and the rhinoviruses that cause colds.

In **minus-strand RNA viruses**, each virion carries only minus-strand RNA that is complementary to the plus-strand RNA needed for protein synthesis (Figure 4.32b). Once in a cell, this RNA serves as a template to produce plus-strand RNA. As in the plus-strand RNA viruses, RNA is transcribed into complementary RNA by the enzyme RNA-dependent RNA polymerase. The newly transcribed RNA is used for two purposes: it is replicated back into many copies of minus-strand RNA, and it codes for the production of proteins such as the capsomeres required for new virion assembly. Rabies virus, introduced in our case, is an example of a minus-strand RNA virus. Others include measles and influenza viruses.

The last group of RNA viruses is the **retroviruses**. By far, the most important human retrovirus is HIV, the causative agent of AIDS. Retroviruses also carry plus-strand RNA in their nucleocapsid, but they do not convert it to minus-strand RNA. Instead, they convert their RNA to DNA in a process called **reverse transcription** (Figure 4.32c). Reverse transcription requires an enzyme called **reverse transcriptase**. The newly synthesized DNA is then transcribed back to many copies of the viral RNA. This RNA serves as the genetic material for new virions and is translated into structural proteins.

Because reverse transcriptase is not used by our cells, it makes an attractive target for antiviral drugs. Ideally, interfering with this enzyme should have no adverse consequences for the host cell. Many of the important currently available anti-HIV medications, such as AZT, act by targeting this enzyme. The topic of antiviral drugs will be more fully explored in Chapter 12.

Not all viral infections cause symptoms or kill host cells

Not all viral infections affect the host in the same way. If the virus undergoes repeated rounds of replication, the infection is termed **acute**; infected cells essentially devote themselves to viral production and ultimately die as a consequence. Some viruses, such as influenza and rabies viruses, participate only in acute infection cycles. The viruses keep reproducing and infecting new cells until either the host immune system eliminates the intruder or the host dies (**Figure 4.33**).

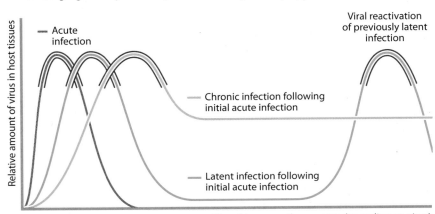

Time highlighted in pink corresponds to when signs and symptoms of disease are apparent.

Time (days, months, or years depending on virus)

Figure 4.33 Types of viral infections. Symptoms of acute infections are associated with that period of time when levels of virus are high. Chronic infections typically start with an acute infection; viral levels then diminish, and viral replication continues at a lower rate. The amount of virus produced during the course of a chronic infection may or may not be sufficient to cause symptoms. Latent infections also typically begin as acute infections. Following the acute infection phase, the virus enters a period of dormancy (the *latent period*), during which it does not replicate. This latent period lasts indefinitely; reactivation does not necessarily occur. If reactivation does occur, there will be a new acute episode.

Chronic infections are characterized by a slow release of viral particles that may or may not kill the host cell; if such infections do result in individual cell death, host cell replication is able to keep up with cell death, and there may be few or no signs and symptoms of disease in an infected individual. A good example is a hepatitis B virus (HBV) infection. A person infected with HBV initially suffers through an acute phase with severe illness. In many individuals, the virus is destroyed by the immune system and the patient returns to the uninfected state. In others, the immune system is able to eventually suppress but not entirely eliminate the virus. Consequently, low levels of virus continue to be produced over a prolonged period of time, sometimes lasting many years.

Some DNA viruses and retroviruses cause **latent** infections. In such infections, there is no immediate production of viral progeny. The viral DNA does not code for viral proteins, nor is it replicated independently of the host DNA. If the host cell divides, however, the viral DNA is duplicated along with the host DNA prior to cell division. Consequently, if a cell infected with a latent virus divides, the two newly produced daughter cells will both be infected with the virus.

Viruses may persist in this latent state for many years or in some cases for the entire life of the host. However, a sudden change in the health of the host or certain environmental factors can cause viral **reactivation**. A reactivated virus is a previously latent virus that has resumed replication. During this time, symptoms may become apparent in an infected individual.

Herpesviruses undergo replication cycles of this type. For instance, human herpesvirus type 1 (HHV-1) is the causative agent of cold sores. When a person is first infected, an acute infection causes cell death, resulting in the oral lesion. An effective immune response eventually destroys replicating viruses. However, the virus will remain latent and may be reactivated at any time. The reappearance of the cold sore is associated with such reactivation. If you are infected with HHV-1, you will remain infected for life, even if you never experience any symptoms after the initial infection. If you get a cold sore a second time in the same spot, it is a consequence of the same initial infection event. This is quite unlike an infection with influenza virus; once you recover from a bout of the flu, the virus is gone. If you suffer through another case of the flu in the future, you can blame it on a second infection.

We are only just beginning to understand the factors that might cause a latent virus to reactivate. Clearly, anything that suppresses the host immune system, such as cancer chemotherapy or HIV infection, makes reactivation more likely. In the case of HHV-1, factors such as exposure to UV light seem to play a role.

Viruses can damage host cells in several ways

Basically, disease-causing viruses cause problems for cells by interfering with their normal functions. Any such interference is known as the virus's **cytopathic effects**. One important cytopathic effect is the ability of many viruses to interfere with the host cell plasma membrane. Recall from Chapter 2 that the plasma membrane both maintains cell integrity and is important in regulating transport into or out of the cell. In virally infected cells, the budding of enveloped viruses can significantly compromise the ability of the plasma membrane to carry out its functions. Additionally, viruses can damage cells simply by interfering with the ability of the cell to carry out its own protein synthesis or by using resources that the cell requires for its own needs. Some viruses prevent the host cell from copying its DNA. Others can initiate a process that ultimately leads to cancer.

Let us return to Edgar Allan Poe and the rabies virus that may have killed him. Oddly, in spite of the extreme symptoms exhibited by rabies victims, examination of infected neurons reveals little damage. There are, however, large, prominent clumps of viral proteins visible in the cytoplasm. These *Negri bodies* are often the only visible sign of infection. Presumably, Negri bodies interfere with normal neuron function, resulting in the neurological symptoms typical of human rabies.

Like animals, plants are susceptible to many viral infections

Unlike animal cells, plant cells are surrounded by a rigid cell wall, and plant viruses are unable to reach the plasma membrane when the cell wall is intact. Consequently, plant viruses infect cells through breaks or wounds in the cell wall. In many cases this damage is caused by insects, bacteria, or fungi.

In other respects, plant viruses are similar to animal viruses. Like animal viruses, they may carry DNA or RNA. Their mode of replication is essentially the same, as is their overall morphology. In one important respect, however, viral infections in plants are quite different. Plant cells are connected to one another via small junctions, which essentially form connections between the cytoplasm of adjacent cells. In effect, all cells linked by these junctions have a common cytoplasm. Progeny of plant viruses can spread to new host cells through these junctions and through the phloem, which functions as the plant's circulatory system. Because of this, plant infections are often systemic. In other words, unlike animal viruses, most of which only affect specific organs or tissues, plant viruses can affect cells throughout the entire organism.

Many important plant diseases are caused by viruses. Plant viruses take a large economic toll in crops such as corn and wheat, but they cause even greater problems in perennial plants such as potatoes. Plants suffering from viral infections are often stunted (**Figure 4.34a**). Irregular coloration often appears on leaves and stems, and green pigments may be lost. Unlike animals, plants rarely recover from a viral infection because they lack the ability to mount a specific immune response. In a few happy cases, viral infection in plants is actually desirable. The attractive variegated color of some tulips is the result of a viral infection, transmitted through the bulbs (Figure 4.34b).

Bacteriophages are viruses that infect bacteria

It may be surprising to learn that even bacteria have their own pathogens to worry about—bacteria-attacking viruses called **bacteriophages** or **phages** for short. Maybe the English satirist Jonathan Swift was prescient when he wrote in 1733, "Big fleas have little fleas upon their backs to bite them, and little fleas have lesser fleas and so ad infinitum."

The most intensively studied phages are those that infect *E. coli*. A specific phage known as T4 provides an illustrative example. A T4 virion has a complex structure, with an icosahedral head containing DNA and a hollow, helical tail (**Figure 4.35**). The tail region is associated with other characteristic structures called the tail fibers, tail pins, and base plate.

The initial step in the infection cycle of bacteria involves the contact and adherence to the bacterial surface by the phage (**Figure 4.36**). This process, called **adsorption**, cannot occur on just any bacterial cell. To successfully adhere, molecules on the phage tail and tail fibers must match specific molecules on the bacterial surface that serve as receptors. A bacterium lacking these molecules is resistant to infection.

Figure 4.34 Plant viruses. (a) Tobacco mosaic virus. The leaf on the right is from a tobacco plant infected with tobacco mosaic virus. Note its stunted and discolored appearance compared with the leaf from an uninfected plant on the left. (b) The attractive variegated color in some tulips is the result of a viral infection.

(a)

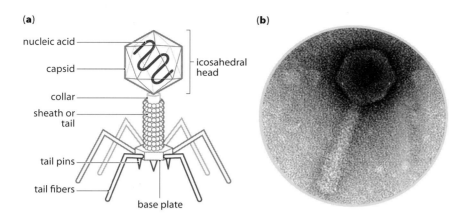

nucleic acid

capsid

icosahedral head

collar

sheath or tail

tail pins

tail fibers

base plate

(b)

Figure 4.35 Structure of T4, a representative bacteriophage.
(a) Note the complex structure, including the icosohedral head, encompassing the nucleic acid, and the helical tail, which makes initial contact with the host cell. (b) An electron micrograph of a bacteriophage.

Once the T4 phage has attached, it injects its DNA into the bacterium in a step called **penetration**. This involves the contraction of the helical sheath, which forces the hollow tube into the cell cytoplasm, much like a microscopic syringe. In the process, the viral DNA is released into the cell's interior. The viral capsid does not enter the cell. It remains as an empty shell, attached to the cell exterior.

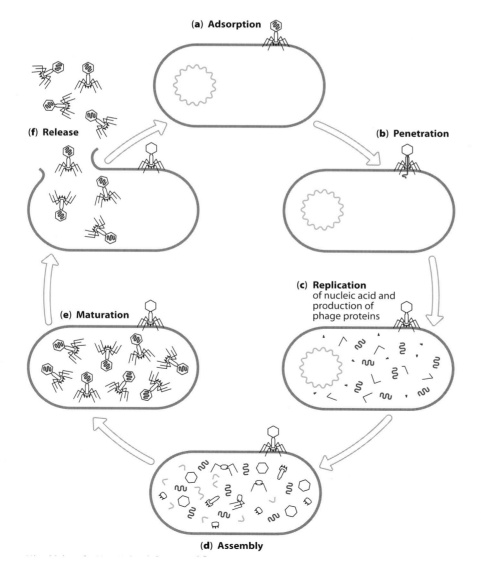

(a) Adsorption

(f) Release

(b) Penetration

(e) Maturation

(c) Replication of nucleic acid and production of phage proteins

(d) Assembly

Figure 4.36 Replicative cycle of bacteriophage T4. This bacteriophage is able to infect *E. coli* cells. (a) Adsorption: the phage makes contact and adheres to the bacterial surface. (b) Penetration: the helical sheath contracts, forcing the viral DNA into the cell cytoplasm. (c) Replication involves both replication of the viral DNA and inhibition of host-cell activity. (d) Assembly of new phage particles. (e) Maturation of phages into newly produced infective viral particles. (f) Lysis of the host cell and release of newly produced, mature phages.

Figure 4.37 Lysogenic cycle of bacteriophage lambda. Once phage DNA is released into the cytoplasm of the host cell, it incorporates into the host DNA. During the lysogenic cycle, no new progeny phages are produced. If the bacterial cell replicates, the phage DNA, called the *prophage*, is replicated along with the host DNA. Under appropriate conditions, prophages in the lysogenic cycle may return to the lytic cycle.

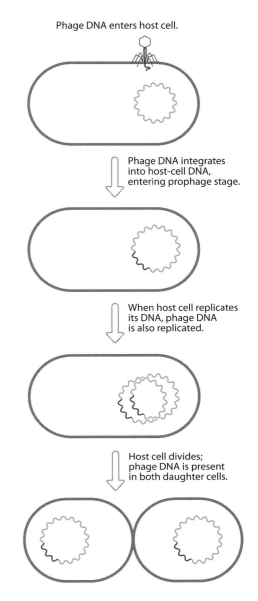

Phage DNA enters host cell.

Phage DNA integrates into host-cell DNA, entering prophage stage.

When host cell replicates its DNA, phage DNA is also replicated.

Host cell divides; phage DNA is present in both daughter cells.

Once T4 DNA has entered the host cell, the metabolic activity of the cell is blocked and the cell begins producing proteins coded for by the viral DNA. These include proteins that block normal host-cell activity, enzymes required for viral replication, and structural proteins needed to construct new capsids. The viral DNA is also repeatedly copied. All energy required for these processes is provided by the host cell.

As new viral DNA and structural proteins are produced, they spontaneously assemble into new virions. Up to several hundred new phages may ultimately be produced. Eventually, the host cell becomes so crowded with virions that it bursts, releasing the progeny phages. These newly released phages may then contact and infect other susceptible cells.

The T4 replicative cycle we have just described is termed a **lytic cycle**, because it results in bacterial cell lysis and death. Not all phages, however, undergo a lytic cycle. Infection of *E. coli* with phage lambda, for instance, does not necessarily cause lysis. After releasing its DNA into a susceptible bacterium, lambda does not immediately reproduce. Rather it enters an inactive period called the **prophage** state (**Figure 4.37**). In the prophage state, the viral DNA is actually incorporated into the bacterial DNA. If the bacterium itself divides, the viral DNA will also replicate along with it, but otherwise, the infected cell remains largely unaffected. This state of viral inactivity inside an infected cell is called **lysogeny**. However, under certain conditions, the virus may begin to replicate, thus returning to the lytic cycle. Ultraviolet light or exposure to certain chemicals is known to promote a return to the lytic cycle in phage lambda.

Phages can influence the numbers of bacteria and even the diseases that they cause

Because they do not infect animals or plants, one might get the idea that phages have little importance for humans. However, bacteriophages affect us in several important direct and indirect ways. First, the numbers of phages in the environment are almost inconceivable. A single milliliter of seawater may contain close to 50 million of them, and virologists have found similar numbers in soil. According to some estimates, in the oceans alone, up to 40% of all bacteria may be destroyed each day by phages. Consequently, they may influence the entire world's food supply by limiting the numbers of bacteria available for other organisms to eat. The vast numbers of killed bacteria no doubt impact nutrient cycling in the ocean and other environments.

Of perhaps more immediate concern to humans is the effect lysogenic phages have on the virulence of some bacteria. For example, cholera is caused by *Vibrio cholerae*, a Gram-negative species. This often deadly disease is caused by a toxin produced by the bacteria that results in massive diarrhea (**Figure 4.38**). However, the bacterium cannot make the toxin by itself. Rather, lysogenic phage genes inside the cell instruct the host to create the toxin. Bacteria that are uninfected with these phages do not cause disease.

Figure 4.38 Cholera. This 1912 painting by the French painter Jean-Loup Charmet depicts the horror associated with cholera epidemics. Only bacteria infected with the proper lysogenic phage produce the cholera toxin, responsible for disease.

Finally, because they destroy specific bacteria, some scientists have suggested that bacteriophages could be utilized therapeutically to treat bacterial infections. This idea was first proposed in the early twentieth century. Spotty results, however, combined with the development of powerful antibiotics dampened interest in "phage therapy." However, in recent years, with the development of antibiotic resistance by many bacteria, some researchers are taking a fresh look at this approach. We will return to this topic in Chapter 12.

Prions

CASE: LAST LAUGH FOR THE "LAUGHING DEATH"

Up until the late 1950s, 2–3% of the Fore people of New Guinea mysteriously died of the "laughing death" each year. The disease, also called kuru, started with increased clumsiness, headaches, and a tendency to giggle inappropriately. Within a few months, victims could no longer walk. Total incapacitation, followed by death, occurred within a year. Curiously, the disease affected women and young children almost exclusively (**Figure 4.39**). Older boys and men were spared. The reason for this was unknown until 1957, when Carleton Gajdusek, of the U.S. National Institutes of Health, determined how the disease was transmitted. The Fore people practiced ritual cannibalism as part of their burial ceremony. When a member of the tribe died, it was customary for female family members to prepare the body and eat parts of the deceased's brain. Small children also consumed brain tissue, but older boys and adult men did not participate. Gajdusek postulated that an infectious agent was transmitted through the brain tissue, explaining why only those who had consumed such tissue were affected. Kuru has now been eradicated, as cannibalism has not been practiced since the 1960s.

1. What exactly causes kuru?
2. How is this infectious agent unlike any other that we have thus far discussed?
3. What other diseases are caused by similar agents?

Figure 4.39 A young Fore child infected with kuru. This photo was taken in the early 1950s, before measures to eradicate kuru were initiated.

Kuru may be gone, but other diseases such as Creutzfeldt–Jakob disease (CJD) and bovine spongiform encephalopathy (BSE), also known as "mad cow" disease, are still with us. These diseases all cause neurological symptoms, and they are all uniformly fatal. They also are all caused by highly unusual infectious agents called **prions**.

Prions are infectious proteins. Their very existence was not proposed until 1982, and it is only since the 1990s that most of the scientific community has accepted the notion that a protein could cause infection. The concept of prions was hotly contested because their existence violates a basic tenet of biology—that only nucleic acids can replicate themselves and serve as genetic material.

But prions do not replicate in the traditional sense. Prions seem to be abnormally folded forms of a protein that already exists in the brain. The best-studied prion disease is scrapie, which affects sheep. It is named for the behavior of infected animals, which includes their scraping themselves bare against fence posts or other solid objects. On the surface of the sheep's brain is a protein called PrP^c. Occasionally this protein folds up incorrectly. It is then called PrP^{Sc} (**Figure 4.40a**). This aberrant protein is able to interact with normal PrP^c proteins, forcing them to refold in the abnormal three-dimensional shape (Figure 4.40b). In other words, PrP^{Sc} acts like a mold or template, which forces PrP^c to take on the disease-causing shape. Eventually, the PrP^{Sc} proteins build up to such an extent that they begin to form abnormal plaque on the surface of brain cells. Whether or not it is these plaque deposits that are the direct cause of the disease remains to be seen. Other prion diseases, including kuru and CJD, are also believed to be caused by

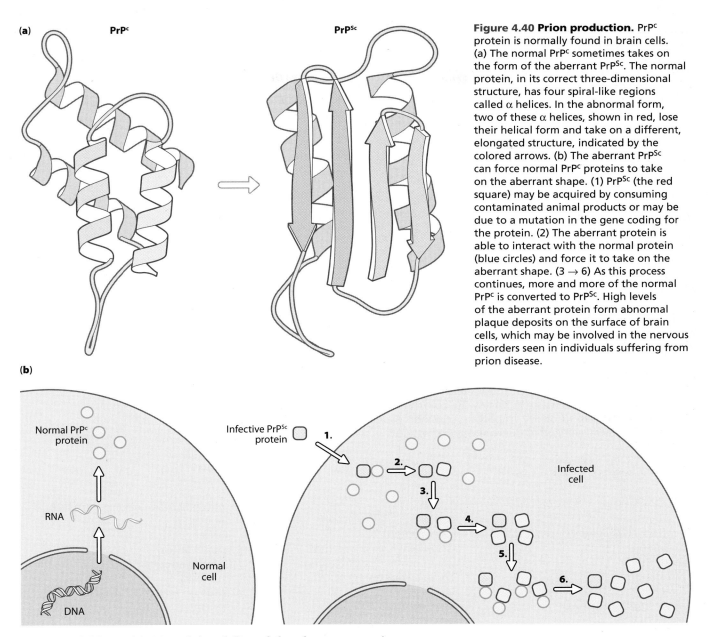

(a) PrP^c   PrP^{Sc}

(b)

Normal PrP^c protein

Infective PrP^{Sc} protein

1.

2.

Infected cell

3.

RNA

4.

Normal cell

5.

DNA

6.

Figure 4.40 Prion production. PrP^c protein is normally found in brain cells. (a) The normal PrP^c sometimes takes on the form of the aberrant PrP^{Sc}. The normal protein, in its correct three-dimensional structure, has four spiral-like regions called α helices. In the abnormal form, two of these α helices, shown in red, lose their helical form and take on a different, elongated structure, indicated by the colored arrows. (b) The aberrant PrP^{Sc} can force normal PrP^c proteins to take on the aberrant shape. (1) PrP^{Sc} (the red square) may be acquired by consuming contaminated animal products or may be due to a mutation in the gene coding for the protein. (2) The aberrant protein is able to interact with the normal protein (blue circles) and force it to take on the aberrant shape. (3 → 6) As this process continues, more and more of the normal PrP^c is converted to PrP^{Sc}. High levels of the aberrant protein form abnormal plaque deposits on the surface of brain cells, which may be involved in the nervous disorders seen in individuals suffering from prion disease.

improper folding of PrP^c and the ability of the aberrant protein to cause normal proteins to adopt the abnormal shape.

Prion disease can be caused by mutation of the gene coding for PrP^c, but it can also be transmitted between individuals. Creutzfeldt–Jakob disease, for instance, can be transmitted via transplants or surgical instruments. Ninety percent of all prion disease in humans has been identified as CJD. Typical age of onset is approximately 65, and it is characterized by progressive dementia. Like all known prion diseases, it is always fatal.

Bovine spongiform encephalopathy (BSE) is a prion disease of cattle. The disease is named for the brain's "spongy" appearance in affected cows. It is also popularly called mad cow disease because of the erratic behavior of affected cattle.

It is believed that the disease was spread by the previously common practice of adding bone meal to cattle feed. Bone meal often has nerve tissue attached to it, and such tissue may contain infectious prion particles. The disease came to the attention of the general public when it suddenly appeared in British cattle in 1980. Since that time there have been more than 160,000 cases. Millions of cattle were slaughtered and the use of animal products in cattle feed was banned in an effort to control the epidemic. Many were concerned that humans who consumed meat from infected cows might develop a similar disease. There was also much speculation that CJD was simply a human form of mad cow disease. This is probably not the case, but disturbingly, in 1996, a new form of CJD called variant CJD (vCJD) was first reported in Great Britain. Unlike CJD, vCJD has an average age of onset of 27. Furthermore, the disease kills its victims in just over a year, as opposed to about 3 years for CJD. So far, about 100 people have died of vCJD.

Could vCJD be the human equivalent of mad cow disease, and did British victims become infected by eating contaminated meat? That remains to be seen, but certainly the sudden appearance vCJD in 1996 is consistent with such a link.

Coming Up Next...

The stage is set. We now know "who" the microorganisms are, and we are starting to understand what they *do*. But before we continue to investigate the activities of this remarkable cast of characters, we will investigate what they *did*. In the next chapter, we will consider microorganisms with a historian's eye, focusing on the astounding ways in which they have influenced human affairs throughout history. You may be surprised to learn that, in many cases, empires have fallen and societies have been altered, all because of microorganisms. We will also review the colorful history of microbiology itself, looking at some of the important individuals and discoveries that have advanced the science of microbiology to where it is today.

Key Terms

acute infection	fusion	obligate intracellular parasite
adsorption	genus	oocyst
Aquifex/Hydrogenobacter	glycoprotein	order
archaeon	Gram-positive bacterium	penetration
assembly	green nonsulfur bacterium	phylogenetics
autotroph	green sulfur bacterium	phylum
bacterium	helix	plus (+) strand RNA virus
bacteriophage (phage)	heterotroph	prion
basidiocarp	host range	prophage
budding	hypha	proteobacterium
capsid	icosahedron	radioresistant coccus
capsomere	intermediate host	reactivation
Chlamydiae	kingdom	release
chronic infection	latent infection	retrovirus
class	lysis	reverse transcriptase
complex viral structure	lysogeny	reverse transcription
cyanobacterium	lytic cycle	RNA-dependent RNA polymerase
cyst	methanogen	saprophyte
cytopathic effect	minus (–) strand RNA	species
decomposer	minus strand RNA virus	specificity
definitive host	mold	spirochete
domain	mutation	strict anaerobe
envelope	mutualism	taxonomy
eukarya	mycelium	thermophile
family	mycorrhizal association	trophozoite
filamentous fungus	mycosis	uncoating
flavobacterium	nonenveloped virus	vector
fleshy fungus	nucleocapsid	virion

Concept Questions

1. If your dog has fleas, a likely culprit is the dog flea, otherwise known as ctenocephalides canis. Rewrite the scientific name of the dog flea in proper form.

2. Lobsters and crabs are in the class Crustacea. Grasshoppers are in the class Insecta, as are fruit flies and mosquitoes. Both fruit flies and mosquitoes are in the order Diptera. All of these animals are in the phylum Arthropoda. Construct a phylogenetic tree that shows the evolutionary relatedness of these five types of arthropods.

3. As we saw in this chapter, there is more than one lineage of bacteria that is photosynthetic. But these groups are not necessarily closely related. On the basis of this information, and by looking at the position of the photosynthetic bacteria relative to other bacteria as illustrated in Figure 4.6, does it appear that the process of photosynthesis evolved once or more than once?

4. Suppose that bacteria of some sort were contaminating a spacecraft on the launch pad and that the spacecraft was sent into orbit. When the spacecraft returns to earth, amazingly, some bacteria are still alive. Considering the various lineages of bacteria discussed in this chapter, which would you suspect might survive their journey into space?

5. Why is penicillin ineffective against any eukaryotic microorganism that might be infecting a human?

6. *Pneumocystis carinii* is a fungal pathogen that can cause serious respiratory disease in AIDS patients. It was at one time classified as a protozoan. Speculate about the sort of evidence that may have resulted in the reclassification of this organism.

7. Suppose that, for some reason, all fungi on the planet instantly went extinct. What would be the environmental consequences?

8. What is fundamentally different about the manner in which viruses reproduce compared with other forms of life?

9. Referring back to Chapter 1, in which the properties of living things were discussed, make an argument that (a) viruses are not living things or (b) viruses are living things.

10. A new anti-HIV drug called Enfuvirtide binds to and interferes with the glycoproteins found in the viral envelope. Which stage of the viral replicative cycle is Enfuvirtide interrupting?

11. Would RNA-dependent RNA polymerase make a reasonable target for drugs active against RNA plus- and minus-strand viruses? Why or why not?

12. If you know that a particular virus is enveloped, what does this information alone suggest about its replicative cycle?

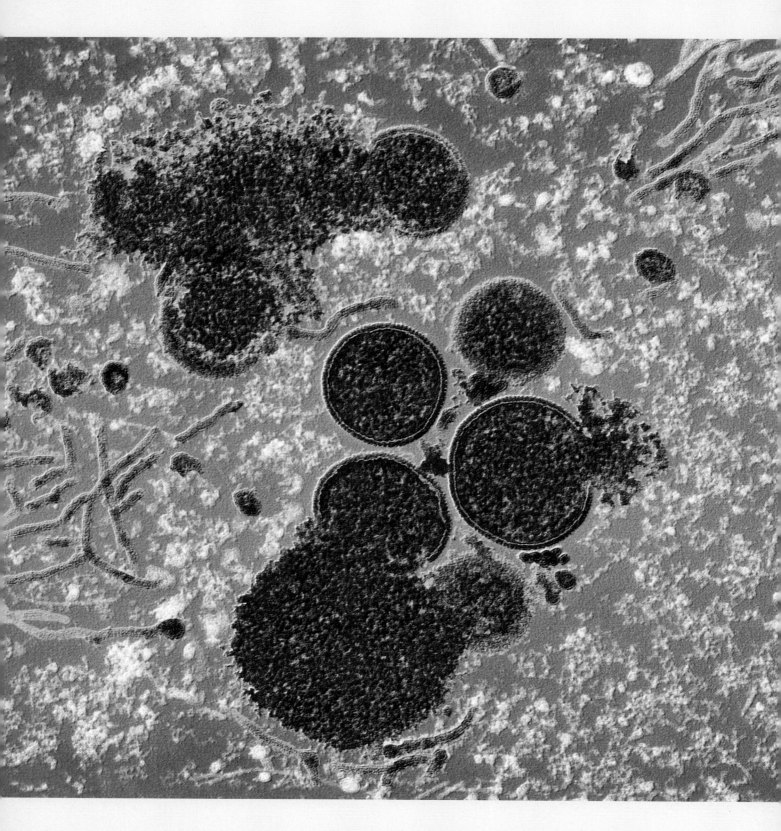

Vaccinia virus causes cowpox and is used to vaccinate against smallpox, a related but more deadly human disease.

Chapter 5

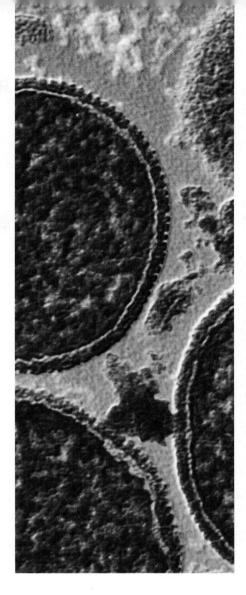

The Microbiology of History and the History of Microbiology

Most history examines the political, sociological, military, or economic events that have defined our past. But throughout our existence, unseen and often underappreciated microscopic participants have helped to shape the human experience. In many of history's important wars, for example, far more combatants and civilians have been killed by microorganisms than by bullets and bombs. In the American Civil War, just to name one, 140,414 Union soldiers died on the battlefield or of battle-inflicted wounds. Disease, however, was responsible for 224,097 Union army deaths, many due to typhoid. Similar statistics are not available for the Confederate army, but the death toll due to typhoid was believed to be even higher for the South. It was not until the 20th century that mankind's ability to kill in battle exceeded that of microorganisms. Infectious disease may have hastened the fall of Rome and the abandonment of feudalism in medieval England and contributed to Napoleon's decision to sell French territories in North America to the United States.

In this chapter we will first explore these and other examples of how microorganisms have altered the course of history. This is not to suggest that microbes have played the only or even the principal role in determining past events. To do so would clearly be absurd. However, in many cases, our understanding of why certain important historical events occurred as they did can be improved by also considering them from the perspective of the microbiologist (**Figure 5.1**). Furthermore, by thinking about the manner in which history has been altered by microorganisms, we gain added insight into the biology of the organisms themselves.

We will then turn our attention to the history of microbiology itself. The study of microorganisms has a rich and vital history, and we will focus on some of the important people and key discoveries that have led to the development of modern microbiology.

If Microbiologists Wrote History

CASE: THE SPANISH CONQUEST OF MEXICO

In the early 1500s, the Native American population in what is now Mexico is estimated to have been between 15 and 30 million people. Much of the land was under the control of the Aztec empire. The Aztecs developed a sophisticated society, comparable to the great civilizations of ancient Europe and Asia. In 1519, the Spanish leader Hernán Cortés landed on the Yucatan Peninsula with about 500 soldiers. Captivated by stories of the Aztec's wealth and the splendor of their capital of Tenochtitlan (now Mexico City), he began to march inland, intent on conquest (**Figure 5.2**). Cortés exploited several advantages as he moved toward the capital. He obtained the support of many other native tribes by promising to relieve

Figure 5.1 *The Triumph of Death.* This painting by the Flemish artist Pieter Brueghel (1525–1569) depicts the impact of plague on society. During Brueghel's life, Europe was still recovering from the devastation of the "black death," caused by bubonic plague, almost 200 years earlier.

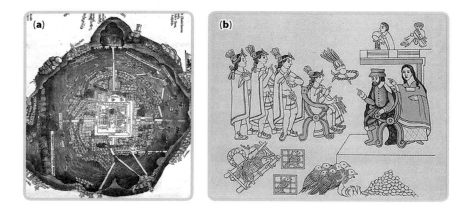

Figure 5.2 The Spanish arrive in Mexico.
(a) A map of Tenochtitlan, based on Cortés' memory of the city, and published in 1524. (b) Initial meeting between Cortés and the Aztec leader, Montezuma, as depicted in *Lienzo de Tlaxcala*.

them of Aztec domination. According to Aztec legend, the god Quetzalcoatl was prophesized to return from across the eastern ocean, and the unusual appearance of the white, bearded soldiers in their metal armor convinced many that the Spanish must be the fulfillment of that prophecy. If the newcomers truly were Quetzalcoatl and his minions, as divine figures they would be invincible. The Spanish also had horses and weapons such as muskets and cannons that were unknown in the Americas. Yet after initially appeasing the Spanish, the Aztecs fought fiercely once the Spanish had occupied Tenochtitlan and their intent became obvious. The Aztecs expelled the invaders from the capital but failed to destroy them. Eventually the Spanish returned, and began an extended siege of Tenochtitlan. During this time, a smallpox epidemic raged among the Aztecs, severely crippling their ability to continue the fight. The Spanish eventually conquered the Aztecs, forcing them into servitude. The death toll among the native peoples was staggering. Within 50 years, the native population in Mexico had shrunk to 3 million, and by 1600 it was under 2 million, a mere 5–10% of the pre-conquest population. Many of these deaths were due to smallpox.

1. Why did smallpox exact such a heavy toll on Native Americans but not the Spanish?
2. Why were the Spanish aided by smallpox but not decimated by "American diseases" to which the native peoples were less vulnerable?
3. How was the conquest of the New World different from the European experience in Africa?
4. Are there other examples of diseases influencing the outcome of armed conflict? What characteristics do the involved microorganisms have in common?

As you have no doubt guessed, the answer to the first part of question 4 above is emphatically "yes." Unseen microbial allies and enemies have turned the tide in many of mankind's battles. Before we explore further the role of smallpox in the conquest of the Americas, let us go back to even earlier times and attempt to understand the part played by infectious organisms in some of history's seminal events.

Disease influenced important events in ancient Greece and Rome

The Peloponnesian War. Following the defeat of the invading Persians in the epic naval battle of Salamis in 480 BC, the power of Athens was at its height (**Figure 5.3**). Thus began the "Golden Age," an unparalleled period of creativity and advances in philosophy, architecture, and the arts.

This period of extraordinary enlightenment, however, was all too brief. In 431 BC, war broke out between Athens and the rival Greek city state of

Figure 5.3 Ancient Greece at the start of the Peloponnesian War. Sparta's ultimate victory is thought to be in part attributable to an unknown disease that decimated Athens, starting in 430 BC.

Sparta. The conflict, known as the Peloponnesian War, initially was a stand-off. Sparta had a strong army, but no navy. Athens, with a powerful navy but a weak army, fought a defensive war on land and an offensive war at sea. Although the Athenians were crowded behind their city fortifications, they might have lasted indefinitely, since the city was connected to a port and they could bring in supplies by ship. Crowded conditions, however, increase the likelihood of epidemic disease. Dense aggregations of people often coincide with poor sanitation, which makes the transmission of many pathogens, especially those transmitted in contaminated food or water, more likely. Other forms of transmission, including airborne transmission, are enhanced when people are crammed together in close quarters. Rodents, which carry many diseases transmissible to people, also thrive in congested urban environments. In some respects, unless proper preventative measures are taken, a crowded city with poor sanitation is an epidemic waiting to happen, and in Athens in 430 BC that is exactly what happened. A plague, perhaps arriving in Athens via a trading vessel from Egypt, broke out and killed between one-third and two-thirds of the city's population.

The nature of this disease remains unknown. The historian Thucydides describes the illness as striking rich and poor alike. The disease was characterized by rapid onset, high fever, and extreme thirst. Eventually, pustules formed on the skin of victims. Thucydides noted that those who recovered from the illness were those best suited to care for the sick, because they somehow became resistant to a second infection following their recovery. This is one of the first descriptions of acquired immunity, an important aspect of our immune defense against invading microbes and a topic we will explore more fully in Chapter 11.

Many experts feel that the disease described by Thucydides was an unusually virulent form of scarlet fever, caused by the Gram-positive coccus *Streptococcus pyogenes*. Other authorities have implicated smallpox, measles, or bubonic plague. It remains possible that the plague of Athens was caused by a disease that no longer exists or by one that has changed so much that it is now unrecognizable. Regardless of the cause, the plague undoubtedly contributed to the eventual defeat of Athens and the end of

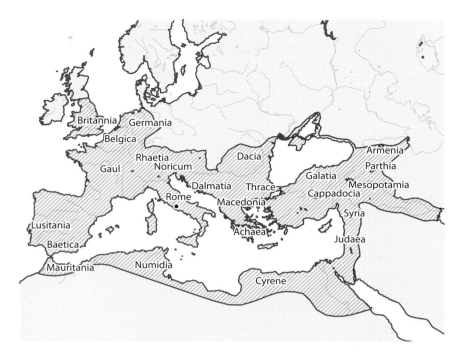

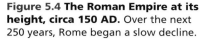

Figure 5.4 The Roman Empire at its height, circa 150 AD. Over the next 250 years, Rome began a slow decline.

the Golden Age. With the loss of so much of its population, Athens was never able to strike a decisive blow. The war dragged on for 27 years before Sparta ultimately prevailed in 404 BC.

The fall of the Roman Empire. In the late second century, the Roman Empire stretched from the British Isles to North Africa and from Portugal to Arabia (**Figure 5.4**). But over the next few centuries Rome and its empire began a long and slow decline, greatly hastened by the sack of Rome by the Visigoths in 410 AD. The fall of Rome is certainly one of the most important events in the history of Western civilization. Historians have debated for years why Rome collapsed, and the reasons are undoubtedly complex. Certainly, centuries of pressure from tribes such as the Goths, Vandals, and Franks along the northern frontier were important. The Roman economic system and its inability to sustain itself over the vast distances in the Roman Empire are often cited as an important reason for Rome's decline. The general decadence of the wealthy and luxury-seeking Romans is also considered part of the story behind the breakdown of the Roman army and its fighting spirit. Indeed, by the late fourth century AD, Germanic tribesmen rather than Roman citizens formed the bulk of the Roman army in the West.

Why did the Romans allow the security of their empire to fall increasingly into the hands of non-Romans? While "moral decay" may have been a factor, it is also probable that a decline in Rome's population played at least a part in the changing makeup of the Roman legions. Certainly some of the reasons for the population decline were disease-related. Starting in the first century BC, epidemic malaria became a regular part of Roman life. The problem was especially bad in the agricultural districts surrounding the capital where mosquitoes thrived. Mosquitoes and other arthropods that transmit pathogens between humans are known as **vectors**, and the diseases they transmit are called **vector-borne diseases**. Malaria is an important example. The problem became so bad that many cultivated areas around Rome were actually abandoned, resulting in a reduction in food supply.

Other epidemics may have contributed to Rome's slow decline. These included the "Great Plague of Cyprian," which began in 251 AD. The cause of this plague is unknown. Cyprian, the Christian bishop of Carthage, described symptoms such as explosive diarrhea, high fever, and gangrene of the hands and feet. The plague lasted at least 19 years, with higher mortality than any previous epidemic known. The disease moved quickly across the empire and was said to be spread not only by contact with infected individuals but also via any articles used by the sick, including clothing. The plague ebbed each summer, with fresh outbreaks each fall. Over half of those who contracted the mystery illness died.

In the late fourth century AD, the Roman Empire was already divided into Western and Eastern Empires. After the Visigoths sacked Rome in 410, the Western Empire was weakened and disorganized. Those in the Eastern Empire dreamed of recapturing the West and this dream almost became reality under the leadership of the Emperor Justinian. With the goal of resurrecting the grandeur of a united empire, Justinian attacked the former Western Empire in 533 AD. He recaptured Sicily as well as portions of the Italian peninsula, including the capital at Rome. After retaking parts of Spain, he made bold plans to lead his forces back into Gaul and even Britain.

In 540 AD, disaster struck. Bubonic plague broke out in North Africa and quickly spread to the rest of the known world, reaching the Eastern capital of Byzantium between 541 and 542 AD. The outbreak, now known as the "Justinian Plague," may have been the worst epidemic that has ever assaulted mankind. Up to 5000 people a day died in the Eastern Empire. So high was the death rate that graves could not be dug in time. Whole buildings were filled with corpses and set ablaze. Ships were loaded with the dead, rowed out to sea, and cast adrift.

Descriptions of the disease allow its clear diagnosis. Victims experienced high fever, combined with swollen glands or "buboes" in the armpits and groin. Often the buboes ruptured, resulting in gangrenous lesions and causing excruciating pain. Death typically followed within a week. All ages were vulnerable, but men were seemingly at higher risk than women. The plague would strike an area and then recede, only to return again. No town or village was spared. Many cities were wiped out or abandoned, as were agricultural areas. The entire empire descended into panic and confusion.

We can never know to what extent the plague quashed Justinian's ambitions and led to the long, gradual decline of the Eastern Roman Empire. But it seems clear that, at the very least, these events were hastened by this disease and that bubonic plague assisted in irrevocably changing the history of the Western world.

The Black Death altered European society forever

In 1346 the Tatars from Central Asia attacked a Genoese trading colony on the shores of the Black Sea. During the siege, bubonic plague broke out among the attackers. In an early example of biological warfare, the Tatars catapulted bodies of dead plague victims over the walls into the city. Although the Tatars did not understand it, plague is another example of a vector-borne disease. The vector in this case is not mosquitoes but fleas. Fleas abandon a dead host and search for new, living hosts to colonize. This is apparently what happened in the trading colony. Plague soon appeared among the Genoese, some of whom managed to escape the beleaguered colony via ship. A few days after their ship arrived in the home port of Genoa, the first cases of plague in the city were reported.

Thus began the notorious bubonic plague epidemic known as the "Black Death" (**Figure 5.5**). While the Italian merchants had brought the disease to Europe, the Tatars and their Mongol allies, in whom the epidemic began, spread plague north into Russia, and east into India and China. The

Figure 5.5 Doctor in protective clothing during the "Black Death." The beaklike structure contained herbs, which were somehow thought to decontaminate the "bad air" that was believed to cause plague.

devastation caused by this worldwide epidemic, also known as a **pandemic**, is hard to comprehend. Although the exact numbers are not known, in Europe alone, it is estimated that 24 million people may have died from the plague, representing over 25% of the total population at the time.

A mortality rate of such stupendous proportions had major social and political ramifications. Deaths from plague destabilized the economies of major European powers, and even temporarily halted the Hundred Years War between France and England. In Siena, local Catholic authorities were hoping to attract the papacy to the city by tripling the size of Siena's cathedral. With the outbreak of bubonic plague, these plans were permanently shelved. The fate of the Greenland Vikings is also intriguing. Ships from Norway probably carried the infection to the Greenland colonies, which had been founded by Erik the Red. Plague may have weakened the colonists to such an extent that they were either conquered or absorbed by the indigenous Inuit people. Greenland was then forgotten for over 200 years until it was "rediscovered" in 1585. Because the Viking colonists in Greenland were known to have visited the Canadian coast, the extinction of the Greenland settlements perhaps permanently altered the history of North America.

The impact on the social fabric of society was particularly acute in England. The epidemic first spread to the British Isles in the summer of 1348. It reached London in November of that year and spread to Ireland in 1349. By 1350 it had arrived in Scotland. At the time, England was still largely an agricultural society, governed by a feudal system of indentured servitude. Prior to the outbreak of plague, England's growing population resulted in a regular supply of cheap labor. The feudal lords were able to keep wages extremely low and restrict the movement of peasants to the cities or other places they might seek better financial opportunity. But with bubonic plague came a severe manpower shortage. Peasants could now extract higher wages, and were allowed greater freedom of movement to meet labor needs. As they gained unprecedented economic independence, the feudal system began to weaken. Within 150 years it had entirely disappeared.

European explorers imported many diseases into the new world

Rarely in the history of humanity is the role of infectious microorganisms more clear than in the conquest of the Americas (**Figure 5.6**). Cortés, introduced earlier in our case study, was not the only conquistador who was assisted by lethal pathogens. Once established in Mexico, smallpox spread rapidly to South America. Millions of Incas in Peru, including the reigning emperor, died. When Francisco Pizarro arrived with a small band of Spanish soldiers, the Inca empire was in political and social upheaval. Pizarro was thus able to usurp control without significant military opposition.

Smallpox was not the only disease that Europeans unwittingly introduced into the New World. Measles, influenza, and malaria were also absent in the Americas before Columbus. Native Americans, unlike the Europeans, were highly susceptible to these diseases. There are several interrelated reasons for this difference in susceptibility. These imported diseases originated in Europe and Asia. Europeans already had a long history, stretching back perhaps thousands of years, interacting with the microorganisms causing these diseases. As we will explore more fully in later chapters, all mammals, including humans, eventually evolve some measure of resistance to regularly encountered diseases. Because viruses such as smallpox were absent in the Americas before the arrival of the Spanish, no such opportunity was afforded to the Native Americans. Therefore, although the Spanish could still contract and transmit diseases such as smallpox, they were far less likely than the native populace to develop serious illness. The Aztecs watched as their own people died in large numbers while the Spanish remained relatively disease-free.

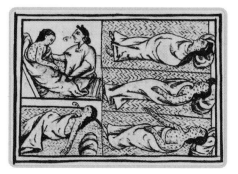

Figure 5.6 Smallpox in the Americas.
A drawing of Aztec smallpox victims, from the *Florentine Codex*.

But this raises an interesting question: Why, as we asked in our case study, were the Spanish not decimated by "American diseases" to which the native population was resistant? The conquest of the Americas occurred in a breathtakingly short period of time, in large part because there were few such diseases. This was certainly not the case when Europeans attempted to colonize Africa. A plethora of indigenous African maladies lay in ambush to waylay potential invaders. Consequently, while the Spanish conquest of Mexico and Peru was completed in a few short years, it was not until the 20th century that the European colonization of Africa was complete.

To answer this question, it is necessary to consider some of the qualities shared by diseases such as smallpox, measles, and influenza. All of these diseases are **acute** in nature. Once a host is infected with an acute disease, the pathogen replicates rapidly, and symptoms of disease appear quickly. Furthermore, there are only two possible outcomes of this host–pathogen interaction: within a short period of time the host either recovers or dies. If the host recovers, he or she will typically be immune to the same pathogen, at least for a period of time. We will explore how such "immunological memory" is developed in Chapter 11.

Because agents of acute disease either rapidly kill their host or are themselves destroyed, they must be transmitted to new hosts quickly and efficiently if they are to survive. Smallpox virus, for instance, is transmitted through the air. A single cough or sneeze by an infected individual can infect many other potentially susceptible hosts. The situation is different for **chronic diseases** (**Table 5.1**). These are conditions that last indefinitely, sometimes for years or decades. The host immune system is able to control but not eliminate the disease-causing microorganism. For a pathogen such as this, there is far less need for rapid transmission. It can survive in its current host indefinitely and has a relatively long period of time in which to reach new hosts.

Acute diseases need large populations of potential hosts. Consequently, they are often called "crowd diseases." Remember that each new host either recovers quickly, and is subsequently immune, or dies. The pathogen, therefore, needs a large supply of susceptible hosts to persist. When an acute disease is introduced into a small isolated population, it quickly "uses up" all potential hosts and then dies out itself. Studies have shown, for instance,

CHARACTERISTIC	ACUTE DISEASE	CHRONIC DISEASE
Replication of infectious agent in the host	Very rapid	Generally less rapid
Disease course	Short incubation; rapid onset of symptoms; quick resolution (immunity or death of host)	Longer incubation; slow onset of symptoms; no quick resolution (may persist months, years, or decades)
Immune response	Generally rapid; result is complete elimination of the pathogen, with resistance to subsequent infection by the same pathogen	Incomplete; although the pathogen may be contained, it is not eliminated
Transmission	Rapid and efficient transmission to new susceptible hosts is required for pathogen survival	Rapid and efficient transmission is not required for pathogen survival
Host population size	Large population of susceptible hosts is required for pathogen survival	Large, susceptible host population is not required for pathogen survival
Example	Smallpox, measles	Tuberculosis

Table 5.1 Characteristics of acute and chronic diseases. Acute diseases generally begin with a rapid onset of symptoms, and the disease resolves rapidly, with either elimination of the pathogen or death of the host. Chronic diseases last indefinitely and may cause symptoms for long periods of time.

Figure 5.7 Animal husbandry has a long tradition in the Middle East. Cattle are thought to have been domesticated in the valley of the Tigris and Euphrates Rivers, in what is present-day Iraq, approximately 8000 years ago. This sculpture of a Mesopotamian king with the body of a cow illustrates the importance that cattle had in early Mesopotamian culture.

that measles virus will die out in populations under 1 million. When measles was accidentally introduced into the Faroe Islands in 1781, a severe epidemic broke out. The Faroes are an isolated group of islands in the North Atlantic, and after the virus spread through the entire population it went extinct. The islands then remained measles-free until an infected individual from Denmark brought the virus back to the islands by ship in 1846. In large populations, new hosts in the form of newborns, immigrants, or people in whom immunity has worn off ensure a constant supply of susceptible individuals, allowing the virus to persist indefinitely.

Molecular analysis of pathogens causing acute disease has confirmed that many of them are most closely related to pathogens causing similar "crowd disease" in animals. Smallpox virus and measles virus, for instance, are most closely related to viruses causing acute diseases in cattle. Most authorities believe that in the 8000 years since cattle were domesticated in western Asia, these animal viruses made the jump from cattle to humans (**Figure 5.7**). Other acute disease agents are similarly related to viruses found in domestic animals such as pigs and chickens that had been raised in large numbers for thousands of years in Europe and Asia.

But in the Americas there was no similar long-standing tradition of animal husbandry. The only animals that had been domesticated prior to European contact were dogs, turkeys, guinea pigs, llamas, and ducks. Thus, the opportunities for animal-to-human transmission were greatly reduced. This is not to say that the Americas were disease-free. But the relatively infrequent or even nonexistent contact between Native Americans and large herds of animals resulted in far fewer human acute illnesses.

In this manner, Cortés and other conquerors were aided by a one-sided exchange of deadly microbes. Perhaps a dozen or more major infectious diseases were introduced into the New World by Europeans. With the possible exception of syphilis, the origins of which are still debated, no similar lethal disease was transferred from the Americas back to Europe.

The rapidity of America's conquest stands in stark contrast to the European experience in Africa. In spite of numerous efforts, attempts to colonize or even to explore Africa through the 18th and 19th centuries often met with dismal failure. In 1816, for instance, the exploration of the Congo River by Captain James Tuckey was cut short when almost all members of his party were struck by "an intense, remittent fever and black vomit in some cases." In 1841, Captain H. D. Trotter set out to explore the Niger River with 145 Europeans and 133 Africans. After they had traveled 100 miles upstream, severe fever broke out. Among the Europeans, 130 became seriously ill, and 50 died. All of the Africans remained healthy.

Native Africans did not have large herds of domestic animals, but unlike the situation in Mexico and South America, the Africans lived in close proximity to enormous herds of wild grazing animals. These animals served as reservoirs for many agents of disease to which the animals themselves, as well as native Africans, were relatively insusceptible. Sleeping sickness, yellow fever, and other diseases took a heavy toll on both Europeans and their animals. Until the cause and prevention of such ailments was understood, little progress in the colonization of Africa was possible. Indeed, the tsetse fly, which carries the protozoans that cause sleeping sickness, has been called "Africa's greatest game warden." Until it became possible to control these flies, large segments of Africa's most fertile lands were "off limits" to the colonists for raising domestic livestock because these non-African animals were so susceptible to sleeping sickness.

Several important events of the 19th century owe their outcomes to microorganisms

The 19th century is especially illustrative of the relationship between infectious disease and human history. During this time period, national armies of unprecedented size swept back and forth across continents, providing the large crowds necessary for acute disease as well as a means for introducing them into new populations. Cities, too, were expanding rapidly and the crowded, unsanitary conditions in these urban centers permitted many diseases to flourish. Furthermore, the 19th century saw a dramatic improvement in communication and record-keeping, allowing us to better chronicle events of the day. It is probably no accident that, as we will see later in this chapter, the late 19th century gave birth to the modern science of microbiology, as scientists struggled to make sense of disease. Our next few examples are all well-documented instances of the impact infectious disease has had on human affairs during this time period, setting the stage for the "Golden Age" of microbiology.

Napoleon's microbial Waterloo. Napoleon is one of the major historical figures of the 19th century. Decisions made by the French emperor doubled the size of the United States and changed the political landscape of Europe. In at least two of his most monumental endeavors, disease loomed large.

At the end of the 18th century, France controlled an enormous tract of land in North America, stretching from the Mississippi River to the Rocky Mountains and from the Gulf of Mexico to Canada. The French also possessed several Caribbean islands, including the colony of Saint-Domingue (now Haiti), where large sugar plantations were established.

In 1801, after more than a decade of revolt and turmoil in Haiti, Napoleon sent over 20,000 troops to the colony. Yet it was not long before most of these soldiers were laid low by yellow fever. Well over half eventually died. These heavy losses influenced Napoleon's decision not to risk even more French soldiers in defense of French territory in North America. Napoleon reasoned that these troops could be better used extending his empire in Europe. In 1803, he sold the vast Louisiana Territory to the United States (**Figure 5.8**).

In the early summer of 1812, Napoleon turned his attention east and invaded Russia. Initially his plans went well, but as his army of over 500,000 men crossed through Poland, typhus broke out among his troops. Typhus is caused by the bacterium *Rickettsia prowazekii* and is transmitted by body lice (**Figure 5.9**). Typhus most often strikes when people are crowed together under unsanitary conditions, as failure to bathe or to wash clothing allows the lice to flourish. History is filled with examples of armies being plagued with typhus because soldiers often find themselves in situations that are ideal for typhus transmission.

By the time Napoleon's army reached Moscow on September 14, only 100,000 of his troops remained. Typhus, combined with starvation of the ill-equipped soldiers, accounted for most of the deaths. Ultimately, Napoleon was forced to withdraw from Russia, defeated by typhus, hunger, and the Russian winter. By the end of the year approximately 20,000 French troops staggered into Vilnius on the Baltic Sea. Although reinforcements arrived, the spirit of the French army had been destroyed. Fewer than 1000 of the once mighty army were ever again fit for duty. Napoleon's dream of conquering Russia had ended for good.

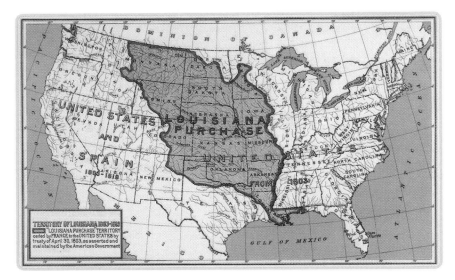

Figure 5.8 The United States, 1803. Territory purchased by the United States from France as the Louisiana Purchase is highlighted in red.

The Irish potato famine. The development of agriculture, starting in about 7000 BC, constituted one of the greatest revolutions in the history of mankind. Farming allowed people to live in large, settled communities for the first time and gave them some measure of control over their food supply. Such progress, however, comes at a cost. Agriculture carries with it certain inherent risks. Oftentimes, when one crop is especially well suited to a particular region or when there is an overwhelming preference for a specific crop, there is a tendency to cultivate only this one plant species. The result is a **monoculture**—an agricultural area dominated by a single crop. Depending on monocultures is a dangerous gamble. During favorable conditions, a farmer may count on bumper harvests. If crop pests such as insects, plant viruses, or fungi invade, however, the results can be disastrous. The pest can increase in numbers rapidly as it feeds on its favored food source and quickly spreads through and destroys the entire crop. If the population depends too heavily on this monoculture for sustenance, the result is famine. Such devastation is rare in natural ecosystems consisting of many plant species. It is unlikely that any plant pest can attack more than a few plant types; a mix of crops can severely curtail the damage to the overall environment caused by any one pest.

The Irish potato famine provides a startling example of how a crop disease can affect a society. It also warns against relying too heavily on a single food source. It is rumored that the potato was first introduced into Ireland by Sir Walter Raleigh in about 1588 when he planted potatoes in his garden. The crop thrived in Ireland, and by the late 1600s potatoes had become a staple in the Irish diet. In 1846 disaster struck in the form of a plant disease called potato blight (**Figure 5.10**). The disease is caused by *Phytophthora infestans*. These organisms appear funguslike but may actually be more closely related to algae. The disease spread rapidly, resulting in almost total crop failure. Because the Irish were so heavily dependent on this single crop, the consequences were severe. Within only a few months the entire economy collapsed, and eventually, 750,000 to a million of Ireland's approximately 8 million inhabitants starved to death or died of diseases related to malnutrition. Many more emigrated from Ireland to other countries.

Today the population of Ireland is over 4 million, still well below the prefamine total. The large Irish communities that are currently found in cities such as Chicago and Boston owe their existence in some measure to the funguslike organism that precipitated the disaster.

The Panama Canal. Before 1914, the only way to travel between the Atlantic and Pacific Oceans was via a long and hazardous trip around Cape Horn or through the Straits of Magellan at the tip of South America. As early as the

Figure 5.9 The human body louse, *Pediculus humanus humanus*. Typhus is a bacterial, vector-borne disease transmitted by body lice. These lice thrive in crowded conditions where people are unable to bathe or wash clothes regularly. Consequently, soldiers through the mid-20th century were especially plagued by typhus. Note the prominent front legs, with which the louse holds tightly onto clothing.

Figure 5.10 Potato infected with *Phytophthora infestans*. The brown streaks growing through the potato are caused by the infection. Though often described as a fungus, this plant pathogen, the cause of the Irish potato famine, is probably more closely related to algae.

16th century, the Spanish proposed building a canal across the Isthmus of Panama to solve this problem, but no major attempt to complete the work was made until the late 19th century. In 1879 the Frenchman Ferdinand de Lesseps secured government backing for what he called "la grande entreprise." The project would be the biggest financial undertaking of all time and would enhance French prestige and influence tremendously.

Yet it was not to be. Mosquitoes thrived in the lakes and swamps through which the canal was to pass, and the French suffered enormously from two mosquito-borne diseases—malaria and yellow fever. In 1882, a year after digging began, approximately 20% of the French workforce had died. By 1883, the death toll averaged 200 each month. When 17 newly graduated French engineers arrived in 1885, only one remained alive 1 month later. As word of the horrendous death rate filtered back to France, it became increasingly difficult to find employees willing to risk death by disease in the Panamanian jungle. By 1889 the project was abandoned.

It was not until 1897 that the English scientist Ronald Ross obtained evidence that mosquitoes transmitted the malaria protozoan *Plasmodium*. Although the Cuban Carlos Finlay had suggested that mosquitoes transmitted yellow fever in 1881, this idea was not confirmed until the 1890s.

This newly discovered information about the transmission of malaria and yellow fever proved invaluable when, in 1904, the American president Theodore Roosevelt decided to tackle the challenge of linking the Pacific and Atlantic (**Figure 5.11**). William Gorgas, who had helped to control mosquito-transmitted diseases in Havana, was put in charge of health and sanitation in Panama. Gorgas employed the same techniques that he had used successfully in Cuba. Buildings for workers were made mosquito-proof with fine-gauge copper mesh. Standing water was drained wherever possible. Water bodies that could not be drained were sprayed regularly with kerosene to kill mosquito larvae. When digging began in 1907, yellow fever was already under control, and deaths from malaria were significantly lower. When the canal was completed in 1914, the annual death rate among American workers in Panama from all causes was 6 in 1000. In comparison, the death rate in New York and London was about 14 in 1000.

Gorgas became a hero. He was appointed Surgeon General in the United States and was knighted by the King of England. Meanwhile, the United States gained unprecedented influence in Central America that continues until the present day.

Certain episodes in both World Wars I and II were influenced by microorganisms

A solution to a gunpowder shortage. Microbes are not only important because of the diseases that they cause. Many common food items are made with the help of microorganisms. Industry relies on microbial assistance for many common industrial processes. Here we describe one such industrial use of microorganisms, which in a small way influenced the outcome of the First World War.

As World War I wore on, the British Navy began to suffer from a shortage of cordite. This compound was used on naval vessels as a smokeless gunpowder. With cordite, a ship could fire its weaponry without billows of smoke giving away its position. Acetone is a carbon compound required for the

Figure 5.11 Theodore Roosevelt confronts "yellow jack." A cartoon from the early 1900s, first published in *Harper's Weekly*, depicting President Roosevelt and Uncle Sam squaring off with yellow jack, as yellow fever was often called. Because of recent discoveries about the transmission of yellow fever, Roosevelt knew that successful construction of the Panama Canal required that disease control measures first be implemented.

Figure 5.12 Chaim Weizmann and David Lloyd George. (a) Weizmann was born in what is now Belarus in 1884. After studies in Switzerland, he moved to Britain, where he developed a technique for the synthesis of acetone, using bacteria. (b) David Lloyd George, after a successful military career, went on to become Britain's first, and to date only, prime minister of Welsh ancestry.

production of cordite, and it can ordinarily be cheaply produced by heating wood. Acetone was obtained at that time as a by-product of charcoal production. So great was the demand for cordite, however, that supplies of acetone became severely limited. To meet the demand, Britain would have had to reduce much of its forests to charcoal.

In an effort to develop alternative means to produce acetone, David Lloyd George, Director of Research for Explosives, traveled to Manchester, where he met with Professor William Perkin, a leading organic chemist. While there, he also met Perkin's student, Chaim Weizmann, who was interested in fermentation (**Figure 5.12**). Weizmann had already discovered that *Clostridium acetobutylicum*, a Gram-positive bacterium, produced a mixture of butyl alcohol and acetone from sugar. When Lloyd George heard of this, he immediately dispatched Weizmann to London to consult with Winston Churchill, who was then First Lord of the Admiralty.

Weizmann explained to Churchill that the bacterium could produce the required acetone in abundance, if a large enough fermentation and distillation plant could be found. Churchill used his authority to commandeer a gin factory and have it converted to acetone synthesis. Weizmann was appointed chief scientist of the Nicholson Gin Company.

The cordite shortage was abated. Acetone production was eventually shifted to Canada, where an abundance of corn ensured a constant supply of sugar for the fermentation process. Lloyd George went on to become Prime Minister of Great Britain, while Weizmann eventually became president of the World Zionist organisation, which was seeking the creation of a Jewish homeland in the Middle East. Weizmann sought the Prime Minister's support, and in 1917 Britain issued the Balfour Declaration, instrumental in the establishment of such a homeland and the eventual founding of the State of Israel. Weizmann was elected as Israel's first president in 1948.

The ruse that spared two Polish villages. A little-known event in World War II illustrates how ingenious people can be when it comes to survival under difficult circumstances. The incident also offers yet another example of how a microorganism can impact human affairs in unexpected ways.

Normally when the immune system is confronted with a particular pathogen, proteins called **antibodies** are produced that react with and neutralize the microorganism. Antibodies are usually very specific—each type of antibody binds to only one type of microbe. In some cases, however, antibodies against one microorganism will bind to a second organism. When an antibody against one microorganism is able to also bind to a second microbe, the antibody is said to be **cross-reactive**.

During the time of the First World War, Polish doctors discovered that antibodies against *Rickettsia prowazekii*, the causative agent of epidemic typhus, were cross-reactive with *Proteus vulgaris* OX19, a harmless strain of Gram-negative bacteria. They used this phenomenon to develop a reliable test for typhus (**Figure 5.13**). Blood was drawn from a suspected typhus patient and mixed on a glass slide with *Proteus* OX19 cells. If the patient was infected with typhus, anti-*Rickettsia* antibodies present in his or her blood would bind to the *Proteus* and cause easily observed clumping on the slide.

World War II began in 1939 when the Germans invaded Poland. In the Polish villages of Rozvadow and Zbydniowie, about 200 km southwest of Warsaw, two Polish physicians decided to try a clever deception to spare their villages from the Nazi onslaught. Drs. Eugeniusz Lazowski and Stanislav Matulewicz began injecting people in their villages with *Proteus* OX19. These villagers remained healthy but produced anti-*Proteus* antibodies. The Polish doctors took blood samples from these individuals and sent them to the German occupation authorities for testing. When the tests came back positive, the Germans became convinced that a typhus epidemic was raging in the region. The Germans were so afraid of typhus that they declared an "epidemic zone" in the area, and German oppression in at least one part of occupied Poland was dramatically decreased.

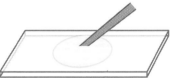

1. Take blood sample from suspected typhus patient.

2. Mix blood sample on glass slide with *Proteus* OX19.

3. Examine slide for evidence of cross-reactivity between anti-*Rickettsia* antibodies and *Proteus* OX19.

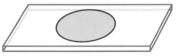

No anti-*Rickettsia* antibodies in patient's blood

Without infection no cross-reactivity takes place between anti-*Rickettsia* antibodies and *Proteus* OX19.

Patient does not have typhus.

Anti-*Rickettsia* antibodies in patient's blood

Antibody cross-reactivity causes clumping of *Proteus* OX19.

Patient has typhus.

Figure 5.13 The Weil–Felix test for typhus. If a patient is infected with *R. prowazekii*, he or she will have antibodies to this bacterial species circulating in the blood. These antibodies will also bind to *Proteus* OX19, a harmless soil bacterium. Because of this cross-reactivity, a diagnosis of typhus can be easily confirmed by mixing the patient's blood serum with *Proteus* OX19. Clumping of the *Proteus* bacteria indicates that the patient is positive for typhus. A lack of clumping rules out a typhus infection.

Some infectious disease is caused by nonmicroorganisms

Our emphasis in this text is, of course, microorganisms—living things ordinarily too small to see with the unaided eye. In addition to prokaryotes, viruses, fungi, and protozoa, humans are susceptible to infection with a variety of much larger *worms*, known as **helminths**. Helminths represent an important category of nonmicroscopic **parasites**. A parasite is any organism, microscopic or not, that lives in or on another organism, from which it obtains a place to live and reproduce and a source of nutrients. Our final example of how history can be shaped by disease involves such a helminth parasite. Specifically we will discuss how an outbreak of schistosomiasis may have influenced the geopolitical landscape in the early days of the Cold War.

Following the defeat of Japan in 1945, the island of Taiwan reverted to Chinese administration. Yet the late 1940s were a time of great unrest in China. The Chinese Communists, led by Mao Zedong, were at war with the Kuomintang led by General Chiang Kai-shek. As the civil war progressed, the Kuomintang nationalists were forced to retreat. Finally, in late 1949 Chiang and his government abandoned the Chinese mainland and retreated to the island of Taiwan, establishing a new capital in Taipei.

The Chinese Communists began immediate preparations to invade Taiwan. Thousands of troops moved into southern China to prepare for an attack across the Taiwan Strait. Because invasion of the island would require an amphibious assault, the Communist troops were taught to swim in the rivers and irrigation canals in the region. Many of these water bodies harbored schistosome-infected snails. As we learned in Chapter 4 (see p. 74), schistosome worms develop as larvae in specific freshwater snails. When larval development is complete, the worms leave the snails and penetrate directly through the skin of their human hosts, where they mature to adulthood. The Chinese soldiers were infected in large numbers, and an epidemic of acute schistosomiasis broke out. Between 30,000 and 50,000 soldiers are believed to have been affected. Although we may never know to what extent schistosomiasis influenced the thinking of the Communist authorities, the invasion was postponed. Meanwhile, a U.S. naval fleet arrived off the coast of Taiwan to protect the island against Communist invasion. With the threat of American involvement now looming large, plans to invade Taiwan were shelved indefinitely. In 1954, the United States and Taiwan signed a mutual defense treaty, assuring the survival of the Nationalist government.

The Science of Microbiology: A Brief History

CASE: THE ASSASSINATION OF PRESIDENT GARFIELD

On July 2, 1881, only 6 months into his term, President James A. Garfield was shot twice by an assailant in Washington, D.C. The first shot caused a superficial wound in the president's arm. The second entered Garfield's chest. The immediate concern of doctors was to find the second bullet, and even before Garfield was moved they began digging in the wound. In the late nineteenth century, any bullet wound was extremely serious, and if the bullet could not be extracted at once, the prognosis for recovery was poor. Garfield himself realized this, and when an attending physician tried to comfort him, the President replied, "I thank you doctor, but I am a dead man." Two unsuccessful operations to find the bullet were attempted. Alexander Graham Bell, the inventor of the telephone, was summoned to look for the bullet with a crude metal detection device, but he too was unsuccessful (Figure 5.14). Finally, 80 days after he was shot, Garfield died. Unfortunately, if the incident had occurred only a few years later, or if doctors attending the stricken president had utilized recently discovered principles of infection control, Garfield might have survived. There is a sad

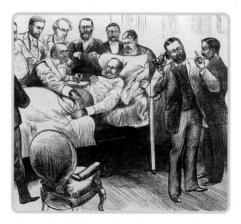

Figure 5.14 Alexander Graham Bell and President James Garfield. An illustration from *Harper's Magazine*, 1881, showing Alexander Graham Bell looking for a bullet in President Garfield with his "magnetic induction" device.

grain of truth to a statement made by the assassin, Charles Guiteau, before he was hung. Upon learning that Garfield had died, Guiteau replied, "The doctors killed him. I simply shot him."

1. What discoveries in the emerging field of microbiology might have saved the life of President Garfield?

Garfield almost certainly died of a bacterial infection. Antibiotics to treat such infections were still a half century in the future, and even simple infection control procedures such as hand washing were rarely if ever utilized. The doctors that first probed President Garfield's wounds did so with unsterilized instruments and with fingers still soiled from work on other patients earlier in the day. During the two surgeries, no attempt would have been made to prevent infection of surgical incisions.

The emerging science of microbiology led to vast improvements in the treatment of wounds and the prevention of infection during surgery. To understand how, we need to examine the origins of microbiology itself (**Table 5.2**).

Microorganisms were first discovered in the 17th century

In 1674 Anton van Leeuwenhoek first observed pond water with a simple microscope he had constructed (**Figure 5.15**). It had been known for some time that lenses could be used to magnify objects, but van Leeuwenhoek was the first to combine precision lens grinding with an intense curiosity about things nobody had thought to look at before. And what he saw astounded him: an entire universe of previously unknown living things, invisible to the naked eye. Van Leeuwenhoek not only saw but also carefully described numerous bacteria and protozoa. When he looked at other samples such as small particles of soil and even his own feces, he continued to find more of what he called "animalcules." While looking at organisms from teeth scrapings in his mouth, he noted that they were apparently killed if he drank hot tea before taking a sample. His findings were made public through a series of letters he sent to the Royal Society of London until his death in 1723.

Van Leeuwenhoek's discoveries started an intense scientific debate that lasted nearly 200 years. Where did these microscopic organisms come from? Many believed that they could somehow arise spontaneously from nonliving substances. Nine years before van Leeuwenhoek's first descriptions of microorganisms, the theory of spontaneous generation had been disproved for animals by the Italian Francesco Redi (**Figure 5.16**). Many people, however, refused to accept Redi's conclusions, and for the newly discovered microorganisms in particular, spontaneous generation still had many adherents.

The science of microbiology was born in the second half of the 19th century

Microbiology truly emerged as a distinct scientific discipline after 1850. One of the first important breakthroughs occurred in 1859 when Louis Pasteur conducted a simple yet elegant experiment, which finally settled the question of spontaneous generation in regard to microorganisms. In his experiment, Pasteur demonstrated that a liquid broth could not give rise to microorganisms spontaneously (**Figure 5.17**). Pasteur believed that the microorganisms apparently arising "spontaneously" in a liquid broth were merely carried into the broth through the air. To test this idea, Pasteur sterilized meat broths by boiling them in flasks. Half of his flasks had standard straight necks, while the other half had necks bent into an S-shape. Both types of flasks were open to the air. Microorganisms soon appeared in the straight-necked flasks, but the S-necked flasks remained sterile. Pasteur

1674	Van Leeuwenhoek views "animalcules" with his microscopes
1721	Lady Montague learns of smallpox variolation, introducing the technique into Europe
1796	Jenner develops an anti-smallpox vaccine
1847	Semmelweis introduces hand washing into Vienna hospital, reducing incidence of "childbed fever"
1854	Nightingale introduces modern nursing techniques, including routine sanitation
1857	Pasteur demonstrates fermentation in wine
1858	Pasteur disproves spontaneous generation
1864	Pasteur develops "pasteurization"
1865	Lister develops aseptic technique
1876	Koch develops "germ theory of disease" and identifies causative agent for anthrax
1881	Koch develops pure culture technique
1882	Hesse suggests agar as bacterial medium; Koch identifies causative agent for tuberculosis
1884	Koch's postulates formally established; Gram stain introduced
1885	Pasteur develops concept of attenuation and develops rabies vaccine
1892	Ivanowski describes "filterable agent" causing tobacco mosaic disease
	Smith and Kliborne describe the first vector-borne disease: Texas red water fever, transmitted by ticks
1894	Yersin identifies bacteria responsible for plague
1896	Ross demonstrates mosquito transmission for malaria
1898	Loeffler and Forch describe filterable agent for hoof-and-mouth disease
1900	Reed identifies role for mosquitoes in yellow fever transmission
1908	Ehrlich develops salversan for treatment of syphilis
1929	Fleming discovers penicillin
1933	Electron microscope developed
1935	First sulfa drugs; viruses first crystallized
1944	DNA identified as the genetic material
1946	Transfer of DNA between bacterial cells first described
1953	Watson and Crick publish their paper on the structure of DNA
1960	Gene regulation in bacteria described
1973	First recombinant DNA
1979	First pharmaceutical (insulin) produced in bacteria via recombinant DNA technology
1986	First approval of a vaccine (hepatitis B) created with recombinant DNA technology
1995	First entire genome, the bacterium *Haemophilus influenzae*, sequenced
	First "designer drug" against a microorganism, protease inhibitors used against HIV, introduced
2006	Introduction of the vaccine against cervical cancer, caused by the human papillomavirus

Table 5.2 Some key highlights in the history of microbiology. A timeline showing a few of the major milestones in microbiology over the last 350 years. The period between about 1850 and 1910 is often called the "golden age" of microbiology, because of the unprecedented number of breakthroughs that occurred during this time.

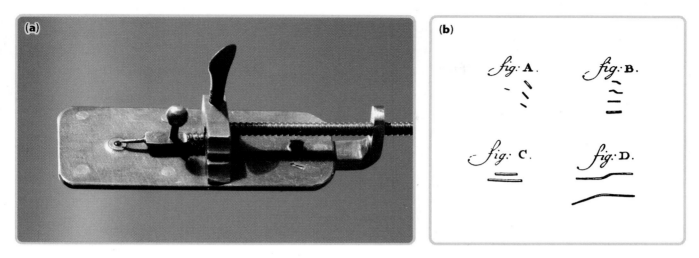

Figure 5.15 **Anton Van Leeuwenhoek was the first to describe microorganisms.** (a) One of van Leeuwenhoek's early microscopes. (b) Drawings of "animalcules" from one of van Leeuwenhoek's notebooks.

reasoned that although both types of flasks were open to the air, microorganisms were unable to enter the S-necked flasks because they were trapped in the bend of the neck. When Pasteur snapped off the necks of these flasks, microorganisms quickly appeared in these broths as well. These results made it clear that organisms could not be created spontaneously, even if air entered the flasks freely. Life could only arise from previous life. The notion of spontaneous generation was discredited for good.

Pasteur helped to advance the science of microbiology in numerous ways, making him, by any measure, one of the giants of 19th-century science (**Figure 5.18**). At about the same time that he was disproving the idea of spontaneous generation, Pasteur was making important contributions to the science of winemaking as well. The wine industry has always been an important part of the French economy. Prior to the mid-1800s, however, wine was too often spoiled when it mysteriously turned sour.

Meat is exposed in open container. Flies enter and lay eggs on meat, which hatch into maggots.

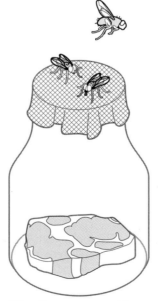

When flies are prevented from reaching meat by fine gauze, no maggots appear in meat.

Figure 5.16 **Redi's experiment refuting spontaneous generation.** The experiment, conducted in 1665, showed that maggots developed on rotting meat only if flies had access to the meat first. When flies were unable to reach the meat, no maggots developed.

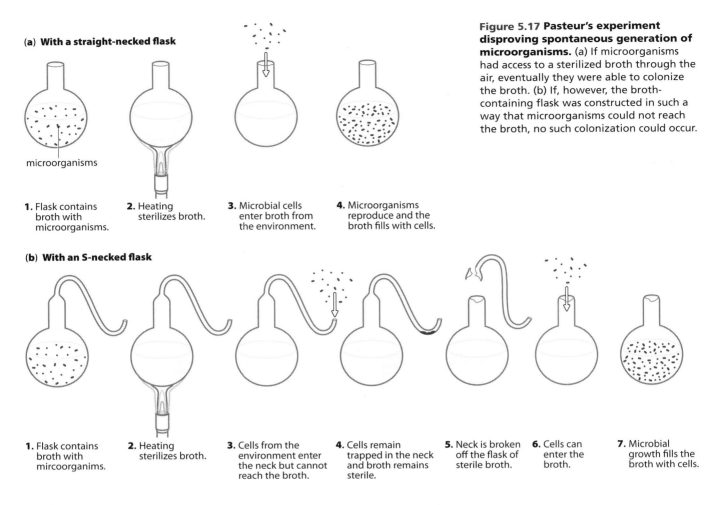

(a) With a straight-necked flask

microorganisms

1. Flask contains broth with microorganisms.
2. Heating sterilizes broth.
3. Microbial cells enter broth from the environment.
4. Microorganisms reproduce and the broth fills with cells.

(b) With an S-necked flask

1. Flask contains broth with mircoorganims.
2. Heating sterilizes broth.
3. Cells from the environment enter the neck but cannot reach the broth.
4. Cells remain trapped in the neck and broth remains sterile.
5. Neck is broken off the flask of sterile broth.
6. Cells can enter the broth.
7. Microbial growth fills the broth with cells.

Figure 5.17 Pasteur's experiment disproving spontaneous generation of microorganisms. (a) If microorganisms had access to a sterilized broth through the air, eventually they were able to colonize the broth. (b) If, however, the broth-containing flask was constructed in such a way that microorganisms could not reach the broth, no such colonization could occur.

In 1857 a group of French wine merchants asked Pasteur to investigate the problem. At that time, it was not clear how sugars in wine or other alcoholic beverages were converted into alcohol. Pasteur found that yeast, a microscopic fungus, released alcohol as a metabolic waste product in the absence of oxygen in a process called **fermentation**. Furthermore, he discovered that if certain bacteria had contaminated the wine, these bacteria competed with the yeast for sugar, converting it into acetic acid (the key ingredient in vinegar). It was this acid that gave the spoiled product its sour taste.

Pasteur found that he could eliminate these bacteria by temporarily heating the wine. This process, now called **pasteurization**, is still routinely used to remove unwanted microorganisms from many beverages, including beer, milk, and fruit juice. This link between microorganisms and food spoilage suggested that microorganisms might play a role in other important processes as well, including disease.

The "germ theory of disease" was advanced by the understanding of the need for hospital sanitation

In the second century AD, the Greek physician Galen attributed good health to a proper balance of the body's four elemental fluids or "humors." When these four fluids—blood, phlegm, black bile, and yellow bile—were not present in the correct proportions, disease resulted. Galen's ideas influenced medical thought for hundreds of years. Others believed that various illnesses were caused by foul odors, by vapors emanating from swamps, or by evil spirits. Up through the mid-19th century, few had even considered the idea that organisms too small to see could cause certain diseases.

Figure 5.18 Louis Pasteur (1822–1895). Pasteur is regarded as one of the founders of modern microbiology. His contributions include the first vaccine for rabies and the development of pasteurization, still used to remove potentially harmful microorganisms from many food products.

Things began to change in the mid-19th century. In 1847, the Austrian physician Ignaz Semmelweis (**Figure 5.19**) made an important discovery that paved the way for the **germ theory of disease**—the concept that microorganisms could cause disease. Semmelweis was concerned by the large numbers of women who died shortly following childbirth in Vienna's largest hospital. All too frequently, women in the hospital delivered healthy babies, only to die within a few days of what was called "childbed fever." Curiously, only women who were admitted to the hospital's "First Clinic" developed the illness. When the First Clinic was full, obstetric patients were admitted to the Second Clinic, and these women did not contract childbed fever. Semmelweis noted that the two clinics were the same except for one thing: in the First Clinic, babies were delivered by medical students. The Second Clinic was staffed by midwives.

Semmelweis got a clue to the mystery when a medical student cut his finger during an autopsy and died of an illness identical to childbed fever. Medical students in the hospital performed autopsies in the morning and later delivered babies in the First Clinic obstetric ward. Semmelweis reasoned that the students picked up something deadly on their hands while they performed autopsies and carried it to the new mothers in the First Clinic.

In 1847 hand washing was not routine in hospitals, but Semmelweis ordered the students to wash their hands with a chlorine solution before entering the clinic. The number of childbed fever cases in First Clinic immediately plummeted. Unfortunately, Semmelweis' discovery was ridiculed and the practices he instituted were not widely adopted. It is now known that childbed fever is caused by the bacterium *Streptococcus pyogenes*, which can easily be carried on the hands between patients.

Yet Semmelweis' work did not go entirely unnoticed. During the Crimean War of 1854–1856, thousands of lives were saved due to the groundbreaking work of Florence Nightingale (**Figure 5.20**), widely recognized as the founder of modern nursing. Nightingale introduced the basic concepts of hospital cleanliness to British military hospitals and insisted that filthy clothes and bandages be regularly replaced or sanitized. Her Nightingale School of Nursing, founded after her return to England, was the first school in the world to specifically train nurses and educate them about the importance of sanitation in the hospital environment.

The English surgeon Joseph Lister (**Figure 5.21**) was also familiar with Semmelweis as well as with Pasteur's studies linking wine spoilage to bacteria. Furthermore, Pasteur had also demonstrated that protozoan pathogens could cause disease in silkworms. Could microorganisms cause disease in humans as well? Lister knew that carbolic acid, also called phenol, killed bacteria. He reasoned that if bacteria did in fact cause human disease, treating wounds and bandages with phenol might reduce such disease. He successfully tested his ideas in 1865. As word of Lister's **aseptic technique** spread, it gradually became accepted around the world.

Unfortunately, in 1881 neither routine hand washing, first championed by Semmelweis, nor the use of aseptic technique as developed by Lister was commonplace in the United States. Had these practices been more readily adopted in this country, it is likely that the story of President Garfield would have had a happier ending.

Figure 5.19 Ignaz Semmelweis (1818–1865). When Semmelweis proposed that childbed fever was caused by a common agent transmitted on the hands of attending doctors, he was ridiculed by others in the medical profession. He was eventually dismissed by the hospital, whereupon he suffered a mental breakdown. Today, however, he is regarded as the "savior of mothers" in Austria.

Figure 5.20 Florence Nightingale (1820–1910). Called "the lady with the lamp," Nightingale is regarded as the founder of modern nursing, training hundreds in sanitary technique at her Nightingale School of Nursing, the world's first such institution. She was also an accomplished writer and helped to found the modern discipline of epidemiology.

The development of pure culture technique allowed rapid advances in microbiology

The definitive proof that a specific bacterium caused a particular disease was finally provided by the German microbiologist Robert Koch in 1876 (**Figure 5.22**). Crucial to this endeavor was the development of **pure culture technique**. Koch realized that if a specific microorganism was causing disease, any host would also harbor many other microorganisms that may have nothing to do with the disease in question. Unless he could isolate these various microorganisms, he would have no means to determine exactly *which* species was actually the pathogen. Koch found that he could take samples from a host and spread them out on solid nutrients such as potato slices. Different microbial species grew into different shaped and colored colonies. He then inferred that each of these colonies had arisen from a single bacterial cell and consequently represented a pure culture. Cells could then be extracted from each of these colonies for further testing.

Koch's use of potato slices, however, was crude and did not support the growth of all microorganisms. He next experimented with gelatin as a growth medium, but gelatin, which melts at temperatures required for maximum bacterial growth, turned out to be a poor substitute. Fortunately, Fannie Hesse, the wife of one of Koch's co-workers, suggested agar, which she regularly used as a solidifying agent in her home fruit preserves. Agar turned out to be an ideal growth medium and substantially assisted Koch in his isolation of pathogenic bacteria.

Figure 5.21 Joseph Lister (1827–1912). Lister, a surgeon at The Glasgow Royal Infirmary in Scotland, found that infection of surgical wounds could be prevented if wounds and bandages were treated with phenol.

Koch was the first to link a specific microorganism to a particular disease

Koch was interested in determining the cause of anthrax, a serious disease of domestic livestock. He inspected the blood of animals that had recently died of anthrax and found characteristic rod-shaped bacteria. When he isolated these bacteria and then injected them into healthy animals, the animals developed anthrax. Koch found bacteria in the blood of these animals as well. He isolated these bacteria and compared them with the bacteria he had isolated initially. They were identical. Koch thus confirmed that this organism, later named *Bacillus anthracis*, was the causative agent of anthrax.

On the basis of his work with anthrax, Koch developed a protocol for demonstrating that a particular disease is caused by a specific microorganism. These investigative steps, known as **Koch's postulates (Figure 5.23)**, led to remarkable advances in the field of medical microbiology. Koch himself used his postulates to identify *Mycobacterium tuberculosis* as the causative agent of tuberculosis, and by 1900 the causes of many diseases, including typhus, diphtheria, and gonorrhea, had been identified. At the close of the 19th century the causative agent of bubonic plague was finally discovered as well. Alexander Yersin, a Swiss scientist, traveled to China to track down the source of this perennial scourge of humankind. Using pure culture techniques, he isolated bacteria from the inflamed lymph nodes or buboes of plague victims and then injected these bacteria into rats. Not only did the

Figure 5.22 Robert Koch (1843–1910). Along with Pasteur, Koch is regarded as one of the founding fathers of microbiology. He was the first to show that a bacterial species (*Bacillus anthracis*) could cause a specific disease (anthrax), and he was awarded the Nobel Prize for his work demonstrating that tuberculosis was caused by the bacterium *Mycobacterium tuberculosis*. The protocol that he developed to link a specific microbe with a particular disease, called Koch's postulates, is still in use today.

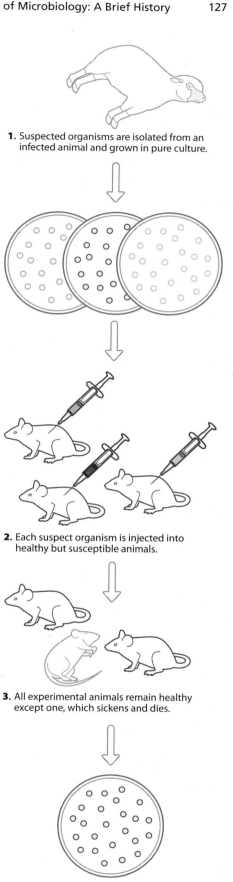

Figure 5.23 A demonstration of Koch's postulates. Developed by Robert Koch, Koch's postulates are used to demonstrate that a specific microorganism causes a particular disease. All four postulates, numbered 1–4 in the figure, must be satisfied to prove a definitive link between the disease and the microbe.

1. Suspected organisms are isolated from an infected animal and grown in pure culture.

2. Each suspect organism is injected into healthy but susceptible animals.

3. All experimental animals remain healthy except one, which sickens and dies.

4. Organisms isolated from the dead experimental animal are confirmed as being one of the original suspect organisms.

rats develop disease, but Yersin was able to isolate more of the same bacteria from the rats, satisfying Koch's postulates. This bacterial species was subsequently named *Yersinia pestis* in honor of its discoverer. Over 100 years later, Koch's postulates are still valuable and are still employed to track down the cause of apparently new diseases, a topic we will return to in Chapter 13.

The germ theory led to major advances in disease control

Once the link between microorganisms and specific diseases was firmly established, it became possible to develop ways to control these microorganisms. Such control was realized in various ways. Improved hospital care reduced the likelihood of infection. The pioneering work of individuals such as Semmelweis, Nightingale, and Lister was crucial in this regard. People also began to appreciate the necessity of having clean food and water. Before the acceptance of the germ theory, drinking water was frequently contaminated with raw sewage (**Figure 5.24**). One could argue that safe drinking water and indoor plumbing have done more to reduce the incidence of infectious disease than any other single innovation.

Nevertheless, not all diseases can be controlled simply by providing safe food and water supplies. In the late 1800s, evidence was beginning to accumulate that insects and other arthropods could transmit disease as well. In 1878 it was demonstrated that elephantiasis, caused by the accumulation of specific parasitic worms in lymph nodes, was transmitted through the bite of an infected mosquito. In 1892, Theobald Smith and Frederick Kilbourne determined that Texas red water fever, a serious disease of cattle, was caused by a protozoan parasite that was transmitted by ticks, and in 1896, Ronald Ross, an officer in the British Army, proved that malaria was mosquito-transmitted.

The demonstration that yellow fever was a mosquito-borne disease is particularly illustrative. As we have seen, until the role of mosquitoes was clearly elucidated, control of the disease was essentially impossible. You will recall that the Cuban Carlos Finlay was the first to provide a scientific argument for the hypothesis that yellow fever was transmitted via the bite of infected mosquitoes. In 1898, with the outbreak of the Spanish–American War, the U.S. Army became especially interested in yellow fever. Consequently, the Yellow Fever Commission, headed by Walter Reed, went to Cuba to determine the source of this disease. To test Finlay's mosquito transmission hypothesis, the commission carefully reared mosquitoes in the laboratory to rule out any previous exposure to humans. The mosquitoes were then allowed to feed on patients with yellow fever. Subsequently, these same mosquitoes were permitted to feed on human volunteers, including commission members James Carroll and Jesse Lazear. Both developed yellow fever, and while Carroll survived the ordeal, Lazear vomited blood, became delirious, and died 12 days following infection. Although Lazear paid with his life, the way was now paved for the control of yellow fever, which, as discussed previously, contributed to the construction of the Panama Canal.

Even though yellow fever could now be managed by controlling mosquitoes, it would still be a number of years before the exact infectious agent causing this disease would be discovered. Yellow fever is caused by the yellow fever virus. As discussed in Chapter 4, viruses are very small: too small, in fact, to be seen by even the best microscopes available at the turn of the

20th century. The first indication that there might be something smaller than bacteria that could cause disease came in 1892. The Russian Dmitri Ivanowski was trying to determine the cause of mosaic disease in tobacco plants. He found that whatever caused the disease was so small that it could pass through filters fine enough to prevent the passage of all known bacteria. When he injected healthy plants with the filtered fluid, they too developed mosaic disease. In 1898 a similar filterable agent was found to cause hoof-and-mouth disease in cattle by the Germans Friedrich Loeffler and Paul Forsch. In the 1930s it finally became clear that these were not merely unusually small bacteria but were a fundamentally different type of microbe. Wendell Stanley showed that these mysterious "viruses" ("virus" from the Latin for "poison") were so simple that they could be crystallized like a chemical compound. In the 1940s, the development of the electron microscope gave microbiologists their first look at viruses.

Figure 5.24 London's water. A cartoon from an 1827 London newspaper describing the Thames River, the source of the city's water, as a "monster soup."

With the development of the first vaccines, the field of immunology was born

Until recently, smallpox was one of the great scourges of humanity. Yet over 1000 years ago, medical practitioners in China and India inoculated healthy people with material taken from smallpox pustules. Generally the inoculated individual developed only a very mild case of smallpox, and he or she would be immune to future infection with the disfiguring and often lethal form of the disease. This practice was not without risk. Occasionally the inoculated person developed a full-blown case of virulent smallpox. The procedure, known as **variolation**, was unknown in the West until 1721. In that year, Lady Montague, wife of the British Ambassador to the Ottoman Empire, observed the practice and became an enthusiastic supporter. She had her own children inoculated and introduced variolation to England.

Although variolation became popular in England, many people remained unprotected, and in the 1790s a smallpox epidemic erupted in west England. Dr. Edward Jenner, a young physician, was dispatched by the Royal Society of Medicine to the scene of the outbreak. In the course of his investigations, Jenner learned that the young girls who milked cows were unafraid of contracting smallpox. Among these milkmaids it was common knowledge that although they developed a mild condition called cowpox on their hands, they rarely contracted smallpox. Jenner wondered if infection with cowpox somehow protected against the far more serious smallpox. In 1796 he inoculated fluid from a cowpox lesion into the skin of an 8-year-old boy (**Figure 5.25**). The boy developed cowpox as expected. He then inoculated the boy with fluid from a smallpox lesion. The boy remained healthy. Inducing immunity to smallpox in this way became known as vaccination. The word "vaccination" derives from the Latin word "vacca," meaning "cow."

Louis Pasteur is credited with the quotation "chance favors the prepared mind," and he demonstrated this to good effect in the mid-1880s. Fresh from his assistance to the French wine industry, Pasteur was asked by agricultural officials in 1885 to investigate a serious disease in chickens. Using the postulates developed by Koch, Pasteur identified the bacteria causing fowl cholera. He then began work on a vaccine for the disease. At one point he mistakenly left some of the bacteria unattended while he went on a short holiday. When he returned to work, he inoculated chickens with these same bacterial cultures, and found that while the chickens became ill, they did not die. He attributed this to the fact that the cultures were old. But when he inoculated these same chickens with fresh bacteria, the birds developed no illness whatsoever. Pasteur correctly reasoned that the weakened bacteria stimulated resistance to a later challenge with the more virulent bacteria. Nowadays, microorganisms are often purposely weakened for use as vaccines against unweakened, still virulent strains. The actual process of weakening microbes is called **attenuation**. Pasteur had discovered the concept

Figure 5.25 The first smallpox vaccine.
(a) In this painting, Edward Jenner is inoculating a young boy with material from a smallpox lesion. The boy, who had earlier been inoculated with cowpox, failed to develop smallpox. (b) Cartoon by the British satirist James Gillray suggests the fear felt by many about the new technique of smallpox vaccination. The cartoon implies the vaccine, made from cowpox, will turn people into part-human and part-cow.

of an attenuated vaccine, in part because of his earlier error. He went on to develop similar attenuated vaccines against anthrax and rabies.

Antimicrobial drugs are a 20th century innovation

Killing microorganisms is easy. The trick is to do so without harming the patient. Early attempts at antimicrobial therapy often did as much harm as good because substances were used that were toxic to both host and microbial cells.

In 1906 Paul Ehrlich, a German chemist, found that certain histological dyes would stain bacterial but not animal cells. Ehrlich reasoned that if these chemicals bound only to bacterial cells, there must be fundamental differences between the two cell types. Furthermore, he thought that if these dyes could be made in a toxic form, the dyes might bind to and kill bacterial cells while leaving animal cells unaffected. He thus began the search for what he called his "magic bullet"—a compound that would selectively kill only bacterial cells.

Ehrlich's quest for such a magic bullet bore fruit in 1908 when he discovered salvarsan, a compound that was effective against syphilis. Unfortunately, it would be almost 30 years before a second reliable antimicrobial became readily available. In 1935 researchers at the German company I.G. Farben discovered that the reddish dye prontosil could be used effectively to treat streptococcal infections. The chemist Gerhard Domagk played the leading role in the drug's development. Further work demonstrated that prontosil was converted in the human host to sulfanilamide, the first sulfa drug. There are now hundreds of sulfa derivatives, and they are still frequently used today.

Sulfa drugs are considered to be synthetic drugs because they are organic chemicals manufactured in a laboratory. **Antibiotics**, on the other hand are antibacterial compounds that are produced by microorganisms. The first true antibiotic was discovered in 1929. In another example of how mistakes can sometimes result in good fortune, Alexander Fleming saw that his *Staphylococcus* cultures had been contaminated with mold. As discussed in Chapter 1 (p. 10), instead of merely disposing of his apparently ruined cultures, Fleming noticed that the bacteria failed to grow in an area close to the contaminant. Subsequent analysis showed that the mold, *Penicillium notatum*, was releasing a compound that was toxic to the bacterial cells.

Fleming called the substance "penicillin." Although he understood that it might be useful for combating bacterial infections, penicillin proved difficult to purify in large quantities. It was not until the 1940s that production problems were overcome and penicillin became widely available for clinical use.

Into the modern era

In the last 70 years microbiology has continued to advance, sometimes at an astonishing pace. In the late 1940s and early 1950s, biologists began to understand such fundamental concepts as the structure of DNA, the mechanism of enzyme activity, and the basis of metabolism. As it became increasingly understood that these processes were essentially the same in all living things, scientists began to use microorganisms as **model organisms**. In other words, important concepts first discovered in microorganisms were found to have broad application in other living things as well. Microorganisms are especially well suited to serve as model organisms because they are relatively easy to rear in the lab and they reproduce so quickly. Consequently, large numbers necessary for study can be grown in a short time. Furthermore, as we saw in Chapter 3, their cell structure is simple compared with that of eukaryotes, making it easier to examine fundamental biological processes.

In 1944 DNA was established as the genetic material, and in 1946 it was shown that DNA could be transferred between bacterial cells, a process we will explore more fully in Chapter 6. In 1953, the American James Watson and the Englishman Francis Crick published their model for the structure of DNA. In the 1960s we began to understand exactly how proteins were produced and how the activity of DNA was regulated, again through the use of microbial model systems. As the 1970s began it was demonstrated that pieces of animal DNA could be inserted into bacterial DNA, resulting in hybrid, **recombinant DNA**. The stage was now set for the ongoing "molecular revolution"—our ability not only to decode the genetic material but also to manipulate it. In Chapter 14 we will discuss the Human Genome Project, the production of pharmaceuticals by bacteria, the creation of pest-resistant agricultural crops, and other ramifications of this revolution.

Coming Up Next...

After our detour into microbiology's past, we return to the present. We have already commented on the remarkable diversity of microorganisms and other living things. In the next chapter we will explain the biological basis of that diversity. We will also provide a scientific explanation for the common observation that offspring tend to resemble their parents. In other words, in Chapter 6 we turn our attention to genetics—the study of heredity. In many ways, genetics can be equated to the study of DNA—the genetic material. Consequently, we will take a close look at this most remarkable of molecules, upon which life itself is based. Although we will focus on microorganisms, we will find, as we recently learned in our discussion of model organisms, that basic genetic processes are similar in all living things.

Key Terms

acute disease	fermentation	parasite
antibiotic	germ theory of disease	pasteurization
antibody	helminth	pure culture technique
aseptic technique	Koch's postulates	recombinant DNA
attenuation	model organism	variolation
chronic disease	monoculture	vector-borne disease
cross-reactive antibody	pandemic	vector

Concept Questions

1. Alternative history has become a quite popular genre for fiction writers (e.g., a fictional history in which the Confederacy won the Civil War). Suppose you wished to write an alternative fiction novel in which ancient Romans had access to modern antibiotics. Provide a short, one-paragraph summary of your novel's plot.

2. What is the difference between an acute and a chronic disease? Why are small isolated populations at less risk of acute disease, unless such diseases are introduced from the outside?

3. In writing to his wife on November 29, 1812, Marshal Ney, top commander of Napoleon's army, said "General Famine and General Winter, not Russian bullets, have conquered the Grand Army." Why should he have included "General Louse and General Typhus"?

4. During the French attempt to construct the Panama Canal, bedposts were often placed in small bowls of water, to prevent ants from crawling up the bedposts and into the beds. How might this practice have contributed to the French failure to complete the canal?

5. Suppose you were working in Vienna's Central Hospital, along with Ignaz Semmelweis, in the mid-1800s. Semmelweis shares with you his ideas regarding the cause of childbed fever. He asks you to conduct an experiment in the hospital to see if he is correct. How would you proceed?

6. In this chapter we learned about the experiments of Redi and Pasteur that disproved the notion of spontaneous generation. Design another experiment that would likewise support the conclusions of Redi and Pasteur, that life can come only from preexisting life.

7. One of the necessary conditions, when using Koch's postulates to demonstrate that a particular microorganism causes a specific disease, is that it must be possible to grow the microorganism in pure culture. Why is this?

8. What makes microorganisms especially useful as model organisms?

9. Why is the late 19th century often referred to as the Golden Age of microbiology?

Vibrio cholerae exchange DNA through a process known as conjugation.

Chapter 6

Microbial Genetics

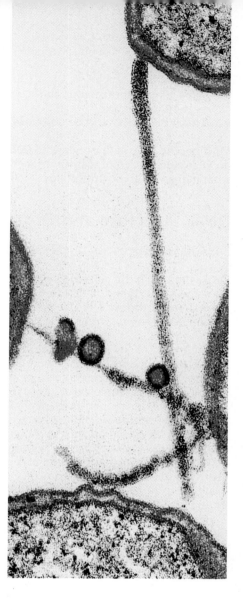

Anyone can see that even members of the same species differ in their appearance. All dogs, for instance, are easily recognizable as dogs, but nobody is likely to confuse a Pekingese and a Dalmatian. Microorganisms are no less variable. Why do some strains of *Escherichia coli* cause serious disease, while most others live peacefully in your intestines without causing problems? Some bacteria have extracellular structures such as fimbriae and capsules. Others of the same species do not. What explains this difference? How do organisms that were originally susceptible to a particular antimicrobial drug become resistant? Why do all living things, microbial or otherwise, tend to resemble their parents or siblings more than they do unrelated individuals?

The answers to these questions ultimately come down to genetics, the study of biological inheritance. Most of the variation that we observe between organisms and most of the similarities between parents and offspring can be explained in terms of differences and similarities in their genetic material. With the exception of the RNA viruses discussed in Chapter 4, the genetic material is deoxyribonucleic acid (DNA).

Since the dawn of agriculture and animal domestication, roughly 12,000 years ago, humans have been indirectly manipulating the DNA of plants and animals through breeding and agricultural practices. However, little was understood about how these manipulations of nature worked until the mid-1800s when Gregor Mendel, an Austrian monk, established important principles explaining how the characteristics of pea plants were transmitted from one generation to the next. Mendel and other early geneticists experimented with inherited and easily recognizable traits such as vine length in peas, wing shape in fruit flies, and coat color in mice. Although these genetic pioneers were able to determine that certain traits were passed from generation to generation, they still did not understand the underlying mechanisms responsible for inheritance. It was not until the mid-20th century that scientists discovered that the genetic material was DNA, and in 1953 James Watson and Francis Crick published their seminal paper on DNA's structure. Since then, our understanding of this remarkable molecule has exploded, and although many questions remain, we have now reached the point where we can manipulate an organism's DNA through the powerful techniques of genetic engineering.

In this chapter we turn our attention to DNA and its role as the genetic material. Although we will focus on prokaryotes, our task is simplified by the fact that DNA carries out the same basic functions in all living things. Specifically, we will consider the following:
- How DNA encodes genetic information
- How genetic information results in specific characteristics
- How DNA's activity is controlled

- How genetic information in the form of DNA is passed from one generation to the next
- How a cell's DNA can occasionally change

Most people have at least heard of DNA and know that it has something to do with genetics. Here we will discover exactly what DNA is and precisely what it does that is so important.

DNA: Structure and Organization

DNA is an information molecule

DNA is found in the cells of all living things, and in all of these cells, DNA does the same thing: it determines which proteins a particular cell can make. In other words, DNA, as an information molecule, *encodes* the information necessary to produce various proteins.

In Chapter 2 (p. 40) we began to understand how DNA accomplishes this task. DNA is composed of individual building blocks called **nucleotides** (**Figure 6.1a**). Part of each nucleotide is composed of a nitrogen-containing base, and each base may be of one of four types: adenine, thymine, guanine, and cytosine. Consequently there are four types of nucleotides. A strand of DNA may be composed of thousands or millions of nucleotides, and nucleotides with different bases may be aligned in any possible order. Much of an organism's DNA is organized into **genes** (Figure 6.1b). A gene is a sequence of DNA nucleotides that codes for a single protein. In any gene, the sequence of nucleotide bases specifies the exact amino acid sequence needed to make a specific protein. The nucleotides therefore represent a code, utilized for protein production.

(a)

(b)

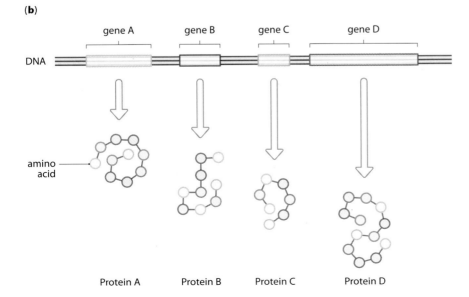

Figure 6.1 Deoxyribonucleic acid (DNA). (a) A strand of DNA consists of many nucleotides linked together. Nucleotides found in DNA contain one of four bases: adenine (A), thymine (T), guanine (G), and cytosine (C). A DNA strand may be composed of thousands or millions of these nucleotides, and they may occur in any order. (b) Genes code for proteins. Each gene contains the genetic information, encoded in the sequence of nucleotides, for the production of a protein. The nucleotides "spell out" which of the 20 amino acids (indicated by the colored circles in the protein) is required at a particular point in the protein. A small protein may consist of a few dozen amino acids. A large protein may be composed of 1000 or more amino acids.

An easy way to think about this is to picture the four types of DNA nucleotides as letters in a genetic alphabet. These "letters" spell out "words," each of which represents one of the 20 amino acids found in proteins. If, for example, we know that a particular protein consists of 300 amino acids (a fairly typical size), we know that the gene that codes for this protein must contain 300 genetic words, each composed of DNA nucleotides. A single bacterial cell may have several thousand genes, indicating that it can produce several thousand proteins.

Although DNA is composed of genes and contains the exact instructions for protein production, DNA does not actually link the amino acids together to form the protein. Remember that DNA is an information storage molecule; it is like a cook book with thousands of "recipes" for different proteins. But it is not the cook. As we will see, other molecules have the responsibility of reading these protein recipes and assembling the specified amino acids, per the instructions in the DNA.

The structure of DNA is the key to how it functions

To fully appreciate how DNA serves as the genetic material, we first need to become familiar with its unique structure. To begin, let's take a closer look at an individual nucleotide (**Figure 6.2**). Each nucleotide consists of three parts: a five-carbon sugar called deoxyribose, a phosphate, and a nitrogen-containing base. The sugar and the phosphate are identical in all nucleotides. Differences in the associated base make a nucleotide either A, T, G, or C.

Notice that in Figure 6.2b, each of the five carbons in the deoxyribose sugar has a specific number. The phosphate is always attached to carbon number 5, and the base (adenine, thymine, guanine, or cytosine) is always attached to carbon number 1. In a strand of DNA, nucleotides are linked together into long chains (**Figure 6.3**). The phosphate on each nucleotide (attached to carbon 5) is linked to carbon 3 on the adjacent nucleotide. The end of the DNA molecule on which a phosphate is found is termed the 5' ("5-prime") end. The other end, with an unlinked carbon number 3, is called the 3' ("3-prime") end. A one-letter code is used to describe the nucleotides in the DNA strand. In Figure 6.3 the code in this short DNA strand is written G-A-T-C (the code is always read from the 5' end). Furthermore, Figure 6.3 also illustrates the fact that the four nucleotide bases are not identical in terms of size. Thymine and cytosine are smaller, single-ringed molecules called **pyrimidines**. Larger, double-ringed adenine and guanine are called **purines**.

We have now examined the structure of a DNA strand, but in reality, a complete molecule of DNA is usually composed of two strands. In other words, DNA is a double-stranded molecule (**Figure 6.4**). The sugars and phosphates on each strand form a backbone with the bases pointing inward. The two strands are held together by hydrogen bonds that form between bases on opposing strands. The shape and chemical properties of the bases are such that the base pairing between opposite strands is highly specific: T pairs with A, and C pairs with G. You might think of the bases as pieces of a jigsaw puzzle to help visualize this point. The structure of A is such that it

(a)

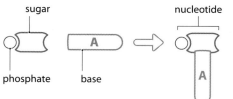

Figure 6.2 DNA is composed of nucleotide building blocks.
(a) Simplified diagram of an individual nucleotide, in this case, A. All nucleotides contain identical phosphates and sugars, but the base of any one may be adenine, thymine, guanine, or cytosine.
(b) A more detailed diagram, depicting the same nucleotide. Note that the carbons on the deoxyribose sugar are numbered 1–5. The base is always attached to carbon number 1, and the phosphate is always attached to carbon number 5.

(b)

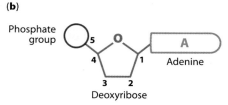

Structure

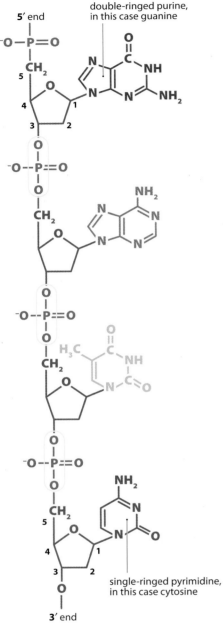

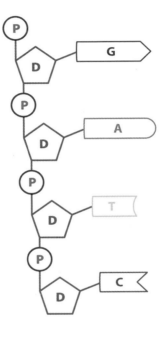

Schematic

Figure 6.3 DNA is composed of nucleotides. A short length of one DNA strand shown both in detail and in a simplified, schematic form. The number 3 and number 5 carbons on the sugar link each nucleotide to other nucleotides in the DNA strand (the linkages are highlighted in yellow). Reading in the 5′ to 3′ direction, the sequence depicted here is GATC. Note that the bases on a G or A nucleotide are double-ringed structures called *purines*. The bases on a T or C nucleotide, on the other hand, consist of a single ring. These smaller, single-ring structures are known as *pyrimidines*.

typically can form hydrogen bonds only with T. Likewise, the specific shape and chemical properties of G and C ensure that they can hydrogen-bond only to each other. Note that, as Figure 6.4 shows, a T-A pair is held together by two hydrogen bonds, whereas a C-G pair forms three hydrogen bonds. Furthermore, the two strands are positioned opposite each other in an inverted orientation, known as **antiparallel**; one strand runs in the 5′ to 3′

direction, while the opposing strand is in the 3' to 5' direction. The two DNA strands wind around each other in a spiral configuration. Because there are two such strands and because a spiral configuration is known as a *helix*, DNA is said to adopt a structure known as a **double helix**.

When Watson and Crick developed their model for DNA structure, they reasoned that the four bases could occur in any order on a single DNA strand and that they could function like letters in a word. The linear sequence of the bases spells out the genetic instructions encoded in the DNA. Because the four bases can occur in any order, the amount of information that can be encoded in DNA is virtually limitless. Furthermore, if the sequence is known on one of the two DNA strands, the sequence on the opposing strand can instantly be inferred because of the specificity of base pairing. As we will see, this last observation about DNA's structure provided a powerful clue as to how the genetic material is replicated.

DNA is found on chromosomes

Storing DNA within a cell is no trivial matter. The DNA of a single bacterium, for example, if stretched out, is approximately 1000 times longer than the cell itself (**Figure 6.5a**). Clearly, such a large molecule cannot be dispersed at random throughout the cell. Rather, it must be neatly packaged in a manner that allows it not only to fit within the cell but also to carry out its functions. In both prokaryotic and eukaryotic cells, DNA is organized into discrete units called **chromosomes**. A chromosome consists of both DNA and associated proteins. These proteins provide a "scaffolding" around which the DNA coils (Figure 6.5b). Such packaging allows the DNA to be condensed many times. In a typical bacterial cell such as *E. coli*, for instance, the condensed DNA takes up only about 10% of the cell's total volume.

In prokaryotic cells, which lack a membrane-bound nucleus, the chromosome is found in the cell's cytoplasm. In eukaryotic cells, chromosomes are located within the nucleus. Prokaryotic and eukaryotic chromosomes differ in several other notable ways. First, a prokaryote's DNA is usually circular (**Figure 6.6a**). This circular DNA, associated with its scaffolding proteins, forms the circular prokaryotic chromosome. In eukaryotes, linear DNA and its associated scaffolding proteins form linear chromosomes (Figure 6.6b).

Second, although there are exceptions, a prokaryotic cell generally has one chromosome. This means that *all* of a prokaryote's genes are found on this single chromosome. In contrast, the DNA of eukaryotes is organized into a number of chromosomes. The exact number is species-specific. Human cells contain 46 chromosomes. The cells of fruit flies, horses, and onions contain 8, 64, and 16 chromosomes respectively. Eukaryotic chromosomes often come in pairs, so the 46 chromosomes of humans consist of 23 pairs. Each member of the pair carries the same genes, meaning that a eukaryotic cell usually has two copies of each gene: one on each chromosome in the pair. Prokaryotes, with only one chromosome, usually have only a single copy of each gene. There are, however, examples of bacteria with multiple copies of a particular gene on the same chromosome.

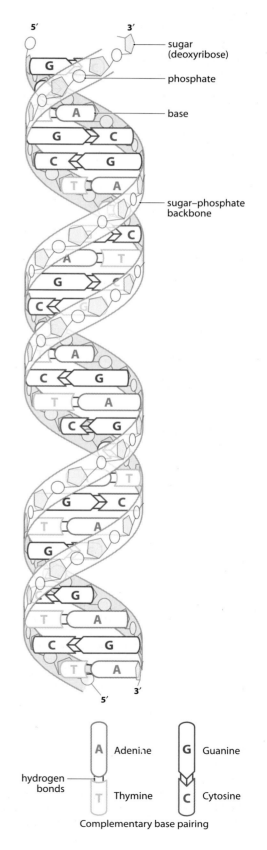

Figure 6.4 The DNA double helix. A DNA molecule is composed of two DNA strands held together by hydrogen bonds (in red) between the paired bases. The 5' and 3' ends of each strand are indicated. Note that the two complementary strands are antiparallel to each other. Below the DNA molecule, the specific complementary base pairing between adenine and thymine and between guanine and cytosine is illustrated.

(a)

(b)

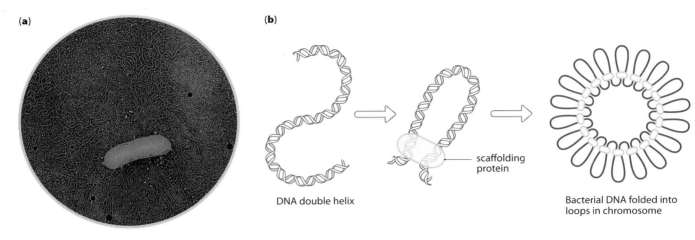

DNA double helix

scaffolding
protein

Bacterial DNA folded into
loops in chromosome

Figure 6.5 The prokaryotic chromosome. (a) In this photograph, DNA, seen here as a mass of squiggly lines, is emerging from a bacterial cell in which the membrane has ruptured. (b) DNA wraps around proteins, forming loops, which help to compact DNA within the cell.

DNA Function: Replication

Each DNA molecule can be accurately copied into two new DNA molecules

In this section we will describe the process in which DNA, the genetic information, is replicated. In brief, the two strands of the DNA double helix separate from each other at a precise point (**Figure 6.7**). Each of these two original, "parent" DNA strands is used as a template for the synthesis of a new, complementary "daughter" strand. If a short section of the parent strand consists of the nucleotides ATTAGC, the specified nucleotides on the daughter strand will be the complementary nucleotides TAATCG. Finally, when the entire nucleotide sequence of each parent strand is paired with complementary nucleotides on a daughter strand, DNA replication is complete. There are now two complete molecules of DNA, each consisting of one original parent strand and one newly synthesized daughter strand. An enzyme called *DNA polymerase* is required for DNA replication. This enzyme moves along the two parental strands as they separate, "reading" the nucleotide sequence and synthesizing the daughter strands accordingly.

CASE: DNA: FORM SUGGESTS FUNCTION

The short, 900-word paper published in the British journal *Nature* by James Watson and Francis Crick on the structure of DNA in April 1953 is one of the most important publications in the history of biology. With the molecule's structure now elucidated, it became clear how DNA was able to code for characteristics and how genetic information was transmitted from generation to generation. The structure of DNA also provided important clues as to how it could be duplicated prior to cell division. Indeed, in a masterpiece of understatement, Watson and Crick concluded their paper with the following sentence: "It has not escaped our notice that

(a)

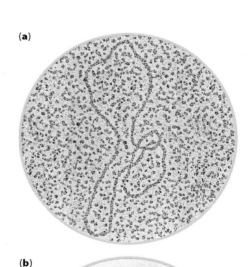

(b)

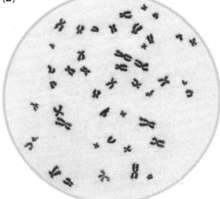

Figure 6.6 Organization of eukaryotic and prokaryotic DNA. (a) Prokaryotic DNA is organized into a circular chromosome. This photograph, taken through an electron microscope, shows a bacterial plasmid (see Chapter 3, p. 59), represented as a long, twisted loop. A bacterial plasmid such as this may contain fewer than a dozen genes. Larger bacterial chromosomes, also circular, typically carry a few thousand genes. (b) Eukaryotic DNA is organized into linear chromosomes. The human chromosomes seen here are an example.

the specific pairing we have postulated immediately suggests a possible copying mechanism for the genetic material."
1. **What did Watson and Crick mean by this statement?**

DNA replication refers to the process in which a DNA molecule is copied into two identical DNA molecules. Such replication is critical. When cells divide, each newly formed cell must receive a complete set of genetic information if it is to produce proteins and carry out its various functions. As the quote in our case suggests, the secret to DNA replication lies in the specific base pairing of the double helix.

When DNA is replicated, the original or **parental DNA** is copied into two identical **daughter DNA** molecules. For this to happen, the double-stranded parental DNA molecule unwinds and separates into two single strands (**Figure 6.8**). Each of these parental strands is then used as a template for the synthesis of new daughter strands. In other words, the base sequence on each parental strand dictates the complementary sequence on the newly synthesized daughter strands. If the specific base on the parental strand is guanine, then, as the growing daughter strand is synthesized, a nucleotide containing cytosine will be added opposite this guanine. If the next base on the parental strand is a thymine, the next nucleotide added to the daughter strand will contain an adenine. This continues until every nucleotide on the parental strands has been successfully paired. This is exactly what Watson and Crick alluded to in our case's quotation.

The overall process of DNA replication is mediated by more than a dozen enzymes and other proteins. One of these, **DNA polymerase**, helps add the new nucleotides in the growing daughter strands (see Figure 6.8). Exact points where nucleotides are added to the growing daughter strands are called **replication forks** (**Figure 6.9**). As the parental DNA continues to unwind, nucleotides are added individually at a progressing replication fork. On one of the parent strands nucleotides are added one after another, until the daughter strand is complete. Along the other parent strand the nucleotides are added until they form a short fragment of nucleotides. A series of these fragments are joined together at a later time. Each new DNA molecule will consist of one strand of original parental DNA and one strand of new daughter DNA (**Figure 6.10**).

DNA polymerase is very accurate but occasionally it makes a mistake. For example, if a base on the parental strand is cytosine, DNA polymerase might erroneously add an opposing adenine instead of guanine. Yet the truly remarkable thing about DNA polymerase is how rarely such mistakes occur. For prokaryotes in general, the DNA might be copied perfectly thousands or millions of times before a single error takes place. This level of accuracy is an important feature of DNA replication. As we will discuss later in this chapter, even when it makes a mistake, DNA polymerase is often able to detect the error and correct it. Between its high precision and its capacity to correct many of the mistakes that sneak though, DNA replication is accurate indeed.

Figure 6.7 An overview of DNA replication. The original double-stranded DNA molecule to be replicated is known as the parent DNA. Although the two strands of a DNA molecule wind around each other in a double helix, the strands are shown as two straight lines in this diagram for simplicity. As the two parental strands are separating, an enzyme called DNA polymerase moves along them toward the progressing separation point and uses the sequence found on the parental strand as a template for the synthesis of a new daughter strand. The two new DNA molecules produced are termed the daughter DNA. Each of these two, identical molecules consists of one original parent strand (in gray) and one newly synthesized daughter strand (in red).

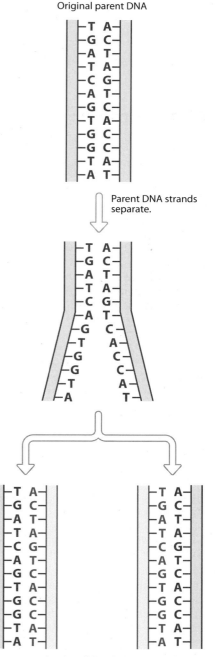

Original parent DNA

Parent DNA strands separate.

Two new molecules of daughter DNA result, each consisting of one of the parent strands and one newly synthesized daughter strand.

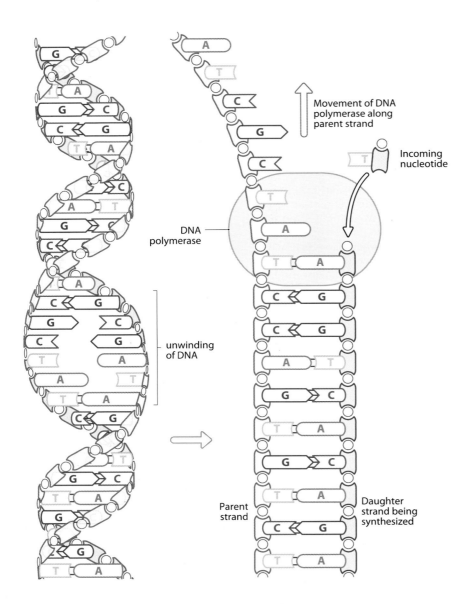

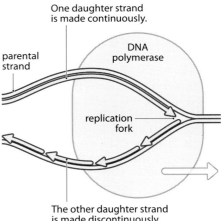

Figure 6.8 DNA replication. DNA replication begins when the DNA double helix unwinds and the bases separate. As the strands separate, DNA polymerase moves along the resulting single parental strands (in brown), using its nucleotide sequence as a template for the synthesis of the daughter strands (in red). The diagram shows the process as it takes place along one parental strand; a similar process is simultaneously occurring opposite the other parental strand.

Figure 6.9 A replication fork. A precise point where nucleotides are added to the growing daughter strands (in red) by DNA polymerase is called a *replication fork*. One daughter strand is synthesized continuously into a single long DNA molecule, while the other daughter strand is synthesized in a series of short fragments, which are later joined together. The replication fork continues to move forward as the parental DNA (in gray) unwinds ahead of the DNA polymerase. The open arrow shows the direction in which DNA polymerase moves.

DNA replication begins at a site called the origin

The **origin** is the point on the DNA where the two parental strands first separate and DNA replication begins (**Figure 6.11**). Replication in both prokaryotes and eukaryotes begins when specific proteins recognize and bind to the origin. The origin typically consists of about 250 base pairs. Although the exact sequence can vary, it is generally a location that is rich in A-T pairs. Recall that A and T are held together by two hydrogen bonds and G-C pairs are maintained by three such bonds. The fewer hydrogen bonds in A-T pairs enable strand separation to occur more easily.

In prokaryotes the DNA is generally a circular molecule (**Figure 6.12a**). The parental strands separate at a single origin, and DNA polymerase binds at each of two replication forks moving in opposite directions around the circular, parental DNA. Because the DNA is a closed loop, the two forks ultimately meet when DNA replication is completed. In eukaryotes, alternatively, DNA is linear. Moreover, eukaryotic DNA has multiple origins (Figure 6.12c). This means that replication is occurring simultaneously at several points. Two replication forks form at each of these origins and move in opposite directions.

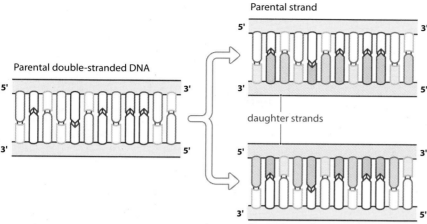

Parental strand

Parental double-stranded DNA

daughter strands

Parental strand

Figure 6.10 Parental DNA serves as a template during replication. Each newly replicated daughter DNA molecule consists of one original parent strand and one newly synthesized daughter strand. Observe that the base pairs on each new, double-stranded daughter molecule are identical to those of the original, double-stranded parental molecule.

DNA Function: Genes to Proteins

What precisely is the connection between a series of nucleotides and a characteristic such as your cat's black fur or whether or not a strain of *Staphylococcus aureus* is resistant to penicillin? How exactly *does* a specific gene influence a particular characteristic? In this section we will answer this question by describing the actual manner in which DNA's code is interpreted by cells.

A cell's characteristics are largely determined by the proteins that the cell produces

It may be surprising to learn that DNA, as an information molecule, primarily codes for proteins and that a fundamental difference between different organisms is simply the proteins that they can make and when they make them. After all, cells also contain many nonprotein components and these also vary in different organisms. Recall from Chapter 2, however, that many proteins are enzymes, and enzymes catalyze biological reactions, including the synthesis of molecules such as carbohydrates and lipids. Although DNA does not code for specific nonproteins directly, it *does* code for the enzymes that synthesize these molecules. In this way, a cell's DNA determines *all* of the processes that cell can carry out.

As we have recently learned, each protein produced by a cell is encoded in a gene within that cell's DNA. Furthermore, we have seen that a gene is a linear sequence of nucleotides, and the sequence of bases in these nucleotides encodes the information necessary to produce the specified protein (see Figure 6.1b). Consequently, genes determine exactly what proteins a cell can synthesize, and proteins are ultimately responsible for the characteristics of a cell. The sum total of an organism's genes is called its **genome**. The genome of a bacterial species may consist of approximately 4000–5000 genes, whereas the human genome is composed of about 25,000–30,000 genes. Because each gene codes for a single protein, geneticists regard the gene as the basic unit of heredity.

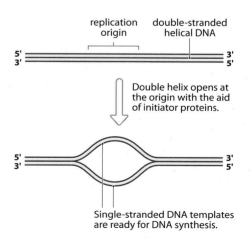

replication origin   double-stranded helical DNA

Double helix opens at the origin with the aid of initiator proteins.

Single-stranded DNA templates are ready for DNA synthesis.

Figure 6.11 Replication starts at an origin. Specific proteins are able to recognize the base sequence at the origin and separate the parental DNA strands at that point. The now single-stranded parental DNA can serve as a template for synthesis of the daughter strands.

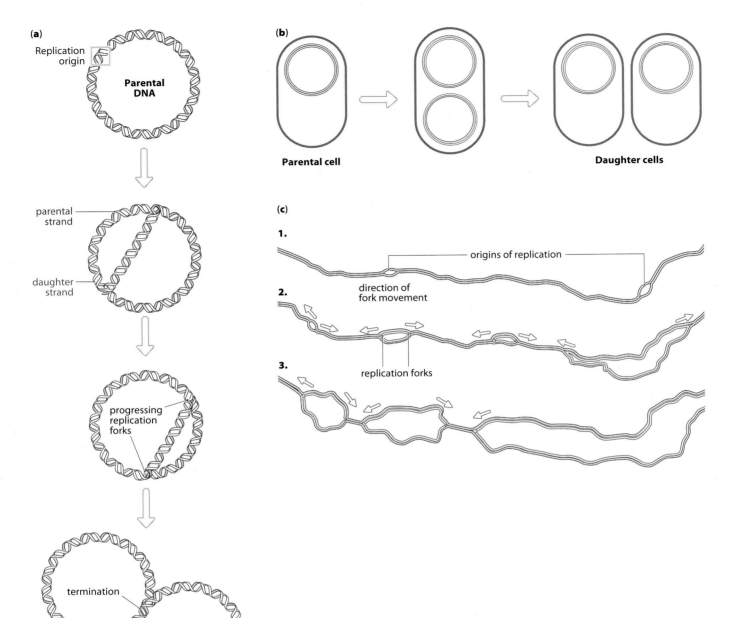

(a)

Replication origin

Parental DNA

parental strand

daughter strand

progressing replication forks

termination

Daughter DNA

(b)

Parental cell

Daughter cells

(c)

1.

origins of replication

2.

direction of fork movement

3.

replication forks

Figure 6.12 Replication at prokaryotic and eukaryotic origins. (a) In prokaryotes, the circular DNA double helix is opened at a single origin (highlighted in green). Two replication forks are formed and move in opposite directions. DNA polymerase binds at each replication fork, producing the daughter strands (in red). When the two replication forks meet, DNA replication terminates, and the newly formed DNA molecules separate. (b) Following DNA replication, the parental cell divides and each new daughter cell receives one complete DNA molecule. (c) In eukaryotes there are multiple origins along the length of the linear DNA molecule. As the DNA strands separate at each origin, two replication forks are formed. The replication forks at each origin move in opposite directions, until the entire parental DNA molecule is replicated.

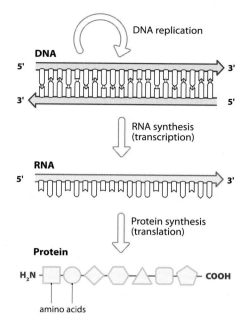

Figure 6.13 Gene expression. In addition to its ability to replicate, DNA encodes genetic information. The flow of genetic information from DNA to RNA (transcription) and from RNA to protein (translation) occurs in all living cells. This flow of information is so critical to our understanding of life that it is often referred to as the *central dogma* of genetics.

It is important to state exactly what it means when we say that a gene "codes" for a protein. Recall that although DNA contains information specifying protein structure, DNA itself does not link the required amino acids together. Rather, a gene's encrypted instructions must be "read" by other molecules and ultimately translated into a series of amino acids forming the specified protein. The conversion of a gene's nucleotide sequence into a protein is called **gene expression**.

We will now turn our attention to how DNA directs protein synthesis. Such synthesis occurs in two basic steps. The first step is called **transcription**. In this step, the DNA in a specific gene is used to guide the synthesis of a different nucleic acid called RNA (see Chapter 2, p. 41). The genetic information in the gene has now been encoded in the molecule of RNA. In the second step, **translation**, the DNA code that has been encoded or "transcribed" into the RNA is used to link the specified amino acids in the proper order to form a protein (**Figure 6.13**). Like DNA replication, protein synthesis is similar in all cells, but we will highlight a few of the important differences in the ways that bacteria, archaea, and eukaryotes transcribe and translate DNA's genetic blueprint.

In transcription, a gene's DNA is used to produce complementary RNA

In the first step of protein synthesis, transcription, the genetic information encoded in the nucleotides of a gene is transferred to RNA through complementary base pairing. Briefly, a specific enzyme, RNA polymerase, recognizes the beginning of a gene, and once it binds at this site, the enzyme unwinds the DNA sequence to be transcribed. It next assembles complementary RNA nucleotides into a molecule of *messenger RNA* (**Figure 6.14**). If a short series of nucleotides in the DNA is GCGCT, for example, the corresponding nucleotides strung together by RNA polymerase will be CGCGA. When RNA polymerase reaches the end of the gene, indicated by a special *termination site*, the RNA polymerase detaches and the newly synthesized messenger RNA is released.

CASE: CHRISTMAS TREES AND TRANSCRIPTION

If you take electron-microscopic images of DNA in cells that are actively producing protein, you will likely see structures that look very much like Christmas trees. There are central fibers corresponding to the "trunk" and attached "branches," which are short at the top and longer at the bottom of the "tree" (**Figure 6.15**). If such cells are treated with an enzyme that digests DNA but not RNA (a DNase), the trunks of the trees disappear, leaving only the branches. If similar cells are treated with an enzyme that digests RNA only (an RNase) the trunks remain but the branches disappear.
1. What is the nature of these "Christmas trees"?
2. What is the explanation for the results observed following treatment with DNase or RNase?

Like DNA, RNA is composed of a series of nucleotides (**Figure 6.16**). Unlike DNA, however, which is a double-stranded molecule, RNA is typically single-stranded, and in place of the deoxyribose sugar that is an essential part of DNA, a slightly different sugar, ribose, is found in RNA. Furthermore, RNA

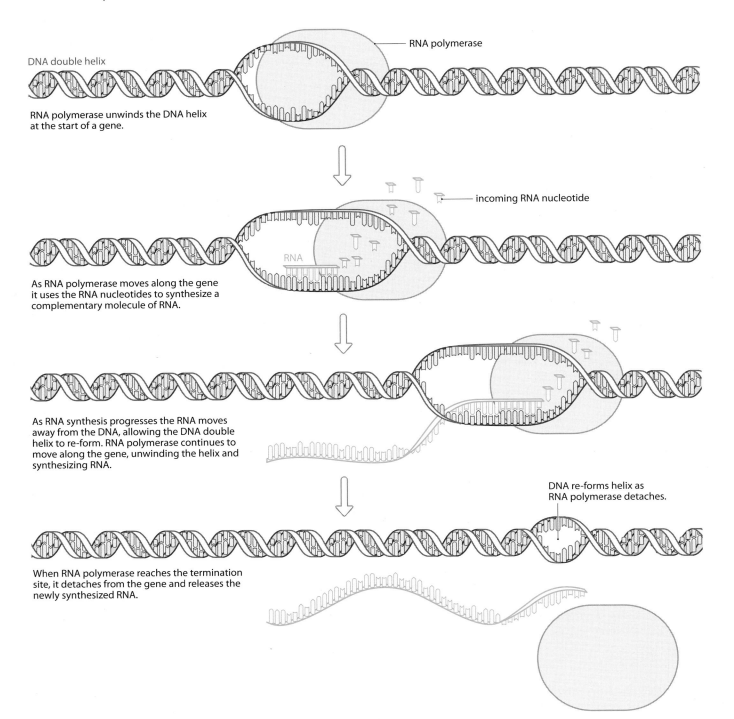

DNA double helix

RNA polymerase

RNA polymerase unwinds the DNA helix at the start of a gene.

incoming RNA nucleotide

RNA

As RNA polymerase moves along the gene it uses the RNA nucleotides to synthesize a complementary molecule of RNA.

As RNA synthesis progresses the RNA moves away from the DNA, allowing the DNA double helix to re-form. RNA polymerase continues to move along the gene, unwinding the helix and synthesizing RNA.

DNA re-forms helix as RNA polymerase detaches.

When RNA polymerase reaches the termination site, it detaches from the gene and releases the newly synthesized RNA.

does not contain the base thymine. In its place RNA uses a different single-ringed pyrimidine called uracil. The other three bases found in DNA (guanine, cytosine, and adenine) are also found in RNA.

During the first step in protein synthesis, transcription, the code in DNA guides the synthesis of a complementary sequence of RNA nucleotides. For example, if a short segment of DNA consists of the bases GGCATC, the complementary bases on the newly synthesized RNA will be CCGUAG. Note that adenine in the DNA is transcribed into uracil in the RNA instead of thymine. However, not all the DNA in an entire DNA molecule is transcribed at once. Genes are transcribed either individually or in small groups as needed, meaning that each RNA molecule is transcribed from only a small portion of the entire DNA molecule.

Figure 6.14 An overview of transcription. RNA polymerase recognizes the beginning of a gene, and once it binds at this site, it separates the DNA strands. RNA polymerase then moves along the gene, separating the DNA strands as it goes. Using the nucleotide sequence of one of the DNA strands as a template, RNA polymerase synthesizes a complementary molecule of RNA. When RNA polymerase reaches the termination site, it releases the RNA and detaches from the DNA, allowing the DNA to regain its original double helix structure.

(a)

Figure 6.15 Treelike appearance of genes being transcribed. (a) The "branches" (in blue) are RNA molecules being transcribed from the DNA (in red). Shorter branches indicate an earlier point in the transcription of each of the three genes illustrated here. Longer branches, representing longer RNA molecules, are nearing completion of transcription. (b) An actual electron-microscopic image of transcription "Christmas trees." Experiments similar to the one described in the case, conducted in 1970, offered powerful clues regarding RNA's role in gene expression.

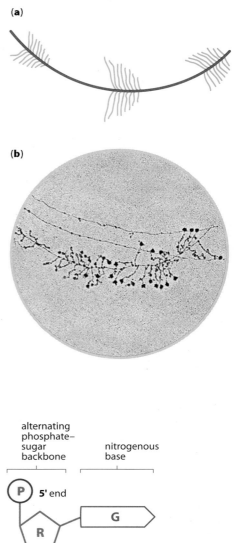

(b)

In bacteria, transcription of a particular gene begins when a specific short base sequence found near the beginning of a gene is recognized and bound by an enzyme called **RNA polymerase**. This base sequence, called a **promoter**, essentially acts as a "start" signal (**Figure 6.17a**).

A given promoter typically controls the transcription of only one or a small number of genes. Once RNA polymerase binds the promoter, it unwinds the DNA to expose the DNA nucleotide bases. Incoming RNA nucleotides that are complementary to the DNA are added, forming a growing RNA chain. In this way, the sequence of bases in the RNA is dictated by the base sequence in the DNA. Transcription continues until RNA polymerase reaches a series of DNA nucleotides called the **termination sequence**. At this point the RNA is released and RNA polymerase detaches from the DNA, allowing the DNA double helix to re-form. Because only a few genes at most are transcribed into one RNA molecule, RNA is far shorter than DNA.

For a particular gene, only one DNA strand (the **coding strand**) is transcribed (see Figure 6.17b). How does RNA polymerase distinguish between the coding and the opposite, noncoding strand? Recall that transcription begins at a specific promoter region. These promoters are found only on the coding strand. The noncoding strand, with a sequence complementary to the promoter, is not recognized by RNA polymerase.

The "Christmas trees" described in our case are actually genes "caught in the act" of transcription (see Figure 6.15). Each gene is being transcribed by many RNA polymerase enzymes at once. Some of the RNA polymerases are almost finished transcribing the gene and are approaching the termination sequence. Their attached RNA molecules represent the longer branches at the bottom of the tree. At the top of the tree, other RNA polymerases are just starting out at the beginning of the gene. Since they have not yet transcribed many nucleotides, their attached RNA molecules are shorter and represent the shorter branches. DNase digests the DNA trunk of our trees, leaving only the branches. RNase has the opposite effect; the trunk remains, stripped of its RNA branches.

Transcription is much the same for eukaryotes. One important difference, however, is the manner in which a gene's promoter is recognized. Recall that, in bacteria, RNA polymerase recognizes and binds to promoter regions to initiate transcription. In eukaryotes, promoters are first bound by a complex of proteins called **transcription factors**. RNA polymerase recognizes and binds to these proteins rather than directly to the promoter. Interestingly, archaea also use transcription factors and are consequently more like eukaryotes than they are like bacteria in this regard.

The genes of bacteria are also less complex than those of eukaryotes. Once a coding sequence begins, it continues uninterrupted until the termination sequence. Eukaryotic genes were originally assumed to adhere to this simple pattern, but in the late 1970s came the remarkable discovery that many eukaryotic genes were interrupted by intervening noncoding sequences called **introns** (**Figure 6.18**). The coding regions that are actually translated into protein products are known as **exons**. The number of introns found in

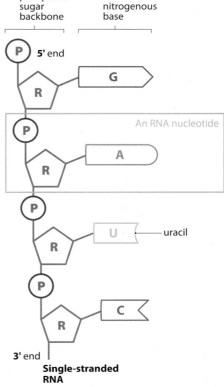

Figure 6.16 RNA structure differs slightly from that of DNA. RNA contains the sugar ribose and the base uracil in place of thymine. It is typically single-stranded.

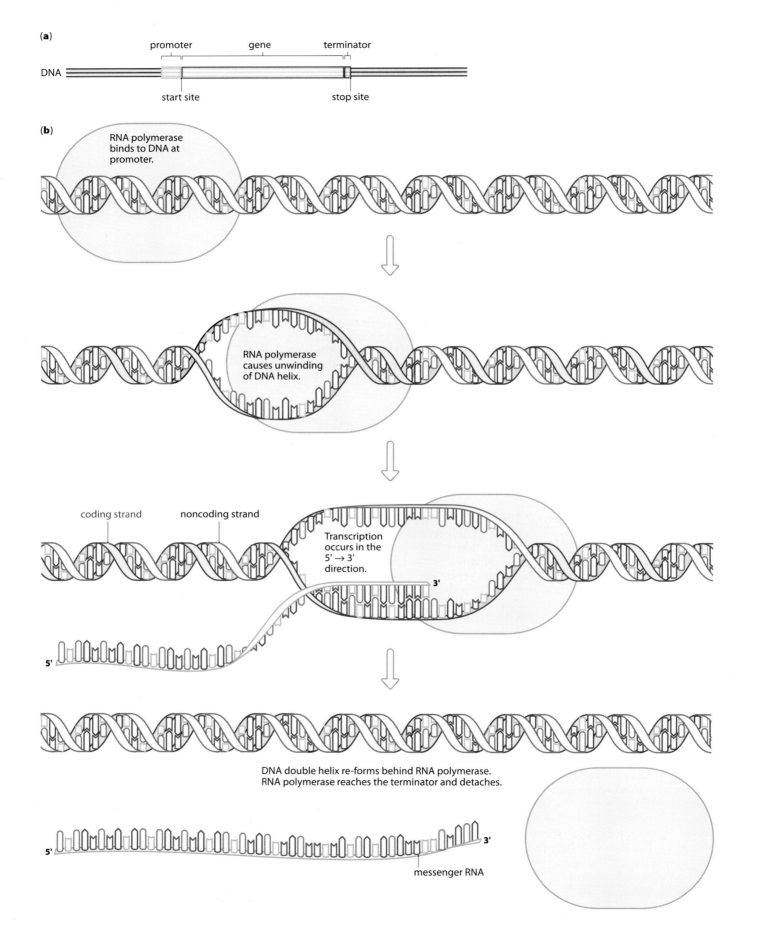

(a)

promoter   gene   terminator

DNA

start site   stop site

(b)

RNA polymerase binds to DNA at promoter.

RNA polymerase causes unwinding of DNA helix.

coding strand   noncoding strand

Transcription occurs in the 5' → 3' direction.

3'

5'

DNA double helix re-forms behind RNA polymerase.
RNA polymerase reaches the terminator and detaches.

5'   3'

messenger RNA

Figure 6.17 Transcription (*left*). (a) A length of DNA showing the respective locations of a promoter, a gene, and a termination site. (b) The process of transcription begins when RNA polymerase binds the promoter of a gene. RNA polymerase unwinds the DNA and, using the DNA as a template, moves along the DNA, adding the appropriate, complementary RNA nucleotides. As transcription continues, the DNA double helix re-forms behind the RNA polymerase. Eventually, RNA polymerase arrives at the termination site, indicating the end of the gene. The newly formed RNA is released and the RNA polymerase detaches. Only one of the two DNA strands (the coding strand) is transcribed by RNA polymerase.

a eukaryotic gene can range from zero to close to 100. Some genes in archaea are also thought to contain introns.

To help visualize this concept, consider the sentence "THE RAT SAW THE CAT." If this sentence were equivalent to a genetic message in bacterial DNA, the same message in a eukaryote might read "THE XQXY RAT SAW VRVNGT THE CAT." The nonsense letters represent two introns, interspersed between three exons. Both introns and exons are initially transcribed, but following eukaryotic transcription, introns are removed from mRNA in a process called **RNA splicing**. The *reason* eukaryotic genes contain introns is still open to question.

Amino acids are assembled into protein during translation

In the second step of protein synthesis, translation, the mRNA formed during transcription is used to specify the amino acids that are linked together to form a protein. Briefly, a structure called a ribosome binds at the end of the mRNA molecule (**Figure 6.19**). The ribosome moves along the mRNA, reading the nucleotides and translating the information in the nucleotides into a series of amino acids. Each set of three nucleotides specifies a particular amino acid. For example, if the first three nucleotides read by the ribosome contain the bases AUG, a particular amino acid, methionine, will be introduced at this point. If the next three nucleotide bases are CGU, a different amino acid, in this case arginine, will next be linked to the growing protein chain. This process continues as the ribosome progresses along the mRNA, until it reaches a three-base sequence at the end of the mRNA that does not specify any amino acid. At this point, the ribosome detaches from the mRNA, the newly completed protein is released, and protein synthesis is complete.

CASE: PROTEIN SYNTHESIS: DEAD IN ITS TRACKS

When she begins to experience vague symptoms including coughing, tightness in the chest, and fatigue, Leslie decides to start taking some amoxicillin that is left over from an old prescription. After a week, since

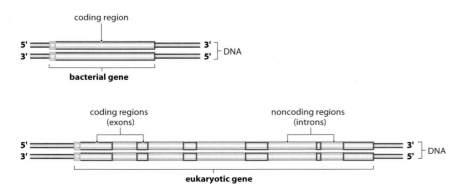

Figure 6.18 Bacterial and eukaryotic genes are organized differently. Bacterial genes consist of a single stretch of uninterrupted nucleotides that encodes the amino acid sequence of a protein. In contrast, many eukaryotic genes are interrupted by noncoding sequences called *introns*. The nucleotide sequences in eukaryotic DNA that are ultimately translated into a protein are called *exons*.

she has not improved, she visits the doctor who, following an examination, prescribes a different antibiotic—erythromycin. The doctor tells Leslie that he suspects that she has a case of "atypical" or "walking pneumonia," caused by the bacterium *Mycoplasma pneumoniae*. He also tells her that the amoxicillin that she took has no effect on this particular bacterial species and he uses the incident to discourage her from such self-treatment in the future. Within a few days of starting erythromycin treatment, Leslie feels much better, and by the end of the 10-day prescription, her symptoms have completely resolved.

1. How did the erythromycin cure Leslie?
2. Why didn't the amoxicillin work?

Cells produce not just one but a number of different RNA types, several of which are needed for the translation of RNA into protein. The RNA that encodes DNA's specific instructions for the production of a protein is called **messenger RNA (mRNA)**. As the name implies, mRNA carries DNA's message to the site in the cell where amino acids are actually assembled into protein.

The mRNA sequence of bases specifies the precise amino acids that make up a protein. These bases are recognized in groups of three, called **codons**, and each three-base codon specifies a particular amino acid. For example, UCU (transcribed from AGA in the DNA) codes for the amino acid serine (Ser), while GGA (transcribed from CCT) codes for glycine (Gly). **Figure 6.20** provides the complete genetic code. Note that of the 20 different amino acids, only tryptophan (Trp) and methionine (Met) can be specified by only a single codon. Other amino acids use as many as six different codons. Three

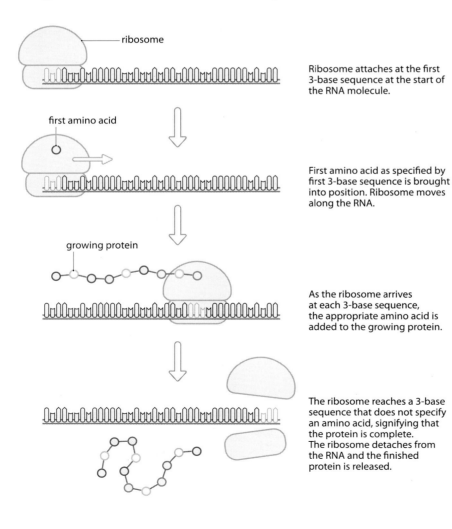

ribosome

Ribosome attaches at the first 3-base sequence at the start of the RNA molecule.

first amino acid

First amino acid as specified by first 3-base sequence is brought into position. Ribosome moves along the RNA.

growing protein

As the ribosome arrives at each 3-base sequence, the appropriate amino acid is added to the growing protein.

The ribosome reaches a 3-base sequence that does not specify an amino acid, signifying that the protein is complete. The ribosome detaches from the RNA and the finished protein is released.

Figure 6.19 An overview of translation. Translation begins when a ribosome binds to the beginning of the RNA nucleotide sequence. Each set of three nucleotide bases on the RNA specifies a particular amino acid. The ribosome moves along the RNA molecule reading the nucleotides, three at a time, and adding the amino acids in the order specified by RNA. Finally, the ribosome reaches a three-base sequence that does not specify an amino acid, indicating that the protein is complete. At this point, the ribosome disassociates from the RNA, and the now-completed protein is released.

Ala	Arg	Asp	Asn	Cys	Glu	Gln	Gly	His	Ile	Leu	Lys	Met	Phe	Pro	Ser	Thr	Trp	Tyr	Val	Stop
	AGA									UUA					AGC					
	AGG									UUG					AGU					
GCA	CGA						GGA			CUA				CCA	UCA	ACA				UAA
GCC	CGC						GGC		AUA	CUC				CCC	UCC	ACC			GUA	UAG
GCG	CGG	GAC	AAC	UGC	GAA	CAA	GGG	CAC	AUC	CUG	AAA		UUC	CCG	UCG	ACG		UAC	GUC	UGA
GCU	CGU	GAU	AAU	UGU	GAG	CAG	GGU	CAU	AUU	CUU	AAG	AUG	UUU	CCU	UCU	ACU	UGG	UAU	GUU	
Ala	Arg	Asp	Asn	Cys	Glu	Gln	Gly	His	Ile	Leu	Lys	Met	Phe	Pro	Ser	Thr	Trp	Tyr	Val	Stop
A	R	D	N	C	E	Q	G	H	I	L	K	M	F	P	S	T	W	Y	V	

of the possible codons do not specify any amino acid. Rather they function as "stop" codons, to indicate when a protein has been completed.

Amino acids are linked together into a growing protein on **ribosomes** (see Chapter 3, p. 59). Each ribosome is a complex assembly of many proteins along with a second type of ribonucleic acid, **ribosomal RNA (rRNA)**. A ribosome attaches to the end of an mRNA molecule to initiate translation. The ribosome then moves toward the other end of the mRNA, one codon at a time, reading the codons from one end to the other. At each codon, the appropriate amino acid is brought into position by a third type of ribonucleic acid, **transfer RNA (tRNA)** (**Figure 6.21**). Each tRNA has a three-base sequence called the **anticodon** that is complementary to a specific codon on the mRNA. When a ribosome arrives at a particular codon, a tRNA bearing the appropriate amino acid temporarily binds the codon via its anticodon. A peptide bond is then formed between the newly arriving amino acid and those amino acids already incorporated into the growing protein chain. The previously arriving tRNA is then released. In this way, codons are read sequentially along the mRNA until the ribosome arrives at a stop codon indicating that the protein is complete. At this point, the ribosome releases its protein and detaches from the mRNA (**Figure 6.22**).

The erythromycin that Leslie took to treat her infection acts by blocking translation of bacterial mRNA. Specifically, erythromycin binds to the bacterial ribosome and prevents it from moving along the mRNA molecule. Unable to translate mRNA, the bacteria fail to produce protein, and reproduction grinds to a halt. The bacteria are then generally eliminated by the immune system within a short time. Many other antibiotics interfere with bacterial translation in different ways, and such drugs might be used as alternatives to erythromycin to treat a case of atypical pneumonia such as Leslie's. But amoxicillin will not work because, as a penicillin-type drug, it acts by inhibiting the synthesis of new cell wall material (see Chapter 3, p. 54). As you may recall, *M. pneumoniae* is unusual because it lacks a cell wall. With no cell wall synthesis to inhibit, amoxicillin is useless against *M. pneumoniae*. We will examine antibiotics and when to use (or not use) them in Chapter 12.

Figure 6.20 The genetic code. The three-base sequences represent the 64 different codons, specifying the 20 different amino acids, each of which is given in both its three-letter and one-letter abbreviation. Most amino acids are represented by more than one codon, with some using up to six different codons. Three codons do not specify any amino acids. Rather, they function as "stop" codons, signaling the end of the protein-coding sequence.

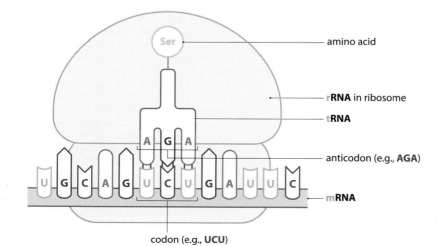

codon (e.g., **UCU**)

amino acid

rRNA in ribosome

tRNA

anticodon (e.g., **AGA**)

mRNA

Figure 6.21 Ribosomal RNA (rRNA), transfer RNA (tRNA), and messenger RNA (mRNA). Ribosomes, the site of amino acid assembly into proteins, contain both protein and rRNA. Transfer RNA (tRNA) carries the correct amino acid (serine in this example) into a growing protein chain, by binding temporarily to specific codons on the mRNA. Note the position of the anticodon, complementary to a specific codon on the mRNA.

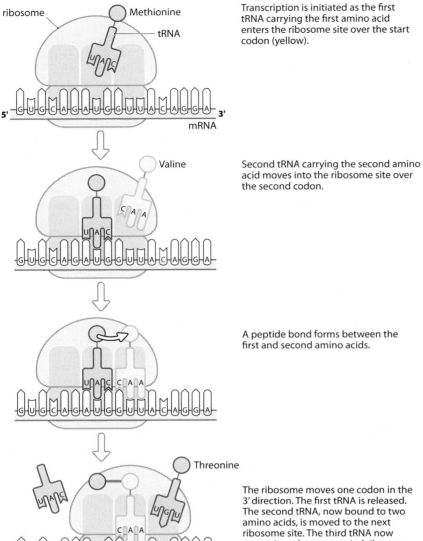

Transcription is initiated as the first tRNA carrying the first amino acid enters the ribosome site over the start codon (yellow).

Second tRNA carrying the second amino acid moves into the ribosome site over the second codon.

A peptide bond forms between the first and second amino acids.

The ribosome moves one codon in the 3' direction. The first tRNA is released. The second tRNA, now bound to two amino acids, is moved to the next ribosome site. The third tRNA now moves into the unoccupied ribosome site over the the third codon.

A second peptide bond is formed linking the first two amino acids to the third amino acid. All three amino acids are now bound to the third tRNA.

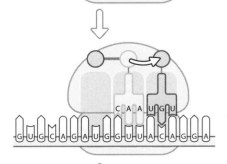

Ribosome moves to next codon, releasing the second tRNA. The next tRNA carrying the next amino acid now moves into the vacant site. This process will be continually repeated until a stop codon is reached. At that point the now completed protein will be released and ribosomal subunits will detach from the mRNA.

Figure 6.22 The process of translation. Translation begins when the ribosome binds to the 5' end of the mRNA. Note that the first codon (termed the "start codon", yellow) on the mRNA codes for the amino acid methionine (Met). The first codon is almost always AUG, coding for Met. The first tRNA, with an anticodon complementary to the first codon, brings the first amino acid (Met) into position. The second tRNA, bearing the amino acid specified by the second codon, likewise binds, and a peptide bond forms between the first two amino acids. The ribosome then moves, one codon at a time, and the correct amino acid is brought into the growing peptide chain by the appropriate tRNA. Ultimately, the ribosome reaches a stop codon, causing both the ribosome and the now-completed peptide to be released.

Like transcription, translation is very similar in bacteria, archaea, and eukaryotes, but we will mention two important differences. First, in eukaryotic cells, transcription occurs in the nucleus. Before translation can occur, a molecule of mRNA must pass through pores in the nuclear membrane to reach the ribosomes in the cytoplasm (see Chapter 3, p. 64). In bacteria and archaea, there is of course no nuclear membrane. This means that, in a prokaryotic cell, ribosomes can associate with the end of an mRNA before transcription is even completed. In other words, the two processes of transcription and translation, completely separate and discrete in eukaryotes, are overlapping in prokaryotes (**Figure 6.23**). This is part of the reason why proteins are produced more quickly in most prokaryotic cells, and it is another example of how the elegant simplicity of prokaryotes often gives them an edge in terms of efficiency.

Another difference with great practical significance involves the ribosome. Although they fulfill the same function, ribosomes in eukaryotes contain more protein and rRNA than bacterial ribosomes. Their mass is therefore correspondingly greater. This difference may *sound* trivial, but the relative safety of drugs such as the erythromycin used in our case is based on this fact. Although erythromycin can bind to and interfere with bacterial ribosomes, the antibiotic is unable to bind the larger, more massive ribosomes found in eukaryotes. Because human or other animal ribosomes are unaffected, these drugs can be used to control bacterial growth without unacceptable side effects. The ribosomes of archaea are more like those of eukaryotes than those of bacteria—another reminder of how different the two domains of prokaryotes are from each other.

Gene activity is often carefully regulated

Arriving home on a cold winter night, your first action might be to flip on some lights and turn up the thermostat on your heater. Later you might cook your dinner or watch TV. These uses of energy might be considered essential or at least enjoyable. Suppose, however, that the next morning you left home quickly, forgetting to turn off lights and leaving the coffee maker and perhaps the radio on. In this case you are clearly wasting energy and you will pay the price when the utility bills arrive.

Cells must make the same kinds of decisions regarding their energy budgets. At certain times, specific proteins are required and must be synthesized, in spite of the energy needed to make them. As conditions change and less of a particular protein is needed, cells need a way to repress protein production in order to conserve energy. Then, should the protein be needed later, cells must be able to switch protein synthesis back on. Those that cannot regulate protein synthesis in this way pay their own price in terms of wasted energy, inefficiency, and reduced growth. Genes that are turned on and off as conditions warrant are called **regulated genes**. Other, nonregulated genes are always being expressed. Such genes code for proteins that the cell needs all the time, and there are no situations in which it makes sense to shut them off. Nonregulated genes are called **constitutive genes**. Examples include genes coding for crucial structural proteins that are always needed in constant, large amounts.

We will now turn our attention to the manner in which the expression of regulated genes is controlled. In prokaryotes, many such genes are organized into gene clusters, called **operons**. Operons are groups of genes the products of which are all involved in the same overall process, even though each one codes for its own protein (**Figure 6.24**). An operon also contains control regions that regulate gene activity. Genes are either activated or repressed as a unit—they are either all "on" or all "off." The genes coding for lactose-digesting enzymes are organized in an operon called the *lac* operon. Several different proteins, each coded for by a different gene in the operon, are required if a bacterium is going to digest lactose. There is never a time

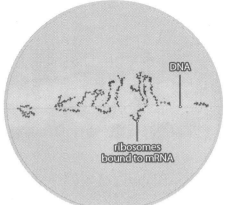

Figure 6.23 In prokaryotes, transcription and translation are overlapping. Because prokaryotes lack a nuclear membrane, translation of an mRNA message can begin even before transcription of that message is completed. This electron micrograph shows a gene being transcribed from left to right. The darkly stained bodies are ribosomes attached to mRNA. The mRNA becomes progressively longer as transcription occurs, allowing greater numbers of ribosomes to bind and initiate translation.

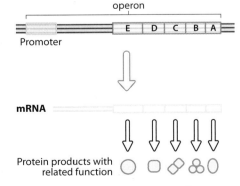

Figure 6.24 Operons are gene clusters whose expression is controlled as a unit. The letters A–E identify different genes in an operon, all of which code for proteins involved in a related process. They therefore all use a single promoter and are activated or inhibited as a unit.

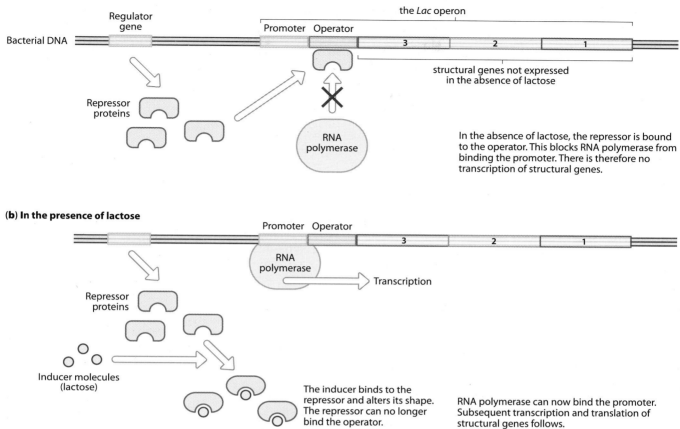

(a) In the absence of lactose

In the absence of lactose, the repressor is bound to the operator. This blocks RNA polymerase from binding the promoter. There is therefore no transcription of structural genes.

(b) In the presence of lactose

The inducer binds to the repressor and alters its shape. The repressor can no longer bind the operator.

RNA polymerase can now bind the promoter. Subsequent transcription and translation of structural genes follows.

Figure 6.25 The *lac* operon. To digest lactose, three different proteins are required. The genes coding for these proteins (the *structural genes*) are numbered 1–3. (a) When lactose is absent from the environment, the repressor is bound to the operator, preventing transcription of the structural genes by RNA polymerase. (b) In the presence of lactose, the repressor is removed from the operator by lactose itself, which functions as the inducer. This permits RNA polymerase to bind the promoter and begin transcription of the genes.

when a cell would want only *some* of these genes active. It is therefore efficient to regulate them as a group. Operons are common in bacteria but are not found in eukaryotes, where most genes are regulated individually.

Consider the conditions under which the genes of the *lac* operon should be expressed or repressed. When lactose is present in the environment, it is to a cell's advantage to digest it. Bacteria respond to the presence of lactose by activating the genes of the *lac* operon. In the absence of lactose, expression of these genes is energetically wasteful, and these same genes are repressed.

Figure 6.25 illustrates a portion of bacterial DNA showing the three protein-coding genes involved in lactose digestion. Adjacent to these genes are crucial control regions. The **promoter**, as described previously, is a DNA base sequence recognized by RNA polymerase. The **operator** is found between the promoter and the three protein-coding genes. Protein-coding genes are often called *structural genes*, to distinguish them from control regions of the operon. An additional regulator gene is found at some distance from the control regions. It codes for a large protein called the **repressor**, which is able to bind to the operator in the absence of lactose.

When lactose is absent from the environment, the repressor binds tightly to the operator (see Figure 6.25a). With the repressor in place, RNA polymerase is unable to bind the promoter. Consequently, no transcription of the genes in the operon can occur. Alternatively, if the environment changes

and lactose becomes available, some lactose enters the cell. The repressor protein has a binding site on its surface for lactose. The binding of lactose causes the shape of the repressor to change slightly. In its new shape, the repressor cannot bind the operator (see Figure 6.25b), and this allows RNA polymerase to bind the promoter. Transcription of the genes ensues, and following translation, lactose digestion can begin. Because it is the presence of lactose itself that activates the *lac* operon, lactose is termed the **inducer**.

Note that this mechanism allows the cells to produce lactose-digesting enzymes only under the proper environmental conditions, minimizing energy waste. The *lac* operon will be repeatedly induced or inactivated in response to changing amounts of lactose in the environment. Many other operons function in a similar manner. Operons such as these are termed **inducible**, because the substance to be digested actually activates or induces transcription.

Other operons utilize a slightly different type of regulation called **repression**. In this case, the operon is turned *off* when a crucial substance is available in the environment, as opposed to the inducible *lac* operon, where we saw that the operon was turned *on* in the presence of lactose. A good example is provided by those bacteria that are able to make their own tryptophan (Trp), a necessary amino acid. Those bacteria with an operon called the *trp* operon can synthesize their own tryptophan. However, when a particular bacterial cell has sufficient tryptophan, it is in the cell's interest to suspend tryptophan synthesis until levels of the amino acid again fall to low levels. Consequently, abundant tryptophan acts not to induce but to inhibit the operon.

Repressible operons such as the *trp* operon have similar structure and work in a similar way to the *lac* operon, with one big difference: RNA polymerase can bind the promoter only when tryptophan is *not* present and the cell needs to make more of the amino acid. Unlike lactose, which *removes* the repressor protein from the operator, tryptophan, when it is present, binds the repressor and *allows* the repressor to bind. Once available tryptophan is used up and there is none left to bind the repressor, the repressor leaves the operator. RNA polymerase can once again transcribe the genes in the operon, resulting in increased production of the needed amino acid.

Like prokaryotes, eukaryotic organisms also regulate gene expression. Eukaryotes, however, utilize a large number of mechanisms to activate or inhibit their genes, many of which are still poorly understood. One important mechanism depends upon how tightly the DNA on a chromosome is wound around its scaffolding proteins. Inactive genes tend to coil more tightly around these proteins, making the DNA much denser at these points. Genes that are actively being expressed are wound in a looser configuration, and as inactive genes become active, the DNA making up these genes becomes less dense. Another exciting, recent discovery is that RNA, long thought to function only in transcription and translation, has been found to play an important role in gene regulation. Small RNA molecules called *microRNAs* now seem to be crucial in determining when eukaryotic genes are expressed. This is certainly one of the most important biological discoveries in recent years and is certain to be a major research focus for the foreseeable future.

The environment influences the nature of genetically determined characteristics

The specific genetic makeup of an individual for a particular trait is termed that individual's **genotype** for the trait. The expression of that genotype, which results in the outward appearance of the individual, is called the individual's **phenotype**.

To illustrate the difference between genotype and phenotype, consider one genetic trait in humans—freckles. Imagine two people with the same genotype. In other words they have the exact same genetic information for this trait. Even these two individuals, however, may end up somewhat different—one person may have more numerous and/or more prominent freckles. If so, their phenotype is different. Various factors might explain this difference. Exposure to sunlight, for one thing, can cause more pronounced freckling, demonstrating the impact that environmental factors can have on genetically based characteristics. This example highlights an important concept in genetics: genes for a characteristic (the genotype) strongly influence, but do not wholly shape, the final observed outcome (the phenotype). An individual's *potential* for a certain characteristic is determined by genotype. The actual outcome of that potential, the phenotype, is influenced by a variety of other, often environmental factors. This important principle applies to all living things, from amoebas to zebras.

Sources of Genetic Variation

In May 2007 came the troubling news that a man infected with an untreatable form of tuberculosis boarded a jet in Europe and flew to North America, possibly exposing other passengers to this dangerous strain of *Mycobacterium tuberculosis*. Fortunately, none of the other people on this flight became infected, but amid all the anxiety and news coverage, a fundamental question remained. Some strains of *M. tuberculosis* are easily treatable, while others are not. Why, if they are all members of the same species, do they differ in this way?

We have already learned part of the answer to this question. All organisms, even those of the same species, vary in terms of their genetic material. When it comes to genetics, variety is indeed the spice of life. In this section we will find out where that variety comes from.

Both asexual and sexual organisms are able to generate new genetic combinations

Genetic information is passed from parents to offspring when an organism reproduces. Reproduction in bacteria is a relatively simple affair; a single-celled bacterial cell divides into two new cells. The original, reproducing cell is called the **parent cell**. The newly formed progeny cells are called **daughter cells**.

Each daughter cell will contain an exact copy of the parent cell's DNA. The DNA of a parent cell is copied prior to actual cell division and each daughter cell receives a single copy as the cell divides (**Figure 6.26**). This type of reproduction, in which a single parent cell gives rise to two identical daughter cells, each receiving the same genetic information, is called **asexual reproduction**.

Reproduction in many multicellular organisms, including many plants and most animals, is a more intricate process. This **sexual reproduction** involves the mixing of DNA from two parent organisms to create offspring with a novel combination of genetic information. Unlike the simple division of a parent cell in asexual reproduction, sexual reproduction involves the union of specialized sex cells called **gametes**—one female and one male. In many familiar organisms, the female and male gametes are called the **egg** and the **sperm**, respectively. In most cases, each gamete contains exactly half of the parent organism's DNA. When an egg and a sperm combine, the resulting offspring obtains a full complement of DNA, half from each parent. Because the offspring has genetic information from both the male and female parents, it is genetically distinct from either parent. This ability to generate new combinations of genetic material is called genetic **recombination**.

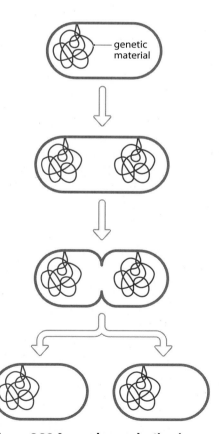

Figure 6.26 Asexual reproduction in bacteria. Following replication of the genetic material into two copies, the bacterial cell divides, forming two new individual bacteria.

genetic material

Figure 6.27 Potential for genetic variability in sexually reproducing organisms. In this example, only three genes are considered. Each parent has two copies of each gene. **A** and a, **B** and b, and **C** and c represent different alleles for each of these three genes. Each parent donates only one of each gene pair to each gamete. Consequently, each parent can form eight possible different gametes. Any female gamete can potentially be fertilized by any male gamete, resulting in 64 combinations. Suppose the egg (highlighted in red) containing **A**, b, and **C** is fertilized by the sperm (highlighted in blue) containing **A**, **B**, and c. The result is a fertilized egg with the genotype **AA**, b**B**, and **C**c. In typical sexually reproducing organisms with thousands of genes, there are almost limitless new gene combinations.

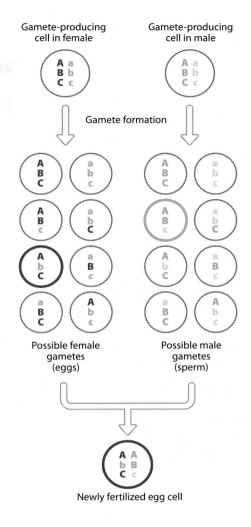

We have learned that the gene is the basic unit of heredity. And even though all members of a given species have the *same* genes, they do not necessarily have the same *forms* of those genes. For example, all humans have genes that code for eye color, but the specific form of those genes determines whether the eyes will be brown or blue. Such alternative forms of the same genes are called **alleles** of that gene. In most cases, a eukaryotic cell carries two copies of each gene—one obtained from the female parent and one from the male. The two gene copies may or may not be the same allele. Prokaryotes, on the other hand usually have only a single copy of each gene.

Sexual reproduction provides a mechanism to greatly increase the amount of genetic variability seen in a population. During gamete formation, each egg or sperm receives only one copy of each gene, which explains why each gamete ends up with only 50% of the genetic material. Exactly *which* of the two copies a specific gamete receives is, however, largely a random process. That means that different gametes end up with different combinations of the parents' alleles (**Figure 6.27**). This is somewhat analogous to a shuffling of the alleles with the formation of each gamete, and this process increases genetic variation enormously. During sexual reproduction, a male gamete unites with a female gamete, in both of which the parent's alleles were scrambled. The result is that no two offspring are ever likely to be identical. This random assortment of alleles during gamete formation accounts for much of the enormous genetic variability observed in sexually reproducing organisms.

While sexual reproduction has an obvious, built-in mechanism for generating genetic variation, this is not the case for asexual reproduction. In Figure 6.26 we saw that asexual reproduction should normally result in offspring with genetic material identical to that of the parent cell. Yet prokaryotes also display considerable genetic variability. As we will now see, compared with most eukaryotes, bacteria and archaea have an impressive number of tricks up their sleeves to recombine their genes in new and interesting ways.

CASE: GRIFFITH'S TRANSFORMING FACTOR

In the 1920s the English microbiologist Frederick Griffith was studying *Streptococcus pneumoniae*, a bacterial species that sometimes causes pneumonia. Griffith knew that only those *S. pneumoniae* strains with a polysaccharide capsule exterior to their cell wall could cause disease. Strains lacking capsules were harmless. Griffith wanted to understand the role of the capsule in the disease process. In the course of his investigations he performed the following experiment (**Figure 6.28**). He first inoculated mice with bacteria lacking the capsule. As he expected, these mice remained disease-free. When he inoculated mice with encapsulated bacteria, the mice sickened and died. Griffith next killed encapsulated bacteria by exposing them to heat. When mice were exposed to these dead cells, the animals remained healthy. None of these results were particularly surprising. But when Griffith exposed mice to a mixture of killed encapsulated bacteria and living *S. pneumoniae* lacking capsules,

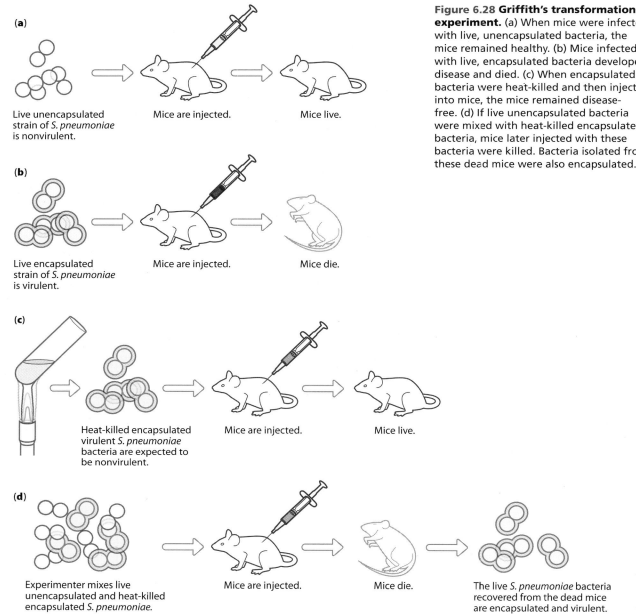

(a)

Live unencapsulated strain of *S. pneumoniae* is nonvirulent.

Mice are injected.

Mice live.

(b)

Live encapsulated strain of *S. pneumoniae* is virulent.

Mice are injected.

Mice die.

(c)

Heat-killed encapsulated virulent *S. pneumoniae* bacteria are expected to be nonvirulent.

Mice are injected.

Mice live.

(d)

Experimenter mixes live unencapsulated and heat-killed encapsulated *S. pneumoniae*.

Mice are injected.

Mice die.

The live *S. pneumoniae* bacteria recovered from the dead mice are encapsulated and virulent.

Figure 6.28 Griffith's transformation experiment. (a) When mice were infected with live, unencapsulated bacteria, the mice remained healthy. (b) Mice infected with live, encapsulated bacteria developed disease and died. (c) When encapsulated bacteria were heat-killed and then injected into mice, the mice remained disease-free. (d) If live unencapsulated bacteria were mixed with heat-killed encapsulated bacteria, mice later injected with these bacteria were killed. Bacteria isolated from these dead mice were also encapsulated.

he was astounded when the mice developed pneumonia and died. When he cultured bacteria from these dead mice, he was further amazed to find live, encapsulated bacteria. Griffith hypothesized that a "transforming factor" of some sort had been released by the dead encapsulated bacteria and that this factor had been absorbed by the living, previously harmless bacteria. The transfer of this factor permitted the living cells to now produce the capsule and cause disease.

1. What was Griffith's transforming factor?
2. How did this factor pass from dead to living bacteria, and once it had passed, how did it permit the unencapsulated bacteria to start forming capsules?
3. How common is this phenomenon? How is it similar to or different from other mechanisms of genetic exchange in prokaryotes?

During Griffith's lifetime, the nature of the genetic material remained elusive. It was already established that DNA was present in cells, but its function was unknown. Indeed, most researchers at the time believed that genes were composed of protein rather than nucleic acid. It was not until the 1940s that Griffith's "transforming factor" was identified as DNA and the central importance of DNA as the genetic material became firmly established.

Although Griffith was unable to fully understand his observations, we now know that DNA from dead cells had been transferred to living cells, resulting in the phenotypic change. This process, now called **transformation**, is just one of the mechanisms by which bacteria swap genetic information.

Transformation. Transformation occurs when DNA released by dead bacteria is taken up by a living bacterium and incorporated into its chromosome. When genes transferred in this manner are different from those found in the recipient bacterium, the genotype of the recipient is altered or *transformed* (**Figure 6.29**). In the case of Griffith's experiment, some of the living unencapsulated cells absorbed genes that code for capsule production from the heat-killed encapsulated bacteria. After incorporating these genes, not only could the living cells produce capsules themselves, but when they divided, their progeny were also encapsulated, disease-causing bacteria. This explains why Griffith was able to culture encapsulated cells from his dead mice, even though these mice had not been exposed to living encapsulated organisms.

Streptococcus pneumoniae can undergo transformation, in part, because it has receptors on its surface that can bind DNA fragments. Many other groups of bacteria are thought to be incapable of transformation. Some species such as *E. coli* that do not normally undergo transformation can be induced to do so in the laboratory. Such artificial transformation is an important technique used in genetic engineering. We will return to this topic in Chapter 14.

Transduction. In Chapter 4 we learned that even bacteria can be infected with viruses. These bacterial viruses, called bacteriophages, normally replicate their own genetic information within the bacterial cell, frequently killing their bacterial host in the process. When bacteriophages are actively replicating, the bacteriophage DNA codes for an enzyme that cleaves the bacterial chromosome into many small fragments.

Occasionally, during assembly of new viral particles, a small amount of bacterial DNA is incorporated into the phage head with the viral nucleic acid. The resulting virion will be unable to replicate. If it is released, however, it may still be able to inject its DNA, including the mistakenly acquired bacterial DNA, into a new bacterial cell. Occasionally, genes transported in this manner become incorporated into the chromosome of the recipient bacterial cell.

In this process, known as **transduction**, the bacteriophage has essentially acted as a device to shuttle bacterial genes between donor and recipient cells (**Figure 6.30**). Like transformation, transduction may alter the genotype and phenotype of the recipient bacterium. Transduction is a common form of genetic exchange between bacteria, and it occurs in many species.

Conjugation. In addition to the main bacterial chromosome, some cells have an additional circular loop of DNA called a **plasmid**. Like the chromosome, the plasmid carries genes that code for specific proteins, although the number of genes is small. A typical plasmid may even code for fewer than a dozen proteins, while a bacterial chromosome typically codes for a few thousand. Plasmids are usually able to replicate independently of the main chromosome, and cells may have more than one of these structures.

Many bacteria with plasmids can transfer these plasmids to other cells. This process of genetic exchange is called **conjugation**. The best-studied example of such conjugation involves the plasmid found in *E. coli*. Cells that carry

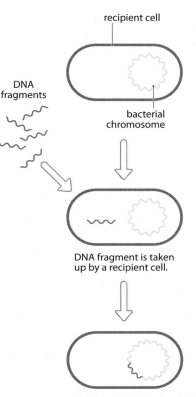

DNA fragment is taken up by a recipient cell.

The fragment is incorporated into the cell's chromosome. The recipient cell is transformed.

Figure 6.29 Transformation in bacteria. Donor DNA fragments may be taken up by certain bacterial species. If the donor DNA is incorporated into the chromosome of the recipient cell, the recipient becomes transformed.

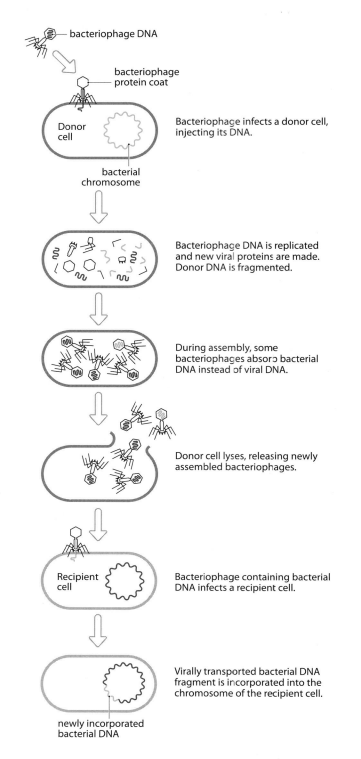

bacteriophage DNA

bacteriophage protein coat

Donor cell

bacterial chromosome

Bacteriophage infects a donor cell, injecting its DNA.

Bacteriophage DNA is replicated and new viral proteins are made. Donor DNA is fragmented.

During assembly, some bacteriophages absorb bacterial DNA instead of viral DNA.

Donor cell lyses, releasing newly assembled bacteriophages.

Recipient cell

Bacteriophage containing bacterial DNA infects a recipient cell.

Virally transported bacterial DNA fragment is incorporated into the chromosome of the recipient cell.

newly incorporated bacterial DNA

Figure 6.30 Transduction in bacteria. A bacterial virus or bacteriophage mistakenly incorporates a fragment of bacterial DNA into its capsid during assembly. If the bacteriophage is released and infects a second bacterial cell, the bacterial DNA it contains is transferred to the recipient cell, where it may be incorporated into the recipient cell's chromosome.

the plasmid are termed **F+ cells** (fertility-positive) while those lacking it are called **F− cells**. A gene on the plasmid allows F+ cells to produce a structure called the **sex pilus** (**Figure 6.31a**). The sex pilus, constructed of protein, physically connects the F+ and F− cells and pulls them together. This contact induces the F+ cell to replicate its plasmid. The two strands of plasmid DNA separate, and one of the resulting single strands is transferred to the F− cell. The single strand of plasmid DNA in both cells now give rise to their respective complementary strands. Essentially, the F− cell is now converted into an F+ cell. Such a cell has been genetically altered and is now capable of producing any new products coded for by genes on the plasmid.

In conjugation, unlike transformation and transduction, the donor and recipient cells must come into actual physical contact. Furthermore, conjugating cells must ordinarily be of opposite types, with only F⁺ cells serving as donors and only F⁻ cells acting as recipients.

Genetic recombination in prokaryotes has great significance for humans

It might seem as if microbial gene exchange has little practical importance for humans. However, the consequences of such seemingly trivial events can be monumental. To illustrate, consider a modern hospital where large amounts of antibiotics are being used. Within the hospital environment, there are often many strains of bacteria that develop resistance to one or more antibiotic. Without gene transfer, if a particular bacterial strain became drug-resistant, such resistance would remain confined to that strain. Because of mechanisms such as those discussed in this section, however, resistance genes can be exchanged between different strains and even different species of bacteria. As a result of such wholesale genetic transfer, strains of bacteria resistant to all antibiotics can arise. It is an unfortunate fact that people often develop serious illness and sometimes die from infections that were previously easy to treat. Drug resistance (to be explored in more detail in Chapter 12) is not the only dangerous characteristic that bacteria can swap by gene transfer. For example, in 2006, *E. coli* bacteria grabbed headlines when many people became ill after consuming *E. coli*-contaminated spinach. Symptoms of infection included bloody diarrhea and kidney failure. Several children who contracted the illness died as a result. It is currently thought that normally benign *E. coli* bacteria changed into the infamous "killer *E. coli*" (*E. coli* O157:H7) when they obtained pathogenic genes by transduction.

Conjugation, in particular, may be an especially effective way for bacteria to exchange dangerous genes. The genes most likely to be transferred are often those that are most worrisome in terms of public health and disease prevention. To understand why, consider that plasmid genes, while perhaps useful, are generally not essential to survival. Clearly this must be true, or cells such as the F⁻ *E. coli* previously described would be unable to survive. Examples of such nonessential genes often include genes for **virulence factors**—characteristics that make bacteria more virulent, or likely to cause disease.

Plasmids may also carry genes for antibiotic resistance, and such genes can be transferred via conjugation. Obviously drug resistance easily qualifies as an important virulence factor. And it is no coincidence that plasmids carrying resistance genes are most commonly found in hospitals and other places where antibiotic use is most widespread.

Not all microbial gene exchange is cause for alarm. Soil bacteria able to degrade environmental pollutants have transferred this capability to other bacteria, which are then also able to break down dangerous chemical

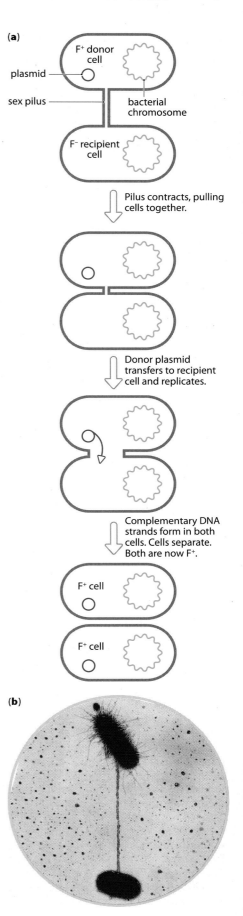

(a)

F⁺ donor cell

plasmid

sex pilus

bacterial chromosome

F⁻ recipient cell

Pilus contracts, pulling cells together.

Donor plasmid transfers to recipient cell and replicates.

Complementary DNA strands form in both cells. Cells separate. Both are now F⁺.

F⁺ cell

F⁺ cell

(b)

Figure 6.31 Bacterial conjugation. (a) The process of conjugation as it might occur between an F⁺ donor *E. coli* cell and an F⁻ recipient cell. Following conjugation, the F⁻ cell is converted to an F⁺ cell and may now serve as a donor. (b) Notice that the upper cell (the F⁺ cell) has numerous fimbriae. If the ability to produce these structures is encoded by a plasmid gene, the F⁻ cell below may obtain it and consequently undergo a phenotypic change. Because fimbriae help a cell to adhere to surfaces inside a host, cells with fimbriae are sometimes more virulent than those without fimbriae. Thus, in this photograph, the bottom cell may be in the process of acquiring a new virulence factor.

contaminants. These newly created bacteria may eventually be used to degrade pollutants in a process called bioremediation. We will examine the topic of bioremediation in Chapter 16.

We have now seen that genes can exist in different forms, known as alleles, and that both prokaryotes and eukaryotes have various mechanisms to mix and match these alleles, generating novel genetic combinations. Yet one big question regarding genetic variation remains—how do various forms of a gene arise in the first place? In the next section we will provide this important missing piece to the puzzle of genetic variability.

Mutations are the original source of genetic variation

CASE: END OF THE LINE FOR A LAST-LINE DEFENSE?

Vancomycin has been a powerful antibiotic against Gram-positive bacterial infections since the 1950s. It has been especially valuable against those bacteria that are resistant to other commonly used antibiotics. For some bacterial strains, such as certain strains of *Staphylococcus aureus* that are resistant to a variety of drugs, vancomycin is the only drug that is still effective. The antibiotic works by disrupting assembly of the bacterial cell wall.

Since the 1980s, however, many strains of bacteria have acquired resistance to vancomycin as well. In these bacteria there is a mutation that changes the composition of the cell wall. Vancomycin cannot attach to the modified cell wall in the mutant strain. Many patients infected with vancomycin-resistant bacteria are essentially untreatable, because these bacteria are often resistant to other antibiotics as well.

1. **What exactly is a mutation?**
2. **What type of mutation would cause the vancomycin resistance seen in the case?**
3. **How and why do mutations occur?**

Nothing in this world *always* goes right. In our discussion of DNA replication and gene expression, we described things as they are *supposed* to happen. Although the mechanisms we have discussed are remarkably faithful, they do not occur perfectly every time. Damage to or mistakes in DNA base sequences can occur, and such changes in the genetic material are called **mutations**. Most mistakes are immediately repaired by various mechanisms and therefore do not persist. We will discuss some of those repair mechanisms later in this section.

Many mutations have no effect on an organism's phenotype, and cells with such mutations look and behave normally. Other mutations may alter a cell's appearance or ability to function. If a mutation is severe enough it may be lethal, and in rare situations, as in our case, a mutation can even provide a new advantage. In other words, the consequences of a change in the base sequence of a gene can span the gamut from deadly to neutral to beneficial. Of course if a mutation kills the cell there will be no transmission to progeny cells. Other nonlethal mutations may become a permanent part of the cell's genetic material and will thereafter be transmitted to subsequent generations. In this section we will consider some of the ways that DNA can be altered and why it is that the effects of mutations on phenotype can vary so greatly.

As we have stated, many mutations have no effect whatsoever on phenotype. Although the base sequence of a particular gene is changed, the protein coded for by this gene remains the same. To understand how this is possible, recall that a gene's DNA code is transcribed into mRNA, which is translated one codon at a time into protein. Imagine that a particular codon, transcribed from normal, nonmutated DNA *was* UU**U**. Further imagine

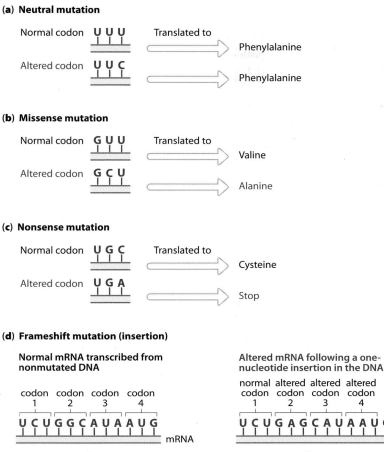

(a) Neutral mutation

Normal codon   U U U   Translated to ⟹ Phenylalanine

Altered codon   U U C   ⟹ Phenylalanine

(b) Missense mutation

Normal codon   G U U   Translated to ⟹ Valine

Altered codon   G C U   ⟹ Alanine

(c) Nonsense mutation

Normal codon   U G C   Translated to ⟹ Cysteine

Altered codon   U G A   ⟹ Stop

(d) Frameshift mutation (insertion)

Normal mRNA transcribed from
nonmutated DNA

Altered mRNA following a one-
nucleotide insertion in the DNA

normal altered altered altered
codon   codon   codon   codon
  1   2   3   4

codon codon codon codon
  1   2   3   4

U C U G G C A U A A U G   mRNA

U C U G A G C A U A A U G

Translation

Translation

Glycine   Methionine

Serine   Isoleucine

Glutamic acid   Asparagine

Serine   Histidine

Figure 6.32 **Types of mutations.** (a) In a neutral mutation, the amino acid sequence in a protein is unchanged. (b) A missense mutation alters the amino acid sequence of a protein. Such mutations, depending on how they affect protein function, may range from lethal to harmful to beneficial. (c) Nonsense mutations, in which a readable codon is converted into a stop codon, are almost always highly detrimental. Neutral, missense, and nonsense mutations are caused when DNA polymerase makes an error during DNA replication, inserting an incorrect nucleotide on the daughter strand. (d) An example of a frameshift mutation, in this case, an insertion.

that, due to a mutation in the DNA, the codon that results is *now* UU**C** (**Figure 6.32a**). This would be caused if an adenine, at the appropriate spot in the DNA, were erroneously replaced by a guanine.

In the genetic code (see Figure 6.20), you will see that both the UUC and UUU codons code for the same amino acid, phenylalanine. Consequently the resulting protein will be unchanged. Clearly, because the specified protein is unaltered, such a mutation has no effect, either detrimental or beneficial, on phenotype. Mutations of this sort are often called **neutral mutations**.

Other mutations can result in a change in a protein's amino acid sequence, and such a change can be harmful or beneficial. For example, returning to Figure 6.32b, consider a mutation that changes a normal codon reading G**U**U into G**C**U. While GUU codes for valine (see Figure 6.20), GCU codes for alanine. Consequently, as a result of this mutation, the translated protein will have a different amino acid sequence. If the mutated form of the protein works less efficiently than the nonmutated form, the mutation is likely to be harmful. If the mutated protein is completely unable to carry out its specific function, the mutation may be lethal, especially if the protein carries out an essential task. Rarely, the mutated protein actually works *more* efficiently or is able to do something new that the normal protein cannot. If this happens, we may observe a rare, beneficial mutation. Mutations such as these, which change the amino acid sequence of a protein, whether harmful or beneficial, are often called **missense mutations**, because they change the *sense* of the DNA.

A particularly harmful mutation can occur when a readable codon is converted into a stop codon. In the example provided in Figure 6.32c, note that the codon UGC normally codes for cysteine. If, due to a mutation in the DNA, the codon is now UGA, the normal codon has been converted into a stop codon. During translation, protein synthesis will be prematurely terminated at this point, resulting in an incomplete protein. Clearly such mutations will almost always have a severe negative impact on phenotype. A mutation such as this, in which a readable codon is converted into a stop codon, is called a **nonsense mutation**.

Harmful mutations also occur when an extra base is either mistakenly inserted into or deleted from newly synthesized daughter DNA by DNA polymerase during DNA replication. Such a mutation is called a **frameshift mutation** (see Figure 6.32d). Frameshift mutations generally have a very severe impact on the resulting protein, because many amino acids are changed in the protein chain. If a base is inserted or deleted, all codons after the mutation are altered. In other words, during translation, the codons are "shifted" by one nucleotide, altering all the remaining amino acids in the protein.

The mutation in our case that resulted in resistance to vancomycin is almost certainly a missense mutation, because only missense mutations can alter the amino acid sequence in this manner. This particular mutation is an example of a rare beneficial mutation. In a patient treated with vancomycin, such mutant bacterial cells have an enormous advantage over sensitive cells. Because they continue to replicate while sensitive cells are being destroyed, resistant cells quickly come to predominate.

Mutations are caused by many factors

You may be wondering what causes mutations to occur in the first place. Remember that enzymes such as DNA polymerase do not work perfectly every time. When DNA is replicated, there is a chance that a mistake will occur and a mutation may result. These random errors are called **spontaneous mutations** and they tend to be rare. A given gene might mutate only once for every several million times it is replicated, although certain genes tend to be more or less prone to mutation.

Exposure to certain chemicals and some forms of radiation can increase the frequency of mutations significantly, often up to 1000 times. Such agents are called **mutagens**.

Chemical mutagens can alter DNA in various ways, increasing the likelihood of replication errors. Nitrous acid (HNO_2), for instance, removes amino groups ($-NH_2$) from nitrogenous bases. This may cause the base to subsequently base-pair incorrectly during DNA replication. For example, if adenine loses its amino group, it is likely to base-pair with cytosine rather than thymine (**Figure 6.33**).

Figure 6.33 Nitrous acid, an example of a chemical mutagen. Exposure to nitrous acid (HNO_2) may cause deamination of nitrogenous bases. In this example adenine has been deaminated. It subsequently base-pairs with cytosine rather than thymine. When this occurs during DNA replication, it can lead to a neutral, missense, or nonsense mutation.

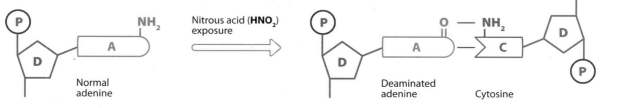

Deaminated adenine pairs with cytosine instead of thymine.

Figure 6.34 Formation of a thymine dimer. UV light can break the hydrogen bonds between thymine and adenine. Two adjacent thymine bases, disrupted in this way, may then covalently bond to each other, forming a dimer.

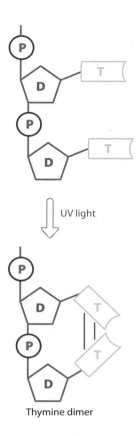

Another group of chemical mutagens are the **intercalating agents**. These three-ringed molecules are about the same size as a pair of DNA nucleotides. They cause problems by inserting themselves at a replication fork during DNA synthesis. As the nucleotides are pushed apart, an extra base may be added to or left out of the new DNA strand, resulting in a frameshift mutation. The mold *Aspergillus flavus* produces aflatoxin, which can induce frameshifts in this manner. Because *A. flavus* often contaminates peanuts, this food is sometimes associated with aflatoxin-induced mutations.

Radiation can also damage DNA and thus act as a mutagen. Gamma radiation and X rays are powerful forms of radiation that can break bonds in molecules. The remaining atoms or molecules often form **free radicals**. These are highly reactive compounds that latch onto and damage other molecules including DNA. Such damage can cause errors in DNA replication or even cause breaks between a nucleotide sugar and its base.

Ultraviolet (UV) radiation can break the hydrogen bonds between thymine and adenine nucleotides. If two such thymines are adjacent to each other on the same DNA strand, they may then covalently bond with each other. Two thymines covalently linked together in this way are called a **thymine dimer** (**Figure 6.34**). The dimer distorts the shape of the DNA molecule, preventing further DNA replication. Ultraviolet radiation can damage the DNA of human skin cells, and it is for this reason that we are often warned about the danger of overexposure to sunlight. Unlike gamma radiation and X rays, it cannot penetrate below the skin's surface and consequently poses no risk to deeper tissues. Most microorganisms, however, are easily penetrated by UV radiation, making UV light an effective means of control. Ultraviolet lighting is sometimes installed in places such as hospital operating rooms where contaminating microorganisms can be a serious problem. In other situations, such as prisons, where unnatural crowding provides increased opportunities of disease transmission, UV lights can help to minimize such risks. To prevent human DNA damage, however, such lights are turned on only when people are not actually in the area to be decontaminated.

The first bacterium to develop resistance to vancomycin may have been briefly exposed to a mutagen similar to those described. Alternatively, this now-major medical problem may have started out as a random spontaneous mutation in a bacterium living in the intestine of a hospitalized patient. We will never know. The consequences, however, of this seemingly minuscule event will be with us for years to come.

Many mutations are repaired before they can affect phenotype

Part of the reason we see so few mutations is that the vast majority of them are repaired before they can have an impact. Both prokaryotic and eukaryotic cells have a sizable repertoire of repair enzymes that guard against mutations. Only those few mutations that manage to slip by these repair mechanisms persist.

In our discussion of DNA replication, we mentioned that not only does DNA polymerase incorporate new nucleotides during DNA synthesis, but also it can "check its work" and repair any mistakes. This self-checking capacity, present in all cells, is called **proofreading**. If DNA polymerase inserts the wrong nucleotide, the enzyme is often unable to proceed because the

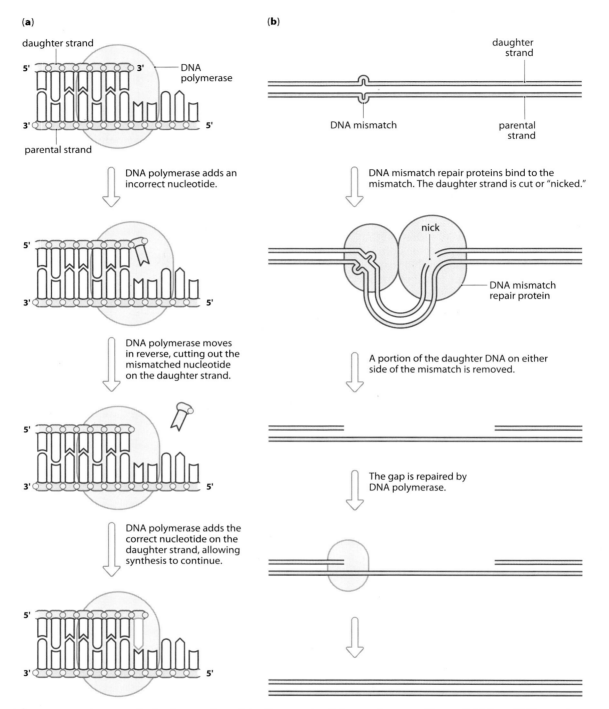

(a)

daughter strand

5'   3' —— DNA polymerase

3'   5'

parental strand

↓ DNA polymerase adds an incorrect nucleotide.

5'

3'   5'

↓ DNA polymerase moves in reverse, cutting out the mismatched nucleotide on the daughter strand.

5'

3'   5'

↓ DNA polymerase adds the correct nucleotide on the daughter strand, allowing synthesis to continue.

5'

3'   5'

(b)

daughter strand

DNA mismatch   parental strand

↓ DNA mismatch repair proteins bind to the mismatch. The daughter strand is cut or "nicked."

nick

DNA mismatch repair protein

↓ A portion of the daughter DNA on either side of the mismatch is removed.

↓ The gap is repaired by DNA polymerase.

↓

aberrant base cannot properly hydrogen-bond to the nucleotide on the template strand (**Figure 6.35a**). DNA polymerase then actually moves in reverse, cutting out the last nucleotide that it added. Following removal of the incorrect nucleotide, the enzyme gets a second chance to insert the proper nucleotide into the growing daughter strand.

One way that dimers can be repaired is by **DNA mismatch repair enzymes**, which recognize and cut the damaged strand and remove a small section of DNA on both sides of the dimer in the process. A molecule of DNA polymerase replaces the removed bases, restoring the DNA to its mutation-free state. This same mechanism can be used to remove and replace mismatched bases that were able to avoid repair by DNA proofreading. Such **mismatch repair** thus functions as a backup to the proofreading ability of DNA polymerase (see Figure 6.35b).

Figure 6.35 Two DNA repair mechanisms. (a) Proofreading by DNA polymerase. If an incorrect nucleotide is added to a growing strand, the DNA polymerase can cleave it from the strand and replace it with the correct nucleotide before continuing. (b) Mismatch repair. DNA repair enzymes can correct errors that occur during DNA replication.

Mutations are the raw material of evolution

Although mutations are rare, random events, they are the *only* way that new genetic information can initially come into being. If mutations *never* occurred, all members of a species would not only have the same genes, they would also have precisely the same *alleles* of those genes; they would be genetically identical. Genetic recombination, whether sexual or asexual, would accomplish nothing, since new gene combinations would not be possible. If the environment changed in any way, as it does for bacteria such as *S. aureus* when vancomycin is being used, there would be no "lucky few" who just happened to have the right form of a gene to survive and thrive in the face of this changing environment.

In other words, there would be no **evolution**—the genetically based change in the phenotype of a population over successive generations. Such change can happen over time precisely because of genetic variability in a population. Because of such variation, under any specific environmental conditions, some individuals are likely to be better adapted for survival than others. Those whose genetic makeup makes them more suited to a particular environment are more likely to survive and tend to produce more offspring. These offspring inherit those genetic traits that result in better adaptation, and over many generations, the population as a whole comes to take on the phenotype of the better-adapted individuals. The diversity of living things we see today, both microbial and nonmicrobial, owes its existence to this never-ending interplay between genes and the environment. We will revisit the topic of evolution, especially as it applies to microorganisms, in Chapter 8.

Coming Up Next...

In this chapter we have explored the genetic material and how it functions. One of the most important aspects of gene expression is metabolism, and the means by which cells go about using energy. In Chapter 7 we will investigate metabolism in detail, focusing on how a cell's metabolism impacts where it lives, how it grows, and in some cases, how it interacts with humans and other animals. Understanding metabolism will help us appreciate how, among other things, bacteria are used to make yogurt, why hydrogen peroxide is a good antiseptic to keep in your medicine chest, and why rotting fish stink. Furthermore, while some microorganisms can divide at the accelerated rate of every half hour or so, others take hours or days to reproduce. The difference often comes down to how they use energy. Understanding the fundamental metabolic requirements of different microorganisms often pays big dividends when we wish to inhibit the growth of those organisms that cause problems or spur on the growth of those that are beneficial.

Key Terms

allele	egg	intercalating agent
anticodon	evolution	intron
antiparallel orientation	exon	messenger RNA (mRNA)
asexual reproduction	F⁻ cell	mismatch repair
chromosome	F⁺ cell	missense mutation
coding strand	frameshift mutation	mutagen
codon	free radical	mutation
conjugation	gamete	neutral mutation
constitutive gene	gene	nonsense mutation
daughter cell	gene expression	nucleotide
daughter DNA	genome	operator
DNA mismatch repair enzyme	genotype	operon
DNA polymerase	inducer	origin
double helix	inducible operon	parent cell

parental DNA	repression	termination sequence
phenotype	repressor	thymine dimer
plasmid	ribosomal RNA (rRNA)	transcription
promoter	ribosome	transcription factor
proofreading	RNA polymerase	transduction
purine	RNA splicing	transfer RNA (tRNA)
pyrimidine	sex pilus	transformation
recombination	sexual reproduction	translation
regulated gene	sperm	virulence factor
replication fork	spontaneous mutation	

Concept Questions

1. If a particular DNA molecule consisted of 28% guanine-containing nucleotides, what percentages of nucleotides would contain adenine, cytosine, and thymine?

2. DNA is described as having an antiparallel structure. What exactly does this mean?

3. In this chapter, we learned about the "proofreading" ability of DNA polymerase. But as we learned in Chapter 4, some viruses rely on RNA rather than DNA as their genetic material. Furthermore, when these viruses replicate their RNA genome, they rely on other enzymes, such as RNA-dependent RNA polymerase, that do not have proofreading ability. On the basis of this information, how do you think the mutation rate in RNA viruses compares with that in DNA viruses?

4. A stretch of DNA has the following sequence:

5′-GTAATCGTAGCTTAC-3′

What would the complementary DNA strand be (be sure to label the 5′ and 3′ ends, remembering that nucleic acids are always drawn in the 5′ to 3′ direction).

5. A stretch of mRNA has the following sequence:

5′-AGUUCUAGGCCC-3′

Using Figure 6.20, translate this sequence into the proper amino acids.

6. Suppose that, in a particular strain of bacteria, the *lac* operon is mutated in such a way that it can never be repressed. In other words, it is permanently in the induced state. What problems would you expect this to cause for the bacteria?

7. A particular species of bacteria has an operon called the *ara* operon. When it is active, enzymes are produced that allow the bacteria to digest the sugar arabinose. Does this sound like an inducible system (like the *lac* operon) or a repressible system (like the *trp* operon)? Under what environmental conditions would you expect the operon to be actively transcribed?

8. A strain of *E. coli* is resistant to tetracycline. You mix some of these *E. coli* with *Klebsiella* bacteria, which are sensitive to this antibiotic. You later find that some of the *Klebsiella* cells are now resistant to tetracycline. How might you determine if the *Klebsiella* acquired their resistance through transduction, transformation, or conjugation?

9. A key protein in influenza virus is the HA protein, which allows the viral particles to bind onto appropriate host cells. At the crucial point where the HA binds to the host-cell receptor, the HA amino acid sequence is Ser-Glu-Thr-Glu. Serine (Ser) has a polar R group. Glutamic acid (Glu) has an acidic R group. Threonine (Thr) has a polar R group.

You analyze two mutant strains of influenza virus. In strain 1, this crucial four-amino-acid sequence has been altered to Ser-Asp-Thr-Glu. The amino acid aspartic acid (Asp) has an acidic R group. In strain 2, the sequence has been mutated to Ala-Glu-Thr-Glu. The amino acid alanine (Ala) has a nonpolar R group. What type of mutations are these (neutral, missense, or nonsense)? In which of the two strains is the mutation more likely to negatively impact the ability of the virus to replicate?

10. A new form of prokaryote is discovered that has its own, unique genetic code. Unlike any other form of life, this new organism uses only a single type of codon for each amino acid. Compared with other organisms, do you think that harmful mutations would be more or less common in this newly discovered prokaryote?

11. In a large colony of bacteria, all individual cells have identical genotypes. You remove members of this colony and place them in different environmental conditions. For example, some are placed at relatively high temperature, while others are placed at relatively low temperature. Some are placed in a nutrient-rich environment, while others are placed in a nutrient-poor environment. After several generations in these new environments, you inspect the cells in each. Based on what you have learned about genetics in this chapter, do you expect them to all have the same phenotype? Why or why not?

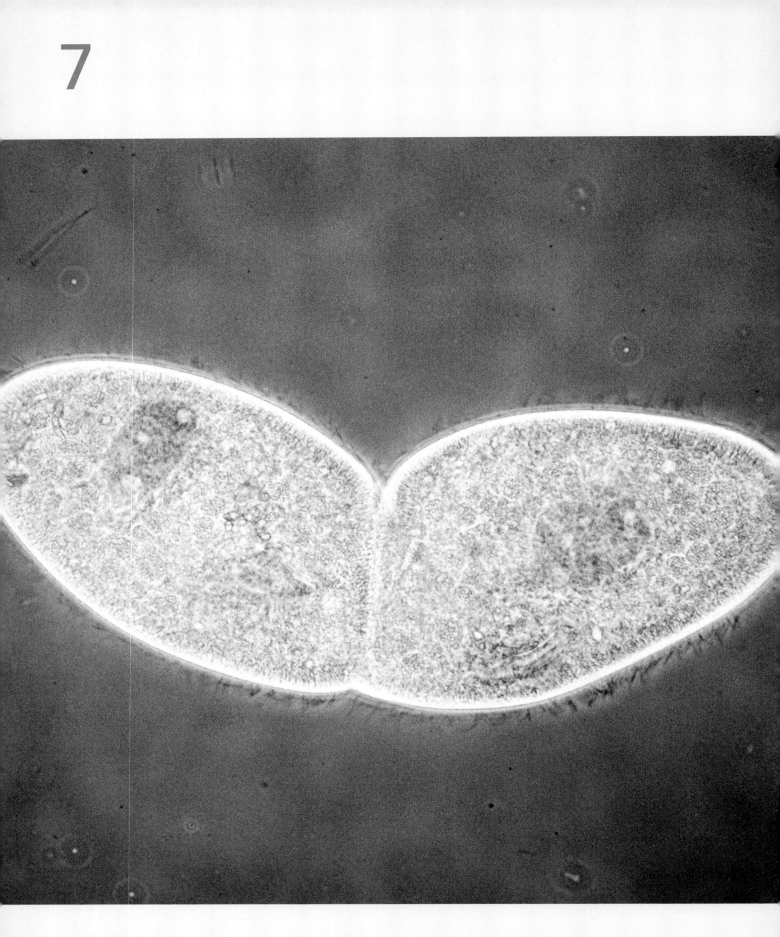

Paramecium caudatum replicate through binary fission.

Chapter 7

Metabolism and Growth

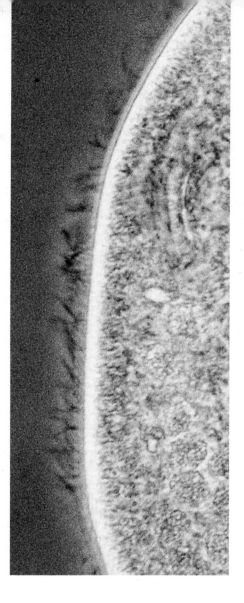

On the evening of March 20, 1966, Washtenaw County, Michigan was the scene of one of the best documented UFO sightings of all time. Local residents observed a number of "flying saucers" hovering above and even landing in a local swamp. Several photographs were taken before the mysterious objects vanished into the night sky (**Figure 7.1**).

In the 1960s, the U.S. Air Force still administered "Project Blue Book," in which it investigated purported UFO sightings. As part of this project, J. Allen Hynek arrived on the scene to look into the mystery. Although UFO buffs contend to this day that Hynek, who later coined the phrase "close encounters of the third kind," was pressured by his superiors, the investigator's conclusion was that the sightings were the result of "swamp gas"—a term that has subsequently often been used to dismiss UFO sightings.

Washtenaw County may or may not have been visited by extraterrestrials, but one thing is certain: countless microorganisms thrive in the oxygen-depleted, organically rich water of swamps. Many of these microbes produce swamp gas—waste products such as methane gas that can sometimes produce eerie glows. And such swamp gas may be responsible for at least some of the odd things, including UFOs, that people sometimes report seeing in swamps.

There is nothing very unusual about what these microorganisms are doing. All cells, microbial or otherwise, break down molecules to provide the energy needed for survival and growth. As cells digest these molecules, waste products are released. In the case of swamp-dwelling prokaryotes that waste is swamp gas. As we will see, other organisms release different wastes, depending on exactly how they go about the business of breaking down large molecules.

Once cells digest these molecules, the energy they obtain can then be used in other reactions that *require* rather than *release* energy. For instance, all cells must construct larger biological molecules such as proteins or nucleic acids from smaller building blocks such as amino acids or nucleotides, and such synthesis requires an input of energy. When we discuss *all* the ways that cells use energy, both how they get it and how they use it, we are discussing **metabolism**.

This definition of metabolism means that for many of us, we must change the way we think about this term. Often, in daily conversation, we use "metabolism" only as it applies to weight gain or the rate at which we burn calories (for example, "Brenda can eat anything she wants, because she's got a rapid metabolism"). But that is only *part* of what metabolism is all about. Metabolism is the sum total of our energy budget. It is somewhat analogous to how we use money. We don't just "earn it" (by eating); sometimes we "save it" in the form of polysaccharides and fats (when we

Figure 7.1 Alien visitors or microbial metabolism? March 20, 1966, Washtenaw County, Michigan.

"earn" more then we need), and we also need to "spend" it (to synthesize needed cell components).

Metabolism is similar in all living things

One important point should be emphasized because it simplifies our task in this chapter considerably. The basics of metabolism are not very different, whether we are talking about archaea growing in a hot spring, a lilac bush growing in your garden, or a rabbit sneaking in at night to eat the lilac. You may recall from Chapter 1, in listing the properties that distinguish life from nonlife, that all organisms need an energy source, which they convert to other usable forms of energy. This means that in a broad general sense, what applies to one type of living thing applies to others as well. This is not to say all living things are identical in every detail. To highlight just one difference, while animals, fungi, and many microorganisms have to eat or absorb organic compounds, plants and some microorganisms make all their own biological molecules. The next time you feed your house plants, look closely at the ingredients in the "plant food"; you will see no amino acids, lipids, or sugars listed. Plants make all these things themselves as long as they are supplied with a few crucial raw materials such as carbon dioxide and certain minerals.

In this chapter, we will start by discussing the basics of metabolism, especially as they apply to microorganisms. We will then examine some of the metabolic differences that distinguish microorganisms from each other. These differences explain not only why some microorganisms are the basis of alien sightings but also why cheddar cheese is different from Gouda, why rotting fish stinks, or how beer is made (**Figure 7.2**). Another reason for studying metabolism is that it is linked closely to growth. By understanding how microorganisms utilize energy, we will be ready to reflect on why various cells reproduce at very different rates, or how a change in the environment might cause a change in a particular organism's growth rate.

Figure 7.2 Courtesy of microbial metabolism. The cowboy in this wall painting in Tijuana, Mexico, may not know it, but the beer he is enjoying relies on single-celled fungi (yeast), which break down the sugars in grains to obtain energy. When the yeast cells are denied oxygen, they release ethyl alcohol as a metabolic waste.

Basic Concepts

Energy released from food molecules is used for other processes that require energy

When a cell digests any biological molecule—a long chain of sugars (a polysaccharide), for instance—energy is released. In this instance, the polysaccharide contains more energy than the simpler sugars that remain. Such a process is an example of an **exergonic** ("energy-out") reaction. What happens to the released energy that is no longer part of the sugars? Much of it is lost as heat or, in some interesting cases, as light (**Figure 7.3**). Some, however, may be captured by cells for energy-requiring reactions. Reactions that need an input of energy to occur are called **endergonic** ("energy-in") reactions.

Figure 7.3 Light emitted by bacterial metabolism. This mysterious sea creature, called a pyrosome, is related to the common sea squirt. Individuals can be up to several feet in length. Light-emitting bacteria in their tissues cause them to pulsate eerily at night. The light is due to the release of energy from exergonic reactions carried out by the bacteria. It is believed that the intermittent glow of the bacteria attracts food for the pyrosome. It might also start a war. In an odd historical note, it has been suggested by some that the 1964 Gulf of Tonkin torpedo attack on an American ship, which led to American military involvement in Vietnam, was actually pyrosome flashes.

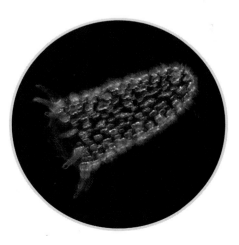

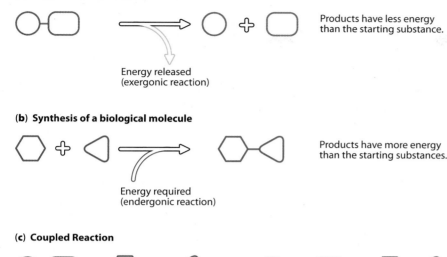

(a) Digestion of a biological molecule

Products have less energy
than the starting substance.

Energy released
(exergonic reaction)

(b) Synthesis of a biological molecule

Products have more energy
than the starting substances.

Energy required
(endergonic reaction)

(c) Coupled Reaction

The energy required by the endergonic reaction is provided
by the energy released by the exergonic reaction.

Figure 7.4 Exergonic, endergonic, and coupled reactions. (a) When a biological molecule is digested, energy is released. The bonds of the larger molecule have more total energy than the bonds of all the smaller substances that remain. Such reactions, in which energy is released, are termed *exergonic*. (b) Synthesis reactions require an *input* of energy, because the bonds of the initial starting substances have less total energy than those of the newly synthesized, larger molecule. Reactions that require an energy input are called *endergonic* reactions. (c) In a *coupled* reaction, an endergonic and an exergonic reaction occur simultaneously. The exergonic reaction provides the energy necessary for the endergonic reaction to proceed.

Exergonic and endergonic reactions are intimately linked in living things. To visualize this, think of a waterwheel in a river. For the wheel to move, an input of energy is required. In other words, the turning of the wheel is endergonic. The energy is provided by the energy released as the water flows downstream. Likewise, the synthesis of simple sugars into a polysaccharide is endergonic. As the sugars are being linked together to form the polysaccharide, another molecule is being digested simultaneously to provide the necessary energy. Put simply, the endergonic reaction needs the exergonic reaction as an energy source. When reactions are combined in this manner, they are known as **coupled reactions** (**Figure 7.4**).

Cells convert the energy in biological molecules into ATP

For most reactions that need an energy source, that energy is supplied by **adenosine triphosphate** or **ATP**. You can think of ATP as a cell's energy currency. Much of this chapter will be devoted to investigating how we "earn" ATP. In a nutshell, cells try to capture energy released from exergonic reactions, so they can use it to make ATP. Cells attempt to make as much ATP as they can, to keep their energy "bank account" topped up as much as possible. The stored ATP can then be spent as needed to power endergonic reactions such as the synthesis of a protein or a large polysaccharide.

ATP is composed of three component parts: a molecule of the sugar ribose covalently bound to a molecule of adenine and three phosphate groups (**Figure 7.5**). The phosphates represent the "business end" of the molecule; they provide the large amount of energy found in ATP. When one of these phosphates is removed, a large amount of energy is released, which is available for use in endergonic processes. After loss of a phosphate, the remaining molecule, with only two phosphates left, is called ADP (adenosine diphosphate). To have enough energy on hand, cells must replenish their ATP stock by adding phosphates back on to ADP.

Why is ATP so energy-rich? Each phosphate group carries a negative charge and the three negative charges strongly repel each other. This makes the bonds between the phosphates unstable. The chemical energy in these bonds is released when a phosphate is cleaved, much as the energy in a compressed spring is liberated when the spring is released.

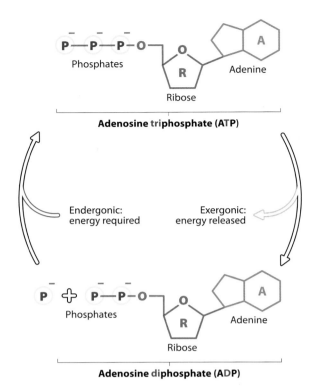

Adenosine triphosphate (ATP)

Phosphates · Adenine · Ribose · R · O · A

Endergonic: energy required

Exergonic: energy released

Adenosine diphosphate (ADP)

Phosphates · Adenine · Ribose · R · O · A

Figure 7.5 Interconversion of ATP and ADP. ATP stores large amounts of chemical energy in the bonds linking the phosphate groups. When the bond between the second and the outermost phosphate group is broken, energy is released as the terminal phosphate separates from the remaining molecule of ADP. The synthesis of additional ATP from ADP and phosphate is an endergonic (energy-requiring) process.

As mentioned, cells need a lot of ATP, and cells that are more active metabolically have even greater ATP demands. Of course, making ATP *requires* rather than *releases* energy, and cells go to a lot of trouble to make sure they do not run out. Returning to the similarity of metabolism in all living things, microorganisms, plants, and animals alike usually convert the energy in organic molecules into ATP, although there is variation in how they do it. Some living things, for example, need oxygen to synthesize ATP. Others make their ATP in the absence of oxygen, and for some of these, oxygen is a deadly poison. Microorganisms fall into all of these categories. Furthermore, not all microbes use organic molecules as an energy source for ATP production. Some use inorganic compounds as an initial energy source, while some use light for this purpose. A full discussion of *all* the diversity found in microbial metabolism is well beyond the scope of this text, but it is worth remembering that "alternative energy" is par for the course for many microorganisms.

Energy is transferred from one molecule to another via oxidation and reduction

How is the energy in biological molecules converted into ATP? Chemical energy is actually a property of electrons. Recall, however, that not all electrons in a molecule have the same amount of energy. As discussed in Chapter 2, when electrons are held more tightly by the nucleus, they are more stable and therefore have less energy. Electrons held less tightly by the nucleus are less stable and consequently have more energy. Energy is released when electrons at higher energy are removed from one atom or molecule and transferred to a different atom or molecule to which they are bound more tightly and thus have less energy. The atom or molecule losing electrons is said to be **oxidized**, while the one gaining electrons is said to be **reduced** (**Figure 7.6**). Oxidation and reduction do not happen independently; if one atom or molecule is losing electrons, another is simultaneously gaining them. We refer to the paired oxidation and reduction as a **redox reaction**. Cells break down food molecules through a series of such reactions.

Something similar happens, for instance, when we burn fuel such as a piece of wood (**Figure 7.7a**). Wood is composed of long-chain organic molecules containing many carbon–hydrogen bonds, and the electrons forming these C–H bonds are at relatively high energy. As these bonds are broken, the electrons are transferred to oxygen, where they are more stable and at low energy; the carbon atoms in these compounds are oxidized and the oxygen is reduced. Although the compounds in wood are not themselves hydrocarbons, we can illustrate this process by looking at methane, the simplest hydrocarbon. Note in Figure 7.7b that when a methane molecule is fully oxidized, the result is no longer a hydrocarbon. Ultimately all that remains is carbon dioxide; no C–H bonds remain at all. Meanwhile, additional oxygen is reduced and converted into water. The energy that was lost by the electrons during their transfer from the hydrocarbon to oxygen is released as light and heat.

As we see in the wood-burning example, when organic molecules are oxidized they often lose hydrogen atoms, while molecules being reduced often gain them. When an electron is transferred, it often picks up a proton (H^+) at the same time. In biological reactions these protons are freely available in water. Consequently, when a redox reaction occurs, the net effect is that one molecule loses a hydrogen atom while a different molecule gains one. It is easy to describe what happens when a biological molecule such as a sugar or fat is oxidized or reduced; oxidation results in the loss of carbon–hydrogen bonds (Figure 7.7b). Reduction, on the other hand, results in the gain of carbon–hydrogen bonds.

When wood or a hydrocarbon fuel burns, a very large amount of energy is released in a single step—the transferred electrons go from very high to very low energy and the result is fire. But such an explosive release of energy would never work in cells. When cells oxidize biological molecules, they do so in a large number of small steps. Not only would a single enormous release of energy be inefficient, it would no doubt kill the cell. By releasing the energy gradually, the cell can use it efficiently, while remaining at temperatures that are compatible with life.

To visualize this more easily, think about the way that octane (a hydrocarbon) in gasoline is used to power an automobile. In the engine, the octane is oxidized by oxygen slowly and in small increments. In this way, the energy that is released can be efficiently utilized to move the car. You could, of course, release exactly the same amount of energy from a tank of gas, simply by throwing a lighted match into the gas tank, oxidizing all octane at once.

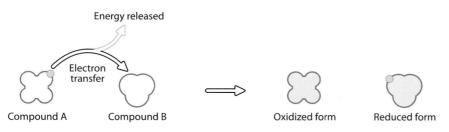

Energy released

Electron transfer

Compound A   Compound B   Oxidized form   Reduced form

Electron is held less tightly by compound A and is therefore at higher energy.

Electron is now more bound more tightly to compound B and is therefore at lower energy.

Figure 7.6 Oxidation–reduction reactions. In this example, the compound that loses one or more electrons (compound A) is being oxidized, while the compound gaining those electrons (compound B) is reduced. Because the electron is held less tightly by compound A, it is less stable and at higher energy. Following transfer to compound B, the electron is bound more tightly and therefore is more stable and at lower energy. Thus, this is an exergonic reaction, with energy being released by the electron during its transfer from compound A to compound B.

(a)

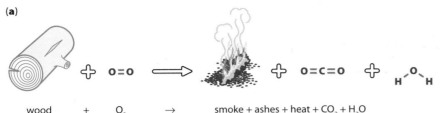

wood   + O_2 →   smoke + ashes + heat + CO_2 + H_2O

(b)

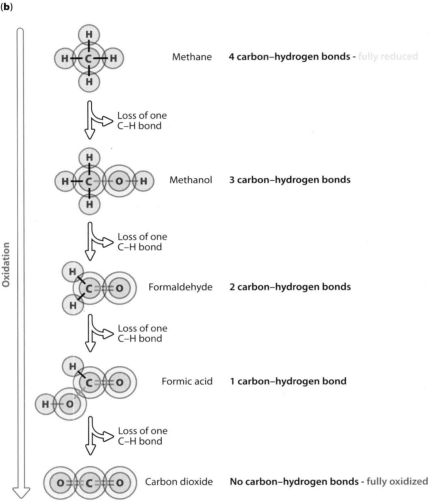

Methane   **4 carbon–hydrogen bonds -** fully reduced

Loss of one
C–H bond

Methanol   **3 carbon–hydrogen bonds**

Loss of one
C–H bond

Formaldehyde   **2 carbon–hydrogen bonds**

Loss of one
C–H bond

Formic acid   **1 carbon–hydrogen bond**

Loss of one
C–H bond

Carbon dioxide   **No carbon–hydrogen bonds -** fully oxidized

Oxidation

Figure 7.7 Oxidation of carbon fuels.
(a) When electrons are transferred to oxygen from the carbon atoms in a fuel such as wood, energy is released in the form of light and heat. Carbon dioxide is the completely oxidized form of carbon, while water is the reduced form of oxygen. (b) Often when biological molecules are oxidized, they lose carbon–hydrogen bonds. When they are reduced, they gain carbon–hydrogen bonds. In this example, methane (CH_4) is fully reduced, because a carbon atom can form a maximum of four bonds. Here, methane is being progressively oxidized as C–H bonds are replaced by bonds between carbon and oxygen. In its fully oxidized form (CO_2), no C–H bonds remain.

This certainly would move the car, but hardly in an efficient way or in a manner that would permit the survival of either the car or the driver.

Oxygen is not the only molecule that can accept electrons during redox reactions. In living things several other molecules act as electron acceptors when food molecules are being oxidized. **Nicotinamide adenine dinucleotide** or **NAD⁺** is an important example (**Figure 7.8**). This molecule is present in one of two states; in its oxidized state, it exists as NAD⁺. When reduced, NAD⁺ picks up two electrons from the molecule being oxidized, as well as a single proton, to become NADH. A second proton is released as an H⁺ ion. NADH can later be oxidized with a different molecule being reduced, releasing still more energy. The now oxidized NAD⁺ is again ready to accept another pair of electrons. Thus, NADH acts as a sort of "electron carrier," first accepting and later donating these electrons. As we will see, the energy released as these electrons are passed along provides the energy needed for a cell to produce its ATP.

Cell Respiration

When they hear the term respiration, most people immediately think of breathing. Furthermore, most people know that we breathe to get enough oxygen, which of course is absolutely necessary for us to survive. Far fewer, however, know exactly *why* we need oxygen. We have already alluded to the answer. Many living things, including animals and many microorganisms, use oxygen to *oxidize* biological molecules in order to obtain the energy these biological molecules contain. Not all organisms, however, use oxygen for this purpose. As we will see, many microorganisms use other molecules precisely the way we use oxygen—to oxidize biological molecules. No matter which molecules are used, the slow incremental breakdown of biological molecules necessary for ATP production is called **cell respiration**. Essentially, breathing is simply the way animals get the oxygen needed to produce ATP. And without ATP, life is impossible.

Carbohydrates are the primary source of energy used in cell respiration

Although many different biological molecules may be digested to meet energy needs, carbohydrates are generally the molecule of first choice. Most commonly, glucose is the starting material of cell respiration. Because glucose is so important for cell respiration, cells tend to store up as much of it as possible, generally by stringing many glucose molecules together into long polysaccharides.

(a)

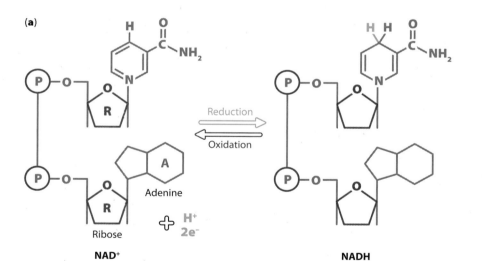

(b)

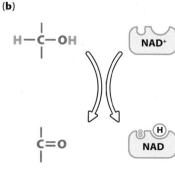

Oxidation of molecule 1   Reduction of NAD$^+$

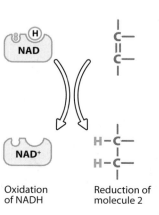

Oxidation of NADH   Reduction of molecule 2

Figure 7.8 Nicotinamide adenine dinucleotide. (a) This important electron carrier can exist in either oxidized (NAD$^+$) or reduced (NADH) state. Note the position of the additional carbon–hydrogen bond (in blue) in NADH. (b) When NAD$^+$ is reduced, a biological molecule (molecule 1) is oxidized, losing a carbon–hydrogen bond. The now reduced NADH may itself become oxidized, reducing a different molecule (molecule 2), which consequently gains carbon–hydrogen bonds.

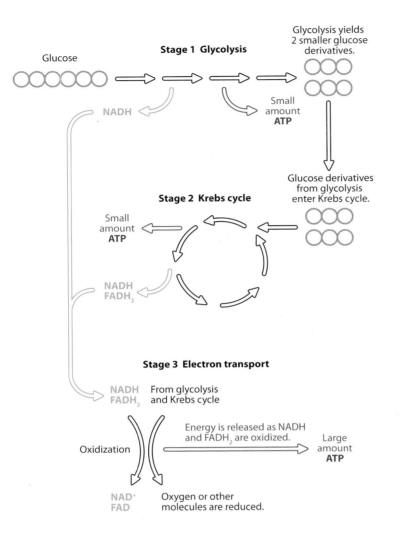

Stage 1 Glycolysis

Glucose

Glycolysis yields
2 smaller glucose
derivatives.

NADH

Small
amount
ATP

Glucose derivatives
from glycolysis
enter Krebs cycle.

Stage 2 Krebs cycle

Small
amount
ATP

NADH
FADH$_2$

Stage 3 Electron transport

NADH
FADH$_2$

From glycolysis
and Krebs cycle

Oxidization

Energy is released as NADH
and FADH$_2$ are oxidized.

Large
amount
ATP

NAD$^+$
FAD

Oxygen or other
molecules are reduced.

Figure 7.9 Overview of cell respiration. Cell respiration occurs in three stages. In stage 1, glycolysis, glucose (a 6-carbon compound, represented by six gray circles) is partially oxidized to two smaller glucose derivatives (two 3-carbon compounds, represented by three gray circles). As glucose is oxidized, NAD$^+$ is reduced to NADH. The oxidation of glucose is completed in stage 2, the Krebs cycle, as glucose derivatives from glycolysis are oxidized in a series of steps. Each oxidation results in the reduction of NAD$^+$ or a similar molecule called FAD, forming NADH and FADH$_2$. Small amounts of ATP are made in these first two stages, but most ATP is synthesized when NADH and FADH$_2$ formed in glycolysis and the Krebs cycle are oxidized in stage 3, electron transport. Energy released from these oxidations provides the energy needed for the synthesis of ATP in electron transport.

Depending on the organism in question and the environmental conditions in which that organism finds itself, glucose may be completely or only partially broken down. If glucose can be completely digested, all carbon–hydrogen bonds are oxidized and the released energy is used for ATP production. If glucose is only partially broken down, some carbon–hydrogen bonds remain. Because less energy is released, ATP yield is lower when glucose is only partly digested. First we will consider the complete digestion of glucose. We will then turn our attention to those situations in which glucose digestion can only proceed to a certain intermediate point and no further.

Cell respiration occurs in three stages

Respiration, the oxidation of biological molecules and the conversion of the released energy into ATP, takes place in a large number of small steps. These steps, however, can be grouped into three principal stages (**Figure 7.9**). In the first stage, **glycolysis**, a molecule of glucose is partially digested into two smaller molecules and some of the released energy is captured as NADH. In the second stage, the **Krebs cycle**, the oxidation of these smaller molecules is completed, and more energy-carrying molecules such as NADH are formed. In both glycolysis and the Krebs cycle, a small amount of ATP is also made, but most ATP is made in the third and final stage, called **electron transport**. Here, all the energy-carrying molecules formed in the first two stages (mostly NADH) are oxidized, molecules such as oxygen are reduced, and the released energy is used to produce the bulk of a cell's ATP.

In glycolysis, glucose is partially oxidized, forming two smaller molecules called pyruvate

Glycolysis, the first of the three stages in cell respiration, is where glucose oxidation begins (see Figure 7.9). By the end of glycolysis, glucose has been split into two molecules of pyruvate. These pyruvates will then be the starting material for the Krebs cycle. A small amount of ATP is generated in glycolysis, and some of the electrons originally with the glucose are transferred to NAD^+, forming NADH. We will now consider these events in more detail.

Glycolysis is common to prokaryotic and eukaryotic cells alike. Glycolysis literally means "the splitting of glucose," which accurately describes what happens. The important steps of glycolysis are presented in **Figure 7.10**. Observe that a molecule of glucose (a 6-carbon molecule) is broken down in a series of steps, into two molecules of **pyruvate** (a 3-carbon molecule). Along the way, several notable events occur. First, early in glycolysis there are energy-requiring (endergonic) steps, coupled to ATP hydrolysis, which provides the energy necessary for them to occur. Although the goal of cell respiration is to *make* ATP, a small amount of ATP must be *consumed* in these early steps.

The second significant event occurs when the 6-carbon compound is cleaved into two 3-carbon compounds. Each of these 3-carbon compounds will ultimately be converted into pyruvate, the end product of glycolysis. Before that happens, however, each of these 3-carbon molecules is oxidized. Recall that when one molecule is oxidized, a different molecule is reduced. In this case, the molecule being reduced is NAD^+, which is converted to NADH as a result. These NADH molecules, along with others that will be produced later, will ultimately be used in the final stage of cell respiration, where most ATP production takes place.

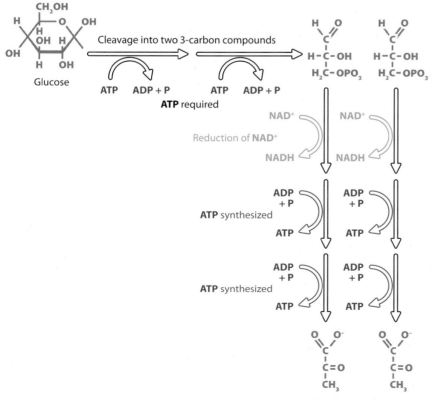

Figure 7.10 Glycolysis. Before the digestion of glucose can begin, a small amount of ATP must be invested to initiate the process. Glucose (containing 6 carbon atoms) is subsequently cleaved into two molecules, each with three carbons. As each of these 3-carbon compounds is oxidized, a molecule of NAD^+ is reduced to NADH. The now-oxidized three-carbon compounds continue through a series of steps, ultimately giving rise to two molecules of pyruvate at the end of glycolysis. Highly exergonic steps in glycolysis are coupled to the synthesis of ATP from ADP and P (an endergonic process), yielding a small amount of ATP.

Some ATP, however, is directly produced in glycolysis. Two steps in particular are highly exergonic (see Figure 7.10). The energy that is released at these steps is coupled to the synthesis of ATP from ADP and phosphate.

To summarize the first stage of cell respiration, each molecule of glucose that has entered glycolysis has been broken down into two molecules of pyruvate. NADH molecules were formed when intermediate compounds in the pathway were oxidized, and exergonic steps were coupled to the synthesis of a small amount of ATP. The pyruvates left over at the end of glycolysis will next be oxidized in stage 2 of cell respiration: the Krebs cycle.

In the Krebs cycle, pyruvate from glycolysis is completely oxidized and the energy released is transferred as electrons to NAD$^+$ and FAD

In the second stage of cell respiration, the Krebs cycle, the oxidation of pyruvate formed at the end of glycolysis is completed in a series of small steps (see Figure 7.9). At a number of these steps, energy-carrying compounds such as NAD$^+$ and a similar compound, FAD, are reduced to NADH and FADH$_2$. These energy carriers, along with those produced in glycolysis, will provide the energy for ATP synthesis in the third and final stage of cell respiration. We will now take a more in-depth look at what takes place in the Krebs cycle.

At the end of glycolysis, two molecules of pyruvate remain from each molecule of glucose. An examination of pyruvate structure (see Figure 7.10) should convince you that there are still a number of carbon–hydrogen bonds available for oxidation. These bonds are oxidized in the next stage, the Krebs cycle.

Figure 7.11 summarizes the Krebs cycle. Observe that before the Krebs cycle actually begins, pyruvate is converted from a 3-carbon to a 2-carbon molecule. The third carbon is released as CO$_2$, and a molecule of NAD$^+$ is reduced to NADH.

Next, observe that the newly formed 2-carbon compound joins an already existing 4-carbon compound to form a 6-carbon compound. As illustrated in Figure 7.11, this 6-carbon compound is progressively oxidized in a series of steps back to the 4-carbon compound, losing an additional two carbons in the form of CO$_2$ along the way. The fact that the starting 4-carbon compound is ultimately regenerated is why the process is referred to as a *cycle*. This newly regenerated 4-carbon compound is now ready to join another 2-carbon compound to begin another round of the cycle.

As shown in Figure 7.11, several key events occur during the cycle. First, as we have mentioned, two molecules of CO$_2$ are released. By the end of the Krebs cycle, all of the carbon atoms that were originally found in pyruvate have been released as CO$_2$. The hydrogens, however, that were originally bound to carbon were transferred to NAD$^+$ in a series of redox reactions. In one step, a molecule similar to NAD$^+$ called FAD is reduced instead, forming a molecule of FADH$_2$. FAD has the same function as NAD$^+$: it serves as an electron carrier, becoming reduced to FADH$_2$ as another molecule is being oxidized. Additionally, note that in one highly exergonic step of the Krebs cycle a molecule of ATP is made, in a manner similar to what we observed in glycolysis. Finally, remember that at the end of glycolysis, two molecules of pyruvate remained from the initial glucose. Each of these two pyruvates is completely oxidized in the Krebs cycle, meaning that for each molecule of glucose, the Krebs cycle takes place twice.

To summarize, by the end of glycolysis and the Krebs cycle, the initial glucose has been completely oxidized. The carbon atoms originally present in

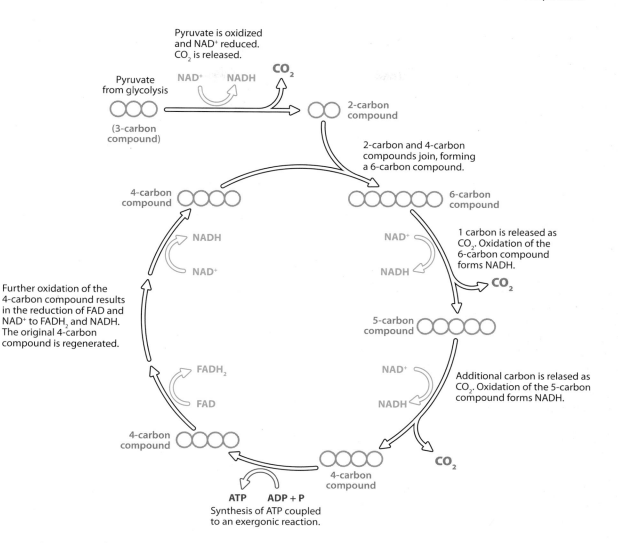

Figure 7.11 The Krebs cycle. Gray circles represent carbon atoms. Pyruvate is first converted to a 2-carbon compound, which enters the Krebs cycle. This 2-carbon molecule joins a preexisting 4-carbon compound to form a 6-carbon compound. This compound is subsequently oxidized in a series of steps back to the 4-carbon compound. Carbons are released in the form of CO_2, and NAD^+ and FAD are reduced to NADH and $FADH_2$ respectively. ATP is generated by coupling its synthesis to an exergonic step of the cycle.

the glucose have been released as CO_2. A small amount of ATP has been formed in coupled reactions, but much of the energy that *was* in the carbon–hydrogen bonds of the glucose molecule is *now* present in the reduced forms of NAD^+ and FAD, namely, NADH and $FADH_2$. These electron carriers will now in turn be oxidized in the third and final stage of cell respiration. As NADH and $FADH_2$ are oxidized, the energy released will be used to produce most of the cell's ATP.

NADH and $FADH_2$ are oxidized in electron transport, providing the energy for ATP synthesis

Most ATP is synthesized in electron transport, the third and final stage of cell respiration. Briefly, the NADH and $FADH_2$ produced in glycolysis and the Krebs cycle are oxidized. The released electrons are passed between a series of electron carriers and are ultimately passed to oxygen or other compounds, which represent the final reduced waste of cell respiration (see

Figure 7.9). The energy released as electrons pass from carrier to carrier is used to pump protons across a membrane. As these protons build up on one side of the membrane, they provide a source of potential energy. The protons then flow back to their original location through an enzyme complex. Like water flowing over a dam that is used to generate electricity, the flow of protons through this enzyme complex is utilized to synthesize ATP. Now we will examine the details of this remarkable process.

A large number of NADH and a smaller number of $FADH_2$ units were produced in glycolysis and the Krebs cycle, due to the step-by-step oxidation of glucose. In electron transport, these reduced molecules are oxidized, and the electron pair released from each NADH or $FADH_2$ is passed through a series of electron carriers known collectively as the **electron transport chain**. The components of the electron transport chain are embedded in membranes: the plasma membrane of prokaryotes, and the inner mitochondrial membrane of eukaryotes (**Figure 7.12**).

As we see in Figure 7.12, the first components of the electron transport chain oxidize NADH or $FADH_2$, converting either of these electron carriers back to NAD^+ or FAD, which can now be reused in glycolysis or the Krebs cycle. Electrons are subsequently passed through the entire chain in a series of redox reactions that ends when the electrons are ultimately passed to a **final electron acceptor**. The final electron acceptor, now reduced by the electrons that have passed down the chain, is released as a waste product. As electrons are passed from one component of the chain to the next, these electrons lose energy. Thus, the redox reactions of the electron transport chain are exothermic. As we will shortly see, this energy release is used to generate the lion's share of a cell's ATP.

The components of the electron transport chain fall into only a few categories of molecules. Some are membrane-bound proteins, which contain associated molecules derived from B vitamins. Other protein components contain metallic ions such as iron. Still others are nonprotein, hydrophobic molecules, some of which are derived from other vitamins.

Different organisms use different molecules in their electron transport chains. Bacteria, archaea, and eukaryotes all typically differ in their electron transport components. Within the domain Bacteria alone, there is significant variety. In fact, the identification of pathogenic bacteria in a diagnostic laboratory is sometimes based on the identification of electron transport components. *Pseudomonas* bacteria, for example, have an electron chain component called cytochrome oxidase, which is lacking in many other species. A simple test for the presence of this compound can help to distinguish *Pseudomonas* from other organisms.

In many prokaryotes, and even more commonly in eukaryotes, oxygen is the final electron acceptor at the end of electron transport. As oxygen accepts the electrons that were released by NADH or $FADH_2$ at the beginning of the electron transport chain, the oxygen is reduced to H_2O. By the time the electrons unite with oxygen, they have lost a great deal of their energy. The bonds between the hydrogens and the oxygen are very stable and at low energy. Water, in this case, represents a waste product. This is the reason that so many living things need oxygen to survive; we need it as a final electron acceptor in electron transport. Without oxygen to remove electrons at the end of the electron transport chain, respiration grinds to a halt. Organisms that utilize oxygen in this way are relying on **aerobic respiration**.

CASE: SOMETHING'S FISHY IN THE FRIDGE

Anya loves seafood, and with company coming over for dinner she goes to the fish market and purchases several fresh halibut filets. After buying the

fish, she runs several errands, and by the time she arrives home, over an hour has passed. When she unpacks the fish in the kitchen, Anya notices that it smells funny. Pressing it to her nose, she now cannot avoid the fact that her planned entrée absolutely stinks, emitting a strong "fishy" odor. Confused and angry, she discards the fish in the trash. The halibut was odor-free and fresh when she selected it in the fish market. In any event it looks like fish is off the menu this evening.

1. What happened to the fish? Why does unfrozen fish so quickly turn "fishy"?
2. Why don't beef and chicken start to smell bad unless they've been left unfrozen much longer?

We are so used to thinking of oxygen as essential for life that it can be surprising to learn that many microorganisms do just fine without it. These **anaerobic** microorganisms rely on final electron acceptors other than oxygen. Some anaerobes never utilize oxygen. Others use oxygen when it is available but readily switch to a different final electron acceptor when oxygen is absent.

Examples of final electron acceptors used by anaerobes include sulfate (SO_4^{2-}) and nitrate (NO_3^-). When sulfate accepts electrons at the end of electron transport, the now-reduced molecule, hydrogen sulfide (H_2S), is released as a waste. Some organisms use carbon dioxide as a final electron acceptor, reducing it to methane (CH_4). Methane is a "greenhouse gas," so-called because of its ability to trap heat in the atmosphere. Consequently *methanogens* ("generators of methane") are thought to be involved in the problem of global warming. Methane is also the primary component of

(a)

NADH or FADH$_2$ is oxidized by components of the electron transport chain. Electrons originally released by NADH or FADH$_2$ are passed through the chain in a series of redox reactions.

Energy released by electrons as they pass through components of the chain is used to pump protons across a membrane.

(b)

Pumping of protons results in proton gradient with high proton concentration on one side of the membrane and low concentration on the other.

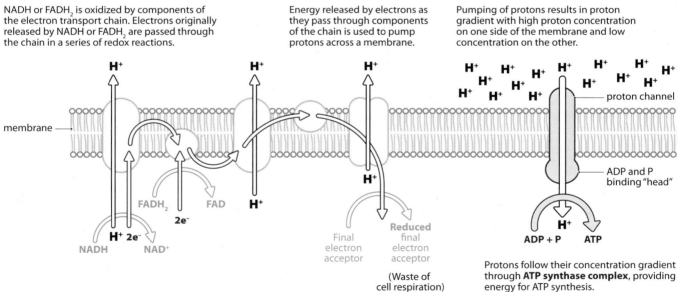

Protons follow their concentration gradient through **ATP synthase complex**, providing energy for ATP synthesis.

Figure 7.12 The electron transport chain. (a) In prokaryotes, the components of electron transport are embedded in the plasma membrane. In eukaryotes, electron transport components are found in the inner mitochondrial membrane. NADH or FADH2, formed in glycolysis and the Krebs cycle, is oxidized by the first components of the chain. Following this oxidation, electrons are passed between components of the chain in a series of redox reactions. Ultimately the electrons are passed to a final electron acceptor, which, in its reduced form, is a waste of cell respiration. As electrons are passed from component to component, the energy lost by the electrons is used to pump protons across the membrane (from the cytoplasm to the extracellular environment in prokaryotes, and from the mitochondrial matrix to the inner membrane space in eukaryotes). (b) The buildup of protons supplies energy for ATP synthesis. The protons flow back to either the cytoplasm (prokaryotes) or the mitochondrial matrix (eukaryotes) through the ATP synthase proton channel. As they do so, these protons release energy, which causes the "head" portion of the ATP synthase to spin rapidly. This spinning provides the force to join ADP to ATP, forming ATP.

swamp gas. Some would argue that fantastic sightings of aliens, swamp creatures, and the like have more to do with methanogens simply busy making ATP than with anything else.

And Anya's ruined fish dinner? The flesh of saltwater fish contains a substance called trimethylamine oxide. This substance is odorless, and many species of anaerobic bacteria use it as a final electron acceptor in electron transport. Once it accepts electrons, however, the now-reduced trimethylamine is anything but odorless. It has the strong "fishy" smell that we associate with fish that is not fresh. Because animals such as cows and chickens do not have trimethylamine oxide in their flesh, beef and chicken will not develop this unpleasant odor. Any bacterial growth via anaerobic respiration is enough to make fish smell bad. If fish is very fresh or frozen there will be no foul smell, because it contains very few bacteria. That is why, when purchasing seafood, it is important to get it straight home and into the refrigerator.

Protons that are pumped across a membrane in electron transport provide an energy source for ATP synthesis

The yield of ATP obtained directly in glycolysis and the Krebs cycle is modest. Small amounts of ATP were produced in these earlier stages when ATP synthesis (an endergonic reaction) was coupled to exergonic reactions. This relatively small ATP harvest represents only about 4% of the energy found in a molecule of glucose—and so, by the end of the Krebs cycle, only about 4% of glucose's energy has been converted into a form that a cell can directly use to meet its own energy needs.

Although we have discussed how the energy in electrons is slowly released in electron transport, we still have yet to explain how this exergonic release of energy translates into ATP production. In Figure 7.12a, observe that as electrons pass through the various components of the chain, the energy released by these electrons is used to pump protons (H^+) across the membrane. This results in a buildup of protons on one side of the membrane (outside the plasma membrane in prokaryotes, outside the inner mitochondrial membrane in eukaryotes). This unequal distribution of protons (high concentration of H^+ on one side of the membrane and low on the other) is referred to as a **proton gradient** (Figure 7.12b). The energy from the electrons has thus passed to the proton gradient, where it is temporarily stored.

The protons that have been pumped across the membrane are unable to simply diffuse back. They can, however, flow down their concentration gradient back through an enzyme complex called **ATP synthase** (see Figure 7.12b). This remarkable enzyme complex consists of two important and related parts. First, there is a channel-forming portion, which permits protons to move across the membrane. Second there is a head portion that binds onto ADP and phosphate. As the protons that have built up on one side of the membrane stream back across the membrane through the ATP synthase channel, they release energy. This energy causes the ADP and phosphate-binding head to spin rapidly. This rotation provides the force to link ADP and phosphate, forming ATP, much as the spinning of a windmill provides energy to pump water or perform other tasks.

It is estimated that, in aerobic respiration, each glucose molecule digested in glycolysis and the Krebs cycle (**Figure 7.13**) generates enough NADH and $FADH_2$ to make approximately 35 ATP molecules in electron transport. Earlier we stated that the ATP synthesized in glycolysis and the Krebs cycle represented about 4% of the energy originally contained in a molecule of glucose. With the addition of ATP produced in electron transport, about 40% of glucose's energy has now been converted into ATP. The amount of

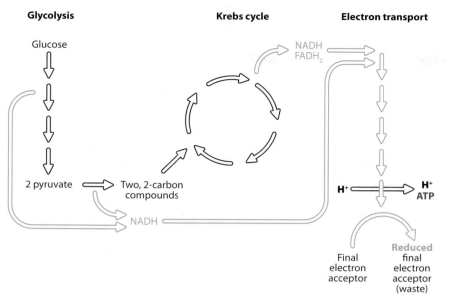

Glycolysis   **Krebs cycle**   **Electron transport**

Figure 7.13 Summation of cell respiration. Glucose is progressively oxidized, first in glycolysis and then in the Krebs cycle, resulting in reduction of NAD^+ and FAD to form NADH and $FADH_2$. These electron carriers are then oxidized in electron transport. The energy lost by these electrons as they are ultimately passed to a final electron acceptor is used to synthesize the bulk of a cell's ATP.

ATP produced in anaerobic respiration is variable, but since there are usually fewer carriers in the electron transport chain, there are fewer protons pumped, and ATP yield is often correspondingly less than it is for aerobic respiration. Because they produce less ATP, many anaerobes tend to grow slowly relative to aerobes.

Now that we have discussed the three stages of cell respiration, we should have a good idea about how cells convert energy in the carbon–hydrogen bonds of glucose into ATP. Remember, however, that sometimes cells are unable to completely break glucose down. Instead, glucose is only partially digested. We will now turn our attention to those situations in which cells "call it quits" before cell respiration is complete.

Incomplete glucose oxidation results in fermentation

The next time you are gasping for air after a particularly strenuous workout, try thinking about what is happening in your cells to take your mind off your discomfort. In a nutshell, your muscle cells are running out of their final electron acceptor—oxygen. That means that respiration can proceed only through glycolysis before it stops. To get back to aerobic respiration and its greatly increased ATP yield, your body responds with increased breathing and heart rate, while you sit down to catch your breath. Many prokaryotes, however, can survive indefinitely on glycolysis alone. It should not be surprising that, when doing so, such microorganisms grow slowly, owing to their relatively paltry ATP production.

If cells are relying solely on glycolysis to meet their ATP needs, even for a short period of time, what happens to the end product, pyruvate? You might guess that it is released as a waste, but this is not the case. Remember that glycolysis produces a small amount of NADH as well as pyruvate (see page 177). In the *presence* of the final electron acceptor, this NADH is oxidized in electron transport, resulting in additional ATP synthesis. *Without* the final electron acceptor, this cannot occur. Not only is the NADH of no further value, but the cell is now in danger of running out of NAD^+. If the supply of

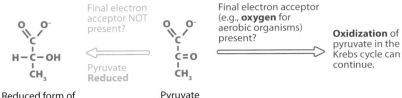

Figure 7.14 The fate of pyruvate. If the necessary final electron acceptor is available, pyruvate will be further oxidized, releasing additional energy to be used in ATP synthesis. If it is unavailable, pyruvate must be reduced in order to regenerate NAD⁺, which is required in glycolysis. The reduced form of pyruvate (lactate, in this example) is released as a waste.

NAD^+ gets too low, even glycolysis stops. Consequently, in the absence of its final electron acceptor, the cell regenerates NAD^+ by oxidizing NADH and reducing pyruvate. The now-reduced form of pyruvate is released as a waste. This process is known as **fermentation**, and the waste that is released is the **fermentation waste product** (**Figure 7.14**).

For many cells, fermentation is something like a "safety net." Although ATP yield is limited to the low quantities made in glycolysis, you can think of fermentation as a way that such cells "get by" until the final electron acceptor again becomes available. The length of time that cells can persist on such skimpy ATP production depends on how much ATP they need to survive. Most large animals are unable to survive like this for long, because their demands for ATP are too high. Deprived of their final electron acceptor (oxygen) for more than a few minutes, they die. Many microorganisms, however, can persist on fermentation alone indefinitely.

The fermentation waste product varies according to cell type. In an animal's muscle cells, pyruvate is reduced to lactate (see Figure 7.14). Yeast cells growing in anaerobic conditions reduce pyruvate to ethyl alcohol, with CO_2 gas also being released. It is these yeast fermentation wastes upon which baking and brewing depend (**Figure 7.15**). Other microorganisms release a wide variety of fermentation waste products, many of which are useful in food preparation. Pickles and sauerkraut, for instance, are the result when bacteria are used to ferment the sugars found in cucumbers and cabbage, respectively. Different cheeses have different tastes because the various microorganisms used to ferment milk release different fermentation wastes. Other microbial fermentation wastes such as acetone (see Chapter 5, p. 117) are utilized in industrial processes. In Chapters 15 and 16 we will take a closer look at the role of microbial fermentation in the production of many familiar foods and industrial products.

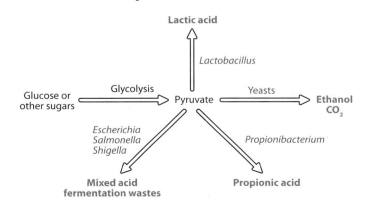

Figure 7.15 Microbial fermentation waste products. Depending on the microorganism, pyruvate may be reduced to a variety of final fermentation wastes. Many organisms release a combination of waste products, and many fermentation wastes are utilized in the production of various food products.

Molecules other than glucose can be used to generate ATP

CASE: NAME THAT BACTERIUM

Sydney, a microbiology student, is trying to identify an unknown bacterial species as part of a laboratory assignment. Using a series of tests, he is able to narrow his choices down to two species: *Citrobacter freundii* **and** *Proteus vulgaris.* **In his laboratory manual, Sydney notes that** *P. vulgaris* **can digest the amino acid phenylalanine, but** *C. freundii* **cannot. He therefore decides to inoculate his unknown species onto a phenylalanine slant. After allowing the bacteria to grow for 24 hours at 35°C, he adds ferric chloride to his tube and observes that a greenish color quickly forms (Figure 7.16). This allows Sydney to deduce that his unknown bacterial species can digest phenylalanine and therefore must be** *P. vulgaris.*
1. **What is a phenylalanine test and how does it work?**
2. **Why can only some species digest phenylalanine?**

Although our discussion of cell respiration has focused on glucose, cells can meet their energy needs by means of a variety of other organic molecules, including other sugars, fats, and proteins. As biological molecules are digested into their component parts, these smaller molecules may enter the cell respiration pathway at various points. Fats, for example, are energy-rich molecules with many carbon–hydrogen bonds available for oxidation. During digestion, fats are first hydrolyzed into glycerol and fatty acids by a group of enzymes collectively called **lipases**. Glycerol and fatty acids are then converted into other intermediate compounds in the cell respiration pathway (**Figure 7.17**). Proteins are first digested into amino acids by enzymes known as **proteases**. Depending on the amino acid in question, it may enter the respiratory pathway at various points.

It is worth remembering that not all cells can use all substances. Only those with the required enzymes are able to digest a specific type of biological molecule. Sydney was able to take advantage of this fact to successfully complete his laboratory assignment. *P. vulgaris* has an enzyme called phenylalanine deaminase, but *C. freundii* does not. The phenylalanine slant

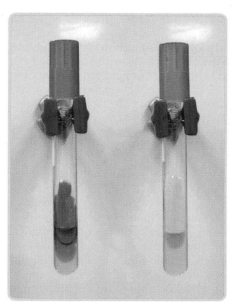

Figure 7.16 The phenylalanine deaminase test. The slant contains the amino acid phenylalanine. Bacteria are streaked onto the slant. Bacteria that are able to digest the amino acid can do so because they produce a specific enzyme called phenylalanine deaminase which permits phenylalanine digestion. An acidic waste is produced as the amino acid is digested. Ferric chloride added to the slant reacts with the acid, producing a green color. If bacteria are negative for the enzyme, no acid waste is produced in the slant. Added ferric chloride remains yellow.

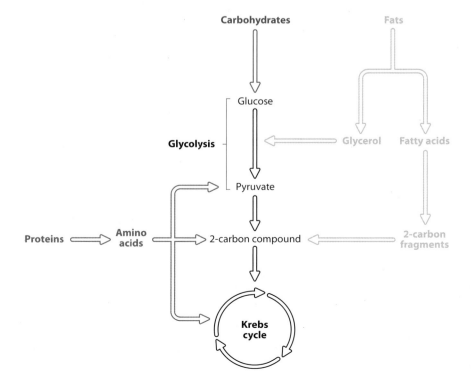

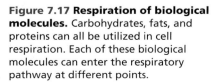

Figure 7.17 Respiration of biological molecules. Carbohydrates, fats, and proteins can all be utilized in cell respiration. Each of these biological molecules can enter the respiratory pathway at different points.

that Sydney used is prepared with a medium enriched with this amino acid. When a species such as *P. vulgaris* is streaked onto the medium and allowed to grow, the bacterium digests the amino acid to help meet its energy needs. Phenylalanine deaminase converts the amino acid to an acid. Ferric chloride binds to this acid, producing the green color (see Figure 7.16). *C. freundii*, on the other hand, does not produce phenylalanine deaminase and therefore cannot digest phenylalanine. If Sydney's unknown organism had been this species, the phenylalanine in the slant would have remained undigested, and when he added the ferric chloride it would have retained its original yellow color.

Autotrophs: the Self-feeders

Metabolism includes both the harvesting of energy from biological molecules, such as carbohydrates and fats, and the use of this energy by the cell to carry out various life processes. To obtain these energy-yielding molecules, all living things fall into two broad categories. **Heterotrophs** ("different feeders") must either eat or absorb such molecules. All animals and fungi, along with many microorganisms, fall into this group. Green plants, however, as well as numerous prokaryotes, are able to make their own biological molecules, as long as they are supplied with a few crucial resources such as carbon dioxide and nitrogen. These "self-feeders" are called **autotrophs**.

The synthesis of biological molecules by autotrophs is endergonic. It therefore requires energy in the form of ATP. Autotrophs get this energy in one of two ways. **Photoautotrophs** convert the energy in light to ATP. **Chemoautotrophs** use the chemical energy in certain molecules to accomplish this same task. We will next briefly consider these two categories of self-feeding organisms.

In photosynthesis, energy in sunlight is used to produce biological molecules

Photoautotrophs—green plants and some microorganisms—have a unique skill: they can create biological molecules out of carbon dioxide and hydrogen-containing compounds, using sunlight as an energy source. This ability is called **photosynthesis**.

Not all photosynthesis happens in precisely the same way, but for green plants and cyanobacteria, photosynthesis is in many respects the opposite of aerobic cell respiration (**Figure 7.18**). We have already seen that, in aerobic respiration, glucose ($C_6H_{12}O_6$) is digested and the carbon atoms that *were* part of the glucose are released as carbon dioxide. Oxygen serves as the final electron acceptor, and once it has accepted electrons it is released as water. Because this process is exergonic, released energy is used to make ATP.

In photosynthesis, on the other hand, glucose is being *made* rather than digested (see Figure 7.18). The raw materials needed to make glucose are carbon, oxygen, and hydrogen atoms. The carbon and oxygen come from carbon dioxide, while the hydrogen atoms are supplied by water. When hydrogen atoms are stripped away from water for use in glucose construction, the remaining oxygen is released as a waste. In photosynthesis, energy is *required* rather than released, because sugar production is endergonic. Sunlight provides the initial energy source for glucose synthesis.

Photosynthesis is basically a two-step process (**Figure 7.19**). In the first step, a series of reactions called the **light reactions**, the energy in sunlight is used to produce ATP. Additionally, a molecule of **NADP$^+$** (nicotinamide adenine dinucleotide phosphate) is reduced. The resulting reduced molecule,

Aerobic cell respiration

Final electron
acceptor for aerobic
respiration is oxygen.

Reduced form
of the final electron
acceptor is water.

$$C_6H_{12}O_6 \; + \; 6O_2 \longrightarrow 6CO_2 \; + \; 6H_2O \; + \; \text{Energy (ATP)}$$

Glucose

Carbon dioxide
waste remains
from glucose after
it is completely
oxidized.

Energy is harvested
from the exergonic
(energy-releasing)
digestion of glucose.

Photosynthesis

$$\text{Energy (sunlight)} \; + \; 6H_2O \; + \; 6CO_2 \longrightarrow C_6H_{12}O_6 \; + \; 6O_2$$

Sunlight supplies
energy required for
the endergonic
(energy-in) synthesis
of sugar.

Water is the
source of
hydrogen
atoms
for sugar
synthesis.

Carbon dioxide
is the source of
carbon atoms
for sugar
synthesis.

Glucose is
the end
product of
photosynthesis.

Oxygen is the
photosynthetic
waste formed
when hydrogen
is split from
oxygen in water.

Figure 7.18 Summary equations for aerobic cell respiration and photosynthesis. Aerobic cell respiration and photosynthesis in green plants and some bacteria are largely opposites of each other. In cell respiration, glucose is digested and the released energy is used for ATP synthesis. In aerobic respiration, oxygen serves as the final electron acceptor and is ultimately reduced to water. When glucose is fully oxidized and all carbon–hydrogen bonds are broken, remaining carbon is released as carbon dioxide. In photosynthesis, glucose is produced rather than digested, and energy is required, rather than released. Carbon dioxide and water provide the raw materials for glucose synthesis.

NADPH, serves as a source of hydrogen with which to construct the sugar molecule. NADP$^+$ and NADPH are structurally very similar to NAD$^+$ and NADH (see page 175) and they serve a similar function: the transfer of electrons from one molecule to another. In step 2 of photosynthesis, ATP and NADPH made in the light reactions are utilized to synthesize sugar. Usually, this occurs in a set of reactions called the **Calvin cycle**. In the Calvin cycle, CO_2 absorbed from the environment is reduced in a series of steps by NADPH to produce sugar. ATP provides the energy for this endergonic process.

The light reactions depend on the presence of the green pigment chlorophyll. The light-absorbing chlorophyll molecules form complexes, which are embedded into membranes. In cyanobacteria, these membranes are called **thylakoids** (**Figure 7.20a**). In other prokaryotes, chlorophyll may be found in the plasma membrane. In eukaryotes, chlorophyll complexes are embedded in membranes found inside organelles called **chloroplasts** (Figure 7.20b; also see Chapter 3, p. 67).

Light striking a chlorophyll complex excites electrons in the chlorophyll, elevating them to high energy (**Figure 7.21**). These energized electrons are then captured by components of an electron transport chain. As electrons travel down the chain between the chlorophyll complexes, protons are pumped across the membrane. The return of these protons through ATP synthase results in the production of ATP in a manner similar to what was described for cell respiration. At the end of the electron transport chain, the electrons reduce NADP$^+$ to form NADPH. In green plants and cyanobacteria, light also is used to split a molecule of water. Electrons extracted from the water replace those lost by chlorophyll at the beginning of the light reactions. Oxygen (O_2) is released at this "water-splitting" step. In other photosynthetic prokaryotes, other molecules, such as hydrogen gas (H_2) or

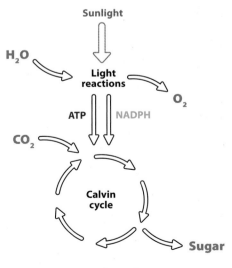

Figure 7.19 Overview of photosynthesis. In the light reactions, energy in sunlight is used to generate ATP, and NADPH is formed. These products are then used in the Calvin cycle, along with atmospheric CO$_2$, to synthesize sugar.

Figure 7.20 Role of membranes in light reactions. Light reactions require the light-absorbing green pigment chlorophyll. Chlorophyll is found embedded in cell membranes. (a) In cyanobacteria, these membranes are called *thylakoids*, and the chlorophyll in these membranes gives the cells their green color. In many other photosynthetic prokaryotes, chlorophyll is embedded in the plasma membrane. (b) In eukaryotic plants and algae, chlorophyll is found in the membranes of organelles called *chloroplasts*, seen here in these plant cells as the small green structures within the cell.

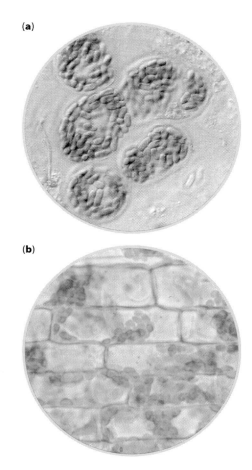

(a)

(b)

hydrogen sulfide (H_2S), are split to provide a source of electrons. These photosynthetic microbes do not release oxygen.

The ATP and NADPH produced in the light reactions are then used to produce biological molecules, most commonly in the Calvin cycle. The principal event in the Calvin cycle is **carbon fixation**—the incorporation of atmospheric CO_2 into organic molecules. The Calvin cycle occurs in a series of steps, as illustrated in **Figure 7.22**. The resulting product is glucose or other biological molecules that are required by the cell. The Calvin cycle occurs in the cytoplasm of prokaryotes or inside the chloroplasts of eukaryotes.

Chemoautotrophs use chemical energy the way that photoautotrophs use solar energy

Chemoautotrophs, like photoautotrophs, make their own biological molecules. They also use CO_2 as a carbon source, but instead of light energy, chemoautotrophs use reduced compounds such as ammonia (NH_3), methane (CH_4), or hydrogen sulfide (H_2S) as an energy source. The energy released when these compounds are oxidized is ultimately converted to ATP, which can be used to synthesize biological molecules. We have recently described ammonia, methane, and hydrogen sulfide as wastes released by anaerobic bacteria. Chemoautotrophs use these waste products and thus play crucial roles in nutrient cycling and other fundamental ecological processes. We will revisit the topic of how microorganisms help to recycle nutrients in the environment in Chapter 9.

Microbial Metabolism and Growth

You may be wondering why, having completed our discussion of metabolism, we are now moving on to microbial growth. The reason is simple. The rate at which microbial populations grow is in large part dependent on metabolism. Those microorganisms that are most efficient at producing and using energy tend to grow the fastest. We have already mentioned, for instance, that aerobic organisms usually produce more ATP per molecule of

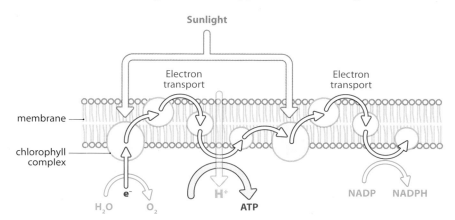

Figure 7.21 Light reactions in green plants and cyanobacteria. Light striking chlorophyll complexes energizes electrons within those complexes. These high-energy electrons are then passed through a series of electron carriers in an electron transport chain. As these electrons are shuttled along the chain, they lose energy, and as seen in cell respiration, this energy is used to pump protons across the membrane. As these electrons pass back across the membrane through an ATP synthase complex, ATP is generated, as in cell respiration. The final electron acceptor in the light reaction electron transport chain is NADP. As it accepts these electrons, it is reduced to NADPH. The electrons that reduce NADP to NADPH, originally lost by the chlorophyll complexes, are replaced by electrons removed from water, and oxygen is released as a waste. The products of the light reactions (ATP and NADPH) will be used in the Calvin cycle in the synthesis of sugar. Other photosynthetic microorganisms rely on molecules other than water to provide electrons and consequently do not generate oxygen.

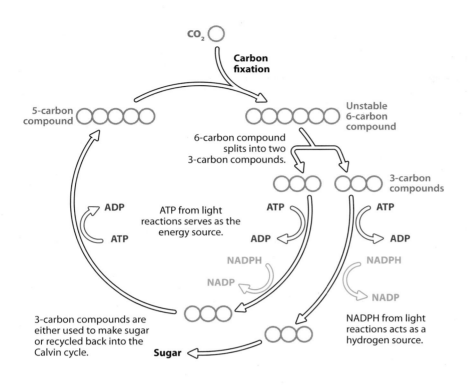

Figure 7.22 The Calvin cycle. In the Calvin cycle, carbon dioxide is fixed into sugar molecules, using NADPH and ATP, made in the light reactions, as sources of hydrogen atoms and energy. Carbon dioxide from the environment is first is linked to a 5-carbon compound (carbon fixation), forming an unstable 6-carbon compound that immediately splits into two 3-carbon compounds. The resulting 3-carbon compounds either remain in the cycle or are used to synthesize sugar.

glucose than anaerobic organisms. Consequently, as a general rule, aerobic microorganisms increase in numbers much faster than anaerobes do.

But things are not quite that elementary. Environmental variables such as nutrient availability, pH, temperature, and oxygen levels impact metabolic activity, and as metabolism changes, we expect the growth rate to change. Because different species have different requirements for such factors, different species respond in various ways to changes in the environment.

Why should we care about any of this? As we have seen throughout this text, some microorganisms, such as certain pathogens, pose grave risks to our health and well-being. Others cause less critical problems ranging from food spoilage to underarm odor to unsightly buildups on toilet bowls and kitchen drains. Many other microorganisms are more benign and provide a variety of useful services. Such organisms help us fend off attack by pathogens, produce food and industrial products, and maintain a healthy environment. Knowing which conditions permit optimal metabolic efficiency and thus maximum growth for a given microorganism allows us to either inhibit growth of the "bad guys" or spur on the growth of beneficial microbial species.

Microbial growth refers to population growth

When we talk about microbial growth, we usually mean population growth rather than an increase in size. We are referring to an increase in the *number* of cells, not the size of individual cells. Although individual microbial cells do in fact get bigger, this growth is of little significance compared with the astronomical increases in population size that microorganisms can achieve under favorable conditions.

Reproduction, and consequently increase in numbers, is generally a straightforward affair in prokaryotes. Most bacteria and archaea divide by **binary fission**—the simple division of an original parental cell into two new cells (**Figure 7.23**). This process begins with the cell's elongation and the replication of its single chromosome. The replicated chromosome becomes attached to the plasma membrane, and new membrane is formed between the two chromosome attachment sites, moving the two chromosomes apart. Subsequently, new plasma membrane and cell wall material forms along the cell's midsection, dividing the cytoplasm into two sections, each with its own chromosome. The separation of these two sections gives rise to two new daughter cells.

Binary fission can quickly give rise to an enormous population, because it can result in exponential growth. As long as conditions remain favorable, the population can double every time the cells in that population divide. A single cell divides in two, the division of two cells results in four, four give rise to eight, and so forth. The time between divisions is called the **generation time**. Some species have a generation time as short as 20 minutes. Others may require 24 hours or more. Obviously, the length of the generation time has important implications for how fast the population grows.

Consider the consequences of exponential growth by binary fission. Suppose we are measuring population size in *Escherichia coli*, which under ideal laboratory conditions can divide every 20 minutes or so. If we start with a liquid broth containing 100 cells per milliliter, in 1 hour (three generation times) we would have 800 cells per milliliter (**Table 7.1**). After 2 hours the population would be 6400 cells per milliliter, and in just 5 hours we would have over 3 million cells per milliliter. Of course, cells are rarely fortunate enough to experience ideal growth conditions for long, and population increase of this sort cannot occur forever. As the population continues to increase, the growth conditions deteriorate as resources such as nutrients, oxygen, and space are used up. Nevertheless, it is worth emphasizing that, given the opportunity, prokaryotic organisms are capable of astonishing feats of population increase.

As you might guess, a graph showing the number of generations and the number of cells in each generation quickly becomes unwieldy because of the very rapid rate of increase (**Figure 7.24**). Consequently, growth curves

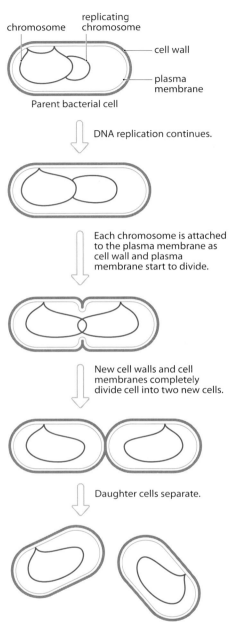

Figure 7.23 **Binary fission.** A single bacterium gives rise to two daughter cells. Each daughter cell, in turn, can give rise to two further daughter cells, and so forth.

TIME	GENERATION NUMBER	CELLS PER MILLILITER
0 min	0	100
20 min	1	200
40 min	2	400
60 min	3	800
80 min	4	1600
100 min	5	3200
120 min	6	6400
140 min	7	12,800
160 min	8	25,600
180 min	9	51,200
200 min	10	102,400
220 min	11	204,800
240 min	12	409,600
260 min	13	819,200
280 min	14	1,638,400
300 min	15	3,276,800

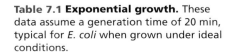

Table 7.1 **Exponential growth.** These data assume a generation time of 20 min, typical for *E. coli* when grown under ideal conditions.

are often plotted in terms of logarithms (powers of 10) over a given period of time or number of generations. A logarithmic value of 1 is equal to 10 (10^1 or 10×1). A logarithmic value of 2 is equal to 100 (10^2 or 10×10) while a logarithmic value of 3 is equal to 1000 (10^3 or $10 \times 10 \times 10$), and so on. By plotting time versus the logarithm of cell numbers, we obtain a more manageable straight line as the cells grow exponentially (**Figure 7.25**).

Why a straight line instead of a stair-step pattern? Put another way, if all cells waited exactly 20 minutes to divide, wouldn't there be 20-minute periods of no growth, followed by a sudden doubling in the number of cells (see Figure 7.25)? Such a growth pattern would reflect **synchronous growth**—a situation in which all the cells in a population divide at the same time. Bacterial cultures may approximate synchronous growth initially, but different cells very quickly get out of synchrony. Some are dividing while others have just divided. Some take a little longer to divide while others take somewhat less time. The result is a smooth ascent in population size, and the generation time for the population is actually the average generation time for all the cells in the population.

Oxygen, temperature, nutrient levels, and other environmental factors all influence microbial growth rate

In the mid-1990s, Diana Duyser made herself a grilled cheese sandwich. As she went to take a bite, she noticed a strange image on the sandwich. To the Florida woman, the image looked like the Virgin Mary. She kept the sandwich in a plastic box for 10 years, and mysteriously no mold ever grew, proof to some this was indeed a miracle of divine intervention. The story drew pilgrims and widespread media attention. In 2004, Ms. Duyser sold the sandwich on eBay for $28,000.

We may never know if the lack of fungal growth was due to spiritual or physical causes. Fungi, however, like all living things, require not only adequate nutrition but also a hospitable environment where conditions such as temperature, moisture, and pH are compatible with survival, growth, and reproduction. Throughout this text we have seen numerous examples of how microorganisms may differ in terms of what constitutes a favorable environment. Earlier in this chapter, for example, we learned how some microorganisms depend on oxygen as a final electron acceptor in electron transport, while others use compounds such as sulfate or nitrate.

In the case of Ms. Duyser's sandwich, microbiologists have speculated that any one of several factors might have prevented the fungal cells from sprouting. These include preservatives in the bread or the sandwich's dried-out condition. The acidity or calcium in the cheese may have helped to inhibit fungal growth. Whatever the cause, divine or otherwise, this example nicely illustrates that in the absence of necessary conditions living things are unable to thrive.

In this section we will take a closer look at how oxygen and other crucial physical and nutritional factors can affect the metabolism, and consequently the growth, of different microorganisms. As we will see, knowing a particular organism's environmental requirements allows us to predict where it lives, and in some cases how it may interact with humans or other living things.

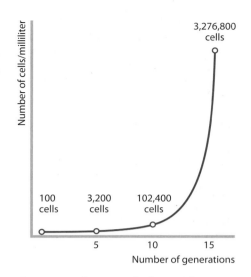

Figure 7.24 Astronomical growth potential of microbes. The graph depicts the potential for exponential growth over a number of generations.

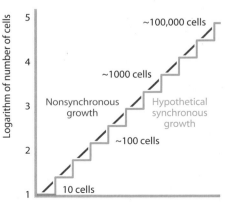

Figure 7.25 Logarithmic microbial growth. If all cells in a growing culture divided at exactly the same time, the result would be synchronous growth and the graph of growth over time would resemble a stair-step pattern. Cells in a growing culture, however, very quickly divide out of synchrony, resulting in a smooth ascent in population size.

CASE: CANINE FIRST AID

Jill has quite a menagerie of pets and she regularly has to deal with various minor animal injuries. So when Johnnie, her rooster, spikes Jessie, her collie, on the muzzle with one of his long leg spurs, Jill knows what to do; she reaches for the hydrogen peroxide and thoroughly floods Jessie's puncture wound with the liquid. When she applies the hydrogen peroxide, Jill sees that, as always, heavy fizzing occurs at the site of the wound. She repeats this process several times a day for the next week, carefully inspecting the puncture for signs of infection. The injury heals without complication, and Jessie is none the worse for wear, although he has a new respect for roosters.

1. What is hydrogen peroxide, and how does it reduce the risk of infection?
2. Why is hydrogen peroxide especially useful for treating puncture wounds?
3. What causes the fizzing associated with hydrogen peroxide use?

Oxygen. Oxygen can affect growth in different ways, depending on the species we are considering. Our discussion of aerobic respiration demonstrated that atmospheric oxygen is important for many forms of life. Indeed, most multicellular organisms and a significant number of microbial species are unable to survive without it. Organisms that require oxygen are called **obligate aerobes**. They need oxygen because it is the *only* final electron acceptor suitable for electron transport. They are, therefore, "obligated" to live where oxygen is present.

For other microorganisms, oxygen is important because it's so deadly. It consequently becomes something to be avoided rather than sought out. Why exactly must some species steer clear of oxygen at all costs?

When oxygen takes part in biological reactions, it can be converted into a number of compounds that have a high likelihood of removing electrons from other biological molecules. In this respect, these reactive oxygen compounds are powerful *oxidizing agents*, which can damage other cellular components. Examples of such oxygen compounds include the superoxide ion (O_2^-) and hydrogen peroxide (H_2O_2). Cells that utilize oxygen in cellular respiration must be able to deal effectively and extremely quickly with these highly reactive and damaging compounds. To do so, aerobic cells have developed enzymes that are able to detoxify them. For example, the conversion of hydrogen peroxide into harmless oxygen and water takes place spontaneously; but cells require an enzyme called catalase to speed up the decomposition so that H_2O_2 causes no damage. Catalase converts two molecules of hydrogen peroxide into two molecules of water and one molecule of oxygen:

$$2H_2O_2 \rightarrow 2H_2O + O_2$$

Aerobic microorganisms come equipped with enzymes such as catalase to deal with reactive oxygen compounds such as hydrogen peroxide, but many anaerobes lack them. These are the **obligate anaerobes**—they are obligated to *avoid* oxygen, because if oxygen is present, reactive oxygen compounds might be formed, and without detoxifying enzymes, reactive oxygen is lethal. Since oxygen is not an option as a final electron acceptor, obligate anaerobes rely on compounds such as sulfate (SO_4^{2-}) and nitrate (NO_3^-), and they tend to thrive in low-oxygen environments, such as lake bottoms or soil. The Michigan swamp we described at the beginning of this chapter is a perfect environment for obligate anaerobes, because there is little oxygen in swamp water and because decaying plant material releases other, alternative final electron acceptors.

Because many anaerobes produce little or no catalase, hydrogen peroxide can be used as an antiseptic to help keep wounds infection-free. This is especially true for a puncture wound like Jessie's, because puncture wounds create ideal oxygen-free environments for obligate anaerobes. Because obligate anaerobes introduced into the wound lack catalase, the hydrogen peroxide used to irrigate the injury is likely to kill them, preventing infection. The fizzing that Jill observed is mostly due to the catalase that is present in Jessie's own cells. The enzyme converts some of the hydrogen peroxide to oxygen and water, and the oxygen bubbles that are liberated cause the fizz (**Figure 7.26**).

The oxygen requirements of other microbial species are not so cut and dried, and metabolic "switch hitters" abound. As previously mentioned, some microorganisms will use different final electron acceptors, depending on availability. A large proportion of microorganisms, including many familiar groups such as *Staphylococcus* and *E. coli*, are **facultative anaerobes** (**Figure 7.27**). They readily utilize oxygen for aerobic respiration when oxygen is available, and they usually possess enzymes such as catalase. In the absence of oxygen, facultative anaerobes rely on fermentation alone or anaerobic respiration to meet their energetic needs.

Microbial diversity in terms of oxygen usage does not stop there. **Aerotolerant anaerobes** can partially break down toxic oxygen compounds, but they do not use oxygen for cell respiration. Consequently, oxygen has no effect, either positive or negative on their growth. They simply ignore it, and they grow equally well in environments with or without oxygen (see Figure 7.27). Finally, there are the **microaerophiles** that require a small amount of oxygen but are unable to grow at normal atmospheric oxygen concentrations.

Temperature. Like oxygen, temperature has a major impact on metabolism and growth. Furthermore, a comfortable temperature for one microorganism can be freezing or sweltering for another. As discussed in Chapter 4, some microorganisms are capable of growth at extremely high temperatures and are termed **thermophiles** ("heat-loving"). Others, called **psychrophiles** ("cold-loving"), grow only at low temperatures. Many microorganisms grow best at temperatures close to animal body temperature. Because they prefer intermediate or mid-range temperatures, they are called **mesophiles**. Most microorganisms grow over a range of temperatures, allowing the microbiologist to determine a **minimum growth temperature** (the lowest temperature allowing growth) and a **maximum growth**

Figure 7.26 Catalase in action. The obvious fizzing that is observed when hydrogen peroxide is used as an antiseptic is caused by the release of oxygen. In this example, the hydrogen peroxide is being broken down by catalase-positive bacteria. If catalase were placed in contact with a catalase-negative, obligate anaerobe, no fizzing would be observed.

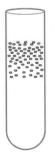 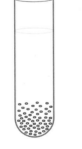 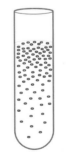 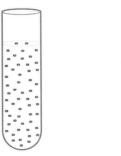

Obligate aerobes require oxygen and are catalase-positive.

Obligate anaerobes cannot tolerate oxygen and are catalase-negative.

Facultative anaerobes grow best with oxygen but can also grow anaerobically. They are catalase-positive.

Aerotolerant anaerobes tolerate but do not use oxygen. They are catalase-positive.

Microaerophiles use small amounts of oxygen. They produce small amounts of catalase.

Figure 7.27 Oxygen use by prokaryotes. All organisms able to survive in the presence of oxygen have the catalase enzyme necessary to help detoxify reactive oxygen compounds. A lack of catalase and other detoxifying enzymes is the reason why obligate anaerobes cannot tolerate oxygen. Oxygen concentration is high near the surface of the agar tube and low near the bottom. The location of microbial growth reflects the manner in which each type of organism responds to oxygen. Obligate aerobes grow near the surface, where oxygen concentration is the highest, while obligate anaerobes grow only at the bottom of the tube. Facultative anaerobes grow best near the surface but are able to grow at lower concentrations of oxygen as well. The growth of aerotolerant anaerobes is independent of oxygen concentration, while microaerophiles grow best slightly below the surface, where oxygen concentration is somewhat less.

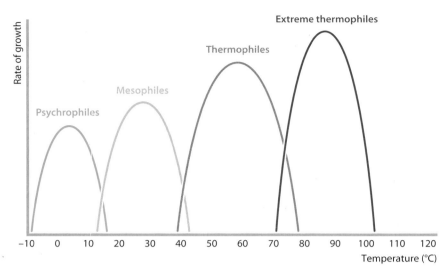

Figure 7.28 Microbial growth rates in response to temperature. Note that for each of the four representative microbial types, there is a minimum growth temperature, below which growth cannot occur, and a maximum growth temperature, above which growth cannot occur. At the minimum and maximum growth temperatures, growth rate is slow, and the generation time is relatively long. Each type of organism also has an optimum growth temperature, at which growth is most rapid with the shortest generation time.

temperature (the highest temperature allowing growth). Population growth will be highest at some intermediate value (the **optimum growth temperature**) between these two extremes (**Figure 7.28**).

It is often relatively easy to predict where one would find thermophiles, psychrophiles, or mesophiles. Clearly if you were to sample a hot spring for microorganisms, you would expect to find thermophiles, while cold ocean water is a habitat of choice for psychrophiles. Most of the organisms found in an animal's body, whether pathogenic or normal flora, are mesophiles.

These temperature preference categories are not rigidly defined, and many microorganisms are somewhat harder to pigeonhole. For example, food is refrigerated to discourage the growth of mesophiles that can lead to food spoilage. Clearly, however, food often does spoil even in refrigerators, meaning that some microorganisms are able to grow, even at these chilly temperatures. Perhaps surprisingly, however, most of the common "food spoilers" are not psychrophilic. Although they are able to grow at temperatures close to freezing, they grow better between 20 and 40°C and are often called **psychrotolerant**—they "tolerate" the cold, even though they prefer warmer temperatures. Among mesophiles that infect humans there is often variability in the optimum growth temperature, sometimes with interesting consequences. *Mycobacterium leprae*, the causative agent of leprosy, for example, typically infects fingers, ears, feet, and other cooler regions of the human body. It is unable to grow well at higher body temperatures found elsewhere on the host. *Treponema pallidum*, which causes syphilis, is especially sensitive to any deviation from its preferred temperature. Indeed, for many years syphilis was actually treated by inducing a high fever to inhibit growth of the bacterium. This was accomplished by deliberately infecting patients with malaria, which causes body temperature to skyrocket.

A variety of factors interact to determine a microbe's temperature preference. To a large degree, the maximum growth temperature is determined by the temperature at which a cell's enzymes begin to denature (unfold) and lose their ability to function. Lipid composition in the plasma membrane is also different in psychrophiles and mesophiles. You may recall from Chapter 2 that saturated fatty acids form straight hydrocarbon chains, while unsaturated fatty acids have twists and kinks formed by carbon–carbon double bonds. This allows saturated fatty acids to be packed together much more tightly. The looser packing associated with unsaturated fatty acids makes them correspondingly more liquid. Consequently, psychrophiles incorporate a large proportion of unsaturated fatty acids in their membrane to help keep their membranes from freezing. Thermophiles, on the other hand, rely on densely packed saturated fatty acids to help prevent their membranes from melting.

pH. In Chapter 2, we learned that pH is a measure of the concentration of H^+ ions or acidity, and the lower the pH, the greater the acidity. Like temperature and oxygen, the environmental pH can have a substantial impact on microbial growth. Most species grow best when the environment is neither very acidic nor very basic: in other words, at a pH close to neutrality (pH = 7.0). There are, however, some interesting exceptions. In Chapter 2, for example, we encountered the case involving *Acidithiobacillus ferrooxidans*, thriving in the highly acidic leachings of a coal mine. **Acidophiles** such as *A. ferrooxidans* grow best at very low pH. Many fungi are acidophiles, which explains why microbial contaminants of acidic foods such as tomatoes and oranges are usually fungal (**Figure 7.29**). Alkaliphiles, on the other hand, thrive in basic environments, such as soda lakes, with pH values above 8.5. As with temperature, some species defy easy categorization. *Helicobacter pylori*, for example, thrives in the stomach of many people. Because the stomach is so acidic, you might guess that *H. pylori* is an acidophile. But is it? As it grows on the stomach lining, this bacterial species produces and excretes the enzyme urease, which converts urea in the stomach into ammonia. Ammonia is basic, and it neutralizes the stomach acid in the immediate vicinity of the bacteria. In effect, the bacteria create a small "island" of neutrality in a sea of acidity. This bacterial species has gained notoriety in recent years for another reason. Traditionally, stomach ulcers were believed to be caused by an oversecretion of stomach acid. In the 1980s, however, an Australian physician found he could successfully treat many stomach ulcers with antibiotics. This led him to the remarkable conclusion that most stomach ulcers were caused not by acid, but by bacteria. Although this notion was originally the subject of skepticism and ridicule, it is now clear that *H. pylori* is the culprit in the majority of stomach ulcer cases. Antibiotics are now routinely employed to treat such ulcers.

Salt. Before the advent of refrigeration, salt was routinely used as a food preservative (**Figure 7.30**), because it is so good at inhibiting microbial growth. Most microorganisms are unable to grow in the presence of high salt concentration, because the salty environment draws water out of the cells, causing them to dehydrate. Yet some microorganisms, the **halophiles** ("salt-loving"), not only survive but thrive in extremely saline habitats. In Chapter 4, for example, we encountered halobacteria, a group of archaea that prefer a salt concentration many times higher than that found in the sea. In some habitats such as the Great Salt Lake in Utah, halophiles account for essentially all the aquatic microbial life.

Nutritional factors. In addition to a favorable physical environment, microorganisms, like all living things, must obtain certain crucial nutrients if they are to grow efficiently. Carbon-based biological molecules are needed by

Figure 7.29 pH as a factor in microbial growth. An acidophilic fungus growing on a tomato.

Figure 7.30 Salted cod drying at Smith's Wharf, Halifax, Nova Scotia, circa 1950. Salt has routinely been used to prevent microbial growth in food products for thousands of years.

most cells to produce ATP as previously described, and many carbon-containing compounds serve as building blocks for cellular components. Microorganisms also need phosphorus for the synthesis of nucleic acids and membrane phospholipids. Protein synthesis depends on a reliable supply of nitrogen. Certain vitamins are needed because they form important parts of enzymes. Small amounts of certain minerals or "trace elements" such as zinc or magnesium also interact with proteins to form a complete enzyme. Complexes in the electron transport chain depend upon small amounts of iron or other metallic ions. If these necessary ions are not available, ATP production, and consequently growth, is curtailed.

Certain pathogenic microorganisms take drastic measures if the trace elements they require are not present in necessary amounts. *Corynebacterium diphtheriae*, for instance, lives in the throats of many people, often without causing problems. In others it produces a powerful protein toxin that results in the disease diphtheria. Why do these bacteria, which can sometimes live peacefully in your throat, suddenly turn vicious? In the presence of abundant iron the diphtheria-causing toxin is not produced. The toxin is coded for by an operon (see Chapter 6). Operon expression is usually repressed, because the operator is bound by a repressor protein. And to bind to the operator, the repressor must first bind to iron. When iron is present, some iron binds the repressor, and the iron/repressor complex binds the operator, silencing the operon. When iron is lacking, the repressor cannot bind the operator, and the toxin genes are transcribed and translated. The result is diphtheria. As cells in the upper respiratory tract are killed by the toxin, the iron in these host cells is liberated, giving the bacterial cells the iron they need to survive.

Because various microorganisms have particular nutritional requirements, a microbiologist trying to rear them in the laboratory will need to provide different "diets" to different species. Microorganisms that cannot make many of the biological compounds required for growth are called **fastidious**. Growing fastidious organisms in the laboratory can be tricky, because their nutritional requirements are complex. *Lactobacillus* is a common example. To grow these bacteria in the laboratory, a special **chemically defined medium**, containing specific amounts of purines, pyrimidines, and various vitamins and amino acids, is required. **Nonfastidious** organisms, on the other hand, are easier to grow. Because they are capable of making almost all the biological molecules that they require, they can be grown on **minimal medium**, containing little more than a carbon source and a few other inorganic minerals and salts. Many species, including most medically important bacteria, are grown on **complex medium**, containing a variety of nutrients such as yeast extracts or predigested proteins. Even if a bacterium is capable of synthesizing some of these compounds on its own, growth is accelerated when these biological molecules are provided.

Environmental and nutritional factors, coupled with microbial diversity in terms of metabolism, show just how complicated it can be to coax microorganisms to grow (or not grow) as we would like. Nevertheless, whether or not microorganisms are behaving the way we want, there are certain predictable features to their growth. We will conclude this chapter by looking at the sequence of events as a microbial population grows, levels out, and ultimately decreases.

Microbial populations pass through a sequence of phases called the growth curve

When we are sick with food poisoning or a bacterial respiratory infection, a reduction in bacterial growth, or at least a slowing of population increase, is clearly in our best interest. In other situations, when bacteria are churning out a useful product at our behest, we would like them to grow as quickly as possible. To encourage microorganisms to grow (or not grow) in a manner that suits us, it is useful to understand the phases of growth that all microbial populations go through.

The growth of a bacterial population typically goes through four characteristic phases: the lag, log, stationary, and decline phases (**Figure 7.31**). To describe these phases, consider a small number of cells that have just been inoculated onto fresh medium in the laboratory. For the purposes of illustration, assume that initially all conditions are conducive to optimal growth.

Lag phase. For a time there will be little if any change in the numbers of cells on our growth medium. During this **lag phase**, although the population is not increasing, the cells are preparing to divide. The length of the lag phase depends on a variety of factors, including the normal generation time for the species in question, as well as the environmental conditions the cells were experiencing before they were placed on fresh medium. If the fresh medium contains the same nutrients that the original medium contained, the lag period will be relatively short. If the newer medium contains different nutrients, the lag period will increase in length, because the bacteria cannot begin growth until they have synthesized the enzymes needed to digest these new nutrients.

Log phase. Eventually the cells will begin to divide and enter a period of exponential or logarithmic growth called the **log phase** (see Figure 7.31). Cells in log phase are metabolically at their most active, and they divide at their maximal rate. This is not to say that no cells are dying during this time, but during log phase, the production of new cells by binary fission greatly outpaces any cell death.

Stationary phase. As previously mentioned, rapid growth cannot continue indefinitely. As cell number increases during the log phase, resources such as nutrients become depleted. Cells often produce waste products that begin to build up, inhibiting growth, and in some cases cells simply fill up all the available space. Consequently, growth begins to slow and the death rate of cells begins to increase. At some point the rate of new cell increase approximates the rate at which cells are dying, and we say that the population has entered the **stationary phase**, a period during which the population remains more or less constant (see Figure 7.31).

Decline phase. Although fewer new cells are produced during the stationary phase, those cells that are produced continue to use resources and produce wastes. Consequently, the conditions for growth deteriorate to such a point that the number of dying cells outnumbers any production of new cells, and the population begins to decrease (see Figure 7.31). Bacterial populations at this stage of growth are said to have entered the **decline phase**. The population may be drastically reduced to only a small number of cells, or it may die out entirely.

As the environment changes, microbial metabolism and therefore the growth curve changes in response

CASE: MAKING YOGURT

Sheri and Jim have always liked cooking and experimenting in the kitchen, so when Sheri sees a recipe for yogurt, they decide to give it a try. They first heat whole milk to about 170°F to reduce the numbers of unwanted bacteria. They then let the milk cool and pour it into a dozen cups. To each cup they add exactly 1 tablespoon of powdered milk and 1 teaspoon of plain yogurt containing active *Lactobacillus acidophilus* cultures (**Figure 7.32**), completely mixing all ingredients. The directions state to carefully

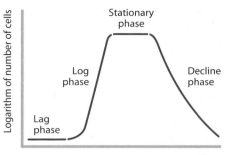

Figure 7.31 Four phases of microbial growth. Bacterial populations typically pass though four characteristic phases in their growth. Initially, there is a period of no growth called the *lag phase*, during which bacterial cells are preparing to divide. During the *log phase*, cell division is at its maximum, with the production of new cells exceeding cell death. As conditions deteriorate, growth slows, and eventually reproduction and death are approximately equal. This corresponds to the *stationary phase*—a time when the population is neither increasing nor decreasing. Finally, populations enter the *decline phase*, during which cell death exceeds the production of new cells.

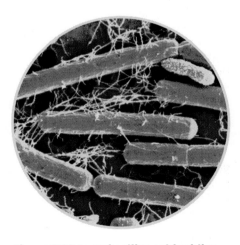

Figure 7.32 *Lactobacillus acidophilus*. This facultative anaerobe can grow in either the presence or absence of oxygen. When oxygen is limited, the reduction of pyruvate during fermentation results in the production of lactic acid, which is released as a waste product. When these bacteria are used in the production of yogurt, the released lactic acid causes milk proteins to denature, giving yogurt its firm texture. It also gives unsweetened yogurt its characteristic sour taste.

seal each cup completely with plastic wrap, but after sealing the first eight cups, they run out of wrap. They place all 12 cups, however, including the uncovered ones, in the oven on the lowest possible setting (120°F). After about 10 hours, they check the progress of their yogurt. The covered cups have all solidified, and Sheri and Jim, after adding some sugar and fresh fruit, enjoy real homemade yogurt. The cups that remained uncovered are still fairly liquid.

1. Why did only the sealed cups produce yogurt?
2. In which of the cups, the sealed or the unsealed ones, was bacterial metabolism more efficient? Which of the cups contained more bacterial cells? If growth curves could be made for the bacteria in sealed and unsealed cups, how would they differ?
3. After each having a cup of yogurt, Jim and Sheri place the remaining cups into the refrigerator. How would this alter the growth curves for the bacteria in the yogurt?

Although microbial populations all go through the same basic stages of growth, the growth curves for different species will appear somewhat different. Consider, for instance, the log phases of an aerobe growing in the presence of oxygen and an anaerobe growing in the absence of oxygen. Generally speaking, we would expect that the rate of increase would be significantly faster for the aerobe for the simple reason that aerobic respiration usually results in a greater ATP yield than anaerobic respiration does (**Figure 7.33a, b**). Consequently, reproduction for the aerobe is typically faster.

The shape of the growth curve will be impacted by the other factors as well. The lag phase will become longer and the log phase will rise more slowly as physical and nutritional conditions become less ideal. Cells growing under less than ideal conditions may enter the stationary phase sooner, and the reduction in population during the decline phase may be steeper. Furthermore, the shape of the curve may change as conditions change. Imagine, for example a thermophile with a minimum growth temperature of 40°C, a maximum growth temperature of 75°C, and an optimum growth temperature of 65°C. If we prepare a culture of these cells, incubate them at 45°C, and then plot their growth curve, we will see that after a prolonged lag phase the cells start to increase in number slowly (see Figure 7.33c). If we then change the incubation temperature to 65°C, we should note that the log phase becomes considerably steeper, rising more quickly.

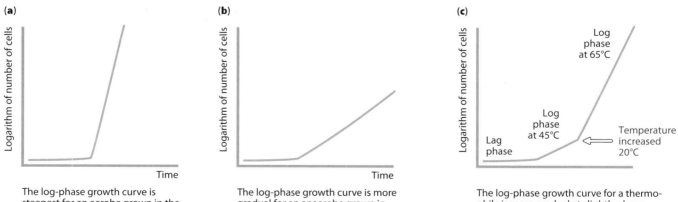

(a)

Logarithm of number of cells / Time

The log-phase growth curve is steepest for an aerobe grown in the presence of oxygen.

(b)

Logarithm of number of cells / Time

The log-phase growth curve is more gradual for an anaerobe grown in the absence of oxygen.

(c)

Logarithm of number of cells

Log phase at 65°C

Log phase at 45°C

Lag phase

Temperature increased 20°C

The log-phase growth curve for a thermophile is more gradual at slightly above minimum growth temperature and steeper at optimum growth temperature.

Figure 7.33 Effect of metabolism and growth conditions on rate of growth. Logarithmic-phase growth for (a) an aerobe growing in the presence of oxygen and (b) an anaerobe growing in the absence of oxygen. The faster growth rate for the aerobe reflects the higher ATP yield achieved through aerobic respiration. (c) Effect of temperature on logarithmic growth. For this thermophile, 45°C is just above the minimum growth temperature. Consequently growth is slow. At the optimum growth temperature (65°C) the rate of growth is significantly higher.

(a)

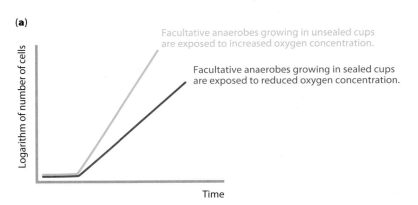

(b)

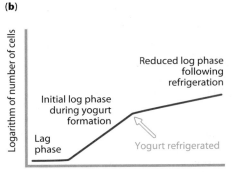

Figure 7.34 *L. acidophilus* growing under different conditions. (a) As a facultative anaerobe, *L. acidophilus* can grow in either the presence or absence of oxygen, but growth is faster when oxygen is present. The bacteria in the sealed cups are exposed to less oxygen, and therefore are depending to a greater extent on fermentation than those bacteria in the unsealed cups. (b) When placed in a colder environment, the growth rate of this psychrotolerant species slows.

The *L. acidophilus* used by Jim and Sheri to make yogurt provides an excellent way to illustrate some of the key points we have made regarding metabolism and its effect on growth. This bacterial species is a facultative anaerobe; it can grow in either the presence or absence of oxygen, but metabolism will be more efficient and growth will be faster when oxygen is available (**Figure 7.34a**). To make proper yogurt, anaerobic conditions are required. Without oxygen, the bacteria are forced to rely on glycolysis alone to meet their ATP needs. At the end of glycolysis, pyruvate is reduced to lactic acid via fermentation, which lowers the pH of the milk, causing the milk proteins to denature. In their denatured state, these proteins take on a solid form, giving yogurt its firm texture. The reason the bacteria initially added in the recipe were thoroughly mixed and the reason the cups were tightly sealed was to provide this anaerobic environment. Because the bacteria in the unsealed cups were exposed to increased oxygen, many of the cells in these cups were able to respire aerobically. Less fermentation resulted in less lactic acid, preventing the milk in the uncovered cups from solidifying properly. Of course, if we prepared growth curves for the sealed and unsealed cups, we would detect a longer lag period and a less steep log period in the sealed cups that ultimately produced yogurt. The unsealed cups would undoubtedly contain more bacterial cells.

Although *L. acidophilus* is mesophilic in terms of its temperature preferences, it is somewhat psychrotolerant. Consequently, when the yogurt was placed in the refrigerator, although log phase may have continued, the slope of the rise would be considerably reduced (see Figure 7.34b). Growth might continue slowly until other factors, perhaps the digestion of all remaining sugars, moved the bacteria into the stationary phase.

Yogurt making is just one situation in which we might want to influence microbial metabolism and growth. As we continue through this text we will encounter many other situations in which we may wish a population of microorganisms to be in one part of the growth curve or another. Certainly there is usually an urgency to move disease-causing pathogens into the decline phase. But as we will see in Chapter 12, we do not always want even pathogens to decline in number as steeply as possible. For other beneficial organisms it may be in our interest to keep them in the log phase as long as we can. As we have seen, much of our ability to do this depends on knowing how a particular species makes and uses energy.

Coming Up Next...

In the last two chapters we have reviewed some of the fundamental processes applicable to all microorganisms. Genetics and metabolism are not always easy topics to grasp, but some familiarity with these subjects is crucial if we are to truly understand how microorganisms live and how we interact with them. Now that we know a good deal about what microorganisms are and what they do, in the next chapter we will explore how they got that way: we will turn our attention to microbial evolution.

We will start by exploring the origins of life itself and how a barren, sterile Earth ultimately gave rise to the first cells. Next, we will consider the mechanism of evolution and how this process has allowed those first pioneering cells to diversify into the almost limitless number of different living things that surround us. Furthermore, an understanding of how living things evolve permits us not only to understand but even to predict phenomena such as bacterial drug resistance and changing levels of pathogen virulence. And finally, we will investigate a few of the astounding ways in which microorganisms have influenced the evolutionary history of humans and other host organisms. It may or may not be true that "man is what he eats," but he certainly has in some ways become what he is thanks to the microbes with whom he shares the planet.

Key Terms

acidophile	facultative anaerobe	NADPH
adenosine triphosphate (ATP)	fastidious microorganism	nicotinamide adenine
aerobic respiration	fermentation	dinucleotide (NAD$^+$)
aerotolerant anaerobe	fermentation waste product	nonfastidious microorganism
anaerobic microorganism	final electron acceptor	obligate aerobe
ATP synthase	generation time	obligate anaerobe
autotroph	glycolysis	optimum growth temperature
binary fission	halophile	oxidation
Calvin cycle	heterotroph	photoautotroph
carbon fixation	Krebs cycle	photosynthesis
cell respiration	lag phase	protease
chemically defined medium	light reaction	proton gradient
chemoautotroph	lipase	psychrophile
chloroplast	log phase	psychrotolerant microorganism
complex medium	maximum growth temperature	pyruvate
coupled reactions	mesophile	redox reaction
decline phase	metabolism	reduction
electron transport	microaerophile	stationary phase
electron transport chain	minimal medium	synchronous growth
endergonic reaction	minimum growth temperature	thermophile
exergonic reaction	NADP$^+$	thylakoid

Concept Questions

1. Electrons are transferred from molecule A to molecule B.
   a. Which molecule has been reduced?
   b. Were the electrons at higher energy with A or B?
   c. Were the electrons held more tightly when they were with A or with B?
   d. With which molecule, A or B, are the electrons more stable?
   e. Is this an endergonic or exergonic reaction?

2. What exactly is meant when we say that, in electron transport, oxygen is the final electron acceptor?

3. Twenty bacterial cells of a psychrophilic species find their way into your milk carton while it's out of the refrigerator. When you return the milk to the refrigerator, the cells begin to grow with a generation time of 2 hours. How many cells, in theory, will be in your milk carton the next time you go to the refrigerator, 10 hours later?

4. A facultative anaerobe suffers a mutation, resulting in its inability to produce catalase. How might this change the environmental conditions under which this organism could survive?

5. Why does oxygen kill obligate anaerobes?

6. Flask A contains yeast in a glucose broth. The flask is exposed to the atmosphere. Flask B also contains yeast in a glucose broth. The flask is tightly sealed, creating an anaerobic environment. Answer the following, considering that these yeasts are facultative anaerobes:
   a. In which flask would there be more cell division over a period of 48 hours?

b. In which flask would ethanol be found after 48 hrs?

c. In which flask would the generation time be shorter?

d. In which flask would the yeast be able to produce greater amounts of ATP?

e. In which flask would pyruvate get reduced? In which would it be oxidized?

f. Draw the growth curves for the cells in each flask. How would they differ and why?

7. Draw growth curves for the following:

a. An obligate anaerobe is placed in optimal environmental conditions. At time x, oxygen is introduced.

b. A facultative anaerobe is placed in anaerobic conditions. At time x, oxygen is introduced.

c. An obligate aerobe is placed in optimal environmental conditions. At time x, all oxygen is used up. Conditions are now anaerobic.

d. You're at a picnic and your potato salad is contaminated with a mesophilic species. After the picnic you take the remaining potato salad home and place it in the refrigerator at time x.

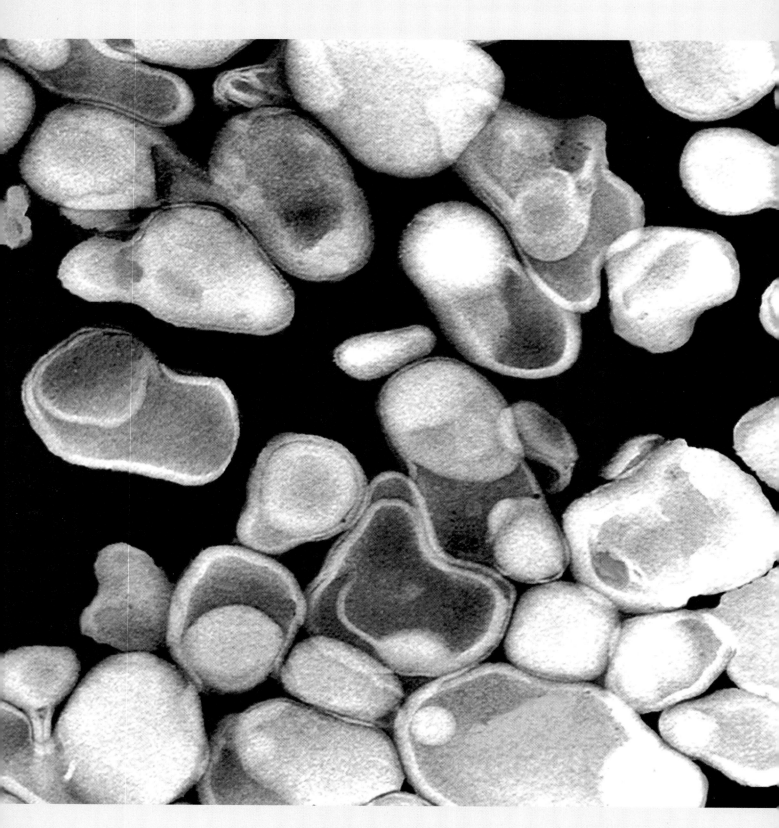

Liposomes may have formed the first cells by incorporating self-replicating RNA.

Chapter 8

Microbial Evolution: The Origin and Diversity of Life

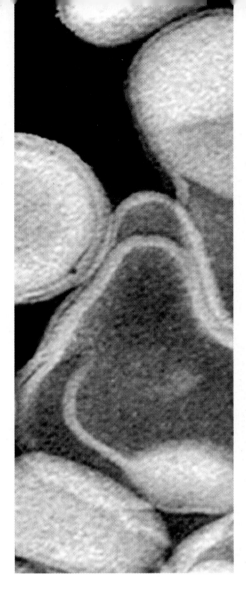

In the last chapter we learned that microorganisms vary considerably in how they respond to environmental conditions. Some, for example, require oxygen, while others are killed by it. The reason for this, as we saw, was that some organisms have enzymes to detoxify dangerous oxygen compounds while others do not. But *why* would organisms develop in such different ways? When we ask about the origin of such a difference, we are going straight to the heart of one of science's most fundamental questions: what accounts for the remarkable variety of living things that exists today?

The short answer is that ever since life on our planet began, it has been subject to the rules of biological evolution, and today, after more than 3.5 billion years, we see the result: an almost limitless assortment of organisms, both microbial and nonmicrobial. Furthermore, whether we are considering *Escherichia coli* or elephants, the rules governing evolutionary change are the same. In any population of organisms there is genetic variability, and in any given environment some members of the population will be more likely to reproduce than others. Given enough time, those that are better adapted to a particular set of environmental conditions come to predominate because they are better at passing on their genes to the next generation. Those less well adapted will eventually perish. No two environments, however, are identical, and any environment also changes over time. This means that genetically determined biological characteristics that result in reproductive success in one environment do not necessarily guarantee success elsewhere or forever. Under different conditions, different genetic makeups (genotypes) are favored, and this never-ending interplay of environment and genotypes has given rise to life's diversity.

As we contemplate the myriad living things around us, it is worth remembering that every one is the product of eons of evolution, and each has its own evolutionary story to tell. In this chapter we will try to make sense of those stories, particularly as they apply to microorganisms. And as with all good stories, the best place to start is at the beginning—in this case, the origin of life on Earth.

How Life Began

Conditions on the early Earth were very different than they are today

Scientists believe that our planet was formed approximately 4.5–4.6 billion years ago (BYA) (**Figure 8.1**). According to accepted theory, our solar system was formed when dust-sized particles accumulated around the sun. As these particles collided, they often stuck together due to electrical attraction. Over time these growing masses attracted still more particles due to

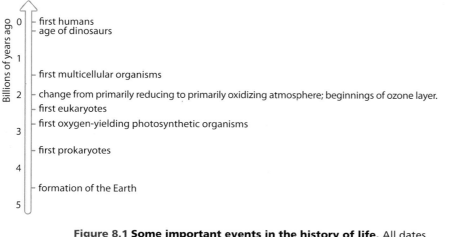

Billions of years ago

0 — first humans
— age of dinosaurs

1

— first multicellular organisms

2 — change from primarily reducing to primarily oxidizing atmosphere; beginnings of ozone layer.
— first eukaryotes
— first oxygen-yielding photosynthetic organisms

3

— first prokaryotes

4

— formation of the Earth

5

Figure 8.1 Some important events in the history of life. All dates are approximate. Major episodes in the origin of living things will be revisited throughout this chapter.

their increasing gravity. Ultimately these protoplanets developed into the planets of our solar system.

The young Earth was certainly a vastly different place than the Earth of today. First of all, it was far hotter. Since its formation, the Earth has steadily cooled, losing the heat that was first generated when it was formed. Second, the early Earth's atmosphere contained very little oxygen (O_2). Today, oxygen makes up 21% of our atmosphere. Some O_2 was added to the atmosphere by geologic processes, but oxygen did not begin to approach current levels for another 2 billion years or so. At this time, certain photosynthetic organisms evolved the ability to use water as an electron source (see Chapter 7, p. 186), releasing oxygen as a waste product in the process. Consequently, the early atmosphere of our planet was a *reducing* atmosphere, as opposed to our current *oxidizing* atmosphere. This fact may have been crucial for the initial development of life. Recall from Chapter 7 that when molecules are more likely to be reduced, the synthesis of biological molecules is more likely. In an oxidizing environment, digestion is more likely because molecules tend to be oxidized. The low levels of free oxygen therefore may have set the stage for the sorts of reactions that would ultimately give rise to the first biological molecules.

Hydrogen gas (H_2) was especially abundant in Earth's early, reducing atmosphere. Other reduced molecules containing hydrogen such as methane (CH_4), ammonia (NH_3), and water vapor (H_2O) were also present. These molecules provided the building blocks for the beginnings of life itself.

The first biological molecules were formed from nonbiological precursors

In Chapter 5 we discussed some of the discoveries that disproved the notion of spontaneous generation. Experiments by scientists such as Redi and Pasteur demonstrated that life could not arise from nonliving material. In the 1920s, however, it was suggested that in the primitive Earth's reducing atmosphere, inorganic molecules *could* assemble into the biological molecules that are the precursors of life. Reduced gases in the atmosphere could donate electrons, meaning that the redox reactions necessary for synthesis could take place. This no longer occurs in our current oxidizing atmosphere, because atmospheric oxygen is such an effective oxidizing agent.

Even in a reducing environment, an energy source was needed to synthesize large biological molecules from smaller molecules such as those present

Figure 8.2 Energy for abiotic synthesis. If biological molecules formed in the Earth's primitive reducing atmosphere, lightning strikes may have provided the energy necessary for synthesis.

in the primitive Earth's atmosphere. Scientists have suggested that UV radiation or lightning could serve in this capacity (**Figure 8.2**). The lack of significant free oxygen meant that no ozone layer would have been present to absorb UV radiation. Consequently, the intensity of such radiation reaching the Earth's surface was far higher than it is today.

Thus, it was hypothesized that although spontaneous generation is unlikely today, the unique conditions found on the early Earth meant that the scene was set for **abiotic synthesis**—the building of biological molecules from inorganic precursors.

The hypothesis of abiotic synthesis was tested in 1953 in a brilliant experiment conducted by Stanley Miller, who was then a graduate student at the University of Chicago. Miller actually recreated the Earth's primitive reducing atmosphere in an experimental vessel (**Figure 8.3**). The apparatus consisted of two chambers connected by glass tubing. In one of the chambers, Miller created a miniature "ocean" of liquid water. A different chamber consisted of a primitive "atmosphere," complete with reduced methane (CH_4), ammonia (NH_3), and hydrogen (H_2) gases. The ocean was boiled with a heating source, such that the water vapor passed into the atmosphere, where in addition to mixing with other atmospheric gases, the molecules were subjected to artificial "lightning" in the form of electrical discharge from an electrode. Within a day, the mixture began to change color, taking on a reddish hue. After a week, Miller analyzed the chemical composition of the molecules in his apparatus. Amazingly, he found numerous organic compounds, including several amino acids. The possibility of abiotic synthesis in an environment simulating that of the primitive Earth had been confirmed.

Others have proposed that the composition of the primitive Earth's atmosphere was less important than Miller's work suggests. As an alternative hypothesis, it has been proposed that early biochemical reactions first occurred in deep sea vents, where hot water and minerals, along with reduced compounds such as methane and hydrogen sulfide, are released into the ocean (**Figure 8.4a**). According to this hypothesis, as reduced compounds rose up through the Earth's crust, they provided the raw materials necessary for the synthesis of biological molecules. The superheated water

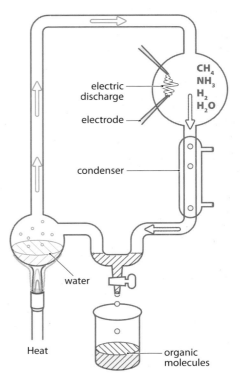

Figure 8.3 Stanley Miller's "origin of life" experiment. Water vapor was passed through a primitive "reducing atmosphere" consisting of NH_3, CH_4, and H_2, into which energy, in the form of electric sparks, was discharged. Within a few days, amino acids and other organic molecules were formed.

Figure 8.4 Alternative explanations for abiotic synthesis.
(a) Deep ocean, hydrothermal vent called a "black smoker." The black color is due to the large amounts of minerals such as iron and carbon that are dissolved as the superheated water leaves the vent. Biological molecules may have first been synthesized at sites such as this. (b) Piece of the Murchison meteorite, which landed in Australia in 1969. Traces of water and amino acids were found on the meteorite, leading some to speculate that Earth was originally seeded with organic molecules formed in space and delivered to Earth on meteorites.

supplied an energy source. Interestingly, these hydrothermal vents provide a suitable environment for many microorganisms today. Many of these bacteria and archaea are extreme thermophiles, able to thrive in the hot water. Others have noted that molecules such as hydrogen cyanide and formaldehyde form in gaseous clouds in space and that these molecules can react to form other more complex biological molecules. It is possible that Earth was originally "seeded" with these organic molecules by meteorites that bombarded the Earth shortly after its formation (Figure 8.4b).

Regardless of where the process took place, however, it now appears likely that under certain specific conditions biological molecules, the building blocks of life, can indeed form from abiotic beginnings.

Genetic information may have originally been encoded in RNA instead of DNA

The "building blocks of life" are not the same as life itself. An essential attribute of living things is their ability to reproduce. Reproduction requires some form of genetic material to transmit information from one generation to the next, and for all modern prokaryotic and eukaryotic organisms, this material is DNA. In our primitive mix of amino acids and other organic molecules it is therefore vital to find a molecule acting as the genetic material, able to both duplicate itself and transmit genetic information. Yet if DNA itself initially served as the genetic material, we are presented with a "chicken or egg" type conundrum. Although DNA can replicate itself, such replication occurs only when necessary enzymes are available. Enzymes, however, are proteins, coded for by DNA. So which came first, the DNA that codes for the enzymes or the enzymes needed to copy the DNA?

There was no good answer to this question until the 1980s, when researchers discovered enzymes made of RNA, which they called **ribozymes**. This discovery led to the "RNA world hypothesis": the idea that the current "DNA world" was preceded by an "RNA world" in which RNA, rather than DNA, served as the genetic material.

Why RNA? Prior to the discovery of ribozymes, RNA was thought to be involved only in conveying DNA's message first into mRNA and ultimately into a protein product. Ribozymes, however, first discovered in tiny ciliates called *Tetrahymena*, demonstrated that RNA could also act as an enzyme, catalyzing certain reactions. In other words, the discovery of ribozymes overturned the long-held assumption that only proteins could serve as enzymes. It has subsequently been found that ribozymes can also catalyze the synthesis of new RNA (**Figure 8.5**).

RNA can also encode genetic information in its sequence of nucleotide bases, and it can serve as a template for replication. On the other hand, DNA can store information, but it cannot catalyze the replication of that information. Proteins can serve as enzyme catalysts but they cannot store information. RNA can do both, and proponents of the RNA world hypothesis suggest that RNA served as the initial genetic material *and* that it catalyzed its own reproduction.

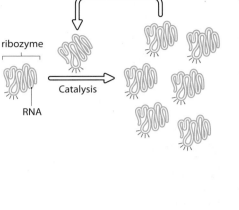

Figure 8.5 RNA is able to catalyze its own synthesis. Although most enzymes are protein-based, ribozymes are composed of RNA. Some ribozymes are involved with RNA synthesis. In this figure, the red lines represent the active site of the ribozyme. If RNA could serve as the initial molecule for encoding genetic information, *and* if it could catalyze its own synthesis, the dilemma of how the first living things were able to replicate may be solved.

The first cells required a membrane and genetic material

Two important things are needed before we can start to consider any assemblages of biological molecules to be bona fide cells. First, a barrier between the inside and the outside of the cell in the form of a membrane is required. Second, a cell needs genetic information that can replicate and code for essential proteins. We do not know precisely what the first cells looked like, but we do know that, when they are mixed with water, phospholipids assemble into small, spherical compartments (**Figure 8.6a**). These compartments, called **liposomes**, are thought to have formed the first cells when they incorporated self-replicating RNA (Figure 8.6b). Liposomes are sometimes able to grow by engulfing other liposomes, and because their membrane is selectively permeable, they are able to absorb and release certain materials. The enclosed RNA would have allowed these primordial cells to replicate and continue to evolve.

We can never know for sure how the transition to a DNA world took place, but it is clear that the earliest RNA-based cells would have had an extremely simple metabolism. Compared with proteins, RNA is an inefficient catalyst, and early in life's history, RNA probably catalyzed only a few absolutely critical reactions. As proteins began to appear in cells, it is likely that they took over since they are more efficient catalysts, leaving RNA to mainly encode information (**Figure 8.7**).

DNA may have arrived on the scene as cells gradually became more complex and needed to store increasing amounts of information. RNA may have provided the template for the first DNA molecules, but DNA is inherently far more stable than RNA. DNA is therefore better suited for the transfer of large amounts of information from generation to generation. With more efficient information storage and replication, cells utilizing DNA would eventually replace those relying on RNA. RNA was then available to assume its current task of serving as an intermediate between DNA and protein.

Consequently, somewhere at an early stage in the evolution of life, the transfer of information from DNA to RNA to protein became universal. The

Figure 8.6 The first cells. (a) Phospholipid bilayers form spontaneously when phospholipids are placed in water. They then readily close in on themselves to formed sealed compartments. The closed structure is stable because it avoids the exposure of the hydrophobic hydrocarbon tails to water. Such phospholipid spheres, known as *liposomes*, may have constituted the first plasma membranes. (b) The earliest cells may have consisted of liposomes surrounding self-replicating RNA.

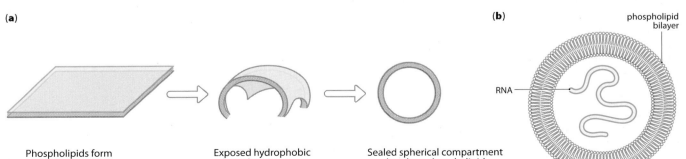

(a)

Phospholipids form a planar bilayer when exposed to water.

Exposed hydrophobic edges close in on themselves.

Sealed spherical compartment results when phospholipid bilayer closing completes.

(b)

phospholipid bilayer

RNA

precise time when this happened has not been determined, but it is worth noting that Bacteria, Archaea, and Eukarya all rely on DNA as their genetic material. This fact strongly supports the idea that any transition from an RNA world to a DNA world would have occurred before these three domains of life diverged.

The first prokaryotes are thought to have arisen approximately 3.5 billion years ago

Not surprisingly, the point at which simple, primordial cells qualified as authentic living things is fuzzy. We can guess that the time frame is enormous and that the first living organisms were prokaryotes that arrived some time around 3.5 BYA (**Figure 8.8**). Evidence to support this has come from the recovery of microbial fossils in layered sedimentary rock formations called **stromatolites** (**Figure 8.9**). Many of these structures contain the fossils of filamentous or rod-shaped microorganisms that resemble organisms that are still alive today. These early microbial pioneers were most likely obligate anaerobes and, as discussed in Chapter 4, were probably well-adapted to withstand the intense UV radiation and high temperatures that characterized the Earth's environment at that time.

Certain important metabolic pathways evolved in a defined sequence

In Chapter 7 we reviewed important metabolic processes such as glycolysis, cell respiration, and photosynthesis. Most researchers agree that glycolysis was probably the first of these metabolic pathways to evolve. Powerful evidence for this is provided by the fact that, unlike photosynthesis or specific types of cell respiration, glycolysis occurs in all cells. This suggests that glycolysis evolved before the diversification of cells into the three domains of life—Bacteria, Archaea, and Eukarya. Later, other less universal metabolic pathways arose in each domain independently.

Now that cells had a mechanism to harvest the energy in biological molecules such as glucose, it obviously would be advantageous for a cell to develop the capacity to produce its own biological molecules, reducing reliance on environmental sources. Consequently, photosynthesis is thought to have evolved sometime after glycolysis and probably after anaerobic cell respiration. The earliest photosynthetic organisms did not use water as a source of electrons and consequently did not release oxygen. The release of oxygen as a photosynthetic by-product (see Chapter 7, p. 187) is believed to have started about 2.7 BYA (**Figure 8.10a**), and the organisms responsible may have been the distant ancestors of modern-day photosynthetic cyanobacteria (Figure 8.10b). Starting at that time, oxygen began its steady buildup

RNA-based systems

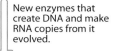

RNAs that could direct protein synthesis evolved from simpler RNAs.

RNA- and protein-based systems

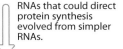

New enzymes that create DNA and make RNA copies from it evolved.

Present-day cells

Figure 8.7 Transition from RNA world to DNA. According to this hypothesis, RNA combined genetic and catalytic functions in the earliest cells. Proteins are more efficient catalysts and eventually took on most enzymatic functions. DNA, more stable than RNA, took over the encoding of genetic information as greater amounts of information were required by increasingly complex cells. RNA remained, functioning primarily as a go-between in protein synthesis, while retaining catalytic functions for a few reactions.

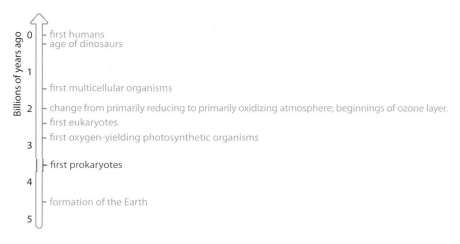

Figure 8.8 Earliest cellular life. The first prokaryotic cells formed approximately 3.5 billion years ago.

Figure 8.9 The first prokaryotes. (a) The oldest known stromatolites to date have been found in Sharks Bay in western Australia. The oldest stromatolite fossils are estimated to be approximately 3.5 billion years old. (b) Fossilized ancient bacterium from northern Australia.

in the atmosphere, and by about 2 BYA atmospheric oxygen was common enough that iron-bearing rocks began to undergo oxidation, forming iron oxide. Iron oxide, commonly called rust, is red in color, and the presence of "red beds," reddish-colored iron-rich rocks that were exposed to the atmosphere 2 BYA, attests to the presence of greatly increased levels of oxygen at that time (**Figure 8.11**).

This buildup of atmospheric oxygen (O_2) changed the environment in profound ways. When O_2 is bombarded by UV radiation, for example, it can be converted into ozone (O_3). Ozone can absorb UV radiation, and as levels of ozone increased, the Earth's atmosphere gained an *ozone shield* to protect living things against UV radiation. In Chapter 6 we discussed how such radiation acts as a powerful mutagen, and before the development of the ozone layer, most living things could survive only under rocks, in the oceans, or in other habitats protected from direct radiation from the sun. Following the development of an ozone layer, early microorganisms could spread out over the surface of the Earth, safe from the damaging effects of UV radiation.

Furthermore, recall from Chapter 7 that oxygen can form powerful and sometimes lethal oxidizing compounds such as superoxide ions (O_2^-). Consequently, for many anaerobes, oxygen is highly toxic. Before oxygen levels reached high concentration in the atmosphere, all organisms would have been anaerobic, relying on glycolysis and/or anaerobic cell respiration to meet their energy needs. So for cells living 2 BYA the choice was simple: either they could avoid oxygen by living in anaerobic environments only or they could become extinct. Undoubtedly the obligate anaerobes alive today are the descendents of those that dealt with the problem of oxygen by avoiding it. Other primitive prokaryotic lineages perished.

As an alternative, organisms could evolve ways to detoxify oxygen. Some organisms developed the capacity to reduce oxygen to water through a primitive form of electron transport. As we saw in Chapter 7, reducing oxygen in this manner releases energy, and some organisms were able to capitalize on this released energy by using it to produce additional ATP. In other

Figure 8.10 The first photosynthetic organisms. (a) Oxygen-yielding photosynthesis is believed to have evolved about 2.7 BYA. (b) The first photosynthetic prokaryotes that released oxygen may be related to modern-day cyanobacteria, seen as the long slender green filaments in the photograph. These bacteria, like green plants, use water as a source of electrons and release oxygen as a waste. Other photosynthetic prokaryotes rely on different electron sources and do not release oxygen.

(a)

Billions of years ago

0 – first humans / age of dinosaurs
1 – first multicellular organisms
2 – change from primarily reducing to primarily oxidizing atmosphere; beginnings of ozone layer / first eukaryotes
3 – first oxygen-yielding photosynthetic organisms / first prokaryotes
4 – formation of the Earth
5

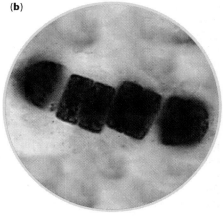

Figure 8.11 Red beds. Uluru, a well-known rock formation in the Northern Territory of Australia. Iron oxide, formed when iron deposits in the rock were oxidized by atmospheric oxygen, gives Uluru its red color. These rocks were formed about 2 BYA, indicating the approximate time when the atmosphere changed from reducing to oxidizing. Rock strata older than 2 billion years that were not exposed to the atmosphere show no such oxidation.

words, the electron transport chain that provides the lion's share of ATP for aerobic organisms first evolved as an oxygen detoxification system. By a happy coincidence, those organisms fortunate enough to develop this strategy received an additional benefit—aerobic respiration.

Eukaryotes evolved from prokaryotic ancestors

CASE: CELLS WITHIN CELLS

As amazing and perhaps as creepy as it sounds, even your own cells are not completely your own. As mentioned briefly in Chapter 3, some parts of eukaryotic cells have bacterial ancestors. In the late 19th century, the idea was first proposed that at least some of the organelles in eukaryotic cells originated from bacteria that had entered into a symbiotic relationship with the eukaryotic cell. Over time, the two cell types developed such a high degree of interdependence that independent life was no longer an option for either cell. This explanation for the origin of organelles, known as *endosymbiosis,* was largely considered to be a harebrained idea until the 1980s. At that time, some scientists began to take a fresh look at endosymbiosis, and as they compiled evidence to support it, endosymbiosis gradually gained credibility. Today, most biologists agree that endosymbiosis is the most likely explanation for the origin of at least two important organelles.

1. **Which organelles were most likely the result of endosymbiosis?**
2. **What evidence exists to support endosymbiosis?**
3. **In what way would endosymbiosis be advantageous to both partners in the symbiotic relationship?**

Following the discovery of fossil prokaryotes in the mid-20th century, it became apparent that there was an abundant fossil record of early prokaryotic life. Yet nothing that looks like a eukaryotic cell appears in the fossil record until just over 2 BYA (**Figure 8.12**). For the first 1.5–2 billion years of their existence, prokaryotes had the planet to themselves.

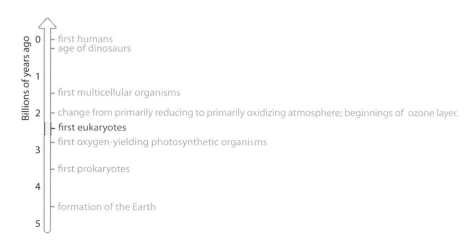

Figure 8.12 The first eukaryotic cells. The earliest fossils that show clear evidence of a membrane-bound nucleus and other distinguishing features of eukaryotic cells first appeared a little more than 2 BYA.

The first eukaryotes were probably structurally simple single-celled organisms, resembling prokaryotes, except for the fact that they had certain internal membrane systems such as the nuclear membrane. The nuclear and other internal membranes such as the endoplasmic reticulum and the Golgi apparatus are thought to have evolved from infoldings of the plasma membrane (**Figure 8.13**). As discussed in Chapter 3, such compartmentalization of the early eukaryotic cell allowed increased intracellular specialization and meant that the cell could carry out more metabolic processes at the same time.

Other organelles, such as mitochondria and chloroplasts, were still lacking. These organelles were later acquired by **endosymbiosis** (see Figure 8.13). *Symbiosis* refers to a special relationship between two organisms, which is often mutually beneficial. Mitochondria are believed to have evolved when aerobic prokaryotes developed a symbiotic relationship with primitive anaerobic eukaryotic cells. Later, some eukaryotes acquired chloroplasts in a similar manner, allowing them to carry out photosynthesis.

It is generally assumed that mitochondria were acquired before chloroplasts, because of their presence in virtually all eukaryotes (**Figure 8.14**). Later, after eukaryotes had begun to diversify, photosynthetic bacteria were acquired by the branch of eukaryotes that ultimately gave rise to plants.

There is no shortage of evidence to support the theory of endosymbiosis:
- Symbiotic relationships between prokaryotic and eukaryotic cells can still be found today. For example, some photosynthetic cyanobacteria reside inside the cells of some unicellular eukaryotes.
- Similarities exist between prokaryotic cells and modern mitochondria and chloroplasts. Like prokaryotes, these organelles can reproduce by binary fission, independently of host-cell division. Both mitochondria and chloroplasts have their own DNA, and like prokaryotic DNA, the chromosomes found in these organelles are circular.
- Using their own DNA, both mitochondria and chloroplasts code for some of their own proteins. Translation of these proteins occurs within the organelle on smaller prokaryotic-like ribosomes, instead of the larger ribosomes found in the cytoplasm of eukaryotic cells.
- Comparison of gene sequences from mitochondrial DNA and certain aerobic bacteria confirms that mitochondria and bacteria are related. Similar evidence confirms the relationship between chloroplasts and photosynthetic cyanobacteria.

The advantages enjoyed by primitive eukaryotes with mitochondria and primitive plant cells with both mitochondria and chloroplasts were straightforward. As we learned in Chapter 7, aerobic respiration results in higher ATP yields than anaerobic respiration. Eukaryotes harboring aerobic bacterial partners could now partake of this ATP bonus. Early plant cells that acquired photosynthetic bacteria via endosymbiosis no longer needed to depend on an outside source of food molecules. As autotrophs, they could now synthesize their own. And for the once-independent bacteria, now

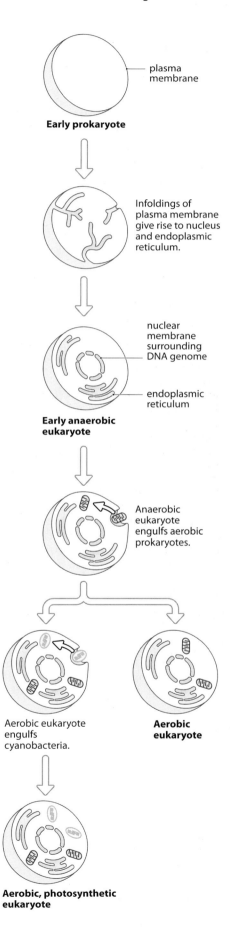

Figure 8.13 Origin of eukaryotes. The first eukaryotes were probably formed when infolding of the plasma membrane gave rise to membrane-bound organelles, permitting increased subcellular specialization. These early eukaryotes would have been anaerobic and nonphotosynthetic. The first aerobic eukaryotes were the result of endosymbiosis with aerobic prokaryotes. These symbiotic aerobes ultimately gave rise to mitochondria.

Some of these now-aerobic eukaryotes subsequently developed a second endosymbiotic relationship with photosynthetic bacteria. Increasing interdependence between these bacteria and the eukaryotic cells resulted in the evolution of chloroplasts and eukaryotic cells that were both aerobic and photosynthetic.

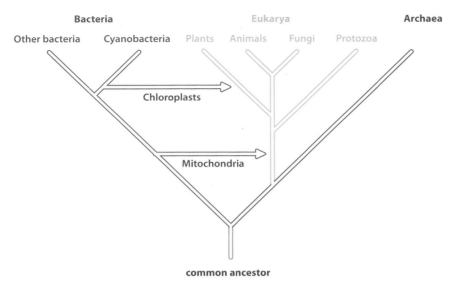

Figure 8.14 Cell evolution. Present-day cells evolved from a common prokaryotic ancestor along three major lines of descent, giving rise to domains Bacteria, Eukarya, and Archaea. Eukarya acquired mitochondria prior to the differentiation of the domain into its various groups. Only the group that ultimately gave rise to plants acquired chloroplasts.

living cozily inside eukaryotic cells, there was less risk of predation as well as the benefits of more stable environmental conditions.

Multicellular life arose from colonies of unicellular eukaryotes

Some time around 1.5 billion years ago, the first multicellular organisms evolved (**Figure 8.15**). To this day, some unicellular eukaryotes form colonies of cells, which resemble a transition stage between unicellular and multicellular modes of life (**Figure 8.16**). As cells within such colonies became increasingly specialized, the capacity for independent life as a single cell was lost, and true multicellular organisms first appeared. As cells continued to specialize, they consequently became increasingly diverse, giving rise to the many types of cells currently seen in modern multicellular animals, fungi, and plants.

Multicellularity confers several advantages. First of all, as an organism increases in terms of cell number, it becomes larger and is therefore less likely to be consumed by another organism. Larger organisms are also better able to maintain suitable internal conditions. A single cell, for example,

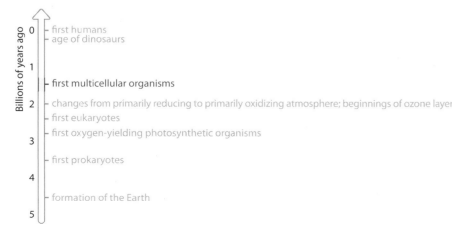

Figure 8.15 The first multicellular organisms. Eukaryotic organisms consisting of many cells first appear in the fossil record about 1.5 BYA.

has little ability to maintain a particular temperature. With size, however, heat gain or heat loss is reduced, and a multicellular organism is to some degree buffered against the vagaries of the environment. Furthermore, multicellular organisms have less to fear from genetic mutations. A single serious mutation can spell doom for a unicellular organism. In a large multicellular organism, a mutation that causes the death of any single cell will probably go unnoticed.

But clearly, unicellular microorganisms must be doing something right, or else they would no longer be so abundant. We have already made note of several ways in which prokaryotes have the edge on eukaryotes. In Chapter 3 we discussed the advantage that small size alone confers, while in Chapter 6 we observed that protein synthesis occurs more rapidly in prokaryotic cells. These advantages mean that unicellular prokaryotes are able to reproduce at an accelerated rate, and when environmental conditions are to their liking, microbial populations can skyrocket. The take-home message is that no one mode of existence, be it prokaryotic or eukaryotic, single-celled or multicellular, is inherently superior. All come with advantages and disadvantages, and the bottom line is always the ability to reproduce and pass on genetic information. Because of variable environmental conditions, a successful evolutionary strategy in one environment may be considerably less successful in a different environment. Thus the process of evolution relies on the environment to serve as a sort of filter, permitting the survival of the best adapted, while poorly adapted organisms ultimately face extinction. The result is the almost limitless variety of life that surrounds us.

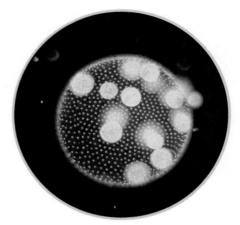

Figure 8.16 *Volvox:* a colonial green alga. A *Volvox* colony consists of a gelatinous matrix forming a hollow sphere, in which many individual cells are embedded. Such colonies, common in pond water, frequently reach sizes of up to 1 mm, making them visible to the naked eye. Colonial organisms such as *Volvox* may represent a transition state between unicellular and multicellular organisms.

The origin of viruses is obscure

So far our discussion has focused on cells. We know less about how viruses first evolved. More than likely, the earliest viruses were **bacteriophages**— viruses of bacteria that emerged during the time when prokaryotes were the only living things on the planet. As eukaryotes evolved, they also became targets for these parasites. But *how* viruses originated is still open to considerable speculation. Currently, there are three principal hypotheses to explain viral origins.

The **regressive hypothesis** postulates that viruses started out as prokaryotic microorganisms that gradually lost most of their genetic information and ultimately became dependent on host cells for most of their metabolic needs. According to this idea, an intracellular parasite could have become more and more dependent on its host cell, losing the ability to synthesize various biological molecules until it retained little except those genes needed for replication and transmission to new host cells. Eventually, this now-obligate parasite evolved into nothing more than a few genes within a protein coat. The regressive hypothesis therefore proposes an origin for viruses something like endosymbiosis, except that instead of evolving into intracellular organelles, viral ancestors evolved into parasitic life forms. If this hypothesis is correct, it means that with a bacterium like *Rickettsia* we could be witnessing viral evolution before our very eyes (**Figure 8.17**). *Rickettsia* are unable to survive outside of a host cell, and they have very simple cell walls, lacking peptidoglycan. Have they lost the capacity for independent existence en route to becoming a new virus? Will they one day dispense with their cell walls altogether? We will have to return in a few million years to find out.

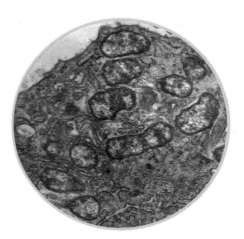

Figure 8.17 *Rickettsia:* an intracellular bacterium. The elongated, darker-staining bodies are *Rickettsia* in the cytoplasm of a eukaryotic cell. These small, intracellular parasites are unusual in that they have lost their capacity for independent life. According to the regressive hypothesis, this bacterial species may be in a transition stage on the road to becoming a virus.

According to the **escaped-gene hypothesis**, fragments of nucleic acid from a host cell escaped and developed the ability to duplicate independently of the cell. These rogue DNA or RNA fragments would have needed to encode the necessary information to replicate themselves, utilizing resources provided by the host cells. The original source of this independently replicating nucleic acid may have been plasmid DNA or **transposons**. Transposons are short sequences of DNA that can move from one site on a chromosome to another site. They are found in both prokaryotes and eukaryotes. Evidence for the escaped-gene hypothesis has come from studying a specific viral transposon in retroviruses (see Chapter 4, p. 97). The transposon was found to contain base sequences similar to sequences found in the host, suggesting that it originally *was* part of the host's genetic material.

The third hypothesis proposes that viruses evolved alongside cellular forms of life and ultimately developed into obligate parasites of these early cells. While some forms of primitive life developed the cell membranes and metabolic capabilities previously discussed, others merely surrounded their genome with a protein coat, giving rise to the first viruses. As these primordial viruses coevolved with cells, they became increasingly dependant on them. Thus the **coevolved hypothesis** suggests that life initially headed off in two separate directions. One resulted in increasingly complex cells while the other maintained the simple, streamlined structure that we now recognize as viruses.

As yet, there has been no way to determine which of these three hypotheses is the most plausible. It is quite possible that viruses evolved more than once and that no one hypothesis explains the origin of all viruses.

Prions may have arisen from abnormal host proteins

As for prions, their origins are even murkier than those of viruses. You will recall from Chapter 4 that prions are infectious proteins that have adopted an aberrant shape and that they are able to force normal forms of the protein to take on the altered conformation.

A clue to the origin of prions may stem from the fact that all known prion diseases in animals affect protein folding in the brain only. Recent research has suggested that certain proteins in the brain ordinarily change shape naturally as part of memory storage. That means that if the prion protein (PrP^{Sc}; see Chapter 4, p. 102) shows up, either due to gene mutation or infection, the brain already has many shape-shifting proteins that are vulnerable to attack. Because this sort of protein "flip-flopping" does not occur naturally outside the brain, proteins in other parts of the body are not susceptible to PrP^{Sc}, thus explaining why prion diseases are exclusively neurological. In other words, it is possible that prions are occasionally formed by normal brain proteins that are already prone to adopt different shapes, because of their function in memory formation. If they take on the shape of PrP^{Sc}, the result is Creutzfeldt–Jakob or another related prion disease.

Exobiology is the search for extraterrestrial life

If life could develop on Earth, might it not evolve on other worlds as well? This is one of the compelling questions asked by exobiologists—scientists who search for clues of life beyond our planet.

At the present time, only Earth is known to support life. Yet microorganisms live in practically every conceivable habitat on Earth, including deep sea hydrothermal vents where the water is above the boiling point, acidic environments where the pH is comparable to battery acid, or even inside rocks buried far beneath the Earth's surface. If life can exist in these extreme environments, some exobiologists believe that it might also exist on other

Figure 8.18 **Life on other planets?** One candidate location for extraterrestrial life is Europa, a moon of Jupiter. Europa is covered with ice, and examination of the moon indicates that large blocks of ice have moved, suggesting an underlying ocean. Conditions on cold, anaerobic Europa are thought to be similar to certain environments on Earth where prokaryotic life is abundant.

planets, where similar conditions are found. These scientists often study extremophiles on Earth for insight as to what extraterrestrial life might be like.

Because of the ubiquity of microbial life in Earth's most extreme habitats, it is likely that if life is found on other worlds, it will also be microbial, rather than the sorts of large menacing creatures popular in science fiction movies. Two places in particular, Mars and Europa, a moon orbiting Jupiter, are potentially places where evidence of extraterrestrial life may be encountered (**Figure 8.18**). Both of these celestial bodies appear to have or have had water, which is essential for all life as we understand it.

The announcement in 1996 that a meteorite found in Antarctica bore microbial fossils caused an immediate sensation (**Figure 8.19**). Analysis of the meteorite indicated that it came from Mars, because upon heating, it emitted gases unique to the Martian atmosphere. The "fossils" on the meteorite certainly resembled small microorganisms; however, subsequent analysis suggested that they were formed by geological rather than biological processes.

Evolution: Explaining Life's Diversity

"Nothing in biology makes sense, except in the light of evolution." This statement by the famed geneticist Theodosius Dobzhansky highlights the preeminence of evolutionary theory in our understanding of modern biology. Indeed, evolution is more than just *a* core principle of biology; it is *the* core principle. Here we will investigate why evolution is central to biology and what exactly Dobzhansky's quotation means.

We have already had ample opportunity in this book to acknowledge the great diversity of life on our planet. But within this bewildering variety of living things, there is a unity. Whether we are discussing *Streptococcus* bacteria, single-celled yeast, a eucalyptus tree, or a hippopotamus, we should remember that all life ultimately shares a common ancestor; we are all related. The same evolutionary forces are at work on all living things, and an understanding of how organisms evolve allows us to comprehend the living world in all its manifestations and complexity. Without evolution, the tree of life is nothing more than a thicket of unconnected branches. With it, the tree becomes a comprehensible road map, detailing the twists, turns, dead ends, and about-faces that have occurred over life's 3.5 billion year history.

The theory of evolution also provides us with a tool to explain new biological developments, many of which have direct relevance for the microbiologist. Why are antibiotics that used to eliminate bacterial infections now less effective or even useless? Why do some infectious organisms become less deadly over time, while others do not? With an understanding of how living things evolve, phenomena such as these not only become intelligible, they become predictable.

Figure 8.19 **Alien life found?** In 1996 a meteorite from Mars caused a sensation when it was reported that it bore fossils of "Martian bacteria." The apparent bacterial fossils are now believed to be due to natural geological processes.

Natural selection is the driving force of evolution

Although he was not the first to propose that life evolved over time, Charles Darwin was the first to lay out a detailed mechanism for evolution, supported by voluminous evidence (**Figure 8.20**). Darwin based his theory of evolution on two fundamental principles. First, he proposed that the organisms alive today descended from ancestral species. He called this view of life *descent with modification*. Second, he suggested a mechanism by which such descent could occur. He called this mechanism **natural selection**, and today it is understood that natural selection is the primary driving force behind evolution.

Natural selection can be broken down into a series of steps or observations. The outcome of these steps is evolution—changes in the genetically programmed characteristics of a population over time.

1.   All living things have an enormous potential to reproduce. Under ideal conditions, all populations can grow exponentially if all individuals survive and reproduce.
2.   Environmental resources will eventually become limited, and because reproduction will produce more individuals than the environment can support, not all individuals will survive and reproduce.
3.   Individuals within any population are not identical in terms of their characteristics, and many of the differences between individuals are genetically based.
4.   Those individuals most suited to their environment are the most likely to survive and reproduce. They therefore leave more offspring to the next generation, passing on those genetic traits that are best adapted to the environment.
5.   This unequal capacity of individuals to survive and reproduce leads to a gradual change in the population, as those with favorable characteristics make up an increasingly larger proportion of the population over many generations. Less well-adapted individuals may eventually die out. The environment has thus *selected* the best-adapted genetic traits for survival.

It should be emphasized that natural selection works only on *genetic* variation, causing some genotypes to become more or less common (**Figure 8.21**). The sources of genetic variation (mutation and genetic recombination) were discussed in Chapter 6. Characteristics such as blood type or the ability to make a particular enzyme have a genetic basis and can consequently be selected for or against by natural selection.

Traits that develop during an organism's life, on the other hand, are not subject to natural selection, because if a trait is not genetically based it is not transmitted in reproduction to the next generation. Animals that gain weight because of an abundant food supply do not pass genes for larger size to their offspring. A tree that grows low to the ground because of exposure to strong winds will not transmit a "low-growing" characteristic to the next generation in its seeds.

A few other points about natural selection should be highlighted. First of all, natural selection is limited. It can work only on *existing* genetic variation. Such variation can arise by mutation at any time, but the environment cannot *cause* a specific favorable mutation to occur. Mutations simply occur, either spontaneously or in response to mutagens (see Chapter 6, p. 162). Once they do, natural selection is free to act upon them.

Figure 8.20 Charles Darwin at age 51. In the mid-1800s Charles Darwin laid the groundwork for modern evolutionary theory by developing the concept of natural selection, which he published in *The Origin of Species*.

Cell undergoing a random unfavorable mutation is removed by natural selection.

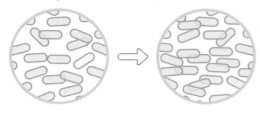

Unadapted cell type fails to reproduce successfully. Best-adapted genotype remains predominant in population.

Figure 8.21 Natural selection in bacteria. When all members of a population are identical, natural selection cannot occur. When genetic variation is introduced, perhaps as a consequence of a random mutation, natural selection may select against or in favor of a mutated individual, depending on how the mutation affects the individual's adaptation to the environment.

Cell undergoing a random favorable mutation survives and reproduces better than predominant genotype.

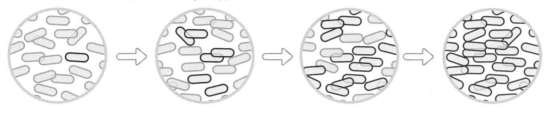

New genotype, favored by natural selection, replaces former genotype as predominant in population.

Furthermore, natural selection cannot result in organisms that are perfectly adapted to their environment. All organisms must be able to do many different things, and a characteristic that is valuable for one task may be less so for another. The same thick, waxy, acid-fast cell wall that protects *Mycobacterium tuberculosis* from damaging chemicals or antibiotics also means that it absorbs nutrients and oxygen very slowly, resulting in a very slow (for bacteria) growth rate (**Figure 8.22a**). Some species of *Staphylococcus* bacteria produce an enzyme called coagulase that causes host proteins to clump up around them, protecting the bacterial cells from host immune system cells. This protective armor, however, comes with a cost. Such bacteria are less able to spread easily to new host tissues (Figure 8.22b). An organism's various characteristics thus represent a compromise between the competing demands that any living thing must deal with.

Humans have guided the evolution of domestic plants and animals for thousands of years, and such **artificial selection** provides powerful evidence for natural selection. Consider, for example, the many breeds of domestic dogs. Over the centuries, humans have directed the evolution of dogs, by selecting those dogs with specific favorable traits and allowing only those individuals to reproduce. As a result, we now have breeds as diverse as pointers, selected for the ability to hunt; boxers, bred as guard dogs; and sheepdogs, selected for herding skills.

As microbiologists, however, we will now zero in on those organisms that are our primary focus and consider some examples of microbial evolution.

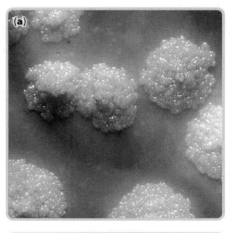

Figure 8.22 Natural selection is frequently a trade-off. All living things are subject to competing demands, and the results of natural selection are a compromise. (a) Colonies of *Mycobacterium tuberculosis* growing on an agar plate. Note the waxy appearance. Although *M. tuberculosis* is well protected from environmental hazards, its thick, waxy cell wall means that it absorbs nutrients and oxygen slowly and consequently has a relatively slow growth rate. (b) *Staphylococcus aureus* is a common cause of skin lesions such as boils. The bacteria secrete coagulase, which causes the coagulation of surrounding host proteins. Although this protects the bacteria from host immune system attacks, it limits their ability to spread throughout the host tissues.

Microorganisms are subject to the laws of natural selection

CASE: VANCOMYCIN RESISTANCE: THE SEQUEL

In Chapter 6 (p. 160) we encountered the case of vancomycin resistance and how some bacteria are now unaffected by this once-powerful antibiotic. You may recall that the antibiotic works by interfering with peptidoglycan synthesis by binding to the amino acids in peptidoglycan (Figure 8.23). Peptidoglycan synthesis is necessary for cell wall formation, and without a properly formed cell wall, bacteria are unlikely to survive. Due to a random mutation, some bacteria use different amino acids in their peptidoglycan. Vancomycin cannot bind to these altered amino acids, rendering bacteria with the mutation impervious to the effects of the antibiotic. Since vancomycin resistance first appeared in the 1980s, the number of resistant bacterial strains has increased, and today vancomycin is useless

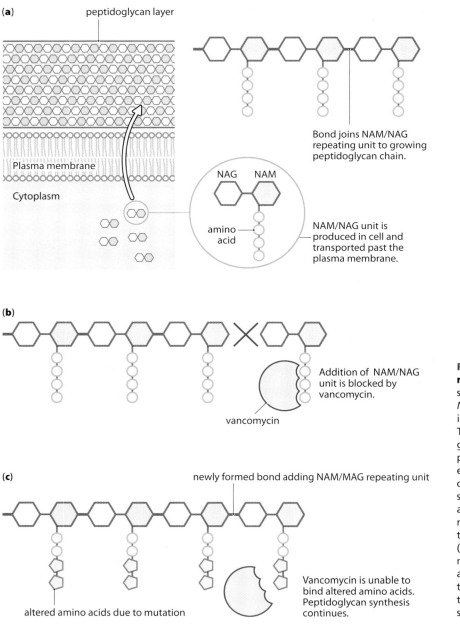

(a) peptidoglycan layer

Plasma membrane

Cytoplasm

Bond joins NAM/NAG repeating unit to growing peptidoglycan chain.

NAG   NAM

amino acid

NAM/NAG unit is produced in cell and transported past the plasma membrane.

(b)

Addition of NAM/NAG unit is blocked by vancomycin.

vancomycin

(c)   newly formed bond adding NAM/MAG repeating unit

altered amino acids due to mutation

Vancomycin is unable to bind altered amino acids. Peptidoglycan synthesis continues.

Figure 8.23 Vancomycin action and resistance. (a) During peptidoglycan synthesis, *N*-acetylmuramic acid (NAM) and *N*-acetylglucosamine (NAG) are assembled into the peptidoglycan repeating unit. The repeating unit is then added to the growing peptidoglycan molecule. Note the presence of the short amino acid chain on each NAM. See Chapter 3, p. 53, for a more complete description of peptidoglycan structure. (b) Vancomycin can bind to the amino acid chain, preventing addition of new peptidoglycan repeating units and thus interfering with cell wall synthesis. (c) In vancomycin-resistant bacteria, a mutation changes some of the amino acids attached to NAM. Vancomycin is unable to bind to these altered amino acids and is therefore unable to interfere with cell wall synthesis.

against certain infections. As if to add insult to injury, in the late 1990s it was announced that some strains of bacteria go beyond mere resistance. Researchers reported that strains of *Enterococcus* were not only resistant to vancomycin, they actually digested it and used it as an energy source. In other words, far from killing these microorganisms, the antibiotic actually facilitates bacterial growth in these strains.

1. How was the initial mutation for vancomycin resistance affected by natural selection?
2. How might resistance to vancomycin be reversed?
3. What allowed some bacteria to actually feed on this antibiotic? Do we expect such strains to become more common as well?

Drug resistance in certain bacteria provides an excellent example of how natural selection is involved in the evolution of specific characteristics. Like all other living things, bacteria that are better able to survive in a given environment tend to produce more progeny. Those that are less well adapted tend to perish over time. The arrival of widely available antibiotics on the scene, starting in the late 1940s, constituted a large shift in a microorganism's environment. Those individual microbes that were genetically able to survive in the presence of such antibiotics suddenly, and quite by chance, gained an enormous advantage over those less able to cope. Bacteria that were able to survive the introduction of antibiotics by whatever means reproduced more successfully and passed on characteristics that made resistance possible to their offspring.

Vancomycin resistance is an example: the altered peptidoglycan did not originally evolve millions of years ago to prevent damage by an antibiotic that would not even exist for eons. If on occasion, a random mutation in a bacterium resulted in the production of the aberrant peptidoglycan, this organism would gain no advantage and would not increase in number relative to bacteria with normal cell wall structure. A single mutated bacterial cell carrying this mutation, without a selective advantage, would not be expected to increase in numbers. In time any cells with the mutated cell wall might have died out entirely.

Now let us fast-forward a few million years to the age of antibiotics. Cells are now exposed to compounds such as vancomycin, and strictly by chance, the occasional mutation changing the cell wall structure still occurs (**Figure 8.24**). This means that if a cell happens to experience this mutation *and* is subjected to the antibiotic, this bacterium suddenly has an enormous advantage. It consequently survives and reproduces, passing on the now-favorable mutation to its progeny. Other cells, those with normal cell wall structure, are not so lucky. In the face of vancomycin exposure, they die and fail to reproduce. Over time, most members of the population will have descended from those cells that were initially able to resist the antibiotic. Cells bearing the mutation will continue to enjoy an advantage over normal cells as long as vancomycin exposure continues. This process, continuing over an extended period, might eventually result in a completely vancomycin-resistant strain of the bacteria.

The key phrase here is "as long as vancomycin exposure continues." Should vancomycin use stop, the advantage enjoyed by the mutated bacteria would disappear. Because they are no longer at a disadvantage, the proportion of

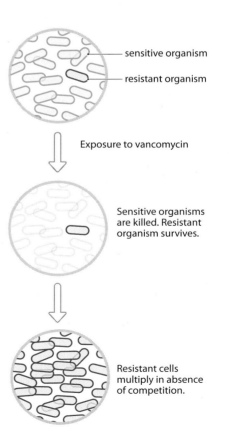

sensitive organism

resistant organism

Exposure to vancomycin

Sensitive organisms are killed. Resistant organism survives.

Resistant cells multiply in absence of competition.

Figure 8.24 Natural selection for vancomycin resistance. In the presence of vancomycin, any cell that experiences a mutation rendering it resistant has a selective advantage. Sensitive cells are killed, but the resistant cell survives and reproduces, ultimately giving rise to a completely resistant strain. This selective advantage would disappear if vancomycin use were discontinued. Any remaining sensitive cells might then again increase in abundance.

normal bacteria in the population might once again begin to rise if, under antibiotic-free conditions, the normal cell wall is in any way superior to the mutated form. Consequently, by discontinuing the use of vancomycin or by using it judiciously, only in those cases where it is absolutely required, the reversal of resistance is possible.

As for the *Enterococcus* that actually digests vancomycin, this was also the result of a random mutation in a strain of bacteria that already had the mutated cell wall type. This second mutation affected an enzyme, permitting it to now use vancomycin as an energy source. Because this strain of *Enterococcus* is unharmed by the antibiotic and actually grows faster in its presence, it could eventually increase in number as well, provided that its exposure to vancomycin continues.

Natural selection can influence the virulence of disease-causing organisms

CASE: A TALE OF TWO COUNTRIES—AUSTRALIA, ENGLAND, RABBITS, AND THE MYXOMA VIRUS

Rabbits are not native to Australia, but European rabbits were introduced there by settlers in the mid-19th century. This seemingly harmless event triggered an ecological nightmare. With no natural enemies, the rabbit population exploded at an astonishing rate, alarming farmers as rabbits completely obliterated millions of acres of farming and grazing lands (Figure 8.25). In 1950, a bold solution was attempted. The myxoma virus, common in South American rabbits, was introduced into the Australian rabbit population. This virus has little effect on South American rabbits, but its impact on Australia's European rabbits was immediate and severe. The virus spread rapidly, and within only a few years, over 99% of all rabbits were dead. But within a few more years, rabbit populations once again began to increase. Research confirmed that not only had the rabbits begun to evolve resistance to myxoma virus but also the virus itself had become less deadly. Interestingly, this was not the case when the myxoma virus was introduced into England. In England, unlike Australia, the virulence of the virus has remained extremely high, and exposed rabbits still rapidly succumb to infection.

1. **Why were South American rabbits unaffected by myxoma virus?**
2. **Why did the myxoma virus, so deadly initially for rabbits in Australia, become progressively less pathogenic?**
3. **Why didn't this reduction in pathogenicity occur in England also? What is different about the situations in Australia and England?**

The capacity of an organism to cause disease is an indication of that organism's **virulence**, and any virulent, disease-causing microorganism is called a **pathogen**. The organism the pathogen infects is called a **host**. Genetics of the pathogen and host play an important role in the host–pathogen interaction, and like other genetically based traits, the host–pathogen relationship is subject to natural selection. It consequently evolves over time.

It is often argued that, over time, pathogenic microbes should become less virulent. After all, if you are a pathogen, you might think it would be a mistake to kill your host. Doing so *sounds* counterproductive in the extreme, because a pathogen that kills quickly has little chance to be transmitted to new hosts and suddenly faces extinction itself. Less deadly pathogens enjoy a selective advantage over more lethal strains of the same species, because the imperative to be transmitted is not so urgent. A more benign pathogen may remain in its host longer and consequently will have more time to reach a new host. Furthermore, the hosts are also evolving and often evolve resistance over time. Individual hosts who have a more vigorous immune response will also be more likely to have offspring and will pass on

Figure 8.25 Rabbits gone wild.
Following introduction into Australia, the population of European rabbits exploded, devastating millions of acres of farm and grazing land. Myxoma virus was introduced to control rabbits, but after initial stunning success, rabbit numbers began to increase once again. In England, the virulence of the virus has persisted.

resistance in the process. As a result, goes the argument, both pathogens and hosts evolve to the point where they reach a sort of evolutionary truce, allowing the survival of both.

We alluded to this principle previously in Chapter 5, when we considered the role of infectious diseases in the conquest of the Americas and in the exploration of Africa. In our case study we observe very much the same thing with myxoma virus in Australia. The virus evolved in South America, and South American rabbits evolved in response to it. After a long coexistence, the rabbit population was relatively unaffected by the virus, as those rabbits that developed resistance reproduced more successfully than nonresistant rabbits. Meanwhile, the virus was also evolving, with less virulent strains of the virus better able to persist. Ultimately, both the virus and the rabbits in South America reached a point in their evolution where mutual survival was possible.

But the myxoma virus was poorly adapted to European rabbits in Australia when it was first introduced, and consequently the virulence of the virus was high. At the same time, rabbits in Australia had never been exposed to myxoma virus, so there had never been any selection for resistance. The result was the initial sky-high death rate for rabbits in Australia. This situation benefited neither the rabbits nor the virus. Once again, because of random genetic variability in both the rabbit and viral populations, natural selection went to work. Those few rabbits fortunate to have some resistance to myxoma virus were the only ones to survive and reproduce. Those rare viral particles that were genetically less virulent in European rabbits also fared better than more virulent strains, because they allowed an infected rabbit to live long enough for the virus to get to a new rabbit.

It may be comforting to think that natural selection always promotes reduced virulence over time. Yet the English experience with myxoma virus demonstrates that it does not always work that way. Remember, when it comes to natural selection, the environment is all-important. No trait, including lower virulence, is necessarily always beneficial or always detrimental; environmental conditions set these ground rules. So what is the difference between Australia and England? In Australia, myxoma virus is transmitted between rabbits by mosquitoes (**Figure 8.26**). When a mosquito feeds on an infected rabbit, it picks up viral particles, which can then be transmitted to a different rabbit the next time the mosquito feeds. In England, the virus is transmitted by fleas. And while most people probably consider both mosquitoes and fleas to be a similar sort of nuisance, there is actually a big difference between them. Adult mosquitoes die when the weather turns cold. Fleas survive the winter months.

During the Australian winter, there are no mosquitoes and consequently there is no viral transmission. The selection pressure on the virus for reduced virulence was therefore enormous. Virulent strains that killed their rabbit hosts quickly died out themselves, because once their host died, they were essentially stuck with no way to get to a new rabbit. Killing their host before they were transmitted meant these virulent viral strains effectively cut their own throat. Less virulent strains that permitted rabbit survival, at least until the next mosquito season in the spring, had a large selective edge and ultimately came to predominate. In England the situation was quite different. Myxoma virus was transmitted year-round, because fleas were always available. There was never any difficulty getting to the next rabbit, so for a virus,

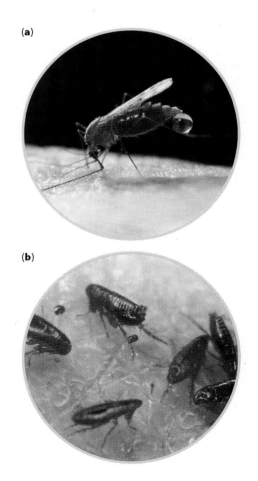

(a)

(b)

Figure 8.26 Myxoma virus transmission. (a) In Australia, myxoma virus is transmitted by mosquitoes, which do not survive the winter months. Consequently, viral transmission ceases during the winter. (b) In England, the virus is transmitted through flea bites. Since fleas survive throughout the year, virus transmission likewise is year-round. How does this fact explain the difference in viral virulence in Australia and England?

killing its host was much less of a problem. Consequently, less virulent viral strains did not enjoy a reproductive advantage. With the selection pressure for reduced virulence greatly reduced, the virus retained its ability to quickly kill most infected rabbits.

Biologists now understand that virulence, like any other trait, can be acted on by natural selection and that pathogens will become less virulent over time only if a reduction in virulence comes with a selective advantage. In other situations, becoming less virulent carries no such advantage. Often, as we saw in the case of myxoma virus, the determining factor is ease of transmission. When it is easy to get from one host to the next, less vicious pathogens gain no real edge on their nastier cousins. When, on the other hand, transmission becomes difficult, selection suddenly favors more benign disease organisms, allowing them to persist in their host until they can finally reach a new host.

To illustrate this concept in humans, let's consider the two important bacterial pathogens that cause the diseases typhoid and cholera. Both are transmitted through contaminated water and thus are called water-borne diseases. In parts of the world where sanitation is especially poor, the typhoid bacteria are particularly virulent (**Figure 8.27**). Poor sanitary conditions ensure that the pathogen will have ample opportunity to achieve transmission. Routine sanitation, however, can change this. With safe drinking water and proper disposal of waste, transmission is much less assured. Selective pressure on the bacteria now favors reduced virulence. The now more benign strains can survive indefinitely, until they get a relatively rare chance to reach a new host. Likewise, when a cholera epidemic struck South America in the early 1990s, the virulence of the bacterial pathogen was high in Ecuador, where sanitation was relatively poor. In Chile, with comparatively better sanitation, those who contracted cholera suffered milder symptoms.

Accordingly, improved sanitation provides a double bonus. Not only do far fewer people contract water- or food-borne diseases in the first place, but when they do, they are likely to encounter a less virulent strain of the disease-causing organism. Scientists are now beginning to consider the possibility of directing the evolution of disease organisms by changing the environment in ways that select for reduced lethality. Their thinking is that if we could comprehend the underlying principles that control virulence, perhaps we could manipulate the host–pathogen relationship in ways that cause virulence to decrease. This emerging field of public health is called **Darwinian medicine**.

Microorganisms often influence the evolution of their hosts

Our emphasis in this chapter has been on the evolution of microorganisms. By and large we have not looked closely at how microorganisms can influence the evolution of their hosts. Microorganisms, however, are an important part of *our* environment and consequently act as an important selective force on host evolution. Clearly hosts that had certain adaptations, which best allowed them to fend off microbial attack, had higher rates of survival and contributed more offspring to the next generation. Micro-

Figure 8.27 Transmission of water-borne pathogens and the evolution of virulence. (a) A slum in Latin America. When sanitation is poor, transmission is assured and there is little selection pressure on pathogens to reduce virulence. (b) A low-cost filtration device provides safe drinking water for these Bangladeshi villagers. Improved sanitation results in far fewer transmission opportunities. Less virulent pathogens now gain a selective advantage, because they can persist in their host without killing for a longer period of time and thus have more time to achieve transmission.

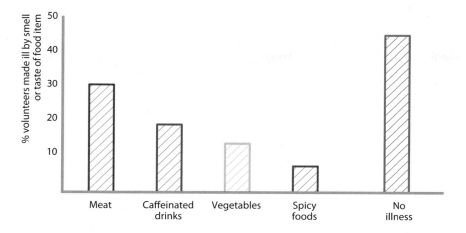

Figure 8.28 Mothers-to-be and morning misery. Researchers surveyed 79,000 pregnant women and found that meat was far more likely to induce morning sickness than any other type of food. Is this an evolved, adaptive response to protect the fetus from potentially dangerous meat-borne pathogens?

organisms have certainly spurred the evolution of the immune system, which has become increasingly complex over evolutionary time. Many other host adaptations, less obviously related to immune defense, may also have been honed by microbial selection pressure. To make this point, we will consider a few surprising examples.

Morning Sickness. As any woman who has been through it understands, pregnancy can have all sorts of unusual effects on a woman's body. One of the oddest is morning sickness. Just when a woman might require good nutrition the most, even the smell of certain foods can make her ill. More specifically, studies have shown that more women feel ill after smelling or tasting meat than any other food type (**Figure 8.28**). Many have suggested that morning sickness evolved as a way to protect a developing fetus from food-borne pathogens, and meat is especially likely to harbor organisms that might otherwise damage a developing baby. *Toxoplasma*, for instance, was discussed in Chapter 4. This protozoan is usually harmless for an adult, but an exposed fetus can develop a variety of problems, including deafness and mental retardation. Accordingly, morning sickness might be realistically viewed as a sensible, if not particularly pleasant, evolutionary strategy to increase the likelihood of a healthy baby. Interestingly, morning sickness generally subsides after the first trimester, when the fetus is much less vulnerable and the benefits of extra nutrition outweigh the risk. In any event, perhaps a mother-to-be can take some solace from the fact that her discomfort is actually a sacrifice, maybe the first sacrifice, on behalf of her baby. Experienced parents know it will not be the last.

Prions and brain evolution. In Chapter 4 we also first learned about kuru, the prion disease once transmitted by cannibalism in New Guinea. You will recall that if a person died of kuru, relatives who consumed portions of the deceased's brain became infected with the prion themselves and subsequently developed the disease. But not everyone who partook in this ritualized cannibalism developed kuru, and recent research has discovered why. Some people had a mutation that changed the shape of certain normal brain proteins, and in this mutated shape, the kuru protein could not bind and convert the normal protein to the kuru protein.

In societies where cannibalism was practiced, having the protective mutation would obviously provide a large selective advantage. Individuals who had the mutation would be expected to have more children and to pass on the mutation in the process. Over time, a larger and larger proportion of the population would carry the protective version of the protein. In other societies where cannibalism was unknown, this mutation would provide no selective advantage. Consequently, if the mutation did occur, it would not necessarily increase in frequency and might ultimately disappear altogether. The investigators found that among New Guinea's Fore people, for whom cannibalism was once a regular part of ritual life, 55% of the

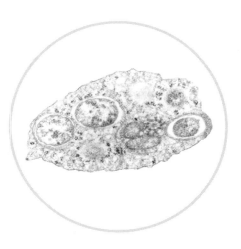

Figure 8.29 Wasps and *Wolbachia*. A wasp cell infected with *Wolbachia* bacteria. The bacteria are the oval-shaped structures within the cytoplasm of the larger, wasp reproductive cell. In some species of wasps, different species of these bacteria are the only things preventing successful mating.

population still carried the mutation. In Japan, with no history of cannibalism, the mutation was found in only 6% of individuals. Other sampled populations had rates of mutation similarly correlated with the historical occurrence of cannibalism.

Bacteria and the formation of insect species. Sometimes infection can result in something as dramatic as the development of new species. As we learned in Chapter 3, in animals, a **species** is a group often defined by its members' ability to mate only with other members. Mating with a different species is almost always unsuccessful.

There are many reasons why mating between different species rarely succeeds. Eggs and sperm from different species are usually incompatible, and even if hybrid offspring are produced, in most cases they will not develop normally. Or, in the case of wasps within the genus *Nasonia*, the inability of different species within the genus to mate successfully is enforced by bacteria. The wasps and bacteria in the genus *Wolbachia* have one of nature's most unusual partnerships. The bacteria live inside the wasp's reproductive cells, but they cause little if any damage (**Figure 8.29**). They do, however, have a huge impact on wasp reproduction.

Two closely related wasp species, *Nasonia giraulti* and *Nasonia longicornis*, are infected with different *Wolbachia* species. Investigators have found that the two wasp species were normally unable to mate successfully. Interestingly, if the wasps are treated with antibiotics, thus wiping out all bacteria in both wasp species, mating between the two species is successful. In other words, in *Nasonia*, the interspecies barrier to reproduction was completely due to the bacteria. The two types of wasps may actually provide us with a glimpse of new species in the process of forming, and it is the bacteria that started the wasps on separate evolutionary roads. Should the now reproductively isolated wasps develop other genetic differences between them, the final species barrier may one day be complete.

CASE: ALICE IN WONDERLAND

In Lewis Carroll's *Through the Looking Glass*, Alice passes through a mirror into a magical world where everything seems to be backward. Alice enters an enchanted garden where she meets the Red Queen, who tells her about the Red Queen's Race. Alice is mystified to learn that in this race the runners never actually move, and she observes, "In our country, when you run, you generally get somewhere." The Red Queen responds "Now, here, you see, it takes all the running you can do, to keep in the same place."
1. **What does the Red Queen's statement have to do with the way hosts evolve in response to infectious organisms?**

The evolution of sex. Why is there sex? This question may sound surprising or even ridiculous at first, but consider the fact that most organisms on Earth get along just fine without it. Prokaryotes do not have males and females, and even many eukaryotes dispense with separate sexes. Many plants and animals reproduce primarily or exclusively asexually. Some species of lizards, for instance are all females. Offspring are produced when eggs, unfertilized by sperm, simply start developing into new lizards on their own. And compared with asexual reproduction, sexual reproduction (see Chapter 6, p. 154) is slow and inefficient. After all, 10 asexually reproducing females can produce many more babies than five males and five females can.

One large advantage of sexual reproduction, however, is that it might permit host populations to survive in the face of a never-ending onslaught by pathogenic microorganisms. According to the **Red Queen hypothesis**, the continual selection pressure imposed by pathogens on their hosts favors genetic diversity in the host. And one prominent way to obtain genetic diversity is through sexual reproduction.

In asexual reproduction, all offspring may have the same genetic makeup as their parent. They are essentially clones of each other. In sexually reproducing organisms, on the other hand, each offspring represents a unique combination of genes from both the mother and the father. The genetic deck is, so to speak, shuffled with each reproductive event. And such diversity might help prevent extinction at the hands of microorganisms, because microorganisms are not able to infect and colonize all members of a host population equally. Some individuals, purely by luck, have genetically determined characteristics that make them a less appealing target for a specific pathogen. Individuals with these characteristics may in time come to be especially numerous in the population, because they are less susceptible to the pathogen in question.

Pathogens, however, are also subject to natural selection. There is intense selection pressure on the pathogens to colonize the most abundant host genotypes. Once pathogens colonize these hosts, however, the selective advantage previously enjoyed by these hosts is gone. Their abundance in the population begins to decline, setting the stage for a decline in the abundance of those pathogens that specialize on them. Meanwhile, however, in a genetically diverse, sexually reproducing population of hosts, a different host genotype is now the most resistant and begins to increase in abundance.

In this way, sexually reproducing hosts are genetic "moving targets" for pathogens. There is always a portion of the host population that is relatively unaffected by the pathogen. Both the host and pathogen undergo repeated cycles of selection, with increasing and decreasing abundance of different genotypes, ensuring the survival of both host and pathogen (**Figure 8.30**). In asexually reproducing organisms, this is far less likely. The largely clonal nature of such populations means that all members are very similar genetically. If a pathogen becomes able to colonize a particular clonal population, the entire population becomes an easy target for infection and may be exterminated.

But like Alice on her trip to Wonderland, in spite of constant natural selection and evolution, the best that sexual organisms can hope for is to stay in place. The eternal assault and counterassault between hosts and pathogens results in a never-ending race, with neither party ultimately winning, just as the Red Queen indicated. Of course, no one would argue that there are not other pressures besides pathogens that might favor genetic diversity and thus sexual reproduction. Yet it is worth remembering that, though unseen, microorganisms are an enormous part of our environment and in many ways have helped to shape us into what we are today.

Coming Up Next...

In earlier chapters we have described the diversity seen in the microbial world, and in this chapter we have explored how that diversity came to be. In the next chapter we will turn our attention to the consequences of that diversity in terms of ecological processes. After a review of some basic concepts we will consider the specific ecological roles of microorganisms in different habitats, such as terrestrial and aquatic habitats, and extreme environments. We will move on to discuss the great biogeochemical cycles that sustain all life on this planet and how microorganisms do more than their

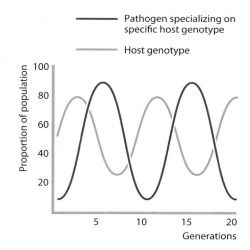

Figure 8.30 Running hard to stay in the same place: the Red Queen hypothesis. Pathogens that specialize on the most common host genotype have an advantage, simply because their host is most abundant. This gives a selective advantage to relatively rare host genotypes, which are relatively unaffected by pathogens. As numbers of the more abundant hosts decline because of high infection rates, the previously rare host type may come to predominate. There is now selection pressure on the pathogen to take advantage of this newly abundant host type. The result is a never-ending cycle of population peaks and crashes for hosts with a particular genotype and pathogens that specialize on that genotype. Sex may have evolved, at least in part, to introduce genetic variability into host populations, so that a pathogen cannot drive a specific host type to extinction.

share to recycle crucial substances such as nitrogen, phosphorus, and carbon. Finally we explore some of the interesting ways that microorganisms interact directly with each other and with nonmicrobial organisms and why these interactions are crucial to a healthy ecosystem. In Chapter 4 we stated that life itself would not be possible on Earth without microorganisms. In Chapter 9, we will find out why.

Key Terms

abiotic synthesis	escaped-gene hypothesis	regressive hypothesis
artificial selection	host	ribozyme
bacteriophage	liposome	species
coevolved hypothesis	natural selection	stromatolite
Darwinian medicine	pathogen	transposon
endosymbiosis	Red Queen hypothesis	virulence

Concept Questions

1. In this chapter we have reviewed the sequence of events that are thought to have occurred in the history of living things. Could they have occurred in a different order? Could animals, for example, or other multicellular eukaryotes have evolved first, with prokaryotes coming later?

2. The specific mutations that lead to antibiotic resistance have no doubt been occurring in bacteria for millions of years. Yet until the mid-20th century, bacteria that experienced these mutations did not increase in numbers. Over the last half century, however, such bacteria often thrive and come to predominate. Why?

3. The entire theory of evolution by natural selection can in some sense be summed up in the phrase "mutation proposes, selection disposes." What exactly does this mean?

4. You are an exobiologist, looking for a planet on which life may be in its earliest incipient stages. What would be the sorts of physical characteristics that you would look for on such a planet?

5. How would the history of life on this planet be different if photosynthesis had never evolved? What would life be like if only glycolysis and photosynthesis had evolved, but aerobic respiration had not?

6. Explain how the discovery of ribozymes helps to resolve the issue of which came first, DNA or proteins.

7. After reviewing the section in this chapter about the origin of the first cells, describe why it appears crucial that these cells evolved in water.

8. You describe the theory of endosymbiosis to a friend, who asks an intriguing question: "Why is it more likely that bacteria gave rise to mitochondria and chloroplasts instead of the other way around?" In other words, why isn't it just as likely that today's bacteria started out as mitochondria and chloroplasts that were in some way ejected from eukaryotic cells? What is your answer?

9. Imagine a hypothetical and highly unrealistic situation involving two isolated human populations of the same size. Initially, rates of HIV infection are the same in both populations. In one population, however (population A), a well-planned educational and prevention program to reduce HIV transmission is implemented. Almost all individuals are tested and subsequently given information as to how to best protect themselves and others. In the second population (population B) no such programs are implemented, and many people remain ignorant about HIV and how to protect themselves. If you compared these populations after 20 years, how might the virulence of the virus have changed in the two populations?

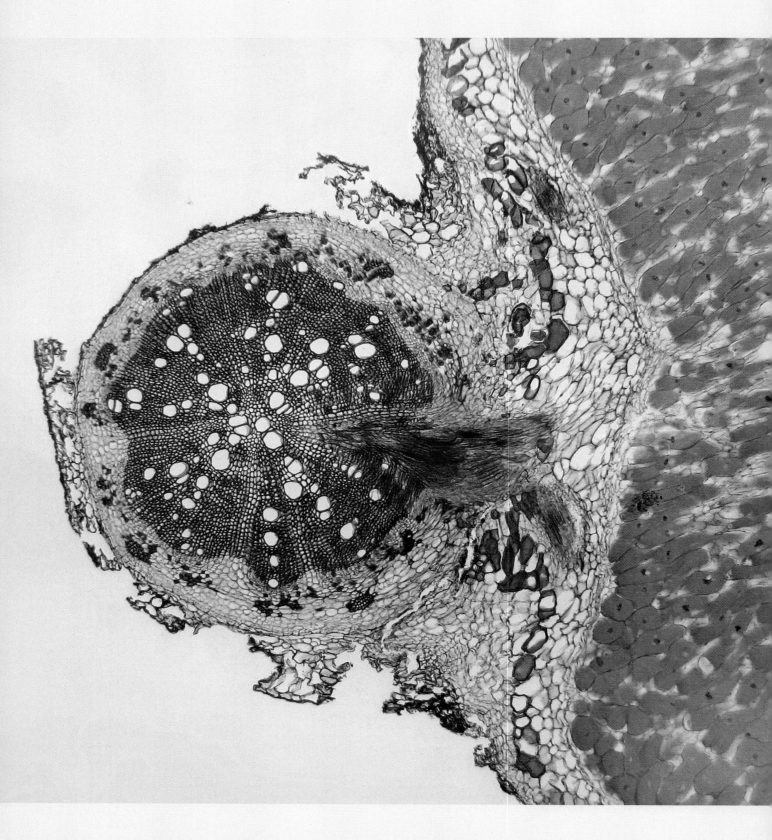

Soybean root nodules host nitrogen-fixing bacteria in a mutually beneficial relationship.

Chapter 9

An Ecologist's Guide to Microbiology

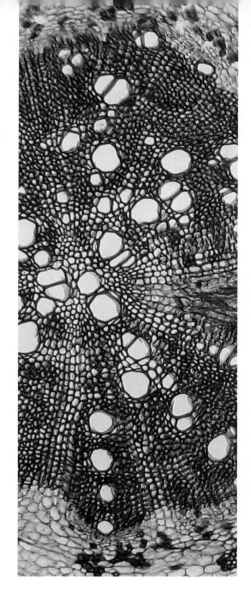

Not many people have heard of siboglinids, commonly known as "tube worms." Fewer still know much about them. These mysterious creatures often inhabit the deepest parts of the ocean, near hydrothermal vents on the sea floor (**Figure 9.1**). As we learned in Chapter 8, extremely hot water enriched with inorganic compounds gushes from the vents, supporting a rich microbial community. The vent-dwelling anaerobes, which live in total darkness and without oxygen, form part of a bizarre biological community that is independent from our infinitely more familiar world, where energy initially comes from the sun.

But there is nothing microbial about siboglinids. On the contrary, these invertebrates can grow to over 8 ft in length, with a circumference of more than 12 in. You might guess that such enormous creatures have appetites to match their size. Yet tube worms completely lack a mouth, a stomach, intestines, or any other digestive structures. Instead, they have a spongy organ called a trophosome that is teeming with billions of bacteria. These microorganisms are chemoautotrophs (see Chapter 7, p. 188). They obtain energy by oxidizing sulfur compounds released from the hydrothermal vents and, in doing so, satisfy both their own and the worms' nutritional needs.

Obviously, for tube worms, these bacteria are crucial to their survival. The role that microorganisms play in other life processes is not always so apparent. When we think about interactions between living things, most of us tend to focus on the organisms that we can see. We observe rabbits and deer eating plants, and we know that meat eaters such as coyotes and wildcats consume the plant eaters. We notice insects such as bees as they pollinate plants and enable them to reproduce, and we may witness vultures feeding on dead animals. All the while, however, microbes, unseen and perhaps underappreciated, are quietly at work breaking down the waste products of other living things and increasing soil fertility. Microorganisms form the base of much of our food supply, and they produce much of the oxygen on which aerobes such as ourselves depend. In fact, all ecosystems rely on microorganisms if they are to survive and remain healthy.

In this chapter we will focus on exactly what microorganisms do that is so critical to a healthy environment. After reviewing some basic ecological concepts, we will explore the roles of microorganisms in various habitats including soil, freshwater, and marine environments. Next we will learn how important elements such as carbon and nitrogen are cycled through the environment and how microorganisms drive these recycling processes. Finally, we will consider some of the interesting ways that microorganisms compete, prey upon, or otherwise directly interact with each other and with nonmicroorganisms. We will see that all life is indeed interlinked. No part of a healthy ecosystem can be altered without affecting every other part. We will now consider some of the smallest, but in certain ways some of the most important, cogs in the highly integrated system that make up the interdependent web of living things.

Figure 9.1 Siboglinids inhabiting a deep sea vent. Siboglinids, commonly called "tube worms," live in mutually beneficial symbiosis with bacteria in hydrothermal vent communities. Members of this community are dependent, either directly or indirectly, on reduced sulfur compounds that are released through the vent. Such vents are often found at depths of over 3000 m beneath the ocean surface.

Basic Ecological Principles

Ecology is the study of how living things interact with each other and the environment

People often use the word "ecology" when they are discussing environmental issues such as pollution or habitat preservation. To the scientist, this word has a more precise definition. **Ecology** is defined as the study of the relationship of organisms to each other and to their environment. The ecologist may therefore be concerned with who eats whom, which organisms compete with each other for the same resources, or which ones interact in ways that benefit both organisms. Such biological interactions, known as **biotic** interactions, are influenced by numerous physical factors such as soil chemistry, temperature, or availability of water. These physical factors, called **abiotic** factors, are important in determining which organisms are found in any given environment, and in many subtle and not so subtle ways they influence the nature of the biotic interactions. Biotic–abiotic interplay, however, is not a one-way street. Living things also alter abiotic conditions. To cite two examples, consider the difference plants can make on the air temperature and levels of sunlight in their immediate vicinity or how a herd of cattle can alter the soil chemistry in a pasture. It is no coincidence that mushrooms, which require high levels of soil nitrogen to thrive, are so abundant where there is a lot of cow manure (**Figure 9.2**).

All the members of a particular species that live in a defined geographic area comprise a **population** of that species. The various populations within a defined region make up the ecological **community**. The hydrothermal vent community that includes both tube worms and their symbiotic bacteria also includes any other populations living near the hydrothermal vent. When we discuss both the community and the abiotic characteristics of a specific area, we are referring to an **ecosystem**. A freshwater lake, a large desert, and a mountain range are examples of ecosystems.

Energy and nutrients are passed between organisms in an ecosystem

The manner in which energy and nutrients flow through any ecosystem is a key feature of that system. Often, the movement of nutrients or energy through an ecosystem is depicted as a **food chain**, which is the "path" of food consumption between the various members of a community (**Figure 9.3a**). At the bottom of the chain are the **producers**—those crucial organisms that first convert inorganic carbon into biological molecules through the process of carbon fixation. Many producers are photosynthetic organisms, but as we discussed in Chapter 7 (p. 186), others rely on chemical energy to synthesize biological molecules. Producers, whether they use solar or chemical energy, are the only organisms able to introduce biological molecules into a food chain, and consequently all life depends on them. Producers are consumed by **primary consumers**, also called **herbivores** (plant eaters), which in turn are consumed by **secondary consumers**, known as **carnivores** (meat eaters). Many food chains include tertiary or quaternary level consumers that feed on those lower in the feeding hierarchy (see Figure 9.3a). The position of an organism in a food chain is called that organism's **trophic level**. Thus producers occupy the first trophic level, while herbivores occupy the second level. Secondary and tertiary consumers occupy higher trophic levels in the food chain.

Figure 9.2 Biotic factors can affect abiotic conditions. Fungi require high amounts of soil nitrogen. Levels of this crucial nutrient may be especially high in cow pastures, where nitrogen is continually introduced into the environment in cow manure.

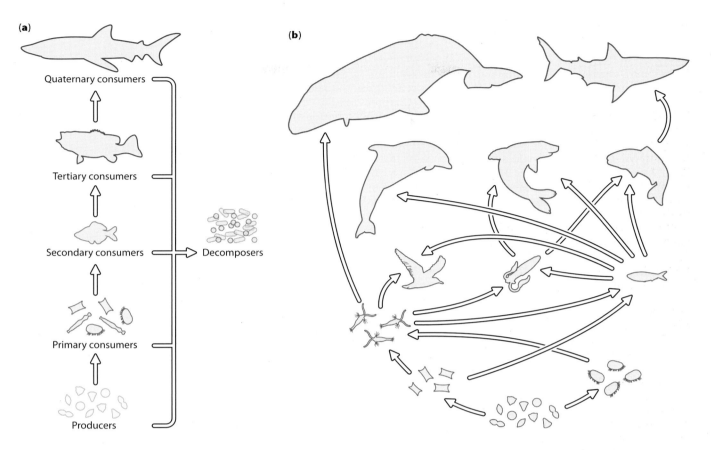

Figure 9.3 Trophic interactions. (a) Typical marine food chain. A food chain is a simple way of depicting feeding interactions between organisms. (b) Marine food web. A food web is a more accurate way to present such interactions in most ecological communities. Decomposers feed at all levels in a food web.

In all but the most basic ecosystems, however, the model of a food chain is probably a gross oversimplification, because many organisms are not restricted to a single location on the food chain. Many animals, for example, are **omnivores**, which eat both plants and other animals. Depending on what an omnivore is eating, it may be a primary, secondary, or even higher level of consumer. Accordingly, a more complex pathway, illustrating how different species in an ecosystem may feed at different trophic levels, is a more accurate way to depict feeding relationships. Such a depiction is called a **food web** (see Figure 9.3b).

Some microorganisms are producers, and some, especially fungi, act as consumers. A few bacteria and many protozoa are predatory and consequently act as consumers as well. But it is as **decomposers** that microorganisms truly take center stage. Decomposers absorb organic materials from dead organisms, including animals, plants, and other microorganisms. Decomposers play a vital role in all ecosystems, because they degrade organic materials into inorganic substances, which are then available to producers. They are absolutely crucial to a healthy ecosystem. As discussed in Chapter 4, a single gram of soil might contain thousands of species of prokaryotes and fungi, many of whom are active as decomposers. Aquatic habitats also teem with microbial recyclers. In certain places such as very cold or very dry environments, decomposers are either scarce or slow-growing. Part of the reason that plants grow slowly in such environments is that with reduced decomposer activity, there are fewer nutrients available in the soil to support plant growth.

Nutrients are recycled in an ecosystem, whereas energy is not

Energy and nutrients do not move through ecosystems in the same way. Nutrients such as iron or phosphorus can be recycled through the environment indefinitely, but energy cannot. Recall from Chapter 7 (p. 173) that, for living things, energy initially resides in the carbon–hydrogen bonds of biological molecules. The conversion of energy from these molecules into a different chemical form such as adenosine triphosphate (ATP) is far from 100% efficient. Much of the energy in a food molecule is lost in the form of heat. Consequently, as energy moves up through a food web, there is less and less of it available at each higher level. When an herbivore eats a plant, as much as 90% of the energy in the plant material is lost as heat. When a carnivore eats the herbivore, most of the calories are once again lost. Put another way, it might take 100 lb of grass to produce a pound of antelope and 100 lb of antelope to produce a pound of lion. This explains why, if you could determine the mass of all organisms in an ecosystem, the collective mass of producers would far outweigh that of consumers. The collective mass of any population or community in a particular location is known as the **biomass**. In any ecosystem, the biomass of producers is greater than that of primary consumers, which is correspondingly greater than that of secondary consumers. By the time energy is transferred to secondary consumers, so much of that energy has been lost that very little is available for higher-level consumers. Only the most productive ecosystems, those where producers are especially prolific, can support a small number of tertiary or quaternary consumers (**Figure 9.4**).

Microbial Ecology

Now we focus our attention more specifically on the ecology of microorganisms. In many ways, microbial ecology is no different than the ecology of other organisms, and most of the same basic principles apply. But microorganisms are unique in that they live in *all* ecosystems. Not only are they found in those places we commonly associate with plants and animals, such as the ocean, lakes, rivers, and soil, but they also live in extreme environments, where few if any larger organisms are found. They colonize the digestive tracts of animals, the roots of plants, and essentially every other living surface. Wherever nonmicrobial life is found, microorganisms are

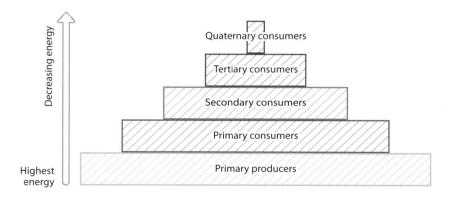

Figure 9.4 Amount of available energy at the various levels of a food chain. Because there is less and less energy available at progressively higher levels of the chain, the numbers and/or biomass of organisms at each trophic level likewise declines. This relationship between biomass and trophic level is depicted here as an *energy pyramid*.

there as well. As previously mentioned, their fundamental role in nutrient recycling and their phenomenal growth rate (see Chapter 7, p. 190) set them apart from other living things. As we saw in Chapter 8, microorganisms did just fine without nonmicroorganisms for more than 1.5 billion years. The reverse could not occur. Without microbes, we would not be here.

Microorganisms live in microenvironments

Because microorganisms are so small, the environment in which they live can be described as a **microenvironment**—the area immediately surrounding a microorganism. Biochemical reactions carried out by microorganisms have a profound impact on the abiotic conditions of the microenviroment. As waste products are released, or as nearby nutrients are absorbed, chemical parameters such as pH or the concentration of ions can be significantly altered.

At first, it is difficult to fathom how small these microenvironments really are. As an example, consider a typical bacterial cell with a length of about 4 μm. If you measured environmental conditions 4 mm (4000 μm) away from the cell, it would be similar to measuring the conditions more than several *kilometers* distant from a large mammal. Over the seemingly small distance of 4 mm, conditions can vary dramatically. One spot, for instance, might be highly acidic, while just a millimeter or two away, basic conditions may prevail. To the casual observer, it might *appear* that acidophiles (acid loving) and nonacidophiles are living side by side, but in reality they occupy vastly different microenvironments. In a single soil particle one might find aerobes living on the particle surface, where oxygen levels are high, while only a short distance away, within the soil particle, a lack of oxygen allows strict anaerobes to grow (**Figure 9.5**).

Environmental conditions affect the growth rate of microorganisms

In Chapter 7 we learned that although microorganisms grow under a variety of conditions, they tend to grow best under one specific set of conditions. For example, typical bacteria that live in the human intestine might divide about twice per day. If, however, these same cells are cultured in the laboratory under optimal conditions, they may replicate in 30 minutes or less. Similarly, soil bacteria usually grow at a rate that is far below what might be observed under controlled laboratory conditions.

Why is this? First, in the laboratory microorganisms are often grown in pure culture (a culture of just one species). This is almost never the case in their natural environment. A single gram of soil might contain thousands of species, and competition between these species can slow growth rate. Second, as discussed in Chapter 7, less-than-ideal abiotic conditions can result in submaximal growth. In many cases, growth rate reflects nutrient availability. Prolonged exponential growth (Chapter 7, p. 190) is actually quite rare in the environment, because microorganisms rapidly exhaust the supply of nutrients. Microbial growth, then, often surges when nutrients are abundant and stalls when those nutrients are depleted. Consequently, many microorganisms are adapted to a life of "boom or bust." When resources are abundant, growth rate accelerates and may approach a maximum. When these resources are used up, growth slows or even stops as cells survive on stored biological compounds. To better visualize this concept, as microbial ecologists we must "think tiny." A single dead insect or a rotting grass blade might represent a food bonanza for microorganisms, resulting in a rapid growth spurt. Once this bounty is used up, it is time again to hunker down and wait for the next nutritional windfall.

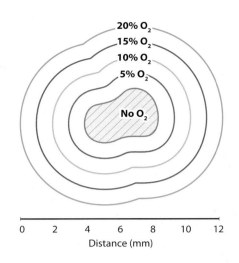

Figure 9.5 Microenvironments in a soil particle. Even a single soil particle that measures only several millimeters across may contain many microenviroments. This diagram depicts how just one abiotic factor, oxygen (O_2), can vary in different parts of the soil particle, creating different microenvironments for different microorganisms.

Many microorganisms live in biofilms

Speaking microbiologically, what do your tongue and a corroded water pipe have in common? They are both likely to be coated with a slimy layer of microorganisms called a **biofilm**—a microbial community in which the cells are encased in a layer of polysaccharide. Although many bacteria live freely, separate from other individuals, they more commonly live as biofilms. The dental plaque that forms on teeth, which we discussed in Chapter 3, is an example of a biofilm, as is the slippery layer found on rocks in a stream bed (**Figure 9.6**). Less-than-fastidious housekeepers are likely to have biofilms growing as scum that gradually builds up in toilet bowls or as a slimy coating on kitchen drains.

Biofilm formation begins when bacteria stick to a surface with their glycocalyx—the loose polysaccharide-based outer coating discussed in Chapter 3. Other unrelated cells may subsequently attach and begin to reproduce, until the entire surface is covered with a slippery layer of bacteria and polysaccharide. Biofilm growth, however, is not as haphazard as it sounds. Clumps of bacteria tend to grow in mushroomlike structures, supported by short stalks (**Figure 9.7a**). Openings between the stalks allow the circulation of water currents that carry resources such as oxygen or nutrients to the cells.

Depending on the environment, biofilms can become increasingly complex, with different types of organisms forming separate layers (Figure 9.7b). On a rock in a steam, for example, there may be photosynthetic organisms on the surface, with a layer of nonphotosynthetic bacteria just below.

Biofilms are certainly unsightly, but their importance goes far beyond the aesthetic. Persistent infections that seem to resist all antibiotic therapy are often caused by bacteria living in biofilm communities, since many

Figure 9.6 Bacterial growth in biofilms. Bacterial biofilms are found on many surfaces. (a) Biofilms are common on rocks in streams. The sticky sugars in the biofilm allow bacteria to adhere tightly to the rocks. (b) A biofilm is clearly visible in this toilet bowl, as the discolored area below or at the waterline. (c) Electron micrograph of streptococci growing as a biofilm on human teeth. The biofilm makes it less likely that the bacteria will be swept away by saliva or by tooth-brushing.

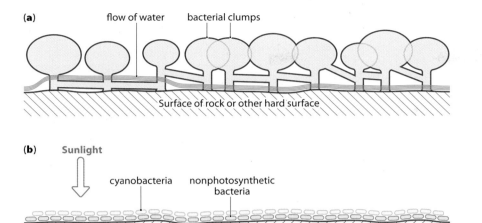

Figure 9.7 Structure of a biofilm. (a) Clumps of bacteria adhere to the surface via short stalks. Water is able to flow between the stalks, permitting efficient nutrient supply and removal of wastes. (b) Different microorganisms may be found in different layers of a biofilm. In this example, photosynthetic cyanobacteria occupy the top layer where they can more readily absorb sunlight. Nonphotosynthetic bacteria form the lower layer.

antibiotics have difficulty penetrating the biofilm. Medical implants, such as artificial joints, are often sites of biofilm development. As many as 10 million infections per year in the United States are caused by biofilms on medical implants or because of invasive procedures with biofilm-tainted devices such as catheters.

Microbial biofilms also cause other, nonmedical problems. When they grow on a ship's hull, for example, they not only cause corrosion but also increase drag on the vessel, leading to higher fuel costs. Biofilms in pipes and drains cause extensive corrosion and block fluid transport. Submerged objects such as docks and oil rigs are often damaged by biofilm growth.

As unsavory as biofilms are, they also have their benefits. As we will see in Chapter 16, **bioremediation**, or the use of bacteria to degrade harmful chemicals, often relies on biofilms.

Microbial Habitats: Here, There, Everywhere

As we have emphasized repeatedly, microbes grow in almost any conceivable habitat. In this section we will review the role of microorganisms in some of their most important habitats, including soil, freshwater, and marine environments.

Soil often harbors rich microbial communities

Soil may appear fairly lifeless to the casual observer, but it actually constitutes a complex ecosystem that is constantly changing due to geological, meteorological, and biological processes. As rocks become weathered over long periods of time, they begin to fragment and crack. Microorganisms that colonize these cracks continue to break down the rock by producing acids and other chemicals. Once a thin layer of particles forms, the first plants may take hold, and as these plants die they contribute their organic material to the developing soil. As the soil continues to form, a top layer, called the topsoil, develops; it contains material rich in organic substances

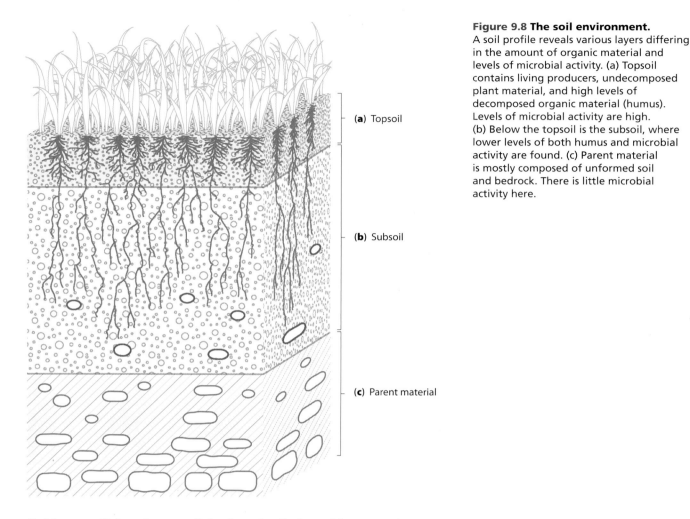

(a) Topsoil

(b) Subsoil

(c) Parent material

Figure 9.8 The soil environment.
A soil profile reveals various layers differing in the amount of organic material and levels of microbial activity. (a) Topsoil contains living producers, undecomposed plant material, and high levels of decomposed organic material (humus). Levels of microbial activity are high. (b) Below the topsoil is the subsoil, where lower levels of both humus and microbial activity are found. (c) Parent material is mostly composed of unformed soil and bedrock. There is little microbial activity here.

called **humus**. Below the topsoil, in the subsoil, the soil becomes increasingly inorganic; less humus is present, and less microbial activity takes place. Below the subsoil is a layer of *parent material* that includes the unweathered bedrock; little organic material or microbial activity is found here (**Figure 9.8**). Oxygen content of the soil also generally decreases with increased depth.

Although the soil teems with microorganisms, their distribution is not uniform. Soil microorganisms typically reproduce only when there is sufficient moisture, and soil moisture is a good indicator of microbial activity. Many soil microorganisms undergo adaptations that allow them to persist during periods of dryness, but reproduction is slow or nonexistent during these times. The endospores produced by some bacteria, as discussed in Chapter 3, are an example of such an adaptation. Soil can be too wet, however. Certain fungi, such as the molds, are obligate aerobes and consequently are typically found in the uppermost layers of soil. If soil becomes waterlogged, however, oxygen levels drop to the point where these fungi cannot grow well. To persist during these less-than-optimal conditions, fungi typically produce reproductive structures called spores as part of their normal life cycle. These fungal spores can survive for long periods of time without growing, until the soil dries and oxygen levels increase.

Temperature, pH, and the availability of inorganic nutrients are other crucial abiotic factors that determine the microbial composition of the soil environment. Most soil bacteria are mesophiles and will consequently grow best between about 20 and 50°C (i.e., between 68 and 122°F). Very high or low pH inhibits the growth of most soil bacteria, but many fungi can tolerate wide fluctuations in soil acidity. In fact, some fungi grow especially well in highly acidic soils, because of reduced competition from bacteria. In many soil ecosystems, low levels of one or more inorganic nutrient, such as iron, place severe restrictions on the growth of some microorganisms.

Although photosynthetic microorganisms such as cyanobacteria and single-celled algae may be found on the surface of soils, most producers in this environment are green plants. As the plants die, microbial decomposers stand ready to return organic molecules to the environment. Plants also contribute to the formation of the **rhizosphere**, the soil environment immediately surrounding the roots (**Figure 9.9**). The microorganisms found in the rhizosphere grow in a biofilm on root hairs, absorbing amino acids and sugars from the plant. This is not a purely one-way relationship, however. The microorganisms return the favor by converting minerals into forms that can be absorbed by the plant. Fungi, for instance, often grow on the surfaces of roots in a mutually beneficial relationship with plants, called **mycorrhizal associations**. Recall that in Chapter 4 (p. 87) we learned that fungi in mycorrhizal associations often supply their plant partners with phosphorus or nitrogen in exchange for sugars. Some bacteria and fungi also produce substances that actually protect the plants from disease-causing soil nematodes or other plant pests.

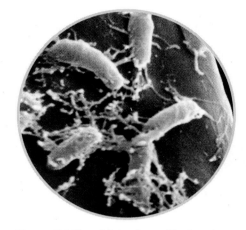

Figure 9.9 The rhizosphere. Plant roots release nutrients that allow growth of soil microorganisms on or near the root surface. In this electron micrograph, bacteria adhere to the roots of a cabbage plant.

Sometimes microorganisms can be found in habitats that seem to stretch the meaning of the *word* habitat—far below the Earth's surface, for example. Microorganisms have reportedly been recovered from drilling sites in northern Texas reaching as deep as 9000 m (30,000 ft), as well as from almost equally impressive depths in Alaska and South Africa.

We actually know very little about what is living in this mysterious subterranean world, in part because it is so difficult to study. It is believed that primary production (the conversion of inorganic molecules into organic molecules) is based on chemical energy released by thermal geological sources and that many of the organisms are no doubt archaea. Other microorganisms have been found actually living inside solid rock. In Antarctica, for example, "dry valleys" are extremely cold and dry, and soil samples are almost microorganism-free (**Figure 9.10**). But this lunar landscape is not as lifeless as one might suppose. Lichens, symbiotic organisms consisting of fungi and algae, live inside porous sandstone. During the brief Antarctic summer, lichens are able to produce organic molecules by photosynthesis, using minute amounts of melted snowfall as a water source. During the rest

Figure 9.10 Life at the bottom of the world. (a) The dry valleys of Antarctica are among the most extreme environments in the world. (b) Life, however, in the form of lichens is able to persist inside porous sandstone rocks, just below the surface on the side of the rock facing the sun. Lichens are symbiotic organisms consisting of fungi and algae. Chlorophyll in the photosynthetic algae gives the lichen its green coloration, seen here as a streak just below the rock surface.

of the year, they exist in an essentially "freeze-dried" state. What this unusual lifestyle lacks in excitement is certainly made up for in life span. Researchers have estimated that some of these lichens are thousands of years old, placing them among the oldest of all Earth's inhabitants.

Many microorganisms are adapted to life in freshwater

Microorganisms abound in freshwater habitats such as lakes, ponds, and rivers. Even the most pristine stream is likely to support abundant microbial life. There are, however, substantial ecological differences between different freshwater habitats such as a large lake and a quickly flowing river, and it should come as no surprise that the microbial community found in different freshwater habitats varies accordingly. Temperature, sunlight, oxygen, and nutrient levels all affect microbial abundance and distribution in freshwater environments.

In terrestrial ecosystems, multicellular green plants are the most important primary producers, but in freshwater, microorganisms take over the leading role. These photosynthetic organisms, mainly algae and cyanobacteria, make up the **phytoplankton**, which provide food for protozoa, small crustaceans, and other small primary consumers—which collectively make up the **zooplankton**. These small consumers are in turn eaten by larger invertebrates, small fish, and other secondary consumers, which provide food for higher-level consumers such as large fish and birds. Meanwhile, bacterial decomposers actively convert dead or excess organic matter back into a form that can be utilized by producers.

CASE: WATER BIRDS AND BOTULISM

The autumn of 2002 was a bad time for birds along the shore of Lake Erie. Over a 2-week period, more than 5500 dead water birds, including ducks, geese, and gulls, were found on beaches between the cities of Buffalo and Dunkirk in New York State. Officials in Ontario, Canada, reported much the same story, and they expressed special concern over the death of approximately 1000 loons, birds that were already uncommon in the area. Because most dying birds are never found, wildlife experts believed that the reported deaths were just the tip of the iceberg. The cause of death was diagnosed as botulism.

1. Why is botulism an especially serious disease for aquatic birds?
2. Are certain types of water bodies more prone to botulism outbreaks?
3. What is the relationship between the time of year and botulism in water birds?

Lakes. Botulism probably kills between 10,000 and 50,000 North American water birds in a typical year. During an especially serious outbreak, that number can rise to over 1 million. The consequences for endangered birds, whose population is already low, can be especially severe. In 1996, a botulism outbreak at the Salton Sea killed an estimated 15% of already threatened brown pelicans in California. In 2002, botulism killed 71 black-faced spoonbills in Taiwan out of a total species population of fewer than 1000 birds. Botulism is caused by the consumption of a toxin produced by *Clostridium botulinum*, and evidently the birds were exposed to this powerful microbial toxin. To understand the link between lakes and this serious wildlife disease, we must first consider some basic principles of lake ecology.

Lakes can be divided into several zones, in which one expects to find different environmental conditions and therefore different microorganisms. The **photic zone** extends from the lake surface to the deepest point of sunlight penetration. Below the photic zone, photosynthesis cannot occur. Dissolved oxygen released as a photosynthetic waste is also highest near the surface,

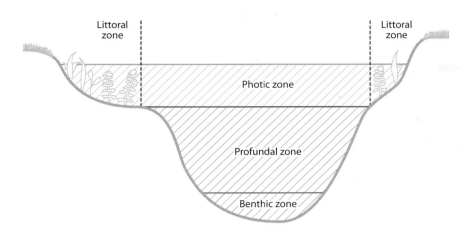

Figure 9.11 Ecological zones in a lake. Each zone differs in terms of abiotic factors such as light, temperature, and oxygen. These differences are reflected in the different microbial communities found in each zone. The *photic zone* is the topmost layer of the lake, through which light penetrates. If the water is shallow enough, the light may strike the lake bottom, a region of the photic zone known as the *littoral zone*. Light does not reach the underlying *profundal zone*. The lake bottom beneath the profundal zone is known as the *benthic zone*.

where the water can also be aerated by wind and wave action. In open water, phytoplankton are the only producers in the photic zone; close to shore in shallower water, multicellular aquatic plants are often abundant and they serve as producers as well. The **littoral zone** (from the Latin word for "shore") is the portion of the photic zone through which light penetrates to the bottom (**Figure 9.11**).

Yet even in the photic zone, oxygen can be the limiting factor for microorganisms and other living things. Oxygen does not dissolve well in water, which means that water can hold only a limited amount of oxygen. Below the photic zone is the **profundal zone**, where oxygen is even more limited. Furthermore, because photosynthesis cannot occur in the profundal zone, consumers are largely dependent on organic material that has sunk from above. Many of the microorganisms living in the profundal zone are facultative anaerobes, with obligate anaerobes taking over in still deeper and even more oxygen-deficient water. The deepest region, called the **benthic zone**, includes the bottom sediments, composed of soil and organic material. Methanogens are often common here, as are other obligate anaerobes such as bacteria in the genus *Clostridium*.

If a temperate (nontropical) lake is deep enough, dissolved oxygen may vary considerably at different depths over the course of the year. During the spring and summer, surface water warms up quickly. Warmer water is less dense and it tends to remain near the surface. Colder water is denser and it tends to sink. The result is a sharp demarcation in temperature known as a **thermocline** between the upper, warmer layers and deeper, colder layers (**Figure 9.12**). Because there is little mixing between the two layers, very little oxygen is available below the thermocline. However, during the autumn, the surface water cools quickly and, as it sinks, it mixes with the deeper water, providing the deep-water habitats with increased oxygen. Unless the lake freezes, such mixing occurs all winter long. Consequently, lake-bottom environments often alternate between being anaerobic in the summer and aerobic in the winter.

The lake's microbial community changes accordingly. This explains the relationship between avian botulism in lakes and time of year. During the warm summer months, a lack of oxygen in bottom sediments provides ideal growth conditions for obligate anaerobes such as *C. botulinum*. Once the weather begins to cool and the thermocline breaks down, two things happen. First, increased oxygen in the benthic zone inhibits *C. botulinum* growth and induces spore formation. Second, mixing of the surface, and bottom waters helps to move the bacterial spores toward the surface, where they may be ingested by birds feeding in the lake. The result can be the type of outbreak described in our case. Deep lakes in temperate regions are especially prone to this problem, because they are most likely to develop a

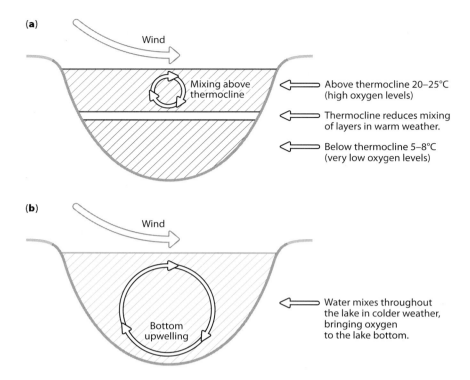

(a)

Wind

Mixing above thermocline

Above thermocline 20–25°C (high oxygen levels)

Thermocline reduces mixing of layers in warm weather.

Below thermocline 5–8°C (very low oxygen levels)

(b)

Wind

Bottom upwelling

Water mixes throughout the lake in colder weather, bringing oxygen to the lake bottom.

Figure 9.12 Seasonal change in a temperate lake. (a) During the summer, the development of the thermocline greatly reduces the mixing of surface and bottom layers. Oxygen levels are very low beneath the thermocline, because the oxygen consumed by organisms beneath the thermocline cannot be replenished. (b) In the autumn, the thermocline breaks down, allowing mixing between these different layers. Upwelling distributes oxygen levels throughout the lake.

seasonal thermocline. The problem can likewise occur in other water bodies, such as rice paddies, in which anaerobic conditions are likely.

Most of the bacteria that live in lakes can grow on relatively low levels of nutrients. Such organisms, which are able to prosper where others would starve, are called **oligotrophs**. They stand in contrast to the **eutrophs**, such as those living in an animal's intestine, that require relatively high nutrient levels. Because oligotrophs can survive on low levels of organic molecules and inorganic nutrients such as nitrogen and phosphorus, they have slow growth rates relative to eutrophs. Yet in most lakes and ponds, fast-growing eutrophs cannot compete with the oligotrophs that are adapted to starvation rations. As we will shortly see, however, when lakes receive a large increase in nutrient material, either artificially or naturally, there are no certainties.

Antarctic lakes and ice. Lake Vida in Antarctica's McMurdo Dry Valley region is frozen over with a thick layer of ice, but 19 m below the ice is a pocket of liquid water that has been isolated for over 2800 years. When researchers took an ice core from a depth of 12 m, they found living microorganisms such as cyanobacteria, essentially surviving in solid ice. These scientists have not yet sampled the liquid water below, which is many times saltier than the ocean and hovers at about −10°C. They will only attempt this sampling once they are confident it can be done without contaminating the pristine water with surface microorganisms. If life is eventually found in this isolated pocket of water, these organisms would essentially represent a microbial time capsule that has been cut off from the rest of the living world for almost 3000 years.

Rivers and Streams. Flowing water habitats are quite different from lakes and ponds, where the water is stationary. First, rivers and streams are generally much shallower, and light can often penetrate to the river or stream bottom. Rapid water flow and water turbulence ensure adequate oxygenation, at least in nonpolluted rivers. Many of the microorganisms living in flowing water are found in biofilms attached to rocks and other hard surfaces, where they come in contact with nutrients that flow past.

In larger rivers, photosynthetic organisms found in the river itself may serve as primary producers. In small streams, on the other hand, most nutrients may first enter the environment as runoff from nearby terrestrial habitats or in the form of leaves or other organic material that directly falls into the water.

Water pollution can lead to severe oxygen depletion

CASE: LAKE ERIE: A NEAR-DEATH EXPERIENCE

In the 1960s Lake Erie was dying. Oxygen levels in the water were declining, and as a result there were large die-offs of fish and other aquatic vertebrates. Excessive growth or "blooms" of algae gave the water a sickly green color, while beaches and shorelines, fouled by masses of rotting algae, emitted an unpleasant odor. Many native fish species disappeared from Lake Erie, to be replaced by species more resistant to the new conditions. Scientists from the University of Manitoba investigated the problem and found that the addition of inorganic compounds, especially phosphorus, was the primary culprit. Such nutrients entered the lake as runoff from agricultural fields, golf courses, or other places that were heavily fertilized. Untreated or partially treated domestic sewage was another major source, particularly because of phosphates used in household detergents. Following the implementation of laws to control water pollution as well as the removal of phosphates from detergents, Lake Erie began to recover. Today, in an ecological success story, the lake is considered to be largely rehabilitated. Nevertheless, such nutrient pollution is still perhaps the number one water quality problem worldwide.

1. What was the exact relationship of the agricultural and domestic runoff to the problems in Lake Erie?
2. How are microorganisms involved in this example of water pollution?
3. How have microbial processes caused the death of animals such as fish and the unpleasant odor of the lake?

We have previously mentioned that many of the microorganisms found in freshwater environments are oligotrophs that are well-adapted to low levels of nutrients. In some circumstances, however, water bodies become inundated with organic material and nutrients. Sometimes this is due to a natural event such as a flood that has washed large amounts of organic matter into the water. More often, however, human activity is to blame. When sewage enters a lake or river, it often carries a heavy load of organic material. Soils in agricultural areas often contain large amounts of phosphate and nitrate fertilizers, which can run off into lakes and streams following rains or irrigation. Phosphates in particular, especially those added to detergents to improve cleaning action prior to the mid-1970s, were an especially important part of the problem in Lake Erie.

The addition of these extra nutrients can wreak havoc on aquatic communities. Warm summer temperatures, combined with the added nutrients, stimulate an excessive growth of algae called an **algal bloom** (**Figure 9.13**). Phosphates are naturally low in such habitats and therefore are often a limiting factor in algal growth. With the addition of phosphate pollution, algae populations in Lake Erie were able to increase dramatically.

Photosynthetic bacteria may run rampant as well. As populations of cyanobacteria explode, the water may take on an unhealthy green color. Populations of aerobic bacteria also soar, as they respond to the oxygen released by photosynthetic plants and microorganisms. As they deplete the oxygen in the surface water, the environment becomes increasingly anaerobic. Ultimately oxygen levels fall too low to support animals such as fish and large invertebrates, which may then die in large numbers. Anaerobic bacteria, on the other hand, are able to thrive in the now oxygen-depleted

Figure 9.13 Lake eutrophication. Water pollution in the form of inorganic nutrients is often the result of runoff containing agricultural fertilizer. Algae, cyanobacteria, and aquatic weeds grow in response to this pollution. Although these photosynthetic organisms produce oxygen, their growth causes a surge in the growth of aerobic bacteria, which ultimately deplete the water of oxygen. The now oxygen-depleted water cannot support the survival of large invertebrates, fish, and other aquatic animals.

environment, and they release metabolic wastes such as hydrogen sulfide that give the water a decidedly unpleasant smell. This addition of excess nutrients to aquatic ecosystems, along with the subsequent ecological problems, is known as **eutrophication**. While stringent water pollution control measures and the removal of phosphates from detergents have resuscitated Lake Erie, the problem of eutrophication is far from solved. As reliance on fertilizers in agriculture continues to increase, the problem of eutrophication is likely to plague us for the foreseeable future.

Although similar in many ways, the marine environment is distinct from freshwater

The Earth is often called "the water planet" because most of its surface is covered with water. And the vast majority of water is found in the ocean. As in freshwater lakes, almost all primary production in the marine environment is due to phytoplankton, consisting of cyanobacteria and algae. These microbial organisms represent the base of the food chain, and they are eaten directly by zooplankton, which serve as primary consumers. As in freshwater habitats, higher levels of consumers include larger invertebrates, fish, and birds. Larger fish, sharks, and marine mammals are found at the top of the marine food chain.

Like lakes, the oceans can be divided into a photic zone, a profundal zone, and a benthic zone. Within the photic zone, we can distinguish between the near-shore littoral zone and the open-ocean **pelagic zone**. Along the shoreline there is also an **intertidal zone**, which is exposed to air at low tide and is covered by the sea at high tide.

While there are similarities, in many respects the marine environment is quite different from freshwater environments. Compared with lakes and rivers, oceans are far less variable in temperature and pH, and they have a much greater concentration of dissolved salts. Seawater is about 3.5% dissolved salts, whereas in freshwater, a typical value is around 0.5%.

The depth to which sunlight can penetrate varies considerably in different parts of the ocean. Depending on the amount of organic and inorganic material in the water, the photic zone may be as shallow as 50 m or as deep as 300 m. Photosynthetic organisms, wave action, and surface turbulence are sources of oxygen. Like lakes, deep sea environments lack both light and oxygen.

Nutrient levels may also range from very low to very high, and often the determining factor is proximity to shore. Near land, runoff from rivers and streams enriches coastal waters, and underwater currents frequently collide with land masses. As these currents are deflected upward from the sea floor, they carry organic material with them. The pelagic zone, on the other hand, has neither coastal runoff nor current-driven upwelling, and consequently it is relatively nutrient-poor.

Because environmental conditions in the ocean are so different from freshwater conditions, it stands to reason that living things, including microorganisms in these two ecosystems, often display very different adaptations. For one thing, marine bacteria are somewhat **halophilic** (salt-loving) and grow best at salt concentrations of approximately 3.5%. Freshwater bacteria have a much lower optimal salt concentration. Furthermore, marine bacteria, unlike those found in freshwater or on land, generally require sodium for growth. Other adaptations vary, according to exactly where in the ocean a particular microorganism is found.

Pelagic and littoral zones. If you have ever spent time on a boat, far from shore, you have no doubt been struck by the deep blue color of the water. By

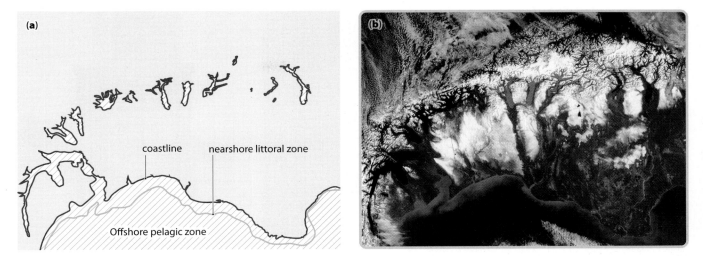

Figure 9.14 Littoral and pelagic waters. (a) Runoff from land introduces relatively large amounts of organic material and inorganic nutrients into the nearshore *littoral zone*. The green color is due to the multitude of tiny marine plants and photosynthetic microbes that grow well in the nutrient-rich water. The *pelagic zone* of the open ocean contains relatively little organic and inorganic material. Consequently, with fewer photosynthetic organisms, the ocean is blue in color. (b) Satellite image of the southern South American coastline. Note the greenish hue close to shore, indicating the presence of many photosynthetic organisms. Further offshore, deeper water reduces upwelling. With fewer nutrients available, photosynthetic organisms are far less abundant.

comparison, close to shore, the sea often takes on a greenish hue (**Figure 9.14**). Perhaps it will come as a surprise to learn that much of the reason for the color difference is microbial. Bereft of the upwelling and runoff found along the coast, nutrient-poor pelagic water contains little organic matter and relatively few microorganisms. In many marine environments, nitrogen is the primary factor limiting biological activity, while in certain areas, such as the North Pacific Ocean, iron is the limiting factor. Understandably, most microorganisms found in pelagic waters are oligotrophs, adapted to low nutrient levels.

The green color typical of more inshore waters reflects the fact that, compared with the pelagic zone, this littoral zone is far richer in nutrients and consequently supports more microorganisms, in particular, large numbers of chlorophyll-containing phytoplankton. It is no coincidence that the world's most productive fisheries are generally littoral or relatively close to shore: the rich and constant supply of nutrients provides ample resources for phytoplankton, which in turn support large numbers of primary and higher-level consumers. However, one does not have to look too hard to be reminded that too much of a good thing can have disastrous consequences. With increased human activity along the coast come increased pollution, unwanted agricultural runoff, and microbial contamination of coastal waters. In many coastal areas, after a rainstorm, the harvest of mussels, oysters, and other shellfish must be suspended for a week or more until the bacteria that have washed into the ocean have had time to die.

The area of the U.S. Gulf Coast near the mouth of the Mississippi River provides a particularly grim example of how excess nutrients can negatively impact coastal ecosystems. Excess industrial and agricultural pollutants entering the Gulf of Mexico from the Mississippi result in eutrophication, characterized by explosive microbial growth and oxygen depletion. Without oxygen, fish and other animals are unable to survive, creating a "dead zone" about the size of Massachusetts (**Figure 9.15**).

"Red tide" is another microbial problem familiar to many coastal dwellers. When excessive nutrients enter nearshore waters due to pollution runoff or increased upwelling, a rapid increase, or bloom, in dinoflagellate numbers

Gulf of Mexico

Figure 9.15 The "dead zone." Runoff of inorganic pollutants such as agricultural wastes, often at river mouths, can create eutrophic dead zones over large areas of sea, such as this one on the U.S. Gulf Coast at the mouth of the Mississippi River.

Figure 9.16 Red tides. (a) Increased nutrients can cause excessive growth, or blooms, of aquatic algae called dinoflagellates. The reddish color of the water is caused by the red pigments in the algal cells. (b) These algae can produce toxins that accumulate in shellfish. If animals, including people, consume contaminated shellfish, there is a serious risk of paralysis caused by shellfish poisoning.

turns the water red. Dinoflagellates are single-celled algae whose reddish pigments are responsible for the discoloration of the seawater (**Figure 9.16a**). Some dinoflagellates produce a powerful neurotoxin that can cause paralysis in many animals. During a bloom, levels of the poisonous substance increase in the bodies of many marine invertebrates. The shellfish are not harmed by the toxin, but as toxin levels in their bodies rise, animals consuming contaminated shellfish are at risk (Figure 9.16b).

Prokaryote numbers are highest near the surface and decline with increasing depth. In the upper 150 m or so, most prokaryotes are bacteria, whereas below that depth, numbers of archaea begin to increase, until in the profundal zone, the numbers of these two types of prokaryotes are about equal. This latter fact came as a surprise. You will recall that originally it was thought that archaea were found almost exclusively in "extreme environments," but researchers have recently found that the sea literally teems with archaea.

Deep sea and benthic zone. For organisms living below the photic zone, there are several major problems that must be overcome: it is dark, it is cold, there is not much to eat, and the pressure is intense. Photosynthesis is not possible, and any primary nutrient production must rely on mechanisms that are not light-dependent (see Chapter 7 for a discussion of chemosynthesis). Organisms that live in the profundal and benthic zones are thus largely dependent on organic material and nutrients that fall from the overlying photic zone. For microorganisms living below 300 m, it can be "slim pickings" indeed. By the time organic material falls to this depth, most of it has already decomposed, and only about 1% of photosynthetically produced material actually makes it to the sea floor in an undigested form. Because there are such low levels of organic input, most microorganisms living at great depths are oligotrophs, able to grow at low nutrient levels.

At depths of more than about 100 m, the water temperature remains at approximately 2–3°C, and microorganisms found here are psychrophiles ("cold-loving" organisms). Furthermore, for every 10 m of water, the pressure increases by 1 atm. (At sea level we experience 1 atm of pressure. One atmosphere equals 14.7 pounds per square inch.) This means that microorganisms living at 1000 m must be able to survive 100 atm of pressure. Such pressure-adapted organisms are called **barotolerant**.

One of the more remarkable discoveries in marine microbiology in recent years reveals a previously unknown ecosystem of astonishing proportions in the rocks beneath the sea bottom. Methanogens (methane-producing archaea) live in inconceivable numbers hundreds of meters beneath the sea floor. The methane that they produce has been estimated to have a greater mass than all the known reserves of oil, coal, and natural gas combined. Methane is a "greenhouse" gas linked to global warming, and it has been suggested that sudden bursts of methane, released from the sea bottom, have altered Earth's climate throughout its history and perhaps even ended ice ages.

So why is this superabundance of methane not being released into our atmosphere now, making current concerns about global warming seem like child's play? Other microorganisms present in more shallow sediments oxidize the methane as an energy source. Thanks to these organisms that break down the methane, our atmosphere is spared the input of an estimated additional 300 million tons of the gas a year.

The combined biomass of all these methane producers and methane oxidizers, living unobtrusively beneath the ocean floor, has been estimated at as much as a third of all life on the planet! Recall that in Chapter 7 we learned that when the atmosphere ultimately became oxygenated due to photosynthesis, living things had to either avoid oxygen or evolve mechanisms such as electron transport to detoxify it. The finding that a sizable proportion of Earth's life continues to pursue a strictly anaerobic existence, far removed from the aerobic world with which we are familiar, suggests that the former option was a popular one.

Cloud-dwelling bacteria may be important in promoting rainfall

Clouds scarcely seem like a promising habitat for any living thing, but some microorganisms are at home even here. Even more surprising is the recently discovered role these microbes have as rainmakers. Before a cloud can produce rain or snow, raindrops or ice particles must form. Tiny particles serve as the necessary nuclei around which the raindrops or snowflakes form. Most such particles are of mineral origin, but in 2008, researchers announced that in many cases bacteria, fungi, or tiny algae can do the job just as well. In some precipitation samples, well over 70% of the nuclei were biological rather than geological. It is thought that when raindrops or snowflakes form around the microbial cell, the airborne microbes are then able to return to earth. The researchers suggest that atmospheric scientists and meteorologists need to consider the implications of these microorganisms as they work to better understand the forces that drive our climate.

Microorganisms and Biogeochemical Cycles

If you had a slice of toast for breakfast this morning, you may find it interesting to consider that the carbon atoms in the toast were once part of CO_2 molecules in the atmosphere, which were fixed into organic molecules during photosynthesis. Likewise, an iron atom that carries oxygen in your blood may have once done the same for a dinosaur, while in between it may have been in the soil, where it helped fertilize a cottonwood tree. Nature is indeed the great recycler, and although the term "biogeochemical cycle" may sound imposing, it simply refers to the movement of chemical elements back and forth between living things and the nonliving environment.

All elements involved in life processes move through ecosystems in cyclical fashion, and in all such cycles, microorganisms play crucial roles. As decomposing microorganisms in soil and water break down organic matter, they release the inorganic building blocks back into the environment, where they can once again enter food chains. In this section we examine some of the most important of these cycles.

Carbon moves between living things and the environment

CASE: IRONING OUT GLOBAL WARMING?

In 2003, Atsushi Tsuda of the University of Tokyo conducted an unusual experiment that he believed might increase fish populations and might even help to reduce the problem of global warming. He and his colleagues distributed 770 pounds of iron-laced powder over 20,000 acres in the North Pacific Ocean. Two weeks later, he found that phytoplankton populations had exploded in the study area. Meanwhile, levels of carbon dioxide in the air immediately over the iron-fertilized water fell as well. Tsuda proposed that using iron in this way could help fisheries by stimulating increased plankton growth. More controversial was his suggestion that the technique could reduce global warming, by removing CO_2 from the atmosphere.

1. **How exactly would the addition of iron to seawater increase phytoplankton populations, and possibly those of fish as well?**
2. **Would iron fertilization work in other marine environments as well?**
3. **Could this technique actually reduce global warming? If so, how?**
4. **Might there be other unforeseen consequences of such iron supplementation?**

Carbon is truly the stuff of life, abundant in all biological molecules such as carbohydrates, lipids, and proteins. In addition to these organic forms, carbon can also be found in the atmosphere as carbon dioxide (CO_2) or in mineral form as calcium carbonate ($CaCO_3$).

Photosynthesis by green plants and cyanobacteria is the gateway by which CO_2 often enters the living world. Photosynthetic organisms are able to fix atmospheric CO_2 in the Calvin cycle, incorporating it into biological molecules (see Chapter 7, p. 188). When animals or other consumers feed on plants or photosynthetic microorganisms, carbon begins its trip through the food chain (**Figure 9.17**). Some of the consumed carbon is released back into the atmosphere as a waste product of cell respiration in the form of CO_2. Other carbon atoms are incorporated into carbohydrates, proteins, and other biological molecules that will remain in a consumer until they are excreted as waste, or until the consumer is eaten by a higher-level consumer. Of course, many consumers die before they are eaten, and when they do, bacteria and fungi, the decomposers, step forward to break down organic material, returning CO_2 to the environment. As discussed in Chapter 7, some chemosynthetic microorganisms are able to fix CO_2 using chemical energy instead of sunlight. Although less important to the overall carbon cycle than photosynthetic carbon fixation, these organisms also contribute to the overall ebb and flow of carbon between the living and nonliving worlds. Fermenting organisms also release CO_2 as a waste product.

An understanding of how carbon moves through a food chain explains the logic behind the idea that adding iron to seawater could actually enhance fishing. Recall that as energy, in the form of carbon–hydrogen bonds, moves up a food chain, much of the energy is lost in the form of heat. Consequently, at each level of a food chain, there is less and less biomass (see Figure 9.4). If the number of primary producers can be increased, however, biomass of consumers, including commercially valuable fish, may increase as well. Of course, this strategy of iron fertilization will work only in those environments where iron is the limiting factor on phytoplankton growth. As mentioned previously, the North Pacific is noteworthy for its limiting factor of iron. Consequently, this was one environment where Tsuda's idea was most likely to work. In the Atlantic Ocean, on the other hand, it is nitrogen rather than iron that limits phytoplankton growth. The impact of added iron would be less because nitrogen would still limit phytoplankton growth.

Carbon is also stored as calcium carbonate ($CaCO_3$) in rocks such as limestone. Limestone itself, although inorganic, is the result of biological activity, as it is secreted by marine organisms such as snails and clams to form their hardened shells and by corals during reef formation. Many microorganisms such as single-celled marine algae and protozoans also secrete $CaCO_3$-containing shells. When these organisms die, their shells accumulate as marine sediments, and when limestone deposits that were originally formed in the ocean are eventually exposed to the atmosphere, they slowly decompose, releasing CO_2 back into the environment.

Figure 9.17 The carbon cycle. Carbon, in the form of CO_2, is removed from the atmosphere by producers, which convert it into biological molecules through photosynthesis. Consumers obtain carbon compounds in their diet and release CO_2 as a respiration waste product. Microorganisms likewise return carbon to the environment as they decompose organic matter. Some CO_2 is used by marine organisms to form calcium carbonate ($CaCO_3$) skeletons. When eventually exposed to the atmosphere, these skeletons can release CO_2 back into the environment. Combustion of fossil fuels adds to the atmospheric CO_2 pool.

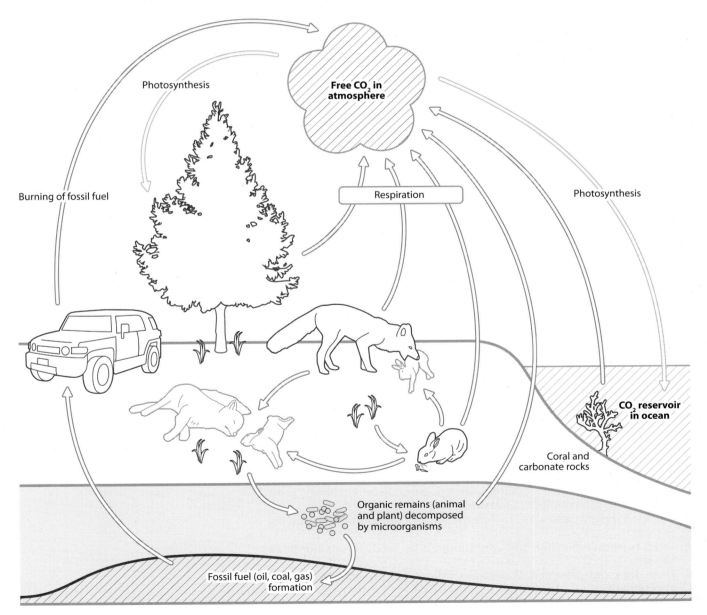

When organic carbon accumulates in anaerobic environments, the combined forces of heat and pressure can eventually convert it into fossil fuels such as petroleum and natural gas. As discussed in Chapter 7, when these hydrocarbon fuels are burned, they are oxidized, releasing CO_2 as a waste. Because of the ever-increasing demand for fossil fuels, CO_2 is being released into the atmosphere faster than it can be fixed by photosynthesis. Atmospheric concentrations of CO_2 therefore continue to increase, upsetting the carbon cycle. This problem is only made worse as forests are cut down, especially in the tropics, reducing rates of photosynthesis and consequently carbon fixation. Because CO_2 is a "greenhouse" gas that traps heat in the atmosphere, many researchers believe that this imbalance in the carbon cycle is an important piece of the global warming problem.

Some of the excess CO_2 is absorbed by the ocean, which stores large amounts in dissolved form. Here we find the reasoning behind the idea that adding iron to seawater could reduce global warming. By increasing phytoplankton biomass, the iron seeding will cause increased amounts of CO_2 to be removed from the atmosphere through carbon fixation. This CO_2 might then remain in the marine environment as the phytoplankton die or are eaten by consumers. As atmospheric CO_2 decreases, the reduction in levels of this greenhouse gas might reduce the rate of global warming.

Tsuda's idea is interesting, but would it actually work? At this point nobody knows, but many experts believe that the amount of iron required and the area of ocean that would need fertilization is far too large to cause any meaningful reduction in atmospheric warming.

Furthermore, even if it did work, nobody knows what the consequence of introducing so much extra CO_2 into the ocean would be. We also know very little about the rate at which the ocean absorbs CO_2 and releases it back into the atmosphere or how increased levels of CO_2 in the ocean would affect other parts of the marine ecosystem. But one thing we know for certain: ecological interactions are complex. When one part is altered, the ramifications spread out in many and often unforeseen directions.

Bacteria convert nitrogen into forms that plants can absorb

CASE: THE KEY TO A BOUNTIFUL HARVEST

Veteran backyard gardeners know that different plant species vary considerably in the amount of fertilizer they require. Garden peas, beans, and other legumes provide a good example. An inexperienced gardener who fertilizes such plants might be surprised to find that although the plants appear healthy and continue to send out extensive leaves and vines, they produce very few pea pods. Usually the principal problem in overfertilization is too much nitrogen. Legumes require very little nitrogen fertilizer or even none at all, and with excess nitrogen, they will funnel their energy into leaf and stem growth at the expense of the flowers, which are what develop into edible pods. Most garden supply stores, however, sell a product called "legume booster," which can be spread on pea seeds prior to planting or simply mixed with the soil. While not a fertilizer, legume booster results in increased pea and bean pod production. Furthermore, many other cool-weather vegetables, such as broccoli and cabbage, grow especially well when planted in proximity to pea plants.

1. What is a legume, and why is overfertilization a problem when growing them?
2. What is legume booster, and why should it be used with pea or bean seeds?
3. How does the presence of legumes stimulate the growth of other vegetables?

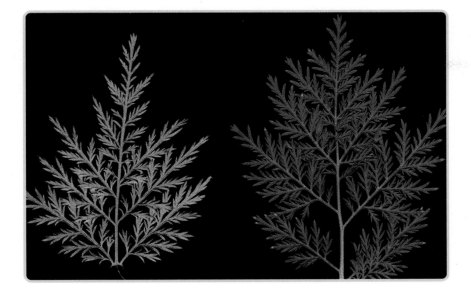

Figure 9.18 Effects of nitrogen on plant growth. Wormwood grown in soils with low (left) and high (right) levels of nitrogen. Note the smaller size and yellowish color of the nitrogen-deficient leaves.

Nitrogen is a major component of biological molecules, especially nucleic acids and proteins (**Figure 9.18**). Yet although pure nitrogen is extremely abundant in the environment, making up 79% of the atmosphere, this vast resource is unavailable to all but a few organisms. Atmospheric nitrogen is in the form of nitrogen gas (N_2), which animals and plants cannot use directly. These multicellular organisms, along with most microorganisms, need to obtain their nitrogen in other forms. Plants absorb nitrates (NO_3^-) or ammonium (NH_4^+), while animals obtain their nitrogen as part of proteins or other biological molecules in their food. How then does atmospheric nitrogen enter food chains? That's where microorganisms take over, acting as the nitrogen conduit between the nonliving and the living worlds.

The major way that nitrogen enters biological systems is through the process of **nitrogen fixation**—the conversion of atmospheric nitrogen into a form that plants can absorb. This process is carried out by certain *nitrogen-fixing bacteria* that live in soil or water (**Figure 9.19**). These bacteria have a unique enzyme that can break the bonds between two nitrogen atoms in atmospheric N_2. They then reduce the individual nitrogen atoms, producing the ammonium ion NH_4^+. Members of the genera *Azotobacter* and *Azospirillum* are examples of nitrogen-fixing bacteria.

Dead organic material also contains a large amount of nitrogen, and much of this is also released into the environment as ammonium ions through **ammonification** (see Figure 9.19). This additional NH_4^+ is formed as bacteria such as *Proteus* and *Clostridium* degrade proteins, nucleic acids, and other nitrogen-containing compounds. Ammonium, released as a waste, is then available to plants.

Not all ammonium, however, is directly absorbed by plants. Much of it is converted to nitrite and nitrate ions, through another microbial process called **nitrification** (see Figure 9.19). First, soil or aquatic bacteria such as *Nitrosomonas* oxidize NH_4^+ as an energy source. Nitrite is released as a waste. Then, a second group of bacteria such as *Nitrobacter* converts NO_2^- (nitrite) to NO_3^- (nitrate), which can be utilized by other bacteria, as well as by fungi and plants. To complete the nitrogen cycle, many common bacteria convert nitrate back to atmospheric nitrogen (N_2).

The role of bacteria in the overall nitrogen cycle is so crucial to plants that some plants go to extremes to make life comfortable for nitrogen-fixing

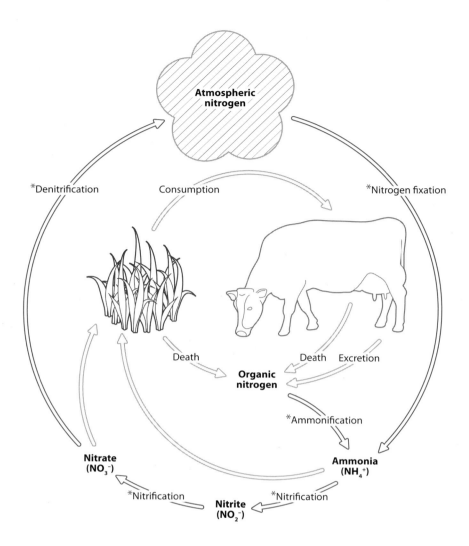

Figure 9.19 The nitrogen cycle. Asterisks indicate microbial processes. Nitrogen from the atmosphere (N_2) is converted into ammonia by nitrogen-fixing bacteria. Some ammonia is then converted by soil bacteria into nitrite and nitrate, which, along with ammonia, are used by plants to produce nitrogen-containing organic molecules such as amino acids and nucleic acids. Organic molecules are then consumed by animals and other consumers. As plants and animals die, organic molecules release ammonia to the environment by the process of *ammonification*. This ammonia can be recycled by bacteria into nitrites and nitrates and again be made available to plants. Certain bacteria convert some nitrate back to atmospheric nitrogen gas in a process called *denitrification*.

microorganisms. Snow peas and other legumes, such as beans and alfalfa, develop symbiotic relationships with these nitrogen-fixers. When the roots of such plants are colonized by these bacteria, the roots start to form special structures called **root nodules** (**Figure 9.20**). These small aggregations of root cells provide not only housing for the bacteria but also nutrients and energy. The bacteria, in turn, supply their hosts with a constant supply of ammonium (NH_4^+), which is used for the synthesis of nitrogen-requiring biological molecules.

Because legumes can utilize nitrogen from the air, they are essentially "self-fertilizing," and if colonized with the proper nitrogen-fixing bacteria, they can thrive even in nitrogen-poor soils. Extra fertilizer is not necessary, and if provided, legumes utilize it for extra growth rather than for flower and pea pod development. If soil does not contain the necessary nodule-stimulating bacteria, the farmer or gardener can actually inoculate the seeds or the soil with them as a means of increasing crop yield. The "legume booster" mentioned in our case is simply a powdered formula containing nitrogen-fixing bacteria.

Other plants such as cabbage and broccoli can benefit by close proximity to legumes. Some of the excess ammonia is released into the soil, where it becomes available to plants that cannot fix nitrogen on their own. Sometimes, entire fields are sown with legumes simply to increase soil fertility. After the legumes are harvested, other crops are better able to grow in the nitrogen-enriched soil.

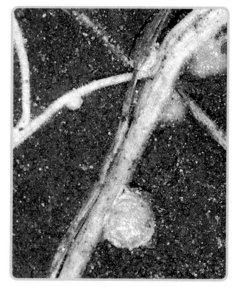

Figure 9.20 Root nodules in a soybean plant. Nitrogen-fixing bacteria such as *Rhizobium* form mutually beneficial relationships with legumes. The bacteria are housed in special structures called *root nodules* where they fix nitrogen for the plant. The root nodules are seen here as swollen areas along the length of the root. In return, the bacteria are supplied with nutrients by the plant.

Sulfur in organic material is returned to the environment by microorganisms

The next time you visit an estuary or mudflat, before you flee in disgust because of the "rotten egg" smell, you might use the experience as olfactory confirmation of yet another essential role that bacteria play in another biogeochemical cycle: the sulfur cycle. Sulfur forms part of several amino acids and it is thus necessary for protein structure.

Accordingly, all living things require sulfur, but like nitrogen, different types of organisms acquire it in different forms. Again like nitrogen, animals must get sulfur from their food. Plants, like many microorganisms, absorb sulfur in the form of sulfate (SO_4^{2-}).

Sulfate is abundant in rocks and water, ready for use by plants (**Figure 9.21**). When these plants die, microorganisms act on organic molecules containing sulfur to produce hydrogen sulfide (H_2S); these are anaerobic microorganisms that use sulfur the way we use oxygen—as a final electron transport acceptor in respiration. This is why mudflats stink. The thick mud in the sediments creates an anaerobic environment, ideal for anaerobic sulfur-reducing bacteria. The H_2S they release as a respiration waste produces the familiar and unpleasant rotten egg smell. These bacteria are therefore simply using the sulfur and releasing hydrogen sulfide in electron transport the way we use oxygen and release water. You might also contemplate why mudflats are black (**Figure 9.22**). The metallic sulfides, formed when H_2S reacts with metals in the mud, are black in color.

The sulfur cycle is completed when other bacteria oxidize the H_2S back to sulfate. As with all oxidations, energy is released and the microorganisms utilize the energy for ATP production (see Chapter 7, p. 188).

Like other cycles, the phosphorus cycle relies on microbial activity

At one time, the tiny Pacific island nation of Nauru was among the richest nations in the world, all thanks to bird droppings. When sea birds eat

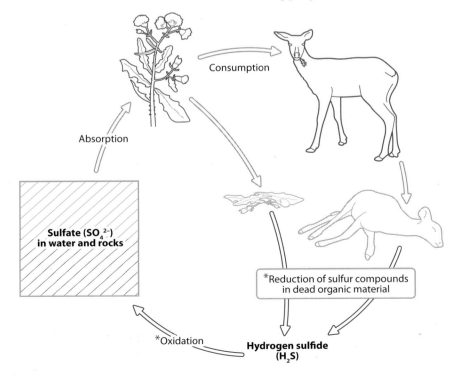

Absorption

Consumption

Sulfate (SO_4^{2-}) in water and rocks

*Reduction of sulfur compounds in dead organic material

*Oxidation

Hydrogen sulfide (H_2S)

Figure 9.21 The sulfur cycle. Asterisks indicate microbial processes. Sulfur, in the form of sulfate (SO_4^{2-}), is absorbed by plants, where it is converted into organic molecules such as amino acids. Sulfur moves through the food chain in the form of these organic molecules, or it is reduced by decomposing bacteria into hydrogen sulfide (H_2S). Sulfur-oxidizing bacteria then convert the hydrogen sulfide back to sulfate.

Figure 9.22 The sulfur cycle in action. The smell of mudflats is caused by hydrogen sulfide released by anaerobic bacteria as a respiration waste product. Hydrogen sulfide reacting with metallic ions in the mud causes the black color of the mudflats.

phosphorus-containing fish, they deposit phosphorus in the form of *guano* on islands where they nest. Such phosphorus deposits are mined for use as fertilizer, and in this regard Nauru was particularly blessed. The phosphorus mines on this island are now exhausted but they serve as a good illustration of the value placed on this essential nutrient, and provide insight into yet another biogeochemical cycle.

Phosphorus, essential for DNA, RNA, ATP, and many other crucial biological molecules, cannot enter the atmosphere and ends up accumulating in the ocean, which explains why marine fish and the sea birds that eat them acquire so much of the nutrient. Except for sea birds, the only other ways for phosphorus to return to the land are through the geological uplifting of ocean floors or the decomposition of organic material. Because the supply of phosphorus is limited in many soils, the demand for phosphorus-enriched fertilizers is high.

Phosphorus in rocks and guano deposits is released as the phosphate ion (PO_4^{3-}) by the dissolving power of bacterial acids (**Figure 9.23**). Sulfuric acid, secreted by bacteria of the genus *Acidithiobacillus*, is an example. Once in the soil or water, phosphate can be absorbed by primary producers such as plants or photosynthetic microorganisms. Here it is incorporated into organic molecules, where it can be passed up the food chain to

Figure 9.23 The phosphorus cycle. Asterisks indicate microbial processes. Phosphorus in rocks or guano is released into the environment as phosphate (PO_4^{3-}), either by erosion or by the dissolving power of bacterial acids. Phosphate then enters the food chain, where it is incorporated into organic molecules such as nucleic acids, ATP, or phospholipids. These organic molecules are returned to the environment by microbial decomposers.

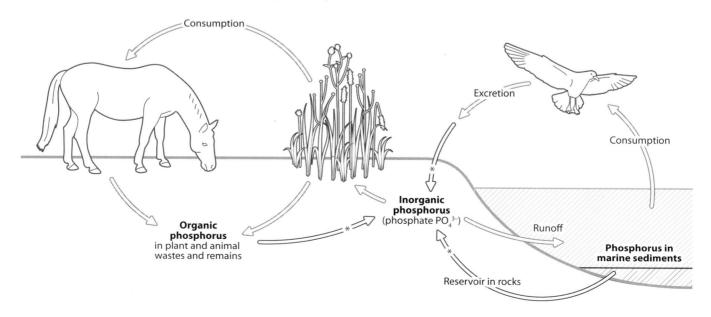

Consumption

Organic **phosphorus** in plant and animal wastes and remains

Inorganic **phosphorus** (phosphate PO_4^{3-})

Excretion

Consumption

Runoff

Phosphorus in marine sediments

Reservoir in rocks

subsequent levels of consumers. Organic phosphate in molecules such as DNA or ATP is returned to the cycle by decomposers. Eventually it returns to the inorganic reservoir in the sea.

Phosphorus is certainly essential for living things, but as in so many cases, too much of it has negative consequences. When humans interfere in the phosphorus cycle, by allowing phosphate-containing fertilizers to run off into lakes and streams, the enriched water bodies can quickly become eutrophic, causing algal blooms, oxygen depletion, and the death of animals. As previously discussed, until phosphate-containing laundry detergents were banned in many places, they were a serious part of this pollution problem.

Ecological Interactions Involving Microorganisms

By now it should be obvious that no living thing, microorganisms included, operates in a vacuum. All life is interconnected, and organisms depend upon and interact with other organisms in a myriad of complex and indirect ways. But not all interactions are so roundabout and subtle. In this final section of the chapter, we will take a closer look at some of the ways microorganisms interact directly with those around them.

Both organisms in a mutualism benefit from the relationship

We opened this chapter with a brief description of tube worms and the bacteria upon which they depend. In return for the nutrients they provide, the microorganisms are rewarded with a suitable habitat in which they are able to thrive. An interaction of this sort, in which both partners benefit, is called **mutualism** (**Figure 9.24**).

Mutualism involving microorganisms is common, and we have already had occasion to describe mutualistic relationships. Recall the case of the sick horse that was given "beneficial bacteria" as part of his treatment, discussed in Chapter 2. The bacteria are able to digest carbohydrates such as cellulose that the horse was unable to digest on his own. The bacteria benefit from a steady stream of nutrients in the horse's intestine and a stable environment. In Chapter 4 we discussed the mutually beneficial relationship between plants and the soil fungi that colonize roots.

Another fascinating example of mutualism involves termites. You may be surprised to learn that termites cannot digest wood on their own; they do not produce the necessary enzymes to break down the cellulose and other complex carbohydrates contained in wood. Without their associated gut microorganisms, termites would pose little threat to homes and other wooden structures.

Some termites are not wood eaters and instead depend on fungi, which the termites actually cultivate in their nests. When termites move to a new nest, they carry fungal spores with them to start fungal gardens in their new homes. These termites "farm" the fungi, which they tend and provide with organic matter in the nest. As the fungi grow, the termites feed upon them. Termite guts are also filled with mutualistic protozoa that can digest cellulose. Many termites also harbor the nitrogen-fixing bacteria *Enterobacter agglomerans* in their digestive tract that supply their termite hosts with a usable form of nitrogen. In return, the bacteria receive a suitable environment and a regular supply of nutrients.

As a final example of mutualism, consider the flashlight fish, at once recognizable because of the luminous organ beneath the eye (**Figure 9.25**). The

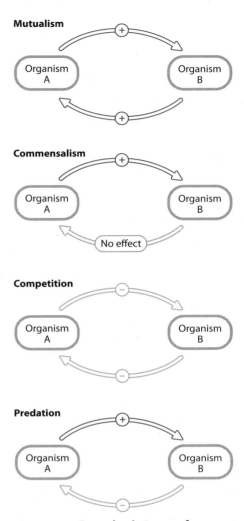

Figure 9.24 Some basic types of microbial interactions. The signs indicate whether the effect on the participating organism (A or B in the figure) is generally positive (+) or negative (–).

Figure 9.25 The flashlight fish. The fish possesses a luminescent organ beneath its eyes. Light is produced by bioluminescent bacteria living in the organ.

light is produced by bioluminescent bacteria and is utilized by the fish for species recognition, prey attraction, and schooling behavior. Flashlight fish can actually control the light emission by covering their light organ with a fold of tissue, similar to an eyelid. The bacteria possess a compound called luciferin, which is oxidized by an enzyme called luciferase. As electrons lose energy during this oxidation, light rather than ATP is produced.

In a commensal relationship one organism is benefited while the other is unaffected

Commensalism differs from mutualism in that while one organism benefits, the other is neither helped nor harmed (see Figure 9.24). In some commensal relationships, one microorganism is able to make a living from the waste product of another. For example, in our previous discussion of nitrification, we learned that bacteria such as *Nitrosomonas* release nitrite as a waste. Other bacteria such as *Nitrobacter* then use the nitrite, converting it to nitrate. While *Nitrobacter* is benefited by this relationship, *Nitrosomonas* is unaffected one way or the other. Another example of a bacterial commensal relationship is when facultative anaerobes living in an animal's intestine use up most of the available oxygen. Under these conditions, obligate anaerobes are now able to grow, without obvious cost or benefit to the facultative anaerobe.

Most of the microorganisms growing on human body surfaces are commensals. For example, bacteria such as *Corynebacterium* growing on your skin obtain nutrients, and you are neither hurt nor assisted by your microbial guests. The same is true of the bacteria that dwell on moist regions of your body, such as the groin and armpits. Many of these bacteria use secreted lipids as an energy source, releasing fatty acids as a waste. Some of these fatty acids readily dissolve in the air and produce a strong, foul smell; they cause body odor. Of course, you might argue that this is not necessarily a commensal relationship, because of the cost the human host can pay in social ostracism. Hence, the popularity of underarm deodorants, many of which contain antimicrobial compounds.

Microorganisms in the same environment may compete for certain resources

CASE: A MICROSCOPIC DOG-EAT-DOG WORLD

As the grand finale of his microbiology course, Dr. Miller expects his students to identify unknown species of bacteria, using techniques that they have learned over the course of the semester. Each student is provided with a liquid broth, into which Dr. Miller has placed one Gram-positive and one Gram-negative species. The student then streaks this mixed broth onto two media. One is a special selective medium on which only Gram-positive bacteria can grow (Figure 9.26); the other is selective for Gram-negative growth only. With the two species now isolated, students can perform a variety of tests on each one to make their identifications. When mixing his samples, Dr. Miller adds 0.9 mL of the Gram-positive species but only 0.1 mL of the Gram-negative species. When mixing one of the unknown samples, however, he is distracted and adds 0.9 mL of both species. This sample is given to Elsie, a student in the class. Elsie streaks her sample onto the two selective media as directed, but following a standard 24-hour incubation period, she is disappointed to find that although her Gram-negative species grew well, the Gram-positive bacteria species did not grow at all. No other students have this problem, and Elsie seeks advice from her instructor about how to proceed.
1. Why didn't Elsie's Gram-positive species grow?

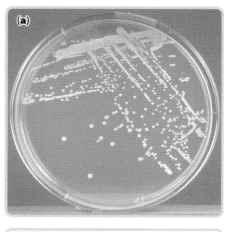

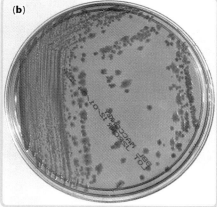

Figure 9.26 Bacterial isolation by use of selective media. (a) Gram-positive bacteria, growing here as whitish colonies, can be isolated from Gram-negative bacteria in a mixed sample by streaking the sample on a colistin–nalidixic acid agar plate. Acids in the agar inhibit the growth of Gram-negative bacteria without affecting Gram-positive species. (b) Similarly, Gram-negative bacteria can be isolated by streaking a mixed sample on a MacConkey plate. Compounds in this selective medium permit Gram-negative growth (pinkish colonies) while inhibiting the growth of Gram-positive organisms.

When two species within the same habitat try to acquire the same resources, **competition** is the result. Such an interaction has a negative impact on both participants (see Figure 9.24). Often, one of the two competing organisms, the one that is better adapted to the habitat in question, will drive the less adapted organisms to extinction. This principle of **competitive exclusion** was first demonstrated in 1934 with two different protozoan populations, forced to compete for the same resources. No matter how the environmental conditions were modified, one species "won" while the other "lost" and disappeared.

If you examine a typical soil or water sample, brimming with many microbial species, you might come to the conclusion that competition among microorganisms is rare, since so many different organisms seem to be peacefully coexisting. Remember, however, the concept of microhabitats from earlier in this chapter and the need to "think small" in microbial ecology. Although you might find dozens of species in a single soil sample, closer inspection reveals that each species occupies its own microenvironment, where it outcompetes all comers.

The reality is that competition in the microbial world is especially fierce, and while competitive exclusion between animal or plant competitors may take weeks, years, or even decades, microbial competition is often resolved within hours. The ability of a microorganism to compete is largely dependent on its growth rate, which in turn is based on its metabolism, and how well it is adapted to any given environment. Because microbial growth rates are so rapid compared with those of multicellular organisms, we do not have to wait long to determine which species reproduces faster. Another important factor influencing microbial interactions is the chemicals released by some organisms that actually interfere with the growth of competitors. As we will see in Chapter 12, humans have taken advantage of many of these antagonistic chemicals, using them in the development of antibiotic drugs.

Dr. Miller knew that, under laboratory conditions, the Gram-negative bacteria in his unknown samples would grow faster and therefore outcompete and eliminate the Gram-positive species, unless the slower Gram-positive bacteria were given a numerical head start. This is why he mixed his samples at a ratio of 9 to 1. But when he mistakenly added equal amounts of both bacteria in Elsie's sample, the Gram-positive organisms didn't stand a chance. Outcompeted by the rapidly growing Gram-negative species, they were eliminated before Elsie had the opportunity to isolate them.

Competition between bacteria probably explains, at least in part, why the bacterial community in the human intestine stays so constant over time. Every day we introduce many new bacterial species into our digestive system along with the food and liquids we consume. Yet the newcomers are rarely able to colonize the intestine, because of competition with well-adapted species that are already colonizing their preferred microhabitats in large numbers.

Some microorganisms are predators

In a predatory interaction, one organism (the predator) consumes another (the prey). Obviously, this benefits the predator, while the impact on the prey is, to put it mildly, negative (see Figure 9.24). Bacteria are often prey, but only a few species are true predators. One of the best studied is *Bdellovibrio*. These bacteria first attach to the cell of their bacterial prey and penetrate into the space between the cell wall and the plasma membrane (**Figure 9.27**). Here *Bdellovibrio* replicates, absorbing proteins, nucleotides, and other biological molecules directly from its host cell. Eventually, the host cell lyses and dies, releasing progeny *Bdellovibrio*. Interestingly, *Bdellovibrio* can prey only on Gram-negative species. Gram-positive

Figure 9.27 *Bdellovibrio*, a bacterial predator. Replication of *Bdellovibrio* occurs after these bacteria (in red) gain access to the space between the cell wall and the plasma membrane of their Gram-negative prey (in blue).

bacteria are safe from attack. Another predatory bacterial species with the colorful name *Vampirococcus* kills its bacterial prey in a different manner. Following attachment, it releases an enzyme that bursts the host cell, releasing its contents.

Many protozoa, especially ciliates and amoebas, are predatory, and bacteria are frequently the victims. Although fungi are generally saprophytic or parasitic (see Chapter 4, p. 86), there are some predators as well. Fungi in the genus *Arthrobotrys*, for example, actually trap nematode worms in the soil, using their hyphae as nooses (**Figure 9.28a**). Once the nematode is stuck in the hyphae, new hyphae grow directly into the worm, where they absorb nutrients directly from the nematode's body cavity.

And we cannot end this chapter without mentioning the fungi that prey on insects. Fungal spores stick to the insect's body as it forages for food. Hyphae produced by the spore penetrate the insect's body, and once inside the fungus grows, consuming the insect from within. After a few days, the now seriously ill insect crawls up to a high spot, where it dies. Shortly thereafter, the fungus literally pushes its way out of the insect's abdomen, spewing new spores in the process (Figure 9.28b). Any nearby insect unlucky enough to be caught in this spore shower will now become infected. Several species of so-called *entomopathogenic* (causing disease in insects) fungi go through this gruesome life cycle. Fortunately for us, vertebrates are not affected. Any of you who were fans of the 1990s television series *The X-Files*, however, may recall an episode in which humans rather than insects were the fungal victims. The writers of this program, while taking a certain poetic license, clearly knew their fungi.

Coming Up Next...

In concluding this chapter with a discussion of ecological interactions, we have purposely avoided one important type: parasitism. In a parasitic relationship, one organism (the parasite) benefits while the other (the host) is harmed, but often in a way that does not necessarily end in death. Because the host may eventually die, however, it is sometimes difficult to distinguish between parasitism and predation. Many microorganisms, including bacteria, viruses, protozoa, and fungi, are parasites of humans and other animals, and we will devote the next two chapters to an exploration of the interactions between animals and microbial parasites. More specifically, we interact with microorganisms, many of which are parasites, every day. Usually we remain healthy, but on occasion we get sick and may even die. The outcome of a parasite–host interaction depends on both the impact of the parasite on the host and the manner in which the host responds to infection. In Chapter 10 we consider the interaction from the parasite's perspective, focusing on how parasites affect the host. In Chapter 11 we will complete the picture by investigating how hosts defend themselves against a never-ending microbial onslaught.

(a)

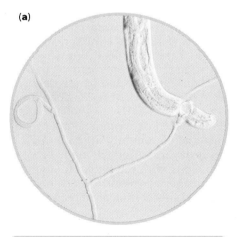

(b)

Figure 9.28 Predatory fungi. (a) Fungi in the genus *Arthrobotrys* live in the soil, where they trap nematode worms. Here, a nematode is trapped in the constricting ring of the fungus. Note the unsprung ring to the left. (b) An entomopathogenic fungus that has burst through the body wall of an infected ant.

Key Terms

abiotic factor	ecosystem	oligotroph
algal bloom	eutrophication	omnivore
ammonification	eutroph	pelagic zone
barotolerant organism	food chain	photic zone
benthic zone	food web	phytoplankton
biofilm	halophilic organism	population
biomass	herbivores	primary consumer
bioremediation	humus	producer
biotic interaction	intertidal zone	profundal zone
carnivore	littoral zone	rhizosphere
community	microenvironment	root nodule
competition	mutualism	secondary consumer
competitive exclusion	mycorrhizal association	thermocline
decomposer	nitrification	trophic level
ecology	nitrogen fixation	zooplankton

Concept Questions

1. Select any habitat and construct a hypothetical food web for that habitat. Indicate which organisms are producers, primary consumers, secondary consumers, tertiary consumers, and decomposers.

2. There is often a striking difference in the rate at which plants grow in tropical versus cold, temperate regions. Explain the role played by microorganisms in this difference.

3. You are hired to write a screenplay for a new movie in which, due to a genetic engineering experiment gone awry, all prokaryotes on Earth are rapidly going extinct. You'd like your screenplay to be as realistic as possible. Describe a possible scenario for what would happen as prokaryotes disappear.

4. If microorganisms are spread out over an agar plate, each individual cell may start to divide, giving rise to a colony of cells. However, cells that were initially well spaced, some distance from other cells, give rise to much larger colonies. Cells that were initially close to neighboring cells give rise to smaller colonies. Explain this observation.

5. Imagine that a large earthquake struck a large American city and that sanitation facilities in the city were badly disrupted. All of a sudden large amounts of raw sewage are pouring into a nearby waterway. What do you think would happen over the next several weeks in the waterway in terms of the population of photosynthetic microorganisms? Of oxygen content in the water? Of fish populations?

6. Although we did not discuss it, oxygen, like other elements crucial to life, cycles between the living and nonliving worlds. On the basis of what you learned about biogeochemical cycles in this chapter and about photosynthesis and aerobic respiration in Chapter 7, draw a simple diagram of how you think oxygen cycles in the environment.

7. Some people have argued that a vegetarian lifestyle is more sustainable because larger numbers of people can be fed on plant material than can be fed on animals that ate the plant material. Is there any truth to this? If so what is the logic?

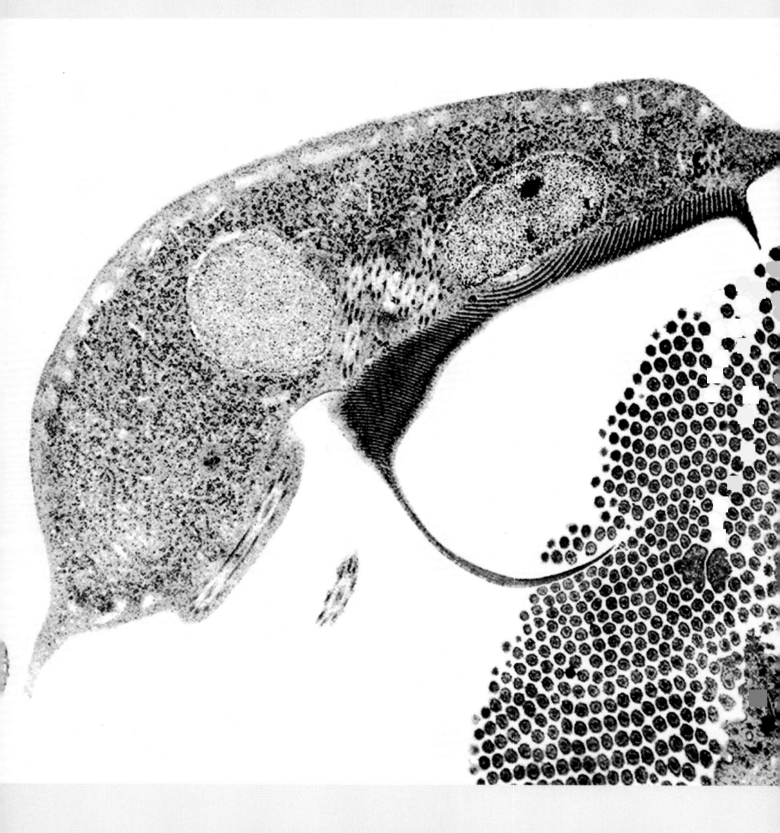

A *Giardia* protozoan (blue) attaches itself to the inner lining of the intestines (green) using an adhesive disk.

Chapter 10

The Nature of Disease: A Pathogen's Perspective

In the 1950s a young biology student named Robert Desowitz told his faculty advisor that he wanted to pursue a graduate degree in parasitology. He specifically wished to study malaria. His adviser counseled against it. After all, infectious disease was on its way out, soon to join the dodo, the passenger pigeon, and other extinct forms of life in the history books. Certainly, after millennia as the primary scourge of civilization, pathogens were in full retreat. New antibiotics, breakthroughs in vaccine development, and steadily improving sanitation all pointed to a future free of infectious disease, and devoting one's professional life to malaria, a disease clearly on the road to oblivion, seemed a poor career move. "Go into cancer research," advised Desowitz' mentor. After all, that is where the future lay—not with some dead-end parasite that might be eradicated before Desowitz was even out of graduate school.

Things did not work out the way the adviser predicted. Half a century later, infectious disease is still with us, and in some respects it is more of a problem then ever. Indeed, it is difficult to pick up a magazine or watch the news without hearing about yet another infectious disease that threatens us. The more spectacular diseases such as AIDS and swine flu usually grab most of the headlines, but other, perhaps less newsworthy maladies such as bacterial dysentery still inflict more than their share of human misery. Desowitz, by the way, did not listen to his adviser and went on to become a respected authority on parasitic disease. With malaria still responsible for over a million deaths each year, students interested in malaria research today would no doubt get very different advice.

In Chapter 8 we began to see *why* infectious microorganisms are such intractable foes. Evolution via natural selection ensures that pathogens remain moving targets, and our efforts to eradicate them are rarely completely successful. Moreover, examples such as the evolution of drug resistance remind us that when control efforts are poorly planned, the situation can become even more serious in the long run.

Yet in spite of the continuing threat, most of the time the majority of us are not sick, and when we do contract an infectious illness we usually recover. But not always. What determines how and when we develop infectious disease? When we do, how serious will the illness be? Will we recover, and if so, will that recovery be rapid or prolonged? To answer these questions, we must understand how pathogens and hosts interact with each other. In this chapter, we will concentrate on what pathogens must do if they are to cause disease, and how and why they actually make us ill. In other words, we will try to view the disease process from the pathogen's point of view. Disease, however, is a two-way street. How we as hosts respond to microbial attack is often just as important or even more important than the attack itself. To fully comprehend the overall disease process, we must also consider the host response, which we will do in Chapter 11.

Basic Principles of Infectious Disease

Contact with microorganisms only rarely results in disease

The simple occurrence of pathogens in or on a host does not mean that illness will follow. Rather, disease is the consequence of a series of increasingly important effects on the host that begins with **contamination**—the mere presence of microorganisms. Contamination does not imply that the microorganisms are causing any problem or that they are increasing in number. It simply means that the contaminated host or inanimate object is not sterile. Indeed, almost all surfaces, living or not, are "contaminated," and in the vast majority of cases the microorganisms involved cause no particular difficulty. In fact, many such organisms provide valuable services to the host. If microorganisms are able to survive and reproduce in or on a host, we say that an **infection** has occurred. Although we are conditioned to view infections as bad things, this is not necessarily true. Many microbial infections have few if any negative consequences. However, if the infection interferes with normal host functions, we are now faced with a **disease**. Depending on the tissues affected and the degree of interference, the disease may range from mild to severe, and the changes in the host caused by the pathogen may be temporary or permanent. For example, once a person recovers from a bout of flu, there are ordinarily no lingering side effects. Recovery is complete. If, however, the eyes of an infant are infected at birth with *Chlamydia trachomatis*, permanent scaring of the cornea can occur (**Figure 10.1**). Most *Streptococcus* infections cause only temporary disease, but occasionally permanent damage can occur to the valves of the heart or to kidney tissue. This condition, known as rheumatic fever, most often occurs when infections with *Streptococcus pyogenes* are not properly treated.

Parasitic microorganisms able to cause disease are known as **pathogens**. As we will see, the capacity to cause disease does not imply that disease will occur every time the pathogen encounters a host. Some pathogens almost always cause disease following infection, while others do so only in rare cases. *Mycobacterium tuberculosis*, the causative agent of tuberculosis, almost always interferes with normal host function to some degree. *Staphylococcus epidermidis*, on the other hand, generally is not considered to be a pathogen, but under certain unusual circumstances it may cause disease. Most infectious agents fall somewhere between these two extreme examples.

Hosts are colonized with normal microbial flora that is usually harmless

CASE: SAYONARA SALMONELLA

Improperly cooked chicken is a frequent source of exposure to the *Salmonella* bacteria that cause food poisoning. Today, however, chicken farmers can ensure that their birds are almost *Salmonella*-free by spraying newly hatched chicks with a product called Preempt. This spray consists of 29 live bacterial species that are part of the chick's normal flora. As the chicks preen they ingest these bacteria, which take up residence in their intestines (Figure 10.2). A single treatment can provide lifelong protection against *Salmonella* infection to almost 99% of treated chicks. Infections with other pathogens, such as *Listeria* and *Campylobacter*, are similarly prevented.
1. What is meant by normal flora?
2. How does spraying the chicks with Preempt prevent later infection with *Salmonella*?

Figure 10.1 Eye damage caused by *Chlamydia trachomatis*. Colonization of the conjunctiva of the eye leads to tissue inflammation. Repeated inflammation due to reinfection can ultimately lead to scarring. If scar tissue accumulates over the cornea, blindness results. This condition, known as trachoma, is one of the most common forms of blindness, affecting millions worldwide each year. Rare in the developed world, trachoma is most common in Africa, Asia, and South America.

Figure 10.2 Acquisition of normal flora. (a) Baby animals begin to acquire their microbial flora at birth. Preening by baby birds is one way bacteria may be introduced into the digestive tract. (b) A baby giraffe enters the world at the Dallas Zoo in August 1984. In giraffes and other mammals, the acquisition of microbial flora begins as the baby exits the birth canal, picking up bacteria present in the mother's reproductive tract.

As previously mentioned, all animals, humans included, are populated by microorganisms that generally cause no health concerns. These commonly encountered, largely benign microbes are called the **normal flora**, and they are typically found on all exposed body surfaces (**Figure 10.3**). Hundreds of species of bacteria and smaller numbers of other microorganisms call the human body home, and their individual numbers can truly be astronomical. It is estimated, for instance, that a typical human body consists of 10^{13}

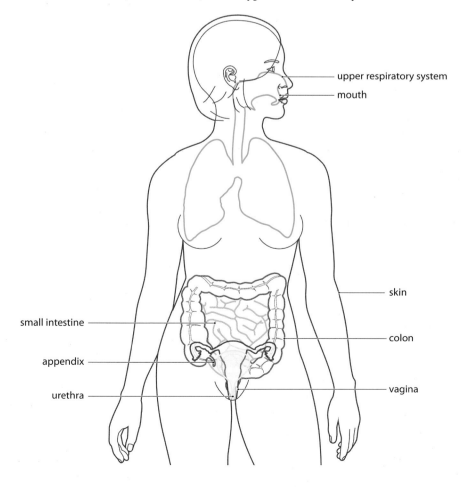

Figure 10.3 Body surfaces colonized by normal microbial flora. All of these surfaces are contiguous with the external environment. Other areas of the body not in direct contact with the exterior should remain sterile.

cells. However, approximately 10^{14} prokaryotes live on your skin, in your mouth, in your intestine and elsewhere. Thus, even within your own body, human cells are outnumbered by microbial cells, by approximately 10 to 1! Fortunately, under normal circumstances, these denizens of our bodies do not cause disease.

The normal flora are different in different host species. The organisms found in Preempt are normal flora for chickens. Humans likewise have their own typical normal flora species, although many of them would also be found in other animals, including chickens. Even within humans there is considerable variation. The normal flora differs between an adult and an infant, as it does between people living in Calgary and Cairo. Even individuals of the same age and in the same geographic area do not have an identical flora. Each of us possesses a unique combination of species and strains. For example, *Neisseria meningitidis* can cause the serious disease meningitis. Yet in 10% of people, this bacterial species lives harmlessly in the throat as part of the normal flora. In others, it may come and go, living as a temporary throat resident. In such an individual, this and other microorganisms that are found only sporadically are considered **transient flora**.

The microorganisms composing the normal flora vary from one body surface to another. On the skin, microorganisms are especially abundant in moister areas such as the groin, the armpits, and between the toes. Most of the normal flora bacteria found here are Gram-positive, as the dryness of the skin prevents its colonization by most Gram-negative species. In the female genital tract, most resident bacteria are Gram-negative. The upper respiratory tract and lower urinary tract, and especially the digestive system, also harbor abundant normal flora. In a normal fecal sample, bacteria make up as much as a third of the weight. Other parts of the body, however, such as the blood, the cerebrospinal fluid, and deep tissues are normally sterile. The presence of any microorganisms in these locations is suggestive of disease. Even normal flora microorganisms can cause disease if they gain access to such sites. This explains, for example, why dentists often prescribe a course of antibiotics following invasive dental work. The mouth supports a rich array of bacteria, which generally are not involved in disease. Serious problems may arise, however, if such microorganisms gain access to the blood due to a dental procedure. When normal flora bacteria cause disease due to unusual circumstances, we say that the host has developed an **opportunistic infection**.

In the above example, the "unusual circumstance" is sudden access to an unaccustomed site. Now, normally harmless organisms have the *opportunity* to cause disease. Likewise, we mentioned earlier that *S. epidermidis* rarely causes disease. This is certainly true when *S. epidermidis* remains on the skin, where it is typically found. But this species also is often found on plastic medical equipment such as catheters, and if such a device is used on a patient without being properly sterilized, a serious infection can result. Gram-negative bacteria such as *Escherichia coli* and *Klebsiella* usually cause few problems in the large intestine, but it is not too unusual for severe alcoholics to develop pneumonia caused by these species. How do these normal flora bacteria of the bowels gain access to the lungs? People suffering from serious alcoholism often pass out, and when they do, they sometimes vomit. If vomit, which may contain microorganisms from the digestive system, is aspirated back into the lungs by an unconscious individual, a lung infection can result.

Normal flora organisms may also become opportunists when the host's immune system is in some way suppressed. This explains why the likelihood of infection increases in patients undergoing cancer chemotherapy. Such therapy often has immunosuppressive side effects, making patients vulnerable to a variety of opportunistic infections. The same is true of patients infected with the human immunodeficiency virus (HIV). For example, respiratory illness caused by the fungus *Pneumocystis carinii* is often the first indication that an individual may be HIV-positive. All people have small numbers of *P. carinii* cells in their respiratory systems, but the fungus is unable to proliferate in those with a normally functioning immune system. In an immunocompromised HIV-positive individual, however, this opportunist is able to reproduce rapidly, causing a life-threatening respiratory disease. The same is true for *Candida albicans*, the fungus responsible for throat infections known as *thrush*. Almost all people have a small number of *C. albicans* organisms in the mouth. In HIV-infected individuals, the decreased immune protection provides an opportunity for the fungus to proliferate (**Figure 10.4**). These last two examples are illustrative in an era when the number of HIV-positive patients continues to rise. You cannot "catch" an opportunistic infection such as *P. carinii* or *C. albicans* from an HIV-positive individual. You are already infected with these organisms yourself, but if your immune system is functioning properly, you stand no risk of developing disease.

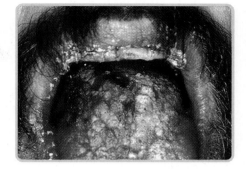

Figure 10.4 An opportunistic infection. A severe case of thrush, caused by an overgrowth of *C. albicans* on the tongue of an HIV-infected patient.

While it is true that normal flora microorganisms cause problems for a host as a source of opportunistic infections, by and large these microbial guests are relatively well behaved, and in many instances they prevent rather than cause disease. One of the characteristics of normal flora organisms is that they are particularly adept at attaching to the surfaces on which they live. Structures such as fimbrae and glycocalyces (see Chapter 3, p. 50) ensure that the normal flora is unlikely to be swept away by intestinal fluid in the digestive system, sweat on the skin, or saliva in the mouth. Many types of normal flora are so well adapted to particular regions of the host's body that they literally cover every square micrometer of space. There simply is no room for latecomers that may be pathogenic but are not necessarily so skilled at attaching to surfaces. Furthermore, many members of the normal flora secrete compounds that actually inhibit the growth of other, competing microorganisms.

Because they are outcompeted by the normal flora, many pathogens, even if they gain access to a host, are unable to colonize. This is exactly how Preempt keeps chickens free of *Salmonella* and other pathogenic bacteria. By establishing a rich normal flora in young chicks, it leaves simply "no room at the inn" if and when *Salmonella* shows up.

We are sometimes vividly reminded of the protective value of our normal flora when we take an antibiotic to treat an infection with pathogenic bacteria. As we will see in Chapter 12, many antibiotics do not discriminate between pathogens and normal flora. When we take such a drug, in addition to destroying the pathogen, the normal flora might be disrupted as well. Such a disruption sets the stage for an opportunistic infection. For example, we recently learned that almost all people have a small number of *C. albicans* in the mouth. In women, this single-celled fungus is also present in the reproductive tract. Normally the growth of this potential pathogen is kept in check by the normal bacterial flora, but if an individual takes antibiotics, normal flora may be reduced in number. This provides an opportunity for *C. albicans*, which is now able to proliferate. The result can be thrush in the mouth or vaginitis in the female reproductive tract.

The Process of Infectious Disease

A pathogen must achieve several objectives if it is to cause disease

Having reviewed some of the basic concepts of relevance to infectious disease, we will now take a closer look at exactly how and why disease occurs. In general, when pathogen meets host, the likelihood of disease depends on variables such as the number of pathogens to which a host is exposed, the overall virulence of the pathogen, and the status of the host's defenses. These factors are all closely interrelated. For example, if a particular patho-

Figure 10.5 Requirements for infectious disease. Even normal flora must meet most of these requirements. Whereas pathogens interfere in some way with normal host function, however, the normal flora ordinarily does not. In this example a bird is serving as the reservoir, and the pathogen is being transmitted via the respiratory route.

gen is highly virulent, fewer pathogens are needed to cause disease. If we know something about these factors in a particular host–pathogen interaction, we can make a reasonable prediction about the likely outcome of that interaction. Yet even the most virulent pathogen is unable to make you ill unless it can accomplish several things, and we will now consider these requirements individually (**Figure 10.5**). To cause disease, a pathogen must:

1. have a place to survive before and after infection (termed a *reservoir*)
2. be transmitted from the reservoir to the host
3. invade the host's body
4. adhere to the host in an appropriate site
5. multiply in or on the host
6. at least initially, evade the host's defenses
7. interfere with normal host functions
8. leave the host and reach either a new host or its reservoir

Reservoirs provide a place for pathogens to persist before and after an infection

CASE: WHAT'S BUGGING KITTY?

Peri, a healthy 12-year-old girl, is thrilled when her parents present her with a new kitten as a birthday gift. While playing with her kitten, Peri is scratched, and a week later she develops a small lesion on her left hand where her kitten scratched her. After another week, she feels quite ill and has developed swollen lymph nodes under her left armpit. Her parents take her to the doctor, who orders blood tests. The results of these tests confirm the doctor's suspicion; Peri is infected with *Bartonella henselae* and has been suffering from cat scratch disease. The doctor orders a course of antibiotics for Peri, during which she makes a full recovery.

1. **Unlike Peri, the kitten shows no signs of illness. Why not?**
2. **What role is the kitten playing in the development of cat scratch disease?**

For many pathogens, following the infection of a susceptible host, there are really only two possible outcomes; the host either eventually recovers or dies. Host survival usually results in at least temporary immunity to further infection by the same microorganism. Either way, this means that if the pathogen is to persist over the long term, it must leave the host at some point and find a new host. Eventually, however, the pathogen is faced with the prospect of a host shortage if most of the hosts are either dead or immune. In that case the pathogen would also cease to exist, because it has no place to survive. Such pathogen extinction is rare, however, because even without a limitless supply of susceptible hosts, pathogens can persist in their **reservoir** until new hosts become available. A reservoir is a place where the pathogen can survive indefinitely without causing serious disease and from which it can infect new susceptible hosts if and when they are available. New hosts typically become available through new births or the waning of immunity in previously infected hosts. Reservoirs are essential for the long-term survival of any pathogen, and all pathogens require at least one.

Animal reservoirs. Many pathogens, such as *B. henselae* infecting Peri's kitten, use **animal reservoirs** (**Table 10.1**). In the case of cat scratch disease, the bacteria cause no particular problem for the cat, because the microorganism and the host are well adapted to each other. Should a human enter

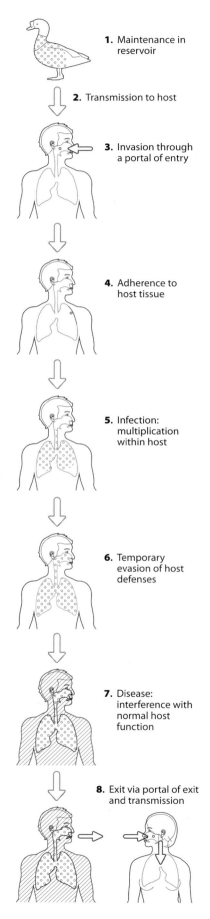

1. Maintenance in reservoir

2. Transmission to host

3. Invasion through a portal of entry

4. Adherence to host tissue

5. Infection: multiplication within host

6. Temporary evasion of host defenses

7. Disease: interference with normal host function

8. Exit via portal of exit and transmission

DISEASE	MICROORGANISM	RESERVOIR	USUAL MODE OF TRANSMISSION	USUAL PORTAL OF ENTRY
Bacterial Diseases				
typhoid	*Salmonella typhi*	human	water or foodborne	digestive system
Legionnaire's disease	*Legionella pneumophila*	environmental	airborne	respiratory system
tuberculosis	*Mycobacterium tuberculosis*	human, animal	airborne	respiratory system
whooping cough	*Bordetella pertussis*	human	airborne	respiratory system
syphilis	*Treponema pallidum*	human	contact	reproductive system
plague	*Yersinia pestis*	animal	vector	flea bite through skin
Viral Diseases				
hantavirus pulmonary syndrome	hantavirus	animal	airborne	respiratory system
measles	measles virus	human	airborne	respiratory system
common cold	many viruses	human	airborne	respiratory system
rabies	rabies virus	animal	contact	animal bite through skin
yellow fever	yellow fever virus	animal	vector	mosquito bite through skin
Protozoan Diseases				
Chagas' disease	*Trypanosoma cruzi*	animal	vector	kissing bug bite through skin
malaria	*Plasmodium* species	human, animal	vector	mosquito bite through skin
giardiasis	*Giardia intestinalis*	animal	food or waterborne	digestive system
Fungal Diseases				
Valley fever	*Coccidioides immitis*	environmental	airborne	respiratory system
athlete's foot	various fungi	environmental	contact	colonization of skin

Table 10.1 Representative infectious diseases. Reservoirs, typical mode of transmission, and portals of entry are indicated.

the disease cycle, cat scratch disease may result, because there is no similar adaptation between humans and *B. henselae*. A human disease such as this, where the pathogen utilizes an animal reservoir, is termed a **zoonosis**. Because no disease is caused in the reservoir, there is no danger of these reservoirs all dying off. Humans, although susceptible, are not essential for the pathogen's long-term survival.

Many similar examples exist. Reptiles often serve as reservoirs for *Salmonella*, and reptile owners sometimes develop *Salmonella* infections after handling their pets. Respiratory disease caused by hantavirus was first described in 1993. The virus is thought, however, to have been present in the Southwest United States for eons, surviving quietly in deer mice, in which it causes no illness whatsoever (**Figure 10.6**). In 2003, monkeypox, an African viral disease, surprisingly showed up in the Midwestern United States. It was eventually determined that the original source of the illness was African rodents, imported into the United States as pets.

Human reservoirs. Other pathogens rely at least in part on **human reservoirs** (see Table 10.1). An example is provided by *N. meningitidis*, described earlier in this chapter, which survives in the throats of some people without causing disease. These individuals serve as a source of infection for susceptible human hosts. Similarly, some people carry *S. pyogenes*, the causative agent of "strep throat," in their throats for decades without becoming ill.

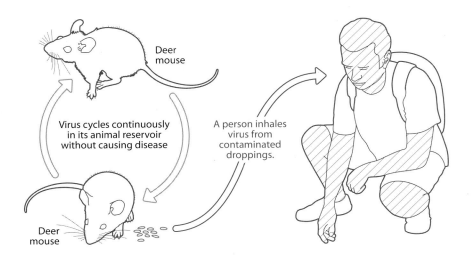

Figure 10.6 Hantavirus: a zoonotic disease. The virus cycles continuously in its animal reservoir, the deer mouse, without causing disease. Virus shed in the mouse's feces or urine can be respired into the lungs of a human, who may then develop a severe, often lethal respiratory illness called hantavirus pulmonary syndrome (HPS). Outbreaks of HPS often seem to follow wet years, which increase vegetation and food for mice. Mice populations then rise, increasing the likelihood of contact between humans and mice.

The infamous case of "Typhoid Mary" is a classic example of a human reservoir from the early 1900s (**Figure 10.7**). Infected with *Salmonella typhi*, Mary herself remained healthy. But because she worked as a cook, and because typhoid can be transmitted through tainted food, people around her often developed this disease. When it became apparent that, in some poorly understood manner, Mary was making those she worked for ill, she was institutionalized for 26 years on a small island in New York's East River, always proclaiming as outrageous the idea that she could make anyone sick. It is now known that when most people contract *S. typhi* they develop the disease but eventually kill off the microorganism and survive. Approximately 3% of humans can act as human reservoirs. They do not develop symptoms of typhoid and carry the bacterium indefinitely, potentially infecting others.

There is no guarantee that a human reservoir will *never* develop illness. Indeed, in many cases, eventually the reservoir also succumbs to disease. This is the case with the human immunodeficiency virus (HIV), which frequently does not cause serious disease until years after infection. During this long symptom-free period, an HIV-positive individual serves as a reservoir for the virus, fully capable of infecting others.

Environmental reservoirs. Some pathogens can survive indefinitely in soil, water, or other **environmental reservoirs** (see Table 10.1). In Chapter 3 we discussed spore-forming organisms such as *Clostridium tetani*, the causative agent of tetanus. This bacterium can survive for decades as a spore in the soil, until it reaches the warm, moist, and anaerobic environment of an animal or human where it can germinate. Similarly, in Chapter 4 we discussed *Coccidioides immitis*, a fungus that lives in the soil of the desert Southwest. When inhaled, it sometimes causes the respiratory ailment Valley fever. Water can serve as a reservoir for many pathogens, including *Vibrio cholerae*, which causes cholera.

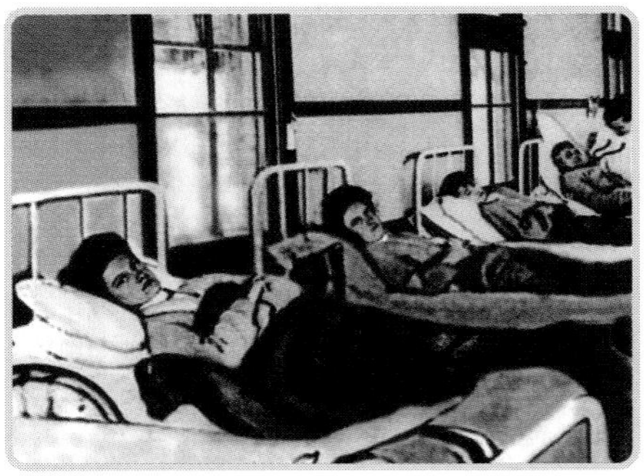

The type of reservoir used by a pathogen has implications for disease control

It is no coincidence that to date the natural transmission of only one important human pathogen, smallpox, has been completely eliminated. Smallpox relies exclusively on human reservoirs; it cannot persist in either the environment or in an animal. Furthermore, a potent anti-smallpox vaccine offers long-term protection against this scourge, and vaccinated individuals cannot serve as reservoirs because any virus is immediately destroyed by the immune system. In the 1950s, international health authorities reasoned

Figure 10.7 Something about Mary. Mary Mallon (in the closest bed) on North Brother Island in New York's East River. She was originally sent to the island when it was realized that she was responsible for cases of typhoid in those she worked for. She was later released on the condition that she not work as a cook, but when a typhoid outbreak began in a hospital, Mallon was found working there under an assumed name. She was shipped back to the island, where she remained until she died in 1938.

Figure 10.8 End of the line for smallpox. (a) Villagers in West Africa line up for smallpox vaccines. Under the supervision of the World Health Organization, smallpox was eradicated thanks to a worldwide vaccination campaign starting in the 1950s. An effective vaccine and the lack of reservoirs other than humans were crucial factors in the success of this effort. (b) This Somali man, who contracted smallpox in 1977, was the last victim of naturally occurring smallpox.

that if enough people could be vaccinated, smallpox virus, without a reservoir, would go extinct. The World Health Organization coordinated an intensive international vaccine effort, and the last victim of naturally occurring smallpox is believed to have been a young man in Somalia, who contracted and later recovered from smallpox in 1977 (**Figure 10.8**). Other diseases that are targeted for eventual eradication, such as polio, likewise depend exclusively on human reservoirs.

When human pathogens utilize animal reservoirs, the problem of control is considerably more problematic. A case in point is provided by efforts to control yellow fever in Cuba and Central America in the early 1900s. In Cuba there were no significant animal reservoirs, and yellow fever was effectively eradicated through mosquito control (see Chapter 5, p. 117). This strategy was never entirely successful in Panama, however, where monkeys serve as important reservoirs and as an ongoing source of infection.

Of course, if all animal reservoirs could be vaccinated, the problem would be solved. Vaccinating wild animals, however, is impractical at best. A more commonly used, if brutal, approach is simply to slaughter animals believed to serve as reservoirs. When "bird flu" first appeared in 1997, the response of officials in Hong Kong was to kill 1.5 million chickens thought to be potential reservoirs. Nevertheless, the virus was able to elude this control effort and has now spread to Vietnam, Thailand, and many other countries. This sort of logic was also behind the attempt in Great Britain to control tuberculosis by killing badgers. Experiments in the lab had shown that badgers, which can carry tuberculosis, can pass this disease to cattle. Humans drinking raw milk from infected cows are subsequently at risk. Consequently, beginning in the mid-1980s, British health authorities, thinking that badgers were serving as important reservoirs, began killing the animals. More recently, however it was demonstrated that although tuberculosis transmission via badgers is possible in the laboratory, it rarely if ever occurred in nature, because cattle and badgers tend to avoid each other. In response to this finding, the "badgering of badgers" was halted in 2003.

Pathogens must reach a new host via one or more modes of transmission

CASE: REVENGE OF THE BLUE DEVILS

Although Duke University has been a basketball powerhouse for years, the Blue Devils are usually not so scary on the football field. But in a game against Florida State in 2000, although Duke lost as expected, the Seminoles hardy got off unscathed. Duke passed more than pigskins that day. Infected Duke players "handed off" Norwalk virus to 11 Florida State players. This was the first known case of transmission of an intestinal virus between sports teams. The original source of the virus among the Duke athletes was traced to tainted turkey sandwiches. The Duke and Florida State players had no contact before the game, but during the contest, several ailing Duke players vomited and continued to play in vomit-stained uniforms. It appears that both aerosol and contact transmission occurred on the field. In any event, it demonstrates yet again that when you deal with the devil, there's hell to pay.

1. What is meant by aerosol and contact transmission?
2. In what other ways are infectious agents transmitted?
3. How long before the game were the tainted sandwiches consumed?

All pathogens need at least one **mode of transmission**—a means of traveling between hosts or between host and reservoir. Modes of transmission can conveniently be categorized into several basic types.

Contact transmission. A variety of microorganisms can move from an infected individual to an uninfected, susceptible host via physical contact (see Table 10.1). Such contact can be direct or indirect. Hand shaking, kissing, sexual contact, and any other form of physical contact that allows the transmission of a pathogen are examples of **direct contact transmission**. In **indirect contact transmission**, an inanimate object acts as an intermediary between an infected and a susceptible individual. Such is the case if you become infected with a cold virus, for example, after using a hand towel or drinking from a glass previously contaminated by an infected individual. The inanimate object that served as the intermediary is called a **fomite**. Other examples of fomites are improperly sterilized catheters in a hospital or needles shared between drug users. Many pathogens, such as HIV and hepatitis C, can be transmitted either sexually or via shared needles and are therefore transmitted by both direct and indirect contact transmission. The Norwalk virus in our case was apparently spread both directly and indirectly. It is likely that, in the football game between Duke and Florida State, vomit on either the skin or the jerseys of the infected Duke players found its way onto the Florida State athletes. Once the transfer had occurred, all a Florida State player needed to do was touch the contaminated material and later rub his face or touch his mouth for the "viral interception" to be complete.

Aerosol or airborne transmission. Many pathogens can move through the air in an aerosol released by an infected individual (see Table 10.1). Such **aerosol transmission** (also called **airborne transmission**) was also implicated as a possible means for the transfer of Norwalk virus. Aerosol transmission simply implies that the infectious agent leaves a host via a cough, a sneeze, or even microscopic droplets released while talking (**Figure 10.9a**). Pathogens that reside in the upper or lower respiratory system, such as cold viruses, influenza virus, or the microorganisms responsible for pneumonia, are transmitted in this fashion. Norwalk virus infects the digestive tract, but in this case, viral particles present in vomit may have been aerosolized and breathed in by opposing players. If some of these viral particles were swallowed, the virus would have arrived at a suitable site for replication in the new host.

Figure 10.9 Modes of transmission.
(a) Airborne transmission can occur via the aerosol spray expelled during a sneeze. (b) Even in places with excellent sanitation, food and waterborne transmission can occur when simple hygienic measures are not followed.

Food- or waterborne transmission. Although Norwalk virus might be transmitted by both contact and aerosol transmission, more typically, transmission of this and other intestinal pathogens is by **food-** or **waterborne transmission**. Microorganisms transmitted in this manner are released through the feces of an infected individual. This contaminated waste must then contact either food or water, whereupon it is ingested by another susceptible individual. *Salmonella*, agents of dysentery, poliovirus, and parasitic amoebas, just to name a few, are transmitted to new hosts in this way. Because of their reliance on transmission by contaminated food or water, these microorganisms are readily controlled through improved sanitation. For this reason, diseases transmitted this way are often associated with undeveloped parts of the world where sanitation is poor. Great strides have been made in many places to control such illness, but even the most developed countries remain susceptible. When animals such as cows and chickens are slaughtered for meat, the meat often becomes tainted with microorganisms that were living in the animal's digestive system. If such meat is improperly handled during preparation, foodborne illness is a real possibility. We are often advised, for example, not to use the same knife to cut up both chicken and salad vegetables. If the knife used on raw chicken is then used to dice vegetables, contamination of the vegetables and subsequent foodborne transmission may occur, even if the chicken itself is well cooked. Occasional outbreaks of hepatitis A are often traced to a single infected food worker who failed to properly wash his or her hands following a trip to the restroom (see Figure 10.9b).

Vector transmission. Transmission of infectious microorganisms by insects or other invertebrates is known as **vector transmission** (see Table 10.1). Vector transmission is generally subdivided into one of two types. **Biological vector transmission** means that the invertebrate vector is an essential part of the life cycle of the microorganism, which replicates and/or undergoes essential developmental changes within the vector (**Figure 10.10a**). In biological transmission, the vector itself is infected with the disease agent. *Plasmodium*, the protozoan responsible for malaria, relies on mosquitoes as a biological vector, as do West Nile virus, yellow fever virus, and many

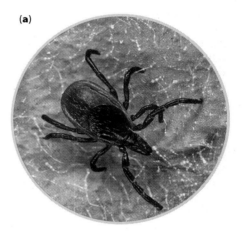

(a)

(b)

Figure 10.10 Vector transmission.
(a) The black-legged tick, *Ixodes scapularis*, is the biological vector of Lyme disease. The tick itself is infected with the pathogen and is an essential part of the pathogen life cycle. (b) Mechanical transmission can occur when pathogens stick to the body of a vector such as this housefly. The pathogen does not develop or reproduce in or on the vector.

Figure 10.11 The principal portals of entry into a human host. All pathogens require a portal of entry into a host. Major portals of entry are illustrated here. Pathogens transmitted by biological vectors are usually introduced directly into the circulatory system as the vector feeds (not shown).

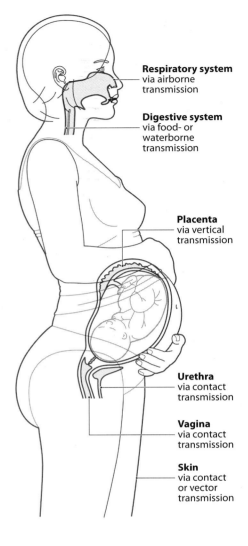

Respiratory system
via airborne transmission

Digestive system
via food- or waterborne transmission

Placenta
via vertical transmission

Urethra
via contact transmission

Vagina
via contact transmission

Skin
via contact or vector transmission

others. *Borrelia burgdorferi*, the bacterium responsible for Lyme disease, is biologically transmitted by ticks, while the protozoan that causes sleeping sickness utilizes tsetse flies.

When an insect serves as an incidental vector in which no pathogen replication or development has occurred, we say that **mechanical vector transmission** has taken place (see Figure 10.10b). Unlike a biological vector, the mechanical vector is not actually infected with the pathogen. The microorganism merely sticks to its body or otherwise "hitches a ride" on the vector. For example, we have already noted that Norwalk virus is generally transmitted via the food- or waterborne route. If virus is present in feces, however, a fly might land on this contaminated waste to feed. When it does so, viral particles may stick to hairs on its body. Later, if this same fly lands on uncovered potato salad at a picnic, viral particles may contaminate the food item, posing a risk to anyone who consumes it. Other typically foodborne pathogens can likewise be transmitted by mechanical vectors in this manner.

Vertical transmission. Some infectious agents are able to pass from a pregnant female to a fetus across the placenta, while others are transmissible through the breast milk. Transmission of pathogens in this manner is known as **vertical transmission**. German measles (rubella), HIV, and syphilis can all be contracted vertically.

Pathogens gain access to the host through a portal of entry

Following successful transmission, the pathogen must be able to invade the host through an appropriate **portal of entry** (**Figure 10.11**). The major portals of entry include surfaces of the respiratory system, the digestive system, or the reproductive system. A break in the skin such as a cut or a burn can be utilized as a portal of entry by many microorganisms. In vertical transmission to a fetus, the placenta is the portal of entry.

Whether or not the pathogens that invade the host will actually establish an infection is dependent in part on the number of invading organisms. The relationship between pathogen numbers and likelihood of infection can be studied experimentally in animals and expressed as the **ID_{50} (infectious dose-50)**: the number of microorganisms that must enter a host to establish an infection in 50% of exposed hosts. As a rule, more virulent microorganisms have lower ID_{50} values than less virulent ones. The value may be as low as 10 microbial cells for *M. tuberculosis* or as high as 100 million for *V. cholerae*. A related value is the **LD_{50} (lethal dose-50)**. The LD_{50} is the number of organisms that must enter a host to kill 50% of the hosts. Again, we expect more virulent organisms to have lower LD_{50} values (**Figure 10.12**).

Once they have entered, pathogens must adhere to the host

Like normal flora, successful pathogens must resist being swept away by host defenses or body fluids. Many pathogens adhere to host surfaces by means of fimbriae or capsules, while others have various proteins on their cell surfaces that can bind to other proteins normally found on host cell membranes (**Figure 10.13**). The specificity of certain pathogens for particular regions in the host's body is often determined by specific binding

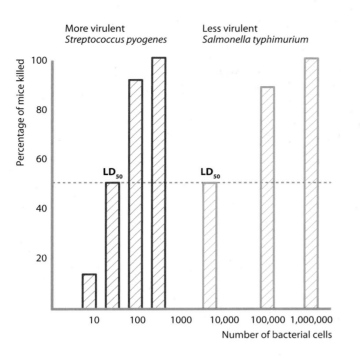

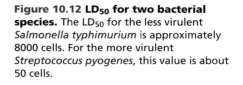

Figure 10.12 LD$_{50}$ for two bacterial species. The LD$_{50}$ for the less virulent *Salmonella typhimurium* is approximately 8000 cells. For the more virulent *Streptococcus pyogenes*, this value is about 50 cells.

between microbial adherence proteins and host proteins present only in certain cell types. For instance, *Neisseria gonorrhoeae* is able to colonize genital areas as well as the throat and the conjunctiva of the eye, because the host proteins recognized and bound by *Neisseria* adherence proteins are found on those host tissues. In Chapter 4 we discussed how viruses attach to cells via their glycoproteins or other surface proteins and how the interaction between these viral proteins and host receptor proteins determines which cell types a specific virus can infect. In the example of the Norwalk virus that infected the football players in our case, these receptors are found on the surface of epithelial cells in the intestine.

Most pathogens must increase in number before they cause disease

Most of the time, the number of organisms to which a host is exposed is too small to cause disease. Before their presence is felt by the host, the pathogens must multiply. The time between the initial exposure and the onset of symptoms is called the **incubation period** (**Figure 10.14**). During the incubation period the number of pathogens is growing, and there is no disease until a given population threshold is reached. Beyond this threshold, the number of pathogens is sufficiently high to cause disease symptoms. If the threshold is not achieved by the pathogen, no illness is seen. Returning to our case, the usual incubation period for Norwalk virus is between 12 and 48 hours. The tainted turkey sandwiches were eaten by the players the day before the game. Over the next 24 hours or so, the players did not feel ill because the number of viral particles infecting them was still below threshold. By game time, the threshold had been reached and the first symptoms appeared.

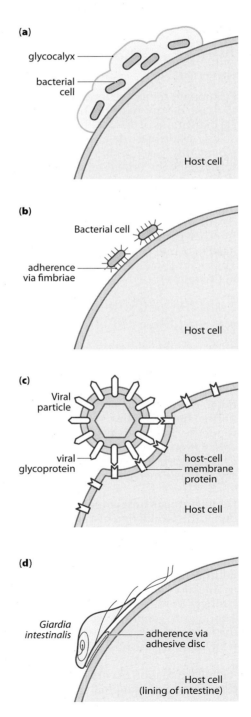

Figure 10.13 Adherence by pathogens. Bacterial pathogens may adhere to host cells in several different ways, including (a) a glycocalyx or (b) fimbriae (see Chapter 3, p. 50, for a complete description of these adherence structures). (c) Viral pathogens often use their envelope glycoproteins (Chapter 4, p. 92) to adhere to membrane proteins on host cells. (d) The protozoan parasite *Giardia intestinalis* uses its ventral adhesive disc to adhere tightly to cells lining the host's intestine.

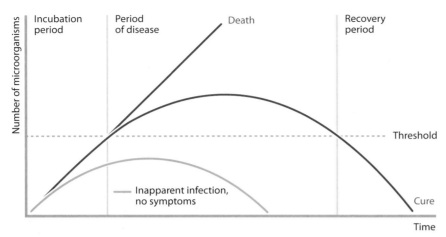

Figure 10.14 Microbial multiplication and disease. Before a patient suffers disease symptoms, a specific threshold of microbial numbers must be reached. The time between infection and the onset of symptoms, a time of microbial multiplication, is the incubation period. As the host fights off the infection, and the number of microorganisms drops below threshold, the patient enters the recovery period. If the host is unable to mount an effective response, microbial numbers may increase until the host dies. If the number of microorganisms never reaches the threshold, the infection will be asymptomatic and inapparent.

Successful pathogens must at least initially evade host defenses

Fortunately for us, the threshold of disease is often *not* reached. Many infections are destroyed by the host's immune system well before there are clinical indications of disease. To cause disease, therefore, the pathogen must be able to evade these defenses, at least until its population rises above threshold. In most cases, even when disease is evident, the immune system eventually gains the upper hand, pathogen numbers start to fall, and we enter the recovery phase, ultimately ending in the complete destruction of the pathogen. At this point the patient is cured (see Figure 10.14). In some cases the immune response is not vigorous or specific enough to completely destroy the pathogen, and the patient may undergo a long, **chronic** illness. In other cases, an overly vigorous or inappropriate immune response is itself to blame for much of the tissue damage observed during an infection. The topic of how hosts defend themselves against pathogens will be explored in Chapter 11.

Bacterial pathogens cause disease in several different ways

So far in our exploration of the disease process, we have discussed several things that a pathogen must do for disease to occur. But by and large, the microorganisms of our normal flora do these same things. What, then, is the fundamental difference between a pathogen and a member of our normal flora? The difference is that pathogens have the capacity to interfere with normal host activities. You will recall that this interference is actually how we define infectious disease. Exactly *how* a bacterial pathogen interferes varies from species to species. In general, however, bacterial capacity to cause disease, known as bacterial **pathogenesis**, can be categorized among several basic mechanisms.

Damage due to host response. The primary symptom in a gonorrheal infection is the production of a large amount of pus in the urethra. Pus is a semi-solid accumulation of dead white blood cells, the material they have engulfed by phagocytosis, and tissue debris. It is a natural consequence, along with swelling, pain, and redness, of inflammation. Inflammation is a natural if sometimes unpleasant component of a normal immune response to infection. Inflammation, however, when it occurs inappropriately, too strongly, or for prolonged periods, can cause damage to host tissues itself. Although *N. gonorrhoeae*, the causative agent of gonorrhea, has stimulated the inflammatory response, this microorganism is not directly responsible for symptoms: the host response is.

A similar situation occurs with *Streptococcus pneumoniae*, which, although it can cause pneumonia, exerts few direct negative effects on host cells. When *S. pneumoniae* infects the lungs, phagocytic cells attempt to engulf and destroy it. This protective response is often ineffective, however, because *S. pneumoniae* is protected by a capsule. The capsule does not stop the phagocytic cells from continuing to try, and as more and more phagocytic cells arrive in the lungs, more and more dead phagocytic cells begin to accumulate. This accumulation of dead cells in the lungs interferes with breathing and causes illness. The symptoms of tuberculosis (**Figure 10.15**) and food poisoning from *Salmonella* are likewise, in large part, a consequence of host response as opposed to a direct effect of the pathogen. In fact, the very term "food poisoning" is a something of a misnomer. Poisons (or toxins, to be discussed shortly) are only part of the reason for the illness caused by *Salmonella*.

CASE: TROUBLE IN PARADISE

Howard and Heidi take a trip to the Galapagos Islands to view the exotic wildlife unique to the archipelago. During the vacation, Howard, who has trouble with acid reflux (the release of stomach acid into the esophagus) and is especially vulnerable to a "nervous stomach" while traveling, regularly takes antacid tablets to relieve his discomfort. Heidi takes no such medication. Following their visit to the islands, they spend a few days in the Ecuadorian coastal resort of Salinas. On their last day they dine at an outdoor café that features ceviche, a marinated raw seafood dish. Although they were warned about raw seafood, they are unable to resist the local delicacy. They fly home the next day, and within 24 hours, Howard is suffering from severe watery diarrhea, muscle cramps, and dizziness. Alarmed, Heidi takes him to the emergency room, where the attending physician finds that Howard also has a weak and rapid pulse and very low blood pressure. When Heidi tells the doctor that they just returned from South America, the doctor suspects cholera. Howard is immediately admitted to the hospital, given intravenous fluids, and started on antibiotics. A stool sample is tested and the diagnosis of cholera is confirmed when large numbers of *Vibrio cholerae* are found. Once treatment begins, Howard's diarrhea decreases, and after 2 days he is released from the hospital.

1. **How exactly does *V. cholerae* cause the symptoms observed in Howard?**
2. **Why didn't Heidi get sick?**

Although poisons may play no part, or only a limited part, in some diseases, for plenty of bacteria, poisons, commonly referred to as **toxins**, are the principal factor responsible for pathogenicity. All toxins act by altering the normal metabolism of host cells in some way, and because they contribute to disease, they are important examples of virulence factors.

There are two basic types of toxins: **exotoxins** and **endotoxins**. Exotoxins are bacterial proteins that are usually secreted into the surrounding environment. They vary considerably in specificity—some act only on certain cell types, while others affect a wide range of tissues or cells. A specific bacterial pathogen may produce one or more exotoxins. Some, such as *S. pneumoniae* discussed above, produce none. Endotoxins consist of components found only in the outer membrane of Gram-negative bacteria. As we will see, they act as important toxins only under certain circumstances.

Exotoxins. The *V. cholerae* contracted by Howard is a Gram-negative bacterial species, transmitted through contaminated food or water. Because ceviche is served raw, and because *V. cholerae* uses seawater as a reservoir, the seafood was the likely source of the infection. Once arriving in the intestine, the bacterium produces a powerful exotoxin that results in Howard's symptoms. Fortunately for him, he received treatment rapidly. Untreated cholera is often lethal as the explosive diarrhea can cause severe dehydration (**Figure 10.16**). Some victims lose up to 25% of their body weight, simply in fluid, within 24 hours.

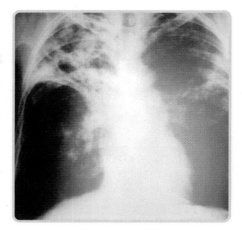

Figure 10.15 Tuberculosis. In this chest X ray, dark areas in the lungs are unaffected areas. White areas reveal damage, primarily the result of the host response to infection with *M. tuberculosis*. The inflammation that occurs in response to infection, while limiting the spread of the bacteria, forms fibrous or calcified lesions. These "tubercles" persist throughout life and are seen as nodules on X rays.

Figure 10.16 A cholera bed. Such beds are sometimes seen in hospitals in countries where cholera is present. The hole allows the patient to remain in bed while experiencing the explosive diarrhea that accompanies a cholera infection.

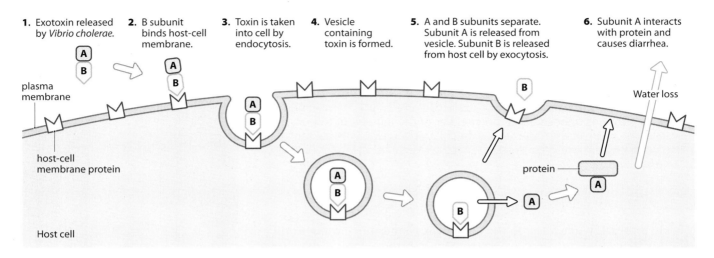

1. Exotoxin released by *Vibrio cholerae*.

2. B subunit binds host-cell membrane.

3. Toxin is taken into cell by endocytosis.

4. Vesicle containing toxin is formed.

5. A and B subunits separate. Subunit A is released from vesicle. Subunit B is released from host cell by exocytosis.

6. Subunit A interacts with protein and causes diarrhea.

plasma membrane

host-cell membrane protein

Host cell

Water loss

protein

Like many exotoxins, cholera toxin consists of an **A subunit** and a **B subunit** (**Figure 10.17**). The B (binding) subunit is what gives cholera toxin its specificity for cells lining the intestine. Like a lock and key, the B subunit protein binds to a specific receptor found only on these host cells. Next, the host cell brings the toxin molecule into the cell by endocytosis. Once inside the host cell, the A and B subunits separate. The A (active) subunit now exerts its toxic effect. Specifically it acts as an enzyme and alters a specific host-cell protein called a G protein. The now enzymatically altered G protein then starts a series of events that causes cells lining the intestine to leak water. It is this loss of water by intestinal cells that explains the massive diarrhea and fluid loss seen in cholera victims like Howard. Note that the bacteria themselves never enter the host cells. Only the secreted toxin does.

The other symptoms that Howard endured were also the result of this loss of water from intestinal cells. As he lost fluid, Howard's blood volume decreased, causing the low blood pressure and weak pulse. *V. cholerae* infections generally respond well to a combination of rehydration and antibiotics if they are treated quickly.

Why didn't Heidi get sick as well? Recall that *V. cholerae* has a relatively high ID$_{50}$. Typically, consumption of 100 million or more bacterial cells is required to ensure an infection in half the people that consume them, since the vast majority of bacteria are destroyed by stomach acid and never reach the intestine. Remember that Howard was taking antacid tablets to combat his nervous stomach. The pH of his stomach was higher (less acidic) than normal, making it easier for the bacteria to survive and reach his intestine. Heidi's stomach remained acidic enough to prevent infection.

Other types of exotoxins have different B subunits, and subsequently bind to different target cells; they also have different A subunits, which means they will interfere with metabolism in a different way, once they enter the target cell. Diphtheria, for instance, is caused by the Gram- positive bacterium *Corynebacterium diphtheriae*. The disease usually starts with sore throat, fever, and fatigue, which if untreated can eventually lead to heart and kidney failure. The B subunit on the diphtheria toxin binds to specific receptors on the membrane of target cells. The specificity of the B subunit explains why certain cells, like those in the heart and kidney, are affected while others are not. Once inside the cell, the A subunit binds and inactivates a protein involved in moving ribosomes along mRNA during translation. In effect, the toxin blocks protein synthesis.

Diphtheria, fortunately, is rare in Europe and North America because of a highly effective vaccine routinely administered to children. However, the breakup of the former Soviet Union in the early 1990s reminds us what can

Figure 10.17 Mechanism of cholera toxin. The B subunit binds specifically to receptors on host cells lining the intestine. Following binding, the toxin is taken into the cell by endocytosis. The A and B subunits separate and the A subunit interacts with cell proteins, initiating a series of events that ends with the host-cell membrane becoming leaky. The water that flows out of intestinal cells and into the intestine accounts for the diarrhea that occurs during a bout of cholera. Other exotoxins have different B subunits, explaining the specificity that exotoxins usually have for certain host-cell types. The activity of the A subunit also is different in different exotoxins.

happen when public heath is neglected. Due to political upheaval, vaccine programs broke down and an epidemic of diphtheria resulted. By the end of the outbreak, 125,000 cases and 4000 deaths were reported from the one-time Soviet republics.

Other well-understood exotoxins include those produced by *Clostridium tetani* and *Clostridium botulinum*, which cause tetanus and botulism, respectively. Both of these bacteria produce toxins with B subunits that bind to motor neurons—nerve cells that carry impulses to muscles. The effects of these two toxins are in some ways opposite. Tetanus toxin prevents the neuron from returning to its relaxed state. Affected neurons continue to send stimulatory signals to muscles, which consequently remain contracted. The result is rigid paralysis or "lockjaw." Botulism toxin, on the other hand, blocks the activity of motor neurons. Because the neurons cannot conduct impulses, associated muscles cannot be stimulated, causing the flaccid paralysis seen in botulism.

Botulism toxin is perhaps the most potent of all biological poisons—a few milligrams would be enough to kill thousands of people. Yet recently botulism toxin has found a more benign niche—as a cosmetic or beauty aid or to treat certain medical conditions. Used under the name BOTOX® Cosmetic, minute amounts of this toxin are injected directly into wrinkles (**Figure 10.18**). The muscles causing the undesired facial lines are unable to contract, and as they relax, the wrinkles disappear. BOTOX® Cosmetic also can be used to treat conditions such as vocal cord paralysis, in which muscles surrounding the larynx contract involuntarily, making speech difficult. Whether used cosmetically or to treat a medical condition, the effect of the BOTOX® Cosmetic is temporary and treatments need to be repeated every 4–6 months.

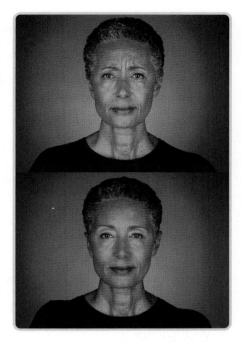

Figure 10.18 Cosmetic use of botulism toxin. A person photographed (top) before and (bottom) after treatment with the BOTOX® Cosmetic.

CASE: A RUPTURED APPENDIX

When Robyn felt severe pain in her lower abdomen, her husband Gary became concerned and took her to the hospital. The doctor who performed the initial examination found her abdomen to be extremely hard and tender. An elevated temperature and low blood pressure suggested impending shock. On the basis of these symptoms, the doctor diagnosed Robyn's condition as a ruptured appendix and immediately began supportive treatment, consisting of IV fluids and large doses of antibiotics. Ten hours later, once her condition had stabilized, Robyn was taken to surgery, where her ruptured appendix was removed. Robyn made a slow but uneventful recovery.

1. What type of bacterial infection is a consequence of a ruptured appendix?
2. How do the involved microorganisms cause Robyn's symptoms?
3. In infections of this type, symptoms often temporarily worsen once antibiotic treatment has begun. Why is this?

Endotoxin. In Chapter 3 (see p. 55), we learned that Gram-negative bacteria have an outer membrane composed in part of lipopolysaccharide (LPS). When Gram-negative bacteria die, their outer membrane degrades and LPS is released. In its released form, LPS acts as an **endotoxin**. Unlike an exotoxin, where each toxic protein has specific and unique effects, all LPS impacts the host in the same way, regardless of the bacterial species that releases it.

The mechanism of endotoxin activity is illustrated in **Figure 10.19**. Certain phagocytic cells such as macrophages have receptors for lipopolysaccharide on their surface. When LPS, released when Gram-negative bacteria die, binds these receptors, a signal is sent, which causes certain genes in the macrophage's DNA to be transcribed and translated. The result is the production of specific proteins. These proteins are released into the blood or

Figure 10.19 Mechanism of endotoxin action. (a) When Gram-negative bacteria die, they release LPS (in red). Certain phagocytic cells called macrophages have receptors for LPS on their cell surface. (b) When LPS binds its receptor, a signal is sent that causes certain genes in the DNA to be transcribed and translated into proteins. (c) These proteins (blue circles) are ultimately released from the cell. Once released, they can travel through the blood, causing the symptoms associated with endotoxin.

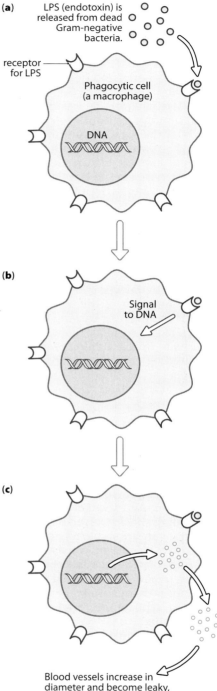

(a) LPS (endotoxin) is released from dead Gram-negative bacteria.

receptor for LPS

Phagocytic cell (a macrophage)

DNA

(b) Signal to DNA

(c)

Blood vessels increase in diameter and become leaky, causing drop in blood pressure. Body temperature rises (fever).

tissue fluid, where they cause blood vessels to increase in diameter and become leaky. These effects can cause blood pressure to drop, resulting in weakness and malaise. Other proteins released by macrophages act to induce a rise in body temperature. In other words, they induce fever. The macrophage receptor for LPS is an example of what are called "Toll-like receptors," an important group of receptors that we are just beginning to understand. We will explore their role in the host–pathogen interaction more fully in Chapter 11.

The symptoms of endotoxin are not too serious when LPS is released by dead bacteria in small amounts. As we have mentioned previously, however, the large intestine is inhabited by trillions of bacteria, most of them Gram-negative normal flora. When the appendix ruptures, these bacteria stream into the normally sterile peritoneal cavity. This opportunistic infection is called peritonitis. As these organisms die, they release endotoxin in large amounts—large enough to cause a severe drop in blood pressure and to cause the additional symptoms experienced by Robyn. If blood pressure falls too far, shock is a real possibility. In such a situation, it is crucial to treat the bacterial infection rapidly with large doses of antibiotics, but the antibiotics themselves often initially make symptoms even worse. As the antibiotics kill large numbers of bacteria, levels of endotoxin in the peritoneal cavity can increase rapidly. During such treatment, therefore, supportive care is especially crucial, and as soon as the patient is stabilized, surgery is required to seal the rupture and prevent further bacterial invasion.

Tissue-damaging enzymes. Some bacteria produce enzymes with pathogenic properties. For example, in Chapter 2 we mentioned streptokinase, an enzyme produced by certain *Streptococcus* species. Streptokinase degrades the protein fibrin, which is common in host connective tissue. By degrading fibrin, the bacteria are better able to spread through the connective tissue, reaching new sites in the host. Some *Streptococcus* species, including *S. pyogenes*, the causative agent of strep throat, produce hemolysins, which lyse red blood cells and release hemoglobin (**Figure 10.20**). The bacteria thus gain access to the iron in the hemoglobin, which they require as a crucial nutrient. In Chapter 8 (p. 217) we described the action of coagulase, produced by some *Staphylococcus* bacteria. Coagulase is in some ways the opposite of streptokinase; instead of degrading fibrin, it causes fibrin to form from a precursor protein called fibrinogen. Fibrin plays a role in clot formation, and the clotting that forms around the *Staphylococcus* bacteria protects them from host immune response.

Other tissue-damaging enzymes produced by different pathogens include hyaluronidase, which degrades the carbohydrate hyaluronic acid that binds many host cells together; and collagenase, which digests collagen, an important structural protein in connective tissue. DNase, as the name implies, degrades host DNA, making the nucleotides available to the bacteria. The infamous "flesh-eating bacteria," which first appeared in the 1990s, are usually a strain of *S. pyogenes* that comes equipped with an impressive arsenal of tissue-degrading enzymes, including streptokinase, hyaluronidase, collagenase, and DNase. An infection with this strain may start with a tiny cut or puncture wound, which quickly spreads over a large area, "eating" flesh in the process (Figure 10.20b). The less colorful, more technical name for this condition is *necrotizing fasciitis*.

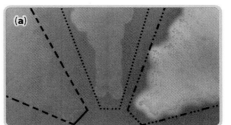

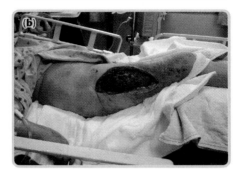

Figure 10.20 Damage caused by bacterial enzymes. (a) Blood agar medium contains sheep red blood cells. When *Enterococcus faecalis* (left) is inoculated onto a blood agar plate, no hemolysis of red blood cells is seen, because this species does not produce the necessary hemolysins. *Streptococcus mitis* (center) causes partial hemolysis, while *S. pyogenes* (right) causes complete hemolysis of red blood cells. The pale areas indicate where red blood cells have been destroyed. (b) Necrotizing fasciitis caused by "flesh-eating bacteria." The term "necrotizing" comes from the Greek *necro,* meaning "dead." Fascia is the connective tissue that binds together internal organs or parts of the body. "Fasciitis" refers to an inflammation of the fascia.

Viruses cause disease by interfering with the normal activities of the of the cells they infect

Viruses must penetrate an appropriate host cell if they are to replicate. Once inside the host cell, most viruses have specific **cytopathic effects (CPE)** that interfere with normal host function. A cytopathic effect is the specific impact that a virus has on a host cell, which may result in host-cell damage or death. Depending on the specific virus, the CPE may or may not actually kill the host cell. Many viruses exert lethal effects when they divert host cell resources and synthetic processes to viral replication. The virus effectively commandeers the host cell, utilizing the cell's nucleic acid and/or protein synthesis machinery for its own replication.

In some viral infections, the host cell is destroyed when cellular lysosomes release their digestive enzymes, and in many others the plasma membrane becomes leaky due to the effect of viral proteins. In other cases, viral proteins found in association with the host-cell plasma membrane target the cell for destruction by the host immune system. In some viral infections, large accumulations of viral proteins and nucleic acids build up in the cytoplasm of an infected cell. These **inclusion bodies** are essentially concentrations of viral "garbage," which will never be assembled into new viral particles, and they may simply get in the way of normal cell processes. Some specific viral infections can be diagnosed from the presence of inclusion bodies. Rabies is a notable example (**Figure 10.21**).

It is now well understood that some forms of cancer are caused by viral infection. Cervical cancer, caused by the human papillomavirus, and liver cancer, caused by the hepatitis B virus, are two examples. Cancer is basically a situation in which cells divide in an uncontrolled manner. Some viruses carry genes that, when inserted into the host DNA, disrupt the normal regulation of cell division. Alternatively, host cells have their own genes that tightly regulate cell division; if these genes are disabled by a virus, the cell may begin unregulated division, ultimately leading to tumor formation. Cancer-causing genes, whether of host or viral origin, are called **oncogenes**.

The point at which the CPE of a given virus becomes noticeable varies considerably (**Figure 10.22**). As we learned in Chapter 4, viral infections can be acute, chronic, or latent. In acute infections, a large number of host cells are rapidly destroyed as the virus undergoes repeated rounds of replication. Symptoms may be severe, but they are generally of relatively short duration. Once the immune system reins in viral reproduction, the infection is destroyed and damaged host cells are replaced by new cells. Influenza virus and the Norwalk virus, discussed earlier in this chapter, cause acute infections of this sort. Some viruses, such as hepatitis B and hepatitis C, replicate slowly over long periods of time, resulting in chronic infections. Other viruses can enter latent stages where all viral reproduction is halted. The period of latency may last years until factors such as a weakened immune system, other infections, or stress allow the virus to begin replicating once again. The varicella zoster virus provides a useful example. When it infects a

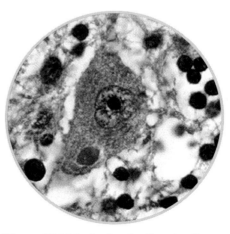

Figure 10.21 Inclusion bodies. A cell from the human brain, taken from a patient who died of rabies. The dark-staining structures in the cell cytoplasm are inclusion bodies. The presence of these inclusion bodies, called Negri bodies, is used to confirm a case of rabies.

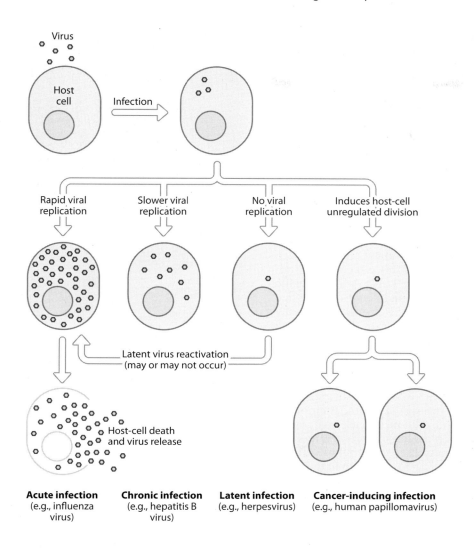

Acute infection (e.g., influenza virus)

Chronic infection (e.g., hepatitis B virus)

Latent infection (e.g., herpesvirus)

Cancer-inducing infection (e.g., human papillomavirus)

Figure 10.22 Some possible outcomes of a viral infection. Depending on the type of virus and the type of cell infected, there can be several possible outcomes when a virus infects a cell. In acute viral infections, the virus replicates rapidly and may kill the cell. Slower viral replication may result in a chronic infection. In some cases, a virus may not replicate at all yet will remain viable within the host cell. Such an infection is called a latent infection. Latency may or may not eventually result in a new acute infection. Some viruses, able to disrupt the normal control of cell division, can cause cancer.

child, it causes the disease typically known as chickenpox. As the infection is brought under control by the host immune system, the virus retreats to nerve cells, where it can remain latent for years. Later in life it may reemerge as shingles.

Different eukaryotic parasites affect their hosts in diverse ways

Protozoa. Protozoan parasites are a varied lot, and the means by which they cause disease are equally variable. The intestinal parasite *Giardia intestinalis* so thoroughly covers the surface of the intestinal wall that it prevents absorption of nutrients and the reabsorption of water, causing diarrhea (see Figure 10.13d). The *Plasmodium* parasites that cause malaria destroy red blood cells and cells in the liver, by entering them and replicating intracellularly (see Figure 4.13d). The trypanosomes that cause sleeping sickness (see Figure 3.21b) are able to repeatedly change the proteins on their cell surface. A different host immune response is generated against each of these changing proteins. This immune response soon becomes of limited value in controlling the parasite, but massive immune response itself can damage host tissues, contributing to the disease. One type of protozoan parasite that you are likely to have heard of is the amoeba. *Entamoeba histolytica*, in particular, is a souvenir of foreign travel that you do not want to pick up (see Figure 4.13b). The parasite is transmitted via the food- or

Figure 10.23 The Salem witch trials. In this painting a young woman is accused of practicing witchcraft. More than 200 individuals were similarly accused in 1692 and 1693, 20 of whom were executed. Some have suggested that witch hysteria was due to ergot poisoning in local residents.

waterborne route and is therefore a risk wherever sanitation is poor. After attaching to cells in the intestine, *E. histolytica* releases proteins that form pores in the target cell membrane. It also releases enzymes that destroy host cells, and in severe cases it may actually digest its way through the intestinal wall. When this occurs, amoebas may spread to other organs of the body, such as the liver and lungs, where they continue to cause tissue damage.

Fungi. Recall from our discussion of basic fungal biology in Chapter 4 that fungi digest nutrients extracellularly and then absorb the digested material. This property forms the basis of much fungal pathology. Parasitic fungi often exert their harmful effects by releasing digestive enzymes that degrade host tissue. As these tissues are digested, the fungi absorb required nutrients and subsequently invade adjacent tissue.

Some fungi also either produce toxins or provoke an allergic response in the host. As an example, *Claviceps purpurea* is a parasite of rye. If infected grain is used to produce bread, anyone consuming the bread may also ingest a fungal toxin called **ergot**. Ergot causes blood vessels to constrict, and if circulation to fingers or toes is cut off, gangrene may result. Ergot also causes mental disturbances including hallucinations. Before the advent of modern milling techniques, a contaminated batch of rye would occasionally cause mass hysteria among those consuming it. Historians even suspect that the Salem witch trials, which occurred in Salem, Massachusetts, in 1692, were instigated by mass hysteria induced by ergot poisoning (**Figure 10.23**).

Pathogens leave the host through a portal of exit

Although it is not necessary for a pathogen to leave its host in order to cause disease, it certainly is necessary for the long-term survival of the pathogenic species. Pathogens leave their host through a **portal of exit**.

For pathogens of the intestinal tract, the usual portal of exit is the anus, although the case involving Norwalk virus demonstrates that at least occasionally, alternative portals of exit are possible for an intestinal pathogen. For airborne pathogens, sexually transmitted pathogens, and those pathogens relying on vectors, the portal of exit is generally the same as the portal of entry. Airborne pathogens, after multiplying in the respiratory system, exit the respiratory system in droplets when the host coughs or sneezes. Sexually transmitted pathogens use genital contact for both entry and exit, and vectored pathogens both enter and leave the blood through the bite of an insect or other arthropod.

In some cases, no portal of exit exists, and the host can be considered a "dead end" from the pathogen's point of view. This is often the case in zoonotic infections, where humans are not the normal or appropriate host. Recall our case involving cat scratch disease. The bacteria in question typically infects cats, but if humans are inadvertently infected, disease can result. Humans, however, provide no portal of exit. Some pathogens that utilize environmental reservoirs also only incidentally infect humans. When they do, although they may cause illness, they are unable to maintain a cycle of transmission because they cannot leave the host. Examples include *C. tetani*, the causative agent of tetanus, and *C. immitis*, which causes Valley fever, both discussed earlier in this chapter. As far as these pathogens are concerned, infecting a human is a mistake—they will never get out alive.

Symptoms of disease often assist the pathogen in its transmission to a new host

CASE: WHO'S HURTING WHOM?

Dr. Adema is an infectious disease specialist, interested in why certain symptoms are observed in connection with specific diseases. In particular, he wants to know if the massive diarrhea associated with two diseases, cholera and shigellosis, is actually a host response, to rid the body of the pathogen, or is due to damage induced by the pathogen, in order to exit the host and find a new one. He assembles four groups of experimental animals. He treats the groups in the following manner:

Group 1: Infected with *V. cholerae* (the bacteria that cause cholera); following infection, the animals are given an antidiarrhea medication.

Group 2: Infected with *V. cholerae*; animals receive no medication.

Group 3: Infected with *Shigella dysenteriae* (the bacteria that cause shigellosis); following infection, the animals are given an antidiarrhea medication.

Group 4: Infected with *S. dysenteriae*; animals receive no medication.

Dr. Adema collects fecal samples from each of his four groups, every day, until the samples no longer contain any pathogen cells. His results are as follows:

Group 1: no *V. cholerae* in feces after 4 days

Group 2: no *V. cholerae* in feces after 11 days

Group 3: no *S. dysenteriae* in feces after 9 days

Group 4: no *S. dysenteriae* in feces after 5 days

1. **What do Dr. Adema's results suggest about the underlying reason for the symptoms seen in cholera and shigellosis?**

Remember that in this chapter we are considering the disease process from the pathogen's perspective. In the next chapter, we will concentrate on how hosts respond to invasion by pathogens, but before we change our point of view, we ask what may sound like an odd question—why do pathogens cause disease? In other words, what do *they* get out of making us ill?

For those pathogens that are poorly adapted to the human host, the answer is often "not much." Remember that many pathogens infect humans only accidentally, and disease is an unfortunate consequence of this host–pathogen interaction that benefits neither party. Some pathogens, on the other hand, are better adapted to humans. Examples include the measles virus, the human influenza virus, or the two bacteria used in Dr. Adema's experiment. Sometimes, symptoms caused by such pathogens represent the host's attempt to eliminate the disease-causing microbes from the body. Diarrhea, for example, is sometimes described as an efficient, if not particularly pleasant, way to get intestinal pathogens out as quickly as possible. When we cough during a cold, many might argue that coughing is the host's attempt to eliminate the pathogens from the body.

This is not, however, necessarily always true. In some cases, symptoms are induced by the pathogen to facilitate its transmission to a new host. A respiratory virus that causes sneezing and coughing is more likely to exit its host and reach a new host than a virus that causes no such symptoms. An intestinal pathogen that makes its host lose large amounts of fluid as diarrhea has a similar edge over more benign cousins. Pathogens that rely on biological vectors might gain a real advantage if they make their host as sick and lethargic as possible; when an appropriate mosquito or other vector lands to take a blood meal, the host is in no condition to swat it away (**Figure 10.24**).

Figure 10.24 Sleeping sickness.
A young woman watches over her comatose husband, who is suffering from the disease. Is the extreme lethargy induced by the parasite a strategy to make it easier for its tsetse fly vector to take a blood meal?

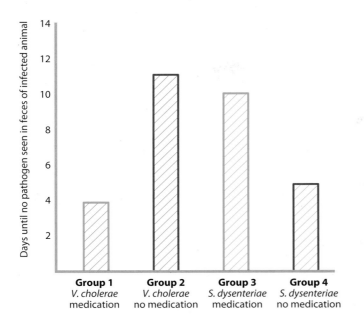

Figure 10.25 Cholera and shigellosis: the reason for symptoms. In *V. cholerae* infections (groups 1 and 2), the bacteria were able to remain in the host longer when hosts did not have their diarrhea controlled with medication. This indicates that continued diarrhea is beneficial for the pathogen and bad for the host, suggesting that it is the pathogen that promotes the diarrhea. In *S. dysenteriae* infections (groups 3 and 4), the opposite results are observed. The bacteria did better in hosts receiving antidiarrhea medication, indicating that diarrhea is a host response designed to rid the body of the pathogenic bacteria.

Dr. Adema's data indicate that things work differently for the two bacteria in question, even though they cause somewhat similar symptoms (**Figure 10.25**). For *Vibrio*, the group taking the antidiarrhea drug (group 1) cleared its infection sooner than the group that received no treatment (group 2). This suggests that stopping diarrhea is beneficial as far as the host is concerned, but harmful from the pathogen's perspective, because it is less able to survive. These data suggest that diarrhea in the case of cholera occurs *in spite* of what is best for the host—the symptom is induced by the microorganism, to facilitate its transmission.

For *Shigella*, the reverse is true; animals receiving antidiarrhea treatment (group 3) actually took longer to rid themselves of the infection, compared with group 4 animals that remained untreated. This suggests that in shigellosis diarrhea is bad for the bacteria because their survival time in the host is reduced. They do better when the diarrhea is stopped. The diarrhea serves a useful purpose from the host's perspective because it more quickly eliminates the infection. In the case of shigellosis, diarrhea might therefore be seen as a host adaptation to help combat infection.

Coming Up Next...

The likelihood of disease in any host–parasite interaction depends on both the effect of the parasite on the host and the response of the host. In this chapter, we have examined the first part of this equation—how *they* affect *us*. In Chapter 11, we will complete this story by considering how *we* respond to *them*. Only by understanding both sides of this dynamic interaction does a complete picture of disease emerge. As we will see in Chapter 11, initially the host attempts to block infections in the first place. Intact skin, stomach acid, and thick layers of mucus in the respiratory and digestive systems, just to cite a few examples, prevent most infectious agents from ever gaining entry to our bodies. But all too often these defenses are breached, and when they are, the immune system swings into action to destroy the invaders. In Chapter 11, we will investigate both the initial barriers to infection and the nature of the immune response, placing us in a position to better understand the answer to the question we posed early in Chapter 10: why do we sometimes get sick, and when we do, why do we usually (but not always) recover?

Key Terms

A subunit	exotoxin	oncogene
aerosol (airborne) transmission	fomite	opportunistic infection
animal reservoir	foodborne transmission	pathogen
B subunit	human reservoir	pathogenesis
biological vector transmission	ID_{50} (infectious dose-50)	portal of entry
chronic illness	inclusion body	portal of exit
contamination	incubation period	toxin
cytopathic effect (CPE)	indirect contact transmission	transient flora
direct contact transmission	infection	vector transmission
disease	LD_{50} (lethal dose-50)	vertical transmission
endotoxin	mechanical vector transmission	waterborne transmission
environmental reservoir	mode of transmission	zoonosis
ergot	normal flora	

Concept Questions

1. You've just been hired by the World Health Organization. Your job is to develop a list of diseases that might realistically be targeted for global eradication. You must also develop a list of diseases for which eradication is most unlikely. What might be the characteristics of disease agents on your two lists?

2. Bubonic plague is caused by the Gram-negative bacterium *Yersinia pestis*. These bacteria can persist in certain rodent populations for long periods of time without causing severe disease. Animals, including humans, generally become infected when a flea feeds first on an infected individual and later on a susceptible individual. Occasionally, if the bacteria spread to the lungs, the patient develops "pneumonic plague," which is spread via airborne transmission. Although cats can develop plague similar to that of humans, dogs are generally resistant to infection.

With the above information, what advice would you give somebody who lives in an area where plague is known to occur as to how to best protect themselves and their pets?

3. Patients admitted to hospitals often acquire new infections during their hospital stays. Careful hand washing by hospital personnel is one of the easiest and most important ways to reduce the numbers of such infections. What do you think the link between hand washing and reduced numbers of hospital-acquired infections is?

4. Imagine two bacterial species. Species A causes disease by virtue of the endotoxin it releases. Species B produces no endotoxin but produces a disease-causing exotoxin. If a doctor has two patients, one infected with species A and one with species B, for which patient will it be easier to diagnose the exact species causing illness, based solely on the symptoms of the patient?

5. Those pathogens that are most virulent often have the most efficient means of transmission. Pathogens with less efficient modes of transmission are often less virulent. What do you think explains this relationship?

6. In Chapter 7 we learned that *Corynebacterium diptheriae* produces a powerful toxin when the environment is low in iron. Would the LD_{50} of this species be higher or lower when iron is abundant in the environment?

7. You are told that the ID_{50} for a particular bacterial species is 1000 cells. Do you think the LD_{50} is greater or less than 1000? Explain.

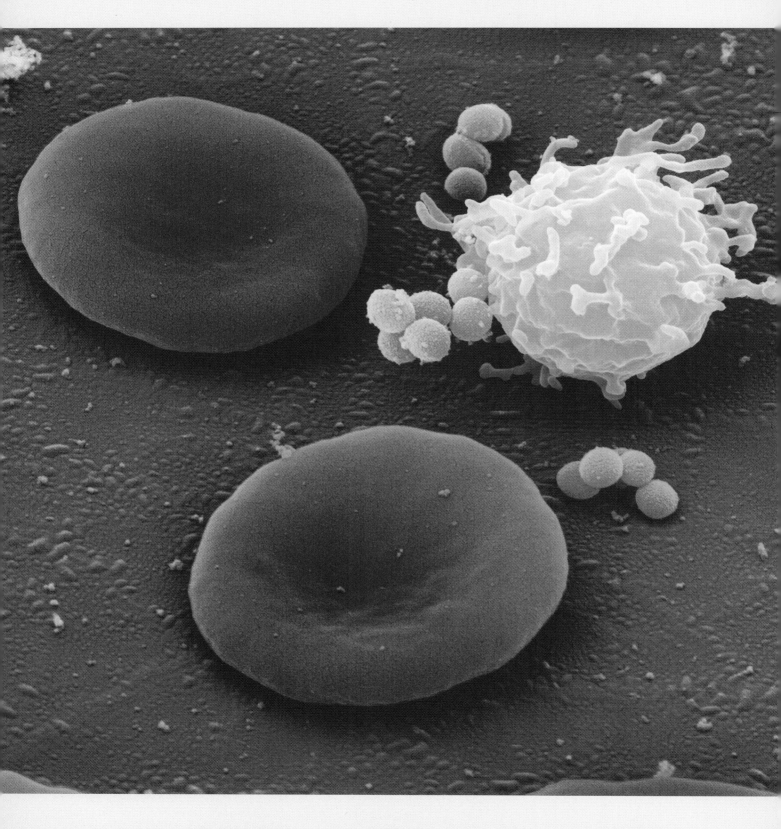

Lymphocytes (white) circulate in the blood along with red blood cells (red) while on patrol for pathogens, such as the *Staphylococcus* bacteria (yellow).

Chapter 11

Host Defense

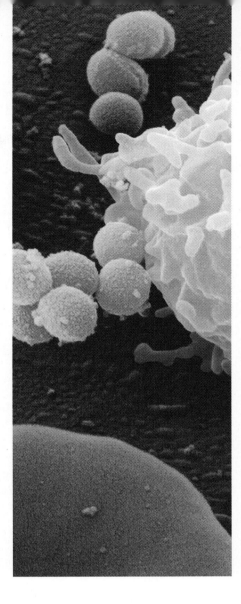

It's a dangerous world out there. Almost anything you touch is covered with microorganisms. Every breath you take and every bite of food you eat is laden with microbes, some of which are potential pathogens. If you dwell excessively on the unseen microbial world seething around you, it can seem as if death and disease are lurking around every corner.

Nevertheless, even though our exposure to potentially deadly microbes is relentless and never-ending, it should be comforting to remember that usually most of us feel fine. Infectious disease is the exception rather than the rule, and often our encounters with pathogens are not much more serious than an occasional cold or bout of diarrhea. Our ability to fend off most pathogens is based on the fact that we do not go through life unprotected. Indeed, animals come equipped with a formidable arsenal of defensive weapons, constantly at the ready to repel microbial invaders.

Infectious disease, of course, can still threaten our well-being. In Chapter 10, as we investigated the host–pathogen interaction from the pathogen's point of view, we began to understand why. However, in any host–pathogen encounter, the likelihood of disease depends on more than just what pathogens do to us. Disease is a two-way street, and how we defend ourselves against microbial attack is equally important (**Figure 11.1**). That defense is what we will now investigate. The scientific study of these defensive mechanisms is called **immunology**.

In other words, having considered in the previous chapter how pathogens can impact us, we will now examine the disease process from *our* perspective—how *we* respond to *them*. After we have done so, we will have the complete answer to the questions posed at the beginning of Chapter 10. Why, when we are exposed to pathogens, do we sometimes *but not always* get sick? And why, if we do get sick, do we usually *but not always* recover?

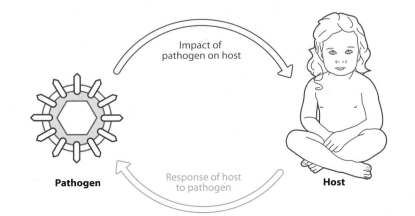

Figure 11.1 The likelihood of disease. When a host encounters a pathogen, there is a possibility of disease in the host. Actual disease depends on both how the pathogen impacts the host (our focus in Chapter 10) and how the host responds to the pathogen (our focus in this chapter). The host's response to potential pathogens is called the immune response, and the study of this response is called *immunology*.

When infection cannot be blocked, first innate immunity and then adaptive immunity is activated

Before we get too far, we need to introduce a few basic concepts. Host defenses, in their simplest form, take a sort of "plan A–plan B" approach (**Figure 11.2**). Plan A is to block invading organisms from infecting us in the first place. As we learned in Chapter 10, infection is one of the crucial steps on the path to disease. To prevent infection, animals are equipped with several types of barriers to entry, which stop most potential invaders before they can colonize us. An example of such a barrier is the skin.

If microorganisms circumvent initial barriers and successfully infect the host, the host switches to plan B—destroy the invading microbes. This immune response to microbial invasion consists of two components. First, elements of our **innate immune system** engage the invader (see Figure 11.2). Innate immunity refers to those components of our defensive repertoire that are inborn. Always in place and always at the ready, they represent our first line of defense. They are not directed specifically against any particular pathogen, and they are not enhanced in response to prior exposure. The activation of phagocytic cells, which engulf and destroy invading pathogens in large numbers, is an example of an innate immune process.

If infection is not rapidly contained, certain elements of innate defense alert and activate the **adaptive immune response**. The adaptive response is a highly directed immune counterattack to specific pathogens. When activated, a variety of cells and molecules are generated in response to infection, and the adaptive immune system has the ability to recognize and eliminate an essentially limitless variety of foreign invaders. The production of antimicrobial proteins called *antibodies*, which bind to and help destroy specific microorganisms, is an example of an adaptive immune response.

Most microorganisms are cleared by innate processes before adaptive immunity is even activated. But if the innate response cannot contain an infection on its own, it activates the adaptive response (see Figure 11.2). Unlike the innate immune response, the adaptive immune response to a particular organism grows stronger with each exposure. This key feature of the adaptive immune response is called **immunological memory**. It is unique to adaptive immunity and is one of its most powerful features.

CASE: INFLUENZA: EXPOSED!

Jeff is an American businessman living in Hong Kong. His work requires frequent travel to and from the United States. In early March, Jeff makes plans to fly to Los Angeles for an important meeting. His flight is scheduled for a midday departure, and that morning he awakens with the vague feeling that something is wrong. He has a headache, feels unusually tired, and cannot escape the feeling that he is coming down with something. He briefly considers postponing his trip. But this business meeting is crucial. He simply has to be there. So he tries to put his symptoms out of his mind and leaves for the airport. As the morning progresses, Jeff's symptoms worsen. Yet he boards the aircraft as planned, hoping that he can relax and will start to feel better. Shortly after takeoff, however, he can no longer deny it; he has the flu. Jeff now has a fever, and he begins to experience the shivering, aching muscles, and malaise that typify an influenza infection. Unfortunately for his fellow passengers, Jeff is still highly infectious to others, and on a crowded airplane there is simply nowhere to hide. As Jeff coughs, infectious influenza viral particles are wafted through the cabin, where they are inhaled by others on the plane. By the time the plane reaches Los Angeles, almost everyone on board has been exposed.

1. Although most people on the flight were exposed, the outcome of this exposure is different in different individuals. Why?

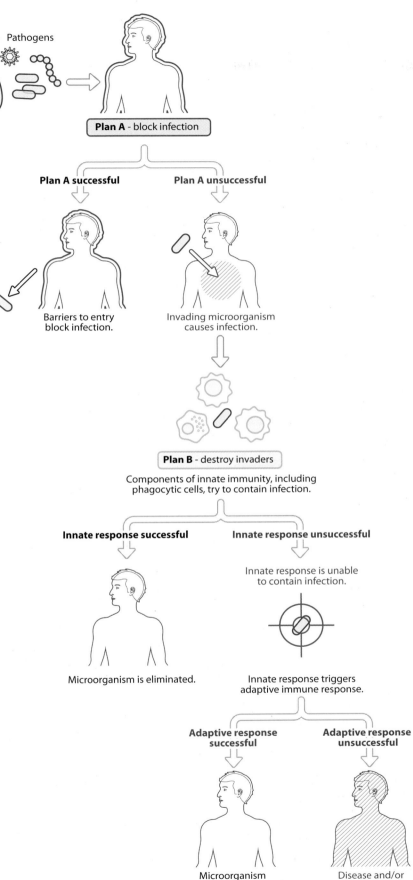

Pathogens

Plan A - block infection

Plan A successful   **Plan A unsuccessful**

Barriers to entry
block infection.

Invading microorganism
causes infection.

Plan B - destroy invaders

Components of innate immunity, including
phagocytic cells, try to contain infection.

Innate response successful   **Innate response unsuccessful**

Innate response is unable
to contain infection.

Microorganism is eliminated.

Innate response triggers
adaptive immune response.

**Adaptive response
successful**   **Adaptive response
unsuccessful**

Microorganism
is eliminated.

Disease and/or
death occurs.

Figure 11.2 The strategy of immune defense. Initially, the host attempts to prevent infection utilizing various "barriers to entry" ("plan A"). Should these barriers fail, "plan B" goes into action: elements of innate immunity try to eliminate the invading microorganisms. If innate mechanisms alone are insufficient, the innate response triggers the activation of adaptive immunity, which mobilizes a highly specific, targeted attack on the infecting microorganism.

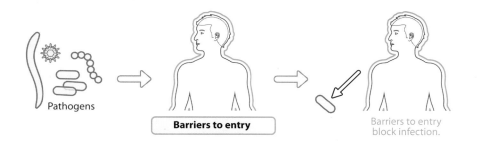

We will spend the remainder of this chapter answering this single question. To do so, we will observe what happens to four representative exposed individuals. The outcome of their exposure will be somewhat different for each of them. The reasons for these differences will be based on how each person responds immunologically. By following these various responses, we will come to understand the highly sophisticated and carefully orchestrated defensive strategies that constitute our immune system.

Barriers to Entry

CASE: FRED

Fred is a 22-year-old exchange student, currently studying in China. He is returning home to the United States for a brief visit. Although he heard Jeff coughing and is somewhat worried about catching whatever his fellow passenger might have, he seems to get off lucky. After his long flight he still feels fine, and after several days he is pleased to realize that he has no symptoms of illness whatsoever. He soon forgets all about the sick passenger on his flight.

1. **Even though Fred was exposed to influenza virus, he did not become ill. Furthermore, he was not even infected by the virus. How did he avoid infection entirely?**

Influenza virus is transmitted via the respiratory route (see Chapter 10, p. 268). Upon reaching the nose or sinuses of a susceptible individual, it may initiate an infection by entering the cells of the upper respiratory tract, where it begins to replicate. But for Fred, plan A worked! His barriers to entry prevented an infection from ever occurring (**Figure 11.3**).

In this example, the primary barrier to entry was the thick layer of mucus coating the membranes of the respiratory system. Any airborne pathogen must cross this mucus layer before it can reach the underlying membranes and subsequently enter otherwise sterile regions of the body. In many cases, including Fred's, pathogens become trapped in this mucus and are thereby prevented from reaching the underlying mucous membranes. Fred was also aided by the cilia lining the respiratory tract (**Figure 11.4**). These cilia move any invading microorganisms back toward the mouth, where they are caught in saliva and either spit out or swallowed. Respiratory infections in general are more common in places where air pollution is a problem, because pollutants can destroy cilia, making it easier for microbes to reach the lungs. Smokers are more prone to respiratory infections for similar reasons. Fred, who neither smokes nor lives in an area with serious air pollution, had the full benefit of these cilia.

In at least one other respect, Fred was lucky. In addition to the cramped quarters and recirculated air on an airplane that enhance airborne transmission, the low humidity on airplanes tends to dry out mucous membranes. As mucus dries, it becomes less effective at blocking access to the underlying membranes, and infection becomes more likely. Perhaps Fred was assisted by drinking lots of water on the flight and thereby avoiding

Figure 11.4 Cilia. Cilia are hairlike extensions of the cell, composed of protein. They help protect the respiratory tract from microbial invasion and are found on many cell types, including the cells lining the respiratory tract. As respiratory tract cilia beat back and forth, they move fluid away from the lungs and back toward the mouth or nose. This movement prevents many invading microorganisms from reaching the lower respiratory tract.

dehydration. Avoiding alcohol may have helped as well, as alcohol hastens the drying of mucous membranes.

Other membranes lining the digestive and urogenital tract are likewise protected by mucus (**Figure 11.5**). In addition to mucus, these surfaces have other barriers that help to thwart invading microbes. Acid produced by the stomach lining destroys many of the microorganisms that might otherwise reach the intestine. Recall from Chapter 10 that when this acidity is disrupted, as it was when the subject of our case took antacid tablets, the chance of infection increases. In the urogenital tract, regular urine flow impedes microorganisms from advancing to the bladder.

Other body surfaces are protected by the skin—one of our most formidable barriers to infection. The skin surface is composed of tightly packed epithelial cells, with an outermost layer of dead cells filled with protein. A thicker region containing blood vessels as well as oil and sweat glands, which bathe the skin with antimicrobial substances such as fatty acids, lies beneath the epithelial layer. Very few microorganisms can penetrate intact skin. The importance of the skin as a barrier to entry becomes evident when its integrity is compromised by burns or injuries. In such cases, the likelihood of infection is greatly increased. In hospital burn units, special measures are routinely taken to ensure that patients remain infection-free.

Mucous membranes and skin, which cover and protect all exposed body surfaces, are examples of **anatomical barriers**—barriers to entry by physical and chemical means. We are also protected by **microbial barriers**. In Chapter 10 (p. 260), we discussed the normal microbial flora and its importance in preventing infection. These well-adapted organisms provide

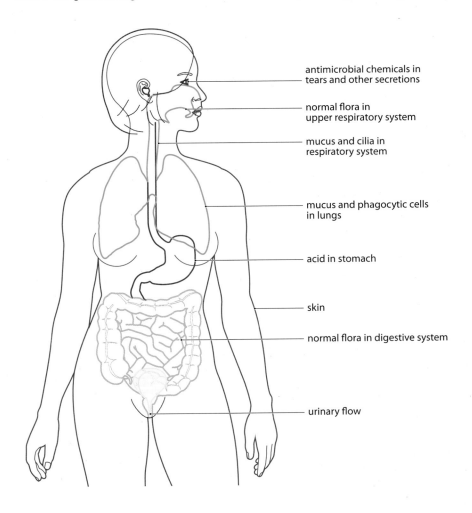

antimicrobial chemicals in tears and other secretions

normal flora in upper respiratory system

mucus and cilia in respiratory system

mucus and phagocytic cells in lungs

acid in stomach

skin

normal flora in digestive system

urinary flow

Figure 11.5 Important barriers to entry. All body surfaces are lined with skin or mucous membranes, which impede microbial colonization. These surfaces are often associated with additional chemical or physical barriers to further hamper microbial entry. Some of the most important barriers are depicted here.

another layer of defense, by producing substances that inhibit the growth of competitors, by monopolizing available nutrients, or by simply taking up all available space. With an intact normal flora, less well adapted pathogens are much less likely to colonize the host.

Certain other barriers to entry can be termed **genetic barriers**. Although you, for example, might contract influenza from an infected person, you cannot infect your cat with this virus. Likewise, your cat cannot infect you with feline leukemia virus. The genetic differences between hosts are simply too great for these pathogens. Even within a particular species, some individuals have greater or lesser genetic resistance to a specific pathogen. Some humans, for instance, are relatively resistant to malaria because their red blood cells lack certain molecules required by the protozoan parasite to enter these cells.

CASE: LAURA

Laura is in her mid-fifties. She is returning home after a 3-week vacation in East Asia. She never really noticed on her return from Hong Kong that one of the other passengers was coughing throughout the flight. Forty-eight hours after arriving home, however, she has developed all the symptoms associated with influenza, including headache, chills, fever, and achy muscles, and she spends most of the next three days in bed. Shortly thereafter the symptoms subside, and although she still feels somewhat weak, Laura makes a full recovery.

1. **Laura has suffered an acute infection of influenza. Although she was ill for several days, the incident ended with the elimination of the pathogen and a complete restoration of health. What occurred immunologically in Laura to account for this pattern of disease?**

The Innate Immune Response

If barriers to entry fail, the pathogen is confronted by elements of innate immunity

As Laura's experience shows us, our barriers to entry, while formidable, are not foolproof. In her case, at least some viral particles reached the epithelium in her upper respiratory tract and then entered these epithelial cells to initiate an infection. Unlike Fred, who got off easily, Laura must now switch to plan B—destroy the invaders. This backup plan begins by confronting the pathogen with a variety of innate defenses.

Think of these innate defenses as built-in security systems that are always on alert for invaders and able to respond to infection quickly. Within a few hours these cellular and molecular "first responders" are on the scene, attempting to eliminate invading microbes. If successful, they will terminate their defensive activities. If they need assistance, they will sound the alarm, activating adaptive immunity.

An innate response often begins when cells at the site of invasion release certain chemicals that summon white blood cells, called **leukocytes**, to the site of infection. In Laura's case, epithelial cells lining her upper respiratory tract released these chemicals as they suffered damage due to viral replication. There are several different types of leukocytes, and the first to arrive are phagocytic cells called **neutrophils** and **macrophages** (**Figure 11.6**). The details of phagocytosis were described previously (see Chapter 3, p. 66). Upon arrival at the site of infection, these phagocytic cells begin to engulf invading microorganisms, generally destroying them in large numbers. **Dendritic cells** are another important type of leukocyte capable of engulfing pathogens. They are vital to a successful immune response and will be

Figure 11.6 Phagocytic cells. The arrival of phagocytic cells at the site of infection is an important early part of the innate immune response. (a) Neutrophils and macrophages are among the most important phagocytic cells participating in an innate immune response. A red blood cell is shown for comparison. (b) A human blood smear containing a neutrophil with its characteristic multilobed nucleus. Note the preponderance of the smaller red blood cells. Neutrophils are the most common type of immune system cell in the blood, and they are generally the first type of phagocytic cells to arrive at the site of infection. (c) Macrophages phagocytosing small polystyrene beads (in blue). In addition to their innate phagocytic activity, macrophages are also involved in the activation of an adaptive immune response.

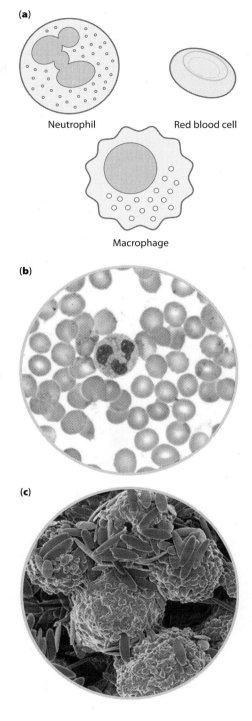

discussed in detail later in this chapter. As we will see, macrophages and dendritic cells form the critical link between innate and adaptive immunity.

Phagocytic cells need a way to recognize various microorganisms that may pose a threat, and one of the ways they do so is through **pattern recognition**—the ability to recognize certain types of molecules that are unique to microbes. The capacity to recognize such molecules, which are not normally found in humans and other animals, is a powerful feature of innate immunity. Pattern recognition means that certain microbial molecules act as "danger signals," indicating that an infection is under way. Macrophages, for example, have special receptors on their surfaces that allow them to recognize microbe-associated molecules. These receptors are called **Toll-like receptors**, or **TLRs** for short, and they are found on several different types of immune system cells, including dendritic cells.

A variety of Toll-like receptors have been discovered in humans. Each type of TLR binds a different microbial product, allowing the immune system cells to identify a variety of different invading microbes as threats (**Figure 11.7**). For example, there is a TLR that binds specifically to bacterial peptidoglycan. Other TLRs bind molecules typically found only on acid-fast bacteria or fungi. Yet another recognizes flagellin, the protein found only in bacterial flagella. When Laura was infected by influenza virus, the TLRs on the arriving macrophages latched onto the viral RNA, indicating to the macrophage that a viral infection was in progress. The macrophages "knew" they were dealing with a virus, because of the binding of viral RNA to the specific TLRs. This allows the macrophages to respond appropriately.

When TLRs and similar receptors are activated on the surface of macrophages and dendritic cells, these cells respond by releasing chemicals messengers called **cytokines** into the surrounding environment (see Figure 11.7). Activation of different TLRs can cause different types of cytokines to be released.

In response to many infections, for example, cytokines are released that promote **inflammation**, a key component of innate immunity. Inflammation refers to a series of events that take place in reaction to tissue damage. We are all familiar with the signs of inflammation. The site of an injury becomes swollen, red, hot, and painful. While unpleasant, this process is part of the normal response to tissue damage and possible microbial invasion. The symptoms we observe are caused when the blood vessels in the affected area increase in diameter, a process called **vasodilation**. They also become leakier than usual, allowing blood serum to leave the blood vessels and enter the tissues. This increased leakiness is referred to as increased **permeability**. Although these changes contribute to the symptoms we observe during an infection, they also allow immune system cells and molecules to likewise leave the blood to confront invading microbes. In Laura, inflammation in the respiratory tract helps to explain symptoms such as her persistent cough.

Figure 11.7 Pattern recognition by Toll-like receptors. At least 10 different Toll-like receptors (TLRs) have been identified in humans. Each type interacts with specific microbial molecules that are typically not found in animals. These molecules thus act as danger signals that an infection is in progress. When a pathogen molecule binds to a particular TLR, a signal is sent to the nucleus. This signal causes the cell to start producing specific chemical messengers called *cytokines*. Leukocytes produce many different cytokines, which regulate and coordinate various immune responses. Different pathogen molecules bind different TLRs, resulting in the release of different cytokines. In this way, the immune response is tailored to the general class of pathogen infecting the host.

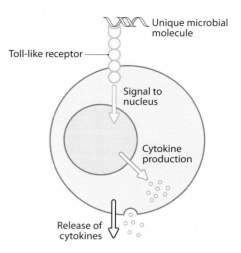

The high fever Laura experienced was also in response to cytokine release. Following her infection, a specific cytokine called interleukin-1 (IL-1) acted on Laura's brain to raise body temperature. Fever is actually a beneficial aspect of the innate immune response, at least to some extent. At elevated body temperature, phagocytic cells are more efficient, and some pathogens, including influenza virus, are less able to reproduce. So even though Laura's fever made her uncomfortable, it helped to limit the spread of the virus in her respiratory system. You may be surprised to learn that even cold-blooded animals, which cannot physiologically regulate their body temperature, will find a way to heat themselves up if they suffer from an infection. Grasshoppers infected with bacteria, for instance, will seek out warmer than normal locations. If prevented from doing so, they are less likely to survive. Tropical fish hobbyists know to raise the temperature of their aquariums slightly if their fish are suffering from an infection. Of course, if fever gets too high it can be dangerous or even life-threatening. Fever, like inflammation, is therefore a good example of an important point: many disease symptoms are not *directly* due to damage caused by the pathogen. Rather, they are often the unpleasant side effects of an immune response. The immune system is very much a double-edged sword. The powerful immunological weapons that keep us safe from most microbial assaults can sometimes cause discomfort, or even serious damage.

The **interferons** are an especially important group of cytokines in viral infections such as Laura's. Their name is based on the fact that they *interfere* with viral replication. These chemical messengers are produced by any cell that becomes infected with a virus. When cells in Laura's respiratory tract became infected with influenza virus, they released interferons, which alerted uninfected cells nearby that a virus was on the loose. The uninfected cells could then take protective measures. In other words, interferons induce an antiviral state in neighboring cells, greatly reducing the ability of the virus to spread (**Figure 11.8**).

When an antiviral state is induced in an uninfected cell, the cell produces special proteins that will shut down further protein synthesis (i.e., translation), in case a viral invasion occurs. Not only is translation of mRNA to proteins inhibited but also the mRNA itself is degraded. Consequently, if a virus infects a cell in which the antiviral state has been induced by interferons, the virus cannot hijack the cell's protein-making machinery to produce the proteins it needs for viral replication. Once the viral infection is cleared and interferon stimulation ceases, surviving cells can once again activate their protein synthesis machinery. Interferons act very quickly and help "hold down the fort" until an adaptive response can be generated. However, like inflammation and fever, they contribute to the symptoms we suffer during an infection. Specifically, many of the "flu-like symptoms" that Laura suffered, including fatigue, headache, and chills, are provoked by interferons.

When they were first discovered, many scientists believed that interferons might be the "magic bullet" with which they could treat any viral infection. These hopes largely turned to disappointment when it was learned that

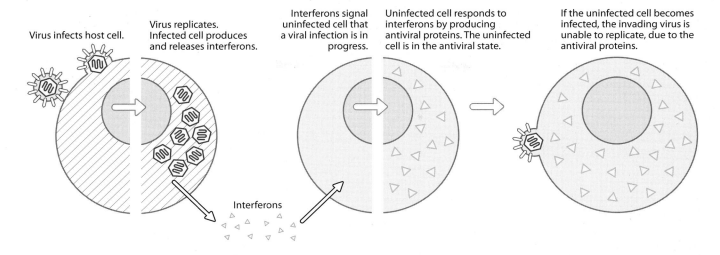

Virus infects host cell.

Virus replicates. Infected cell produces and releases interferons.

Interferons signal uninfected cell that a viral infection is in progress.

Uninfected cell responds to interferons by producing antiviral proteins. The uninfected cell is in the antiviral state.

If the uninfected cell becomes infected, the invading virus is unable to replicate, due to the antiviral proteins.

Interferons

interferons can cause the side effects mentioned above. Additionally, interferons are active in the body only for a very short time. Their short life span makes it difficult to deliver adequate doses of interferons to appropriate parts of the body. Nevertheless, interferons have shown some benefit in the treatment of certain viral infections such as hepatitis C.

When innate mechanisms fail to eliminate an infection, the adaptive immune response is activated

Laura's innate immune defenses were unable to contain the infection on their own (**Figure 11.9**). Consequently, certain leukocytes, specifically macrophages and dendritic cells, sounded the alarm, rousting the cells of the adaptive immune system to action.

Why do innate mechanisms sometimes, but certainly not always, contain an infection on their own? Recall that, in Chapter 10, we learned that the likelihood of disease in any encounter between pathogen and host depends upon the number of pathogens to which a host is exposed, the overall virulence of the pathogens, and the status of host defenses. If we know something about these factors in a particular host–pathogen interaction, we can make a reasonable prediction about the likely outcome of that interaction.

Figure 11.8 Interferons. Interferons induce an antiviral state in uninfected cells. Interferons allow an infected cell to warn its uninfected neighbors that a viral infection is in progress. The uninfected cells respond by taking protective measures, which include the production of antiviral proteins that inhibit protein synthesis by invading viruses. If a virus infects a cell that has produced these antiviral proteins, the virus is less likely to replicate.

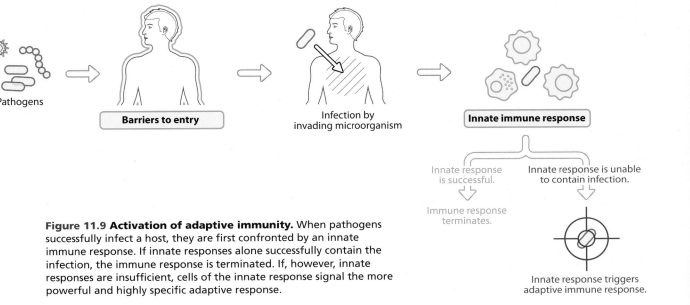

Pathogens

Barriers to entry

Infection by invading microorganism

Innate immune response

Innate response is successful.

Innate response is unable to contain infection.

Immune response terminates.

Innate response triggers adaptive immune response.

Figure 11.9 Activation of adaptive immunity. When pathogens successfully infect a host, they are first confronted by an innate immune response. If innate responses alone successfully contain the infection, the immune response is terminated. If, however, innate responses are insufficient, cells of the innate response signal the more powerful and highly specific adaptive response.

In Laura's case, where innate mechanisms were clearly not enough to prevent illness, it may be that she was exposed to a particularly large dose of infectious virus that simply overwhelmed the capacity of innate defenses. Alternatively, for a variety of reasons, Laura may have been especially vulnerable to viral infection at the time of her flight. Factors such as stress, lack of sleep, or inadequate diet can compromise immune defenses. It is possible that, in Laura's case, a combination of these factors was involved.

Antigen-presenting cells activate those cells responsible for adaptive immunity

When the Toll-like receptors of macrophages and dendritic cells (**Figure 11.10**) are bound by pathogen molecules, these important cells begin not only to release certain chemical messengers (cytokines) but also to actively ingest invading pathogens. Once ingested, by either phagocytosis or other similar processes, the pathogen is digested within the endomembrane system of the cell (see Chapter 3, p. 66). Following digestion, small protein fragments that were part of the pathogen bind onto cellular proteins called **major histocompatibility complex** (**MHC**) proteins. These MHC proteins, bound to pieces of pathogen protein, are then transported to the plasma membrane of the macrophage or dendritic cell and displayed on the surface (**Figure 11.11**). It is as if the cell is saying, "We have been invaded by a pathogen, and here is a piece of that pathogen." These tiny pathogen fragments are called **antigens**. Antigens are pieces of foreign molecules that the adaptive immune system can recognize as part of a specific microorganism. The macrophages and dendritic cells, with these antigens displayed prominently on their surfaces, are serving as **antigen-presenting cells**.

In the case of influenza, the most important antigens are parts of the proteins embedded in the viral envelope (see Chapter 4, p. 91, for a more complete description of the viral envelope). As we will see, once an antigen-presenting cell has these antigens on its surface, it will present them to certain cells responsible for adaptive immunity, informing them what sort of highly targeted response is required. A different invading pathogen, *Streptococcus* bacteria, for instance, would have its own unique antigens, permitting a different adaptive response.

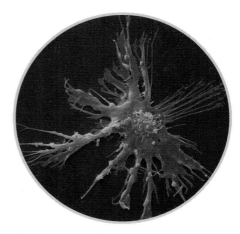

Figure 11.10 A dendritic cell. Dendritic cells, considered to be part of the innate immune response, play an essential role in stimulating the initiation of an adaptive response by acting as *antigen-presenting* cells. The long, fingerlike extensions characteristic of dendritic cells are called *dendrites,* which gives these cells their name.

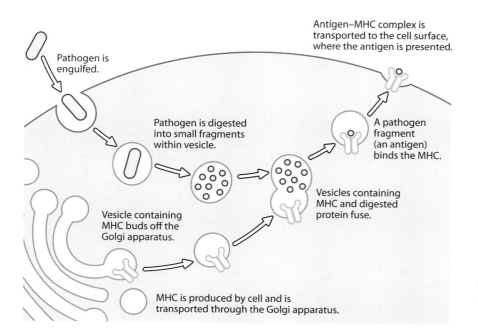

Pathogen is engulfed.

Antigen–MHC complex is transported to the cell surface, where the antigen is presented.

Pathogen is digested into small fragments within vesicle.

A pathogen fragment (an antigen) binds the MHC.

Vesicles containing MHC and digested protein fuse.

Vesicle containing MHC buds off the Golgi apparatus.

MHC is produced by cell and is transported through the Golgi apparatus.

Figure 11.11 Foreign antigens are presented with MHC. When a macrophage or a dendritic cell (an antigen-presenting cell) engulfs a pathogen (in red), the pathogen is degraded by digestive enzymes. Meanwhile, host proteins called *major histocompatibility complex (MHC)* proteins (in green) are produced by the antigen-presenting cell. The MHC proteins move through the cell's endomembrane system, and eventually bud off the Golgi apparatus in a membrane-enclosed vesicle. The vesicles containing the pathogen and the MHC fuse, and small digested fragments of the pathogen bind to the MHC. These small pathogen fragments, recognizable as foreign by the immune system, are known as *antigens*. The antigen–MHC complex is then transported in a vesicle to the cell membrane, where it is prominently *presented*.

Antigen-presenting cells migrate to lymphatic organs to activate adaptive immunity

Once macrophages and dendritic cells are presenting antigen on their surface, they migrate into the **lymphatic system** (**Figure 11.12**). This network of vessels helps to return any fluid that has leaked out of the blood back to the circulatory system. The fluid in the lymphatic system (the **lymph**) passes though increasingly larger lymphatic vessels until it is returned to the blood through a duct, which empties into a large vein near the heart. As the lymph progresses, it passes through lymphatic organs called **lymph nodes**. These small, bean-shaped structures are generally found in places where several smaller lymphatic vessels come together to form larger vessels.

(a)

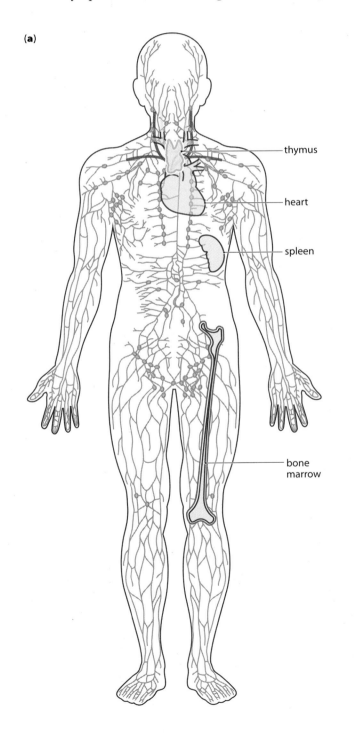

thymus

heart

spleen

bone marrow

(b)

Lymphatic fluid enters the lymph node.

Lymphatic fluid moves through the lymph node packed with immune system cells, including antigen-presenting cells and lymphocytes.

Lymphatic fluid exits the lymph node.

Figure 11.12 The lymphatic system.
(a) The lymphatic system consists of a network of vessels that collect fluid in the tissues and return it to the circulatory system. Lymph nodes are represented by the small colored spots, seen throughout the body, with especially dense clusters in the groin, armpits, and neck. All blood cells are originally formed in the bone marrow. The cells involved in adaptive immunity (the lymphocytes) either remain in the bone marrow or migrate to the thymus to complete their development. The spleen is the site where adaptive immune responses are generated for pathogens in the blood. (b) A lymph node. As lymph travels through the lymphatic vessels, it passes through lymph nodes. Antigen-presenting cells migrate into lymph nodes, where they interact with lymphocytes to initiate an adaptive response.

Lymph nodes also contain large numbers of **lymphocytes**. Lymphocytes are the specific white blood cells that are responsible for adaptive immunity. Lymphocytes, as well as all other blood cells, are initially formed in the red marrow of the bone (see Figure 11.12). As we will see, there are two principal types of lymphocytes, called the **T cells** and the **B cells**. Each cell type plays a specific role in an adaptive response.

Lymphocytes circulate through the blood, but they can also leave the blood and enter the lymphatic system by adhering to capillaries and squeezing between the cells of the capillary walls. Once they have entered the lymphatic system, they may be eventually returned to the blood. But before they get there, they must pass through lymph nodes. If there are no antigen-presenting cells present in a lymph node, they will pass through the node and continue circulating. If, however, antigen-presenting cells (macrophages and dendritic cells) are also in the lymph node, the lymphocytes may be activated to initiate an adaptive response.

The lymph nodes, strategically placed throughout the body, thus concentrate all the components necessary for an adaptive immune response: antigen, antigen-presenting cells, and lymphocytes. In Laura's case, since influenza virus has invaded the epithelium of her upper respiratory system, we would expect the adaptive response to be generated in the lymph nodes closest to the site of infection, in the neck and throat. Once activated, lymphocytes within the node begin to rapidly divide, greatly increasing in number. It is this rapid rise in lymphocyte number that results in the swelling of lymph nodes during infection.

Lymph nodes are not the only lymphatic organ. When microbes reach the blood they are filtered by the **spleen**, instead of lymph nodes (**Figure 11.13**; also see Figure 11.12). The spleen is an organ about the size of a bar of soap, found in the upper left portion of the abdomen. It basically works like a lymph node for blood-borne pathogens, providing a site for the initiation of an adaptive response. Other areas in the body that are particularly prone to invasion and infection such as the intestine are equipped with their own lymphatic organs.

The Adaptive Immune Response

Antigen-presenting cells activate helper T cells to initiate an adaptive response

When antigen-presenting cells arrived in the lymph nodes of Laura's neck and throat, they prominently displayed pieces of the influenza virus on their surface, bound to MHC molecules. They also released certain cytokines that attracted lymphocytes called **helper T cells**. These lymphocytes are just one type of T cell, and like all T cells they migrate from the bone marrow, where they are formed, to the **thymus,** where they complete their development. The thymus is a lobed organ, found just above the heart (**Figure 11.14**). Once they mature, T cells leave the thymus and begin to circulate. As we will see, helper T cells (abbreviated as **Th cells**) are especially crucial in adaptive immunity, because once they are activated by antigen-presenting cells, they coordinate the overall adaptive response.

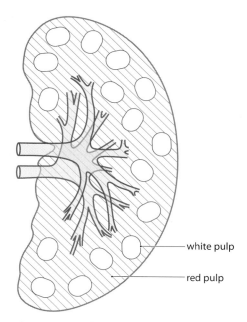

Figure 11.13 The spleen. The spleen consists of red pulp, where old red blood cells are destroyed, and white pulp, where adaptive immune responses to blood-borne pathogens are generated.

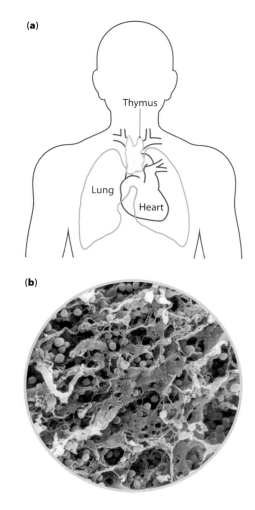

Figure 11.14 The thymus. (a) The lobed thymus is found immediately above the heart. Those lymphocytes that leave the bone marrow and migrate to the thymus develop into T cells. (b) In this image taken with an electron microscope, developing T cells (the small spherical cells) can be seen among the extensive epithelial tissue of the thymus.

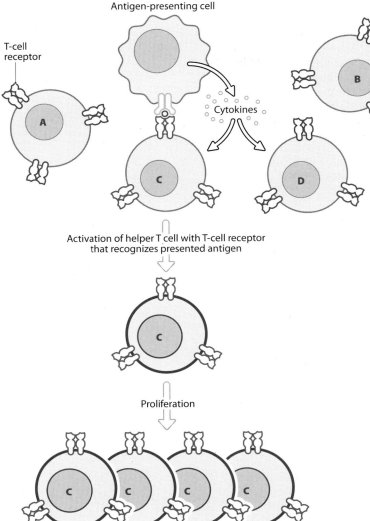

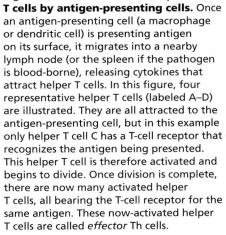

Figure 11.15 Activation of helper T cells by antigen-presenting cells. Once an antigen-presenting cell (a macrophage or dendritic cell) is presenting antigen on its surface, it migrates into a nearby lymph node (or the spleen if the pathogen is blood-borne), releasing cytokines that attract helper T cells. In this figure, four representative helper T cells (labeled A–D) are illustrated. They are all attracted to the antigen-presenting cell, but in this example only helper T cell C has a T-cell receptor that recognizes the antigen being presented. This helper T cell is therefore activated and begins to divide. Once division is complete, there are now many activated helper T cells, all bearing the T-cell receptor for the same antigen. These now-activated helper T cells are called *effector* Th cells.

Helper T cells all have protein complexes on their surface called **T-cell receptors**. The T-cell receptor allows the Th cell to recognize a specific antigen bound to MHC on the surface of an antigen-presenting cell (**Figure 11.15**). An individual helper T cell may have 100,000 or so of these receptors, and they are all identical. But different Th cells have different T-cell receptors, all of which recognize different antigens. One Th cell may have receptors that recognize a certain part of an influenza envelope protein. A different Th cell may have receptors that recognize a different piece of the same protein, while a third Th cell may recognize a piece of a different viral protein. All of these Th cells will be activated by macrophages and dendritic cells in Laura's lymph nodes. Other Th cells have receptors that might recognize antigens from other viruses or from various bacteria or fungi. These Th cells will not be activated in Laura's lymph nodes if she is infected only with influenza, because they will not come in contact with their specific antigen. The Th cells that *do* recognize their influenza antigens, on the other hand, begin to replicate furiously. The newly replicated Th cells all have the same T-cell receptor that was found on the Th cells that were originally activated. In other words, only a relatively small number of Th cells initially recognized their antigens. Now that they have replicated, however, their numbers have increased enormously.

The number of different kinds of Th cells, all with different T-cell receptors on their surfaces, is almost limitless. We even have Th cells with receptors for antigens to which we will probably never be exposed. For example, if you take a trip to Antarctica and turn over a rock exposing yourself to an unusual bacterium of some sort, you probably have T cells that would recognize its antigens. This stupendous diversity is made possible because T-cell receptors are coded for by a collection of genes. Although these genes are finite in number, each receptor is encoded by a different *combination* of these genes. And even with a limited number of genes, different gene combinations allow the assembly of millions of different receptors. The genetics behind this process is complex and beyond the scope of this text. Some of the details remain unclear. However, it is safe to say that, thanks to this remarkable process of gene recombination, almost any invading pathogen will be recognized as a foe to be combated.

Some activated helper T cells activate cytotoxic T cells to initiate a cell-mediated response

The activation of the Th cells that we have described takes a few days. Meanwhile, in spite of innate defenses, the influenza virus that infected Laura has been multiplying, and as the amount of virus surpasses the threshold of disease (see Chapter 10, p. 271), she starts to feel ill. Soon, however, the appropriate Th cells are ready for action and they begin to activate a second type of T cell called a **cytotoxic T cell**. Like Th cells, cytotoxic T cells (Tc cells) develop in the thymus, and like Th cells, different Tc cells come equipped with specific T-cell receptors on their surface. Furthermore, like Th cells, once they are activated, they begin to divide and increase in number.

Cytotoxic T cells are activated when Th cells stimulate them with certain cytokines. Interleukin-2 (IL-2) is an example. The now-active cytotoxic T cells then leave the lymph nodes, in search of infected cells. They can recognize these infected cells because the infected cells display foreign antigens on their surface, bound to MHC (**Figure 11.16a**). By displaying these antigens on their surface, infected cells are essentially broadcasting a message saying, "I'm infected. Kill me before the pathogen can replicate and infect other cells." If a now-activated, patrolling Tc cell interacts with this cell, and if its T-cell receptor matches the displayed antigen, it releases toxic chemicals that result in the death of the infected cells (Figure 11.16b). Uninfected cells do not display foreign antigens on their surface and are unharmed. This targeted cell killing, called the **cell-mediated response**, is similar to performing microscopic amputations; infected host cells are sacrificed to prevent uninfected cells from becoming infected.

The Th or Tc cells that have been activated during a cell-mediated response are collectively referred to as **effector T cells**. Before we go on, it is important to point out that when either Th or Tc cells are first stimulated, the dividing T cells do not *all* develop into active, effector T cells. Some remain as relatively inactive and long-lived **memory T cells**. As we will soon see, these memory cells are a big part of the reason why, if Laura is infected with influenza again in the near future, she will probably remain healthy and blissfully unaware of her second encounter with the same virus.

Helper T cells also activate B cells to initiate a humoral immune response

While some Th cells are busy activating Tc cells, others release different cytokines, such as IL-4 and IL-5, to activate a different type of lymphocyte called a B cell. As opposed to T cells, which migrate to the thymus, B cells remain in the red marrow of the bone, where their development is

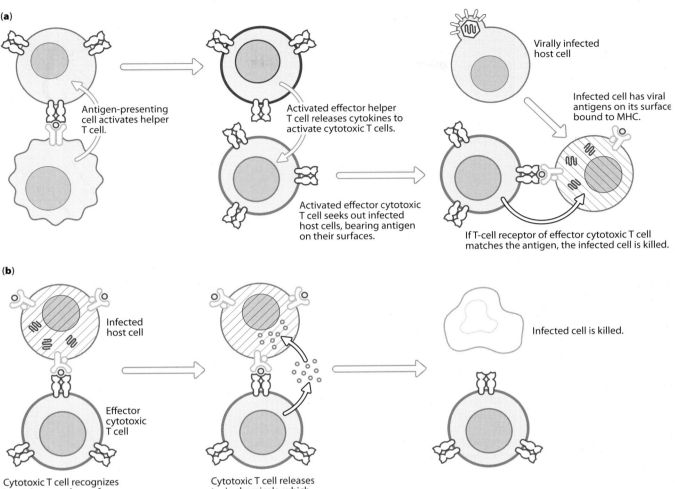

(a)

Antigen-presenting cell activates helper T cell.

Activated effector helper T cell releases cytokines to activate cytotoxic T cells.

Activated effector cytotoxic T cell seeks out infected host cells, bearing antigen on their surfaces.

Virally infected host cell

Infected cell has viral antigens on its surface bound to MHC.

If T-cell receptor of effector cytotoxic T cell matches the antigen, the infected cell is killed.

(b)

Infected host cell

Effector cytotoxic T cell

Cytotoxic T cell recognizes its antigen on the surface of infected cell.

Cytotoxic T cell releases toxic chemicals, which enter the infected cell.

Infected cell is killed.

Figure 11.16 The cell-mediated response. (a) Once helper T cells are activated by antigen-presenting cells, they may release specific cytokines that activate cytotoxic T cells (Tc cells). Cytotoxic T cells also have T-cell receptors on their surface; they can recognize only antigens for which their T-cell receptor is specific. The now-activated, effector Tc cells seek their antigen, bound to MHC, on the surface of host cells. (b) If they recognize their antigen, effector cytotoxic T cells release toxic chemicals that kill the infected cells.

completed. B cells were so named because their site of development was first discovered in chickens where, as in all birds, they complete their development not in the bone marrow but in a structure known as the *bursa*.

Recall that both Th and Tc cells have T-cell receptors on their surface. These receptors recognize specific antigens, when the antigen is bound to MHC on the surface of either an antigen-presenting cell (in the case of Th cells) or an infected cell (in the case of Tc cells). B cells, on the other hand, have **antibodies** in their plasma membranes (**Figure 11.17**). Like T-cell receptors, antibodies are composed of protein. Antibodies can also recognize specific antigens. But unlike T-cell receptors, they bind antigen that is floating freely in the body's fluids. Similar to what we described for T-cell receptors, each B cell has many identical antibodies on its surface, all of which recognize the same antigen. Furthermore, we produce literally billions of different kinds of B cells, all with different antibodies on their surface. This practically limitless variety of antibodies, all recognizing different antigens, ensures that whatever pathogen happens to come along, we will have antibodies that can recognize it. As with T-cell receptors, this diversity is based on the ability to recombine a limited number of antibody genes in an astronomical number of different ways.

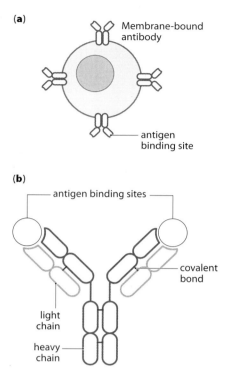

Figure 11.17 B cells and antibodies. (a) Following its development in the bone marrow, a mature B cell expresses antibodies on its membrane. Each antibody has two antigen binding sites, and all antibodies on a particular B cell are specific for the same antigen. Like T-cell receptors on different T cells, different antibodies on different B cells bind different antigens. (b) Antibody structure. Each individual antibody is actually composed of four proteins, held together by covalent bonds. The two smaller proteins (known as *light chains*) are identical, as are the two larger proteins (known as *heavy chains*). Each antibody molecule has two antigen binding sites.

Following B-cell maturation in the bone marrow, the B cells begin to circulate through the blood and lymph. As they circulate, they pass through lymphatic organs such as lymph nodes or the spleen. Foreign antigens also accumulate in these lymphatic organs, much as a drain in a kitchen sink might trap bits of solid material. If a B cell has a surface antibody that matches antigens in the lymph node, the B cell binds this antigen. This antigen–antibody binding and the stimulatory cytokines released by helper T cells provide the two signals that the B cell needs to start dividing (**Figure 11.18a**). After a series of such divisions, the cells have developed into one of two types: **memory B cells** and **plasma cells** (Figure 11.18b). We will discuss memory B cells shortly. Like memory T cells, memory B cells ensure that Laura won't have to worry about influenza again, a least for a while.

Plasma cells are antibody-*secreting* cells. Instead of retaining their antibody on their surface, plasma cells release antibody into the blood or lymph. Soon the body fluids are flooded with antibodies, all of which recognize and bind the same antigen. The production of antibody by B cells is called the **humoral response**. This is the portion of an adaptive response against antigen that is free in the body's fluids, once referred to as "humors."

In a humoral response to influenza, the recognized antigens will again be pieces of the virus, such as pieces of envelope proteins. Once antibodies are locked into place on a viral particle, that particle can no longer bind to appropriate host cells (**Figure 11.19a**) and is therefore unable to penetrate these cells (see Chapter 4, p. 93, to review viral binding to host cells). This capacity to "lock out" viruses from the host cells in which they replicate is called **neutralization**. Bacteria or other pathogens can likewise be neutralized by antibody, if the antibody bound to their surface prevents them from reaching a particular target site in the body. Furthermore, note in Figure 11.19 that antibodies have two binding sites for antigen (the two tips of the Y). Because they can bind two antigens at once, antibodies can form large masses together with antigen. Such clumping is called **agglutination** (Figure 11.19b). Any pathogens bound up in these large agglutinated clumps will never reach their target tissue. They are therefore neutralized and readily available to phagocytes.

Antibodies are useful in other ways, as well. Observe in Figure 11.19c that, in addition to the two antigen binding sites, they also have a binding site (at the base of the Y) for phagocytic cells such as macrophages and neutrophils. Consequently, because they are able to bind antigen at one end and phagocytic cells at the other end, antibodies form a physical link between the phagocytic cell and the antigen. If the antigen is part of a pathogen, the phagocyte can simply engulf both the antibody and the pathogen. When antibodies are functioning in this manner, they are acting as **opsonins**. An opsonin is any molecule that forms a physical connection between a phagocyte and an antigen. Although phagocytes do not require opsonins to phagocytose foreign material, opsonins make their job easier and increase phagocytic efficiency.

Figure 11.18 B-cell activation (*right*). (a) As B cells move through lymph nodes or the spleen, they may recognize and bind their specific antigens. Activated effector helper T cells may also be activated by antigen-presenting cells as previously described. If a particular B cell both binds its antigen and also is stimulated by activating cytokines released by the effector helper T cell, the B cell will be activated. B cells that receive only one of these two stimulatory signals ordinarily remain inactive. (b) Depicted here are five B-cell types (A–E). If they fail to encounter their antigens, these B cells continue to circulate. If a B cell recognizes and binds its specific antigen and receives appropriate stimulation from helper T cells, as B cell type D has done, it starts to divide rapidly and develops into one of two cell types: a plasma cell, or a memory cell. Short-lived plasma cells secrete antibody D into the lymphatic fluid and blood in large amounts. Each antibody has two binding sites for its specific antigen. Memory cells, which retain antibody D on their surface, are long-lived cells that can quickly respond if antigen D reappears in the future.

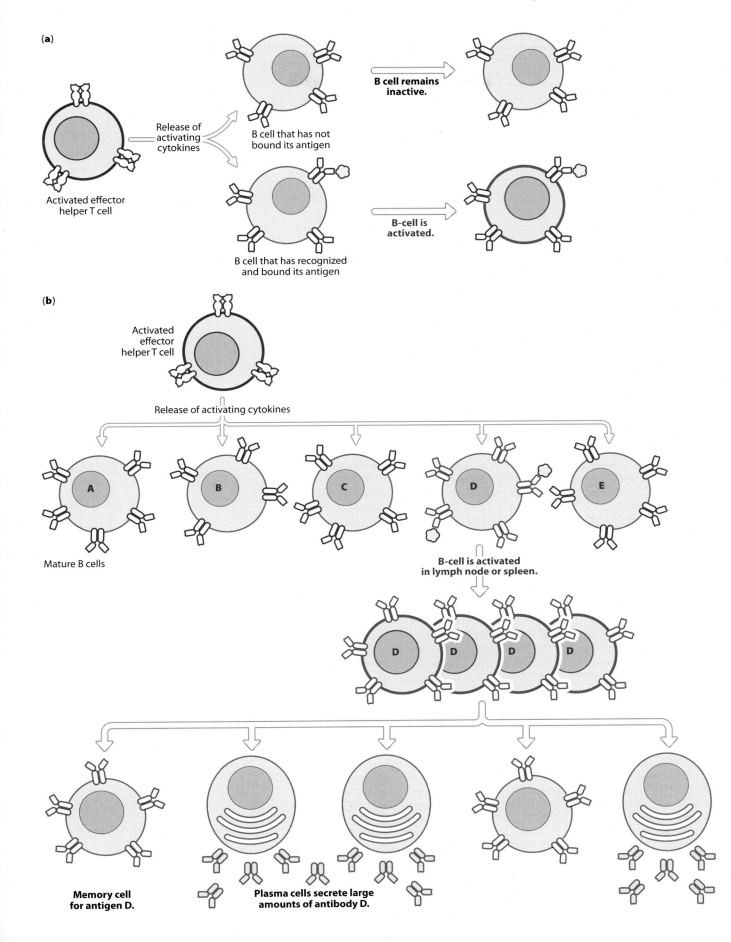

(a)

Activated effector
helper T cell

Release of
activating
cytokines

B cell that has not
bound its antigen

**B cell remains
inactive.**

B cell that has recognized
and bound its antigen

**B-cell is
activated.**

(b)

Activated
effector
helper T cell

Release of activating cytokines

A   B   C   D   E

Mature B cells

**B-cell is activated
in lymph node or spleen.**

D   D   D   D

**Memory cell
for antigen D.**

**Plasma cells secrete large
amounts of antibody D.**

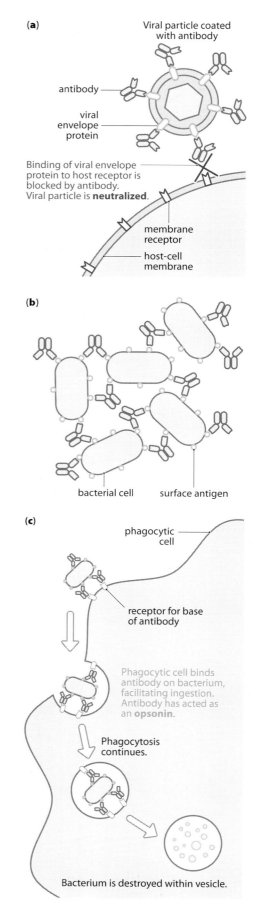

Figure 11.19 Antibody functions. (a) Antibody can coat the surface of viral particles, preventing them from attaching to their cell receptors. This process is called *neutralization*. (b) Because each antibody has two antigen-binding sites, large agglutinated clumps of antigen and antibody often form. These clumps are readily available to phagocytic cells. In this example the antigens (in green) are on the surface of bacterial cells. (c) Antibody can also enhance phagocytosis by acting as an *opsonin* for phagocytic cells, which have receptors on their surface for the "base" of Y-shaped antibodies.

Before we conclude our discussion of antibody function, it is worth noting that when antibodies act as opsonins, they are promoting an innate immune process—phagocytosis. This is a good example of an important point: the separation of immunity into innate and adaptive components is in some ways arbitrary. In a properly functioning immune system, innate and adaptive processes overlap and very much depend on each other.

In a humoral response, several different classes of antibodies may be produced

The antibodies that are secreted into the body's fluids during a humoral response are categorized as one of four basic classes. These are referred to as IgM, IgG, IgA, and IgE. The prefix Ig stands for immunoglobulin, a term that is interchangeable with antibody. Our discussion here will focus on IgM, IgG, and IgA, the three most important antibodies in most humoral responses.

The membrane-bound antibodies on B cells are mostly of the IgM class. A fifth antibody type, IgD, is also present on the membrane. Its overall role in humoral immunity is still not completely clear, and we will not consider it further in this discussion. Once a B cell is activated and matures into a plasma cell, it first produces a secreted form of IgM that is released into the body fluids. Later, the plasma cell may start to produce other antibody types instead, primarily IgG. The antigen specificity of the antibody, however, does not change. In other words, if a plasma cell first secretes IgM and later IgG in response to a particular antigen, both the IgM and the IgG will react with the *same* antigen.

The reason for the switching from IgM to other antibody classes is that the different classes of antibody have somewhat different functions. Therefore, depending on the type of infection, different types of antibodies might be more beneficial (**Figure 11.20**). As stated, the first type of antibody secreted in a humoral response is IgM. When secreted, five IgM antibodies join together to form covalently bound pentamers. With 10 antigen-binding sites, such pentamers are especially effective at agglutinating antigen. IgM is produced only during the first few days of a humoral response. As levels of IgM begin to decline, levels of other antibodies, mostly IgG, start to rise.

IgG is by far the most abundant antibody type produced in a humoral response. IgG can agglutinate antigen, and it enhances phagocytosis as an opsonin. Because it can cross the placenta and enter breast milk, IgG is maternally transferred to a fetus or newborn and thus provides an important source of immunity in infants. In Laura's immune response to influenza, IgG is an important part of the humoral response, because of its ability to bind influenza envelope proteins, neutralizing the virus before it can attach to respiratory cells in the upper respiratory tract.

Individual IgA antibodies bind to each other in pairs, forming dimers. Once secreted, IgA is able to cross from the blood to body fluids such as saliva and intestinal fluid. It is also found in tears and mucous secretions in the respiratory and urogenital tracts. IgA thus plays a special role in immunity on mucous membranes and is particularly valuable in combating pathogens

	Structure		Location	% of total antibody made	Function
IgM	Pentamer		Surface of B cells (as monomer) Blood Lymph	~10%	First secreted antibody in adaptive response Agglutination
IgG	Monomer		Blood Lymph Breast milk Fetal circulation	~80%	Predominant antibody in most humoral responses Neutralization Opsonin Agglutination Newborn immunity
IgA	Dimer		Secretions such as intestinal fluid, saliva, tears, and breast milk	~10%	Protection of mucous membranes

Figure 11.20 Structure and function of the three most important antibody classes. The surface antibody found on mature B cells is of the IgM class. Once a B cell is activated and differentiates into a plasma cell, it begins to secrete IgM into the blood and lymph. It may later begin secreting IgG or IgA. Secreted IgM forms a pentamer, composed of five covalently linked IgM antibodies. IgG is secreted as a single antibody molecule (a monomer). IgA is secreted as a dimer, composed of two individual IgA antibodies. Each antibody type is involved in various aspects of an adaptive immune response.

found in places like the digestive tract. Like IgG, IgA can enter the breast milk and consequently helps to provide immunity for newborns. IgA is also produced during an influenza infection, but it is probably less important in viral neutralization than IgG, as it is produced in substantially lower amounts.

A successful adaptive response culminates in the elimination of the pathogen

Once her bout of influenza began, Laura felt quite ill for several days. The virus, while slowed down by innate defenses, was still able to replicate prior to the activation of a full-fledged adaptive response. The time between initial infection and the start of the adaptive response is called the **inductive period** (**Figure 11.21**). The inductive period thus corresponds to the time that is necessary for T cells and B cells to be activated, to proliferate, and to begin their antimicrobial activity. During the inductive period, with no adaptive response to contend with, the pathogen has something of a free rein. In Laura's cause, as the amount of virus rose above the threshold of disease, symptoms commenced, and she was forced into bed. But after several days, with antibodies now limiting viral spread and cytotoxic T cells

Figure 11.21 The course of infection and immune response. The amount of virus in an infected person is shown in red. Once an individual is infected, the amount of virus in the body begins to rise. For influenza virus, it usually takes about 2 days before viral levels are high enough to cause symptoms. This is the point at which the virus passes the threshold of disease. Other pathogens may take longer or shorter periods of time to reach their thresholds. The pathogen is able to replicate relatively well for the first few days of the infection, because there is no adaptive response to contain it. The amount of time necessary to generate an adaptive response is called the *inductive period*. For simplicity, only the humoral response, measured as the amount of antibody, is shown here (in blue). Activation of the cell-mediated response is similar. Once antibody levels begin to rise, viral levels begin to fall, and the patient begins to recover. Symptoms start to wane as viral levels, contained by adaptive immunity, fall below the threshold of disease.

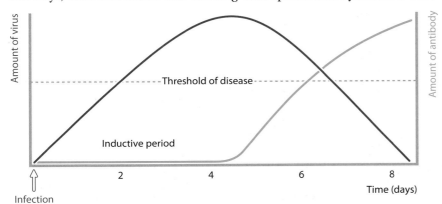

now on their "seek and destroy" mission, levels of virus in Laura's body began to fall, and she began to recover. Gradually, activated lymphocytes began to die off, levels of antibody in the body fluids waned, and the adaptive response came to an end. A recently discovered third group of T cells, called the **regulatory T (T_R) cells**, may also be involved in bringing an adaptive response to an end. Although their exact mode of action is still unclear, it is believed that T_R cells play a role in suppressing the activity of other T cells.

Keep in mind that while the activated lymphocytes do not persist, both T cells and B cells have formed populations of long-lived memory cells. Memory cells can last for years, decades, or even a lifetime, and it is these memory cells that form the basis of immunological memory—the ability of the adaptive immune response to a particular organism to grow stronger with each exposure. We will shortly investigate immunological memory in greater depth.

An adaptive response is not always successful

Unfortunately, not everyone is as lucky as Laura was. Influenza still kills between 30,000 and 50,000 people during a typical year in the United States. During an epidemic, this number can rise considerably. Most fatalities occur among the elderly, in whom the immune response is less vigorous, or in those with underlying respiratory problems, which lessen the effectiveness of barriers to entry and may also impede an effective immune response. In many fatal cases, damage to the respiratory epithelium by influenza virus opens the door to opportunistic bacterial infections, which may result in pneumonia.

Other agents of infectious disease still take their toll in terms of fatalities. As mentioned earlier in this chapter, the likelihood of disease in a host–pathogen encounter depends primarily on three interrelated factors: the number of pathogens to which the host was exposed, the virulence of the pathogens, and/or the status of host defenses. When hosts are exposed to too many pathogens, or if those pathogens are too virulent, even the most robust immune response may not be enough. Additionally, if the host's immune system is in some way compromised, the adaptive response may not be sufficient to protect against even less virulent microorganisms.

CASE: BONNIE

Bonnie is a 30-year-old flight attendant based in Hong Kong. She makes the flight between Hong Kong and Los Angeles approximately every 10 days. She missed one of her flight assignments earlier in the year, in late January, when she was ill with the flu. As Bonnie lay miserably in bed at that time, she became angry with herself for not getting a flu vaccine, which was offered free of charge to all employees of her airline. On this most recent flight, in early March, she interacts with Jeff, the sick passenger, several times, and she can see that he is seriously ill. Bonnie dreads the thought of getting sick again herself, but to her relief, she remains healthy.

1. **Bonnie was infected with influenza virus when Jeff coughed repeatedly in very close proximity to her. In spite of this infection, she developed no signs of disease. Why not?**
2. **How exactly would a vaccine have prevented Bonnie's earlier illness?**

Subsequent exposure to the same pathogen results in a stronger and faster adaptive response

Although Laura and Bonnie were both infected with influenza on their flight to Los Angeles, they had very different experiences; while Laura became quite ill, Bonnie was not even aware that she had been infected. The difference is the immunological memory established in Bonnie when she *was* sick 2 months earlier. Immunological memory is one of the most powerful and distinctive features of the adaptive immune response. It refers to the fact that an adaptive response to a particular pathogen is faster and stronger with each exposure.

When Bonnie became ill in January, the sequence of events was much the same as what we saw in Laura. Innate defenses were unable to contain the infection on their own, and consequently, antigen-presenting cells activated helper T cells in lymphatic organs. Once they were activated, helper T cells activated both Tc cells, to initiate a cell-mediated response, and B cells, to stimulate a humoral response. As we have seen, this series of events takes time. The time between the initial infection and the generation of a full-fledged adaptive response (the inductive period) is usually about 5–7 days. During the inductive period, with neither antibodies nor cytotoxic T cells to impede it, the virus is replicating successfully, and as it increases in numbers, it crosses the threshold of disease and symptoms of illness develop (see Figure 11.21). Once adaptive immunity is in full swing, viral replication slows; and as viral numbers fall back below the disease threshold, the patient starts to recover. This pattern of disease and recovery is typical when the infected individual generates a **primary immune response**—the adaptive response that occurs the first time an individual is infected with a particular pathogen.

Recall that when T cells and B cells are activated and undergo proliferation, some develop into active effector T cells and plasma cells. As the pathogen is destroyed, these relatively short-lived cells die off and the adaptive immune response comes to an end. However, as previously noted, not all the proliferating lymphocytes became effector T cells or plasma cells. Many became long-lived memory T and B cells. These memory cells, some of which may survive for months, years, or even decades, circulate until the same pathogen shows up again, sometime in the future. When it does, memory cells stand ready to quickly differentiate into effector T cells and plasma cells. Rather than a 5–7 day inductive period, within only a day or two, effector Tc cells are destroying virally infected cells, and antibody is present in the blood at high levels (**Figure 11.22**). Although the pathogen

Figure 11.22 Primary and secondary humoral responses. (a) Only the humoral response is shown, for simplicity. The activity of the cell-mediated response (not shown) is similar. During the primary response, the inductive period is about 5–7 days. Initially a small amount of secreted IgM is produced. A larger amount of IgG follows. Later, during a secondary response, memory cells that have persisted since the initial primary response quickly differentiate into plasma cells, resulting in a shorter inductive period and heightened antibody production. (b) The course of infection over this same time period. During the first infection, pathogen numbers cross the disease threshold, and symptoms develop in the infected individual. During the second infection, pathogen numbers stay below the threshold of disease, because the secondary response is so much faster and stronger. Symptoms of disease do not develop. In other words, the infection remains *subclinical*.

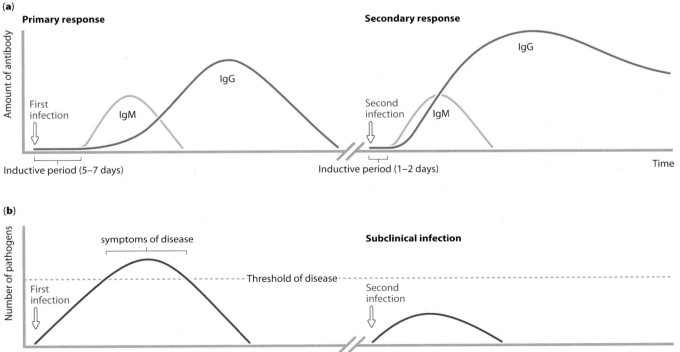

might replicate for a short time, it is quickly confronted by an adaptive response before it passes the disease threshold. Thus, although the pathogen has established an infection, it is stopped dead in its tracks by the presence of memory T and B cells before it can replicate enough to cause symptoms. This is what occurred in Bonnie. During her primary response in January, she became ill and subsequently generated large numbers of memory T and B cells to the antigens on the influenza virus. When she was exposed a second time on the flight to Los Angeles, these memory cells responded so quickly and so strongly that the infection remained **subclinical**—at a population below the threshold of disease. Consequently, she suffered no symptoms. This second, much stronger and faster adaptive response is called a **secondary response**. If Laura, who responded with a primary response to her initial infection, is infected a second time, she too will respond with a secondary response and will probably remain healthy. At worst, she will suffer only mild symptoms. During a secondary response, even greater numbers of new memory cells are generated. This means that if the same pathogen infects for a third time, there will be an even more vigorous adaptive response.

Vaccines induce immunological memory without causing disease

The ability to mount ever-stronger adaptive responses with each exposure is the basis of vaccine therapy. In most vaccines, a primary response is stimulated by exposing the vaccinated individual to either a dead or a weakened form of the pathogen that cannot cause illness. Both sorts of vaccine exist for influenza. If you have been vaccinated against influenza with an injection, you were being exposed to killed virus. Each year these vaccines are reformulated to protect against whichever strains of influenza are most common at the time. Recently, a second type of influenza vaccine has been introduced. Rather than using killed virus (generally called **inactivated virus**), this new vaccine relies on weakened but still viable virus. Although the weakened virus can cause an infection, it replicates poorly and is very unlikely to cause illness. Such a virus is called an **attenuated virus**. The new, attenuated influenza vaccine is inhaled as a nasal mist rather than injected.

Regardless of how vaccines are made, they all work in a similar way. When a vaccinated person is later exposed to a live and virulent form of the pathogen, he or she responds with a secondary response, because memory cells for that pathogen were generated by the exposure to the vaccine.

Sometimes, even a secondary response will not reliably prevent disease. In these cases, more than one vaccine is necessary. Hepatitis B is a good example. When someone is vaccinated against the hepatitis B virus, he or she receives three injections. With the first injection, a primary response is mounted. The second injection some time later is termed the first "booster" because it results in a secondary response, "boosting" the number of memory cells to the viral antigens. The third injection (the second booster) causes a third response (also called a tertiary response), generating even more memory cells. Therefore, if a fully vaccinated individual is infected with the actual hepatitis B virus, the immune system will respond for the fourth time with a quaternary response. So many memory cells would now be ready to respond that the chances of actually developing the disease are extremely slight.

The hepatitis B vaccine does not rely on either inactivated or killed virus and thus represents a third type of vaccine called a **subunit vaccine**. A subunit vaccine does not even contain an entire pathogen. Rather, the vaccine contains only those pieces of protein known to be strong antigens for the pathogen and therefore powerful stimulators of adaptive immunity. In the hepatitis B vaccine, all three injections contain these same viral proteins.

Why might a physician opt for one type of vaccine over another? Generally, an attenuated vaccine results in better immunological memory; although the pathogen is weakened, it *does* result in an actual infection. Consequently, an attenuated vaccine is a closer mimic to what occurs during an infection with the pathogenic microbe. Because the attenuated pathogen replicates, the adaptive immune response is stimulated strongly, resulting in the development of strong immunological memory. A killed pathogen may not persist in the body long enough to fully stimulate adaptive immunity in this way.

You may then wonder why we use inactivated vaccines at all, if attenuated vaccines result in stronger memory and therefore better protection. The answer is straightforward: in some cases, killed vaccines are safer. With an attenuated vaccine, because the pathogen is alive, there is always a slight chance that it will mutate back to its virulent state. In such a case, a vaccine can, on rare occasions, *cause* the disease it was designed to prevent. This problem has not, however, been observed with the attenuated influenza vaccine.

The attenuated polio vaccine, on the other hand, can occasionally revert to virulence. Assuming you have been vaccinated against polio, depending on when and where you were born, you received either the injected Salk vaccine or the liquid Sabin vaccine that you drank in a sugary solution. The Salk vaccine is an inactivated vaccine; it contains killed poliovirus. The Sabin vaccine is an attenuated vaccine, composed of weakened strains of the virus that elicit immunity without causing disease.

Because the virus in the Sabin vaccine is alive, it replicates slowly in the body, and it mimics a natural poliovirus infection, inducing a strong primary response. The Salk vaccine, on the other hand, does not replicate and its persistence in the body is reduced. The memory that is generated in response to the Salk vaccine is consequently somewhat less, which explains why the Salk vaccine is fully protective only after several boosters.

Occasionally, however, the virus in the Sabin vaccine mutates from its weakened form back into the virulent form. This possibility is slight, but about one in every 2.6 million people who get the Sabin vaccine actually develop paralytic polio for just this reason. This may be an acceptable risk if the likelihood of a natural poliovirus infection is great, which accounts for the preferential use of the Sabin vaccine when polio was more common. In recent years, however, the chances of naturally contracting polio in a developed country such as the United States are extremely slight. Consequently, health authorities opt to use the safer, albeit less protective, Salk vaccine.

HIV: A Problem of Immune System Destruction

CASE: OSCAR

Oscar has been HIV-positive for 10 years. He originally sought medical attention when he developed severe respiratory illness, later diagnosed as pneumonia caused by *Pneumocystis carinii*. This opportunistic fungal pathogen does not ordinarily cause disease in healthy individuals. When he was found to be HIV-positive, his physician prescribed appropriate antiviral drugs, and Oscar's health improved dramatically. For the past several years, he has had no serious health problems whatsoever, and except for the drugs he must take daily, Oscar has resumed a normal life. Lately, however, he has become somewhat lax about taking his medications. On this recent trip to Asia, which lasted over a month, he became especially negligent, often going for several days at a time without taking the antiviral drugs. He rationalizes this by thinking to himself that he will "get serious" again about his medications once he returns home.

On the flight to Los Angeles, however, he becomes infected with influenza virus and develops a very severe case of the flu, which requires hospitalization. In addition to serious symptoms of influenza, Oscar develops a case of bacterial pneumonia and he is placed on a respirator in intensive care. With careful monitoring and supportive care, he recovers, but he receives a stern warning from his doctor to take his anti-HIV medications regularly and without fail.

1. Why was Oscar's bout of influenza so severe? Why did he suffer from a bacterial infection in addition to the influenza?
2. Could an influenza vaccine have prevented illness in Oscar?

By any measure, the human immunodeficiency virus (HIV) represents a public health crisis of catastrophic proportions. Over 40 million people are believed to be infected worldwide, and in some parts of the world, notably sub-Saharan Africa and Southern and Southeastern Asia, new infection rates continue to skyrocket (**Figure 11.23**).

Because acquired immune deficiency syndrome (AIDS) is essentially a disease of the immune system, it is appropriate to consider Oscar's experience and how HIV interferes with an effective immune response. Such a discussion will not only reinforce recently reviewed principles but will also allow us to better understand one of the most important world health issues of our time.

Recall that Oscar's first indication that he had a serious health condition was the pneumonia caused by *Pneumocystis carinii*. Almost all cases of pneumocystis pneumonia occur in individuals with severely suppressed immune systems. In fact, prior to the beginning of the AIDS epidemic, the illness was so rare that if a physician in the United States needed to treat it, he or she had to request the drug of choice, pentamidine, from the Centers for Disease Control (CDC) in Atlanta. In a normal year there might be a few requests for the medication. A sudden spike in requests for pentamidine in 1981 was the red flag that alerted health officials that something unusual was going on. Investigations into the mystery revealed that a strange new disease had reared its head, and in June 1981, the CDC's *Morbidity and Mortality Weekly Report* (MMWR) described five previously healthy, homosexual men in Los Angeles who had been diagnosed with pneumocystis

Figure 11.23 The global AIDS epidemic. Estimated worldwide distribution of AIDS cases as of 2007.

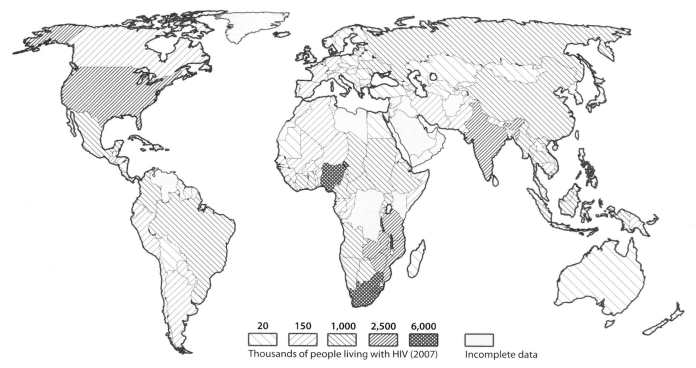

| 20 | 150 | 1,000 | 2,500 | 6,000 |

Thousands of people living with HIV (2007)   Incomplete data

pneumonia (**Figure 11.24**). The report raised the possibility of a link between lifestyle and disease, and on the basis of clinical examination of the five patients, it speculated that the ailment might be caused by reduced immune function.

It was soon realized that the condition was not limited to the homosexual population, and the new disease was given its current name of acquired immune deficiency syndrome or AIDS. In 1983 the responsible pathogen was discovered, a previously unknown human retrovirus (see Chapter 4, p. 97) named the human immunodeficiency virus, or HIV.

HIV is typically transmitted via sexual intercourse, shared blood products, or passage from mothers to infants. As we learned in Chapter 4, viruses recognize their target cells when proteins on the viral surface are able to bind to specific molecules found on the surface of host cells. In the case of HIV, that host molecule is called CD4. The virus recognizes and binds to CD4 by virtue of its envelope proteins (**Figure 11.25**). And crucially, CD4 is found on the surface of helper T cells. Consequently, helper T cells are targeted for infection by this virus.

Once the virus' envelope proteins bind CD4, a series of events begins that allows the virus to enter the host cell's cytoplasm. The virus then releases its genetic material, and viral replication begins. Newly produced progeny virus leaves the infected cell and the cell is killed. Normally T cells survive for years, but after HIV infection, most infected cells die within a day or two.

By examining **Figure 11.26**, it becomes clear how HIV hamstrings an effective immune response. Neither cytotoxic T cells nor B cells have CD4 on their surface. Consequently HIV cannot infect them. But as helper T cells decrease in number due to continuing viral replication, it becomes progressively more difficult to activate either B cells or cytotoxic T cells. Although healthy B and Tc cell populations persist, eventually neither an effective humoral nor cell-mediated response can be mounted. It is at this point that an infected individual becomes susceptible to the opportunistic infections that are the hallmark of AIDS (see Chapter 10, p. 262, for a complete discussion of opportunistic infections).

The good news about HIV is that if an infected person takes his or her medications regularly, the ability of the virus to replicate is inhibited and the number of helper T cells can rebound. In Oscar's experience, as is often the case, helper T cells recover sufficiently to the point where opportunistic infections become far less common. As we will discuss in Chapter 12, however, these anti-HIV drugs do not *cure* the infection; the virus persists at low levels, and if the drug regime is not followed closely, viral replication resumes and helper T cell numbers once again begin to fall. This is what happened to Oscar. Once he became sloppy with his medications, the virus resumed rapid replication and his helper T cells declined in number. When he was exposed to influenza, his adaptive response was therefore less vigorous than normal, and the influenza virus was better able to replicate. Furthermore, in his compromised immune state, and with damage being done to his respiratory system by the rapidly replicating influenza virus, Oscar became susceptible to a bacterial respiratory infection as well. He is lucky to have survived. When people die of influenza, it is often due to such secondary bacterial pneumonia.

For HIV-infected individuals who are controlling their virus with proper medications, an influenza vaccine is certainly a good idea. Only the inactivated vaccine would typically be used on an immunocompromised individual. In such a person, even the slow growth of an attenuated virus might be enough to result in disease symptoms. Of course, if a person has full-blown AIDS, any vaccine would be of limited value, due to the patient's inability to generate protective immunological memory.

Morbidity and Mortality Weekly Report (MMWR) 1981;**30**:250-2 (June 5, 1981)

Pneumocystis Pneumonia Los Angeles

Center for Disease Control

In the period October 1980-May 1981, 5 young men, all active homosexuals, were treated for biopsy-confirmed Pneumocystis carinii pneumonia at 3 different hospitals in Los Angeles, California. Two of the patients died. All 5 patients had laboratory-confirmed previous or current cytomegalovirus (CMV) infection and candidal mucosal infection. Case reports of these patients follow.

Figure 11.24 June 5, 1981. The first report of the new disease that would come to be called AIDS.

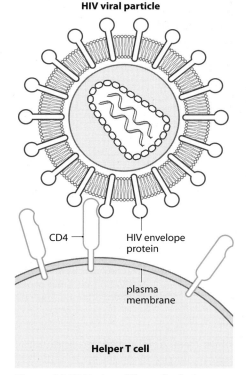

HIV viral particle

CD4

HIV envelope protein

plasma membrane

Helper T cell

Figure 11.25 Recognition of a helper T cell by HIV. The CD4 protein is bound by the HIV envelope protein, allowing the viral particle to recognize a helper T cell. Following binding, the virion is able to enter the host cell's cytoplasm, where it releases its genetic material and begins replicating (see Chapter 4, p. 97, for a review of the retroviral replication cycle).

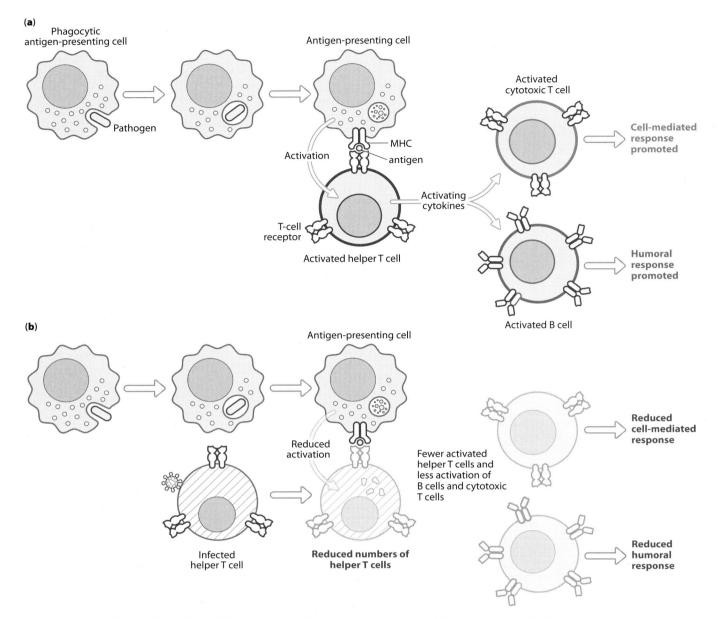

Figure 11.26 How HIV interferes with immune response. (a) In this overview of adaptive immunity, concepts described in this chapter are reviewed. Antigen-presenting cells engulf pathogens and migrate to lymphatic organs, where they activate helper T cells. The activated effector helper T cells activate cytotoxic T cells and B cells, promoting cell-mediated and humoral immunity, respectively. (b) When an individual is infected with HIV, the virus infects and destroys helper T cells, reducing their number. Consequently, although cytotoxic T cells and B cells persist, their activation becomes progressively more difficult as helper T cell numbers continue to decrease. Both cell-mediated and humoral immunity are reduced as a result.

In untreated HIV-infected individuals, the time between infection and severe disease averages approximately 10 years. In the interim there may be few signs of disease, although sometimes there is a short, acute period immediately following infection where the patient may experience fever, rash, and swollen lymph nodes. Note in **Figure 11.27** that this acute phase corresponds to a rapid rise in viral levels and a sharp decline in Th cells.

Why is the period between initial infection and the onset of disease so lengthy in an HIV infection? It is known that the initial infection event stimulates a powerful immune response on the part of the host. This response, both humoral and cell-mediated, keeps the virus in check after its initial

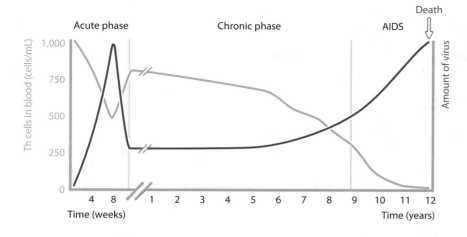

Figure 11.27 Progression of an HIV infection. A normal individual has approximately 1100 helper T cells per milliliter of blood serum. Over the course of an untreated HIV infection, the number of these cells (in blue) rapidly declines during the acute phase, but then rises early in the chronic phase. T cells then begin a long numerical decline. A Th cell count of approximately 200/mL marks the point where a protective adaptive immune response can no longer be initiated and serious opportunistic infections set in. At this point the patient is said to have clinical AIDS. Levels of virus in the blood (in red) rise sharply during the acute phase and then stabilize during the chronic phase. Viral levels slowly climb late in the chronic phase and then rise abruptly, signaling the onset of clinical AIDS. All of these time periods are approximate, and the chronic phase can be extended, perhaps indefinitely, with proper treatment.

burst of replication during the acute phase, and the amount of virus in the blood levels off (see Figure 11.27). Although the patient usually shows few signs of illness during this time, the virus is continuing to replicate while the immune response continues to counterattack. As many as a billion new viral particles are produced daily, which infect and destroy helper T cells. New Th cells continue to be produced, replacing those that have been killed, but eventually, the immune response is unable to keep pace, and the virus gains the upper hand. Helper T cells begin a steady decline, while viral levels surge upwards. Opportunistic pathogens such as *P. carinii* and *Candida albicans*, the cause of a fungal overgrowth in the throat called thrush, now gain the opening they require to establish serious infections. A list of some of the more common and medically significant opportunists affecting AIDS patients is provided in **Table 11.1**.

Host Versus Pathogen: A Summary

So, to rephrase the questions we asked earlier, why *do* we sometimes get sick, and when we do, why do we usually but not always recover? As we have learned in the last two chapters, the answer depends both on how pathogens affect *us* (our theme in Chapter 10) and how *we* respond to *them* (reviewed in this chapter). If a potential pathogen is able to infect us, the likelihood of disease is influenced by interrelated pathogen factors such as the number of infecting organisms, the rate at which the pathogen reproduces, and the number and type of virulence factors that the pathogen possesses.

For example, pathogens with powerful virulence factors generally have a lower threshold of disease (**Figure 11.28**); fewer pathogens are needed before this threshold is crossed and symptoms of disease appear. If a pathogen replicates rapidly and/or the infective dose is large, the pathogen can approach the threshold before an immune response can be generated. Alternatively, low virulence, slow growth, and/or a low infective dose reduce the likelihood of reaching the disease threshold.

Opposing the pathogen is the host's immune response. If the response is fast enough or strong enough to keep the pathogen below the disease threshold, the patient will remain healthy. This is what we expect to happen in a memory response. If the pathogens do cause disease, in most cases, once an adaptive immunity is generated, the immune response will generally gain the upper hand, pathogen numbers fall below the disease threshold, and health is restored. Occasionally, however, because of pathogen or host factors, the host is unable to contain the infection, pathogen numbers continue to rise, and the host succumbs as a result.

MICROORGANISM	OPPORTUNISTIC DISEASE
Fungal	
Pneumocystis carinii	pneumonia
Candida albicans	thrush
Bacterial	
Mycobacterium avium	tuberculosis
Viral	
varicella-zoster	shingles
human herpesvirus type 8	Kaposi's sarcoma
cytomegalovirus	inflammation of retina
Protozoan	
Cryptosporidium species	chronic diarrhea

Table 11.1 Some opportunistic infections seen due to HIV infection. A partial list of some of the more common opportunistic pathogens observed in AIDS patients.

Figure 11.28 The likelihood of disease. Four scenarios are depicted here. Many other scenarios are possible. All three graphs are on the same time scale, and we are making the artificial assumption that the immune response (graph c) is the same for all four scenarios. (a) The course of two infections is illustrated. The more virulent pathogen has a lower threshold of disease, because fewer of these organisms are necessary to cause symptoms. If, as in this example, the immune response to both of these pathogens is the same, the less virulent pathogen is unable to reach its threshold, because it is contained in time by the immune response. Only the more virulent pathogen crosses its threshold and causes illness. (b) Two additional infections are followed. It is assumed that these two pathogens have the same threshold but that they reproduce at different rates. The more rapidly reproducing pathogen reaches the threshold before the inductive period is complete. The more slowly reproducing pathogen is contained by the immune response before it can reach the threshold, and therefore does not cause disease. (c) The adaptive immune response that might occur over this same time period.

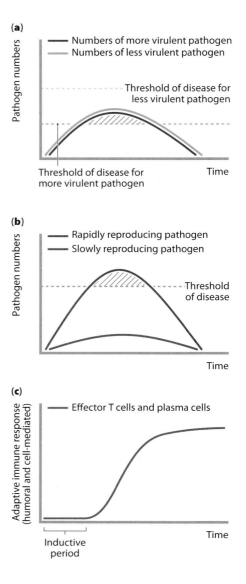

Coming Up Next...

Fortunately, gone are the days when we were unable to reliably and thoughtfully influence microbial growth or, if infected with a pathogen, the best a person could hope for was a successful immune response. We now have numerous methods at our disposal to reduce the risk of infection in the first place or otherwise regulate the growth of microorganisms.

We have mentioned many of these control strategies throughout this text. When we store food in a refrigerator, wipe down a kitchen counter with a disinfectant, or boil a liquid, we are manipulating microbial growth. When we contract an infectious illness, we may avail ourselves of antimicrobial medications. These examples are all designed to limit microbial growth. Yet sometimes, when microorganisms are being cultured for food, industrial, or certain other purposes, stimulation rather than suppression of growth is the goal. We will devote the next chapter to a more thorough consideration of various techniques for controlling the growth of microorganisms, and we will explore some of the factors that influence the choice of a particular control strategy.

Choosing the right strategy is particularly crucial in the matter of antibiotics. Not every antibiotic is appropriate in every situation, and careful evaluation is called for to select the best drug for the job. Furthermore, because these miracles of modern medicine have so often been misused or overused, we now face the frightful reality of antibiotic resistance. In Chapter 12 we will consider how to best use antibiotics, why antibiotic resistance has become a problem, and what might be done to reverse it.

Key Terms

adaptive immune response	dendritic cell	interferon
agglutination	effector T cell	leukocyte
anatomical barrier	genetic barrier	lymph
antibody	helper T cell (Th cell)	lymph node
antigen-presenting cell	humoral response	lymphatic system
antigen	immunological memory	lymphocyte
attenuated virus	immunology	macrophage
B cell	inactivated virus	major histocompatibility complex
cell-mediated response	inductive period	(MHC)
cytokine	inflammation	memory B cell
cytotoxic T cell (Tc cell)	innate immune system	memory T cell

microbial barrier	plasma cell	subunit vaccine
neutralization	primary immune response	T cell
neutrophil	regulatory T cell (Tr) cell	T-cell receptor
opsonin	secondary response	thymus
pattern recognition	spleen	Toll-like receptor (TLR)
permeability	subclinical infection	vasodilation

Concept Questions

1. Meningitis is caused by the bacterium *Neisseria meningitidis*. This microbe is often found in the throat of healthy people, but it sometimes penetrates the epithelium in the throat and invades the nervous system, causing meningitis. In North Africa, there are regular epidemics of meningitis. These outbreaks almost always occur during the dry season. During the rainy season, although the bacteria are still present in the throat and are still transmitted, they are far less likely to penetrate the epithelium and cause disease. Explain this observation, on the basis of what you learned about barriers to entry.

2. Although they make us feel better, medications to reduce inflammation during illness may in some cases actually impede immune function. Why is this?

3. Innate and adaptive immunity are often discussed separately in textbooks as a matter of convenience. Yet this separation is somewhat artificial. Provide two examples of how innate immunity and adaptive immunity depend on each other.

4. People who have had their spleen removed can usually live normal lives, but some, especially children, are placed on prophylactic antibiotics to prevent infection. If they must undergo surgery or even dental work, the dose of the antibiotic is often increased. On the basis of what you learned about the spleen, what do you think is the rationale for the use of prophylactic drugs?

5. In a famous experiment in immunology that demonstrated the concept of antigen presentation, scientists found that they could stimulate antibody production under laboratory conditions in one of the following situations. On the basis of what you have learned about the immune system, in which of the following scenarios do you think antibodies would most likely be produced? Explain your reasoning.
- Scenario A: Macrophages are exposed to antigen and are then mixed with B and T cells.
- Scenario B: B and T cells are exposed to antigen and are then mixed with macrophages.
- Scenario C: Macrophages are exposed to antigen and then mixed with B cells only.
- Scenario D: Macrophages are exposed to antigen and then mixed with T cells only.

6. Imagine that mice have their thymus removed at birth, before the immature T cells in the bone marrow that have migrated to the thymus can mature. What sorts of immune responses would these mice still be capable of? Which types of responses would no longer be possible?

7. Colds are caused by a large number of different viruses. One of the effects of aging is that as we get older we get fewer and fewer colds. Children get frequent colds, but older adults rarely get colds. Explain this phenomenon on the basis of what you have learned about the immune system.

The development of penicillin from the mold *Penicillium* changed medicine forever.

Chapter 12

Control of Microbial Growth

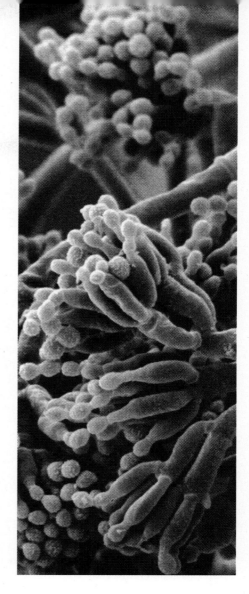

A hospital can be a scary place. In the United States, it has been estimated that up to 10% of all hospital patients acquire a **nosocomial** (from the Greek for "hospital") infection—an infection that is the *result* of a hospital stay. An estimated 2 million patients a year become infected *because* of their hospitalization. Over 70,000 people a year probably die of nosocomial infections, making such infections the eighth leading cause of death in America. The story is much the same in countries such as France, Italy, and Great Britain.

Why are hospitals such hotbeds of infection? Hospitals house large numbers of people who are already sick, many of whom have weakened immune systems or damaged barriers to entry. Medical staff move from patient to patient, providing an easy way for pathogens to spread. Even normal flora on the hands of a staff member can prove dangerous if mistakenly introduced into a wound or a surgical incision.

Many nosocomial infections are preventable with basic control measures. Some of these measures are so simple that they are often overlooked or performed haphazardly. Proper hand washing is probably the single most important procedure that can reduce the transfer of microorganisms between individuals in hospitals and therefore decrease the likelihood of nosocomial infections. Many infections are caused when medical devices such as catheters are not sterilized properly prior to their use. Other problems, however, are more complex. Antibiotic resistance, for example, is an especially thorny issue in the hospital environment, and many infections that could once be easily cured simply no longer respond to the best available drugs.

There is still much, however, that can be done to control microorganisms in the hospital, the home, the laboratory, and elsewhere, and in this chapter we will investigate many of the strategies, as well as the problems, associated with such control. We will also see that the control of microbial growth does not only mean *limiting* their growth. In some situations, such as in the production of many foods and industrial products, microbial growth is something we wish to encourage, and we will also explore how the growth of useful microbes can be facilitated.

But as the example of nosocomial infections should make clear, we rarely if ever truly "control" microorganisms. The best we can really hope for in our relationship with microbes is coexistence—encouraging the growth of some, while trying to keep the dangerous ones at bay. Let us consider some of the ways that coexistence can be achieved. For convenience, the strategies employed can be divided into physical and chemical methods.

Physical and Chemical Means of Control

Killing microorganisms requires no great insight or cleverness; we do it all the time without a moment's thought. We boil, incinerate, or dehydrate microorganisms, or douse them in lethal chemicals as we clean our houses, cook our food, or perform other ordinary tasks. The challenge is to select the right technique for the right situation. High temperature will kill bacteria, but if your goal is to sterilize plastic tubing on a piece of medical equipment, excessive heat may damage the plastic. If you want to be certain that milk is microbe-free, boiling the milk will certainly do the trick, but the milk will not taste very good. When selecting a method to control microbial growth, we need to consider where and what we want to control. Different tasks require different strategies.

Physical control of microorganisms involves manipulation of specific environmental factors such as temperature

We can control microorganisms by a variety of physical methods, including heat, cold, filtration, and radiation (**Table 12.1**). Some of these techniques have been utilized for millennia. Ancient people used salt and drying to preserve food. Romans knew that burning corpses during epidemics helped limit the spread of disease, and Greek armies learned to boil water before drinking it while on the march. Here we will describe some of these physical methods and consider when their use is appropriate.

Temperature. In Chapter 7 we learned that all microorganisms have preferred temperatures at which they grow best. The also have minimum and maximum growth temperatures, below or above which they are unable to grow. Temperature, accordingly, is one of the most direct and inexpensive ways to encourage or inhibit microbial growth. When the production of large numbers of microorganisms in the quickest possible time is the goal, as it often is in the food industry or in a research laboratory, the optimum growth temperature for the species in question is selected. Recall from Chapter 7 (see Figure 7.29) that microorganisms differ in their temperature

METHOD	COMMON USES	EFFECTIVENESS
Physical Methods		
oven	glassware	100%
autoclave	glassware, fabrics, liquids, medical instruments	100%
boiling	liquids	kills bacteria and viruses; does not destroy all spores
flash pasteurization	beverages	kills most pathogens
filtration	heat-sensitive liquids	removes bacteria
UV light	air and surfaces	kills bacteria with direct exposure
Chemical Methods		
iodine	antiseptic	kills bacteria
alcohol	skin and surfaces	removes bacteria and some viruses
soap	antiseptic	removes bacteria
bleach	disinfectant	kills most bacteria and viruses

Table 12.1 Some representative methods for controlling microbial growth.

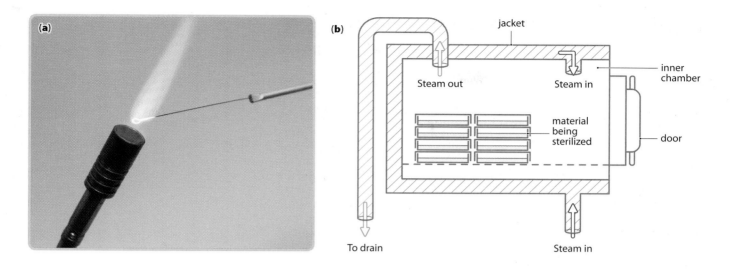

Figure 12.1 Heat as a sterilization technique. (a) Sterilization of an inoculating loop. Incineration destroys all microorganisms that may be contaminating an inoculating loop. The loop is routinely used in microbiology laboratories to inoculate media. (b) An autoclave. Steam pumped into the inner chamber both heats the chamber and increases its pressure, usually to 121°C and 15 pounds per square inch of pressure. Proper autoclaving will completely sterilize any material in the inner chamber. Following autoclaving (usually after 15 minutes), steam is allowed to escape through an outlet valve.

preferences and that the optimum temperature is quite different for psychrophiles, mesophiles, and thermophiles. Likewise, if high or low temperatures are to be used to limit or halt microbial growth, the particular temperature limits of the species of interest must be considered.

Heat is an especially effective means of **cidal** control; the suffix "-cidal" indicates that microorganisms are killed. Bactericidal agents kill bacteria, and fungicidal agents kill fungi. The term "germicidal" ("germ killing") is used to describe the general killing of microorganisms. Unlike certain other methods such as UV radiation, which kill only those microorganisms living on a directly exposed surface, heat can penetrate an object and kill microorganisms living throughout. Heat kills microorganisms in large part because of its ability to denature proteins (see Chapter 2, p. 36).

Heat is often used as a way to **sterilize** certain objects or materials. Sterilization refers to the complete elimination of all organisms. Either dry or moist heat can be used in this way (see Table 12.1). Students in a microbiology laboratory quickly learn that passing a metal inoculating loop through a flame is a fast and effective way to eliminate any organisms on the loop (**Figure 12.1a**). Likewise, an oven can easily be used to sterilize glassware and other heat-tolerant materials. Dry-heat sterilization, however, usually requires considerable time, especially for heat-resistant organisms or endospores. Moist heat penetrates more quickly and is effective at lower temperatures. Often, in laboratory or medical settings, an **autoclave** is used to sterilize materials with moist heat (Figure 12.1b). An autoclave works on the same principle as a pressure cooker. It consists of a metal container that can be filled with pressurized steam. As the steam penetrates the objects placed inside the autoclave, microorganisms are killed. Autoclaves are completely effective in sterilizing a wide range of materials, including less heat-resistant objects such as towels, which would be destroyed by the higher temperatures necessary for dry-heat sterilization. Autoclaves may also be used to sterilize liquids. Although boiling can eliminate microorganisms as well, endospores and some thermophiles can survive prolonged periods of time in boiling water. Autoclaves work better because at high pressure, the temperature in an autoclave rises considerably above the boiling point, even above the temperature necessary to kill endospores. However, when in doubt, boiling is a useful way to make sure certain foods and water are safe. A camper who takes water from a stream and boils it for 15–20 minutes has little to fear from drinking the water.

Some liquids, however, are damaged by boiling. Yet products that we would not boil such as a quart of orange juice or a bottle of wine are still safe to drink because they have been pasteurized. Pasteurized beverages, however, are not microorganism-free. **Pasteurization**, the temporary heating of liquids (see Chapter 5, p. 124), destroys pathogens. Some highly heat-tolerant species, however, survive. How does pasteurization target those organisms that are the most worrisome to us?

During *flash pasteurization*, liquids are generally exposed to a temperature of 72°C for 15 seconds (see Table 12.1). In *batch pasteurization* the liquid is subjected to a temperature of 65°C for 30 minutes. Because it causes less flavor change, flash pasteurization is usually preferable. Both of these processes kill mesophiles, which are most likely to be pathogenic. Surviving thermophiles are far less likely to be involved in disease processes, simply because body temperature is too cool for them to thrive. Therefore, even pasteurized milk contains many microorganisms, but they are not the ones that pose a threat to our health. They can, however, cause milk to spoil, which explains why even milk in an unopened container eventually goes bad. When a longer shelf life for milk is required, the modern technique of "ultrapasteurization" might be used. In this process, the liquid is exposed to a temperature of 134°C for only 1–2 seconds. Not only does ultrapasteurized milk keep for up to 3 months, it can also be stored without refrigeration—an important consideration in parts of the world where a refrigerator is still considered a luxury.

Cold temperatures alone generally do not kill microorganisms. Rather, when the temperature dips below the minimum growth temperature, microorganisms are unable to reproduce. Such control measures that inhibit microbial growth are indicated by the suffix "-static" (**Figure 12.2**). When food is placed in a refrigerator, the food remains fresh for longer periods of time because the growth of most microorganisms, including most pathogens, is held in check. Obviously, anyone who has hauled some long-forgotten casserole or piece of cheese out of a corner of the refrigerator, only to find it spoiled, has learned that cold does not eliminate all growth. The organisms able to grow at such low temperatures usually pose little disease risk, but *Listeria monocytogenes* is an exception. This Gram-positive bacterium is able to grow at refrigerator temperatures, especially in dairy products. Most infections are asymptomatic, but occasionally the pathogen can invade the central nervous system, where it can cause meningitis.

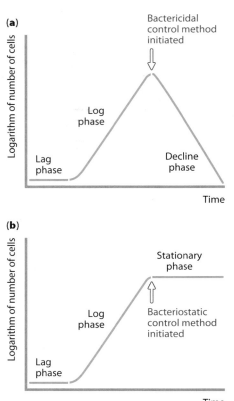

Figure 12.2 Growth curves for bactericidal and bacteriostatic control methods. (a) Bactericidal control methods kill bacteria and move the growth curve immediately into the decline phase. (b) Bacteriostatic control inhibits further cell replication, moving the curve into the stationary phase. For a review of the microbial growth curve, see Chapter 7, (p. 196).

CASE: COLD, HARD FACTS ABOUT COLD-FILTERED BEER

Anyone who has walked down the beer aisle at the local supermarket is familiar with the large variety of choices that confronts the consumer. In addition to the usual brands, many major breweries also market a "cold-filtered" product as well. Certainly the television commercials for cold-filtered beer make it appear especially refreshing. But is cold-filtered beer in any way better tasting than the more conventionally produced product?
1. What exactly is cold-filtered beer and why is it made this way?
2. Does the cold-filtering process actually result in a superior product?

Filtration. Some liquids cannot tolerate even the subboiling temperatures of pasteurization. For example, a biologist studying a protein in solution might need to keep the solution cold to avoid any change in the protein's three-dimensional structure but may still need to eliminate any bacterial contaminants. In this case, the solution might be filtered (see Table 12.1). The most common filters used in laboratories for heat-sensitive liquids are membranes made of nitrocellulose (a modified form of cellulose) that have a pore size of about 0.45 micrometer (μm). This is small enough to remove bacteria, although some viruses are able to pass through pores of this size. In fact, viruses were first discovered when it was found that some sort of infectious agent was able to pass through filters that could retain bacteria. Filtering is still used by virologists today to ensure that viral cultures are bacteria-free.

Beverages can also be filtered, when pasteurization is thought to adversely affect flavor. This is the rationale behind cold-filtered beer. Most beer, like most commercially sold beverages, is pasteurized to ensure its safety. If you believe the advertisements, however, pasteurization damages the flavor of beer. If beer is filtered, on the other hand, bacteria can be removed while keeping the beer at low temperature, which is supposed to result in superior taste. We'll let the consumer judge whether or not this is really the case.

Radiation. In Chapter 6 (p. 163) we saw how radiation can damage genetic material and thus act as a mutagen. It is this ability to introduce mutations that explains how radiation can be used to control microbial growth. Gamma radiation (ionizing radiation) causes water to split into highly reactive molecules such as OH^- and toxic oxygen ions such as O_2^-. Ultraviolet radiation (nonionizing radiation) causes the formation of thymine dimers that inhibit proper DNA replication. Both of these forms of radiation have applications in microbial control.

Gamma radiation can be used to kill microorganisms in food products, although this practice remains controversial. It is widely used to sterilize medical equipment such as plastic instruments, vaccine preparations, drugs, and other materials that cannot be filtered or subjected to heat. It may also be used to eliminate any microbial contaminants from tissue intended for transplantation such as heart valves.

Although sterilization is rarely achieved with ultraviolet (UV) lamps, microbial numbers on surfaces or in the air can be dramatically reduced. Such lamps have value wherever the likelihood of microbial pathogens is either high or especially worrisome. For instance, UV radiation can be used in hospital operating rooms to reduce the likelihood that someone undergoing surgery will be exposed to airborne microorganisms that might otherwise settle into an open surgical incision. Because UV light can also damage our DNA, and because it poses a risk to eyesight, such lights are applied to the room environment only between surgeries—not when people are actually in the operating room. Other similar situations where UV lamps may be used in this way include nursing homes, child care facilities, prisons, and food preparation areas. UV radiation is also commonly employed in water treatment facilities, where it is used as a means of water purification (**Figure 12.3**).

Drying. The next time you eat a piece of beef jerky or a slice of salted fish, you might consider that these products have been preserved with methods

Figure 12.3 Controlling microorganisms with ultraviolet light. At this water treatment plant ultraviolet light is being used to kill microorganisms in drinking water. As discussed in Chapter 6 (p. 163), UV light results in the formation of thymine dimers in the DNA of microorganisms, thereby inhibiting DNA replication and, consequently, microbial growth.

Figure 12.4 Drying is an age-old way of preserving fish and other food products. In Arctic villages, fish is preserved in this manner to provide food throughout the winter.

that date back to antiquity. A newer but similar process is freeze-drying, a technique that produces the Stroganoff you might take on a backpacking trip.

Native Americans have dried meats for hundreds of years as a way of preserving them (**Figure 12.4**). The practice of using salt to draw water out of foods has an equally ancient pedigree. In fruits, sugar may be used in place of salt to remove water. Whether water is removed by simple evaporation or through salt or sugar preservation, the consequences for microorganisms are the same; they are unable to reproduce in the absence of water. Few microorganisms are actually killed, however, and should water become available, growth will begin again.

Freeze-dried foods are likewise preserved when water is removed from a food item. After being placed in a chamber under a partial vacuum, the food is quickly frozen at temperatures of about −75°C while water is removed by vacuum pressure.

Chemical methods can control microorganisms on living and nonliving material

Many chemical compounds can also be used to control microbial growth (see Table 12.1). Like physical methods, different chemicals are appropriate in different situations. Although bleach, for instance, is bactericidal, you certainly would not use it as a mouthwash or on an open wound to reduce the risk of infection. Agents such as bleach, appropriate for use on nonliving material, are termed **disinfectants**. To control microorganisms on living tissue, less damaging **antiseptics** such as iodine and topical sprays such as Bactine® may be used. In this section we will briefly discuss some common disinfectants and antiseptics used to control microorganisms in the environment. In the next section, we will investigate how **chemotherapeutic agents** or antimicrobial drugs are used to combat microorganisms inside the body of a human or animal host.

A variety of chemicals have antiseptic and disinfectant properties

Iodine and chlorine. Just about everyone has at one time swabbed an iodine solution onto a scrape. Likewise, campers know that iodine tablets can be added to water to ensure its safety. Perhaps surprisingly, even though iodine is commonly used as an antiseptic, its exact mode of activity is not clearly understood. A likely mechanism is that it binds to the amino acids of enzymes and other proteins, inhibiting their activity.

Chlorine gas, compressed into a liquid form, is often used to protect drinking water and swimming pools from potentially harmful microorganisms. Chlorine exerts its germicidal activity by forming an acid when added to water (see Chapter 2, p. 25). Chlorine, in the form of sodium hypochlorite (NaOCl), is also the active ingredient in bleach, which can be used as a household disinfectant, or, in small amounts (2–4 drops per liter), to ensure the safety of drinking water.

Alcohol. Great for disinfecting a thermometer, telephone mouthpieces, or other small objects, alcohols kill microorganisms by denaturing proteins and disrupting lipid bilayers in plasma membranes. They are effective against bacteria, fungi, and enveloped viruses. Nonenveloped viruses and endospores are not destroyed. Alcohol is also used to remove microorganisms from surfaces, and from the skin before an injection is given. In spite of what you have seen in movies, where the grizzled old cowboy soaks a bullet wound in whiskey to prevent infection, alcohol is a poor wound antiseptic because it causes host proteins to congeal. Beneath this coagulated layer of proteins, bacteria can continue to reproduce.

Phenols. In Chapter 5, we learned how the English physician Joseph Lister in the late 1800s first treated wounds and bandages with phenol to control infection. Phenolic compounds kill microorganisms by interacting with and denaturing proteins. They are not commonly used as antiseptics today, but they remain one of the active ingredients in Listerine®, named in honor of Lister. When the makers of this product advertise that their mouthwash "kills the germs that cause bad breath," the phenolics are the reason why. Furthermore, every time you spray Lysol® on your toilet bowl you are using a similar phenolic compound called *para*-cresol to eliminate bacteria and other microorganisms.

CASE: WASH YOUR HANDS FIRST!

In her microbiology laboratory class, Becky has heard her instructor say it over and over again: before beginning work, wash your hands carefully to prevent contamination. Becky understands the reasons for this. Normal flora bacteria are common on the hands. Some of these bacteria, if you work without first removing them, are likely to come off on your cultures. Then, when you perform tests on these cultures, you cannot be sure whether the results you see are caused by the bacteria in your cultures, as you intended, or by contaminating organisms from your hands. Becky wonders, however, if this is really such a problem. To find out, she obtains three agar media plates and performs the following experiment; on plate 1, she presses her right index finger onto the agar, without washing her hands. She then gives her hands a "routine" washing— the way someone might before sitting down to eat. She inoculates plate 2 with her middle finger. To conclude, she next washes her hands vigorously for 12 minutes, using lots of soap, hot water, and a scrub brush. She then presses her ring finger onto plate 3.

After a 24-hour incubation at 35°C, she examines her plates. There must be some mistake! As expected, plate 3 has few growing bacterial colonies but plate 1 actually has less bacterial growth than plate 2 (**Figure 12.5**). What happened?

1. What explains Becky's results?

Soaps and detergents. What do a bar of soap and laundry detergent have in common? They both get rid of dirt and they both reduce microbial numbers. Furthermore, they both do it in the same way. We have learned that microorganisms often have a variety of structures, such as capsules and fimbriae, to help adhere to surfaces such as skin and clothing. Soaps and

Plate number 1.

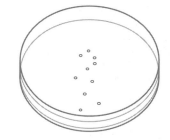

Plate number 2.

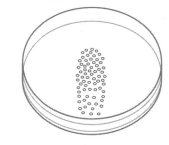

Plate number 3.

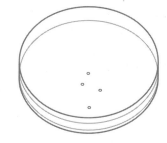

Figure 12.5 Hand washing and bacterial numbers. Plate numbers correspond to plates described in the case. Each red dot represents one bacterial colony.

Figure 12.6 Effect of heavy metals on bacterial growth. The early-20th-century cent and dime were composed mainly of copper and silver, respectively. The clear zones around these two coins are "zones of inhibition" where the bacteria are unable to survive because these metals are present. The metal found in the nickel has little negative effect on the bacterial species in this culture.

detergents disrupt this adherence. They do so by inserting themselves between the microorganisms and the surface to which they are adhering. At this point, the microorganisms are washed away with water. Because molecules of the active ingredients of soaps and detergents have both hydrophobic and hydrophilic regions (see Chapter 2, p. 25), they can also kill some microorganisms by penetrating their plasma membranes, causing leakage and ultimately cell death.

In light of this information, perhaps Becky's results make sense after all. The normal skin flora, adhering tightly to her skin, did not easily come off Becky's dry finger onto plate 1 (see Figure 12.5). The routine washing she performed before pressing her middle finger onto plate 2 was enough to loosen many of the bacteria on her skin, meaning lots of cells were freed up to inoculate the plate. The final, thorough scrubbing she gave her hands detached and washed away almost all of the cells before she inoculated plate 3.

So does this mean that when we briefly wash up before a meal, we are doing more harm than good? No. Normal washing with soap *does* remove dirt and some bacteria. And remember, these are normal flora bacteria. You are already exposed to them all the time, so the risk from them is minimal. But Becky's results do show that a routine wash is not good enough before working in a microbiology laboratory, and it is certainly not sufficient in a hospital setting. Especially before performing surgery, a brisk 15-minute scrub with soap containing bactericidal chemicals is required. Without such precautions, there is a serious risk that normal hand flora could be introduced into surgical wounds, causing an opportunistic infection.

Heavy metals. Salts of heavy metals latch onto sulfhydryl groups (–SH) on certain amino acids, thereby interfering with proteins and killing microbial cells (**Figure 12.6**). Unfortunately, they do the same to eukaryotic cells, which explains why heavy metal compounds are rarely used today as antiseptics. Before the development of antibiotics, however, silver salts were commonly used to prevent infection, especially in the eyes of newborns where trachoma could result in blindness. Mercuric chloride was once widely used as an antiseptic, but due to its toxicity it is no longer employed.

Antimicrobial Chemotherapy

For a real-life horror story, ask your grandparents or others of their generation about microbial disease before the mid-1940s. Pneumonia and bacterial sepsis are still causes for serious concern, but before World War II, infections such as these were downright terrifying. Likewise, there was no such thing as "minor surgery." Procedures that are considered fairly routine today could be life-threatening in the pre-antibiotic era. During any invasive procedure, there is a reasonable likelihood that bacteria will find their way into what *should* be sterile body regions. Today, prophylactic antibiotics prior to surgery greatly reduce that risk, and if an infection should occur, doctors have an arsenal of powerful drugs to combat it. Before antibiotics became available, besides trying to reduce symptoms and hope for a successful immune response, there was little that could be done to treat an infection.

There *was* and *is*, however, plenty that can be done to prevent infection in the first place. So before we delve into a discussion of antimicrobial drugs, consider **Figure 12.7**, which illustrates changes in human survival since the mid-1800s. As we can see, the largest improvement in survivorship occurred prior to the mid-1930s—before antibiotics were even available. During that time, a better understanding of the "germ theory of disease" was accompanied by the introduction of aseptic techniques in medicine. Likewise, improved sanitation greatly reduced the transmission of food and water-borne illness. The smaller, but by no means insignificant, increase up through the mid-1950s can be attributed in large part to the development of antibiotics. You will note that the most recent breakthroughs in medicine, since about the mid-1980s, have had only a modest impact on average human life expectancy. The contribution of antibiotics has been great, but that of good plumbing has been much greater.

Nevertheless, the development of antimicrobial drugs must be considered among the milestones of scientific achievement, and the history of their development is a fascinating tale of foresight, inquiry, persistence, and simple luck. But of course, even dumb luck is of little value unless the lucky individual has the presence of mind to recognize and act upon luck when he or she sees it. As the famed golfer Lee Trevino said, "The more I practice, the luckier I get." Like golf, the history of drug development is replete with people who obviously got plenty of practice.

An ideal antimicrobial drug inhibits microorganisms without harming the host

In the late 1800s, Paul Ehrlich, a German physician, became intrigued with the way certain dyes used to prepare cells for observation under the microscope stained some cells but not others. When he observed that some of these dyes stained bacterial cells but not animal cells, he had one of those "Eureka!" moments that scientists live for: the differential staining properties of the cells must mean that the cells were fundamentally different in structure. Consequently, he thought, it might be possible to develop toxic dyes that would bind to and kill bacterial cells without adhering to and harming animal cells. Such a compound might therefore be able to zero in on and destroy bacterial infections, without undue side effects on the host. Ehrlich's search for what he called the "magic bullet" had begun.

Ehrlich decided to concentrate on *Treponema pallidum*, the bacteria that cause syphilis, as his first target. He initially developed an arsenic-based compound that could easily kill *T. pallidum* but was disappointingly toxic for human cells as well. He then began to systematically alter the structure of his arsenic compound until finally in 1910, after 605 attempts, he came up with a derivative that was highly effective in treating infected laboratory animals. The drug, which he called "salvarsan" (from the words "salvation" and "arsenic"), was subsequently used to treat human syphilis patients. Its use, however, was eventually discontinued because its **selective toxicity**—its ability to kill microorganisms without harming human cells—was far from absolute. Salvarsan was still toxic to humans and even occasionally killed patients. It did, however, offer the possibility of a cure in cases that were previously considered beyond hope, and it demonstrated that selective toxicity was a realistic goal. Pathogens could in theory be eliminated without causing irreversible harm to the human host.

Almost a quarter of a century passed before a truly selectively toxic magic bullet would become a reality. In 1932, another German researcher, the chemist Gerhard Domagk, found that prontosil, a reddish dye, could be used to cure ordinarily fatal bacterial infections in mice. Oddly, prontosil had no effect *in vitro* (under laboratory conditions), when it was added to bacterial cultures. It was effective only *in vivo*, meaning when it was actually administered to living, infected animals. It was later determined that an enzyme in an animal's body breaks the prontosil down into a smaller molecule called **sulfanilamide**, and it is sulfanilamide, not prontosil itself, that

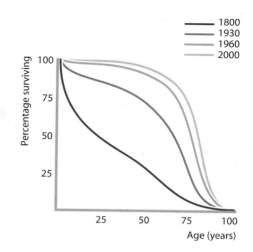

Figure 12.7 Human survival over the last 200 years. The graph depicts the approximate percentage of individuals surviving at any given age at four points in time. The large increase in survivorship between 1800 and 1930 is largely due to improved sanitation. The significant increase between 1930 and 1960 is in large measure due to the introduction of antibiotics. The increase in survivorship since 1960 has been relatively small.

Figure 12.8 Structures of *para*-aminobenzoic acid, an essential bacterial nutrient, and sulfanilamide. Bacteria convert *para*-aminobenzoic acid (PABA) to folic acid, which is required for DNA replication. The conversion requires a specific enzyme. Because PABA and sulfanilamide have similar structures, the enzyme will bind sulfanilamide instead of PABA if the drug is present at high enough concentrations. The enzyme cannot, however, convert sulfanilamide to folic acid, and the bacteria experience a shortage of this essential nutrient, inhibiting their ability to reproduce. Humans obtain their folic acid in the diet and thus do not depend on this enzyme. Consequently, sulfanilamide can be safely used with minimal side effects.

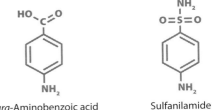

para-Aminobenzoic acid   Sulfanilamide

has antibacterial activity. In a lab culture, this conversion from prontosil to sulfanilamide did not occur. Sulfanilamide was the first of the so-called sulfa drugs, many of which are still used today to treat a wide variety of bacterial infections.

Sulfa drugs work because they have structures very similar to a crucial bacterial nutrient, *para*-aminobenzoic acid, often abbreviated as PABA (**Figure 12.8**). Bacteria enzymatically convert PABA to folic acid, a substance essential for both eukaryotic and prokaryotic cells. But the enzyme that acts on PABA cannot distinguish between PABA and the sulfa compounds. Consequently, if sufficient sulfa is present, the enzyme binds sulfa instead of PABA, and because sulfa cannot be converted to folic acid, the bacteria are unable to replicate. Side effects from sulfa drugs are generally limited, because humans lack the targeted enzyme. We cannot convert PABA to folic acid and must obtain all of our folic acid in our diet.

Sulfa drugs are synthetic chemicals, and therefore they are not technically **antibiotics**, which are a category of specific antibacterial agents that originally came from natural microbial sources themselves. In Chapter 5, we described the discovery of the first true antibiotic when, in 1928, Alexander Fleming noticed that some of his bacterial cultures had become accidentally contaminated with mold. You will recall that a chemical, which Fleming named "penicillin," was being produced by the mold, inhibiting the growth of bacteria in the cultures.

Fleming became frustrated by his inability to purify penicillin, and he did not continue to study it for long. A decade later, however, other British scientists successfully purified penicillin, and in 1941 the antibiotic was first tested on a human patient. Penicillin proved to be extremely useful in treating previously untreatable infections, and with World War II wreaking havoc on Europe, production facilities were set up in America. The development of penicillin changed medicine forever. The antibiotic was so effective that it must have seemed to be a true miracle (**Figure 12.9**). Gone were the days when physicians were more or less helpless when confronted with serious, potentially lethal infections.

Now that penicillin had been developed, the floodgate was opened. Next came streptomycin, an antibiotic isolated from a soil bacterium called *Streptomyces griseus*. With thousands of microorganisms now being screened for useful compounds, the list quickly expanded to include many of the antibiotics we know and use today. The era of antibiotics had begun. Doctors now had an arsenal of drugs from which to choose and were able to select the most appropriate drug for a given situation. But of course, there can be too much of a good thing. A somewhat cavalier attitude developed, and misuse of antibiotics was common. It wasn't long before the first signs of drug resistance appeared, a topic we will explore in more detail later in this chapter.

Figure 12.9 Penicillin: the miracle drug. An advertisement from the mid-1940s for penicillin as a cure for gonorrhea.

A drug's mode and speed of action, the type of infection being treated, the potential for side effects, and the likelihood of drug resistance all influence drug selection

With so many drugs to choose from, how does the physician decide which one to use? Selecting the right drug for the right situation is a complex enterprise, influenced by many factors. Does the patient have any known drug allergies? Has the infectious organism been specifically identified? In which organ or tissue is the infection? These and other factors must be considered before one can decide upon the best antibiotic. Sometimes the deciding factor is not even a medical one. Some drugs for instance, are more expensive than others, and for someone without health insurance, cost may dictate drug choice. For doctors in undeveloped rural areas where electrical supply is uncertain, it may be necessary to opt for drugs that do not require refrigeration. However, here we will focus on some of the important *purely medical* criteria for drug selection.

Speed of drug action. The sulfa drugs previously described, while effective, do not really begin to inhibit bacterial growth for up to a few hours. Most bacterial cells have at least some surplus supply of folic acid. As long as this supply remains, the bacteria can continue to divide. After a number of divisions, the supply of folic acid is exhausted. Sulfa drugs can then exert their influence by preventing further folic acid from being made. Erythromycin, on the other hand, is an extremely fast-acting antibiotic. By binding to and inhibiting the activity of the bacterial ribosome, erythromycin quickly brings bacterial protein synthesis to a halt. This is an especially attractive feature if the bacteria are producing a protein toxin. Erythromycin stops such toxin production almost immediately. This does not imply, however, that the faster drug is always preferable. For example, the physician might consider whether the patient's condition is rapidly deteriorating. If so, clearly speed is of the essence, and the rapidity with which the drug acts may be of overriding importance. In other situations where the patient is relatively stable, or for chronic infections, speed is much less of an issue. In such cases, doctors have the luxury to select a slower-acting drug if it is superior for other reasons.

Bactericidal versus bacteriostatic drugs. Bactericidal antibiotics kill bacteria. Penicillin, for example, by preventing dividing bacteria from properly synthesizing new cell wall material, causes cells to rupture as they absorb surplus water from the environment. **Bacteriostatic** drugs are more subtle; they merely inhibit bacterial cell division and therefore hold the numbers of microorganisms in check. The host immune system is then able to gain the upper hand more easily and destroy the bacteria in question. Both sulfa drugs and erythromycin are bacteriostatic.

Generally speaking, a bactericidal drug is preferable because of its more direct impact. Activity of bacteriostatic drugs relies on the continual presence of the drug until the infection is cleared, because if medication is stopped, the bacteria may begin to reproduce again. With bactericidal drugs, an effect is seen even before an immune response begins. Right from the start, bacterial numbers begin to fall, and as the number of bacteria decreases, symptoms in the patient may start to wane. Of course, in immunocompromised individuals, such as cancer patients undergoing chemotherapy or AIDS patients, bactericidal drugs are essential.

But in some situations, a bacteriostatic drug may be preferable. When Gram-negative bacteria, for example, invade normally sterile body regions, as they might after rupture of the appendix, it might be necessary to use bacteriostatic drugs exclusively. As you may remember from Chapter 10 (see p. 275), when Gram-negative bacteria die, they release toxic components of their outer membrane known as endotoxins. If such bacteria are killed quickly and in large numbers, as might happen if a bactericidal antibiotic is used, large amounts of endotoxin are suddenly released, and the patient may go into shock. It is much safer in such situations to use a bacteriostatic compound. The bacteria are then held in check while the patient's immune response eliminates the bacteria slowly. Because endotoxin is then released slowly, and because it is toxic only in large amounts, the likelihood of serious complications such as shock is greatly reduced.

CASE: A MYSTERY ILLNESS

In the spring of 1994, an unusual epidemic of unknown origin struck the Southwestern United States. In May, although the cause of the disease was still unknown, the New Mexico State Department of Health sent information to every physician in the state with recommendations on how to confront the illness. The following is an adaptation of that letter:

May 28, 1994
Dear Doctor:
    We are investigating a cluster of an unusual illness with high mortality that has occurred primarily in the northwest part of the state. The illness starts with flu-like symptoms, progressing to severe respiratory distress. To date, all viral and bacterial cultures are negative. *Mycoplasma, Chlamydia, Ehrlichia,* and *Rickettsia* are among the microorganisms being considered as possible causes. As the cause remains unknown, it is difficult to recommend antimicrobial treatment or prophylaxis. However as a preliminary precaution, we recommend the following:
    Prophylaxis, for persons who have had close contact with a known case: for adults, tetracycline for 7–10 days; for children, chloramphenicol for 7–10 days.
    Therapy, for acutely ill persons: a broad-spectrum antibiotic plus erythromycin.

1. What is the rationale behind the drug recommendations that were made?
2. What is a broad-spectrum antibiotic?

Broad- versus narrow-spectrum antibiotics. May 1994 was a scary time in Arizona and New Mexico. People were dying and nobody knew why. Most of those affected were young, previously healthy adults in the Four Corners area, where Utah, Colorado, Arizona, and New Mexico meet. Not long after the above letter was sent, a diagnosis was finally made. The mysterious disease was being caused by a previously unknown type of hantavirus, contracted from deer mice (**Figure 12.10**). As it turned out, none of the recommended antibiotics had any effect against the viral pathogen. Nevertheless, the recommendations made good sense in light of the fact that when they were made, the cause was still unknown. Let's find out why.

Some antibiotics can be used to combat a wide range of microorganisms. Such drugs are termed **broad-spectrum**, because they are effective against a broad range of bacterial types. Tetracycline, for instance, can be used to treat both Gram-positive and Gram-negative bacterial infections (**Table 12.2**). It is even effective against intracellular bacteria. Penicillin, on the other hand, is relatively **narrow-spectrum**, because it is primarily active against Gram-positive bacteria only. In other words, broad-spectrum drugs take a shotgun approach, while narrow-spectrum drugs are more highly targeted to specific types of infections.

In most cases, doctors will opt for a narrow-spectrum drug if possible, because broad-spectrum drugs, by definition, will kill other species besides those causing illness. In Chapter 10, for instance, we learned about the important beneficial roles played by the normal flora. Broad-spectrum antibiotics inhibit normal flora as well as pathogens, which explains some of the side effects associated with broad-spectrum drugs. Those who have

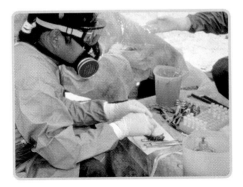

Figure 12.10 The source of the mystery illness. Researchers collect a blood sample from a deer mouse possibly infected with hantavirus. Respirators protect against the accidental inhalation of this airborne pathogen.

BROADEST SPECTRUM	tetracycline	Effective against both Gram-positive and Gram-negative bacteria, including intracellular organisms such as *Chlamydia*.
↑	sulfa drugs	Effective against Gram-positive and Gram-negative species.
	streptomycin	Effective against Gram-negative and acid-fast bacteria.
	erythromycin	Effective against Gram-positive bacteria and bacteria without cell walls (*Mycoplasma*).
↓	penicillins	Original penicillin G is mainly active against Gram-positive bacteria; newer penicillins have broader range, active against many Gram-negative species.
NARROWEST SPECTRUM	isoniazid	Effective against acid-fast bacteria only.

Table 12.2 Some representative drugs with their spectrum of activity indicated.

taken such an antibiotic may find that their tongues develop a coated appearance, as normal flora living on the tongue are destroyed, allowing fungal opportunists the chance to proliferate. Women on a broad-spectrum antibiotic are more likely to suffer from vaginal yeast infections for the same reason, and others taking a broad-spectrum drug may experience intestinal irregularities such as diarrhea because the gut flora is disrupted.

In some situations, on the other hand, a broad-spectrum drug is preferable. This is clearly the case when disease is caused by a pathogen that has yet to be identified. If the pathogen responsible for an illness is unknown, a broad-spectrum antibiotic is more likely to prove effective, simply because such a drug is active against a wider range of potential bacterial causes. Before doctors knew the exact cause of New Mexico's unexplained respiratory disease, it therefore made good sense to use the broadest-spectrum drug possible, such as tetracycline, both for prophylaxis and for therapy. Chloramphenicol was recommended for children instead of tetracycline, because, as we will describe shortly, tetracycline has undesirable side effects in children.

You will notice that erythromycin was also recommended for therapy. Although this antibiotic is not particularly broad-spectrum, as previously discussed it *is* fast. Because bacteria such as *Mycoplasma* were being considered as possible causes, and because erythromycin is effective against some of these candidate pathogens, the use of this rapidly acting drug seemed like a wise precaution in such acutely ill patients.

Broad-spectrum drugs are also often called for when antibiotics are used prophylactically to protect individuals against infections that may occur in the future. This is especially necessary in the case of immunocompromised individuals who may be exposed to any number of potentially opportunistic species.

Site of infection. Some drugs work better in certain parts of the body. For example, streptomycin works well only in parts of the body where oxygen is present. Lung infections are therefore responsive to streptomycin, whereas intestinal infections, infections of the blood, or infections in other relatively anaerobic sites are not. A different drug, metronidazole, which inhibits ATP synthesis, is oxidized to an inactive form if oxygen levels are high. This drug, therefore, is used only to treat infections in relatively anaerobic parts of the body. Nitrofurantoin is used almost exclusively for urinary tract infections.

Nitrofurantoin is rapidly filtered by the kidneys and quickly ends up in the urine in high amounts. Consequently this drug can never achieve sufficiently high levels in other body regions to be effective. It is ideal, however, for bladder or other urinary tract infections, because this is exactly where the drug will become concentrated.

Side effects. Obviously, even the best antibiotic cannot be used if it causes unacceptable side effects in a patient. We have already mentioned how some antibiotics, especially those that are broad-spectrum, may cause problems by disrupting the normal flora. Some individuals are allergic to specific antibiotics, and clearly this rules out the use of the allergy-provoking drug, no matter how appropriate it may otherwise be. Finally, for some antibiotics, the goal of "specific toxicity" is only partly achieved. Such drugs must be used judiciously and only in specific situations. Streptomycin can cause problems such as hearing loss and kidney damage. For this reason, streptomycin is no longer commonly used.

Drug resistance. In Chapter 6 (see p. 160), we discussed the genetic basis of microbial drug resistance. In Chapter 8 (see p. 218), we considered the situations in which drug resistance is likely to become a significant problem. We will conclude this chapter with a more in-depth look at the problem of drug resistance and what, if anything, can be done about it. For the present, it is sufficient to say that resistant microorganisms have greatly undermined our ability to treat many infections and that when resistance crops up, the options available to a physician become limited. In many cases, doctors are forced to use their second- or third-choice antibiotics, because the pathogen in question is resistant to what would ordinarily be the drug of choice. For example, we mentioned that because streptomycin can have important side effects, its use has decreased significantly in recent years. However, in some cases, doctors are forced to use streptomycin to treat tuberculosis because the patient is infected with strains of bacteria that are resistant to other, more effective antibiotics that cause fewer side effects.

Antibiotics work by interfering with specific bacterial structures or enzymes

Now that we have considered some of the factors that enter into a doctor's decision regarding antibiotic selection, we will look at a few of the basic types of these drugs. A thorough description of all antibiotics is beyond the scope of this text. Instead, we will describe how antibiotics work in general and provide representative examples. We will discuss how these examples can be used in light of the criteria for antibiotic selection that we have just reviewed.

To meet the goal of selective toxicity, most antibiotics work by targeting a feature of the microbial cell that is fundamentally different from the host's eukaryotic cells. As discussed previously, sulfa drugs, for instance, achieve this by interfering with an enzyme that is unique to bacteria. Because humans lack the enzyme, suppressing the enzyme's activity poses no particular risk to our cells. Likewise, in Chapter 3 (p. 54) we described how penicillin acts. Briefly, when bacterial cells divide, penicillin interferes with the synthesis of new pentaglycine bridges between different layers of peptidoglycan in the bacterial cell wall (see Figure 3.12c). Without these bridges, the integrity of the cell wall is severely compromised, and the cell is likely to burst as it gains water from its environment. Animal cells have no peptidoglycan to begin with and consequently are unaffected by penicillin. Bacterial cells offer up a variety of such targets (**Figure 12.11**). Whereas penicillin and some other antibiotics inhibit cell wall synthesis, antiribosomal drugs interfere with bacterial protein synthesis. Some antimicrobials target bacterial DNA, while others, such as sulfa drugs, interfere with prokaryotic enzyme activity.

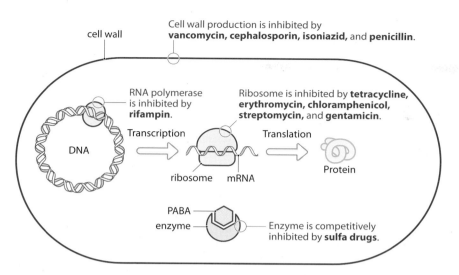

cell wall

Cell wall production is inhibited by **vancomycin, cephalosporin, isoniazid,** and **penicillin.**

RNA polymerase is inhibited by **rifampin.**

Ribosome is inhibited by **tetracycline, erythromycin, chloramphenicol, streptomycin,** and **gentamicin.**

DNA

Transcription

ribosome   mRNA

Translation

Protein

PABA enzyme

Enzyme is competitively inhibited by **sulfa drugs.**

Figure 12.11 Representative targets for antibacterial drug activity. Most antibiotics work by interfering with some cellular process in bacteria that is fundamentally different or absent in eukaryotic cells. Many antibiotics either damage or inhibit the synthesis of cell wall material. Others interfere with protein synthesis by binding to bacterial ribosomes. The larger ribosomes typical of eukaryotic cells are unaffected. Other drugs such as sulfa interfere with unique bacterial enzymes or interfere in some manner with DNA replication or transcription. Although eukaryotic DNA is similar to prokaryotic DNA, eukaryotes surround their DNA with a nuclear membrane, protecting it from otherwise damaging antimicrobial drugs.

Cell wall inhibitors. "Penicillin" actually refers to a large class of antibiotics that may be collectively called the penicillins. The first commercially available penicillin, termed penicillin G, is still with us, but over the years a variety of other, modified penicillins have been developed. All members of this antibiotic class work the same way, but because of slight structural modifications they all have unique characteristics, which might make them more or less useful in specific situations. A few of these compounds are diagrammed in **Figure 12.12.** Notice, for example, that penicillin G is a narrow-spectrum drug, primarily useful against Gram-positive bacteria. It is also acid-sensitive, which means that it must be administered by injection. If taken orally, penicillin G would be destroyed by stomach acid. Some bacteria, those that are resistant to penicillin, produce an enzyme called penicillinase that can destroy penicillin. Penicillin G is especially sensitive to penicillinase and is therefore not recommended when drug resistance is a problem.

Compare the properties of penicillin G and those of ampicillin. Modifications in the structure of ampicillin make it more broad-spectrum than penicillin G, and because it is acid-stable, it can be taken by mouth. Like penicillin G, it is vulnerable to penicillinase. If drug resistance is a concern but a penicillin-like drug is still called for, a physician may opt to use oxacillin instead. Structural modifications in oxacillin greatly reduce the ability of penicillinase to destroy it. Cephalosporins are a separate class of antibiotics, although their structure and mode of action make them similar to the penicillins. Like the penicillins, these antibiotics are available in many different forms. They are in general active against both Gram-positive and Gram-negative bacteria and are resistant to penicillinase. They are thus useful against bacteria that produce this enzyme, although some bacteria secrete a different enzyme, rendering them resistant to cephalosporin as well.

Although it is structurally unrelated to penicillin and cephalosporin, vancomycin also inhibits cell wall synthesis. Its mechanism of action has already been described (see Chapter 8, p. 218, and Figure 8.24). Vancomycin is a narrow-spectrum antibiotic, highly active against certain Gram-positive species. Because it has toxic properties, vancomycin must be used with care, but it is especially valuable against strains of *Staphylococcus aureus*

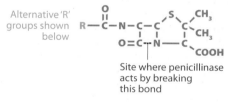

Alternative 'R' groups shown below

Site where penicillinase acts by breaking this bond

Penicillin G

Highly active against Gram-positive bacteria.
Sensitive to acid.
Sensititve to penicillinase.

Ampicillin

More active than penicillin G against Gram-negative bacteria.
Acid stable.
Sensitive to penicillinase.

Oxacillin

Acid stable.
Resistant to penicillinase.

Cephalosporin

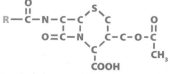

Similar mode of action to penicillins, but a separate class of antibiotic.

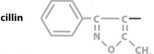

Figure 12.12 Representative cell wall inhibitors. Three examples of penicillins are shown. All have the same basic structure but differ in their R group (diagrammed in blue). Different R groups explain the different characteristics of each penicillin. Cephalosporin is an unrelated drug that has a mode of action similar to the penicillins. It too varies at the R group (in blue), and there are many cephalosporins, all with slightly different properties.

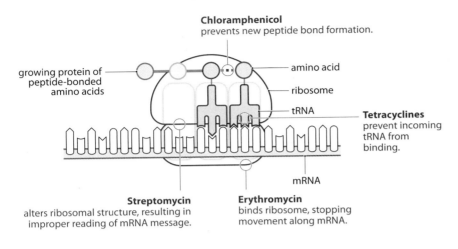

Chloramphenicol prevents new peptide bond formation.

growing protein of peptide-bonded amino acids

amino acid

ribosome

tRNA

Tetracyclines prevent incoming tRNA from binding.

mRNA

Streptomycin alters ribosomal structure, resulting in improper reading of mRNA message.

Erythromycin binds ribosome, stopping movement along mRNA.

Figure 12.13 Representative antiribosomal drugs. Bacterial ribosomes, because they are smaller and of somewhat different structure than those of eukaryotes, make useful targets for antibiotics. A variety of antibiotics interfere with the ribosome's ability to translate messenger RNA into proteins, but the manner in which they interfere varies according to the antibiotic. The specific mode of action for several representative antiribosomal drugs is illustrated here.

that are resistant to most of the penicillins. In some cases, vancomycin has become the last option when patients are infected with certain strains that are resistant to all other antibiotics. As described in Chapter 8, this makes the appearance of vancomycin-resistant *Staphylococcus* especially worrisome.

Isoniazid is a very narrow-spectrum cell wall inhibitor that targets *Mycobacterium*, including *Mycobacterium tuberculosis*. As we learned in Chapter 3 (p. 56), *Mycobacterium* has an acid-fast cell wall, rich in a long-chain molecule called mycolic acid. Isoniazid inhibits mycolic acid synthesis. Because non-acid-fast bacteria do not include mycolic acid in their cell walls, this antibiotic has no effect against them.

Antiribosomal drugs. Recall from Chapter 3 (p. 59) that bacterial ribosomes are smaller and contain less protein and ribosomal RNA than eukaryotic ribosomes. Many antibiotics achieve selective toxicity because they can interfere with the smaller bacterial but not larger eukaryotic ribosomes. These drugs comprise a diverse group and are not all closely related. They are grouped together here simply as a convenience because they all attack the same bacterial target.

As mentioned earlier, erythromycin is an example of such an antibiotic (**Figure 12.13**). By binding to the bacterial ribosome, it prevents movement of the ribosome along mRNA and thereby blocks translation. Because it cannot penetrate the outer membrane of Gram-negative bacteria, erythromycin is mainly used against Gram-positive species. It is, however, highly effective against *Mycoplasma pneumoniae*, the species that causes atypical or "walking" pneumonia. Penetrating the cell wall is not an issue for this unusual species, because, as you may recall from Chapter 3, *Mycoplasma* have no cell wall. As previously mentioned, erythromycin is a fast-acting bacteriostatic drug, especially useful for quickly shutting down production of protein toxins. If its use is discontinued prematurely, however, the bacteria may begin to reproduce again, resulting in a return of symptoms.

Tetracyclines are another group of bacteriostatic antiribosomal drugs that are active against a large array of bacterial species. They work by blocking the binding between transfer RNA and the mRNA–ribosomal complex (see Figure 12.13). Tetracycline is not recommended for children, since it can cause a discoloration of the teeth in younger people. Chloramphenicol is also a broad-spectrum bacteriostatic antiribosomal antibiotic. It works by blocking the formation of peptide bonds between amino acids during translation. Recall that this antibiotic was suggested as a replacement for tetracycline for children in our recent case, describing an unknown respiratory illness in the American Southwest.

Streptomycin and gentamicin also inhibit translation, but they are bactericidal rather than bacteriostatic, because they cause irreversible damage to bacterial ribosomes. Specifically, they alter the structure of the ribosome in such a way that codons on the mRNA are misread, and translation occurs incorrectly (see Figure 12.13). These antibiotics are mainly active against Gram-negative bacteria, although streptomycin is sometimes used to treat tuberculosis when the strain of bacteria is resistant to less toxic, first-choice drugs such as isoniazid. Neither streptomycin nor gentamicin can be used to treat infections in anaerobic parts of the body, such as puncture wounds, because they are active only in their oxidized forms. Consequently they are used only to treat infections that occur in well-oxygenated parts of the body. Because of toxic side effects that can result in hearing loss and kidney damage, these antibiotics must be used with caution.

Antibiotics targeting DNA. The list of drugs in this class of antibiotics is relatively short. A moment's thought should tell you why; bacterial DNA is essentially the same in structure as eukaryotic DNA. Selective toxicity is therefore more difficult to achieve, because anything damaging bacterial DNA might harm host DNA as well. But that does not mean that such selectivity is impossible. Rifampin, for example, binds to and inhibits the activity of bacterial RNA polymerase and consequently blocks transcription. Since it does not bind eukaryotic RNA polymerase, it can be used safely. Its use can be disconcerting, however, because as it is secreted from the host, the urine and even sweat and tears take on a reddish hue. It is especially useful against tuberculosis, because it is able to penetrate macrophages in which *M. tuberculosis* often resides. Rifampin can also penetrate the connective tissue nodules known as "tubercles" that form around infected macrophages in the lungs and elsewhere in tuberculosis patients.

Selective toxicity, while possible, is harder to achieve against eukaryotic pathogens

Compared with the antibiotics used against bacteria, a physician in need of a drug to treat a protozoan or fungal infection has relatively limited options. Why is the list of drugs that fight eukaryotic pathogens so short? Simply because they *are* eukaryotic. Fungal and protozoan cells are so similar to ours that there are far fewer unique targets to attack. Selective toxicity is consequently harder to achieve. Ribosomes, for instance, are the same in human, fungal, and protozoan cells, meaning that using antiribosomal drugs against eukaryotic pathogens is not an option.

The hands of a doctor confronting such a pathogen are not completely tied. For certain fungal infections, for instance, a group of drugs called the "polyenes" provide a treatment option. Polyenes disrupt the fungal plasma membrane. Fungi incorporate a sterol called ergosterol in their plasma membrane to increase membrane stability and flexibility, much as animal cells use cholesterol (see Chapter 3, p. 63). Polyenes bind egosterol better than they bind cholesterol and therefore can generally be used safely to treat fungal infections, at least when those infections are on the skin. Systemic fungal infections are much harder to control. Polyenes can also damage host cells and therefore are considered toxic. Amphotericin B, for example is used to treat systemic fungal infections, but it is used only when the situation is considered life-threatening. Because of its toxic properties, it must be administered at relatively low doses for a long period of time. Side effects can include fever, vomiting, headache, anorexia, and low blood pressure. It is not surprising that this antifungal drug has the nickname "Ampho the Terrible." Other drugs, the azoles, interfere with ergosterol synthesis by inactivating a fungal enzyme used to produce this sterol compound. Azoles are typically found in topical creams used to treat conditions such as athlete's foot.

Protozoa too, while more difficult to treat than bacteria, have a few Achilles' heels that can be exploited in the event of an infection. In fact, the Chinese have known for centuries that malaria could be treated with certain plant extracts, while the Inca in Peru used the bark from cinchona trees for the same purpose. The active ingredient in cinchona turned out to be quinine (**Figure 12.14**). When you see old photos of British colonialists in British East Africa or India, lounging on a veranda with a gin and tonic, now you know part of the reason why this particular cocktail was so popular. The quinine in the tonic actually provided some protection against malaria.

World War II played an unusual role in spurring on the development of new antimalarial drugs. Early in the war, Japanese capture of cinchona plantations in Indonesia created a quinine shortage in the United States. This stimulated a flurry of research, culminating in the development of several new effective drugs, including chloroquine, which is still a workhorse in malaria treatment. Malaria parasites reproduce inside human red blood cells. The hemoglobin in these cells is a crucial nutrient for these parasites, but chloroquine prevents hemoglobin digestion. Unfortunately, resistance to chloroquine and other antimalarial drugs is now widespread.

Metronidazole, also known by the trade name Flagyl, is a drug widely used for amoebic *Giardia*, and other intestinal protozoan infections. Metronidazole inhibits ATP synthesis in anaerobic environments such as the gut. Because it is not well absorbed across the intestinal lining, it is relatively safe to use. If taken with alcohol, however, it can cause vomiting and headache, as well as possible dangerous low blood pressure.

Antiviral drugs must interfere with a particular step in the viral replicative cycle

The development of drugs to combat viral infections has special problems all its own. As we learned in Chapter 4 (p. 93), viral replication is tied intimately to the host. In many cases, the virus simply infects a host cell with its genetic material. Replication depends mainly on the machinery of the host cell, including enzymes and ribosomes. Attacking any of these components means attacking the host cell itself. Furthermore, when viruses are outside the host cell, they are metabolically inactive for the most part. There is little or nothing to attack because the virus is not undergoing replication. Viruses replicate only inside an appropriate host cell, presenting a further complication. To interfere with viral replication, an antiviral drug must ordinarily be able to penetrate host cells without harming them.

Nevertheless, viruses do have unique features, meaning that, at least in theory, selectively toxic antiviral drugs are possible. In recent years, this possibility has begun to be realized. Although the number of antiviral medications is small compared with the number of antibiotics, that number continues to grow, and for at least some types of viral infections, antiviral therapy has become reasonably commonplace. Several steps in the viral replication cycle are vulnerable to attack (**Figure 12.15**).

The first truly effective antiviral drug to become available was acyclovir, which is still commonly used to treat herpesvirus infections. To understand how acyclovir works, remember that a DNA nucleotide is composed of a phosphate, a sugar, and a base (see Chapter 6, p. 135). After the sugar and base are assembled, phosphates are added by an enzyme called thymidine kinase. The now-complete nucleotide can be used in DNA replication by DNA polymerase. When a herpesvirus infects a human cell, it uses sugar–base compounds of the host and converts them to nucleotides by adding phosphates to the combined sugar and base with its own thymidine kinase. The addition of phosphates to another molecule in this way is called **phosphorylation**. The virus then uses these "stolen" nucleotides to copy its DNA and produce new viral particles.

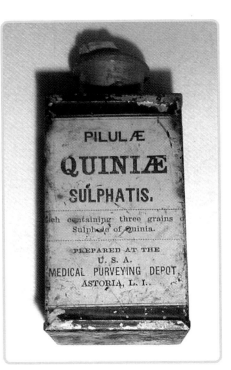

Figure 12.14 Protection against malaria. A quinine tin, from the 1860s. Extracted from the cinchona tree, quinine was widely used during the second half of the 19th century.

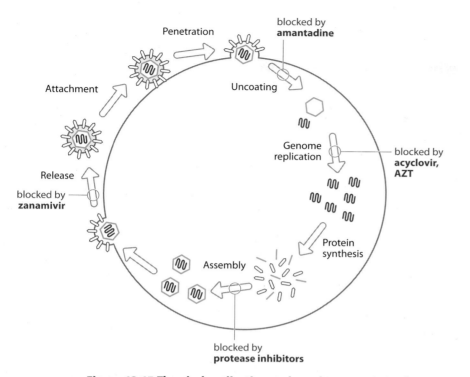

Figure 12.15 The viral replication cycle and some points of intervention. The stages in the cycle where representative antiviral drugs interfere are indicated. Amantadine, used against influenza virus, prevents viral particles from uncoating inside the host cell. Drugs such as acyclovir, used to combat herpesvirus outbreaks, and AZT, an anti-HIV drug, prevent the virus from making copies of its genetic material. Protease inhibitors interfere with HIV at a different point in its replicative cycle—the assembly of new viral particles. Zanamivir, another anti-influenza drug, prevents newly made viral particles from exiting the host cell.

Now look at the structure of acyclovir (**Figure 12.16a**) and observe that the drug resembles a DNA sugar and base that is missing part of the sugar. The herpesvirus thymidine kinase, however, cannot tell the difference between a genuine sugar and the "dummy" acyclovir sugar, and it phosphorylates the acyclovir. DNA polymerase now uses this false nucleotide to attempt replication of the viral DNA, but because part of the sugar is missing, DNA replication cannot continue, and the herpesvirus is unable to make new copies of itself (Figure 12.16c). Human thymidine kinase is not fooled in this way; it will not phosphorylate acyclovir. Consequently, there is no effect on host cells, and selective toxicity is achieved.

Flu patients now have antiviral options as well. Amantadine is thought to interfere with the ability of the influenza virus to shed its protein coat following entry into the host cell (see Figure 12.15). The anti-influenza activity of this drug was actually first noted in 1961, when it was being used to treat Parkinson's disease. Surprisingly, it was observed that people being treated for Parkinson's were less likely to get the flu. Zanamivir is another anti-influenza drug that works by inhibiting the release of newly replicated viral particles from the infected cell (see Figure 12.15).

Although both amantadine and zanamivir can reduce flu symptoms considerably, speed is of the essence in both cases. To make a significant difference in the course of the illness, these drugs must be taken within 24–48 hours of the onset of symptoms. High-risk populations such as the elderly may take amantadine or zanamivir prophylactically to greatly reduce the likelihood of developing the flu in the first place.

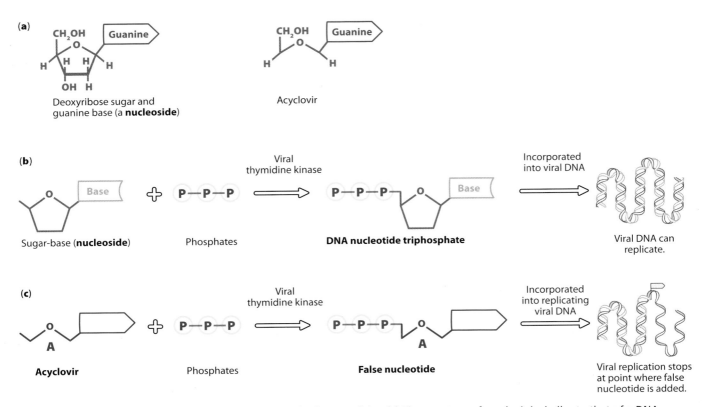

Figure 12.16 How acyclovir stops the herpesvirus. (a) The structure of acyclovir is similar to that of a DNA sugar–base complex with part of the sugar missing. (b) The sugar–base complex must be phosphorylated via the viral enzyme thymidine kinase before it can be incorporated into viral DNA. Initially three phosphates are added. Later, when the nucleotide is incorporated into DNA, two of these phosphates are removed. (c) If acyclovir is phosphorylated instead, this "false nucleotide" may inappropriately be incorporated into the DNA. At this point, viral DNA replication is terminated.

Zanamivir is particularly interesting because, unlike most antimicrobials, it is an example of a "designer drug." The vast majority of antimicrobial drugs, whether for use against bacteria, eukaryotes, or viruses, were basically found by trial and error. Scientists test thousands of compounds for antimicrobial activity, and when they find something that works, they investigate further. Since the mid-1990s, however, scientists have started to construct specific molecules that they think will interfere with a microorganism in a precise manner. In the case of zanamivir, the exact three-dimensional structure of a viral protein crucial in the virus's release from an infected cell was first worked out. Scientists then synthesized the antiviral drug atom by atom, resulting in a molecule with the exact shape and characteristics needed to latch onto the viral protein. Not many designer drugs are currently available, but the success of zanamivir demonstrates that such drugs are possible. Indeed, with advancing technology, we are likely to see the development of truly amazing antimicrobial agents.

The first antiviral drug used to treat HIV-positive patients was zidovudine, also known as AZT. Similar to acyclovir, AZT looks something like, but not exactly the same as, a DNA nucleotide. When the viral enzyme reverse transcriptase converts viral RNA to DNA (see Chapter 4, p. 97), the enzyme may incorporate AZT into the DNA instead of a real DNA nucleotide. If this happens, DNA synthesis cannot continue, and viral replication is halted. Many of the newer **reverse transcriptase inhibitors** work in the same way.

Although reverse transcriptase inhibitors can extend life span and reduce symptoms in HIV-infected individuals, the virus rapidly develops resistance to such drugs. Therefore, these drugs offer no long-term solution by themselves. But the treatment of HIV-positive patients changed dramatically in

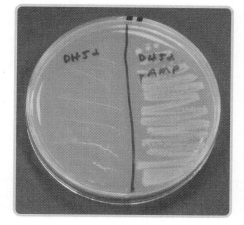

Figure 12.17 Ampicillin-resistant and -susceptible bacteria. The agar used to construct this plate contains ampicillin. A susceptible strain of *E. coli* was streaked on the plate to the left of the blue line. In the presence of the antibiotic, the bacteria failed to survive. To the right of the blue line, resistant *E. coli* display heavy growth, in spite of the ampicillin.

the mid-1990s with the introduction of the **protease inhibitors**. These drugs were the first designer drugs to be successfully developed. They attack the virus at still another step in its replicative cycle (see Figure 12.15). Specifically, protease inhibitors bind to and inhibit a viral enzyme called protease, which cleaves large proteins into the small proteins the virus needs for its assembly (see Chapter 4, p. 94). The use of combination drug "cocktails," usually consisting of two reverse transcriptase inhibitors and one protease inhibitor, has literally given patients with HIV a new lease on life. Patients may remain symptom-free for prolonged periods of time, and the likelihood of resistance is considerably reduced. Some experts have speculated that many patients treated with such drug cocktails may survive and remain relatively healthy indefinitely. Certainly many patients have done so since the mid-1990s, and the hope is that they will continue to do so.

However, these drug combinations do not represent a cure. If the medications are stopped, the virus quickly rebounds. They might mean, however, that HIV is now a manageable disease, somewhat akin to other chronic conditions such as diabetes or heart disease. Only time will tell how effective these and perhaps newer drugs will remain over the long run.

The misuse of antibiotics has led to the problem of drug resistance

CASE: THE IMPORTANCE OF COMPLETING PRESCRIPTIONS

Frank, a healthy third-grader, complained to his school nurse of a sore throat. The nurse found that he had a fever and his mother was called. Frank's mother took him to his pediatrician, who saw that Frank had enlarged lymph nodes, a reddish throat, and swollen tonsils covered with grayish-white patches. The doctor took a throat swab and streaked it on a blood agar plate. Test results confirmed her suspicion of a strep throat infection caused by *Streptococcus pyogenes* (see Figure 10.20). Frank's mother was given a 10-day prescription of ampicillin with instructions to complete the treatment, no matter how Frank felt. Within 72 hours, Frank's sore throat was gone and he felt much better. His mother decided not to give him his remaining medication, saving it to treat either Frank or one of his two sisters the next time one of the children became ill. When the pediatrician heard about this on a later office visit, she strongly voiced her anger.
1. Why was the pediatrician angry?
2. How does the sort of "self-medication" considered by the mother with the remaining antibiotic contribute to drug resistance?
3. Suppose that Frank developed another sore throat soon after the first episode, and so did one of his sisters. How might treatment need to be changed?

Frank was lucky. *S. pyogenes* has so far shown only limited signs of drug resistance. Had he been infected with certain other bacteria, such as certain species of *Staphylococcus* or some strains of *E. coli* (**Figure 12.17**), for instance, he might have paid a heavy price for his mother's inability to follow the doctor's instructions.

Although it has been a major public health issue only in the last few decades (**Figure 12.18**), antibiotic resistance is not new; the first signs began to

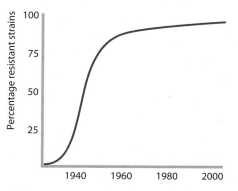

Figure 12.18 Increasing penicillin resistance in *Staphylococcus aureus*. The graph illustrates the proportion of all *S. aureus* strains that have developed resistance to penicillin since the antibiotic's introduction.

appear in the late 1940s, soon after the use of antibiotics became common. We should have seen larger problems looming—after all, the mechanisms behind the emergence of resistance were already well understood. But with most bacterial infections suddenly curable, a complacency developed over our ability to treat infectious disease. With the elimination of such disease seemingly on the horizon, the warning signs were largely ignored.

Antibiotic resistance may work in one of several ways. Tetracycline, for instance, as earlier described, inhibits bacterial protein synthesis. In resistant cells, however, although tetracycline enters the cell, it is rapidly excreted by a protein pump on the bacterial plasma membrane. Consequently, although the drug enters the cell, it is quickly expelled before it reaches its target. Bacteria resistant to erythromycin employ a different strategy; they have ribosomes with a slight structural modification. This modification prevents erythromycin from binding to the ribosome, meaning that protein synthesis in these cells can continue unimpeded. Resistance to penicillins, as previously stated, is due to the production of a bacterial enzyme called penicillinase that cleaves the drug, rendering it inactive (see Figure 12.12). Newer forms of penicillin such as oxacillin have been chemically modified to protect the antibiotic from penicillinase.

Antibiotic resistance, like other genetically determined characteristics, evolves through the process of natural selection. In Chapter 8 we considered the evolution of vancomycin resistance, observing that, like all living things, bacteria that are better adapted to their environment survive and reproduce. Those that are less well adapted tend to perish. With tetracycline resistance, for example, the protein pump that excretes tetracycline out of the resistant bacterial cells did not originally evolve millions of years ago to deal with a drug that would not exist for eons. Instead, this protein transported a different substance such as a metabolic waste out of the cell. Because of genetic variability present in any population, this protein might have had a slightly different structure in different members of the population.

Now we fast-forward a few million years to the age of antibiotics. Cells are now being exposed to compounds such as tetracycline, and strictly by chance, the new drug has a structure similar to the compound normally excreted by the bacteria's transport protein. That means that some bacterial cells, those that happen to transport tetracycline best, are at a huge advantage. Other cells that cannot transport tetracycline are not so lucky. In the face of tetracycline exposure they fail to reproduce, and over time, most members of the population will have descended from those cells that could originally excrete tetracycline. As long as tetracycline exposure continues, any cells undergoing random mutations that enhance their ability to excrete the drug continue to proliferate at the expense of other, less resistant cells. Finally, a completely tetracycline-resistant strain may emerge. The key phrase here is *as long as tetracycline exposure continues*. Once the drug is removed from the environment, resistant bacteria lose their competitive edge and may cease to predominate.

So why is it so crucial that patients like Frank complete antibiotic prescriptions? When Frank became infected, the bacteria causing the infection would be expected to vary genetically in terms of how antibiotic-sensitive they were (**Figure 12.19**). Some cells would be killed easily, while some would be somewhat resistant. The majority of the bacteria would be somewhere between these two extremes.

We saw in Chapter 10 (see p. 271) that the ability of a pathogen to cause disease is in part a function of its population within the host. An important reason that Frank felt so ill was that he was infected with a large number of *S. pyogenes* bacteria. When he started his treatment, many of these bacteria were quickly killed, so it wasn't long before he started feeling better (see

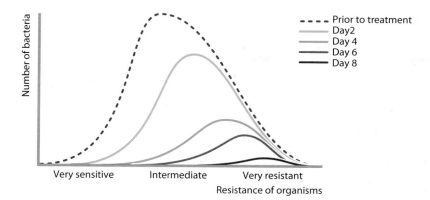

Figure 12.19 Hypothetical changes in bacterial numbers and resistance over a typical 10-day treatment course. Before treatment begins, bacterial numbers are high, causing numerous and possibly severe symptoms. Resistance is mostly intermediate, with only a few very sensitive and very resistant organisms. After 2 days, the patient generally feels much better because bacterial numbers are much lower, but those bacteria that do survive tend to be the more resistant ones. After 3–5 days, the patient may feel well because bacterial numbers continue to be reduced. Once again, however, the few remaining bacteria are increasingly those that have the highest resistance. If treatment is halted prematurely at that time, those now more resistant organisms may once again increase in numbers, causing renewed symptoms. If treatment is continued for the recommended 10 days, all organisms, even the most resistant ones, are destroyed.

Figure 12.19). But those that died first were most sensitive to ampicillin. The most resistant were the last to succumb. Within 72 hours the number of bacteria was so low that Frank felt well. By this time all highly sensitive bacteria, and perhaps most of those with intermediate resistance, had been eliminated. Frank was fortunate because this species, *S. pyogenes*, rarely shows high levels of resistance, and even the most resistant members of this species were still relatively easy to kill. However, had he been infected with a more highly resistant microorganism such as *S. aureus*, the most resistant individuals might still be viable after 72 hours, even if overall numbers had been reduced and Frank felt better. Had he stopped taking his medication early in this case, the most resistant, still living cells might have recommenced reproducing, and very soon Frank might be ill again. This time, however, *all* of the bacteria would have increased resistance. If he then infected one of his sisters, she would also be infected with this more resistant strain. The pediatrician may then have had to either switch antibiotics and/or increase the dosage of ampicillin to deal with these more resistant bacteria.

Such misuse of antibiotics is an important reason for the drug resistance that currently confronts us. The 10-day prescription would ordinarily eliminate the entire infection, including the more resistant bacteria. By stopping treatment prematurely, Frank's mother could have incrementally increased the resistance of the *S. pyogenes*, by giving only the most resistant cells the chance to reproduce. Multiply this by millions of patients not fully complying with their doctor's instructions, and the result is the evolution of highly resistant strains and ineffective antibiotics.

Antibiotic resistance is also promoted when such drugs are used inappropriately to treat nonbacterial conditions. Although Frank's sore throat was caused by bacteria, most sore throats are due to viral infection. Antibiotics, of course, have no effect on viruses, because the targets of these drugs are simply not present in viruses. Nevertheless, antibiotics continue to be misused in situations where they do nothing of benefit. A recent study by the Centers for Disease Control and Prevention (CDC) estimated that in the United States, 30% of all antibiotic prescriptions for ear infections and 50% of those for sore throats are of no value because the illness is nonbacterial. In these situations, although there is no bacterial pathogen, the normal bacterial flora is exposed to the antibiotic. These harmless bacterial may likewise develop resistance, which they can then pass on to a genuine pathogen by processes such as conjugation, as discussed in Chapter 6.

The customary feeding of antibiotics to domestic animals also plays a role in resistance. Since the 1950s, farmers have routinely added antibiotics to animal feed, because it was correctly assumed that conditions for transmission of disease were favorable in closely penned livestock (**Figure 12.20**). The use of prophylactic antibiotics would reduce disease, protecting both the animals and the farmer's investment. Unexpectedly, it was also found

Figure 12.20 Antibiotic use in livestock. Cramped conditions promote transmission of infectious organisms. Prophylactic antibiotics in animal feed can protect the health of domestic animals and, incidentally, can help them gain weight more quickly. Unfortunately, they promote antibiotic resistance as well.

that antibiotics resulted in faster animal growth. It is thought the growth of such animals is somewhat slowed by intestinal bacteria that produce toxins or excessive gas. Because livestock fed antibiotics this way reached marketable size sooner, the use of antibiotics in animal feed soon became widespread.

Over time, the normal flora in these animals develops resistance as well. Humans can become exposed to resistant microorganisms by consuming contaminated dairy products, eggs, or meat. When such food products are undercooked or eaten raw, these bacteria may gain access to human hosts. They may then either cause disease directly, as in the case of *Salmonella*-contaminated chicken, or they may pass their resistance genes on to other bacteria in the human intestine via conjugation, transduction, or transformation (see Chapter 6, p. 155).

New strategies provide options for circumventing antibiotic resistance

With the emergence of multi-drug-resistant microorganisms already a reality, what does the future of antimicrobial therapy hold? Certainly the simplest solution to the problem of resistance is more judicious and selective use of antibiotics. Consider the experience of Hungary, where resistance to penicillin in some *Streptococcus* species rose to 50% in the 1980s because of overuse of the drug. From 1983 to 1992 the use of penicillin in Hungary declined by 50% because doctors found that it was ineffective in too many cases. And as penicillin use went down, so did resistance. By the mid-1990s the proportion of resistant *Streptococcus* dropped to 34%.

The reason for this is simple. When drug use is restricted, the microbial environment is altered. Resistant cells lose their selective advantage and no longer reproduce at an accelerated rate relative to sensitive cells. The example in Hungary demonstrates that at least some resistance can be reversed and previously ineffective drugs can become useful again. The use of antibiotics in animal feed might likewise be restricted, resulting in a reversal of resistance. Currently, more than 50% of all antibiotics employed in the United States are used in animal feed.

Other appealing strategies to fight resistance might involve looking for new antimicrobial compounds against which resistance is unlikely to develop. For example, although insects do not have a typical vertebrate immune system, they have been successfully dealing with microbial pathogens for more than 500 million years. It should therefore not be surprising that insects and other invertebrates have compounds in their blood with powerful antimicrobial properties. Recent studies have found, for instance, that many insects are protected from bacterial infections by blood-borne proteins that perforate the bacterial plasma membrane. Because membrane structure is a fundamental property in all cells, any modifications to it is likely to be detrimental. Consequently, any bacteria that modify their own membrane structure in an attempt to avoid the action of antimicrobial proteins are unlikely to survive. Resistance under such circumstances would not develop.

Vertebrates too, and even other mammals, should be considered as potential sources of new antimicrobial compounds. Take the koala. When a female koala is not reproducing, her pouch contains a rich bacterial flora. When she ovulates, bacterial numbers start to decline, reaching their lowest level just as a baby koala enters the pouch (**Figure 12.21a**). Australian researchers have found that female koalas secrete antimicrobial proteins into their pouch to prevent infection in their babies. Interestingly, these proteins are most effective against aerobic bacteria. This is especially vital for baby koalas, because they are born with an underdeveloped respiratory system and are unusually susceptible to infection by pathogenic aerobes. Such proteins might one day provide the basis for a new narrow-spectrum antibiotic against aerobes in humans.

An intriguing alternative to antibiotics is the use of *phage therapy*. As we learned in Chapter 4, bacteriophages, or "phages," are viruses of bacteria. After replicating in a bacterial cell, they cause the cell to rupture as new progeny phages are released. Phages were described in 1917 by Felix d'Hérelle of the Pasteur Institute, who immediately began to tout their potential as antibacterial agents. Prior to the development of antibiotics, phage therapy was attempted against many diseases including plague, typhoid, and cholera (see Figure 12.21b). Success, however, was spotty, because not enough was understood about the specificity of certain phages for certain bacterial species. With the development of antibiotics, phage therapy was largely forgotten. In light of increasing resistance and a more complete knowledge of phage biology, however, scientists are taking a fresh look at this strategy.

The evolutionary battle between microorganisms and humans is likely to continue for the foreseeable future. But through careful and selective use of antimicrobial drugs, as well as the use of alternative strategies where appropriate, it is a battle that can and must be won. Certainly the alternative, returning to an era where infectious disease was the number one cause of human mortality, is not an acceptable option. New antibiotics will no doubt buy us time, but in the long run our success depends on using drugs with a wisdom that matches our cleverness in creating them.

Coming Up Next...

Will there be a flu epidemic next year? What do health authorities do when people are getting sick and nobody knows why? Why do seemingly new diseases, such as Ebola hemorrhagic fever and severe acute respiratory syndrome (SARS), turn up out of nowhere with frightening regularity? What will be the next new microbial menace to appear, and when will it happen? These are the sorts of questions that are answered by epidemiologists—scientists who study diseases in populations. Epidemiologists are essentially medical detectives. When a mystery crops up—an unexplained outbreak of food poisoning in a large city, for example—epidemiologists get the call. They sift the clues, interview witnesses, and develop lists of potential suspects. Sometimes they nab the perpetrator and sometimes they don't. In the next chapter, we will ride along with epidemiologists as they visit the scene of the most recent microbial "crime" or try to predict the next epidemic.

Figure 12.21 Possible alternatives to traditional antibiotics. (a) Other animals may produce novel antimicrobial peptides. This baby koala faces a reduced risk of infection, due to antimicrobial compounds secreted into the pouch by its mother. (b) Famed Nobel Prize-winning American author Sinclair Lewis. In his novel *Arrowsmith*, a doctor stops a plague epidemic in the Caribbean using phage therapy.

Key Terms

antibiotic	cidal agent	phosphorylation
antiseptic	disinfectant	protease inhibitor
autoclave	*in vitro*	reverse transcriptase inhibitor
bactericidal drug	*in vivo*	selective toxicity
bacteriostatic drug	narrow-spectrum antibiotic	static agent
broad-spectrum antibiotic	nosocomial infection	sterilization
chemotherapeutic agent	pasteurization	sulfanilamide

Concept Questions

1. Pick any microorganism you have learned about in this textbook. List one possible scenario in which you would wish to encourage its growth and one in which you would wish to suppress its growth.

2. You have a common household product, which advertises that it is effective in controlling microbial growth. Describe a simple experiment you could perform to determine whether the effects of this product are cidal or static in nature.

3. Supermarkets now routinely sell "antimicrobial soap," which is simply regular soap with antiseptic compounds added. Are such products worth the expense for routine use in the home?

4. A friend of yours who is taking microbiology becomes somewhat obsessive about hand washing after learning about all the microorganisms to which we are regularly exposed. Even before having lunch, he scrubs his hands for several minutes, as he was taught to do in microbiology laboratory. Explain to him why this is not necessary.

5. As we learned in this chapter, many commonly used antimicrobial drugs were derived from compounds that certain microorganisms themselves produce. Why do you think microorganisms produce such compounds?

6. A new antibiotic is found to be very broad-spectrum and fast-acting. It shows a high degree of selective toxicity. Think of a medical situation in which you might choose to use this drug and a different situation in which you probably would not want to use this particular drug.

7. In Chapter 10 (p. 275), we reviewed the case of Robyn, who had a ruptured appendix. After reviewing her case, if you were a physician in charge of her treatment, which antibiotic might you select and why? Are there any that you would avoid?

8. If it were your job to develop a policy for your state regarding the use of antibiotics in animal feed, what issues might you consider if you were interested in reducing the likelihood of drug resistance, as well as the economic well-being of farmers and ranchers?

9. You are hired by a large hospital to develop a plan for antibiotic use that will decrease the emergence of resistance. What are the components of your plan?

The Ebola virus causes Ebola hemorrhagic fever, a particularly gruesome emergent disease.

Chapter 13

Epidemiology: Who, What, When, Where, and Why?

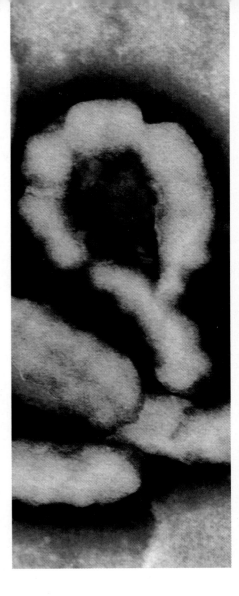

In February 1996, an outbreak of *Salmonella* infections in Denver, Colorado, had health officials scratching their heads. The first 20 patients examined all tested positive for *Salmonella enteritidis* in their feces. This species is often involved in food poisoning outbreaks, and the sudden appearance of the cases suggested that most or all of the affected individuals had been infected in the same way. So public health personnel asked the routine questions. Had they all bought food at the same supermarket? Had they all eaten at the same restaurant? A yes answer to questions such as these would have put authorities on the trail of the outbreak's source. But mysteriously, no common thread among the patients could be found. Meanwhile, the number of cases continued to rise.

Having eliminated the "usual suspects," such as a market where affected people had all purchased chicken from the same tainted batch, health officials considered other, less obvious possibilities. Further questioning finally gave them the break they needed. All patients had recently been to the Denver Zoo. Furthermore, they had all visited the Komodo dragon exhibit. Like many reptiles, Komodo dragons often harbor *Salmonella* in their intestines (**Figure 13.1**).

When the enclosure for these reptiles was searched for *S. enteritidis*, the bacteria were found in large numbers on the bedding of the dragons, in their feces, and on their bodies. There was a wooden barrier around the exhibit, and the giant lizards often raised themselves on their hind legs, putting their front feet on the barrier. Bacteria on the reptile's feet contaminated the barrier, and visitors became infected by touching the barrier and then touching their mouths with their contaminated hands. The mystery was explained, but not before 65 confirmed cases had been reported. All patients had diarrhea, and about half of them were considered severe. Six individuals required hospitalization, and health officials speculated that there may have been several hundred more unreported cases. A higher wall was installed around the enclosure, and the problem was solved.

This case was more unusual than most, but for epidemiologists, it was all in a day's work. **Epidemiology** is the study of how disease occurs in populations. Epidemiologists try to understand the factors that influence a disease's frequency and distribution. They try to predict future problems, and they make recommendations that might prevent disease or limit the number of people affected. When something new or unexplained crops up, epi-

Figure 13.1 The Komodo dragon. Komodo dragons, found on several remote Indonesian islands, are the world's largest lizards. Adults weigh several hundred pounds and commonly reach lengths of over 10 feet. Like other reptiles, they can serve as reservoirs for *Salmonella* bacteria. Komodo dragons were implicated as the source of the mid-1990s food poisoning outbreak in Denver, Colorado.

demiologists try to make sense of it. In this chapter, we will take a closer look at the work of these medical detectives, in order to gain a better understanding of how disease behaves in human populations.

The Birth of Epidemiology

The first modern epidemiological study identified cholera as a waterborne disease

In the mid-1800s, residents in one part of London obtained their drinking water from one of two sources. Some were served by the Southwark and Vauxhall Company, while others relied on water provided by the Lambeth Company. Both of these companies pumped their water from the river Thames. During this time, cholera was still a serious problem in London, and John Snow, a local physician, suspected that contaminated water could transmit the disease.

In 1853, London was struck by a cholera outbreak, and Snow followed the **incidence** of the disease (number of new cases per specified time period) until 1855. When he compared a person's likelihood of contracting cholera and where that person got his or her water, an interesting trend emerged. In the first month and a half of the epidemic, about 3% of Southwark and Vauxhall's customers died of cholera. For customers of Lambeth, the death toll was under 0.4%. Why the difference?

Snow investigated where exactly the two companies got their water. He discovered that Southwark and Vauxhall's water came from a pump near Broad Street—an area where untreated sewage entered the Thames (**Figure 13.2**). The Lambeth Company, on the other hand, drew its water from the river upstream of the city, before it could become contaminated with human waste. Snow concluded that water from the Broad Street pump was the source of the epidemic. The science of modern epidemiology was born.

Figure 13.2 The first epidemiological study. (a) John Snow's map of London, drawn about 1854, showing the area around the Broad Street pump and the incidence of cholera. Water pumps are indicated by ×, and each black spot represents a cholera death. (b) An 1866 drawing depicting "King Cholera" pumping water in London. Thanks to Snow's pioneering work, the mode of transmission for cholera was recognized.

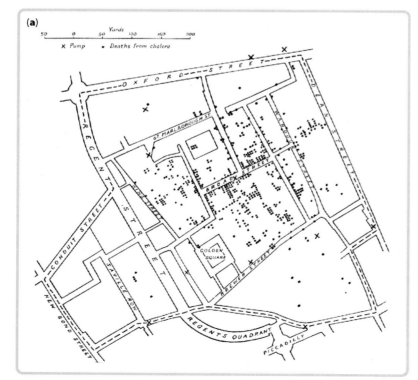

Florence Nightingale found that improved hygiene reduced the likelihood of typhus

Not much later, Florence Nightingale attempted to analyze the factors that put British soldiers at risk for typhus. This bacterial disease, caused by *Rickettsia typhi*, is transmitted by human body lice. Historically, military personnel have been especially at risk, because the lack of regular opportunities for bathing or clothes washing permits the lice to thrive.

Nightingale carefully compared cases of the disease in the civilian and military populations, looking for factors that seemed to link typhus victims in both groups. Unsanitary conditions seemed to be the principal factor that both civilian and military patients had in common. In 1858 she published her report, and on the basis of her work, the British army instituted reforms that led to dramatically lower rates of typhus among its soldiers.

At the time of these early epidemiologic studies, it was still not clear that microorganisms could cause disease. Even though bacteria had been discovered, the "germ theory of disease" (see Chapter 5, p. 124) was still more than 20 years away. Nowadays, of course, we understand the role played by microorganisms in disease, and our understanding greatly facilitates the work of epidemiologists as they attempt to track down possible causes for disease outbreaks. In addition to investigating the cause and source of unknown illnesses, modern epidemiologists also try to predict epidemics and develop organized responses to them.

Epidemics

An **epidemic** is a sudden increase in the number of cases of a specific disease, beyond what is considered to be normal. Of course, "normal" varies from disease to disease. A dozen cases of viral encephalitis in a large city during the summer would be cause for alarm and would probably result in health warnings and canceled outdoor activities. However, during the flu season, if a similar number of people in the same city contracted influenza, health authorities would no doubt breathe a sigh of relief about the surprisingly low number of cases.

Not *all* epidemics are caused by infectious microorganisms. Environmental toxins, for example, might result in epidemics of cancer. Exposure to heavy pollution can cause epidemics of respiratory disease. In this chapter, we will focus on infectious disease epidemics—those clearly caused by a microorganism.

A single contaminated site can give rise to a common source epidemic

CASE: AN OUTBREAK OF HEPATITIS A

In July of 1988, Panama City, Florida, was struck by an outbreak of hepatitis A. Epidemiologists from the Centers for Disease Control and Prevention (CDC) in Atlanta were called in to investigate. It was soon determined that those affected had all consumed raw oysters in Panama City-area restaurants. Dining companions of affected individuals who did not consume raw oysters remained unaffected. Interestingly, some unaffected restaurant patrons had eaten oysters as well. The sale of raw oysters in restaurants was immediately suspended, and the number of new cases waned, with the last new cases reported about 1 month later. A total of 61 cases were eventually reported, making this the largest hepatitis A epidemic in the United States since the early 1970s. The CDC investigators eventually concluded that the source of the epidemic was illegally harvested oysters from the bays surrounding Panama City.

1. **What type of epidemic was this?**
2. **Why was it illegal to harvest oysters near Panama City in the summer? How did this relate to the epidemic?**
3. **Why did some people who ate the oysters remain healthy?**
4. **Why did the number of cases rapidly peak and then decline, with no new cases approximately 1 month following the suspension of raw oyster sales?**

As the name implies, a **common source epidemic** occurs when a large number of individuals become infected from the same original source. The cholera epidemic described in the previous section is an example; the common source was the Broad Street pump. The *Salmonella* outbreak, with which we opened this chapter, was also a common source epidemic; all affected individuals contracted *Salmonella* in the same way. Common source epidemics often occur because of a breakdown in sanitation or due to contamination at a central distribution point for food or water. Not all such outbreaks, however, depend on the consumption of contaminated food or water. When a number of young women all developed *Pseudomonas* infections of the skin in Klamath Falls, Oregon, in 2003, the common source was found to be a body piercing parlor that failed to properly sterilize its equipment.

The hepatitis A outbreak described in our case is yet another example of a common source epidemic. Hepatitis A is contracted via food- and water-borne transmission. Shellfish are sometimes associated with hepatitis A outbreaks because they tend to concentrate the virus in their tissues as they feed. The hepatitis A virus often reaches the ocean in contaminated water runoff from agricultural areas and urban sewage. In Florida and other parts of the Gulf Coast, this problem is often exacerbated in the summer by heavy rains and violent storms. For that reason, in many areas, including Panama City, the harvesting of oysters is banned or restricted in the summer. Although the culprit was bacterial rather than viral, something very similar occurred in New Orleans in the wake of Hurricane Katrina in 2005, when a large number of *Vibrio* infections were reported.

There are several possible reasons why not everyone who consumed the oysters in Panama City became ill. The CDC scientists found that not all oysters carried the virus, and no doubt some restaurant patrons were simply lucky in that they consumed only uninfected oysters. Furthermore, hepatitis A often causes only mild illness, and some infected individuals remain asymptomatic. Consequently, the number of people who were actually infected may have been much greater than the number reported. Finally, a curious fact emerged following the Panama City outbreak. An epidemiological analysis revealed that people who drank liquor or wine while eating contaminated shellfish were significantly less likely to become ill. Beer had no effect. It is believed that the higher alcohol of the spirits or wine interfered with the ability of the virus to adhere to membranes in the intestine, reducing viral penetration of host cells.

When a common source epidemic occurs, there is usually a very rapid rise in the number of individuals affected, all of them becoming ill within a relatively brief period of time (**Figure 13.3**). Once the source is eliminated, cases will continue to be reported for a time period approximately equal to the duration of one incubation period (the time between exposure to the organism and the first onset of symptoms—see Chapter 10, p. 271).

The incubation period for hepatitis A averages about 30 days. In our outbreak caused by contaminated oysters, the number of cases peaked just as the sale of raw oysters in restaurants was stopped, and then began a steady decline. The last few cases cropped up over the next month or so, a length of

Figure 13.3 Common source and host-to-host epidemics. In common source epidemics, there is a steep rise in the number of cases. Once the source is eliminated, the number of cases declines, reaching zero in a time corresponding to one incubation period of the pathogen in question. In host-to-host epidemics there is a slow increase in cases, as the disease agent begins to spread through the population, and a more gradual decline. If an epidemic begins, epidemiologists frequently determine from the shape of the curve what type of epidemic is in progress.

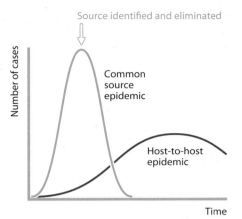

time corresponding to the incubation period. The last patients were infected just before the source of the infection was identified and contained. Of course, when the common source is *not* identified, cases may be reported indefinitely.

Host-to-host epidemics are spread from infected to noninfected individuals

Some infectious agents are propagated through a population when they are transmitted from infected individuals to susceptible ones. In such a situation there is no single source. As the agent spreads through the population there are many potential sources of infection, since every infected individual can serve as a source. This type of epidemic, in which the disease is transmitted from infected to noninfected individuals, is called a **host-to-host epidemic**. Epidemics of this type can start with one infected person, but this individual may infect many other people, each of whom, in turn, can infect many others. An epidemic of the flu is a common example.

Host-to-host epidemics start out slowly (see Figure 13.3). As the infectious agent spreads and the number of potential sources of infection grows, the epidemic picks up steam until the number of cases reaches a peak. The epidemic then slowly wanes, continuing as long as some susceptible individuals become exposed and infected. When confronted with an unexplained epidemic, epidemiologists frequently look at the shape of the curve, as illustrated in Figure 13.3, to determine which type of epidemic they are dealing with.

The *reasons* that a host-to-host epidemic begins are often complex. In the next section we will discuss some of the most important ones.

Epidemics can occur for biological, environmental, and/or social reasons

CASE: CAUGHT RED-HANDED—AND WHITE-BEAKED!

Up through the 1990s there was a curious outbreak of diarrhea in England each May and June. Carol Phillips, a British microbiologist from Northampton, decided to investigate. She knew that the diarrhea was being caused by *Campylobacter jejuni*, Gram-negative bacteria commonly found in many animal reservoirs. Humans become infected when they consume contaminated food or water. Intriguingly, Phillips found that most of the late spring/early summer victims had home milk delivery, with milk and other dairy products delivered to their door early each morning.

1. **What was the link between home milk delivery and infection with** ***C. jejuni***?
2. **Why was this a particular problem in May and June?**

As previously described, common source epidemics usually occur because of a breakdown of some sort in sanitation that suddenly allows transmission to occur. Host-to-host epidemics, on the other hand, can be due to multiple, often interacting factors.

In some epidemics, weather is the key. As an example, consider the fact that, in Bangladesh, the rainy season is the time to be on guard against cholera. A study in 2005 showed that cholera epidemics were related to bacteriophage concentrations in the water. Bacteriophages, as we learned in Chapter 4 (see p. 99), are bacteria-specific viruses. The researchers found that during most of the year, the bacteria-killing viruses hold *Vibrio cholerae*, the cholera-causing bacteria, in check. But during the monsoon season in Southeast Asia, the phages become diluted in swollen bodies of water. With both the bacteria and the phages now greatly dispersed, fewer *V. cholerae* bacteria are killed by the phages. As bacterial numbers rise, the likelihood of an epidemic increases.

The periodic warming of the eastern Pacific Ocean, known as El Niño, has been linked to epidemics of malaria in South America and hantavirus in the Southwestern United States. El Niño causes major changes in prevailing weather patterns; some areas experience drought, while others are deluged with rain. In South America, increased precipitation results in lots of standing water, which is ideal for mosquito breeding. With a boom in mosquito populations, the incidence of malaria, which is transmitted by mosquitoes, rises as well. In the case of hantavirus, most data suggests that the probability of an epidemic is greatest when populations of deer mice, the reservoir for the virus, are at their peak. Deer mouse populations rise and fall with food supply. When winter and spring are relatively wet, as they are during an El Niño event, there is a bonanza of seeds and other plant foods for the mice, and their population skyrockets. With more deer mice, the frequency of human–mouse interaction rises, resulting in increased disease transmission.

Other, often surprising factors sometimes increase opportunities for disease transmission, thereby setting the stage for an epidemic. A recent hepatitis C epidemic in Egypt provides a useful example. Hepatitis C is caused by a blood-borne virus that replicates in liver cells. Once the liver becomes infected, there is a high probability that a chronic condition will develop that can result in significant liver damage. Often, however, symptoms of disease do not appear for decades. Consequently, currently symptomatic hepatitis C patients in Egypt were probably infected years ago.

What was happening in Egypt in the mid to late 20th century that might explain this? Egypt is a hotbed of schistosomiasis infection. As we learned in Chapter 4 (p. 74), schistosomiasis is caused by a parasitic worm called the schistosome. Until the 1980s, this disease was frequently treated in Egypt with an injectable drug. Unfortunately, the syringes used to treat patients were often reused without being properly sterilized; and syringes, if not properly sterilized, are ideal vehicles for hepatitis C transmission. The result was a large number of symptomatic cases many years later. Something similar happened in Japan in the 1960s and 1970s. During World War II, Japanese soldiers were sometimes injected with methamphetamines to "improve their fighting spirit." At that time, no one gave a thought to hepatitis C, and needles were not necessarily sterilized between uses. The result was an unwelcome war "souvenir" for many Japanese war veterans.

Now we return to the odd incidence of diarrhea in England that began late each spring. Dr. Phillips found that most people who became ill had home milk delivery. Furthermore, the delivered milk bottles often had damaged foil caps. She suspected that an animal of some sort, acting as a reservoir, was contaminating the milk, early in the morning before the human residents took the milk inside. To find out, she set up video cameras at four homes to record what happened on the doorsteps in the early morning hours. On 13 occasions, she caught magpies and crows in the act, pecking through the foil-capped bottles to get at the milk for themselves (**Figure 13.4**). When she screened the 13 pecked bottles for *C. jejuni*, nine were positive for the bacteria. Why May and June? Magpies and crows produce young

Figure 13.4 Caught in the act. Birds in England with a taste for dairy products have learned to obtain milk by peeling the foil caps off milk bottles. The pictured bird is a blue tit. Other avian species, including crows and magpies, were behind the *C. jejuni* outbreaks in England each spring.

at that time of year, and Phillips thought the birds drank the milk to regurgitate it back to their babies as food. The problem has since been corrected in most places by replacing the foil with bird-proof caps. But until it was solved, the English, renowned as animal lovers, presumably bore up with their characteristic stiff upper lip, even if it did mean the occasional loose lower bowel.

Just because some factors, such as weather conditions or even thieving animals, make an outbreak more likely, there is no guarantee that an epidemic will occur. One of the most important variables is the immune status of the human population. Next we will consider this piece of the epidemic puzzle.

Epidemics become more likely when fewer people in a population are resistant

In any population, we typically expect that some people will be immune to a particular microorganism, while others will be resistant. As we saw in Chapter 11, immunity follows an initial infection, due to the development of immunologic memory. This immunity may last months, years, or even a lifetime. Individuals who have never been exposed to the microorganism in question remain susceptible.

The proportion of individuals in any population that is immune to a specific pathogen is always changing. When people recover from an infection, they are likely to be immune to subsequent infections with the same pathogen (**Figure 13.5**). Vaccination achieves the same result, without the need for actual disease. Whether individual immunity is generated by infection or vaccination, the population's overall immunity increases. Natural immunity can wane over time, however, and as it does, the proportion of immune individuals in the population declines. This proportion also decreases as babies are born, or when susceptible newcomers enter the population.

When immunity in a population is sufficiently high, an epidemic becomes unlikely or even impossible, because there are too few susceptible people in the population to sustain a large number of cases. Even susceptible individuals are less likely to get sick when many individuals are immune since immune individuals do not transmit the disease to susceptible people. When a high proportion of the population is immune, there are not many routes of transmission available for the pathogen to reach those few

Figure 13.5 Immunity in a population and epidemics. If the proportion of immune individuals in a population to a particular pathogen is low, when that pathogen enters the population, an epidemic is possible because there are so many susceptible potential hosts. If an epidemic begins, as it does in this figure at time zero, immunity in the population starts to rise, as infected people recover and are subsequently immune to a second infection. At some point the proportion of susceptible individuals is too low to permit effective pathogen transmission (approximately week 3 in this example). At that point the number of cases begins to fall and the epidemic wanes. The proportion of immune individuals will continue to slowly rise, reaching its highest level as the last few people become ill at the end of the epidemic (week 5).

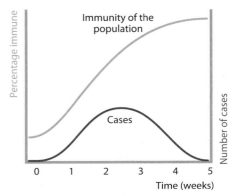

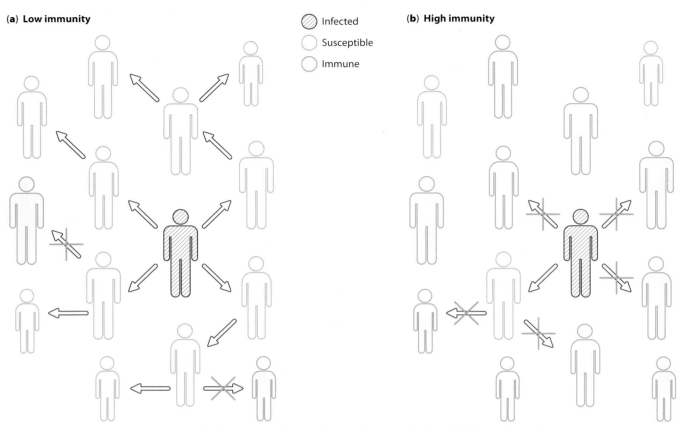

Figure 13.6 Immunity in a population and rate of transmission. (a) If few individuals are immune and an infected individual (in red) enters the population, there are many possible transmission routes to susceptible individuals (in green). Resistant individuals (in blue) are too scarce to block transmission and prevent an epidemic. (b) When most individuals are immune, the few susceptible individuals are protected from transmission by the large number of resistant individuals. Because transmission between infected and susceptible people is unlikely, the number of new cases will be low

susceptible people (**Figure 13.6**). When few individuals are immune, most people are vulnerable and if they become infected they are able to transmit the pathogen to other susceptible individuals. Transmission is consequently relatively easy for the pathogen.

The ever-changing immune status of populations explains why so many host-to-host epidemics occur in cycles (**Figure 13.7**). If an epidemic occurs, many people become ill, and as they recover, become immune. At some point, so much of the population is immune that the number of new cases starts to decline and the epidemic wanes. The proportion of immune individuals will continue to rise as the epidemic subsides, but at a much slower rate, because fewer and fewer people are getting ill and subsequently developing immunity. Once the epidemic has finally ended, immunity in the population is at its peak. Another epidemic is not possible at this point. As time passes, however, and as immunity gradually wanes in previously immune people, or as new, susceptible individuals enter the population, another epidemic is eventually possible. This may take months, years, or decades, depending on the rate at which this decline in the population's immunity occurs. Sporadic cases might occur even when immunity in the population is high, but not the large number of cases associated with an epidemic.

How high must a population's level of immunity be to prevent an epidemic? The approximate level, known as the **epidemic threshold**, varies for different pathogens. With influenza, for example, it is estimated that when approximately 90-95% of the population is immune an epidemic cannot

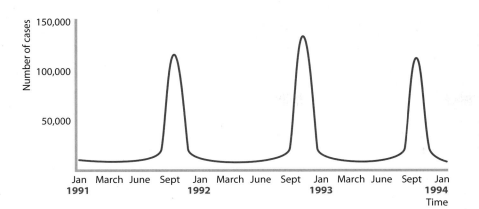

Figure 13.7 Cyclic epidemics of chickenpox. Chickenpox epidemics traditionally occurred at the beginning of the school year in the fall. Before the advent of the chickenpox vaccine, almost all children starting school at age 5 were susceptible. Consequently, each fall a new group of susceptible hosts came together in crowded classrooms, where the chickenpox virus, spread via respiratory transmission, could thrive. A single infected child could start a large epidemic. Once an epidemic began, increasing numbers of children developed immunity, resulting in lower rates of transmission. Finally, viral transmission was so low that the epidemic subsided. Because children who were infected became immune to future infections, another epidemic was not possible until another new group of young students started school again the following fall. Since the introduction of the vaccine in 1995, the number of annual chickenpox cases has significantly declined.

occur. For polio the value is about 70%. In other words, as long as at least 70% of the population is immune, transmission is too poor to permit a polio epidemic. Different threshold values usually reflect, at least in part, how easily a given pathogen is transmitted to new hosts. Pathogens such as the influenza virus, which spread easily, generally have higher threshold values. **Figure 13.8** traces the timing of epidemic cycles for a hypothetical pathogen, with epidemics possible only when the population's immunity is below the epidemic threshold.

For many diseases, public health officials rely on vaccination programs to maintain immunity above threshold and prevent epidemics. As long as the percentage of vaccinated individuals is maintained above threshold, an epidemic is not possible. This of course is an easier task to accomplish for diseases such as polio, with relatively low thresholds. For diseases with high thresholds, it is considerably harder to maintain levels of immunity high enough to prevent outbreaks. The flu, for example, which we saw has a threshold of about 90-95%, is a regular, if unpleasant part of life in many parts of the world, and regular epidemics are to be expected. In the next section, let's take a look at the epidemiology of this relatively common viral disease, which in some respects is anything but common.

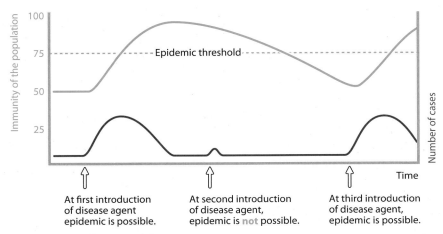

Figure 13.8 Epidemic cycles and immunity in populations. In this hypothetical example the disease in question has an epidemic threshold of 75%. Initially, an epidemic is possible because the proportion of immune individuals is below the threshold. When the disease organism enters the population for the first time, the number of cases and immunity in the population both rise. As the population's immunity rises above the threshold, the number of cases begins to fall. At this point, even if the disease agent is reintroduced, a second epidemic is not possible. Over time, however, a population's immunity will gradually decline because of waning individual immunity or an influx of susceptible individuals. When it eventually falls below the threshold, a second epidemic (third introduction on the graph) is once again possible.

Epidemic outbreaks of influenza occur as the virus changes genetically

CASE: FLU SEASON

One day in early March, Sheila, a healthy 60-year-old woman, returns home feeling achy and feverish. After a fitful night, she goes to the doctor and her suspicions are confirmed—she has the flu. She cannot remember the last time she had the flu, but she clearly recalls the winter of 1957 when it seemed like everyone had the flu. Schools were closed because so many students and teachers were ill. An uncle of hers even died.

Sheila read in the newspaper that this newest strain of the flu arose in China, that it is particularly nasty, and that the chances for a flu epidemic this year are high. She is worried about her elderly parents, who live in her neighborhood and with whom she has frequent contact. Her doctor recommends that they come in right away for a flu shot, even though they had one last year. As for Sheila, the doctor says that she will probably feel better in a week or so but that other than getting plenty of rest and taking pain relievers like aspirin, there is not much that can be done for her.

1. What determines the likelihood of a flu epidemic in any given year?
2. Why are some strains of the flu more dangerous than others? Why was the 1957 strain so severe?
3. Why do Sheila's parents need a flu vaccination when they had one the previous year?

Influenza certainly must make anyone's "top ten" list for major epidemic diseases in Europe, North America, and other developed parts of the world. The disease is no newcomer. An epidemic described by Hippocrates from the 5th century BC is thought to have been influenza, and in the last 900 years or so, there are more than 300 records of flu-like epidemics—an average of one epidemic somewhere in the world approximately every 2.4 years. Furthermore, since the early 1700s, there have been 22 recorded influenza **pandemics**. A pandemic literally means a "worldwide epidemic." The term is used more commonly to imply that an epidemic is spread over a very wide geographic area, not necessarily the entire world. There were three influenza pandemics in the 20th century. Most recently, in June 2009 the World Health Organization declared that "swine flu" had attained pandemic status.

The most deadly influenza pandemic in recent history was the 1918 "Spanish flu," which ultimately affected a large part of the globe and killed an estimated 20 million people worldwide (**Figure 13.9**). In 1957, 70,000 people in the United States alone, including Sheila's uncle, were killed by the "Asian flu." The last pandemic of the 20th century occurred in 1968, caused by the "Hong Kong flu." In years when there is neither an epidemic nor a pandemic, influenza is said to be **endemic**. An endemic disease is one that is always present in a particular geographic area, usually occurring at a low level.

Why is influenza sometimes endemic and sometimes epidemic, and why does it occasionally explode into a bona fide pandemic? To answer this and

Figure 13.9 The Spanish flu pandemic. The Spanish flu was the most serious pandemic of the 20th century. (a) Soldiers of the 505th Service Battalion en route to Europe aboard the USS *President Lincoln*. It is believed that the massive movements of troops and refugees during World War I helped to spread the virus. (b) A streetcar conductor in Seattle checking boarding passengers for masks. During the pandemic, although the cause of the disease was unclear, it was understood that transmission occurred via the respiratory route, prompting the widespread use of masks to limit infection.

the other questions presented in Sheila's case, we first need to review a few details of the structure and genetic makeup of the influenza virus.

There are three types of influenza virus that can affect humans, called influenza A, B, and C. They are related viruses, but they differ in a number of details. Influenza A causes the most severe illness, and it is the only one of the three viruses for which animals are reservoirs. The use of animal reservoirs makes this virus particularly dangerous and it is one of the key reasons that influenza A is the only type of influenza that causes pandemics. Due to its overriding importance in human health, we will confine our present discussion to influenza A.

Influenza virus is a negative-strand RNA virus. This means that it encodes its genome in RNA rather than DNA (for a complete review of RNA viruses, see Chapter 4, p. 96). The viral RNA genome is actually composed of eight segments, each of which carries different genes (**Figure 13.10**). Each newly produced viral particle must have all eight of these segments if it is to replicate.

Two viral genes, located on different segments of the genome, are of particular importance in understanding both the virulence and epidemiology of influenza. These are the genes that code for hemagglutinin (H) and neuraminidase (N) (see Figure 13.10). Hemagglutinin is the viral envelope glycoprotein, which allows an influenza viral particle to latch onto and then enter host cells in the respiratory tract. Neuraminidase is an enzyme found in the viral envelope, which allows newly formed viral particles to leave a host cell. Consequently, both H and N are essential to the completion of the viral replicative cycle (see Chapter 4, p. 92). They are also the principal antigens recognized by the host immune system; when you are infected with influenza, you produce antibodies against the H and N antigens. Memory for these antigens is protective against a second infection by the same strain. Different strains of influenza, however, differ in their H and N antigens, and immunologic memory against one type of H or N does not guarantee protection against other types. There are 13 known variations of hemagglutinin, all of which cause different immune responses. Likewise, there are nine neuraminidase immunological types. The different forms of these antigens are called H1 through H13 and N1 through N9, respectively. Different influenza strains are designated by their specific type of hemagglutinin and neuraminidase. A strain with H4 and N3, for instance, is designated influenza H4N3. If you are infected with H4N3, you will be resistant to further infections with this strain. You would remain susceptible, however, to different strains such as H5N6. You would have partial immunity to strains H4N6 or H5N3.

Remember that influenza has a segmented RNA genome and that the genes for H and N are on separate segments. This segmented RNA genome has two important consequences for influenza epidemiology. First, like all RNA viruses, influenza virus mutates rapidly—much faster than DNA viruses, bacteria, or eukaryotes. Organisms that rely on DNA have repair mechanisms that ensure that many mutations are corrected. These repair mechanisms are lacking in RNA viruses. In Chapter 6, for example, we learned that DNA polymerase has "proofreading" ability; if it makes a mistake, the enzyme can cut out a mismatched nucleotide and replace it. RNA polymerase used by RNA viruses, on the other hand, cannot do this. Consequently, if the enzyme makes mistakes, those mistakes remain in the newly synthesized RNA. With large numbers of mutations constantly being introduced, RNA viruses are constantly changing. Some of these changes might affect proteins that are recognized by a host immune system as antigens. Therefore, if memory is developed against a particular antigen, and if that antigen mutates sufficiently, immunological memory against the original antigen may not be protective against mutated strains of the virus. This phenomenon is known as **antigenic drift**. For influenza virus, changes in both the H

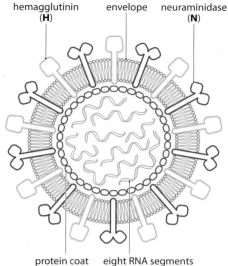

Figure 13.10 Structure of the influenza A virus. The eight RNA segments carry the genes used by the virus for synthesizing various proteins. Among those proteins are the envelope glycoproteins hemagglutinin (H) and neuraminidase (N), which are the primary antigens recognized by the immune system.

and N antigens due to antigenic drift can mean that exposure to one viral strain does not provide protective immunological memory against different strains.

Second, sometimes a cell can be infected by two influenza viruses at the same time. Each of these viruses will replicate all eight of its RNA segments, but when new progeny viruses assemble (see Chapter 4, p. 94), they may get some of the eight segments from one parent virus and some from the other. As long as they obtain all eight segments, the source does not matter, and the virus can replicate successfully. This reshuffling of the genetic material can cause dramatic new viral strains to emerge to which few if any humans are immune. This process, called **antigenic shift**, is the principal reason for the emergence of new viral strains.

For influenza A, antigenic shift is most likely to occur in a chicken, duck, pig, or other animal reservoir. When multiple viral strains infect the same animal, the mixing and matching of the RNA segments can begin, occasionally producing new combinations (**Figure 13.11**). Not all of these newly created strains can easily infect humans. Some can be transmitted from the animal reservoir to the human but cannot be transmitted from human to human. Occasionally, however, the new strain can move not only from animal to human but between humans as well. Furthermore, shift can result in new combinations of the H and N antigens, for which a human population has little or no immunity. For example, if a cell in an animal reservoir is simultaneously infected with influenza H2N2 and H8N8, new viral progeny could be produced that are either H2N8 or H8N2. If a population's immunity to this newly created strain is low or nonexistent, the stage is set for an epidemic or even a pandemic.

Because of antigenic drift and shift, different strains of influenza virus circulate from year to year. This explains why Sheila's doctor recommended a new vaccination for her parents, even though they were vaccinated last year. The previous year's vaccination probably offers little or no protection against this year's strain. Because influenza is a moving target, epidemiologists at the World Health Organization continually monitor the virus in an attempt to predict what strain will be in circulation during the coming year. Such predictions are then used to develop that year's vaccine.

The constantly changing nature of the virus also explains in large part why the virus can vary so much from year to year in severity. When a person is infected with influenza, the severity of symptoms is determined by how well the virus is replicating. In individuals with no established immunity, the virus initially replicates quickly, causing more pronounced symptoms. If there is partial immunity to either H or N, there will be less viral replication and consequently milder symptoms. When Sheila read that this year's strain would be particularly severe, it meant that the viral strain would have hemagglutinin and neuraminidase antigens to which there was little or no immunity. Replication in infected hosts would be rapid, resulting in a more pronounced illness.

The 1918 Spanish flu was caused by an H1N1 strain of the virus. The 1957 Asian flu pandemic that killed Sheila's uncle was caused by an H2N2 strain. In 1968, the Hong Kong flu pandemic was caused by an H3N2 strain. The 1968 pandemic killed fewer people than the 1918 or the 1957 pandemic, partly because in 1968 many people still had partial immunity to the N2 antigen, which had been found on the 1957 strain, just 11 years earlier.

To illustrate just how unpredictable influenza can be, and how the prediction of epidemics remains in part a guessing game, we need look no further than very recent history. Up through 2008, epidemiologists were concerned that the next influenza pandemic might involve the H5N1 strain, known as

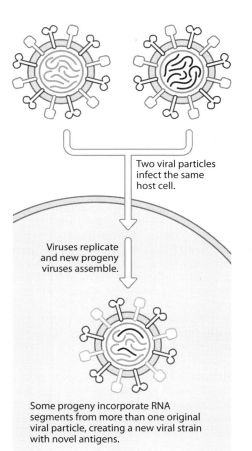

Two viral particles infect the same host cell.

Viruses replicate and new progeny viruses assemble.

Some progeny incorporate RNA segments from more than one original viral particle, creating a new viral strain with novel antigens.

Figure 13.11 Antigenic shift. When influenza viruses infect a cell, they make copies of all eight of their RNA segments. Each new progeny virus must contain all eight segments. Occasionally, when a single cell is co-infected by more than one virus, a progeny virus may obtain RNA segments from more than one original virus. The new "hybrid" virus has undergone *antigenic shift*. Immunity in the population to this new viral strain is probably very low.

avian influenza, or bird flu. The first cases of H5N1 infection in humans occurred in 1997, when 18 people in Hong Kong contracted this strain of influenza A, previously known to infect only birds. Six of the patients died. A second rash of outbreaks in Southeast Asia began in 2003. Since then, the virus has spread across Asia, Europe, and North Africa, killing millions of birds. The number of human cases has remained relatively low because most transmission to humans has been through contact with infected birds (**Figure 13.12**). Human-to-human transmission, while possible, has been infrequent. Nevertheless, the H5N1 virus remains extremely pathogenic, killing up to 60% of those humans who become infected.

With attention focused on avian influenza, public health officials were caught off guard when a new strain of influenza reared its head in March 2009. The virus, commonly called swine flu, was first detected in Mexico and identified as H1N1 in April 2009. Although H1N1 virus was known from the past, most notably from the Spanish flu pandemic, in some respects this is a completely novel form of influenza, created by genetic shift. It is thought to be a reassortment of four known strains of influenza A: one that normally infects humans, one endemic in birds, and two endemic in pigs. The virus is highly transmissible between humans, although so far it seems no more pathogenic than other, more familiar strains of influenza A. In June 2009, as the virus continued to spread around the world, the World Health Organization declared a global pandemic, the first such declaration in half a century. In the first 6 months of the pandemic, there were more than 300,000 reported cases and over 3000 deaths worldwide.

Because this appears to be a new strain of influenza A, public health officials suspect that very few people are immune. Consequently, it is expected that efficient transmission will allow the virus to continue spreading, and the number of human cases will continue to grow. Furthermore, although at present the H1N1 virus does not appear to be unusually virulent, such virulence can change at any moment due to genetic drift or due to a new genetic shift event.

It is difficult to predict exactly what might happen in this pandemic, but should the H1N1 virus change in terms of its virulence or its transmission, it will be up to epidemiologists to detect such changes. In fact, any time there is a suspicious disease outbreak, epidemiologists swing into action. Next we will see exactly what they do when a medical mystery crops up.

Figure 13.12 Avian influenza reservoirs. A marketplace in Vietnam. Many domestic birds, including chickens and ducks, can serve as reservoirs for H5N1, otherwise known as the bird flu.

Investigating Disease Outbreaks

We opened this chapter with a description of a *Salmonella* outbreak in Denver. Although the source of infection in that outbreak was unusual, the pathogen was fairly routine. *S. enteritidis* is known to be transmitted by contaminated food or water, and once epidemiologists knew what they were dealing with, they could zero in on the source. All patients were carefully questioned, as epidemiologists looked for something that they all had in common. Once it was realized that they had all visited the reptile exhibit at the zoo, authorities were able to pinpoint the cause of the problem and implement control measures. Other epidemics that we have described in this chapter would be similarly investigated.

Sometimes, however, outbreaks are more mysterious, and health authorities are unable to quickly identify a specific cause. In that case, the problem becomes trickier and the task of epidemiologists is more complex. Nevertheless, when they are confronted with such a situation, epidemiologists rely on a defined protocol, in which clues and evidence are gathered in a step-by-step manner, to help them understand what is happening. We will next see exactly how this is done, using a recent epidemic as an example.

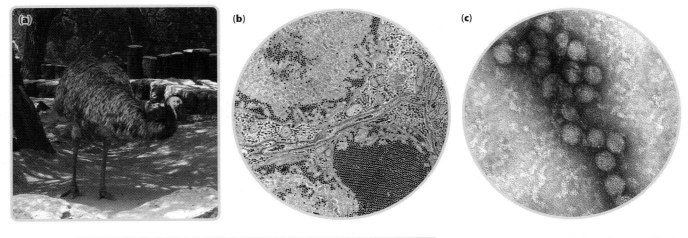

Figure 13.13 Crucial clues in a medical mystery. (a) Healthy emus at the Bronx zoo helped to rule out Eastern equine encephalitis as a possible cause of an epidemic of unknown origin. (b) St. Louis encephalitis virus, first implicated as the epidemic's cause, was the wrong size to be the virus isolated from dead birds from the zoo. The small black spots in the cells' cytoplasm are newly replicated particles. Each of these viral particles is about 50 nm across. (c) Electron micrograph of West Nile virus. Each of these viral particles (seen here as spherical structures in the cytoplasm of an infected cell) is approximately 40 nm across. This is significantly smaller than SLE, and this fact helped rule out SLE as the cause of the 1999 epidemic in New York.

CASE: A NEW BUG ON THE BLOCK

In June 1999, residents of Queens, New York, began reporting that crows with no obvious injuries were dying. On August 23, the city's health department was notified that three elderly patients had symptoms of a neurological condition, suggesting some sort of disease outbreak. By August 31, with six cases now reported, the Centers for Disease Control and Prevention sent epidemiologists to New York to investigate an "encephalitis of unknown origin." Their first guess was that this was a form of insect-transmitted encephalitis virus, most likely St. Louis encephalitis (SLE), in part because SLE uses birds as its reservoir and can cause neurological symptoms in humans. New York City mounted a major mosquito abatement program to control the outbreak, using malathion sprayed from helicopters.

Meanwhile, a veterinarian at the Bronx Zoo had doubts that this was SLE. Many exotic birds at the zoo were dying, but their symptoms did not match those of SLE. Furthermore, SLE does not usually cause such massive mortality, and most birds remain asymptomatic. On the basis of the symptoms and the high mortality, other diseases such as Newcastle disease, avian influenza, and Eastern equine encephalitis (EEE) seemed more likely candidates. Yet if Newcastle disease or avian influenza was the cause, the chickens at the children's zoo should have been the first to die. The chickens, however, were all healthy. Furthermore, the zoo's emus were fine, even though emus are highly vulnerable to EEE (Figure 13.13a).

On September 9, the zoo's veterinarian sent samples from dead birds to the National Veterinary Services Laboratory in Ames, Iowa. The laboratory found virus in the central nervous system of these birds, and when it took electron microscope images, the viral particles were found to be only 40 nm across, too small for SLE (Figure 13.13b). On September 24, the virus was confirmed as West Nile virus (WNV). This was surprising because this virus had never been reported in North America before. On September 30, WNV was isolated from mosquitoes in Queens. By late October, New York City reported 62 cases and seven deaths, and infected mosquitoes, birds, and horses had been found as far away as Connecticut and Maryland.

1. What exactly did epidemiologists do to track down the cause of this unknown outbreak?
2. Where do seemingly new diseases like West Nile virus come from? What factors cause new diseases to appear?

A case definition helps health authorities determine if unusual cases are related

When the first cases of "encephalitis of unknown origin" appeared, health authorities first wanted to be certain that the cases were all related. In a city as large as New York, there will always be patients with unusual neurological symptoms in any given week, and they might all be suffering for different reasons. If so, then there would not be any specific epidemic in progress. However, if all patients had developed encephalitis for the same reason, a red flag would be raised.

To find out whether an epidemic was under way, health authorities first had to develop a **case definition**. A case definition is essentially a list of common symptoms, obtained from doctors who have treated the patients and by questioning the patients themselves. After careful questioning and examination, a case definition for the encephalitis in New York emerged. It included changes in mental status, headache, stiff neck and back, fever, and malaise. With this case definition now in hand, epidemiologists instructed local doctors to be on the lookout for patients with similar symptoms.

Time, place, and personal characteristics of a new disease provide clues to the disease's identity

Now that they were aware of the problem and armed with case definitions, local doctors in New York began to identify new patients who seemingly suffered from the same condition. As the list of patients grew, it became possible to start defining some of the important epidemiological parameters of the outbreak, each of which might provide clues as to what was happening.

One such parameter is the **time characteristic**. The time characteristic is a description of *when* people are getting sick. In the present example, the first cases were appearing in the late summer and early fall (**Figure 13.14**). What are people doing at this time of year? Are they all going to the same places, eating the same foods, or engaging in the same activities? Or are they all being bitten by insects? This later possibility, combined with the case definition suggestive of encephalitis, pointed at candidate pathogens such as SLE or EEE, both of which are mosquito-transmitted viruses.

A second important parameter is the **place characteristic**. The place characteristic describes *where* affected individuals live, or where they were when they became ill. Do patients live in rural or urban settings? Do they all work in a similar sort of environment? In this example, all early patients lived in Queens, New York, and surrounding areas. As more cases were reported, it became obvious that this particular disease could strike in rural areas as well.

It is also necessary to assemble the **personal characteristics**. The personal characteristics portray *who* is getting sick. Are the sick people mainly of one sex, age, or cultural background? If so, epidemiologists would investigate what it is about a particular group of people that places them at elevated risk, in the hope that this will help them uncover the underlying cause of the

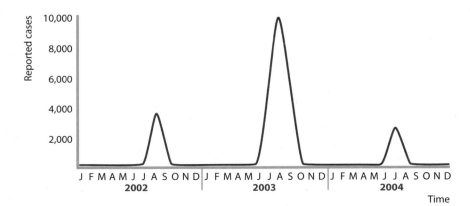

Figure 13.14 Time characteristic for West Nile virus. Numbers of U.S. human West Nile cases for the 2002–2004 seasons. The incidence of West Nile virus mirrors the seasonal abundance of its mosquito vectors, rising sharply in the summer and falling to zero in the winter. Similar time characteristics would be seen for other mosquito-vectored diseases such as Eastern equine and St. Louis encephalitis.

disease. The first WNV patients were all elderly men and women. We now know that younger people are equally likely to become infected with WNV, but in younger individuals, encephalitis is much less likely to develop. Most people infected with WNV remain asymptomatic or develop a much milder flu-like illness. Because so many of these mild or asymptomatic cases went unreported, it is likely that during those few summer months in 1999, many more people than the 62 reported cases were actually infected.

As the number of encephalitis patients grew, time, place, and personal characteristics of the epidemic became increasingly clarified, and epidemiologists could begin to concentrate on possible causes. Certainly the case definition, along with the fact that cases were concentrated in the late summer and early fall, pointed to a viral form of encephalitis transmitted by mosquitoes. Other arthropods such as ticks are also more common in the summer, but since most of the early cases were urban rather than rural, mosquito transmission seemed far more likely. SLE was proposed as a likely cause, but other questions had to be answered before a definite diagnosis could be nailed down. First of all, epidemiologists wanted to know if this disease could be transmitted from infected to susceptible humans. If so, people who spent time with affected individuals, such as family members and caregivers, would be more likely to become sick themselves than people who had no association with the patients. Careful investigation determined that those who spent time with patients were not at elevated risk. This indicated that the disease agent could not pass from human to human and further pointed to SLE, which can infect humans only through the bite of an infected mosquito.

Case–control studies can pinpoint a common risk factor among affected individuals

Once time, place, and personal characteristics are defined, the next step is to develop a **case–control study**. A case–control study is used to determine what factor or factors link the affected individuals and distinguish them from unaffected individuals. In such a study, the affected people are the *cases*. Each case is matched with a *control*—an individual who has not become ill. The matched case and control should be as similar as possible in age, sex, place of residence, and other factors that investigators consider important. The only significant difference should be whether or not they are affected by the disease. Epidemiologists then look for any other difference between the cases and controls that could explain why the cases got sick and the controls did not. It turned out that those who developed encephalitis were far more likely to have spent time outside in the evening than those who remained healthy. Because the mosquitoes that transmit encephalitis viruses tend to feed at night, epidemiologists were further assured that they might be on the right track.

Case–control studies often provide the valuable missing clue that helps epidemiologists crack a difficult mystery. Before anyone understood what Lyme disease was, for instance, the case–control study showed that those who had become ill were far more likely to own household pets. This led investigators to believe that they were dealing with something brought into the home by dogs or cats, perhaps a pathogen transmitted by fleas or ticks. Eventually a newly discovered tick-borne bacterium called *Borrelia burgdorferi* was identified as the causative agent. In the early investigation of AIDS, the case–control study suggested a sexual mode of transmission by revealing that the main factor separating the cases from the control group was having multiple sexual partners. When hantavirus first struck the American Southwest, the case–control study also helped elucidate the disease source. Most of those affected had recently cleaned out buildings such as garages and storage sheds. This knowledge helped investigators zero in on rodent droppings, which are common in such structures.

Although the case–control study in the early West Nile virus epidemic strongly indicated an encephalitis virus such as SLE, as we saw in our description of the early epidemic, the final identification of the virus did not go smoothly. As it turns out, it was the astute veterinarian at the Bronx Zoo who finally set the wheels in proper motion, when she suspected that the cause of the epidemic was perhaps something new. Only after she sent samples to the veterinary diagnostic laboratory in Iowa was it clear that this was not SLE after all. Fortunately, because West Nile virus is also mosquito-borne, control measures instituted for what was thought to be SLE were appropriate for West Nile virus as well. This episode nicely illustrates the importance of communication and the involvement of all interested parties when we are confronted by a new health threat. Veterinarians are not routinely consulted when there is an outbreak of human illness. In this case, their participation was essential.

West Nile virus is just one of a host of new diseases that have sent epidemiologists and other health authorities scrambling in recent years. Over the last few decades, diseases such as Lyme disease, Ebola hemorrhagic fever, acquired immune deficiency syndrome (AIDS), and severe acute respiratory syndrome (SARS) have taken us by surprise. In each instance, epidemiologists proceeded through the same series of steps until the nature of the new epidemic was revealed. In the next section, we will learn where new diseases come from in the first place.

Emergent Diseases

Emergent diseases are new or changing diseases that are increasing in importance. Such diseases often seemingly come out of nowhere on an almost annual basis (**Figure 13.15**), and there are no doubt newer ones lurking in our future. On the other hand, sometimes an old microbial foe, thought to be vanquished, comes back to haunt us. We will next consider where these emergent diseases come from and why they crop up with such alarming frequency.

Figure 13.15 Representative emergent diseases. Geographic locations and dates represent the first report of each disease as a new and emerging or reemerging threat.

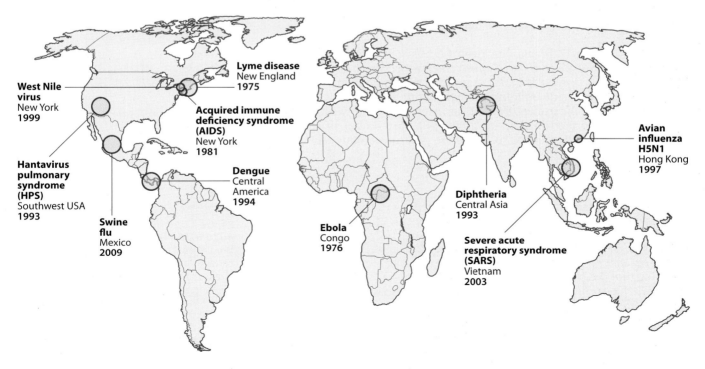

Environmental, biological, behavioral, and social changes can result in emergent diseases

At many points throughout this text we have emphasized that diseases can change as conditions change. Pathogens may become more or less deadly as transmission becomes easier or more difficult. As the environment is altered, reservoirs might become more or less common. Even changes in human behavior can inadvertently alter the disease equation, making some outbreaks more or less likely.

These changes sometimes cause new diseases to appear, or they may cause the resurgence of diseases that were thought to be controlled. Such emergent diseases receive considerable media attention, but they are not strictly a modern problem. Syphilis was an emergent disease of 16th-century Europe, thought to have been transported from the Americas by returning explorers. As we discussed in Chapter 5, yellow fever emerged as a major problem during the first attempt to construct the Panama Canal in the late 1800s. During the mid-20th century, polio, which had been present as an endemic disease for thousands of years, roared to prominence as a serious emerging disease of epidemic proportions (**Figure 13.16**).

At first, the reason for polio's emergence sounds counterintuitive. Polio is transmitted through contaminated food and water. Before improved sanitation became common, most people were exposed to polio as very young children. Although infected children may develop a mild intestinal illness, they rarely develop paralysis, by far the most serious consequence of polio. Newborns are protected by maternal antibodies, and there is some indication that the virus has a more difficult time invading the nervous system of infants. In many parts of the world, however, sanitation vastly improved during the 20th century. Consequently, the number of people exposed to polio as infants declined dramatically. If these individuals contracted the virus later in life, paralysis was a more likely consequence. Thus, while polio was traditionally endemic throughout the world, *epidemic paralytic* polio was to a large degree the *result* of improved sanitation.

Increasing human population and urbanization can also set the stage for the emergence of disease. Even changes in travel and trade influence the distribution of diseases around the world. An excellent example of how such factors contribute to disease emergence is provided by dengue.

Dengue, also known as "breakbone fever" because of the excruciating joint pain it can cause, is caused by a mosquito-transmitted virus. The main mosquito vector, *Aedes aegypti*, is most common in the tropics, where it prefers to feed on humans, inside buildings. Consequently, dengue is described as an urban disease of the tropics. Dengue has been around for years. The disease was first described in the late 1700s, but the number of cases has exploded in the last 25 years or so. There are now an estimated 40 million cases a year worldwide.

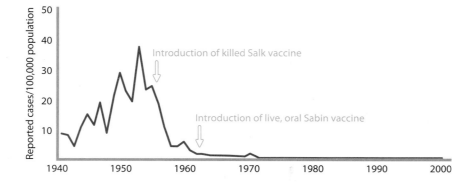

Figure 13.16 Incidence of paralytic polio in the United States during the second half of the 20th century. Paralytic polio was an important emergent disease of the 20th century, with the number of cases reaching a peak in the early 1950s. With the introduction of the Salk vaccine in 1955, the number of cases plummeted. Today, thanks to vaccination, polio has been eradicated from much of the world.

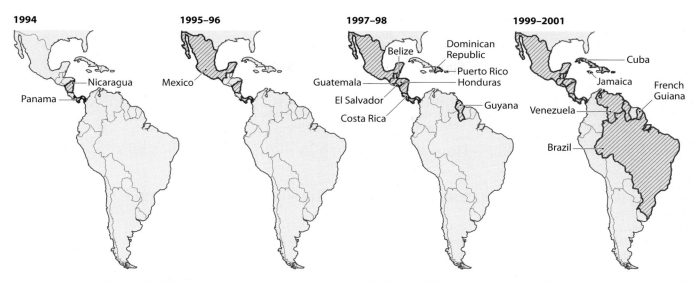

Figure 13.17 Dengue on the march. Since its introduction in 1994, dengue 3, subtype III, has invaded large parts of Latin America. The virus has been reported from those countries highlighted in red.

What accounts for dengue's upsurge? Part of the answer is that in the last 150 years, the human population in the tropics has gone from 2% to 50% urban. As more and more people have moved into cities, previously undeveloped land has become urbanized. More people living in crowded cities is good news for *Aedes* mosquitoes. Not only is there more of their preferred food, but along with the people comes their garbage. Much of this garbage, such as old tires and cans, makes ideal mosquito breeding sites because it holds rainwater. With more people and more mosquitoes, the increase in the number of human dengue cases is not surprising.

Furthermore, dengue, once considered primarily a disease of Asia, has branched out, entrenching itself in the Americas as well. International jet travel, common only since the 1960s, has provided a handy way for dengue and other diseases to reach new destinations. In the case of dengue, if an infected mosquito happens to fly into an open jet door as the plane sits on the runway in Bangkok, that same mosquito might find itself in Miami or Mexico City in a few hours. Once there, if appropriate reservoirs and the necessary mosquito vectors are already in place, the stage may be set for a new dengue outbreak. In fact this has already occurred. **Figure 13.17** shows how one strain of the dengue virus has spread through Latin America since its arrival in Central America in 1994.

This is probably the way that West Nile virus arrived in New York as well. As the name suggests, West Nile virus is endemic in Africa. It is also found in the Middle East and southern Europe. It most likely made its New York debut in 1999 when an infected mosquito crossed the Atlantic in a jetliner. With plenty of the birds it needs as reservoirs, and with abundant mosquitoes present to serve as vectors, West Nile virus has prospered in North America. By 2002 it had reached the West Coast (**Figure 13.18**). In 2003, there were 9862 human cases and 264 deaths. By the start of the 2008 mosquito season, infected mosquitoes, birds, and/or nonhuman mammals had been detected in all 48 of the contiguous states. All states except Maine have recorded human cases.

The importance of factors such as increasing urbanization and international travel points out an important fact about emerging diseases. Human society is changing and the environment is being altered at an unprecedented rate. Consequently, although new diseases have periodically arisen throughout human history, such diseases are emerging today as never before.

Figure 13.18 The spread of West Nile virus in the United States. States in which human cases were reported during a specific year are in red. States without human cases but in which infected mosquitoes, birds, or mammals were detected are indicated in orange. States in gray had no detected West Nile virus activity. Following the first reported cases in 1999, the virus spread rapidly across the United States, reaching the West Coast in 2002. As of 2008, all 48 contiguous states have reported some West Nile virus activity. Maine is the only one of these states in which no human cases have been reported.

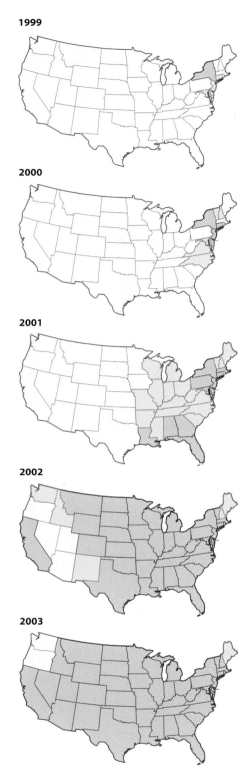

Emergent diseases can be categorized as one of four basic types

Emergent diseases are placed in one of four categories. These can be briefly described as follows:
1. invasion of a new host population by a known pathogen
2. appearance of a completely new, previously unknown disease
3. association of a well-known disease with a new pathogen
4. increased virulence or a renewed problem with a well-known but previously less virulent or well-controlled pathogen.

When a known pathogen is able to reach a new host population and therefore changes its overall distribution, it is considered to be in category 1. Most emerging diseases, including West Nile virus and dengue, are in this category. As described, international jet travel, with no place now more than a few hours away from any other place, plays an important role in emergent diseases of this type.

In June 2003, an outbreak of monkeypox in the midwestern United States reminded us that the world is an ever smaller place. Monkeypox is endemic to Central and West Africa, where it survives naturally in certain rodents. Humans or monkeys can become infected through close contact with such rodents. The monkeypox virus made it to Illinois in a Gambian giant pouched rat that was purchased by an exotic pet dealer in Chicago. The rat was housed with prairie dogs, many of which also became infected. These prairie dogs subsequently infected their new human owners, resulting in several dozen cases of a tropical African disease in decidedly untropical Illinois, Wisconsin, and Indiana (**Figure 13.19**).

Some pathogens exist in the environment undetected for many years. They may occasionally cause illness or death, but if cases are rare or sporadic, such pathogens may remain unrecognized. At some point a convergence of events finally brings a "new" disease to our notice. Such an apparently new disease is placed in category 2. AIDS, first detected in 1981, hantavirus, unknown before 1991, and SARS, which first came to our attention in 2003, are examples.

To illustrate how previously unknown diseases can suddenly emerge, consider Lyme disease, another disease in this category, which was first detected in 1975. The bacteria causing Lyme disease rely on deer and rodents as reservoirs and ticks as vectors. In parts of New England where the outbreak was first detected, the number of deer greatly increased during the latter part of the 20th century, and as more and more people began moving into rural areas, the opportunities for transmission from wildlife to humans increased. Eventually, in the mid-1970s, a cluster of cases affecting children near the town of Lyme, Connecticut, alerted authorities to a disease that had probably been affecting people in low numbers for many years.

Of course, no discussion of previously unknown diseases could be complete without mentioning Ebola hemorrhagic fever, perhaps the most spectacularly gruesome emergent disease of all. This disease, caused by Ebola virus, burst on the scene in 1976 when it caused two outbreaks in Africa. Since then it has occasionally come out of hiding, causing large epidemics in

Figure 13.19 Monkeypox in the midwestern United States.
(a) Blisters typically seen in the course of a monkeypox infection. Other symptoms include fever and headaches. (b) The 2003 outbreak in the United States occurred when infected prairie dogs, purchased as pets, bit their owners. The prairie dogs became infected when they were housed with African rodents serving as reservoirs in a Chicago-area pet store.

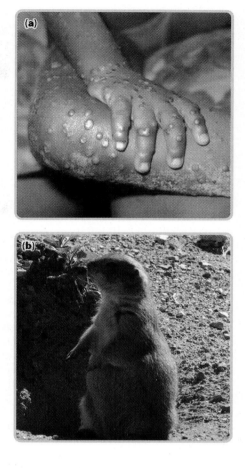

Congo in 1995 and Uganda in 2000 (**Figure 13.20**). Smaller outbreaks have occurred elsewhere in Africa. Ebola can cause up to 90% mortality, although some strains of the virus are apparently somewhat less lethal. A strain of this African virus, found in Reston, Virginia, in 1989 in imported African primates, turned out to be nonpathogenic in humans, although highly virulent in monkeys.

Ebola virus causes massive hemorrhaging in its victims, who essentially die as they "bleed out" from a variety of tissues. Human-to-human transmission can occur by contact with infected material such as blood or contaminated needles. In the outbreak in Reston, there was worrying evidence of airborne transmission. Monkeys housed in rooms that shared ventilation with infected animals developed the disease themselves, in spite of no direct contact. Strains of Ebola that infect humans have thus far shown no signs of airborne transmission.

The reservoir of Ebola virus remains unknown, and we still do not understand how and under what conditions the virus is transmitted from its reservoir to humans. Until these crucial factors are determined, it will be difficult to predict where and when an outbreak might occur. There is still no effective vaccine that has been approved for humans, nor are there any antiviral drugs to treat Ebola patients. Treatment basically focuses on maintaining blood pressure and replacing lost blood and fluids. Containment of epidemics is attempted by isolating affected individuals and otherwise trying to limit transmission.

Category 3 consists of those emergent diseases in which a well-known disease is newly associated with a particular pathogen. Gastric ulcers are an example. Such ulcers were originally believed to be caused by excess stomach acid. Earlier in this text we discussed the discovery of *Helicobacter pylori* and its role in gastric ulcers. Kaposi's sarcoma is another example. This once very rare form of skin cancer is now recognized as an opportunistic infection in AIDS patients, caused by a type of human herpesvirus.

Reemerging diseases are placed in category 4. Reemerging diseases are those that were considered to be under control but have more recently been causing increasingly serious, new problems. Many reemerging diseases are caused by pathogens that have become resistant to drugs. In Chapter 12, we discussed drug resistance and how once easily treated infections have in some cases become untreatable. Multi-drug-resistant *Staphylococcus aureus* and *Mycobacterium tuberculosis* are examples. Malaria in Africa is another reemerging disease. After a few decades of decline, malaria surged again in the 1980s. A World Health Organization report estimated that the number of deaths among children doubled since this time, citing increasing poverty, a breakdown in health services, and drug resistance as important factors.

The above examples demonstrate just how complex the reasons for disease emergence can be. As we have seen, increased travel, changes in weather, and increased contact with animal reservoirs can all tip the scales in favor of emergence. Sometimes even politics is the primary culprit. An epidemic of diphtheria in the former Soviet Union that started in 1993 is widely attributed to the collapse of the government in 1990, because the newly independent nations that emerged from the breakup of the Soviet Union failed to maintain anti-diphtheria vaccination programs. With human society and the environment continuing to change at breakneck speeds, epidemiologists will have plenty to keep them occupied in the foreseeable future.

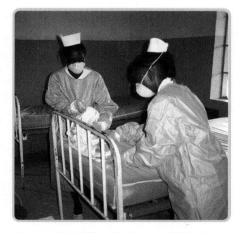

Figure 13.20 Kikwit, Congo, 1995. Two nurses wear protective clothing while changing the bedding in an Ebola hemorrhagic ward in Kikwit. The town in the central African country formerly known as Zaire was the scene of a major Ebola outbreak. There were 315 cases and 244 deaths in the epidemic, a mortality rate of 77%.

Bioterrorism

In 1979, 66 people suddenly died in Sverdlovsk, a Russian city of 1.2 million inhabitants located about 800 miles east of Moscow. Soviet officials reported that the deaths were caused by contaminated meat, but that was just a cover-up. It took almost 15 years before it was revealed that a nearby biological weapons research facility had accidentally released anthrax spores. The victims breathed these spores into their lungs and died of pulmonary anthrax (**Figure 13.21**). More recently, as we learned in Chapter 3 (see p. 60), letters intentionally laced with anthrax spores caused several deaths in the United States in 2001.

It is now an unfortunate sign of the times that the most fearsome emergent diseases may not always arise in nature. With the recent focus on terrorists and radical fringe groups, **bioterrorism** has become part of our vocabulary. Bioterrorism refers to the use of biological agents to kill or otherwise cause fear and mayhem. As with other, naturally occurring emergent diseases, epidemiologists must be able to recognize a bioterrorism event, elucidate its cause, and develop a response strategy.

Compared with other "weapons of mass destruction," such as nuclear weapons, biological agents are cheap and can be highly effective. Sophisticated laboratories are not needed, and unlike a chemical poison, some microorganisms used as weapons can spread from person to person, complicating control efforts. Because only a small quantity of a biological agent is necessary to wreak havoc, transport is less of a problem. In the Sverdlovsk incident, less than 1 gram of anthrax spores was released.

But not any microorganism will do. Spore-forming organisms such as *Bacillus anthracis*, the causative agent of anthrax, are among the most opportune candidates. The resistant endospores can be aerosolized and spread through the environment, where they persist until they happen to infect someone. This bacterium can also be extremely lethal. Untreated pulmonary anthrax, caused when spores are respired into the lungs, has a mortality rate of close to 100%.

A bacterial toxin such as botulinum toxin, produced by *Clostridium botulinum*, would also make an effective bioterrorism tool. Because the lethal dose is so small (less than 2 micrograms), only a tiny amount introduced into a common water or food source could prove catastrophic. Viral agents also pose risks as terrorist weapons. Smallpox is a particular concern, because of its high pathogenicity and its ability to spread through a population rapidly. Although an effective smallpox vaccine was developed many years ago, it is no longer widely used. Many people are consequently susceptible. Although smallpox was eradicated in the wild in the late 1970s, the virus is still maintained in laboratories in the United States and Russia. The fear is that terrorists might somehow gain access to these stocks.

In light of the ongoing terrorist threat, many governments are now ramping up production of vaccines to be used in the event of a bioterrorism incident. In many places response plans are now in place, in order that quick action may be taken to contain any outbreak. In a world where political agendas are often pursued through less than peaceful means, however, bioterrorism remains a serious threat. Understanding the biology of the involved microorganisms is crucial to our protection.

Figure 13.21 Bioterrorism, 2001.
Macrophages (in green) becoming infected with *Bacillus anthracis* (in orange).

Coming Up Next...

A recurrent theme in the first several chapters of this text was that although pathogenic microorganisms grab most of the headlines, the majority of bacteria, viruses, fungi, and protozoa have little if any direct impact on humans. Many are not pathogenic but rather perform vital ecological services or help maintain our health by acting as normal flora. In the last few chapters, however, we have focused on disease, casting microorganisms in a less favorable light. In the interest of fair play, the final three chapters of the text will emphasize some of the more benign aspects of human–microorganism interaction. In Chapter 14, we will concentrate on biotechnology and the vital role that microorganisms play. When diabetics need insulin, or when toxic waste spills must be cleaned up, genetically modified microorganisms more and more frequently lend a hand. In the next chapter, we will find out how they do it.

Key Terms

antigenic drift	emergent disease	incidence
antigenic shift	endemic	pandemic
bioterrorism	epidemic	personal characteristic
case definition	epidemic threshold	place characteristic
case–control study	epidemiology	reemerging disease
common source epidemic	host-to-host epidemic	time characteristic

Concept Questions

1. Consider the following statistics for the number of cases of a particular infectious disease in a given geographic area.

MONTH	NO. OF CASES
January	0
February	4
March	101
April	611
May	2404
June	900
July	313
August	48
September	9
October	0
November	0
December	0

In which month is the proportion of immune individuals the highest? In which month would it be the lowest? Is this proportion increasing or decreasing between February and May? Between May and September? Between October and December? Would you guess that this is a common source epidemic or a host-to-host epidemic? Explain your reasoning.

2. Discuss how the proportion of immune individuals in a population affects the likelihood of epidemics. How can vaccines be used to increase immunity in a population and thereby reduce the likelihood of an epidemic?

3. Imagine that you wish to write a screenplay for a movie in which a new, exceedingly virulent virus emerges from the Brazilian rain forest and within a few years has wiped out a sizable proportion of the human population. You want your movie to be as biologically realistic as possible in terms of factors such as your virus's mode of transmission, its reservoir, etc. You also want as much realism as possible in terms of why and how the virus emerged and how your movie ends. Provide a plot summary for your movie.

4. We have learned that weather is frequently an important contributing factor to when an epidemic will occur. Global warming certainly constitutes an enormous change in our weather. Are there any diseases you have learned about in this text that you think will become more or less likely to occur in epidemics in your country? What about emergent or reemergent diseases? Can you think of diseases that are more likely to emerge where you live due to a warming atmosphere?

5. You are an epidemiologist working for the World Health Organization. If a series of severe cyclones were to hit India, what types of epidemics might you become worried about and why?

6. A number of measles cases are suddenly reported in a large city. At elementary school X, with a student population of 500, 485 students have been vaccinated against measles. At elementary school Y, also with 500 students, only 250 have been vaccinated. Explain why, at elementary school X, the 15 unvaccinated students are less likely to get measles than are the 250 unvaccinated students at school Y.

7. In this chapter we learned about the pioneering work of Florence Nightingale in determining the risk factors for typhus. Knowing what you know now about how a case–control study is designed and used, imagine that you could go back in time and assist Nightingale in developing a case–control study to help elucidate how typhus was transmitted. What would your advice be?

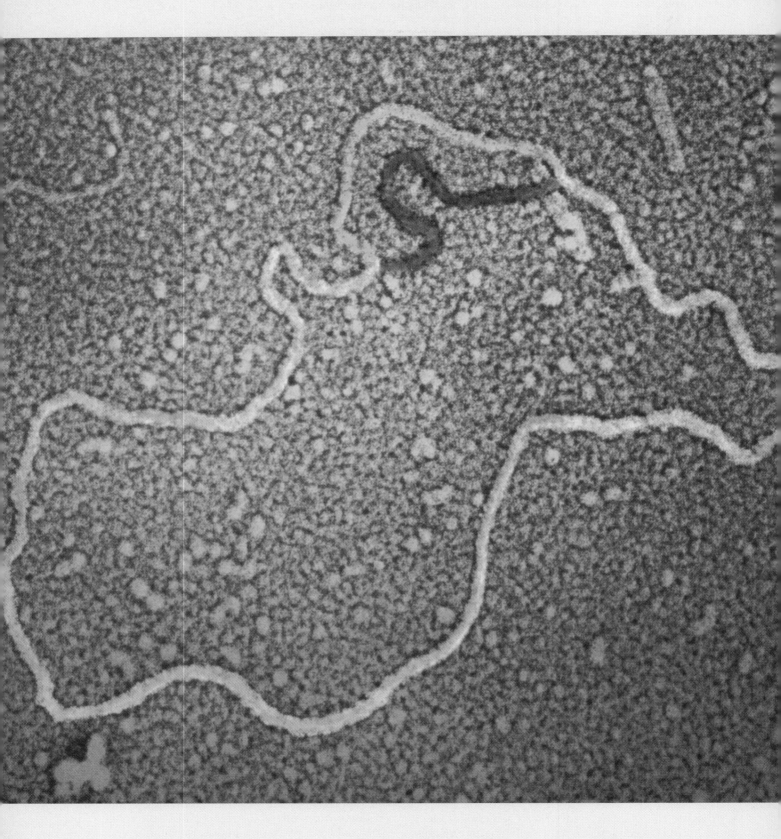

The DNA of this bacterial plasmid has been separated to map a specific gene (blue).

Chapter 14

The Future Is Here: Microorganisms and Biotechnology

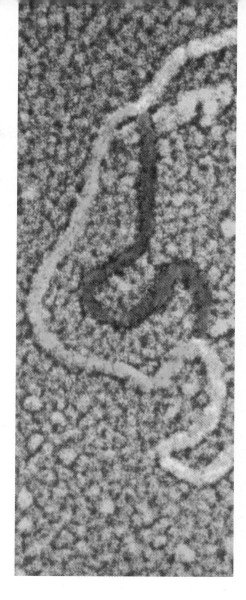

Since the dawn of civilization, humans have shaped the evolution of other organisms according to their whim. Wild animals gave rise to the familiar domestic creatures that we depend on for food and companionship. Food crops such as corn, wheat, and rice have all descended from wild ancestors. Millennia before anyone knew what a gene was, humans began the process of selective breeding, crossing animals and plants that had desirable features with similar organisms, until whole new domestic breeds arose (**Figure 14.1**). These efforts continue today, but in the past few decades techniques have become much more sophisticated. The era of **genetic engineering**, the direct and purposeful manipulation of DNA to create essentially new living things with certain qualities, has arrived.

How did this biological revolution begin? In Chapter 6, we learned that microorganisms are especially adept at exchanging genes. Once scientists understood processes such as conjugation, transformation, and transduction (see p. 156), they began to wonder whether it might be possible to deliberately use these microbial "tricks" to place certain genes exactly where they wanted them. Of course, they had to identify and characterize the various genes of interest before any of this could be attempted. Microorganisms played a role here as well, providing many of the secrets and tools necessary for genetic analysis.

In this chapter we will first investigate how DNA is analyzed and manipulated. Next, we will discuss some of the remarkable consequences and practical applications of this new technology. Finally, we will conclude by considering the debate that swirls around biotechnology. As we have done elsewhere, we will focus primarily on microorganisms. Sometimes the role of microorganisms will be obvious. In other cases, the connections to bacteria, viruses, or fungi will be more subtle. For instance, we will consider how certain genetic diseases can be treated by introducing normal genes into the cells of human patients. At first glance, this topic may seem "nonmicrobial," but these types of genetic techniques and therapies owe their

Figure 14.1 Selective breeding. Humans have been experimenting with DNA since the dawn of time. (a) The oldest known depiction of a rose in Western art, from the palace of Knossos in Crete, around 2000 BC. Modern roses (all members of the genus *Rosa*) are the result of centuries of breeding between wild roses. (b) A poodle and a pug illustrate the range of dog breeds. All dogs, regardless of breed, belong to a single species (*Canis familiaris*).

existence to the microbial world. Most knowledge of genetic engineering depends on lessons and tools provided by microbes.

Some of the topics we will discuss are controversial. Subjects such as genetically modified food, or the creation of **transgenic animals** bearing foreign, introduced genes that nature never intended, evoke strong emotions and opinions. This underscores perhaps the most important reason for this chapter: all of us, scientists and nonscientists alike, are living in the age of biotechnology. Our lives and those of our families are directly affected by much of this technology, and all of us must make decisions about it. Should you buy genetically modified food? Do genetically modified organisms pose a risk to the environment? Should insurance companies or employers have access to genetic information indicating a person's likelihood of developing certain diseases such as cancer? Each of us must come to our own conclusions on issues such as these, but in order to make rational, well-informed decisions, it is necessary to understand the science behind the issues.

The Analysis of DNA

We have had many opportunities throughout this text to refer to specific genes found in microorganisms or their hosts. In Chapter 6 we learned that a gene is a specific sequence of DNA nucleotides that codes for the production of a particular protein. In many cases that specific DNA sequence is known.

You have perhaps wondered how one actually finds a gene and how its sequence can be deduced. Furthermore, you may have heard that scientists now have a record of the entire genome for a number of living things, including humans, and you may have asked yourself how such a monumental task is achieved. We will start by investigating some of the techniques and tools used to accomplish such feats.

Bacterial restriction enzymes have proven useful for cutting DNA at specific sites

We have already learned that even bacteria have their own pathogens. And just as we are protected from potentially dangerous microorganisms by our immune system, bacteria are also armed against bacteria-specific viruses or bacteriophages. One interesting defensive mechanism is a group of proteins called **restriction enzymes**, which literally chop up the DNA of invading bacteriophages. After the discovery of restriction enzymes in the 1970s, biologists began to study the properties of these enzymes and found that many of them did not cut DNA randomly. Instead, they cut only at very specific places. For example, one restriction enzyme, called *Eco*RI, cuts DNA only at the nucleotide sequence GAATTC (**Figure 14.2**). Remember that DNA is a double-stranded molecule and that the two strands are complementary. Therefore, the strand opposite the target base sequence has the complementary sequence CTTAAG. When *Eco*RI finds this sequence of paired bases, it makes identical cuts on both DNA strands, between the terminal G and the adjacent A. This means that the cuts on the two opposing strands are staggered. Because the cut is staggered, the two remaining DNA fragments have short, unpaired, overhanging ends (see Figure 14.2). These are referred to as **sticky ends**, because they readily bond with or "stick" to complementary bases. To date, almost 1000 different restriction enzymes have been identified, all of which cleave DNA at different sites.

The discovery of restriction enzymes gave biologists a powerful new tool that greatly facilitated the study of DNA. For example, some restriction enzymes cut at relatively common base sequences, while others cut DNA at less frequently encountered sequences. This variability gives scientists the

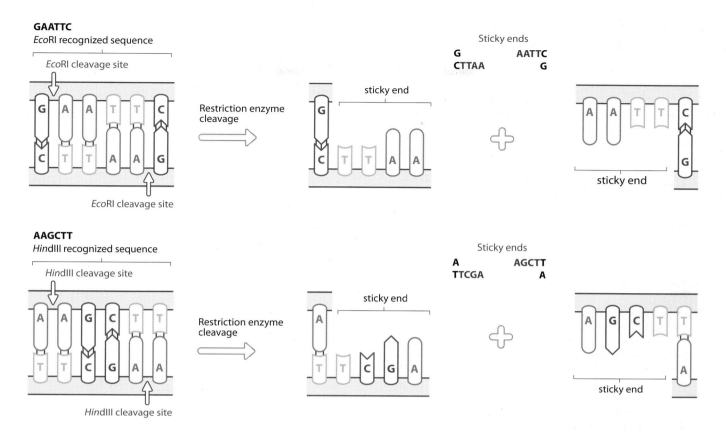

Figure 14.2 Representative restriction enzymes. Many restriction enzymes have very specific cleavage sites. *Eco*RI cuts DNA on both strands only at the sequence GAATTC, always cleaving between the terminal G and A. Likewise, *Hin*dIII cuts between the two terminal A nucleotides at the sequence AAGCTT. Because the same cut is made on both DNA strands, overhanging *sticky ends* that readily bind to complementary bases are produced. The name *Eco*RI indicates that this restriction enzyme comes from *Escherichia coli*. *Hin*dIII comes from *Haemophilus influenzae*.

ability to cut a DNA molecule into any desired length. Restriction enzymes that recognize common sites cleave DNA into a greater number of shorter fragments. Those that recognize less common sites cut DNA into a smaller number of larger fragments (**Figure 14.3a**).

Furthermore, restriction enzymes have proven invaluable when a scientist wishes to insert the DNA from one organism (the donor) into the DNA of another organism (the recipient). If the same restriction enzyme is used to cut the DNA of both the donor and the recipient, complementary sticky ends on the donor and recipient DNA molecules are produced (see Figure 14.3b). If donor and recipient DNA are then mixed, the sticky ends will allow the donor DNA to insert itself into the recipient DNA molecule. This capacity to "cut and paste" DNA from one source into the DNA from another source allows for the production of DNA from two different sources. This type of hybrid DNA is called **recombinant DNA**.

DNA fragments can be separated by gel electrophoresis

Once a large DNA molecule has been cut with restriction enzymes, the variously sized fragments can be separated from each other by the technique of **gel electrophoresis**. As the name suggests, the DNA fragments are placed on a gelatinous material, or gel. The gel contains microscopic pores through which the DNA fragments can pass. An electrical current is run through the gel, and since DNA itself carries negative charges, the DNA fragments move through the gel, away from the negatively charged electrode and

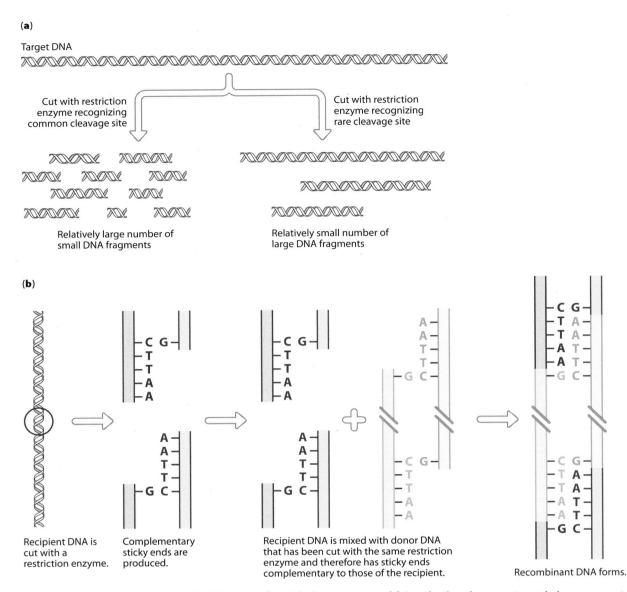

(a)

Target DNA

Cut with restriction enzyme recognizing common cleavage site

Cut with restriction enzyme recognizing rare cleavage site

Relatively large number of small DNA fragments

Relatively small number of large DNA fragments

(b)

Recipient DNA is cut with a restriction enzyme.

Complementary sticky ends are produced.

Recipient DNA is mixed with donor DNA that has been cut with the same restriction enzyme and therefore has sticky ends complementary to those of the recipient.

Recombinant DNA forms.

Figure 14.3 Some valuable uses of restriction enzymes. (a) By selecting the correct restriction enzyme, target DNA of interest can be cut into fragments of different lengths. (b) If both recipient and donor DNA are cut with the same restriction enzyme, they will form complementary sticky ends. When DNA from two sources (from two different bacterial species, for example, or from a prokaryote and a eukaryote) that has been cut in this way is then mixed, donor DNA may insert into recipient DNA, forming *recombinant DNA*. Following insertion, the breaks in the phosphate–sugar backbone are sealed with a different enzyme called ligase that forms new covalent bonds between adjacent nucleotides. The gray breaks in the donor DNA indicate that the donor DNA may be of any specified length.

toward the positively charged electrode (**Figure 14.4a**). Not all fragments move through the pores with equal ease. Smaller fragments move readily, whereas larger fragments are impeded by the fibers of the gel matrix and consequently move more slowly. Therefore, if allowed to run for a given period of time, the shortest fragments migrate the greatest distance, while the largest fragments migrate the shortest distance from the starting point. Each band seen on the gel after staining corresponds to a cluster of DNA molecules of equal length (Figure 14.4b).

Separating DNA fragments in this manner is useful for various reasons, depending on the goals of the researcher. One important reason is to look through or "screen" the DNA fragments to find a particular gene.

Southern blotting can be used to identify a specific gene of interest

CASE: PINPOINTING A CANCER-CAUSING GENE

Breast cancer strikes 185,000 women annually in the United States, accounting for more than 30% of all cancer in women. Approximately 5–10% of breast cancer cases are termed "familial breast cancer," because they are strongly associated with an inherited mutated gene. In most cases, the mutation occurs in either the *BRCA1* or the *BRCA2* gene (from "<u>br</u>east <u>ca</u>ncer"). The exact manner in which the mutated gene increases the likelihood of tumors is still under investigation.

Because breast cancer runs in Mattie's family, her doctor suggests that she may wish to be screened for the mutant *BRCA* genes. If she agrees to such screening, Mattie might then have a better way of assessing her risk of developing cancer. After weighing the pros and cons, Mattie decides to undergo *BRCA* screening, and she submits a DNA sample for analysis.

1. How might such screening be accomplished?
2. If Mattie carries the mutant *BRCA1* or mutant *BRCA2* gene, does this mean she will develop cancer? If she doesn't have the mutation, does this mean she will not develop the disease?
3. How has the potential value of such screening techniques contributed to the controversy over genetic testing?

To find out if Mattie carries a mutant *BRCA* gene, doctors might rely on a technique known as **Southern blotting**, named after Edwin Southern, its inventor. Southern blotting relies on the fact that any DNA molecule, if made single-stranded, will base-pair or *hybridize* with a complementary single-stranded DNA molecule.

The DNA sample from Mattie is first digested with restriction enzymes (**Figure 14.5**), and the resulting DNA fragments are then separated by gel electrophoresis. The DNA fragments are next treated with chemicals that separate them into single strands. Next, as illustrated in Figure 14.5, the single-stranded fragments are transferred to a filter made of nitrocellulose or, more recently, of nylon.

Now that the fragments are single-stranded and have been transferred to the filter, they are exposed to a **gene probe**. The probe is a short, single-stranded length of DNA with a sequence that is complementary to the gene being sought. In the screening of Mattie's DNA, probes of single-stranded DNA with known base sequences, complementary to the base sequences of the BRCA genes, would be necessary. The probe is also labeled, usually with radioactive atoms, making it easy to detect.

Next, the probe and filter are incubated together in a solution for a specified period of time (see Figure 14.5). The solution facilitates the binding of the probe to the target gene. If Mattie carries a defective *BRCA* gene, the probe will hybridize with the gene. If neither gene carries the defect, the probe will

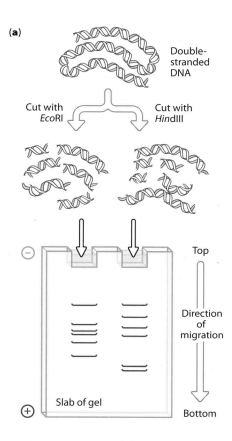

Figure 14.4 Separation of DNA fragments by gel electrophoresis. (a) Following cleavage with restriction enzymes, the fragments within a DNA sample can be separated by loading the sample onto a gel. The DNA sample is placed in a well at the top of the gel, and DNA samples migrate within the gel from the negative electrode toward the positive electrode. The rate of migration is determined by the size of the fragment. Cutting the DNA with different restriction enzymes results in different sets of fragments. (b) Following migration, the DNA is stained with a dye that is visible under ultraviolet light. Each band is composed of DNA fragments of the same length.

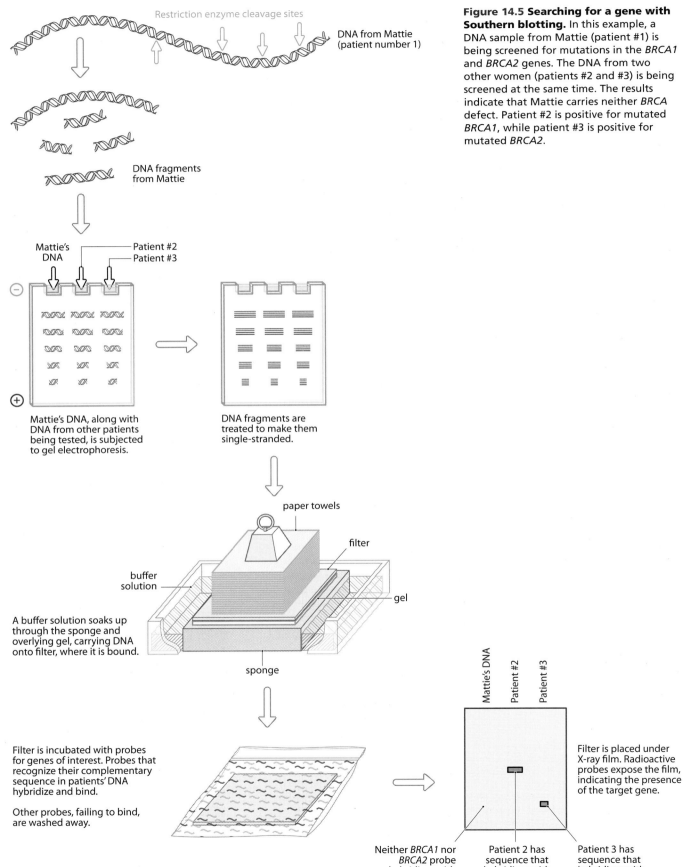

Restriction enzyme cleavage sites

DNA from Mattie (patient number 1)

DNA fragments from Mattie

Mattie's DNA

Patient #2
Patient #3

Mattie's DNA, along with DNA from other patients being tested, is subjected to gel electrophoresis.

DNA fragments are treated to make them single-stranded.

paper towels

filter

buffer solution

gel

A buffer solution soaks up through the sponge and overlying gel, carrying DNA onto filter, where it is bound.

sponge

Filter is incubated with probes for genes of interest. Probes that recognize their complementary sequence in patients' DNA hybridize and bind.

Other probes, failing to bind, are washed away.

Mattie's DNA
Patient #2
Patient #3

Filter is placed under X-ray film. Radioactive probes expose the film, indicating the presence of the target gene.

Neither BRCA1 nor BRCA2 probe hybridizes with Mattie's DNA.

Patient 2 has sequence that hybridizes with BRCA1 probe.

Patient 3 has sequence that hybridizes with BRCA2 probe.

Figure 14.5 Searching for a gene with Southern blotting. In this example, a DNA sample from Mattie (patient #1) is being screened for mutations in the BRCA1 and BRCA2 genes. The DNA from two other women (patients #2 and #3) is being screened at the same time. The results indicate that Mattie carries neither BRCA defect. Patient #2 is positive for mutated BRCA1, while patient #3 is positive for mutated BRCA2.

be unable to bind and will be washed away; no hybridization will occur. To find out if she carries a mutant *BRCA* gene, the technician merely examines the filter. A piece of X-ray film is placed over the filter. The radioactive probe, if present, will expose the film, exactly over the point where it has hybridized. A black band on the film would identify the presence of the mutant gene.

If, on the other hand, Mattie's DNA contains neither of the two mutant genes, the probe will not have hybridized and the radioactivity necessary to expose the film will not be present. The lack of the black bands indicates that the mutated genes are not present and that Mattie can breathe a sigh of relief.

Or can she? Remember that only 5–10% of breast cancers are inherited. Even without a *BRCA* mutation, a woman in the United States has a 1 in 10 chance of developing breast cancer sometime in her life. In Mattie's case, with a family history of breast cancer, the negative test results would be reassuring, since it is likely that her female relatives who developed cancer had a *BRCA* mutation. However, recall an important genetic principle learned in Chapter 6: genes usually provide only the potential for a particular characteristic. Environmental factors often determine whether this potential is realized. Having a mutated *BRCA* gene is no guarantee that cancer will develop, although it makes it much more likely. Similarly, the absence of a *BRCA* mutation does not mean that Mattie is off the hook; it only means that the odds of her developing breast cancer are less than they might have been.

This prognosis is admittedly somewhat ambiguous. And regardless of whether Mattie carries a *BRCA* mutation, she needs to remain vigilant. Controversy arises during discussions of who should have access to such test results. If employers or insurance companies were to obtain this information, might they use it to make hiring or insurance decisions? And if so, would this be a prudent use of valuable information or an outrageous violation of privacy?

In addition to identifying specific human genes, Southern blotting also has more strictly microbial applications. Probes have been developed for genes that are unique to many infectious agents, including *Mycobacterium tuberculosis*, *Salmonella*, and *Escherichia coli*. DNA from the sample to be screened is isolated, separated into single strands, and transferred to a filter. The filter is then incubated with the appropriate probe and observed for the hybridization that would indicate whether a sample is positive for a given microorganism. By use of this technique, the diagnosis of many previously difficult-to-identify infections has become somewhat routine.

Of course, Southern blotting would not be possible if we did not know the sequence of the gene in question, because we would be unable to construct an appropriate probe. There are many other situations in which it is valuable to know the exact order of nucleotide bases in a gene or length of DNA. We will next consider how DNA sequencing is carried out.

Sequencing techniques can be used to reveal the sequence of nucleotides in a DNA sample

To the nonscientist, the ability to determine the order of nucleotides in a molecule of DNA may seem a bit like magic, but the techniques are actually straightforward. To ascertain a gene sequence, we first need to obtain enough of the DNA in question. Once sufficient DNA is on hand, we can determine the actual nucleotide sequence. Thus, the entire business of sequencing a gene is often a two-step process. Here we discuss one way those two steps might be carried out.

Figure 14.6 Exonerating evidence?
Peruvian mummies, over 1000 years old. DNA removed from lung lesions from such mummies indicates that tuberculosis was in the New World for hundreds of years before the arrival of European explorers.

CASE: TB—BC ("BEFORE COLUMBUS," THAT IS)

"In fourteen hundred and ninety two, Columbus sailed the ocean blue." Did he bring tuberculosis with him? In Chapter 5 we learned about the long list of infectious diseases that European explorers introduced into the New World. Tuberculosis was always thought to be on that list. Some experts were doubtful, however, because mummies that predated the arrival of the Spanish in Peru sometimes had lung lesions that looked similar to tuberculosis. To find out if tuberculosis in the Americas predated the arrival of Europeans, researchers obtained lung tissue from a 1000-year-old Peruvian mummy (Figure 14.6). DNA was extracted from lung lesions and subjected to the polymerase chain reaction (PCR). This process is used to make many copies of or "amplify" target DNA (in this case, *M. tuberculosis* DNA). A short segment (approximately 100 base pairs) of bacterial DNA was amplified. When this DNA was sequenced, the researchers found that the DNA collected from the lung lesions contained base sequences identical to DNA from modern *M. tuberculosis*, clear evidence that tuberculosis was present in Peru at least 500 years before the first Spaniard set foot in the New World. Thus, tuberculosis is one illness that cannot be blamed on Columbus.

1. What is the polymerase chain reaction, and how is it used to amplify DNA?
2. Once the DNA from the mummy was amplified, how was it sequenced and compared with DNA from modern *M. tuberculosis*?

Large amounts of specific DNA sequences can be obtained with the polymerase chain reaction

Early attempts to sequence DNA were often hampered by a limited supply of the DNA in question. This problem was brilliantly solved in the 1980s with a new technique known as the **polymerase chain reaction (PCR)**. Starting with a small sample of DNA, PCR allows scientists to quickly copy any genetic sequence in the DNA until they have a sample large enough to sequence.

PCR is carried out in a number of repeated cycles. Each cycle doubles the amount of the DNA of interest (the target gene). If we begin with 100 copies of a particular gene that we wish to amplify, we will have 200 copies after one PCR cycle, 400 copies after two cycles, 800 copies after three cycles, etc.

Because usually 30 or so cycles are run in a PCR reaction, we can get billions of times the original number of copies of the target DNA.

The PCR reaction is prepared in a small *reaction tube*. In this tube, the gene is copied, essentially in a similar manner to the DNA replication discussed in Chapter 6. As with DNA replication, a supply of A, G, T, and C nucleotides and the enzyme DNA polymerase are required to copy the target DNA. But not *any* DNA polymerase will do. As we will see, during each PCR cycle, the DNA is heated to the point at which the two DNA strands separate into single strands. If DNA polymerase from most sources were used, the enzyme would denature when heated and would be unable to copy the target. Consequently DNA polymerase from a thermophilic (heat-loving) bacterium is used. The bacterium, *Thermus aquaticus*, grows only in extremely hot water, and as you therefore might expect, it has an unusually heat-stable DNA polymerase. The enzyme, called "Taq polymerase" (from <u>T</u>hermus <u>aquaticus</u>), is a key component in PCR reactions.

Appropriate **primers**, short DNA sequences of approximately 20 nucleotides that are complementary to the ends of the target DNA (**Figure 14.7**), are also needed. Once the target DNA is separated into single strands, a primer attaches to the end of both strands. The primers essentially act as "start" signals, indicating to the DNA polymerase where the ends of the target gene are located. The DNA polymerase then elongates the primer, synthesizing a new daughter strand of the target DNA. The primers used to amplify a 100-base-pair length of *M. tuberculosis* DNA would be short sequences complementary to the ends of this 100-base-pair region.

Once the reaction tube contains all necessary components, it is placed into a *thermocycler*, a device that can be programmed to alternate among various temperatures. Each cycle of the PCR reaction passes through three temperature stages (see Figure 14.7). The first stage heats the DNA up to the point where it separates into single strands. In the second stage the temperature is lowered, allowing the primers to attach to the ends of the target. Finally, in the third stage, the DNA polymerase (Taq) copies each of the single strands, using the supply of nucleotides added to the reaction mix. All subsequent cycles pass through the same three stages. Because the amount of the target DNA doubles after each cycle, by the end of a PCR procedure, the DNA of interest will have been amplified to literally billions of copies.

The Sanger method can be used to sequence a specific sample of DNA

Following PCR amplification, it is sometimes possible to proceed directly to **DNA sequencing**, in which the exact nucleotide sequence of the target DNA is determined. The principal technique of DNA sequencing is called the Sanger method, named for its discoverer, Frederick Sanger. Although the details of this technique are beyond the scope of this text, DNA sequencing

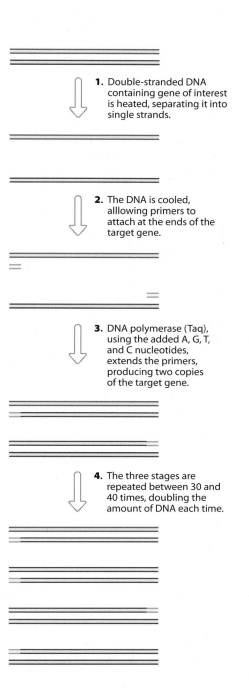

1. Double-stranded DNA containing gene of interest is heated, separating it into single strands.

2. The DNA is cooled, alllowing primers to attach at the ends of the target gene.

3. DNA polymerase (Taq), using the added A, G, T, and C nucleotides, extends the primers, producing two copies of the target gene.

4. The three stages are repeated between 30 and 40 times, doubling the amount of DNA each time.

Figure 14.7 DNA amplification by PCR. Via the polymerase chain reaction (PCR) a particular gene can be copied billions of times. Each reaction is carried out in a reaction tube, into which the DNA to be copied is added. DNA polymerase (Taq), nucleotides, and primers, which are short nucleotide sequences complementary to the ends of the target gene, are also added. Each cycle of a PCR reaction goes through three stages. In the first stage, the reaction mix is heated, causing the DNA to separate into single strands. In the second stage, the reaction is cooled to allow the primers to attach to the ends of the gene. In the third stage, the DNA polymerase uses the added nucleotides to extend the primer and produce new, double-stranded DNA molecules. After each cycle of these three stages, the amount of DNA is doubled.

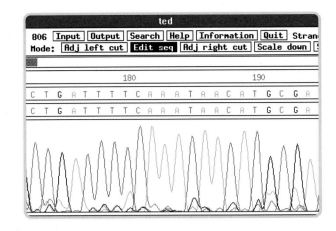

Figure 14.8 Reading a DNA sequence. Following DNA sequencing, the DNA sequence is analyzed by computer. The nucleotide sequence is translated into a *chromatograph,* which corresponds to the DNA sequence of interest. Each color peak (blue = C, red = T, green = A, and black = G) represents one of the four nucleotides in a specific order.

is now routine and is carried out daily at universities and research institutes around the world. Once an actual base sequence for a particular stretch of DNA is obtained, the information is analyzed by computer, yielding a sequential list of the nucleotides in the DNA (**Figure 14.8**).

The ability to rapidly and efficiently sequence genes has opened the door to tremendous research opportunities. As described in Chapter 4, for example, gene sequences are now used in modern taxonomy to deduce evolutionary relationships. In the case of tuberculosis in a Peruvian mummy, we see how the technology was used to answer an interesting question about the presence of an important infectious disease in pre-Columbian America. With advances in technology and improved protocols, sequencing continues to become faster and less expensive. When researchers sequence a new gene, they submit the data to an ever-growing online database called GenBank that now contains billions of sequences. Any newly sequenced gene can be compared with known sequences in the database, and within seconds, a researcher can find out how similar or different it is from the same gene in other organisms. If you sequence a gene from a lung lesion obtained anywhere in the world and want to know whether it is from *M. tuberculosis*, you merely go to GenBank and enter your sequence. In almost no time you find out whether your gene matches *M. tuberculosis* genes in the database. Sequenced genes from other microorganisms can be similarly identified.

For many biologists, the ability to sequence genes has merely whetted their appetites. Perhaps, rather than settling for the sequence of individual genes, it might be possible to sequence the entire genome of an organism. With the entire DNA blueprint in hand, biologists could begin to answer questions they had previously only dreamed about. Whole genomes could be compared to provide an even clearer picture of evolutionary history. One could determine how the genes of a bacterium and a eukaryotic organism differ or how they are the same. Scientists have long wondered about the minimum number of genes necessary to sustain cellular life. Genomes may provide the answer. With the complete genome of a pathogenic microorganism in hand, the time-consuming and difficult task of determining which genes are responsible for virulence might be greatly simplified. In fact, questions such as these are now being investigated routinely, thanks to the modern field of **genomics**—the sequencing and study of entire genomes.

The entire genomes of many organisms have been sequenced

CASE: *E. COLI'S* DARK SIDE

On the surface, *E. coli* might seem to be a bit schizophrenic. We have frequently made reference to this species throughout the text, usually

referring to it as a harmless member of our normal flora. But, in the 1990s, a more malevolent form of the bacteria, known as *E. coli* O157:H7, reared its head. This strain, first detected in contaminated hamburgers, is unusual because it carries genes that code for powerful toxins. An infection can cause severe diarrhea, kidney damage, and in some cases even death.

1. How did normally benign *E. coli* acquire virulence genes for toxin production? How has the study of microbial genomes permitted this question to be answered?
2. Do answers to the above questions suggest any practical applications for the treatment of dangerous *E. coli* infections?

The first genomes to be completely sequenced were understandably some of the smallest. Sanger himself was in charge of the first attempt to sequence an entire genome, a small bacteriophage with a total genome of only 5386 bases. This project was completed in 1976, but it was not until 1995 that a complete bacterial genome was sequenced. It was from a small prokaryote—the 1,830,137 bases that make up the entire genome of *Haemophilus influenzae*.

However, since the mid-1990s the pace of sequencing has been mind-boggling. A year after *H. influenzae*, the genome of the first eukaryote, *Saccharomyces cerevisiae*, or common brewer's yeast, was sequenced. Many animal, plant, and microorganism genomes followed. Finally, in 2001, a "rough draft" of the human genome was published, and things have not slowed down since then. Hundreds of prokaryotic and eukaryotic genomes have now been sequenced, and genome projects for many other living things are in the works (**Table 14.1**).

To sequence an entire genome, the first step is to cut the genome of interest into a large number of random fragments with restriction enzymes. Each of these fragments is randomly inserted into a bacterial chromosome or plasmid, creating recombinant DNA (see Figure 14.3). This recombinant DNA is

SPECIES	GENOME SIZE (MILLIONS OF BASE PAIRS)	ESTIMATED NO. OF GENES
Bacterial		
Haemophilus influenzae	1.83	1743
Vibrio cholerae (causative agent of cholera)	4.0	3885
Escherichia coli	4.6	4288
Fungal		
Saccharomyces cerevisiae	12	6000
Protozoan		
Plasmodium falciparum (an agent of human malaria)	30	6500
Plant		
Arabidopsis thaliana (wild mustard)	119	26,000
Animal		
Drosophila melanogaster (fruit fly)	180	13,000
Mus musculus (house mouse)	2500	30,000
Homo sapiens (human)	3000	25,000

Table 14.1 Some representative completed genomes.

copied many times as the chromosome or plasmid itself replicates. Using normal bacterial DNA replication to make many identical copies of introduced, foreign DNA is an example of **cloning**. We will discuss cloning technology shortly, because it is crucial to the process of genetic engineering.

Next, the fragments are sequenced via the Sanger method. Sophisticated computer programs identify regions on each fragment that overlap with those of other fragments, and the various fragments are ordered into a complete "genetic map." Those fragments that are too large to be sequenced completely are digested into even smaller fragments of about 2000 nucleotide pairs—a manageable length for DNA sequencing. Once all of these smaller fragments are sequenced, they are reassembled by computer, and yet another genome has been completed. There are variations to this strategy, but all genome sequencing basically relies on breaking the genome up into manageable chunks, followed by sequencing and computer-based reassembly.

The potential for DNA analysis and the field of genomics is enormous, and applications for medicine, agriculture, and industry are now starting to be realized. Our case provides an excellent example. To deduce the origins of the *E. coli* virulence factor genes, researchers began by comparing the genomes of virulent and nonvirulent strains. They found several genes present in the virulent bacteria that were absent in the nonvirulent *E. coli*. When these genes were entered into the GenBank database and compared with other genes already in the database, it was found that the novel *E. coli* O157:H7 genes were very similar to genes found in *Shigella*, a genus of bacteria long associated with severe dysentery. The genetic similarity strongly suggested that these shared genes were the ones that code for toxins in both *E. coli* and *Shigella*.

How did the toxin-producing virulence genes get from *Shigella* to *E. coli*? It is currently believed that a bacteriophage shuttled the genes from one species to the other. In other words, the genes were transferred by transduction (see Chapter 6, p. 157). Aside from demonstrating yet again the versatility and diversity of microorganisms, this information has at least one potential practical application. If an antibiotic could be developed that inactivated or interfered with *Shigella* toxin, it should work against pathogenic *E. coli* as well. In the meantime, it is probably a better idea to order your hamburgers medium or well done.

Genomic analysis has led to a better understanding of other diseases as well. For example, examination of the genome of *Borrelia burgdorferi*, the bacterium that causes Lyme disease, reveals that *B. burgdorferi* has an unusually large number of genes that code for surface proteins. It is believed that by regularly changing the proteins on its cell surface, this species is able to evade host response and thereby establish a chronic infection. By better understanding how these genes are switched on and off, it may be possible to interfere with the process, rendering the bacteria more susceptible to immune destruction.

The availability of genomic data for *M. tuberculosis* has similarly suggested new, potentially valuable treatment options for tuberculosis. Genomic analysis of this bacterial scourge shows that *M. tuberculosis* has an unusually large number of genes related to lipid digestion. Whereas a species such as *E. coli* has about 50 such genes, *M. tuberculosis* has 250. This suggests that *M. tuberculosis* may meet an uncommonly large part of its energy demands by metabolizing host lipids. If somehow lipid acquisition could be impeded, bacterial growth would probably slow, and tuberculosis might become an easier disease to treat.

Not all genomic research necessarily deals with pathogens. For example, if bacteria are found living in habitats contaminated with toxic wastes, it is possible that these organisms may be able to digest the wastes and convert them to less harmful substances such as carbon dioxide. If these bacteria are able to digest benzene, dioxin, or other nasty chemicals, they must have genes that code for the enzymes that permit them to do so. Genomic analysis can help identify such genes. Following identification, genes for toxin-digesting enzymes might be transferred to other cells, creating, in effect, genetically modified or "engineered" microbial waste decontaminators. We will look closer at some of these applications later in this chapter. Meanwhile, we will look at how genetic engineering is accomplished.

Genetic Engineering: Whose Gene Is It Anyway?

CASE: BIOLOGICAL BLACKMAIL

In the Clive Cussler thriller *Dragon*, the hero, Dirk Pitt, does battle with a group of evil industrialists bent on world domination. As part of their nefarious scheme, they have genetically engineered special bacteria capable of rapidly digesting petroleum. Their intention is to introduce the bacteria into the United States' Strategic Oil Reserves. Brought to its knees by a lack of oil, America will have no choice but to give in to the demands of the villains. Fortunately, Dirk saves the day, but the creation of such oil-eating bacteria is not as far-fetched as it sounds.

1. How can a gene that codes for the ability to digest petroleum be introduced into bacteria?

In the early 1970s, researchers at Stanford University and the University of California accomplished a breakthrough that has shaken the world ever since: they spliced a gene from a virus into the chromosome of *E. coli* bacteria. The field of recombinant DNA was born. Other accomplishments followed rapidly. Before long, DNA from two different types of bacteria was successfully combined. In 1973 researchers managed to insert a gene from a eukaryote, the African clawed toad (*Xenopus laevis*), into *E. coli* plasmid DNA (**Figure 14.9**).

Since then, molecular biologists have hardly looked back. The initial potential of recombinant DNA is now a reality and has both improved our understanding of basic genetic processes and provided many practical applications. For instance, recombinant DNA technology is now used to produce commercial products and to treat hereditary conditions. One such product is a human protein called AT3, which inhibits blood clot formation and is often used to treat people in whom dangerous blood clotting occurs. To create a relatively inexpensive and renewable source of AT3, the human gene for AT3 was inserted into goat DNA, producing a new type of goat that secreted AT3 human protein in its milk. Animals with genes from at least one other species, like these goats, are called transgenic animals.

In the next section we will explore some of the basic techniques of genetic engineering that make marvels such as transgenic animals possible, and we will also discuss other remarkable uses of this technology.

Genes of interest can be cloned by inserting them into bacteria

Cloning refers to the creation of copies of a cell, a gene, or an entire organism. When a genetic engineer wants to insert a foreign gene into the DNA of a different organism, it is often necessary to first clone the gene to generate a large number of copies. Genes are often cloned by introducing them into bacteria. The bacteria then do all the work, replicating the foreign gene of interest as they replicate their own DNA.

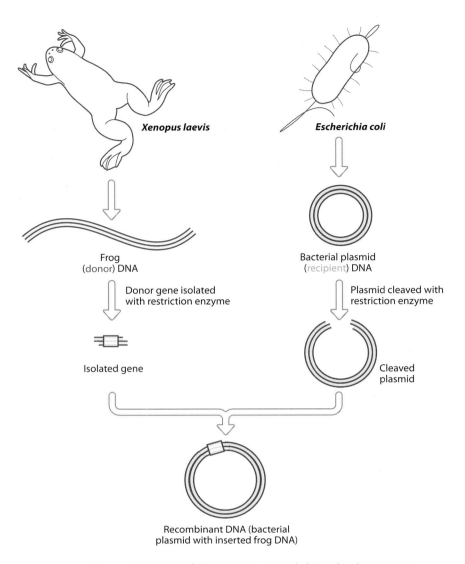

Xenopus laevis

Escherichia coli

Frog
(donor) DNA

Donor gene isolated
with restriction enzyme

Isolated gene

Bacterial plasmid
(recipient) DNA

Plasmid cleaved with
restriction enzyme

Cleaved
plasmid

Recombinant DNA (bacterial
plasmid with inserted frog DNA)

Figure 14.9 Recombinant DNA. An early biotechnology success
involved the splicing of animal DNA into a bacterial plasmid.

In order to get the bacteria to do the work for us, we must incorporate the
gene of interest into the bacterial DNA. Usually the DNA of choice is the
plasmid rather than the main bacterial chromosome, because plasmids,
being much smaller than bacterial chromosomes, are easier to work with.
Recall from Chapter 6 (p. 157) that plasmids are small loops of DNA with
their own origin of replication. They can therefore replicate independently
of the bacterial chromosome, and they are often exchanged by bacteria during conjugation. In genetic engineering, the plasmid serves as a **cloning
vector**, that is, a self-replicating DNA molecule into which foreign DNA can
be inserted.

The first step is to obtain the necessary plasmids and a suitable restriction
enzyme. Both the plasmid and the restriction enzyme must be selected with
care. An appropriate plasmid will usually have two key genes: an antibiotic
resistance gene and a "lethal" gene (**Figure 14.10**). The lethal gene is so-
called because it will kill any bacterial cell in which it is functioning. The
selected restriction enzyme must be able to cut the plasmid within the lethal
gene. As we will see, if any of these requirements are not met, it will be difficult to determine later in the process which bacteria ultimately carry the
recombinant plasmid and which do not.

**Figure 14.10 (*right*) Producing and
identifying recombinant bacteria.**
(a) A foreign gene that is to be cloned (the
"donor DNA") is randomly incorporated
into copies of a replicating bacterial
plasmid. In this example, the bacterial
plasmid has a resistance gene for ampicillin.
There is also a cleavage site for a restriction
enzyme in the lethal gene. (b) After
donor DNA is incorporated into some
plasmids, the plasmids are absorbed by
E. coli cells. The bacteria that contain the
recombinant plasmid survive. Bacteria with
nonrecombinant plasmids are killed by the
lethal gene. Those with no plasmid are
killed by ampicillin.

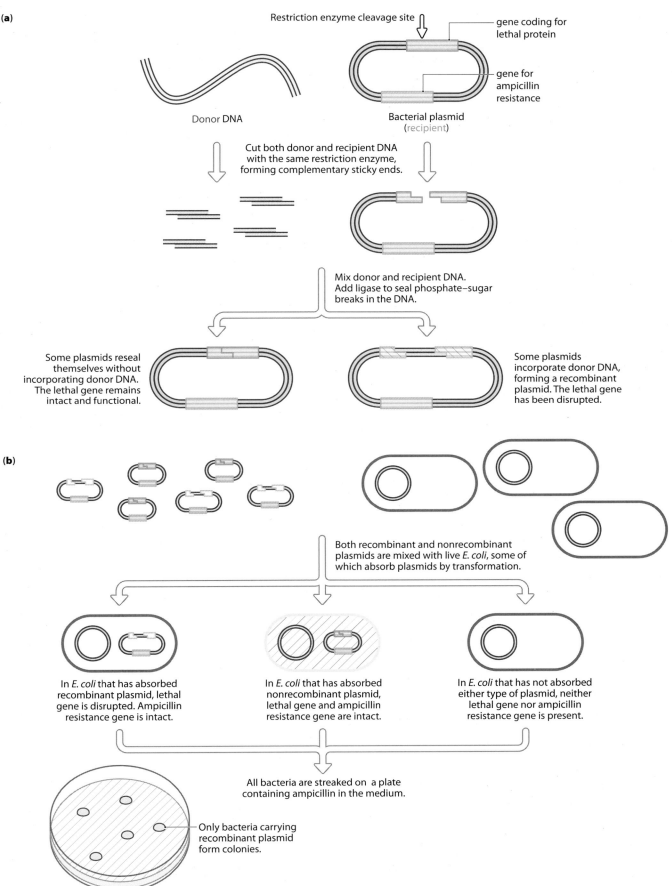

(a)

Restriction enzyme cleavage site

gene coding for lethal protein

gene for ampicillin resistance

Donor DNA

Bacterial plasmid (recipient)

Cut both donor and recipient DNA with the same restriction enzyme, forming complementary sticky ends.

Mix donor and recipient DNA. Add ligase to seal phosphate–sugar breaks in the DNA.

Some plasmids reseal themselves without incorporating donor DNA. The lethal gene remains intact and functional.

Some plasmids incorporate donor DNA, forming a recombinant plasmid. The lethal gene has been disrupted.

(b)

Both recombinant and nonrecombinant plasmids are mixed with live *E. coli*, some of which absorb plasmids by transformation.

In *E. coli* that has absorbed recombinant plasmid, lethal gene is disrupted. Ampicillin resistance gene is intact.

In *E. coli* that has absorbed nonrecombinant plasmid, lethal gene and ampicillin resistance gene are intact.

In *E. coli* that has not absorbed either type of plasmid, neither lethal gene nor ampicillin resistance gene is present.

All bacteria are streaked on a plate containing ampicillin in the medium.

Only bacteria carrying recombinant plasmid form colonies.

To insert the foreign donor gene into the plasmid, the same restriction enzyme is used to cut both the foreign DNA and the plasmid (see Figure 14.10). This treatment produces complementary sticky ends in both the donor DNA and the plasmid. The two types of DNA are then mixed together, and some of the donor DNA pairs with the cut ends of the plasmid. Observe in Figure 14.10 that recombinant plasmids—those that have incorporated the foreign DNA—have had their lethal gene disrupted. In other plasmids, on the other hand, the sticky ends simply rejoin. These plasmids have failed to incorporate the foreign gene and their lethal gene remains intact. At this stage we do not know which plasmids are recombined and which are not.

Next, an enzyme called **DNA ligase** is added. This enzyme binds, or ligates, the breaks in the sugar–phosphate backbone. The result is a sealed, continuous plasmid, which may or may not contain an inserted foreign gene.

Now that the foreign gene has been inserted into the cloning vector (the plasmid), we need to get the cloning vector into a cell where it can replicate. This cell, called the **cloning host**, provides the environment in which many copies of the foreign gene are made. Obviously, a cloning host must be able to accept the cloning vector. It should also replicate quickly and must be able to maintain the foreign gene for many generations. Most often, *E. coli* is the bacterial species of choice. The cells and the plasmids are mixed together, and the bacteria are chemically induced to absorb the plasmid by transformation (see Chapter 6, p. 157). Although some of the bacterial cells will absorb a recombinant plasmid, others will absorb a plasmid without the foreign DNA, while some will fail to absorb plasmids of any type. The problem, now, is to identify the bacterial cells with a recombinant plasmid. The secret to accomplishing this task lies in the lethal and antibiotic resistance genes.

Returning again to Figure 14.10, note that there are now three types of bacterial cells, defined by the plasmids they may have absorbed. First are those cells that absorbed a plasmid lacking donor DNA. They are resistant to the antibiotic but they carry a lethal gene. Second are the cells that did not absorb a plasmid. They do not carry the lethal gene, but they remain sensitive to the antibiotic. Third are the cells that absorbed a recombinant plasmid containing donor DNA. They are resistant to the antibiotic, but the donor DNA has been inserted in the middle of the lethal gene. The lethal gene has therefore been rendered nonfunctional and does not kill the cell.

The cloning hosts are next identified and isolated. This is accomplished by spreading the bacteria over a culture medium containing the antibiotic. Only the bacteria containing the recombinant plasmid are able to grow. Bacteria that absorbed a nonrecombinant plasmid are killed because the functional lethal gene codes for a lethal protein that kills them. Bacteria that did not absorb a plasmid are killed by the antibiotic. Those bacteria that absorbed the recombinant plasmid are protected from these threats and grow into observable colonies that can be maintained indefinitely.

As the colonies of the cloning hosts grow, their recombined plasmids replicate with them. Each of these bacteria can now produce a foreign protein, and with literally billions of bacteria now on hand, this protein is made in large amounts. If you remember the fictional example from the Dirk Pitt novel in our case, this protein might be an enzyme that allowed the genetically engineered bacteria to digest oil. Presumably these bacteria would have been slipped into petroleum storage facilities where they would have reproduced in astronomical numbers, destroying our oil supplies as they replicated. Of course, this is just fiction for the time being. Genetically engineered bacteria, however, are now used routinely to produce commercial or medical products. For example, at one time, the only way to get human growth hormone was to remove it from the pituitary glands of human cadavers. Now, transgenic bacteria with the gene for this human hormone churn it out in substantial quantities.

Plasmids are not the only type of cloning vector available to scientists. Bacterial plasmids cannot carry very much foreign DNA, and so their usefulness for certain purposes is limited. Eukaryotic genes, for instance, which often contain many introns, are often too big for a typical bacterial plasmid. Consequently, other types of cloning vectors are required.

Bacteriophages or bacteria-specific viruses (see Chapter 4, p. 99), for example, make ideal cloning vectors in certain situations. Certain bacteriophages, called phage lambda, are specific for *E. coli* and can be engineered to carry more foreign DNA than a plasmid. Large sections of the phage lambda genome are removed and replaced with the desired donor genes. These recombinant phages are then allowed to infect *E. coli*, inserting their recombined DNA by viral transduction (see Chapter 6, p. 157). In effect, the phage acts as a microscopic syringe to inject foreign DNA into the bacteria. Because so much of the phage DNA has been deleted to make room for the donor DNA, the viral infection does not kill the bacterial cell. Rather, recombined viral DNA is incorporated into the bacterial DNA as in normal transduction (**Figure 14.11**). Once incorporated, donor DNA is replicated along with the bacterial DNA.

DNA from any organism can be maintained in a DNA library

Now that we have reviewed cloning basics, we can back up a little. How do we find the gene we want to clone in the first place? This is not a trivial problem. Suppose, for example, our goal is to place a particular human gene into bacteria, so that the microorganisms can make a human protein. The entire human genome consists of more than 3 billion nucleotides. A typical gene might be 3000 nucleotides long. This means that our gene of interest takes up roughly one-millionth of the genome. Finding it requires a bit of ingenuity.

We have already discussed some of the ways that genes can be identified. Sometimes, PCR can be used to amplify, and therefore to detect, the gene we are looking for. Often, however, PCR will not work because we cannot design primers that will amplify *only* the gene we want. Other sequences will also be replicated, and it may be difficult or impossible to separate out the gene of interest.

Southern blotting is another way to identify genes. If we wish to find a human gene, DNA can be extracted from human cells and then digested into fragments with restriction enzymes. The various fragments are then inserted into a cloning vector and replicated. The cloned fragments are then probed for the gene we seek. This can be a time-consuming process, however, because every single fragment must be probed. In the case of humans, thousands of such fragments must be screened.

The good news is, once a gene has been identified, the cloning hosts that carry it can be maintained indefinitely. A collection of such clones, all carrying different DNA fragments from the same organism, is called a **DNA library**. For example, imagine taking all DNA from the cells of the *Anopheles* mosquitoes that transmit malaria. If we break this DNA into many fragments and clone them in bacteria or phages, we will have a set of cloning hosts with different inserted mosquito DNA fragments. If we have bacteria or phages that contain *all* fragments, our library is complete. We now have a **genomic library** for our mosquito. Anyone wishing to look for a particular gene in mosquitoes can obtain such a library and screen it for the gene of interest. Genomic libraries for many organisms are now available, and more

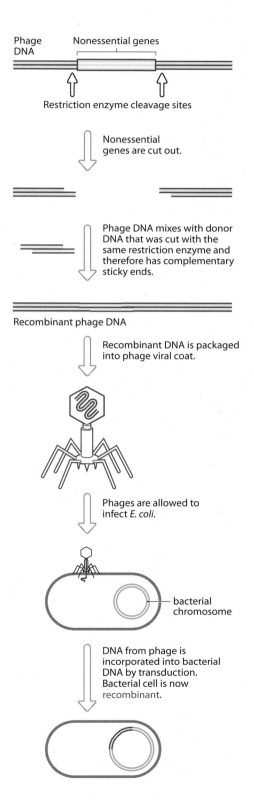

Figure 14.11 Phages as cloning vectors. Phage lambda is a useful tool for inserting donor DNA into *E. coli* DNA.

are becoming available all the time. All the researcher needs to do is order the library he or she needs.

Insights and Applications

Modern recombinant DNA methodology has already provided a great deal of new information about genetic processes. With this greater understanding has come a host of new applications with direct impact on medicine, agriculture, and industry. An entire industry employing thousands of people around the world has sprung up around this new technology, and the pace of discovery gets faster every year. We have truly entered the age of biotechnology. Here we will look at just a few of the new discoveries and applications that are a direct result of our ability to analyze and manipulate DNA.

DNA technology has provided better understanding of genes and how they function

The availability of genomic information has already changed our thinking about many aspects of biology. For example, prior to the publication of the human genome, it was generally stated that humans had about 100,000 genes. We now know that the number of human genes is much smaller, probably between 25,000 and 30,000. More than half of our DNA apparently does not code for protein and has no known function. Another revelation is how similar we humans are. Two unrelated humans differ by only about one nucleotide in a thousand. In other words, your DNA is approximately 99.9% the same as that of any other human.

Our similarity extends well beyond even our own species. Human DNA is more than 98% identical to that of our closest nonhuman relative, the chimpanzee, and 80% to that of the mouse. We even share more than half of our DNA, about 60%, with rice plants.

Examination of microbial genomes has yielded new insights as well. For example, we now know that the smallest bacterial genomes are exactly where you might expect them to be—in intracellular pathogens such as *Mycoplasma pneumoniae* (causative agent of "atypical" or "walking" pneumonia). These are highly fastidious bacteria (see Chapter 7, p. 196) that let the host cell do much of their work. Other bacteria have genomes that are usually about 10 times bigger. Those organisms that are nonfastidious have the largest genomes, reflecting their ability to carry out many metabolic reactions that other, more fastidious bacteria cannot. Comparison of bacterial genomes also confirms a point we made in Chapter 6: that microorganisms are incredibly diverse—certainly as genetically diverse as eukaryotes, in spite of the fact they do not reproduce sexually. On average, between 10% and 20% of the genes in any bacterial species appear to be unique to that species. In many cases, the function of sequenced microbial genes remains to be determined (**Figure 14.12**).

Information obtained as a consequence of genetic technology continues to provide us with a better understanding of biological processes. For example, genome comparison between microorganisms can provide insight into a particular organism's evolution. Many of the basic principles that we have already reviewed in this text have been either confirmed or uncovered through such comparisons. In Chapter 8, for instance, we discussed the theory of endosymbiosis. The mitochondrion and the chloroplast, in particular, are believed to have originated when free-living aerobic and photosynthetic bacteria entered into a symbiosis with anaerobic, nonphotosynthetic primitive eukaryotes. Genome comparisons of the DNA of mitochondria and chloroplasts with bacterial DNA have allowed us to

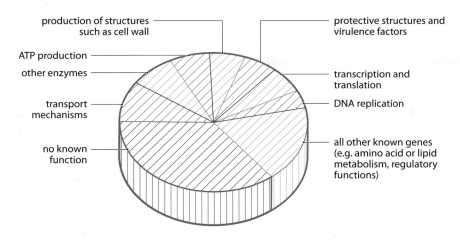

Figure 14.12 Gene function in *Escherichia coli*. *E. coli* has approximately 4200 genes. Of these, well over a third have no known function. The graph indicates the proportion of genes with known function dedicated to specific cellular activities.

determine which bacteria specifically gave rise to these organelles. For mitochondria, sequence similarity is greatest with the ancestor of *Rickettsia prowazekii*, the intercellular pathogen that causes typhus (see Chapter 5, p. 115). Recall that mitochondria have their own DNA, and this mitochondrial DNA shares many similarities with the DNA of *R. prowazekii*.

Ecological adaptations are likewise often better understood following genome analysis. Organisms that live in nutrient-rich environments often have a particularly large number of genes for the metabolism of carbohydrates and other biological molecules. In Chapter 4 (see p. 78) we discussed the bacterium *Deinococcus radiodurans*, which is at the top of the range when it comes to the ability to withstand massive doses of radiation. This species is unusually adept at repairing radiation-induced damage to its DNA, and it was assumed that *D. radiodurans* must have some remarkably unique DNA repair mechanisms. Analysis of its genome demonstrates that this is not the case. What it *does* have is simply *more* of the mechanisms that are already found in other bacteria. For example, most bacteria have one gene called *mutT* that can repair damaged nucleotides. *D. radiodurans* comes equipped with about 20 of these genes.

Might scientists actually "create" living cells in the laboratory one day? Such an idea is not as far-fetched as it sounds, and genome analysis is helping biologists determine exactly what genes would be crucial for any such artificial cells. By analyzing the smallest bacterial genomes—those of *Mycoplasma*, for instance—we are starting to understand which genes constitute the bare minimum necessary for life. It has already been estimated that, to sustain a cellular mode of life, somewhere around 300 proteins would constitute the "rock bottom" for existence.

DNA technology has numerous medical, agricultural, and industrial applications

The practical applications of genetic engineering run the gamut from the ridiculous to the sublime (**Figure 14.13**). At one extreme are "innovations" such as squirt guns that are loaded with a solution containing jellyfish genes that glows in the dark when squirted onto skin. Perhaps equally silly was the company called Transgenic Pets, which had hoped to market cats in which

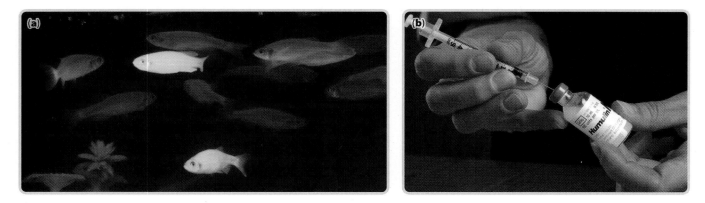

Figure 14.13 Better living through gene engineering? Biotechnology applications range from the whimsical to lifesaving. (a) These fluorescent zebra fish, marketed under the name GloFish®, have been genetically engineered to express a jellyfish gene. (b) Recombinant bacteria now produce pure human insulin for use by diabetics.

the gene responsible for cat allergies had been deleted. At the other end of the spectrum are new medications and vaccines that prevent disease and save lives. We will look at a few of the more important innovations that are already or may soon be available.

Medical applications. Through the late 1970s, insulin was obtained from the pancreases of slaughtered farm animals. This insulin helped a great many diabetics control their disease, but some people developed allergic reactions to the animal protein. In 1979, however, the Eli Lilly Corporation began marketing human insulin produced through genetic engineering (see Figure 14.13b). In the manner previously described, the human gene for insulin production was inserted into a plasmid and then absorbed by bacteria. The result was living microbial insulin factories that churned out pure human insulin in an essentially limitless supply. Other human proteins supplied by recombinant bacteria soon followed. We previously mentioned that similar bacteria now produce human growth hormone, used to treat pituitary deficiencies. Likewise, certain clotting factors, used to treat people with hemophilia, and interleukins, used to modulate immune responses, are now available courtesy of recombinant bacteria. No doubt, more important proteins are on the way.

Other medically valuable proteins are produced not by bacteria but by transgenic animals. The most common way to insert foreign, donor genes into animals to deliver the genes by way of a virus into fertilized eggs or very early embryos (**Figure 14.14**). In other words, the virus acts as a sort of microbial shuttle to carry foreign genetic material into the recipient animal. Alternatively, the donor gene can be injected into eggs or embryos, where it may be incorporated into the host DNA. In either case, the transgenic animal that develops now expresses the foreign gene along with its own genes. When this procedure is used to produce medically important proteins, it is often somewhat amusingly known as "pharming."

The goats we mentioned earlier that produce anticlotting agents are one example (**Figure 14.15**). Other proteins made by "pharmed" animals include human albumin, produced by transgenic cows, which is used to increase blood volume, and a protein made by transgenic sheep that is used to treat hereditary emphysema.

What is the advantage of using transgenic animals over simply letting transgenic bacteria make the protein? One straightforward reason is that, in transgenic animals, the protein is often secreted in milk or semen, where it

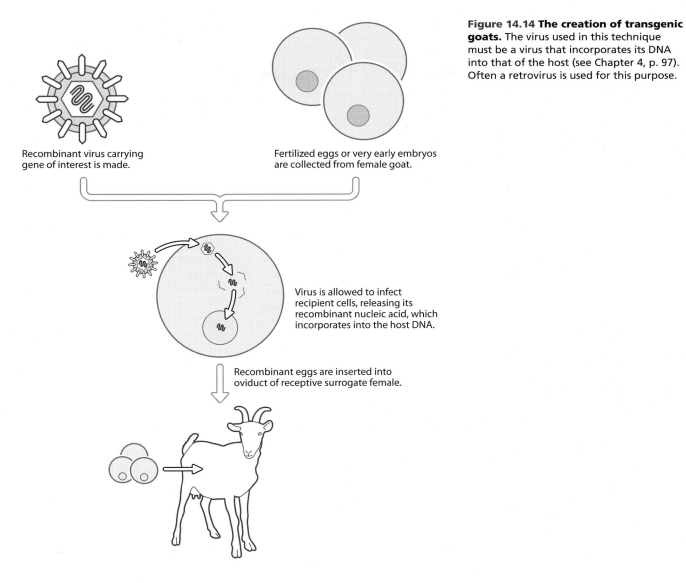

Recombinant virus carrying gene of interest is made.

Fertilized eggs or very early embryos are collected from female goat.

Virus is allowed to infect recipient cells, releasing its recombinant nucleic acid, which incorporates into the host DNA.

Recombinant eggs are inserted into oviduct of receptive surrogate female.

Figure 14.14 The creation of transgenic goats. The virus used in this technique must be a virus that incorporates its DNA into that of the host (see Chapter 4, p. 97). Often a retrovirus is used for this purpose.

is easy to collect. Furthermore, in eukaryotes, many proteins are modified in certain important ways following their production. These modifications take place in the endoplasmic reticulum and Golgi apparatus. Prokaryotes, lacking these organelles, do not necessarily modify proteins in the same way.

Vaccine development has also been greatly impacted by recombinant DNA technology. To cite one example, consider the vaccine that is used to protect chickens and turkeys from two diseases at once: fowlpox and Newcastle disease. To make the vaccine, the fowlpox virus is modified by removing the genes that cause disease symptoms in birds. The genes that code for proteins eliciting immunity (antigenic proteins; see Chapter 11, p. 294), however, are left intact. Meanwhile, the genes that code for immunity-inducing proteins from Newcastle virus are added to the fowlpox nucleic acid. The result is a recombinant virus that could not cause either disease. It could, however, be used as a **recombinant vaccine** to induce immunity against both diseases at once (**Figure 14.16**).

Figure 14.15 Down on the "pharm." These baby transgenic goats will secrete a human anticlotting protein in their milk or semen.

In Chapter 11 we described the **subunit vaccine**, currently used to protect against hepatitis B. Subunit vaccines, unlike traditional vaccines, consist of neither a killed nor an attenuated microorganism. Rather, those microbial proteins that best stimulate an immune response are identified. Once the genes that code for these proteins are pinpointed, they are cloned, resulting in large amounts of the immune-stimulating proteins. These proteins are then utilized as a vaccine to boost immunologic memory. In addition to hepatitis B, subunit vaccines are now in development for a wide range of other pathogens. Not only are these vaccines safer than traditional vaccines, but they can usually be produced much faster as well. Recall, for instance, our discussion of influenza epidemics in Chapter 13. A subunit vaccine against a new strain of influenza might be available on a few months' warning. This is in contrast with the 6–9 months needed to mass-produce traditional influenza vaccines: an important consideration if a new epidemic is looming.

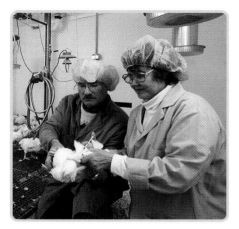

Figure 14.16 A nontraditional vaccine produced through biotechnology. Recombinant vaccines are used to protect poultry against more than one disease at once.

Finally we cannot end our discussion of genetic engineering and vaccines without mentioning "plant vaccines." Imagine simply eating a potato or a banana and developing immunity to cholera as a result. This is not as ridiculous as it sounds. Foreign genes can be introduced into plants, just as they are into animals, to create transgenic fruits or vegetables. If the foreign genes code for antigenic proteins from pathogenic bacteria or viruses, the result is an edible vaccine. Several such vaccines are being developed. They are especially attractive as protection against gastrointestinal pathogens (after all, you *eat* the vaccine, establishing immunity in the gastrointestinal tract) that are particularly a problem in less developed parts of the world. Cholera and typhoid are two examples. A vaccine in a fruit could be transported easily, overcoming some of the logistic problems in the developing world. Additionally, children, who are at especially high risk from diseases of the digestive tract, would hardly put up a fuss if asked to eat a banana.

Because the sequence of the human genome is now available, more and more of the mutant genes responsible for various genetic diseases are being identified. Consequently, it has become possible to treat some of these diseases by introducing normal copies of the gene into affected individuals. The normal gene then produces the necessary protein that the mutant gene was unable to produce. This form of treatment, known as **gene therapy**, was first used successfully to treat a child with a serious immunological deficiency caused by a single mutant gene. The disease is called severe combined immunodeficiency or SCID.

Although gene therapy has been successful in the treatment of SCID and some other conditions, in most cases the technique is still in the experimental stages. Getting the host cells to take up foreign DNA and then to express this DNA in significant amounts still presents a formidable challenge. For example, cystic fibrosis is a genetic disease that takes an especially large toll in North America and Europe. The lungs of cystic fibrosis patients regularly fill with thick, gooey mucus, which interferes with breathing. The gene mutation that is responsible for the disease has been identified. There have been attempts to introduce normal copies of the gene into lung cells, but problems remain. The technique is to splice the normal gene into the genetic material of an adenovirus, a virus that commonly infects the lung cells of humans. The patient inhales the recombinant viruses, and as the virus infects the patient's cells, it introduces the normal human gene along with its own DNA. Thus far, however, it has been difficult to introduce enough copies of the normal gene or to maintain the introduced genes long enough to alter the course of cystic fibrosis.

Nevertheless, gene therapy is still in its infancy and holds great promise in the treatment of various diseases including cancer. There are several ways to introduce foreign genes into a human patient. One such method is illustrated in **Figure 14.17**.

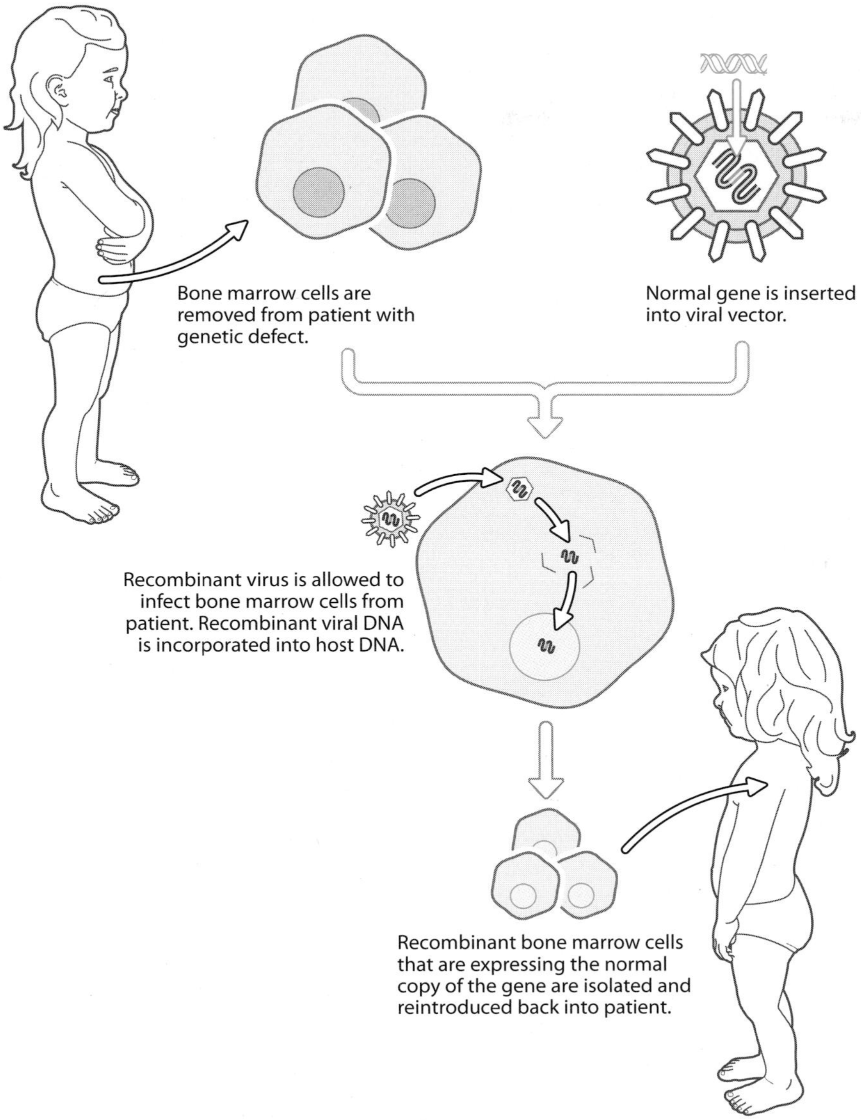

Bone marrow cells are removed from patient with genetic defect.

Normal gene is inserted into viral vector.

Recombinant virus is allowed to infect bone marrow cells from patient. Recombinant viral DNA is incorporated into host DNA.

Recombinant bone marrow cells that are expressing the normal copy of the gene are isolated and reintroduced back into patient.

Figure 14.17 Gene therapy. As in the creation of transgenic animals, recombinant retroviruses or other viruses that insert their DNA into that of the host are used in this protocol. If the reinfused bone marrow cells continue to express the protein coded for by the introduced gene, the treatment may be successful.

CASE: THE MONARCH BUTTERFLY: THE KING IS DEAD OR LONG LIVE THE KING?

In 1999, the monarch butterfly found itself at the center of a biotechnology controversy. These insects lay their eggs on milkweed, and the caterpillars feed exclusively on this plant (Figure 14.18). But concerns about the monarch's survival arose when it was discovered that milkweed growing in the vicinity of genetically modified "Bt corn" was dusted with corn pollen. Monarch caterpillars that were fed pollen from Bt corn in the laboratory died in large numbers. A public outcry about the potential danger of genetically modified crops resulted, and several investigations to further assess the risk to butterfly populations were launched.

1. What is Bt corn? How might it pose a risk to butterflies?
2. What did subsequent investigations show?

Agricultural applications. It may surprise you to learn that about half the soybeans grown in the United States have been genetically altered. The same is true for about a third of the corn and substantial proportions of other crops. These transgenic plants have been created in a manner similar to the transgenic animals discussed earlier. One common technique involves modification of plasmids taken from the bacterium *Agrobacterium*

Figure 14.18 Butterflies at risk? The controversy surrounding genetically modified food crops was highlighted by reports that Bt corn posed serious hazards for monarch butterflies.

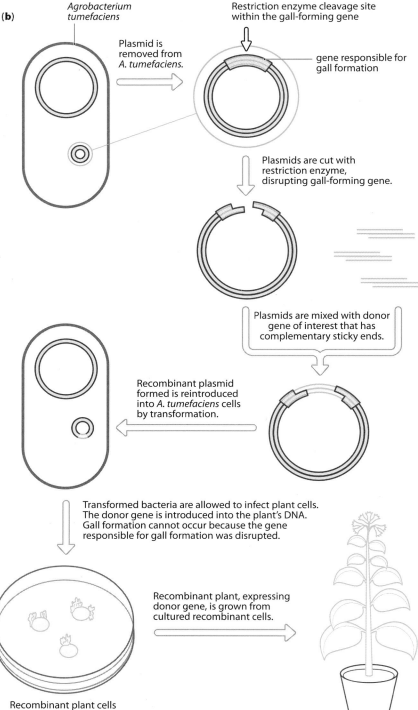

(b) *Agrobacterium tumefaciens*

Plasmid is removed from *A. tumefaciens*.

Restriction enzyme cleavage site within the gall-forming gene

gene responsible for gall formation

Plasmids are cut with restriction enzyme, disrupting gall-forming gene.

Plasmids are mixed with donor gene of interest that has complementary sticky ends.

Recombinant plasmid formed is reintroduced into *A. tumefaciens* cells by transformation.

Transformed bacteria are allowed to infect plant cells. The donor gene is introduced into the plant's DNA. Gall formation cannot occur because the gene responsible for gall formation was disrupted.

Recombinant plant, expressing donor gene, is grown from cultured recombinant cells.

Recombinant plant cells are grown in culture.

tumefaciens. This unusual species infects many plant species, causing the plants to grow tumors commonly called *galls* (**Figure 14.19a**). When *A. tumefaciens* infects a plant, it is able to insert some of its plasmid DNA directly into the plant cells. This bacterial DNA is subsequently incorporated into the plant DNA, where it codes for proteins initiating gall formation. But this bacterium can be made to play a very different role if plasmids are obtained from the bacteria and modified by previously described techniques. Gall-inducing genes are disrupted, and a foreign gene coding for a desirable characteristic is inserted (see Figure 14.19b). The plasmids retain the ability to enter plant cells, but now, instead of gall-forming genes, they

Figure 14.19 Creation of a transgenic plant. (a) A gall, caused by *Agrobacterium tumefaciens*. This bacterium is commonly used to introduce foreign genes into plants, because when it infects plant cells it inserts plasmid DNA into the host plant DNA. (b) A protocol for creating transgenic plants by means of modified *A. tumefaciens* plasmids.

introduce a donor gene, creating a transgenic plant embryo. Some plants will not absorb *A. tumefaciens* DNA, and for them, a "gene gun" can be used to introduce foreign genes of interest. The genes are made to adhere to tiny metal "bullets," which are then fired into the cells with a blast of compressed gas.

Foreign genes that can be introduced in this way may come from other plants that naturally have a desirable characteristic, such as a high oil content or an increased tolerance for drought. Other times the genes do not come from plants at all. For example, the Bt corn in our case carries a bacterial gene from *Bacillus thuringiensis*. The gene codes for the production of a powerful toxin that can kill insects. Corn modified in this manner essentially makes its own insecticide, greatly reducing damage from insect pests. Likewise, a gene from *Salmonella* is currently used in transgenic cotton to make the cotton plants resistant to weed-killing chemicals.

Is all this a good thing? It depends on whom you ask. The genetically modified soybeans are lower in saturated oils, and the altered corn has increased resistance to insect pests. Yet many people are strongly opposed to these genetically modified (GM) foods. Some worry about the consumption of GM foods and about currently unknown effects they might have on health. The European Union has gone so far as to require the placement of labels on all GM foods, identifying them as such.

One major environmental concern is how GM foods might affect wild plants. Many food plants are able to hybridize with native plants, and if such hybridization occurs between GM plants and native plants, new genes might be introduced into wild populations. The consequences are difficult to predict. If weeds, for example, obtain a gene for resistance to herbicides from a resistant GM crop, what will be the result? Our case of potentially endangered butterflies shows that it is not even necessary for GM plants to hybridize with wild plants to cause mayhem. Simply releasing genetically modified plant parts into the environment may be enough to create problems. Certainly the laboratory evidence suggested that Bt corn pollen was likely to harm the butterflies. Such pollen might be carried by the wind to the milkweed plants, where monarch caterpillars could consume it. Subsequent investigations, however, showed that, in the field, monarch caterpillars rarely if ever ingested enough Bt pollen to cause death. This was clearly good news for the butterflies, but controversy over GM crops will no doubt continue. Although capable of great benefit, genetic engineering always includes an element of risk. The risks require careful analysis before any genetically modified product is introduced.

Industrial applications. The next time you slip into a pair of "stonewashed" jeans, you might use the occasion to ponder the marvels of biotechnology and its use of microorganisms (**Figure 14.20**). The cotton used to make jeans is essentially composed of cellulose, a complex carbohydrate found in plant material. Cellulose can be made softer by treating it with cellulases, which are enzymes that digest the cellulose. Some of these enzymes originally came from fungi, while others were first isolated from bacteria that colonize clams and oysters. The genes that code for the commercially valuable enzymes have been introduced into other microorganisms that are easier to rear in the laboratory. These microbes then produce the cellulase enzymes in large quantities. Not only do these enzymes produce a more comfortable pair of pants, but because they are biodegradable, they are also environmentally friendly. Thus, "stonewashing" has nothing to do with stones. Apparently, jean manufacturers did not think "microbially washed" had marketing potential.

Recombinant microorganisms now also play an important role in the new field of **bioremediation**, or the use of biological processes to reduce pollution. Like the bacteria in our earlier case, transgenic microorganisms that

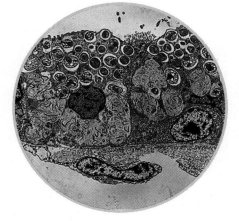

Figure 14.20 Not a stone in sight. Certain clams and other bivalves harbor bacteria that produce the enzyme cellulase (seen here inside the cells of a mussel). The genes for this cellulose-digesting enzyme have been introduced into other microorganisms, which are then utilized in the production of "stonewashed" jeans.

degrade petroleum and other environmental pollutants have been produced. A species of *Flavobacterium*, for instance, actually produces enzymes that degrade organophosphates, a group of chemicals that includes highly toxic nerve gases. The gene that enables these bacteria to digest organophosphates has been introduced into *E. coli*, which can then degrade the toxin in the event of a toxic chemical spill. Should such a chemical be used in an act of bioterrorism, recombinant *E. coli* stand ready to swing into action as important "first responders." We will discuss bioremediation and other industrial uses of microorganisms further in Chapter 16.

Ethics and Safety

Just because we *can* do something, does this mean we *should*? New technology that threatens to change lives has always generated controversy. Protests arose during the Industrial Revolution among those who believed that new mechanized technology posed a serious danger to workers and their way of life. We still debate whether or not the development of nuclear energy was wise. Certainly DNA technology is no exception. From its very inception, genetic engineering has raised safety and ethical issues (**Figure 14.21**). In the 1970s, scientists worried whether recombinant bacteria might escape into the environment, and if so, what problems they might cause. Today we ask ourselves about the safety of GM foods. Are they safe to consume? What ecological havoc might they create? Is the ability to grow more food, perhaps using fewer pesticides, worth the risk?

We now have the ability to screen people for genetic disorders. If such screening is carried out, who should have access to the results? In the future we may have to debate whether or not it would be ethical to create "designer babies" in which genes for desired characteristics—increased intelligence or athletic skill, for example—might be spliced into developing embryos. This ethical dilemma does not begin and end with humans. Is it proper to create new forms of life such as transgenic animals? Is it morally defensible to develop an animal such as a transgenic goat, simply so that it can serve as a living factory for human proteins?

Answers to these questions will come not only from scientists but also from politicians and from the general public. Unfortunately, it is a sign of the times that frequently politicized arguments often generate more heat than light. We live in the "sound bite" era, in which arguments, whether or not they are valid, are used to advance a political or social agenda. The emphasis is too often on winning rather than being right.

Figure 14.21 Genetic engineering: the continuing controversy. A protest against genetically modified food highlights the strong opinions stirred by this issue.

What is the solution? The first step is making sure that the public has the necessary information and genuinely understands the issues under consideration. Only then can we cast votes and make decisions that are based on rational analysis rather than emotion or political agenda. It is certainly true that everyone is entitled to his or her opinion, but opinions mean little when unsupported by knowledge and understanding.

Thus we conclude this chapter very much where we started it. The field of recombinant DNA has enormous potential for good, but it also raises important safety and ethical issues. Each of us, armed with an understanding of what is involved, can help encourage wise policy decisions, allowing this powerful new technology to be used in the most thoughtful, beneficial, and ethical manner.

Coming Up Next...

Ever since the first bread rose, the first wine flowed, and the first piece of tempura was dipped in soy sauce, microorganisms have been involved in making the food we eat. In fact, it would be difficult to get through a meal without consuming *something* that comes to you courtesy of bacteria or fungi. Many of these food items have been around for thousands of years. Others have found a place on our dinner table relatively recently.

In the next chapter we will look closely at the hand that microorganisms have in our diet and how much less interesting that diet would be without the assistance of bacteria and fungi. We will begin by exploring the historical role of microorganisms in food and move on to learn how our microbial partners are used in food production today. In conclusion we will serve up a few selected recipes in which microorganisms are a crucial part of the *plat du jour*, giving you the opportunity to sample your own microbial fare.

Key Terms

bioremediation	gene probe	recombinant vaccine
cloning	gene therapy	restriction enzyme
cloning host	genetic engineering	Southern blotting
cloning vector	genomic library	sticky end
DNA library	genomics	subunit vaccine
DNA ligase	polymerase chain reaction (PCR)	transgenic animal
DNA sequencing	primer	
gel electrophoresis	recombinant DNA	

Concept Questions

1. How would you describe the term "biotechnology" to someone with no scientific background?

2. Explain why the production of "sticky ends" in DNA, when the DNA is cut with a restriction enzyme, is an important reason why such enzymes are valuable when creating recombinant DNA.

3. When performing a Southern blot, why is it necessary to treat the DNA in a way that makes it single-stranded?

4. In this chapter we found that parasitic bacteria, which live intracellularly in host cells, typically have smaller genomes than other bacteria. Why does this observation make sense?

5. Look through a newspaper or magazine, or online to find an article or advertisement for a genetically engineered product. Then provide a brief outline that describes how you think this product may have been produced.

6. We concluded Chapter 13 with a discussion of bioterrorism. Should we be concerned about the use of genetic engineering by terrorists to create more lethal bioterrorism weapons? How might genetic technology be used for such a purpose?

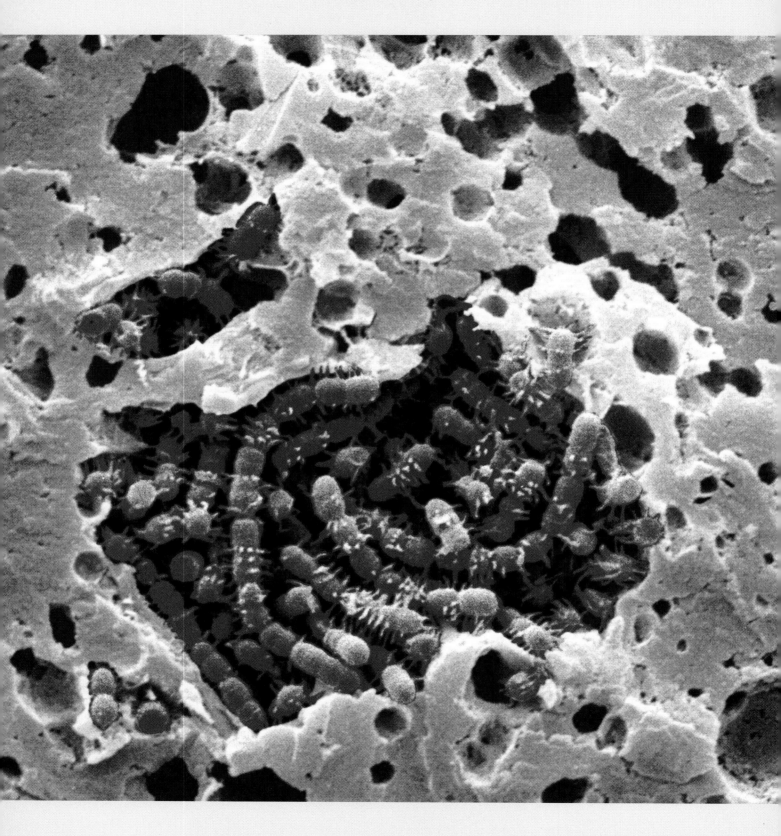

A colony of lactic acid-producing streptococci are used to make feta cheese.

Chapter 15

Guess Who's Coming to Dinner: Microorganisms and Food

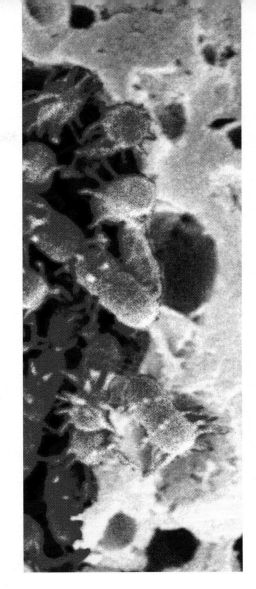

For the true beer connoisseur, there is really no place quite like Belgium, home to some of the world's best brews, and certainly to some of the most unusual. And no trip to Belgium would be complete without sampling the Trappist ales. The Trappists, an order of Catholic monks, have been brewing beer in their monasteries for centuries. But be alert for imitations. While many beers use the name "Trappist," only six authentic Trappist monasteries in the world still produce the genuine article (**Figure 15.1**).

Perhaps the most unusual of the Trappist ales is lambic beer, sometimes called Belgian champagne. This ale is definitely not for the casual beer drinker or anyone who requires the crisp, clean taste of a traditional German-style lager. Lambic beer is decidedly stronger both in alcohol content and in flavor. It strikes many as overly sour. The secret to its unusual flavor is the manner in which it is made.

Most beer is produced when commercial yeast strains, almost invariably members of the genus *Saccharomyces*, anaerobically metabolize the sugars in grain, releasing ethyl alcohol as a fermentation waste product (see Chapter 7, p. 183, for a discussion of fermentation). The production of lambic ales, like that of most beers, begins by adding a conventional *Saccharomyces* strain to a boiled mix of grain, water, and hops. Once the fermentation is under way, the monks open the windows of the brewery and let nature take its course. Wild yeast cells are wafted in on the Belgian breeze, and as they settle in the fermentation vats, they also participate in the fermentation. The result, of course, is somewhat variable, because each type of wild, yeast releases different metabolic waste products, imparting its own

Figure 15.1 Trappist ale. A monk oversees brewing at a Trappist monastery in Belgium. Genuine Trappist ale is made only in Trappist monasteries.

special contribution to the final flavor. The final hearty brew, although initiated with traditional *Saccharomyces* yeast, consequently ends up quite different from what we normally think of as beer. Regular or "straight" lambic beer consists of this simple, wild, fermented beverage. If you find that a bit intimidating, you may wish to sample kriek beer instead. This is a lambic ale that has been flavored with cherries. Alternatively, there are also pêche (peach-flavored); framboise, into which raspberries have been added; and even varieties flavored with apricots and blackcurrants.

Although lambic ale may seem a bit unusual, the use of microorganisms in the production of foods and beverages is commonplace. Consider, for instance, a fairly typical meal. Perhaps you would like to start with a salad. If your favorite dressing is oil and vinegar or blue cheese (or any other dressing containing vinegar or cheese), you will be dining on food items produced by microbes. Likewise, if you select cheese enchiladas, a hearty stew with summer sausage, or an Asian dish flavored with soy sauce as your entrée, microorganisms contributed to the recipe. The same is true for your dinner rolls, or almost any other bread product that might complement your meal. And to drink? Beer, wine, most soft drinks, and even coffee are produced with microbial assistance.

Of course, not *all* microbial activity is so benign when it comes to the food we eat. Throughout this text we have had ample opportunity to discuss food- and waterborne diseases and the manner in which microorganisms cause food to spoil. We have similarly spent plenty of time investigating how foodborne diseases are controlled and treated and how microbially induced food spoilage may be reduced. So enough with the bad news. In this chapter the spotlight will be on microorganisms as partners rather than adversaries. After a bit of history we will look at the ways in which fungi and bacteria are either consumed directly as food or used in the production of everything from pickles to pinot noir to provolone cheese. The production of some nutritional supplements such as vitamins and amino acids also relies on microbial assistance, and we will investigate the part microorganisms play in stocking the shelves of the local health food store. Finally, we will list a few recipes involving microorganisms. Try one, raise a glass to your microscopic culinary colleagues, and say "bon appétit"!

The Beginnings of a Beautiful Friendship

CASE: FLIGHT FROM EGYPT

Microorganisms have contributed to at least some of humanity's dietary needs since ancient times. Even the Bible refers to foods in which microbial processes are crucial. Perhaps one of the most famous biblical quotations that makes such a reference comes in the story of the Israelites and their hasty departure from Egypt.

    "Seven days shall ye eat unleavened bread..."
    Exodus 12:5
1. **What is the connection between the above biblical quote and food microbiology?**

Microbial activity can help to preserve the quality of some foods

Imagine what it would be like to live without refrigeration. Suddenly, eating would become literally a hand-to-mouth affair. Most foods spoil rapidly, especially in warm weather, so for the most part, unless you were willing to simply hold your nose and force down rotten food, you would have to buy, hunt, harvest, or gather your meals almost daily. Or, if you were rich enough, you could buy spices. Not only do spices make food tastier, but many contain compounds that inhibit microbial growth, reducing food spoilage.

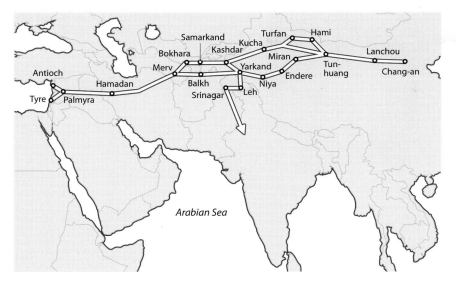

Figure 15.2 The spice trade. This map highlights the main routes of the Silk Road between Europe and Asia, along which many goods, including spices, were transported. The quest for spices was a response in part to their ability to inhibit microbial growth and consequently food spoilage.

Garlic, oregano, thyme, cumin, and capsaicin (found in chili peppers) are just a few commonly used spices that have been found to exert strong antimicrobial activity. Prior to the development of more modern food preservation techniques, spices were one of the few reliable ways to slow down food spoilage. In fact, through much of history, spices were big business, in part for just this reason. Many early explorers and traders were motivated by the high price that spices commanded (**Figure 15.2**). Marco Polo, for instance, set off for China with his father and uncle in search of marketable spices found only in East Asia.

Or, as an alternative, you might actually *encourage* the spoilage of certain foods. As we learned in Chapter 7, many microorganisms are fermenters, able to break down biological molecules to a variety of waste products in the absence of oxygen. In some cases, these wastes serve a valuable purpose; they add flavor to food as well as prevent additional spoilage caused by less savory microbes. Fermented foods have formed an important part of the human diet for thousands of years. Of course those producing the foods had no idea that organisms too small to see were involved; they simply knew that the process worked.

Fermented dairy products and grains have been used for thousands of years

The earliest production and consumption of fermented foods is lost in antiquity. Fermented dairy products have probably been around ever since a horseman in Central Asia filled up a goatskin with milk, tied it to his pack, and hit the trail. Taking a drink at the end of the day, the rider found that the milk, which had sloshed around all day in the hot sun, had turned into a semisolid curd. The first yogurt had been made, all because of the bacteria that had fermented sugars in the milk (see Chapter 7, p. 197). Cheese, another fermented milk product, is believed to have been first made in the Middle East at least 8000 years ago. Some thousands of years later, cheesemaking became both an industry and an art in ancient Greece and Rome. Subsequently, Roman tastes and techniques influenced cheesemaking throughout Europe and in much of the rest of the world. There are hundreds of types of cheese today, and as we will see, the type of cheese produced is in large measure due to the type of microorganisms utilized in the cheesemaking process.

Alcoholic beverages have likewise been a part of the human experience for thousands of years. In 4000 BC, ancient Sumerians were producing wine and beer from fermented grapes or barley (**Figure 15.3a**). Historical records from ancient Egypt show that beer was a taxable item almost 5500 years ago,

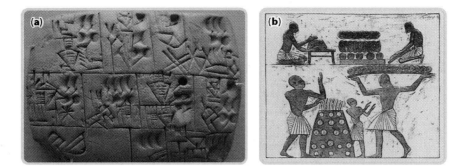

Figure 15.3 Fermented foods in ancient history. (a) The Sumerians were among the first to brew beer. The photograph shows a brewing recipe written in cuneiform script, the earliest known writing system. (b) Egyptian hieroglyphics depicting breadmaking.

and Romans, living in pre-caffeine times, frequently started their day with a flask of wine. In fact, because water was often polluted, wine was the beverage of choice for many early Europeans. By the 11th century, drinking establishments were a fixture across Britain and other parts of Europe.

Bread, which like alcohol relies on yeast fermentation (see Chapter 7, p. 184), also got its start in the Middle East. Ancient Egyptian hieroglyphics clearly depict breadmaking in about 2500 BC (see Figure 15.3b), and large excavated Egyptian bakeries are believed to have supplied thousands of people each day with "the staff of life." The preference of ancient people for bread that had risen ("leavened" bread) as opposed to bread dough that was baked without first rising ("unleavened" bread) is well illustrated by the biblical quote in our case. The Israelite slaves, abruptly freed by the Egyptian Pharaoh, were in such haste to abandon Egypt that they left before allowing their bread to rise. The hard, flat, unleavened matzos eaten by observant Jews at Passover commemorate this event. The New Testament also makes reference to the use of yeast in breadmaking: "Know ye not that a little leaven leaveneth the whole lump?" (I Corinthians 5:6).

While fermented foods have been produced and consumed throughout human history, it was not until the late 1800s that the role of microorganisms was at last understood. As we learned in Chapter 5, at this time scientists, such as Pasteur and Koch, first described the role of microorganisms in processes such as disease and food spoilage. Once early microbiologists understood the importance of microorganisms, they began to work out the relationships between particular food products and specific microorganisms. With the advent of pure culture techniques, the "right microorganism for the job" could be cultivated and introduced into food items, reducing the chance of contamination with unwanted microbial species. The early 20th century witnessed the development of large-scale industrial food production, and microbial fermentation became big business.

Microorganisms and Food Production

As you have may have already deduced, the impact of most microorganisms used in food preparation is indirect. Microbial activity in products such as milk, cheese, and vinegar results in the release of compounds that alter the taste, smell, or consistency of the food. However, a few microorganisms have a more immediate connection to human gastronomy. These are the ones we simply eat "as is." Next we will consider some of the ways that microorganisms themselves are on the menu.

Some fungi and bacteria are consumed directly as food

CASE: WHAT'S FOR TUCKER, MATE?

On his first trip to Australia, Hal arrives in Sydney for a professional conference. On his first morning in town, he goes down to his hotel's buffet for breakfast. Hal picks up a couple of slices of toast, and after finding a table, he notices that along with butter, there are several jams and jellies available as condiments. There is also an unfamiliar dark brown spread that Hal suspects might be something like apple butter. Eager to try something new, he spreads a thick layer on his toast. Yet when he takes a bite of his toast, he quickly realizes that the unknown spread is certainly not apple butter. The strong, salty flavor is almost more than he can take, and after forcing it down, he quickly gulps a glass of orange juice to get rid of the taste. Later, he notices that many people are eating this spread, and they seem to like it. Asking someone at a neighboring table, he finds out that the brown spread is called "Vegemite" and that it is practically an Australian national dish.

1. What exactly is Vegemite, and how is it made?

To many people, the thought of dining on fungi or bacteria sounds revolting. Yet throughout this text we have been reminded that most single-celled organisms pose no threat and are often beneficial. Furthermore, every bite of food or swallow of liquid that we take is laden with microorganisms. So why should it be unappetizing to consume them intentionally? Nevertheless, for some people a food product that is obviously microbial raises a red flag. Perhaps it is a case of "out of sight, out of mind." The bacteria concealed in our lunch that we cannot see and usually do not think about are one thing. Intentionally seeking out microorganisms for lunch is something else altogether.

Yet we readily eat mushrooms, the larger cousins of microscopic fungi. Indeed, some mushrooms are considered to be a true delicacy and often demand a high price. When you include mushrooms in your salad, pizza, or omelet, you are actually eating the "fruiting bodies," which are large reproductive structures of even larger filamentous fungi living under ground (see Chapter 4, p. 86). Each mushroom is formed by a large number of hyphae that together make up the reproductive body, known as the mycelium. Of the more than 300,000 tons of mushrooms consumed in the United States annually, most are grown commercially in temperature- and humidity-controlled buildings, in an organically rich bed of compost (**Figure 15.4a**).

Mushrooms are appreciated for their taste and texture. Although they are not a rich source of vitamins or protein, mushrooms can easily form part of a healthy diet. They are essentially fat- and cholesterol-free, and they provide a good source of dietary fiber. For many people, they are also fun to collect. During the summer months, taking to the woods to harvest morels, chanterelles, and other highly prized varieties is an age-old tradition (see Figure 15.4b). Unfortunately, it can also cause friction between collectors, landowners, and conservationists who decry the "destruction" of mushroom populations at the hands of overzealous mushroom hunters. Part of the problem, however, is based on a general misunderstanding regarding basic mushroom biology. The fungus is not "killed" when a mushroom is

Figure 15.4 The business of mushrooms. (a) White button mushroom production at a California mushroom farm. (b) Chanterelles are especially prized by gourmets and mushroom hunters.

Figure 15.5 As Australian as kangaroos. A Vegemite advertisement from the 1960s. Most Australian children, from a very early age, are familiar with the words to a commercial jingle for this product: "We are happy little Vegemites as bright as bright can be. We all enjoy our Vegemite for breakfast, lunch, and tea. Our mummy says we're growing stronger every single week. Because we love our Vegemite, we all adore our Vegemite, it puts a rose in every cheek!"

picked; it has merely lost a fruiting body. This means that picking mushrooms is more like picking berries from a bush than it is like overfishing a lake. The subterranean fungus that produces the mushroom continues to survive. Indeed, mushroom harvesting likely encourages the production of a new crop of fruiting bodies. On commercial mushroom farms, systematic picking actually increases the production of additional mushrooms.

Perhaps this last point was lost during Oregon's "mushroom war" of 1993. Two people were killed and several others were injured when tempers exploded and heavily armed mushroom rustlers fought it out in the Oregon woods over wild edible fungi. Commenting on the hostilities, a Morrow County Sheriff's deputy observed that "nearly all the mushroom pickers are armed and it's real scary."

So if we are willing to eat mushrooms, why shouldn't we eat smaller, microscopic fungi? In certain places, such as Australia, we already do. Vegemite is a commercial product with a unique recipe, made from an extract of *Saccharomyces cerevisiae* (brewer's yeast) to which vegetable flavorings and spices have been added (**Figure 15.5**). The yeast used to make Vegemite is often left over from beer brewing. Vegemite has a dark reddish-brown color and it is one of the richest sources of essential B vitamins. Although Vegemite has the consistency of peanut butter, it has a flavor all its own. As Hal found out, Vegemite is definitely an "acquired taste." It is very salty, and to the uninitiated, the flavor can be overpowering. Yet for Australians, Vegemite is an essential part of the culture. It is found in more than 90% of homes, and children are brought up on it from the time they are infants. A Vegemite sandwich is the quintessential Aussie snack.

Vegemite and other consumable forms of yeast are examples of **single-cell protein**—bulk microorganisms that are used as a food source for humans or animals. Because of its high vitamin content, yeast can be added to wheat or other grains to enhance nutritional value. When yeasts are grown on high-nitrogen media, they are also fairly high in protein content. Simple powdered yeast can also be added to foods as a nutritional supplement. Too much yeast in the diet, however, may be a problem for some people because of the high nucleic acid content, which can lead to gout.

(a)

Bacteria have been less successful at working their way into the human diet. A few decades ago, cyanobacteria in the genus *Spirulina* were hailed as the answer to food shortages (**Figure 15.6**). It is estimated that 10 tons of protein per acre could be harvested from *Spirulina*, compared with 0.16 tons per acre for wheat and 0.016 tons per acre for beef. *Spirulina*, however, has yet to gain broad acceptance as a food item, although it is available as a nutritional supplement in health food stores. If you really want to sample *Spirulina*, a good place to go is the North African country of Chad. Here, *Spirulina platensis* is collected from the shallow waters of Lake Chad. The bacterial mats are sun-dried and cut into protein-rich cakes called *dihe*.

Figure 15.6 Bacteria as food. (a) *Spirulina* tablets. Although it is a rich source of protein, *Spirulina* has yet to catch on as a mainstream food. It is, however, readily available in health food stores as a dietary supplement. (b) *Spirulina platensis* cells.

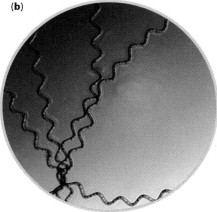

(b)

As previously mentioned, however, it is microbial *metabolic products* rather than the microorganisms themselves that are far and away the principal route by which bacteria and fungi affect our diet. Most of these by-products are the result of fermentation. Next we will take a look at how this basic aspect of microbial metabolism is so crucial to many of the foods we eat.

In the absence of oxygen, some microorganisms undergo fermentation, releasing specific waste products

Fermentation was described in detail in Chapter 7 (p. 183). As we learned, in the absence of oxygen, many cell types continue to produce ATP via glycolysis alone. The final product of glycolysis is pyruvate, but without oxygen, pyruvate cannot be broken down further. Instead, it is reduced to a final waste product in order to oxidize NADH. The now oxidized NADH (NAD$^+$) is required for glycolysis to continue.

As we have already discussed, humans have taken advantage of microbial fermentation for thousands of years, using it to preserve food or enhance its flavor. If fermenting microorganisms are either added or naturally gain access to a food product, the fermentation waste products change both the chemical and textural qualities of the food, extending its shelf life and/or creating new desirable flavors and aromas. We will take a closer look at how some of these foods are made, highlighting the indispensable services provided by fermenting bacteria and fungi.

Many plant products can be fermented into various food items

Bread. Considering its central role in the diet of so many people, bread is a good place to start our survey of microbially produced foods. To make bread, the flour, obtained from a grain such as wheat or rye, serves as a sugar source for the yeast. As the yeast ferments the sugars in the flour, it causes the bread to rise and also contributes to the taste and aroma of the final baked product.

To make bread, a strain of the single-celled fungus *S. cerevisiae* is added to the bread dough (**Figure 15.7**). Surrounded by dough, some of the yeast cells find themselves in an anaerobic environment. Consequently, in order to grow, they must ferment the sugars in the dough.

The pyruvate produced at the end of glycolysis is reduced to ethanol (ethyl alcohol) and carbon dioxide. The CO_2 gas causes the dough to rise, giving the bread its light, fluffy texture. Small amounts of other fermentation wastes contribute to the flavor of the bread. All traces of ethanol evaporate from the bread as it is baked. If yeast is not added to bread dough, the final baked product will remain flat and hard.

Occasionally, other microorganisms are employed to produce more unusual breads, such as sourdough. Sometimes a complex assemblage of microorganisms is called for that contains both yeast and bacteria. In sourdough bread, *Lactobacillus sanfrancisco* contributes to the flavor by releasing lactic acid, giving the bread its characteristic acidic or sour flavor. The easily identifiable flavor of rye bread is due in part to the use of lactic-acid-producing bacteria such as *Lactobacillus bulgaricus*, *Lactobacillus plantarum*, and *Streptococcus thermophilus*.

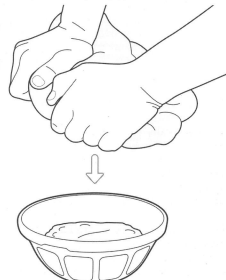

The bread dough is prepared, mixed with yeast, and thoroughly kneaded.

The yeast ferments the sugars in the dough, causing it to rise. CO$_2$ gas, released as fermentation waste, creates gas pockets in the dough, giving it its finished texture.

After it has risen, the bread is shaped into loaves and allowed to go through a second rising. It is then baked. Ethanol produced during fermentation evaporates.

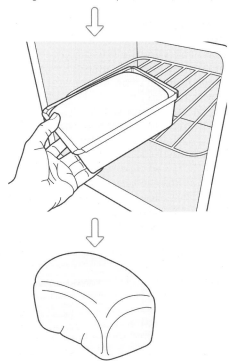

Figure 15.7 How bread is made. Yeast is kneaded into bread dough, where it ferments sugars found in the dough. Carbon dioxide gas, released as a fermentation waste, causes the dough to rise. Ethyl alcohol, also a fermentation waste, evaporates during baking.

CASE: HOMEMADE WINE

Joe has recently taken up the hobby of winemaking and he has already made a variety of wines that he enjoys sharing with family and friends. Of course, the final products vary considerably in character and taste, depending on the type of wine he is making. For one thing, the sweet wines tend to be lower in alcohol content, while the dry wines have a higher final alcohol content. Although Joe has always liked wine, he never noticed this correlation between sweetness and alcohol content before he started making wine himself.

1. **Why is there a relationship between sweetness and alcohol content of homemade wine?**
2. **Why is this trend less valid for commercially made wine?**

Just about any type of plant that has enough carbohydrates in its tissues can by fermented by yeast into some type of alcoholic libation. If you look hard enough, you can find beverages made with fermented plums, oranges, or even dandelions. Typically, however, most alcoholic beverages are produced by grape or grain fermentation.

Wine. To produce wine, grapes are first crushed to obtain the grape juice, known as the **must**, which contains most of the sugars (**Figure 15.8**). The must is usually treated with sulfur dioxide to rid it of any microorganisms that are naturally present on the grapes. If these bacteria and fungi are not eliminated, their own fermentation waste products will end up in the wine, resulting in an unpredictable, and usually undesirable, final product.

Once contaminating microorganisms have been removed, an appropriate strain of *S. cerevisiae* (brewer's yeast) is added. As the yeast ferments the sugars in the must, ethyl alcohol is produced. During fermentation, which usually lasts about four days for red wine and 10 days for white wine, the temperature is maintained at approximately 25°C. Eventually, the yeast cells are killed off by the alcohol waste they produce, resulting in a final alcohol content between 7% and 18%. The now-fermented must is placed in large vats to allow sediments to settle. It is also filtered and flash-pasteurized to kill any remaining microorganisms. The wine can then be stored in casks, and it is during this stage that aging or the development of the wine's final *bouquet*, or distinctive and characteristic fragrance, occurs.

Of course, wines may be red or white, sweet or dry, wonderful or undrinkable. This variety is the result of many factors. Filtered must is always clear or "white" in color, regardless of the type of grape. The grape skin, on the other hand may be "red" or "white," based on the presence or absence of reddish pigments. To make a white wine, either white grapes are used or the skin of red grapes is removed (see Figure 15.8). If a red wine is being made, the grape skin is simply permitted to stay in the must prior to fermentation. During that time, chemical components of the skin seep into the must, imparting the familiar red color.

As Joe observed, the sweetness and alcohol content of wine are actually opposite sides of the same coin. To make a sweet wine, fermentation is stopped relatively early, before the yeast can ferment many of the sugars in the must. The remaining sugars give the wine its sweet taste and the alcohol content is relatively low, because of the short fermentation period. To make a dry wine, on the other hand, fermentation is allowed to go to completion.

Figure 15.8 Production of wine. A general scheme for winemaking is diagrammed. Many details, such as the length of fermentation and aging, are different for red and white wines. Grape juice (the *must*) extracted from the grapes is used as a sugar source for fermentation. Yeast uses the sugars as an energy source and releases ethyl alcohol as a fermentation waste product.

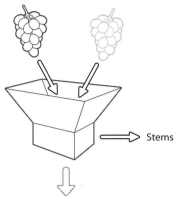

To make red wine, red grapes are destemmed and crushed. To make white wine, either deskinned red grapes or white grapes are used.

Stems

The must is transferred to a fermenting vat. SO₂ (sulfur dioxide) is added to inhibit the growth of contaminating microorganisms. Appropriate strains of yeast are added and fermentation begins.

The fermented must is transferred to a settling vat. It is siphoned several times to separate solid debris from the fermented liquid.

Solids

The wine is transferred to casks for aging. Chemical changes during aging contribute to the wine's final flavor and aroma.

The aged wine is filtered and bottled.

Filtering

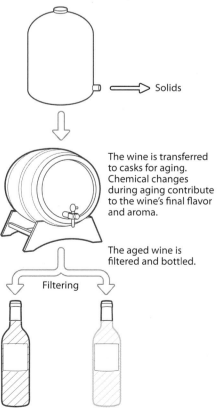

All sugars have been broken down during the extended fermentation period and alcohol content is relatively high. The length of the fermentation period can be regulated by various means. One way is to select a strain of yeast that is either relatively tolerant or intolerant of alcohol. Alternatively, the wine-maker can simply intervene, stopping fermentation at any point. When wine is commercially produced, the alcoholic content can be adjusted "after the fact," simply by adding or removing alcohol.

Champagne and other sparkling wines are produced like still wines, except that fermentation is allowed to continue after the wine is bottled. Sealed inside the bottle, the carbon dioxide produced as a fermentation waste is unable to escape, resulting in the bubbly final beverage.

If all of this sounds fairly simple and straightforward, then why does wine vary so much in quality? Winemaking is both a science and an art. Because so many factors can affect the final product, the making of a truly fine wine is a complex affair. The climate and soil in the grape-growing region can affect the chemical composition of the grapes, which in turn influences the flavor. The precise time at which the grapes are harvested determines their sugar content. The preparation of the must, the strain of yeast used for fermentation, the duration of fermentation, and the length of aging all contribute to the wine's ultimate flavor. One can even get a degree in **oenology**, the science of winemaking, at certain universities.

Beer. Beer brewing also consists of a number of steps (**Figure 15.9**). Grains, most commonly barley, serve as the source of sugar for yeast fermentation. However, unlike grapes, grains cannot be directly used for fermentation, because grains store their sugar in the form of starch. Yeast cells lack the enzymes necessary to metabolize starch, so an additional trick or two is needed to permit yeast fermentation.

First, the barley is **malted**. This simply means that the barley seeds are allowed to sprout, because sprouted barley naturally contains starch-di-gesting enzymes. The sprouted barley, or malt, is then dried and roasted and mixed with water in a step called *mashing*. During mashing, the naturally occurring enzymes in the malt break down the starch, converting it into simple sugars that the yeast cells are able to ferment. Once the starch has been degraded, the remaining solids are removed, and the liquid portion containing the sugars, known as the **wort**, is mixed with hops. Hops are the flower portion of the hop plant, which account for much of the variation of a beer's flavor. Hops have a bitter taste and the brewer adjusts the amount of hops depending on the type of beer desired. The mixture is then boiled. During boiling, the mixture is concentrated, flavor-imparting chemicals are leached out of the hops, and potentially contaminating microorganisms are killed. The liquid wort is then separated from the solid hops, in preparation for fermentation.

Once again, strains of *Saccharomyces* are usually the yeast of choice. The particular species selected depends on what type of beer the brewer is making. To brew a lager-style beer, a "bottom fermenter" such as *Saccharomyces carlsbergensis* is used. The cells of this species settle to the bottom of the fermentation tank. Fermentation by *S. carlsbergensis* takes about 10 or 12 days at a temperature of about 8 or 10°C. On the other hand, if an ale is desired, the brewer selects *S. cerevisiae*, a "top fermenting" yeast. The less dense cells of this species are carried to the surface of the wort by carbon dioxide bubbles. Fermentation takes only about 6 days, and it is carried out at a higher temperature of about 20°C. If you prefer pilsner or lager beer, you are favoring a bottom-fermented beverage (**Figure 15.10**). Those preferring stouts, India pale ale, and the like are opting for a top-fermented product.

Either bottom or top fermentation yields an alcohol content of around 3.5–6%. Because carbon dioxide was allowed to escape during fermenta-

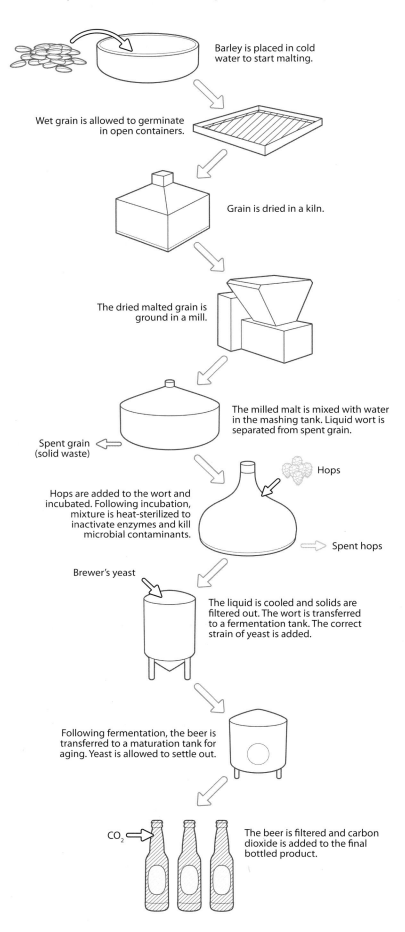

Barley is placed in cold water to start malting.

Wet grain is allowed to germinate in open containers.

Grain is dried in a kiln.

The dried malted grain is ground in a mill.

The milled malt is mixed with water in the mashing tank. Liquid wort is separated from spent grain.

Spent grain (solid waste)

Hops

Hops are added to the wort and incubated. Following incubation, mixture is heat-sterilized to inactivate enzymes and kill microbial contaminants.

Spent hops

Brewer's yeast

The liquid is cooled and solids are filtered out. The wort is transferred to a fermentation tank. The correct strain of yeast is added.

Following fermentation, the beer is transferred to a maturation tank for aging. Yeast is allowed to settle out.

CO_2

The beer is filtered and carbon dioxide is added to the final bottled product.

Figure 15.9 Beer brewing. An overall scheme for the production of beer. One or more of these steps may be somewhat modified, depending on the type of beer being made. In all cases, however, the barley used as a sugar source for fermentation must be *malted* or allowed to sprout. Once sprouted, the barley is dried, ground, and mixed with water in a process called mashing. During mashing, starch is converted to simple sugars, which can be used by yeast for fermentation. Solids are removed as waste, and hops are added to the remaining liquid (known as the *wort*) to impart flavor. The liquid is then heat-sterilized to kill contaminating organisms, and yeast is added. Following fermentation, the liquid is aged and yeast cells are removed. The remaining beer is ready for the addition of carbon dioxide, to provide carbonation, and final bottling.

Figure 15.10 Bottom and top fermentation. Bottom fermentation results in the lighter colored beers seen in the photo. Darker beers are brewed with top fermentation.

tion, CO_2 must be added to produce the beer's familiar carbonation. Remaining yeast cells or other sediments are allowed to settle, and any contaminating microorganisms are removed by either filtration or pasteurization. The beer is now ready for bottling or canning.

Liquor. Making a liquor such as bourbon or vodka is not much different from making beer. The main difference is the source of starch. Corn is the grain of choice when making bourbon, while scotch is primarily fermented from barley. A *single malt* scotch uses only barley. A *blended* scotch contains whiskey from fermented barley blended with another whiskey in which other grains, such as wheat or corn, were used. Rye whiskey, as the name suggests, contains a minimum of 51% rye. Vodka traditionally relied on potatoes as the initial starch source although it can be made from a variety of grains. Rum is made from fermented sugar cane or molasses, and tequila comes from fermented agave. Unlike beer and wine, liquors are **distilled** after fermentation is complete. This simply means that the fermented liquid is boiled, which concentrates the liquid, yielding a final beverage with a higher alcohol content and a higher concentration of the chemicals responsible for specific flavors.

Other fermented plant products. What do pickles, sauerkraut, and Greek olives have in common? Along with many other, sometimes less familiar foods, they are all examples of fermented fruits or vegetables (**Figure 15.11**). Unlike most of the fermentation we have discussed thus far, the fermentation of vegetables, or *pickling*, does not usually depend on adding a specific starter culture for fermentation. Rather, most pickling relies on bacteria that are naturally found on the surface of these plant products. Under the correct conditions, these bacteria produce a lactic acid fermentation waste that both prevents spoilage and gives the food its unique flavor.

Sauerkraut, which literally means "sour cabbage" in German, is made from shredded cabbage. The cabbage is placed in an anaerobic fermentation chamber, along with bacteria found naturally on the cabbage itself. Salt is added to eliminate Gram-negative bacteria, thus allowing the lactic acid-producing Gram-positive species to thrive. Some of the important fermenting species are *Leuconostoc mesenteroides*, *Lactobacillus plantarum*, and *Lactobacillus brevis*. As fermentation continues, the increased lactic acid production causes the pH to drop, eventually killing off the fermenting organisms. By the time the pH drops to about 3.5, all microorganisms have been killed, and the sauerkraut is ready to eat. The low pH ensures that this food product is slow to spoil, because few contaminating microorganisms can tolerate an environment that is so acidic.

Pickles result from cucumber fermentation. Cucumbers are placed in a brine (i.e., salt solution) and allowed to ferment. The natural bacterial flora

Figure 15.11 Fermented vegetables found in different cultures. (a) Kimchi, popular in Korea, is composed of cabbage and other vegetables, fermented by lactic acid-producing bacteria. (b) Traditional Hawaiian method of pounding taro root into poi. The mashed material is then allowed to ferment to yield the final product.

of the cucumbers is sometimes used, but the brine can also be inoculated with specific fermenting species. Different styles of pickle can be produced by adding dill, garlic, or other flavorings to the fermentation vat. Olives, as well as other fruits and vegetables, are cured in a similar way.

Flavoring agents. Several microbial products have earned a spot in our kitchen as flavor enhancers or as vital parts of many recipes. Vinegar and soy sauce are two of the most commonly used. Additionally, the flavor of many commercially produced foods is at least in part due to citric acid, also produced by microorganisms.

Vinegar is an acetic acid solution used on everything from salad to spaghetti. It has probably been around for thousands of years, ever since the first cask of wine was spoiled by contaminating bacteria. In fact, the name "vinegar" comes from the French *vin* ("wine") and *aigre* ("sour"). The acetic acid is produced by bacteria of the genus *Acetobacter*. These strictly aerobic, Gram-negative bacilli convert the ethyl alcohol produced by yeast fermentation into acetic acid by oxidizing the alcohol. In other words, the bacteria used to make vinegar are not themselves fermenters. Rather they convert a fermentation product, ethyl alcohol, into a different molecule with a distinctly different taste.

Ancient Greeks and Romans made vinegar simply by introducing air into wine casks. Today, one vinegar-making technique involves allowing the bacteria to grow on wood shavings as a biofilm (**Figure 15.12**). A fermented liquid containing alcohol is sprayed onto the bacterially coated wood, and as the alcohol seeps through the shavings, the bacteria oxidize it into acetic acid. Different types of vinegar are made by selecting different sources of alcohol. Any fermented liquid that contains ethyl alcohol is suitable; some very common sources are fermented apple juice (to make cider vinegar), beer (malt vinegar), or wine (wine vinegar).

Soy sauce is an important part of many Asian dishes. To make it, a mixture of cooked soybeans and roasted wheat is inoculated with a mold of the genus *Aspergillus*. The fungus digests the proteins and carbohydrates in the soybeans and wheat, resulting in a solution full of simple sugars and amino acids. The mixture is then mixed with a salty brine, and salt-tolerant bacteria and yeasts are introduced. After up to a year of fermentation, the liquid soy sauce is removed and is ready for use.

If you think the price of gasoline is high, be thankful that a soda still costs only a dollar or so. It would cost a lot more without microbially produced

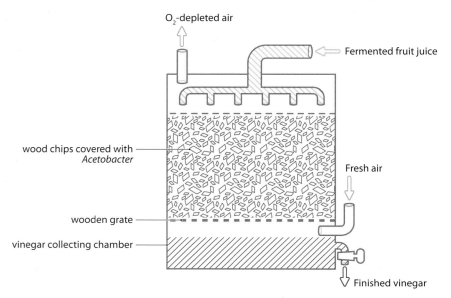

Figure 15.12 A vinegar generation tank. Aerobic bacteria of the genus *Acetobacter* convert ethyl alcohol into acetic acid (vinegar) in the presence of oxygen. Although vinegar production starts with a fermentation waste (ethyl alcohol), the bacteria involved in vinegar generation are not themselves fermenters. The bacteria are allowed to grow on wood chips as biofilms inside the generation tank. The chips are sprayed with a fermented liquid containing ethyl alcohol, and as the liquid percolates through the wood chips, the bacteria growing on the chips convert the alcohol to vinegar. The finished product is collected at the bottom of the tank. Different fermented liquids can be used to produce different types of vinegar.

citric acid. In addition to soft drinks, citric acid is also used to flavor candy, jams, jellies, ice cream, and other foods. Citric acid was originally obtained from citrus fruit, but in 1923, a much cheaper microbial fermentation technique was developed. Most citric acid is produced by allowing the fungus *Aspergillus niger* to ferment the sugars in molasses. The sucrose in the molasses is first digested to glucose and fructose, which are then metabolized to citric acid. More than 130,000 tons of citric acid are produced annually around the world; most of it is used to enhance the flavor of commercially produced foods.

Fermented milk is the basis of making cheese and yogurt

One of the enjoyable things about international travel is trying out new and exotic foods, and some of the more exotic are made from fermented milk. If in Russia, for example, you may want to sample koumiss, or fermented mare's milk. The same beverage, called *airag* in Mongolia, was popular at the court of Genghis Khan (**Figure 15.13**). If, on the other hand, your travels take you to Kazakhstan, don't miss out on the chance to try fermented camel's milk, or *shubat*. Of course, not all fermented dairy products are so out of the ordinary, and here we will discuss a few of the more familiar ones that can be enjoyed without using your passport.

Yogurt. The production of yogurt was described in Chapter 7 (see p. 197). Briefly, yogurt is made by allowing milk to ferment. The milk is made thicker by either removing water or adding powdered milk. Two types of fermenting bacteria are then added. *Streptococcus thermophilus* releases lactic acid, giving the yogurt its tart taste. The acids also denature proteins in the milk, converting it to a semisolid. *Lactobacillus bulgaricus* releases lactic acid as well, but it also produces aromatic compounds that give yogurt its distinct smell.

Cheese. As mentioned earlier, cheese has been a part of the human diet for thousands of years. Although there are hundreds of varieties, cheeses can easily be classified into one of a few groups based on moisture content (**Table 15.1**). Any type of milk can serve as a starting material, and today there are cheeses made from the milk of cows, goats, sheep, horses, and camels.

The first step in cheese manufacture is the conversion of milk into a semisolid **curd**. Traditionally, this is done by inoculating the milk with lactic acid-producing bacteria. As described above for yogurt, the released acid causes the pH to drop, thereby denaturing milk proteins and causing the curd to develop. Today, when cheese is made commercially, the curd is formed by adding a clotting enzyme called rennin to the milk. Rennin was originally obtained from the stomachs of calves. Today rennin is synthesized by bioengineered microorganisms.

Once the milk has clotted and the curd has formed, it is separated from the remaining liquid, called **whey**. The pH, as well as salt and moisture content of the curd, is adjusted and the curd is physically molded into a desired shape. At this point, some cheeses are essentially ready to eat. Cream cheese and cottage cheese, for instance, are soft, high-moisture cheeses that have a short shelf life. No additional aging is required. Many other cheeses must undergo an additional aging process, in which various bacteria and molds are essential. Appropriate microorganisms are generally allowed to colonize the surface of the curd. Enzymes produced by the organisms seep into the curd, digesting carbohydrates, lipids, and proteins. Various by-products remaining behind give the cheese its unique flavor, aroma, and texture. To make Camembert, a soft cheese, for instance, the mold *Penicillium camemberti* is allowed to coat the surface of the curd. Protease enzymes produced by the mold give the cheese its creamy consistency. Typically, a soft cheese requires only a few weeks of aging. Blue cheese, which is somewhat less soft,

Figure 15.13 Mare's milk for airag. In central Asia, horse milk, used to produce fermented milk products, is often still obtained in the traditional way.

TYPE OF CHEESE	EXAMPLES
soft	brie, Camembert, mozzarella
semisoft	blue, Muenster, Roquefort
hard	Cheddar, Colby, Gouda, Swiss, Edam, feta
very hard	Parmesan

Table 15.1 Cheese variety. A convenient way to classify cheeses is by their moisture content. Softer cheeses, which have a higher moisture content, are aged a relatively short time, while hard cheeses, relatively low in moisture content, may be aged up to a year or more.

takes several months to age properly. The "blue veins" that form are due to the growth of *Penicillium roqueforti*, used to inoculate the curd (see Figure 1.10b).

Hard cheeses may be aged for a year or more. They also owe their particular flavor to aging with specific microorganisms. Swiss cheese, for example, owes its easily identifiable sharp, nutty taste to *Propionibacterium*. Not only do these bacteria alter the cheese's flavor, but trapped carbon dioxide gas released during fermentation forms the characteristic holes.

Certain meat products, including salamis and cured hams, require fermentation

Compared with plant and dairy products, the list of fermented meats is short. Summer sausage, as well as salami and country-cured hams, are examples of meat products involving fermentation. When sausage or salami is made, ground meat is inoculated with commercially available fermenting bacteria. Cured hams rely on fungi that naturally colonize the surface of the ham. For the more adventurous, you might wish to try fermented seal flipper, the next time you find yourself north of the Arctic Circle. While you are at it, sample a fermented fish head (locally known as "stinky head"). But be careful; Alaska has the highest rates of botulism in the United States, for a reason. The anaerobic environments created to encourage fermentation of such local delicacies can also permit the growth of *Clostridium botulinum* (see Chapter 10, p. 275).

Production of Other Foods and Dietary Supplements

Coffee beans are readied for roasting through the use of bacteria

Microbial involvement in other aspects of our diet is more subtle. Coffee, for instance, owes little or nothing of its taste or aroma to microorganisms, but the beans are processed with microbial assistance. Before coffee beans can be roasted, the outer pulplike coating of the bean must be removed. Much of this pulp is composed of the carbohydrate pectin, and although it can be removed in other ways, the bacterium *Erwinia dissolvens* is especially well suited to the task. This species produces a pectin-digesting enzyme, permitting easy pulp removal.

Many common dietary supplements are produced by microorganisms

It is certainly not uncommon for us to supplement both our own diet and that of animals with any number of nutritional aids. Microorganisms figure prominently in the production of some of these supplements. Here we consider a few of the more important ones.

CASE: MICROBIAL METABOLITES FOR A FEATHERED FRIEND

When Judy's scarlet macaw Vladimir suddenly becomes listless and develops diarrhea, she becomes very concerned. After all, Vladimir has been a member of the family for years and he has always been very active and healthy. Judy assumes that Vladimir must have some sort of infection, and she takes him to the veterinarian. The veterinarian begins by giving Vladimir a complete examination. He allows the bird to grip his finger with his claws and notes that Vladimir is very weak. He also notices that one of Vladimir's toes is twisted or curled out in an unnatural manner. When he asks about the bird's diet, Judy responds that he mainly eats a commercial parrot food available at the supermarket. To Judy's surprise, the veterinarian says that he does not think Vladimir has an infection. His diagnosis is a condition called "curly toe," which is treated by increasing the bird's dietary vitamin intake. Riboflavin is especially necessary because very few commercial bird foods contain enough of this vitamin. Judy thanks the veterinarian profusely, and within a week, Vladimir seems to be doing much better. After a few more weeks he appears to be completely recovered.

1. What exactly is riboflavin, and why is it so crucial for birds?
2. How do microorganisms contribute to the commercial production of riboflavin?

Vitamins. Pernicious anemia, characterized by low numbers of red blood cells, is caused by a lack of vitamin B12. Human cells are unable to synthesize this essential vitamin, and consequently, we obtain most of our vitamin B12 from our food. Some bacterial strains, on the other hand, are able to make vitamin B12 in abundance, and some of these organisms are part of our normal intestinal flora. We therefore go through life with our vitamin B12 needs met at least in part by our symbiotic microbial partners. Vitamin B12 is also produced commercially, for both human and animal consumption, and much of this supply is synthesized by bacteria of the genera *Pseudomonas* and *Propionibacterium*. To ensure high levels of vitamin B12 production, these microorganisms are grown on media supplemented with cobalt, which is an essential structural part of the vitamin. Vitamin B12 is regularly added as a supplement to hog and poultry feed, or it can be bought over the counter in any health food store or pharmacy. It is also commonly added to foods such as bread and breakfast cereal.

Vladimir's problem was principally due to a deficiency of riboflavin (vitamin B2), which is also produced commercially by microorganisms. Animals require riboflavin to form enzymes essential for carbohydrate metabolism. In humans, a riboflavin deficiency can cause eye problems or lesion formation on the skin or in the mouth. Domesticated animals such as horses may show similar signs if they are fed a low-quality diet.

Birds are especially vulnerable to riboflavin deficiency. Only a few commercially available bird foods contain enough riboflavin to meet nutritional requirements, especially for growing chicks. In some long-lived birds such as macaws, riboflavin deficiency can take years to develop. Symptoms can then appear quite rapidly. Vladimir's symptoms, including listlessness, diarrhea, and the curled toe, are typical. In serious cases, there is significant neurological damage. Fortunately, as Judy found out, the problem can often be remedied by adding sufficient riboflavin to the diet, although complete recovery may take several months. Like vitamin B12, riboflavin is often added to foods such as bread and cereal. Much of the commercially available riboflavin is synthesized by yeasts such as *Ashbya gossypii*.

Amino acids. Commercially produced amino acids are used as additives to enhance nutritional quality and flavor. The low-calorie sweetener NutraSweet® is also made synthetically from amino acids. The Japanese have led the way in the development of microbial techniques to produce amino acids. Large amounts of lysine, for instance, are produced by bacteria of the genus *Brevibacterium*. Lysine is an example of an *essential* amino acid. Humans are unable to synthesize lysine, and it must therefore be included in our diet. This can be a particular problem for vegetarians, since many of the best lysine sources are animal products. Consequently, lysine is often added to plant-derived foods to improve their nutritional quality. It can also be bought as a supplement in health food stores. Glutamic acid is also produced by bacteria in large quantities. This amino acid is used to make the flavor enhancer monosodium glutamate (MSG), common in Asian cuisine.

Probiotics. In Chapter 10 we reviewed the use of Preempt®, a solution containing normal flora that is sprayed on baby chickens. As the chicks preen, they ingest these bacteria, establishing a healthy intestinal flora that can prevent colonization by pathogens such as *Salmonella* (see Figure 10.2a). Preempt® is a good example of a **probiotic**, a product consisting of living bacteria that promotes overall health.

But probiotics are no longer just for chickens. Indeed, the development of probiotic products for humans is a rapidly growing part of the health food industry. Many of the claims made by probiotic advocates have yet to be substantiated by scientific trials, and if probiotics are in fact beneficial, in many cases it is still not entirely understood how they work. As discussed in Chapter 10, the normal flora can outcompete and inhibit the growth of pathogens. As with vitamin B12, it often synthesizes nutrients that the host itself is unable to synthesize, and it aids in the digestion of certain foods. Some normal flora organisms may have anticancer properties. The microorganisms found in probiotics may do any or all of these things.

Time to Eat: A Few Microbial Recipes

If you would care to try your hand at making some of the foods discussed in this chapter, here is your chance. Some of the following recipes require little or no special equipment. For others, you may need to pick up a few supplies. All of them involve microorganisms in some capacity.

Sauerkraut

Let's start with an easy one. As described earlier, sauerkraut is fermented cabbage. The fermentation relies on Gram-positive bacteria already found on the surface of the cabbage leaves (**Figure 15.14a**). Salt is added both for taste and to inhibit the growth of undesirable Gram-negative species.

Ingredients
cabbage
salt

Preparation
1. Wash the cabbage and cut it into quarters. Remove the core, and shred the cabbage finely with a sharp knife.
2. Place a layer of cabbage in a wide-mouthed jar or crock, sprinkle with salt, and press down firmly. Continue to add additional layers of cabbage and salt until the jar is full.
3. Cover the top with a clean cloth, and place a plate on top. Add a weight to the plate to hold it down. Place the jar in a warm spot to ferment. After a few days, remove the froth on the top of the cabbage, replace the cloth, plate, and weight, and allow to stand for another 3 days. Then repeat the process.
4. Move the jar to a cool spot for about 2 weeks. Then enjoy fresh, old-fashioned homemade kraut. To continue the microbial theme, you may wish to serve with summer sausage and a German lager.

Pickles

Pickles are also easy to make. There are, of course, many different varieties of pickles, and the following is just one example. If you like this recipe, you might try different variations of the brine solution next time. To increase the quantity, you need only adjust the amounts of all ingredients. In this basic recipe, we simply rely on microorganisms normally inhabiting the surface of cucumbers to do the fermenting (see Figure 15.14b).

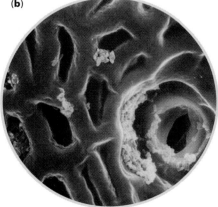

Figure 15.14 Home fermentation.
(a) Homemade sauerkraut is tasty and easy to make. (b) Like sauerkraut, pickles can be made by use of fermenting organisms found naturally on the surface of the plant. The electron micrograph shows *Lactobacillus plantarum* growing naturally on the surface of a cucumber.

Ingredients
1 tablespoon (tbsp) salt per quart of water
1 clove garlic
1 stalk sliced celery
1 sliced carrot
1 chili pepper
1 bay leaf
1 bud dill
1 lb (0.45 kg) washed pickling cucumbers

Preparation
1. Combine all the ingredients except the cucumbers in a pot and boil. Following a brief boiling, allow the pickling solution to cool.
2. Puncture the cucumbers with a fork and add to the pickling solution. Add to wide-mouthed jars and seal tightly.
3. Let stand for 3 weeks, and enjoy.

Bagels

As an example of a homemade bread, you might try bagels, the only type of bread in which the dough is boiled before it is baked (**Figure 15.15**). Bagel making is a bit more complicated than the recipes presented so far, but you still will need no or few special utensils, and all ingredients are readily available. As with most breads, *S. cerevisiae* (baker's yeast), once in the anaerobic environment of the bread dough, will perform the necessary fermentation.

Ingredients
7.6 fluid ounces (225 mL) scalded milk (boiled and hot)
2 ounces (50 g) butter
1 ounce (25 g) sugar
2 teaspoons (tsp) dried baker's yeast
½ tsp salt
1 egg white (medium-sized egg)
14 ounces (400 g) plain flour

Preparation
1. Put the milk, butter, and sugar into a mixing bowl and mix thoroughly. The milk should still be hot enough to melt the butter completely.
2. When the mixture has cooled and is now lukewarm, sprinkle on the yeast and allow to sit for 45 minutes. Then stir in the egg white and salt and add the flour, kneading it into a soft dough. Knead the dough until it is smooth and elastic.
3. Place the dough in a large plastic bowl with a snap-on lid, or seal dough inside the bowl with plastic wrap. Allow the dough to rise until it has doubled in size. This should take about 1 hour. Then knead the dough briefly to remove large gas bubbles.
4. Divide the dough into about 15 lumps, each about the size of a tennis ball. Moisten your finger and cover it with flour. Then drive your finger through each lump to make the bagel hole. Twirl the bagel around your finger on the work surface until the hole is at least one third the diameter of the entire bagel.
5. Place the bagels on a tray and set in a damp place. To prevent drying, you may place a rack over the bagels, and lay a damp towel over the rack. Allow them to rise until they look puffy. This takes only an additional 10 or 15 minutes. Do not let them double in size. Preheat your oven to 390°F or 200°C. While the oven is warming up, prepare a large shallow pan of gently boiling water.
6. Place the bagels in the water a few at a time so that they don't touch each other. They will float about halfway submerged in the water. Leave them for 15 to 20 seconds, then remove with a slotted spoon. Arrange them on a greased baking tray.

Figure 15.15 The bread with the hole. Bagels were supposedly invented in the late 1600s by an Austrian baker. He wanted to honor Austria's king, and because the king was an enthusiastic equestrian, the baker made his gift in the shape of stirrups (German, *bügel*).

7. Bake for about 20 minutes, until they are golden brown and have a hollow sound when you tap them. Allow them to cool, and serve with your favorite topping.
8. Once you've mastered the ABCs of bagel baking, try adding other ingredients such as chopped onions or raisins to the dough after kneading.

Yogurt

Homemade yogurt is easy to prepare, and it makes a delicious and nutritious breakfast or snack. Here are the basics. What you add in terms of fruit flavoring is up to you.

To ferment the lactose sugar in the milk, we will require two types of bacteria, both of which will release lactic acid as a fermentation waste. This acid gives yogurt its tart flavor, and it also converts the liquid milk to the semisolid yogurt as it denatures the milk proteins. To obtain the required *S. thermophilus* and *L. bulgaricus*, we will simply use a small amount of plain yogurt with active cultures as a starter culture. Once fermentation begins, *S. thermophilus* initially digests the lactose. As levels of lactic acid rise and the pH drops, bacterial growth is inhibited. The more acid-tolerant *L. bulgaricus* then takes over, fermenting the remaining milk sugar. As the pH continues to drop, the growth of other contaminating microorganisms that might otherwise spoil the yogurt is inhibited. *L. bulgaricus* also releases certain aromatic chemicals, such as acetaldehyde, that give yogurt its pleasing bouquet.

Ingredients
1 quart (about 1 L) whole milk
2 tbsp natural yogurt with live cultures
1/3 cup nonfat dried milk

Preparation
1. Be sure that all containers and utensils are thoroughly cleaned before use to prevent contamination. The milk must first be sterilized by boiling (remember that even pasteurized milk contains some microorganisms).
2. Add about 1 ounce (20 g) of the milk powder to 3.5 fluid ounces (100 mL) of water. Add this to the whole milk in a cooking pan. Heat the milk to boiling, allowing it to boil for 1 minute (no more), and remove from the heat.
3. Place a kitchen thermometer in the milk and allow the milk to cool until the temperature falls to between 100 and 89°F (38 and 32°C). Pour the cooled milk into ½ quart (½ L) jars. Canning jars with screw-on lids make good yogurt containers.
4. Mix about 0.5 ounce (14 g) of the plain yogurt starter culture into the milk. Cap the jar with its lid. Place the jar in a warm oven (between 100 and 89°F or 38 and 32°C). Do not disturb the jar for 3 hours. Then check every hour for yogurt to form. This may take half a day. If you move or shake the container during the incubation process, the yogurt may not coagulate properly.
5. When the milk has thickened, remove the jar of yogurt from the oven. Taste it, and if a tarter flavor is desired, allow the yogurt to incubate for another hour or so.
6. Add fruit or flavoring, such as vanilla extract, as desired. Once the yogurt is completed, it can be refrigerated. It should be eaten within the next 4–5 days.

Greek feta cheese

Feta is a popular semihard cheese that features prominently in Greek cuisine (**Figure 15.16**). It is also a reasonably easy way to introduce yourself to

Figure 15.16 Feta cheese. Traditional feta cheese, made with goat's milk, is used widely in Greek cooking.

the ancient art of cheesemaking. Although feta cheese can be made with whole cow's milk, here we will stick with tradition, relying on goat's milk.

Cheesemaking requires more ingredients and effort than the recipes provided so far. All ingredients, however, are available from gourmet shops, natural foods outlets, or online cooking supply companies. To ferment the milk sugars we will require *Lactococcus lactis*, most easily obtained as the commercial product Mesophilic-A. The clotting enzyme rennin, available commercially as liquid rennet, and calcium chloride are also necessary.

Ingredients
2 gallons (7.6 L) pasteurized whole goat milk
¼ tsp Mesophilic-A
½ tsp calcium chloride
½ tsp liquid rennet
2 tbsp cheese salt

Preparation
1. Warm the milk to 85°F or 30°C in a large pot. In a separate dish, mix the calcium chloride (½ teaspoon) with 2 tablespoons of cool distilled water. Add this mixture to the milk and stir gently for 30 seconds. Maintain the milk's temperature at 85°F (30°C).
2. Add ¼ teaspoon of Mesophilic-A and stir gently. Then, allow the milk to stand at 85°F (30°C) for 1 hour.
3. Add ¼ teaspoon of liquid rennet to 4 ounces (120 mL) of cool distilled water. Stir in gently. Cover and allow the milk to sit undisturbed for 30 minutes or until the milk forms a solid curd that shows a clean break. If, after 30 minutes, the consistency is more like that of soft yogurt, continue to wait until the solid curd forms.
4. While leaving the curd in its pot, cut the curd into ½-inch (1.25 cm) slices all the way across the pot. Rotate the pot 90° and repeat, as if forming a checkerboard of thin slices.
5. Stir the curd slowly every 10 minutes for 1 hour, gradually heating the pot to 95°F (35°C). After 1 hour of slow cooking, drain the liquid portion (the whey) by pouring the contents of the pot into a colander. Allow to drain for 1 hour. The curd will form back into a solid mass.
6. Cut the curd into small blocks and turn it over in the colander. Allow to drain an additional 30 minutes. Add salt to taste.
7. Pack the curd into a sterilized quart jar. Prepare a brine of 2 cups distilled water and 2 tablespoons cheese salt. Pour the brine over the curd and seal tightly. Refrigerate and allow the jar to sit for at least 2 weeks. The longer it ages, the better it tastes.

Once you are ready to try your cheese, make yourself a Greek salad, complete with other fermented foods such as Greek olives, and a dressing made with balsamic vinegar.

Vegemite sandwich

Last but not least, here is your introduction to this Australian staple (**Figure 15.17**). Although real Aussies love Vegemite, the spread is an acquired taste, and if you are trying it for the first time, it might be a good idea to have plenty to drink on hand, to quickly cleanse your palate of the "distinctive" taste.

Preparation
1. Spread two pieces of your favorite bread with butter or margarine. Either toasted or untoasted bread is appropriate.
2. Cover the buttered bread with a thin layer of Vegemite. Only a small amount is necessary. Some connoisseurs like to mix the Vegemite into the butter before spreading. Put the two slices together and give it a try. If you actually like it, consider yourself "fair dinkum."

Figure 15.17 Down under cuisine. An essential part of Australian culture.

Coming Up Next...

Food production is just one area in which humans have enlisted microbial help. A variety of other industrial processes also rely on bacteria and fungi. In the next and final chapter, we will take a look at the way microbial metabolism is harnessed in a range of industries and aids in the production of everything from detergents to pesticides to pharmaceuticals. Microorganisms can also be part of pollution control, and we will investigate the rapidly expanding field of bioremediation. Some of the microorganisms that participate in industry are products of genetic engineering. Some have been around all along. Both types impact our lives in many and often surprising ways.

Key Terms

curd	probiotic
distillation	single-cell protein
malting	whey
must	wort
oenology	

Concept Questions

1. We learned that the cyanobacterium *Spirulina*, if raised for food as single-cell protein, could produce up to 10 tons of protein per acre. For wheat and beef, the corresponding figures are 0.16 tons per acre and 0.016 tons per acre, respectively. Recalling what you learned in Chapter 9 about the way that energy moves through an ecosystem, explain the above observation.

2. After drinking a portion of a bottle of wine, you recork the bottle and save the rest for another time. A few months later, you pour a glass from this bottle and find it to be undrinkable. What do you think happened microbiologically during the interim?

3. Go back through the chapter and find three particular species of bacteria or fungi used in a food-making process. Are the involved microbes respiring anaerobically or aerobically during the food-making process?

4. In Chapter 7 we discussed the concept of the growth curve as it applies to microorganisms. Imagine that you introduced a population of yeast cells into grape must, for the purpose of making wine. Then imagine how those yeast cells would grow as they fermented the sugars in the must. What factors would cause the population to leave the log phase of growth and enter the stationary phase? The decline phase?

5. In this chapter we discussed the use of "probiotics," or foods containing live microorganisms to provide certain benefits. After having also read Chapter 14, on biotechnology, can you think of any specific way that microorganisms in probiotics could be genetically modified to provided enhanced benefits?

6. In our description of how sauerkraut is made, it is stated that salt is added to eliminate any Gram-negative bacteria. Why do you think it is necessary to eliminate these bacteria?

7. Imagine you are planning a dinner party. Prepare a sample menu highlighting dishes in which microbes played a crucial role. Next to each menu item, specify the microorganism involved and what that microorganism contributed.

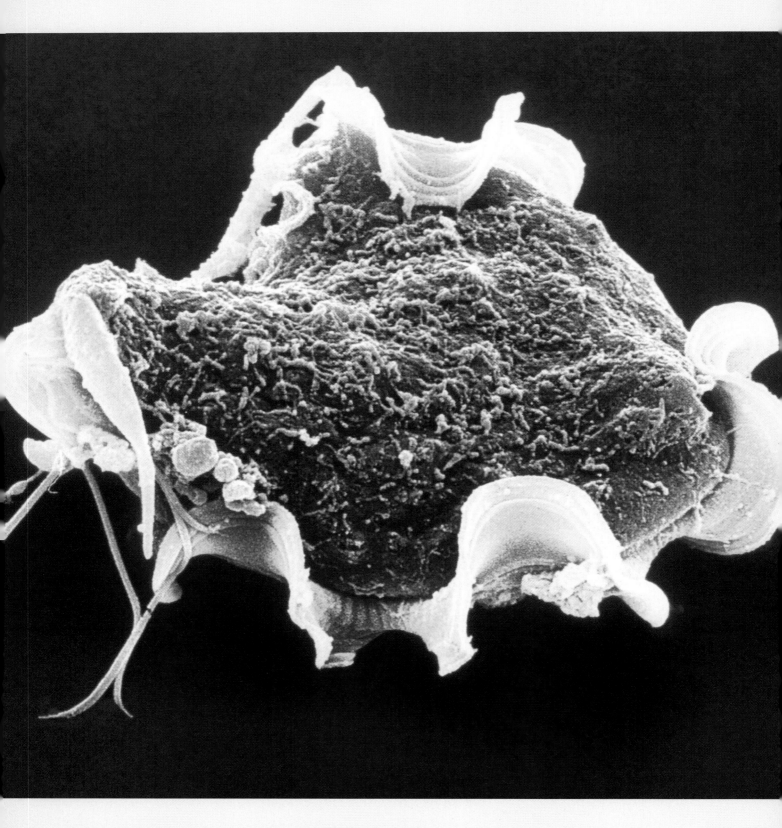

Trichomitopsis protozoa, microbes that inhabit the guts of termites and help digest wood, may one day be used to produce commercially useful enzymes.

Chapter 16

Better Living with Microorganisms: Industrial and Applied Microbiology

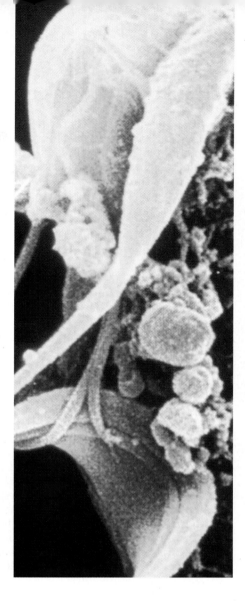

We all know that air and water pollution are harmful for living things. Less obvious is the adverse impact pollution can have on some of our most beloved historical monuments. In large cities around the world, acid rain and urban pollution are compounding the consequences of natural erosion, turning architectural masterpieces into dust. One of the principal problems is the effect that contaminants in the air and rain have on the calcium carbonate found in marble, limestone, and other types of ancient building materials. As pollution dissolves the calcium carbonate, the structures slowly degrade, until previously intricate details erode away and become featureless lumps (**Figure 16.1a**).

One solution to this problem may come from a surprising microbial source. Carlos Rodriguez-Navarro of the University of Granada in Spain has enlisted the help of a common soil bacterium, *Myxococcus xanthus*, to reverse the degradation of Granada's Royal Chapel—the burial spot of Spanish kings and queens (see Figure 16.1b).

Rodriguez-Navarro has found that *M. xanthus* secretes crystals that bind together the grains found in limestone, dolomite, and marble. This bacterial secretion can substitute for the lost or degraded calcium carbonate, and help hold the stones together. When pieces of limestone from the Granada cathedral were placed in a liquid broth containing *M. xanthus*, the crystalline material secreted by the bacteria seeped into the stone, significantly strengthening it within 2 weeks. Just as important, unlike the resins currently used to protect stonework, the bacteria do not clog pores in the stone, which would later trap water, accelerating decomposition. Field testing of the technique will soon begin by either spraying the bacteria on the cathedral or wrapping the buildings in bacteria-impregnated sheets. If the technique works, it might provide a low-cost way to protect other monuments around the world.

This is just one of the ways in which microorganisms offer practical solutions to real-world problems. Microbes are involved in the production of everything from pharmaceuticals to plastics to pesticides. They offer possible solutions to our energy needs, and they can be used to degrade toxic waste or to clean up polluted water. In this chapter we will see how microorganisms are used in industry and survey some of the ways that the industrial use of microorganisms affects our daily lives. We will first look at how microorganisms are utilized in industrial and commercial processes to produce valuable products such as antibiotics, insecticides, or enzymes with industrial applications. We will then turn our attention to some of the ways that microorganisms may be harnessed to clean up toxic sludge, put fuel in our gas tanks, or solve other sticky problems.

Commercial Applications

Industrial microbiology began with the use of microorganisms for food as described in Chapter 15. Once the role of microorganisms in processes such as fermentation was fully understood, it was discovered that many microbial metabolic products had useful, nondietary applications. In Chapter 5, for instance, we discussed how bacterially produced acetone played an important part in World War I. In Chapter 12 we discovered that many of the first antimicrobial drugs came from microorganisms themselves. With the commercialization of these and other products, the modern discipline of industrial microbiology was born. In the past 25 years, the biotechnology revolution, as described in Chapter 14, has opened the door to a host of new potential industrial applications for microorganisms. Yet whether the microorganism in question is a common soil fungus or a genetically engineered bacterium expressing genes that were introduced from another organism, many of the basic considerations for finding, cultivating, and utilizing the organism are the same. We will begin by reviewing a few basic concepts and examining some of the problems that must be dealt with before a microorganism is ready to "go to market."

CASE: UNEARTHING A NEW ANTIBIOTIC

Sam, a student in a microbiology laboratory course, is instructed to find a microorganism in the environment with potential commercial or industrial applications. He decides to screen soil samples taken from his garden for bacteria or fungi that produce new antibiotics. He first prepares bacterial lawns (see Chapter 1, Figure 1.10a) of *Escherichia coli* (a Gram-negative species) and *Staphylococcus epidermidis* (a Gram-positive species). He then places small amounts of soil on each of his media plates. The plates are incubated for 24 hours at 35°C. When Sam inspects his plates, he finds that there are small, clear "halos" without bacterial growth surrounding some of the soil particles on the plates inoculated with *S. epidermidis*. There are no such bacteria-free zones surrounding the soil particles on the plates inoculated with *E. coli* (Figure 16.2). On the basis of these results, Sam concludes that his soil sample contains a microorganism of some type that secretes a compound with antibiotic properties, which is apparently more effective against Gram-positive organisms.

1. **Why would soil microorganisms produce antibiotics?**
2. **Might such an antibiotic be of commercial or medical value? If so, what other factors would have to be considered, and what other steps would have to be followed, before this or any newly discovered, potentially valuable microbial product could be deemed useful?**

Figure 16.1 Bacteria to the rescue? (a) Air pollution and acid rain can wreak havoc on the stonework of historical buildings. (b) Granada's Royal Chapel, or Capilla Real, where a new strategy to control pollution-induced erosion with a common soil bacterium will be put to the test.

Halo that is free of bacterial growth indicates that this soil particle contains microorganisms that secrete an antibiotic effective against *S. epidermidis*.

Bacteria are able to grow up to the soil particles, indicating no antimicrobial activity against *E. coli*.

soil particles

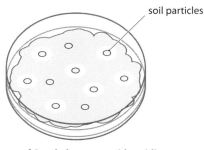

Lawn of *Staphylococcus epidermidis*

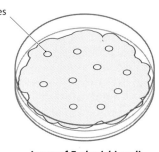

Lawn of *Escherichia coli*

Figure 16.2 Discovering a new antibiotic. Some of the soil particles in the plate on the left are surrounded by zones of inhibition or halos where the bacteria are unable to grow. Microorganisms in these soil particles are secreting an antibiotic that inhibits growth. No such activity is seen in the plate on the right.

Many microorganisms produce metabolites with commercial potential

As we saw in Chapter 7, all living things, including microorganisms, carry on their life processes through numerous metabolic pathways. Each of these pathways consists of a number of steps, in which a starting material is transformed into a series of intermediate compounds and is ultimately converted into one or more final products. These intermediate compounds or final products, produced as a consequence of metabolic activity, are called **metabolites**. Many metabolites have useful applications. Antibiotics are just one important example. In the experiment described in our case, the halos that Sam observed suggest that a soil organism of some sort is releasing a metabolite that is inhibiting the growth of the Gram-positive bacteria. Microorganisms often produce such metabolites because, as we learned in Chapter 9, competition between soil organisms can be especially fierce. Many microbes consequently produce such compounds to inhibit the growth of potential rivals. Other metabolites are likewise produced by microorganisms as a consequence of their normal metabolic activity. The goal of the industrial microbiologist is to identify metabolites that may have practical applications.

To find microbes that produce metabolites with commercial potential, microbiologists have traditionally scrutinized the natural environment, isolating organisms from soil, water, spoiled food, plants, or animals. These organisms are then screened for potentially valuable metabolites, in a manner similar to what Sam did in his hunt for new antibiotics. Although such searching (or, as it is sometimes called, "bioprospecting") has been in progress for years, we have literally only scratched the surface in this regard; most environmental microorganisms remain unclassified and unstudied. Undoubtedly, efforts to find useful microorganisms in the environment will continue because the potential payoff is enormous.

Microbial production of citric acid provides an excellent example of how careful (and sometimes simply lucky) bioprospecting can yield colossal profits. Citric acid, common in many food products (see Chapter 15, p. 406), is also used as a preservative in stored blood and in medicinal ointments. Citric acid is also used in detergents, in which it often replaces phosphates, which are a known water pollutant (see Chapter 9, p. 241). Originally, all citric acid was obtained from citrus fruit, principally from Italy. However, political instability in Italy and the advent of World War I prompted the search for alternative sources of the lucrative compound. In 1917 scientists at Pfizer discovered that *Aspergillus niger*, a common fungal food contaminant, produced citric acid under appropriate environmental conditions. By 1929, techniques had been developed for the large-scale industrial production of citric acid. Suddenly, this valuable industrial product could be produced at a small fraction of the original cost, propelling Pfizer to the forefront of industrial microbiology. The company was well-positioned to use a similar technology in the 1940s to mass-produce the antibiotic penicillin from the fungus *Penicillium notatum* (also known as *Penicillium chrysogenum*). Pfizer's place as one of the world's largest pharmaceutical manufacturers was now assured.

Microorganisms producing promising metabolites must often be subjected to strain improvement

But the odds of bioprospecting success are also long. As Sam's experience demonstrates, finding an organism that produces a useful substance is not necessarily difficult. Finding one with real industrial potential is not so easy. First of all, most potentially useful metabolites are made in only minute amounts. Microorganisms generally tightly regulate their metabolism through processes such as induction and repression, discussed in

Chapter 6 (see p. 151). Strains that overproduce an enzyme or metabolite tend to be selected against because they waste energy by producing substances that are unnecessary. Random mutations ensure that overproducing strains are occasionally found, but as a rule, any strain that has industrial potential requires **strain improvement**—a process of increasing the production of the metabolite in question far beyond what a naturally occurring strain would synthesize. If Sam wishes to pursue his project further, his next task would be to isolate his soil microorganism in pure culture and subsequently subject it to strain improvement.

How are strains improved? There are several strategies that might be employed. One common technique is to subject a promising strain to mutagens such as ultraviolet (UV) light to induce random mutations. Many of these mutations will be of no significance, and some will be harmful to the microorganism. Occasionally, however, a mutation will occur that results in overproduction of a particular metabolite. Such mutations often disrupt regulatory mechanisms that ordinarily limit metabolite production. The microbiologist then uses a process of artificial selection (see Chapter 8, p. 217) to isolate and select for the mutated organisms of interest. The isolated organisms are then grown on fresh medium. Additional random mutations make it likely that among these cells, some will produce even higher levels of the metabolite, and those organisms are once again isolated. Repeated rounds of such selection can result in strains that produce remarkably high levels of a particular substance. For example, the original strains of the *Penicillium* mold, from which penicillin was first discovered, produced about 5 mg/L of the antibiotic. By artificially selecting those strains that produced the most penicillin, scientists raised the yield to 60,000 mg/L—a 12,000-fold increase. Likewise, in Chapter 15, we learned about the mold *Ashbya gossypii*, utilized to produce riboflavin (vitamin B2). Selection in the laboratory for overproduction has resulted in strains that synthesize approximately 20,000 times the amount of the vitamin that the microorganisms need to meet their own metabolic needs.

In some cases, metabolite production can be enhanced by adjusting environmental parameters such as temperature, pH, and oxygen levels. As we learned in Chapter 7, for example, all microorganisms have an optimum temperature at which they grow the fastest. For some metabolites, known as **primary metabolites**, synthesis closely mirrors population growth (**Figure 16.3**). In such cases, simply providing this optimum temperature may result in substantially increased metabolite yields. For other metabolites, called **secondary metabolites**, synthesis occurs only after the microbial population has ended its period of exponential growth and entered the stationary phase (see Chapter 7, p. 197). In this case, the environment may be altered in such a way that entry into stationary phase is encouraged. Sometimes careful regulation of nutrients can alter metabolite synthesis. Penicillin-producing molds, for instance, produce more of the antibiotic when provided with lactose instead of glucose.

Increasingly, the techniques of genetic engineering discussed in Chapter 14 are being utilized to create microorganisms that produce a particular metabolite in abundance. Sometimes, for example, an organism is discovered that, although it produces a potentially valuable metabolite, is difficult to culture and grow in the laboratory. If the gene responsible for the metabolite synthesis can be identified, it may be possible to isolate and clone the

(**a**) **Primary metabolite:** ethyl alcohol

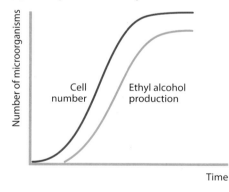

Time

(**b**) **Secondary metabolite:** penicillin

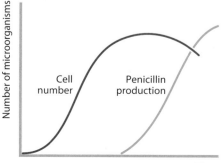

Time

Figure 16.3 Primary and secondary metabolites. (a) Ethyl alcohol (ethanol) is an example of a primary metabolite. Its production by fermenting yeast cells closely mirrors yeast cell population and increases fastest during the yeast's exponential growth phase. As the yeast enter the stationary phase, alcohol production slows. (b) Penicillin is produced by the mold *Penicillium notatum* as a secondary metabolite. Penicillin synthesis does not begin until the mold enters the stationary phase of population growth.

gene, introducing it into a more manageable organism such as *E. coli*. As techniques for the manipulation of DNA become ever more sophisticated, the importance of genetic engineering as it applies to industrial microbiology will continue to grow.

Potentially valuable microbes must also grow well in an industrial setting and must not pose undue risks to humans or the environment

High levels of metabolite synthesis are certainly necessary, but this alone does not guarantee a smooth transition to industrial application. The organism in question also must be able to grow in large-scale cultures. This is not necessarily a simple criterion to satisfy. As we have learned, microorganisms can be quite specific in their environmental requirements, and required conditions are not always easy to duplicate in the laboratory. It is not uncommon, for instance, for bacteria and fungi to form mutualistic relationships with other microorganisms in their environment (see Chapter 9, p. 253). Sometimes these relationships are essential if a given organism is to produce a particular metabolite. One species may produce a desired compound only when it has access to a different compound produced by the symbiotic partner (**Figure 16.4**). Complex microbial assemblages of this sort are very difficult to recreate in a laboratory, and they can greatly complicate the process of developing a strain of commercially viable microorganisms.

Rate of population growth is also an important consideration. Ideally, a microorganism of interest will grow and rapidly produce the metabolite in question. Preferably, the organism will also thrive in a relatively inexpensive liquid medium. In some cases, waste products from one industrial process, such as the liquid whey formed during cheesemaking (see Chapter 15, p. 407), are used to culture microorganisms for an entirely different industrial application.

Finally, an industrial microorganism must be safe, both for humans and for the environment. It is difficult, if not impossible, to prevent the release of industrial organisms into the environment completely. Therefore, any human, animal, or plant pathogen is not acceptable, even if it produces a valuable metabolite. Fortunately, safety standards are easier to satisfy because by the time most industrial strains of microorganisms have been developed, they are so highly specialized and "domesticated" that they are unable to survive in the natural environment (**Figure 16.5**). Selected for rapid and high metabolite production, as well as specific environmental conditions, they are a far cry from the wild strain that was isolated initially. Yet this fact does not mean that we can assume such strains are absolutely safe, and stringent testing and examination of any microorganism is necessary before its use in industrial or other processes can be condoned.

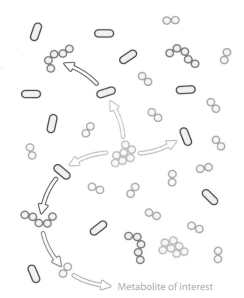

Figure 16.4 Metabolite production in the environment. Microorganisms often produce certain metabolites only when they are in close proximity to other microbial species, because one species produces a metabolic waste used by other species as a substrate for metabolite synthesis. In this example, there are four microbial species indicated by the four colors, living together in their natural environment. Green cells produce a metabolic waste (green arrows) that can by used by red cells. Only those red cells growing close to green cells are able to ultimately produce their own metabolite (red arrows). Orange cells growing nearby absorb the metabolite from the red cells, and produce yet a third metabolite (orange arrows), which in turn is absorbed by blue cells. Only those blue cells absorbing the necessary substrate from orange cells are able to produce the final, commercially valuable metabolite. Such complex relationships are not easy to elucidate and are often extremely difficult to reproduce in a laboratory setting.

"Oh dear! I didn't realize 'in the field' would be like this! We should have stayed in the laboratory."

Figure 16.5 Laboratory-reared microorganisms returned to their original environment. Bacteria that have been subjected to strain improvement are often no longer able to survive well in the environment. Should such bacteria be inadvertently released, they are usually at a competitive disadvantage with naturally occurring wild-type strains.

A defined series of steps are followed to move production from the laboratory to the factory

Suppose we have developed a strain of microorganisms that produces a valuable product in commercially viable quantities. We have determined the environmental parameters that permit optimum growth for our purposes, and we have unambiguously demonstrated that the microorganism in question poses no risks to humans or the environment. Certain problems must still be addressed before commercial production can begin. Metabolic processes that proceed satisfactorily in a few fluid ounces of nutrient broth in a laboratory flask do not necessarily occur efficiently in a 100,000 gallon industrial vessel. To cite just one problem, many metabolic processes of commercial value occur best under aerobic conditions, but proper aeration of a very large volume of liquid nutrient broth is not necessarily a simple task.

The microbial production of industrial metabolites generally takes place inside a device called an **industrial fermentor**. Basically, an industrial fermentor is a large cylinder made of stainless steel. The fermentor is outfitted with equipment to maintain constant temperature and pH, to aerate and adequately mix the liquid medium, to add nutrients as required, and to harvest the metabolite of interest (**Figure 16.6**). Ideally, those conditions that provided optimum metabolite production in a laboratory flask are recreated in the industrial fermentor.

Industrial fermentors can be quite large, holding thousands of gallons. Running and maintaining such a device requires a considerable input of time and money. Consequently, the transfer from the laboratory flask to a

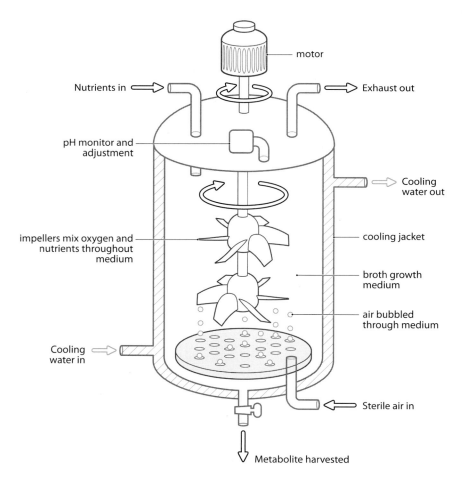

Figure 16.6 An industrial fermentor. The fermentor is constructed so that temperature, aeration, pH, and nutrient levels can be carefully controlled, in order to provide optimum conditions for metabolite harvest.

large industrial fermentor usually involves certain relatively low-cost intermediate steps, in which problems can be addressed and conditions can be adjusted as necessary.

First, a process that appears to have commercial or industrial potential in a small laboratory flask or test tube is carried out in a **laboratory fermentor**, which may hold 5 gallons or so of nutrient broth. Here, various combinations of nutrients, temperature, oxygen, and pH can be tested inexpensively, in order to determine which combination is optimal. If results in the laboratory fermentor are satisfactory, the next step is often to move up to a larger scale of perhaps 500–1000 gallons in a **pilot plant fermentor**. Should things proceed smoothly at this stage, the process is finally transferred to the large-scale industrial fermentor. This sequence of stages is referred to as **scale-up** (**Figure 16.7**).

Once the industrial fermentation process is successfully under way, the metabolite of interest is recovered by filtration, settling, or other methods. Depending on the metabolite, it may now be ready to use or it may require further treatment such as drying or purification. Next we will take a look at some of these valuable substances that come to us thanks to microbial activity.

Many industrially produced microbial metabolites have useful medical applications

Antibiotics. As we learned in Chapter 12, antibiotics were originally isolated from microorganisms themselves, and many are still commercially produced by bacteria and fungi. Soil bacteria of the genus *Streptomyces* have proven to be especially adept at producing valuable antibiotics. Examples include erythromycin, streptomycin, tetracycline, vancomycin, and rifampin. Other antibiotics such as penicillin and cephalosporin are synthesized by fungi.

Commercial antibiotic production is carried out in large fermentors that may hold up to 100,000 gallons. As in other fermentors, oxygen levels, pH, and temperature must be carefully monitored and controlled. The nutrients that are provided usually include a sugary molasses-like substance or a combination of sugars, proteins, and vitamins.

Although thousands of antibiotics from various microorganisms have been described, relatively few actually make it to commercial production. In some cases it has not been possible to produce the antibiotic in sufficient quantities. Other potential antibiotics may be shelved when it is determined that while they work well in a laboratory setting (*in vitro*), they are ineffective in a living animal body (*in vivo*). Others are discarded because they are found to have undesirable side effects on animal hosts. An antibiotic is

Figure 16.7 The scale-up process.
(a) Initial research for commercially valuable products is conducted with standard laboratory equipment, and reactions are carried out in flasks or test tubes. Microbial processes that show promise are next carried out in a sequence of progressively larger fermentors, in which production problems can be worked out. (b) A small-scale laboratory fermentor. (c) A typical pilot plant fermentor. (d) A large-scale industrial fermentor.

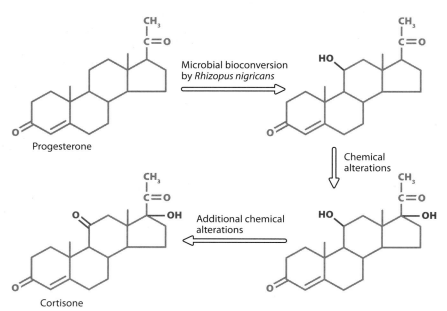

Figure 16.8 Steroid production by bioconversion. Minor chemical alterations by specific bacterial or fungal species can result in the production of medically valuable steroid hormones at a small fraction of the cost and labor of obtaining the hormone from other sources. In this example, the female hormone progesterone is being converted into cortisone, useful for the suppression of inflammation. The first step in the process involves the bioconversion of progesterone to an intermediate compound by the fungal microorganism *Rhizopus nigricans*.

produced and marketed only after research has demonstrated that it is effective against certain microorganisms *in vitro* and that it can be produced in sufficient quantities, and after it has passed stringent testing for effectiveness and safety, first in animals and finally in clinical human trials. Currently, approximately 160 antibiotics are commercially produced.

Despite the difficulties, the search for new antibiotics continues. Researchers regularly screen microorganisms for new antibiotics with increased specificity (an even *narrower spectrum*—see Chapter 12, p. 326), faster action, or other desirable characteristics. Antibiotics with novel mechanisms of action are especially valuable because they provide at least a short-term way to circumvent the resistance that has often emerged against older antibiotics.

So clearly Sam, the student in our case, has a long way to go before he makes his fortune in pharmaceuticals. There is a high probability that the microorganism in his soil sample belongs to a group of fungi or *Streptomyces* that has already been investigated. Even if his organism is previously unknown, its value may be limited, unless the compound it secretes works differently from other antibiotics. Furthermore, the production problems discussed in this section must be overcome.

Steroid hormones. Various steroid hormones, such as cortisone, used to treat pain and inflammation, and progesterone, used to prevent miscarriages, can be synthesized commercially by microorganisms. The process involves providing an appropriate microorganism with a commonly available sterol compound that it can convert into the medically valuable hormone (**Figure 16.8**). Such minor chemical alterations by bacteria and fungi, transforming a common substrate into a valuable product, are known as **bioconversions**.

Microbial bioconversion of steroids into medically useful hormones provides an inexpensive and quick alternative to either chemically synthesiz-

PRODUCT	USE	MICROBIAL SOURCE
pravastatin	lowering cholesterol	*Penicillium citrinum* (a fungus)
cyclosporin	prevents rejection in organ transplant patients	*Tolypocladium inflatum* (a fungus)
ergot alkaloids	induction of labor	*Claviceps purpurea* (a fungus)
mitomycin	anti-cancer drug	*Streptomyces caespitosus* (a bacterium)

Table 16.1 Representative medical products other than antibiotics produced by microorganisms.

ing the hormone in question or recovering it from an animal source. A good example is provided by the production of estrogen, which can be administered to relieve the symptoms of menopause. This female steroid hormone has traditionally been isolated from the urine of pregnant horses. This is a relatively expensive process that is not particularly pleasant for the horses. In fact, to many people, the use of horses in this manner constitutes a serious form of animal abuse. Those concerned with animal welfare might be glad to learn that alternatives produced by microorganisms exist. For example, the estrogen estradiol can be produced synthetically. The production of this estrogen starts with cholesterol, which is converted in several steps to estradiol. Bacteria, specifically *Arthrobacter simplex* and species of *Mycobacterium*, are crucial to the process.

Other medical products. The list of medically useful products synthesized industrially by microorganisms does not end with antibiotics and steroid hormones. Microbial products are used to lower cholesterol, induce labor, fight cancer, and prevent organ rejection in transplant patients. **Table 16.1** lists a few such products.

Industrial microbial metabolites have a wide variety of other, nonmedical uses

Enzymes. Microorganisms produce a large variety of novel enzymes, a number of which have industrial applications. Recall the case in Chapter 2, for example, in which a cardiologist used the bacterial enzyme streptokinase to dissolve a clot in the arteries of a heart attack patient. In Chapter 14 we discussed how stonewashed jeans have more to do with bacterial enzymes and less to do with "stones" than the name implies. Likewise, the next time you tenderize a steak before grilling, you may consider that the tenderizer you sprinkle on the meat may be a bacterial protease enzyme that initiates the process of protein digestion. Such proteases in drain cleaners help to digest clogs caused by hair, which is largely protein. They also lend a hand in laundry detergents, where they digest protein-containing stains, such as blood stains. A grass stain, on the other hand, often is dissolved by bacterial cellulases in the detergent, which digest the cellulose in the plant material into simple sugars that are rinsed away.

These are just a few of the microbial enzymes that find their way into common products. As with the other microbe-based products already discussed, the first step is to find or develop a mutant strain that produces abnormally high levels of the particular enzyme. Increasingly, the genetic techniques discussed in Chapter 14 allow scientists to engineer organisms to suit specific industrial needs. Subsequently, the production is scaled up as previously described, ultimately resulting in quantities sufficient for commercial purposes.

A newer and potentially exciting group of enzymes is the **extremozymes**, produced by thermophilic or other microorganisms normally found in extreme environments. Thermophiles, for example, thrive in very hot water (see Chapter 7, p. 193), and their enzymes are capable of activity at temperatures that would normally inactivate typical enzymes. In Chapter 14, for example, we discussed how the polymerase chain reaction (PCR), so vital in modern genetic analysis and biotechnology, involves a series of high-temperature steps. Most DNA polymerase enzymes would quickly be denatured and would be of no use at these temperatures. Accordingly, PCR relies on a type of DNA polymerase known as Taq polymerase. The enzyme is named after the thermophilic bacterium *Thermus aquaticus*, from which it is obtained.

Many industrial processes that work best at high temperature might benefit from such heat-tolerant enzymes. Likewise, other procedures might require a very cold, acidic, or halophilic environment. Enzymes produced by psychrophilic (cold-loving), acidophilic (acid-loving), or halophilic (salt-loving) bacteria may then be most appropriate.

Biopesticides. If you are fond of gardening but are troubled by insect pests, you might consider enlisting the aid of *Bacillus thuringiensis*. This spore-forming bacterium produces a protein that is highly toxic to many types of insects, particularly the moth larvae that can wreak havoc on the leaves of flowers, tomatoes, and other commonly cultivated plants. A subspecies, *Bacillus thuringiensis israelensis*, is especially effective against mosquito larvae and is often used in mosquito control programs. Spores containing the protein toxin are ingested by the insect, and after the cells lining the gut are destroyed, the toxin enters the insect's blood, causing paralysis and death. The protein is most lethal against insects that have a basic pH in their digestive system. Fortunately, dogs, cats, and other animals, as well as humans, have either a neutral or acidic pH in their digestive system, making the toxin safe to use.

After the bacteria are grown in large numbers in a fermentor, the conditions are altered to induce spore formation. The spores are then collected and mixed with inert compounds, for marketing as a pest control agent that can be dusted on plants. You will find it in almost any garden supply store, where it is often referred to simply as "Bt" (**Figure 16.9**).

Microorganisms as materials. Increasingly, microorganisms are being considered not just for what they produce but as materials themselves in the growing field of *nanotechnology*—technology or the use of materials at an extremely small scale. For example, researchers have converted bacteria into humidity detectors by coating the bacteria with tiny gold particles (**Figure 16.10**). As the environment dries out and humidity falls, the bacteria lose water, moving the gold particles on their surface closer together. Electrodes apply a voltage across the bacteria, and as the gold beads get closer together, an increased current is detected. The scientists found that if the humidity dropped from 20% to 0%, there was a 40× increase in current. These microscopic barometers, which function even if the bacteria die, work best in dry environments. The researchers have even suggested that they might be useful on future space missions to detect humidity on other planets such as Mars. Similar metal-plated microorganisms have been contemplated as semiconductors, magnets, and optical devices, and for use in other nanotechnology applications.

Big Problems, Little Solutions

In addition to the commercial applications we have been discussing, microorganisms can help us solve some of today's thorniest problems. These include water pollution and toxic waste, solid waste reduction, and the ever-worsening energy shortage. Here we will examine some of the ways that microbial partners can help us get out of some of the jams we have created for ourselves.

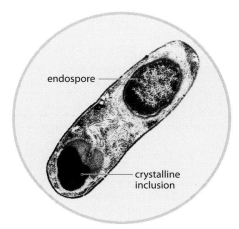

Figure 16.9 *Bacillus thuringiensis*: the microbial source of the biological pesticide Bt. The bacillus has already formed its spore. Such spores contain a protein, which is toxic to many insect pests. *B. thuringiensis* is grown in fermentors, and when the bacteria have reached high numbers, conditions within the fermentor are altered to induce spore formation. The spores can then be sold as Bt, which can be used on plants to reduce insect damage.

Figure 16.10 Microbial humidity sensors. The bacteria in the photograph, *Bacillus cereus*, have been coated with microscopic gold particles. Electrodes can then apply a current, transferred by the gold particles. As the humidity drops, the bacteria lose water, reducing their volume and bringing the gold particles closer together. This is detected as an increase in current.

New strategies are required to combat environmental pollution

CASE: THE SWEET SMELL OF SUCCESS

Sewage treatment plants stink. The rotten-egg smell comes from hydrogen sulfide, released by bacteria as they digest organic material in the sewage. Sewage plants often combat this problem by using devices called chemical scrubbers that filter the hydrogen sulfide through lye and bleach. Although this technique works, it is at best a partial solution, because new environmental problems are created in the process. Researchers at the University of California, Riverside, however, may have come up with a cheaper, more environmentally sustainable strategy—replacing the scrubbers with hydrogen sulfide-digesting bacteria. The sewage is trickled through biofilms of the bacteria, growing on polyurethane foam (Figure 16.11) in a process known as bioremediation. The microorganisms convert the offending hydrogen sulfide into odorless hydrogen sulfate, which is carried away by water seeping over the foam. The technique has already been instituted at several California treatment plants, where it costs about $50,000 to make the switch from scrubbers to bioremediation. Plants then save about $30,000 a year in operating costs.

1. What environmental problems are caused by the use of chemical scrubbers?
2. What exactly is bioremediation, and how does this process help alleviate the environmental problems caused by the scrubbers?
3. Why are the bacteria grown in biofilms on polyurethane foam?

Without question, modern industry and chemical production have improved our standard of living in many ways. But this improvement comes with a cost: increased environmental pollution. While some pollutants are degraded relatively quickly, others can last for years or decades, causing significant environmental disruption. Not only are natural ecosystems damaged, but humans may be at risk as well, if pollutants such as pesticides or heavy metals are concentrated in the food chain (see Chapter 9, p. 230).

Figure 16.11 Microbial biofiltration. (a) At wastewater treatment plants, bacteria can eliminate smelly hydrogen sulfide by percolating the sewage through a tower filled with bacteria-covered foam blocks. Arrows indicate the direction in which the sewage is pumped. (b) Foam blocks on which the bacteria grow as a biofilm. As sewage trickles through the foam, the bacteria convert the hydrogen sulfide into odorless hydrogen sulfate.

Likewise, toxic material that enters ground or surface water may find its way into our drinking water, creating further health hazards.

Since the environmental "dark days" of the mid-20th century, there has been considerable progress, at least in more developed countries, as the problems of water and toxic waste pollution have been recognized and addressed. Important legislation, such as the Clean Water Act, enacted in 1977, ensures that active measures are taken to prevent such pollution and violators are punished. Nevertheless, these problems have not vanished, and as human populations continue to increase, pressure on the environment persists. In the United States, for example, there are thousands of designated hazardous waste sites. Approximately 1200 of these heavily polluted areas have been designated as "Superfund sites" by the U.S. Environmental Protection Agency. These are areas so hazardous that their cleanup, which will cost billions of dollars, is an especially high priority.

Historically, the basic approach to the release of pollutants into water and soil was "out of sight, out of mind." People believed that the environment was so vast that whatever we released into it would eventually be absorbed, diluted, and broken down. Hard lessons taught us that this is not the case, and a number of chemical and physical procedures to remove dangerous materials from water and soil have been subsequently developed. These procedures are often expensive and require constant monitoring. And frequently, they simply transfer the environmental burden elsewhere. Some chemicals, for instance, can be detoxified by incineration, but this increases air pollution. Certain toxins can be concentrated and prevented from entering water, but they must then be stored indefinitely in a secure, leakproof container. For example, the chemical scrubbers used in the treatment of sewage, as described in our case, are highly effective at removing hydrogen sulfide, but they rely on toxic chemicals such as lye or bleach. Were these chemicals to leak into groundwater, there would be serious environmental repercussions. Sewage plants, consequently, must go to great effort and expense to prevent such contamination.

No doubt current technology will continue to play a vital role in environmental protection and the control of pollution, but increasingly, scientists are turning to microorganisms as a way to break down pollutants in an environmentally friendly manner.

Microorganisms are used to digest harmful chemicals through the process of bioremediation

The strategy of using living organisms to degrade environmental contaminants is known as **bioremediation**. This technique is already being used in many ways, including oil spill cleanup, drinking water treatment, toxic chemical degradation, and even the disposal of out-of-date explosives. Bioremediation is often performed by allowing certain microorganisms to grow in a thick biofilm over a solid substrate. A biofilm, as described in Chapter 9 (p. 234), is a growth of microorganisms held together by a secreted carbohydrate matrix. A fluid solution containing contaminants is allowed to seep through the carbohydrate matrix, placing the individual bacteria within the biofilm into direct contact with the fluid. The bacteria then degrade the toxic waste as the fluid percolates through the biofilm. In other words, the biofilms are uniquely constructed to serve as "biological filters," placing the degrading microorganisms into slow and steady contact with toxic materials. Ideally, the fluid leaving the biofilm is contaminant-free.

Not just *any* bacterial species can be used in this way. To break down any toxic chemical, the microorganism in question must produce enzymes capable of such digestion. As we have seen throughout this text, different microorganisms have different enzymes, and therefore different digestive capabilities. It should not be too surprising that microorganisms, which have evolved to live almost everywhere and utilize almost every possible resource, can be found that digest even the most toxic materials.

Often, the most promising species for such toxin digestion are nonfastidious organisms. Recall from Chapter 7 (p. 196) that nonfastidious organisms are those that can digest a very large range of organic molecules, because they are especially well endowed with many different digestive enzymes. *Pseudomonas* is a good example. These nonfastidious bacteria are already commonly used to help clean up oil spills, because some *Pseudomonas* species have enzymes that allow them to break down the complex hydrocarbons found in petroleum. A fastidious species, on the other hand, would likely be a poor choice for bioremediation. Fastidious organisms have very specific nutrient requirements, because they produce relatively few digestive enzymes.

Both the environmental context and the microbe being used determine how bioremediation is conducted

In certain cases, bioremediation is a simple affair because the necessary organisms are already present in the environment, and all that is necessary is to encourage their growth. Certain *Pseudomonas* species, for instance, as previously mentioned, are able to digest petroleum. Following an oil spill, contaminated water can be "fertilized" with nutrients such as nitrogen- and phosphorus-containing compounds to spur on the growth of *Pseudomonas* already found in marine environments. When microbial growth is accelerated in this way, the cleanup of the oil proceeds much faster, with less environmental impact, and at a reduced cost (**Figure 16.12**).

In other situations, bioremediation requires a bit more effort. Sometimes, although a strain of organisms that digests a particular pollutant can be isolated and identified, it either cannot digest enough of the pollutant or it does not survive well in the contaminated site. In such cases, laboratory selection, as described previously, must be used to develop strains with both the capacity to degrade large amounts of a particular pollutant and to survive and grow quickly once released into the environment. As an alternative to isolating and selecting for a useful strain, the genes responsible for degradation in one organism can be spliced into the genetic material of a more manageable species, by use of the techniques described in Chapter 14. Such genetically modified organisms have limited use, however. They are often unlikely to survive for long if released at a contaminated site, and because they are genetically altered, their release into the environment carries substantial risk. With newly created genetically modified organisms, it is not possible to anticipate all of the environmental problems this organism might actually *cause* rather than solve. In most cases, the release of such organisms is illegal. This is not to suggest that such microbes are of *no* value. Genetically modified organisms can be of use in carefully controlled facilities such as water treatment plants, provided suitable precautions are taken to prevent their spread to the environment.

Sometimes, the problem is not so much finding the "right" bacteria as it is finding the correct way to expose these bacteria to the material to be digested. The idea of using bacteria to remove hydrogen sulfide from sewage, as described in our case, has been around for some time. But it has always been considered too inefficient compared with chemical scrubbers. To be degraded, hydrogen sulfide needs to be in contact with bacteria for at least 10 seconds. Until recently, there was no practical and economical way to accomplish this. With the use of polyurethane foam as described in our case, researchers vastly increased the surface area of their biological filter by filling the silo-shaped filtering tower with 4-cm^3 porous-foam blocks (see Figure 16.11b). The bacteria grow as a biofilm throughout the nooks and

Figure 16.12 Bioremediation in progress. Cultures or different oil-degrading bacterial species under evaluation for their bioremediation potential.

crannies of the blocks, and as the sewage trickles through the tower, the microorganisms have ample opportunity to convert the malodorous hydrogen sulfide to less offensive compounds. This strategy may be appealing for other, similar problems, such as the disposal of animal wastes in large-scale agriculture.

Bioremediation can prove valuable in many different settings

Polychlorinated biphenyl compounds (PCBs) are double-ringed compounds containing chlorine atoms that were once used extensively in heavy-duty electrical equipment and as industrial solvents. At the time, it was not understood that these compounds posed a significant environmental hazard. We now know that PCBs accumulate in the fatty tissue of animals that consume them, where they can cause a number of problems. Many are known carcinogens.

PCBs are now banned in most industrialized countries, but many of these compounds can persist for years in the environment, where they continue to work their way into food chains. The Hudson River in New York State was particularly contaminated because of the large electrical equipment plants located along its shores. In the 1980s, however, came the surprising and welcome news that PCB concentrations in the river sediments were much lower than expected. Research subsequently showed that anaerobic bacteria in the sediments were able to break the compounds down to smaller, simpler compounds, which could then be degraded completely by other bacteria. This finding highlighted the potential value of bioremediation for the elimination of these stubborn toxic chemicals. It also demonstrated that specific environmental conditions, in this case an anaerobic environment, were required for successful bioremediation to take place.

Other chlorinated compounds might also provide targets for cleanup via bioremediation. Vinyl chloride, for instance is a cancer-causing agent produced when solvents in metal cleaners and dry-cleaning fluids degrade naturally. In 2003, researchers announced that they had isolated a previously unknown bacterial species that can digest vinyl chloride. The bacterium, known as BAV1 (a strain of *Dehalococcoides*), was discovered living 20 feet beneath the sediments of Lake Huron. Field tests at one vinyl chloride-contaminated site showed that all traces of the pollutant were eliminated in about 6 weeks. This is an exciting finding because vinyl chloride is present in one-third of the high-priority Superfund sites.

Uranium- and plutonium-containing radioactive waste is another daunting problem at many toxic waste sites. One of the difficulties with such wastes is that they are often mixed with other chemicals and are present at low concentration. Bioremediation may one day offer a solution. Recently, a process for concentrating uranium has been developed. The technique relies on bacteria that reduce uranium, making it insoluble in water. The now-insoluble uranium can then be separated out from other, soluble compounds. Additional research focuses on *Deinococcus radiodurans*, a bacterial species with a remarkable tolerance for radiation that we discussed in Chapter 4. Mercury compounds are common in radioactive waste sites, and the goal is to genetically engineer strains of *D. radiodurans* that reduce mercury to a less toxic form. Because of its inherent resistance to radiation, *D. radiodurans* is uniquely qualified to work in such "hot zones."

Military bases, in addition to their obvious purpose, often provide buffers to overdevelopment. In this sense many military bases also serve as a type of wildlife refuge, especially important in light of increasing urbanization and the encroachment upon open space. But often, such sites are heavily contaminated with many toxic compounds or explosives. Until about 1970, one

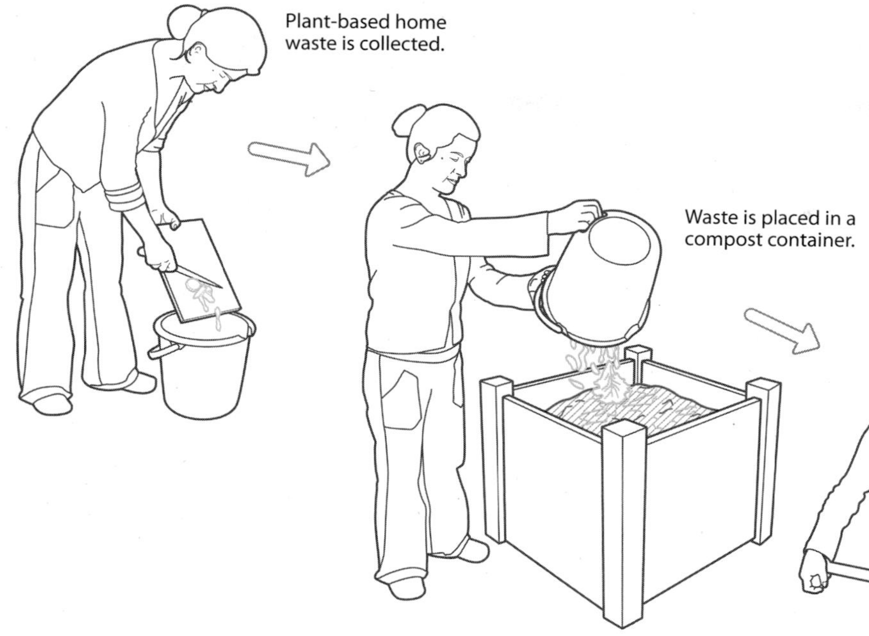

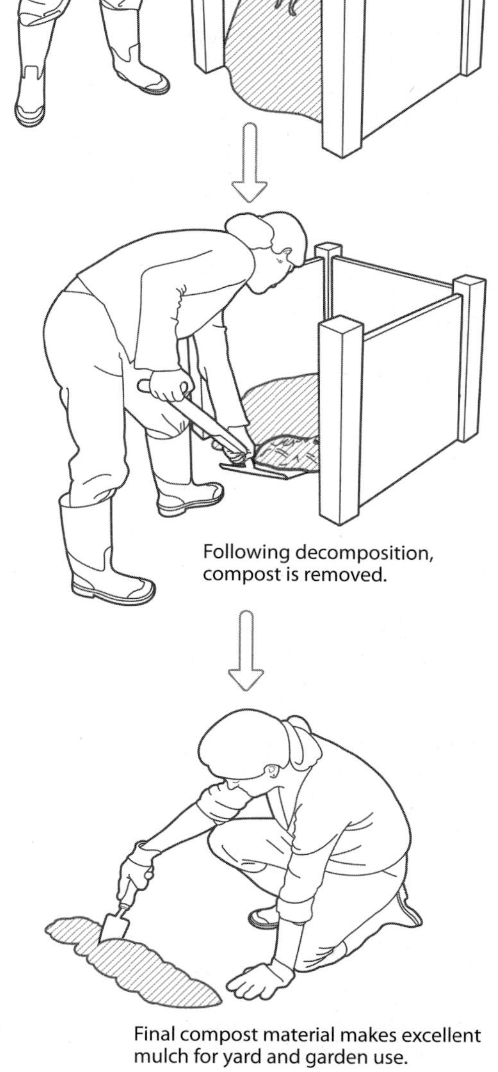

Plant-based home waste is collected.

Waste is placed in a compost container.

Figure 16.13 Backyard composting. Kitchen and yard waste provide the starting material for compost. Such waste is added to a compost container, where regular turning of the compost introduces oxygen. Following decomposition, compost is removed. Such decomposed material can then be used as mulch in the yard or garden.

Turning the compost introduces oxygen.

of the most commonly used explosives was trinitrotoluene (TNT). Although it is not common in modern weapons, tons of TNT still pose a danger of explosion at military instillations and weapons factories around the world. *Clostridium bifermentans*, however, has recently been found to readily digest TNT. The bacteria can adhere to carbon compounds, which are then inoculated into the contaminated soil. Starch is then added to the soil to serve as an energy source for the growing microorganisms. The explosive is then degraded to harmless materials. A number of contaminated sites have already been rendered TNT-free by this technique.

Microorganisms can help reduce solid waste and improve soil quality through composting

CASE: WASTE REDUCTION BEGINS AT HOME

After buying their first home, Bill and Lylette are anxious to start a vegetable garden, something that was never possible in their rented apartments. Although they have little experience, they know that kitchen waste can be turned into a valuable soil amendment by composting. After reading up on the subject, they are somewhat surprised to learn that compositing is a bit more complicated than merely dumping vegetable waste into a pit and waiting for it to decompose. Rather, they note that proper compost must regularly be "turned" either by hand or by a special mechanical device (**Figure 16.13**). Once they begin to compost, they are further surprised to observe that the compost heats up over time, and that if they forget to turn the compost every so often it begins to smell much worse.

1. What exactly is composting?
2. Why does compost have to be periodically turned?
3. Why does the composting material generate heat?
4. What explains Bill and Lylette's observation that compost eventually smells bad if not turned?

Microorganisms can easily be enlisted to quickly reduce the amount of solid organic waste through **composting**. In this technique, familiar to many homeowners, kitchen and yard waste is decomposed by encouraging the growth of certain microorganisms. Not only does this result in less solid

Following decomposition, compost is removed.

Final compost material makes excellent mulch for yard and garden use.

waste, but the degraded products can be used to improve soil quality in gardens or yards. Composting works best when the soil is well oxygenated, as this promotes the growth of aerobic microorganisms. As we learned in Chapter 7, aerobes grow much faster than anaerobes and consequently degrade organic material more quickly. Although anaerobes can eventually degrade organic wastes, their much slower metabolism and reduced growth rate greatly slow the process (**Figure 16.14**).

Consequently, one of the keys to successful composting is to ensure that the waste remains well oxygenated, so that rapid aerobic microbial degradation can occur. As Bill and Lylette learned, this means that the waste must periodically be turned in order to introduce a fresh supply of oxygen (see Figure 16.13). Such oxygenation of the compost also explains why the decomposing material tends to heat up. Within a few days, the temperature of the compost pile rises as aerobic soil microorganisms initiate the decomposition process. As oxygen levels within the composted material fall, the temperature also declines as microbial growth slows. The compost is then turned, oxygen is replenished, and the process repeats until the digestion of the organic waste is complete, typically in a number of weeks. Not only is the bulk of the waste decreased by about two-thirds, but a rich, mulchlike material is produced. As an added bonus, the regular reintroduction of oxygen reduces growth of anaerobic methanogens (see Chapter 7, p. 181), meaning that less polluting and hazardous methane gas is released.

Composting is no longer merely an avocation for the home gardener. In an ever-increasing number of municipalities, residents are able to place leaves, grass clippings, and other yard waste into special receptacles for collection. These wastes may be deposited in long, low piles at the landfill site where they can be regularly oxygenated mechanically (**Figure 16.15**). The benefits of such city-sponsored composting are many. Landfills, now reserved for glass and other non-biodegradable materials, fill up more slowly, reducing the need for yet newer landfills. The final organic product can be used on city-owned parks and golf courses, or it can be made available for home or agricultural use. With less methane production, there is a greatly reduced risk of explosions caused by inadvertent methane combustion in a landfill. Finally, as discussed in Chapter 9, methane is an important greenhouse gas. With their tree trimmings being composted rather than being added to landfills, residents can rest easy at night, knowing that their yard waste is not contributing to global warming.

Plastics may be replaced by biodegradable, microbially produced alternatives

Almost 40 billion tons of plastic are produced each year, and disposing of discarded plastic items is an enormous part of the solid waste problem. Plastic cannot be digested by microbes and is therefore not biodegradable. As new locations for landfills grow increasingly scarce, the search for biodegradable alternatives to plastics has become urgent.

Figure 16.14 Decomposition under anaerobic conditions is a slow process. Because their rate of growth is slow, the ability of anaerobic bacteria to decompose waste can be slow. This newspaper, a copy of *The Daily Mirror*, was buried and therefore persisted under anaerobic conditions for decades. Note the June 16, 1967, publication date.

Figure 16.15 Municipal composting. Composting is increasingly used by cities, both to reduce their solid waste and to provide soil amendments for city parks or for homeowners. In this photo, organic wastes are being oxygenated mechanically.

Research into such alternatives is beginning to bear fruit. Starch-based plastics, for example, use starch molecules to link together other biodegradable molecules. These plasticlike compounds can be digested first by starch-digesting soil bacteria. Other microorganisms next degrade the starch-linked molecules. Bacteria themselves have also been investigated as sources of substitutes for conventional plastics. Some bacteria store carbon in the form of a molecule called poly-β-hydroxyalkane, or PHA (**Figure 16.16a**). When environmental sources of carbon are low, such bacteria, by use of an enzyme called depolymerase, can digest their supply of PHA to meet their metabolic needs. Some bacteria secrete an extracellular form of this enzyme that can degrade any PHA in the immediate environment.

PHA has many of the same properties as plastic, and its use as a biodegradable plastic alternative is currently being investigated (see Figure 16.16b). Providing bacteria with different carbon sources induces them to synthesize different forms of PHA. This means that specific "bioplastics" could be produced for different purposes such as plastic bags, squeeze bottles, or rigid plastic materials such as pipes. When discarded, such products can be degraded by bacteria that secrete the depolymerase enzyme.

A few products are already made from bacterially produced plastic. In 1990, the first such product, shampoo in a bottle made of PHA, was introduced in Europe. The use of PHA has been slow to catch on because, compared with conventional plastics made from petroleum, PHA-based plastics are still quite expensive. With oil prices continuing to rise, however, the time may come when biodegradable plastics become the rule rather than the exception.

Microorganisms may be used to help meet the demand for limited resources

Nonrenewable resources, by definition, are in limited supply. Yet here, too, microorganisms may help extend the supply or in some cases even produce renewable versions of limited materials.

Alternative fuels. Technically speaking, petroleum, natural gas, and other hydrocarbon fuels are renewable. Yet the geological processes necessary to form new hydrocarbon deposits take millions of years. Consequently, for all practical purposes, once we deplete our supply of hydrocarbon fuels they are gone for good. Unless, that is, we increase our use of renewable, biologically produced hydrocarbons.

For example, ethanol is a biological fuel already in widespread use. It is produced by fermenting agricultural products such as corn or sugarcane and then frequently blended with gasoline for use in motor vehicles. Ethanol, however, is not without problems. Ethanol, when made from corn, diverts

Figure 16.16 PHA: a microbial source of biodegradable plastic. (a) The bacterium *Rhodobacter sphaeroides* stores carbon as an energy reserve in the form of poly-β-hydroxyalkane (PHA) in storage granules. The granules are visible in the cytoplasm of the cell in the center of the photo. (b) Because they are biodegradable, PHAs could be an attractive alternative to the petroleum-derived plastics that pollute our environment. The sheet of PHA on the left has not yet been degraded. By six weeks, degradation is almost complete.

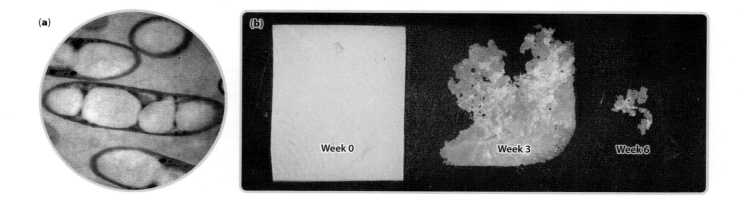

(a)

(b)   Week 0   Week 3   Week 6

an important food item from livestock feed or the dinner table to ethanol production plants. With so much corn being used for fuel, the price we pay for food has risen accordingly. Consequently, there has been renewed interest in "cellulosic ethanol," for which plant wastes such as stalks and leaves are the source instead of edible plant parts such as corn kernels. You will recall from Chapter 2 (p. 32) that cellulose is a long-chain polysaccharide composed of many glucose molecules and it is the principal component of plant cell walls. Although cellulose is energy-rich, very few nonmicrobes can tap this energy because most organisms lack the enzymes necessary to digest cellulose. Many microorganisms, on the other hand, are able to digest cellulose, and if these microbes were used in ethanol fermentation, it might be possible to use plant wastes as a sugar source. In fact, one Canadian company is already doing exactly that, utilizing the fungus *Trichoderma reesei* to convert agricultural waste into ethanol. This fungus has something of a notorious past. In World War II, troops in the Pacific had problems with "jungle rot"—the destruction of clothing and other products made from cotton, which is mostly cellulose. The culprit turned out to be *T. reesei*.

Methane, produced by methanogenic microorganisms, is another potential source of "biofuel." Previously, we have cast methane in a bad light; it acts as a greenhouse gas, and it can cause fires in landfills. But it can also be burned as a hydrocarbon fuel. In many places, methane released from landfills is already being used as a source of inexpensive energy (**Figure 16.17**). Another potential source of methane is the large amount of manure, filled with methanogens, that is an abundant waste product of animal feedlots.

Even petroleum-like substances produced by microorganisms might one day contribute to our overall energy needs. The green alga *Botryococcus braunii*, for example, excretes long-chain hydrocarbon compounds that have many of the properties of oil.

Other, more exotic uses of microorganisms for energy production may await us in the future. For example, bacteria of the genus *Shewanella* are facultative anaerobes. As we learned in Chapter 7 (p. 193), such organisms use oxygen as a final electron acceptor when oxygen is available, and they release the reduced oxygen as water. In the absence of oxygen, a different final electron acceptor is used and consequently a different end product is released. *Shewanella*, however, is unusual because when it is growing under anaerobic conditions in the soil, electrons released at the end of respiration do not reduce a final electron acceptor inside the bacterial cell itself. Rather, the bacteria export the electrons outside their cells, where the electrons attach

Figure 16.17 Methane: an alternative form of energy. Methane gas, released at this Austin, Texas, landfill, is collected and converted into electricity. The structure in the photo is the electrical generator. Methane gas, when used in this way, is an example of a renewable "biofuel." The gas starts to accumulate within months after a landfill is sealed, and gas may be produced at the landfill for 5 years or more.

to metallic ions. In the laboratory, researchers have found that they can substitute an electrode for these naturally occurring ions, effectively turning the bacteria into microscopic electric generators. Such bacteria could be used to create "microbial fuel cells" capable of generating energy. In a "traditional" fuel cell, hydrogen is combined with oxygen to create electricity, with water as a by-product. Microbial fuel cells would involve placing bacteria such as *Shewanella* in an anaerobic environment, along with glucose as a fuel source. The efficiency of such fuel cells ranges from about 60–80%. In other words, between 60% and 80% of the energy available in a glucose molecule's electrons is ultimately transferred to the electrodes by the bacteria. By comparison, the efficiency of a typical internal combustion engine in a modern car is rarely better than 30%. The study of microbial fuel cells is in its infancy, but this emerging technology may one day play a part in helping us to meet our ever-growing demand for energy in an environmentally sustainable way.

Microbial mining. To mine iron, copper, silver, and other metals, it is usually necessary to dig up rocks containing the mineral of interest, crush the rock, and extract the desired metal. Often it is necessary to fuse or melt the raw ore to separate the metal contained within. This process, called smelting, often releases toxic gases or other contaminants that pose a risk both to the miners and to the environment. It is also expensive, which means that mining low-grade ores is not profitable.

In some cases bacteria are able to do the dirty work at a fraction of the cost, making it economically feasible to mine ores that would otherwise be left in the ground. With the help of *Thiobacillus ferrooxidans*, for example, copper can be extracted from low-grade ore that would typically not be economical to mine. This acid-loving, or acidophilic, species lives naturally in the soils surrounding the ore, and it meets its energy needs by oxidizing iron sulfide compounds to sulfuric acid. To spur the growth of the bacteria, an acid solution is sprayed onto piles of the copper-containing ore (**Figure 16.18**). The sulfuric acid produced by the bacteria causes the copper to dissolve, leaching it out of the rock. A similar technique is sometimes used in gold mining.

The use of microorganisms to separate valuable minerals from ores is still limited primarily to organisms that are found naturally in the ore deposits. Through the use of genetic engineering techniques, it may be possible to one day develop strains of bacteria to help remove other minerals from low-grade ores.

Figure 16.18 Microbial mining. An acidic solution is being sprayed onto copper-containing rocks. The acid increases the growth of acidophilic bacteria. Sulfuric acid waste dissolves the copper and leaches it out of the rock.

Microorganisms may be able to stabilize soil, reducing earthquake damage

As we have seen in this chapter, the number of ways in which microorganisms may one day be harnessed to help solve various problems is seemingly limited only by our imagination. We opened this chapter with a description of a novel idea for the restoration of historic buildings. We will conclude with a similar type of microbial stabilization—this time not to combat the wear and tear of pollution but to potentially protect buildings from earthquakes.

When a large earthquake strikes, buildings on sandy soils are at particularly high risk. The earth movement causes the sand particles to shake apart from each other, loosening the soil to the point where a building can easily collapse. Currently such buildings are usually stabilized by oozing cement into the soil beneath them, but the cement can set unevenly, leading to patchy support.

The soil bacterium *Flavobacterium johnsoniae* may provide superior stabilization for threatened buildings. The bacteria grow in a biofilm composed of secreted carbohydrates (see Chapter 9, p. 234). When a liquid culture of *F. johnsoniae* is mixed with sand, researchers have shown that after only a few days, the cohesion of the sand particles almost doubles. In other words, thanks to the bacterial biofilm, the sand is almost twice as solid.

So far the technique has been tried only on a small scale, but if field testing proves satisfactory, the researchers think that bacterial biofilms may become an important component in making buildings earthquake-proof. The same strategy could also be applied to areas prone to landslides.

Coming Up Next...

Who knows? We have reached the end of this book, but the story of microorganisms and how they affect us is far from over. When will the next new epidemic of an emerging disease strike? What will be the next big breakthrough in genetic engineering and what previously unrecognized role for microorganisms in an important ecological process will be elucidated? What newly understood microbial processes will offer novel solutions to the problems of the 21st century, and when will new vaccines or drugs lift the burden of the infectious diseases that threaten us today? The answers to these questions are not a matter of "if" but a matter of "when." Keep your eyes open. The next microbial story is right around the corner, and it will no doubt provide more evidence, if more evidence were needed, that these smallest inhabitants of our planet are among the most remarkable.

Key Terms

bioconversion	metabolite
bioremediation	pilot plant fermentor
composting	primary metabolite
extremozyme	scale-up
industrial fermentor	secondary metabolite
laboratory fermentor	strain improvement

Concept Questions

1. Many people suffer from gas following the consumption of certain fruits and vegetables such as beans, broccoli, and cabbage. For such individuals, a product called Beano® is available. The major ingredient in this product is an enzyme called galactosidase, produced by the fungus *Aspergillus niger*. This enzyme digests the carbohydrates in the plant material that are responsible for much of the gas. Provide a simple description of how you think "scale-up" for the production of Beano may have progressed.

2. We have learned in this chapter that occasionally mutations occur that result in the overproduction of certain metabolites. Suppose that a particular enzyme of commercial interest is produced by a gene in an operon. Furthermore, suppose that this operon functions by induction, similar to the *lac* operon discussed in Chapter 6 (p. 151). After reviewing the mechanism of operon action in Chapter 6, can you think of any specific mutation in an operon that may cause overproduction of an enzyme?

3. Look through the food items, cleaning products, medicine chest, etc., of your home. Which products can you find that are commercial microbial products?

4. In this chapter we have reviewed a number of environmental problems, which may, in part, be solved with microbial solutions. Can you think of an environmental or ecological problem that was *not* specifically addressed in this chapter? Furthermore, can you foresee any way that microorganisms might be involved in addressing the problem that you have thought of?

5. Alcohol, when produced by yeast cells, is an example of a primary metabolite. Penicillin, when produced by mold, is an example of a secondary metabolite. How would the production of alcohol and penicillin differ in terms of the environmental conditions provided for the microorganisms?

6. In this chapter we learned about *B. thuringiensis*, the bacterium that produces the Bt toxin, useful in controlling agricultural insect pests. The gene for the Bt toxin has actually been inserted into many crop plants directly, using the technology discussed in Chapter 14 (p. 398). Farmers who plant such crops sometimes plant non-Bt-producing crops in adjacent fields to their Bt crops, as a way to reduce the evolution of Bt resistance in the insect pests. How might this strategy work?

7. Look closely at the garbage in the wastebaskets in your home. Then make a ballpark estimate about the percentage of the garbage that could be degraded by microorganisms through composting.

8. A microbiologist who is also a big sports fan makes the statement that "without microorganisms there would be no steroid problem in sports." Do you think there is any truth at all to this statement? If the use of microorganisms in bioconversion, as discussed in this chapter, were somehow impossible, would the steroid problem in sports go away? Be reduced? Become more serious?

9. We have learned that that certain bacterial species are unusually resistant to radiation because of their highly efficient DNA repair mechanisms. Can you think of any potential commercial or industrial applications for such bacteria?

Glossary

A subunit the portion of a bacterial exotoxin that interferes with normal host cell function.

abiotic factor a physical environmental parameter such as temperature or oxygen concentration that affects a living organism.

abiotic synthesis the proposition that life began through the formation of biological molecules from simple precursor compounds.

acid a compound that, when placed in water, causes the concentration of protons (H^+) to increase.

acid-fast bacteria bacteria with a characteristic cell wall type, consisting of long-chain fatty acids interlaced with peptidoglycan, from which stains cannot be removed by washing in acid.

acid-fast stain a staining technique allowing acid-fast and non-acid-fast bacteria to be distinguished.

acidophile an organism that grows best in conditions of low pH.

activation energy the energy needed to initiate a chemical reaction.

active site the part of an enzyme that temporarily binds the substrate in a biochemical reaction. This binding alters the shape of the substrate, facilitating its conversion to product.

acute disease a disease in which symptoms develop rapidly but last for only a short time.

adaptive immune response type of response made by the immune system that is directed specifically against the infecting pathogen and will not protect against any other pathogen.

adenosine triphosphate (ATP) a molecule containing chemical bonds whose hydrolysis releases large amounts of energy, used by cells as an energy source for activities in which energy is required.

adsorption the attachment of a bacteriophage to the surface of a bacterial cell.

aerobic respiration cell respiration in which oxygen is the final acceptor of electrons produced by the oxidation of "fuel" molecules in energy generation.

aerosol transmission *see* airborne transmission.

aerotaxis movement toward (positive aerotaxis) or away from (negative aerotaxis) oxygen.

aerotolerant anaerobe an organism that does not use oxygen but is not harmed by its presence.

agglutination the clumping of large aggregations of antibody and antigen.

airborne transmission also called aerosol transmission. Transmission of pathogens in small droplets through the air from an infected host to an uninfected, susceptible host.

algal bloom a sudden explosion of growth in a population of aquatic algae.

allele a specific form of a particular gene.

ammonification the process by which decomposers release ammonium ions into the environment, through the degradation of nitrogen-containing organic molecules.

anaerobic respiration cell respiration in which a molecule other than oxygen is used as the final acceptor of electrons produced by the oxidation of "fuel" molecules in energy generation.

antibiotic an anti-bacterial agent, either synthetic or produced naturally, by fungi or bacteria.

antibody an immune system protein produced by plasma cells in an adaptive immune response, which is able to neutralize the pathogen or otherwise help to control the infection.

anticodon a three-base sequence in a tRNA molecule, which permits the tRNA to interact temporarily with the correct mRNA codon during translation.

antigen a piece of a molecule that induces an adaptive immune response.

antigen-presenting cell (APC) a type of leukocyte that engulfs foreign microbes and presents fragments of their proteins as antigens to T cells.

antigenic drift a minor genetic change in the antigenic makeup of certain viruses, due to random mutations.

antigenic shift a major change in the antigenic makeup of viruses with a segmented genome, caused when gene segments from different strains of the virus combine to create a new strain.

antiseptic a chemical that can be used to eliminate microorganisms from living tissue without harming it.

Apicomplexa a group of mostly nonmotile protozoa, characterized by an intracellular structure known as the apical complex.

applied science the use of basic science to solve a specific practical problem.

Archaea one of the two domains of prokaryotic microorganisms; the other is the Bacteria. Members of this domain are called archaea (singular: archaeon).

artificial selection the purposeful direction of animal or plant breeding, in order to develop breeds with certain beneficial characteristics.

aseptic technique laboratory or medical processes carried out in a prescribed manner intended to prevent contamination by unwanted microorganisms.

asexual reproduction reproduction from a single parent by budding or cell division.

assembly of viruses, the production of new virus particles by the association of newly produced viral proteins and nucleic acid.

atom the most basic chemically recognizable unit of an element. It can be divided into smaller subatomic particles but is then no longer distinguishable as the specific element.

atomic mass the total number of protons and neutrons present in the nucleus of a given atom.

atomic number the number of protons present in the nucleus of a given atom.

ATP synthase an enzyme complex through which protons in higher concentration on one side of a membrane can return to the other side of the membrane. Energy released as the protons move down their concentration gradient is used by the enzyme to synthesize ATP from ADP and phosphate.

attenuation the process of weakening a pathogen to the point at which, although it can still infect a host, it can no longer cause disease.

autoclave a device in which objects can be sterilized though a combination of moist heat and high pressure.

autotroph an organism that can synthesize organic molecules by using inorganic precursors.

B subunit the portion of a bacterial exotoxin that allows the exotoxin molecule to bind to and enter specific host cells.

bacillus general name for any bacterium with a rod-like shape (plural: bacilli). *Bacillus* is the name of a genus of rod-shaped bacteria.

Bacteria one of the two domains of prokaryotic microorganisms; the other is the Archaea. Members of this domain are called bacteria (singular: bacterium).

bactericidal describes a procedure, chemical, drug, or other process that kills bacteria.

bacteriophage a virus that infects bacteria.

bacteriostatic describes a procedure, chemical, drug, or other process that inhibits bacterial replication.

barotolerance the ability to withstand high pressure.

barrier to entry a physical, chemical, genetic, or microbial deterrent to infection by a pathogen in an animal host.

base a compound that, when placed in water, causes the concentration of protons (H^+) to decrease.

basic science scientific inquiry focused on revealing the manner in which fundamental processes in nature proceed.

basidiocarp the spore-producing reproductive structure produced by fungi of the Basidiomycota.

B cell a type of lymphocyte that, upon activation, matures into an antibody-secreting plasma cell.

benthic zone the deepest zone in an aquatic environment, including the bottom sediments.

binary fission mode of reproduction of prokaryotic cells by division into two daughter cells.

bioconversion the change of one compound into another by microbial action.

biofilm a microbial community in which the cells are encased in a layer of polysaccharide.

biological vector transmission transmission of a pathogen to the host in which it causes disease via an invertebrate vector, in which the pathogen replicates and/or develops.

biomass the collective mass of all organisms in an ecosystem.

bioremediation the use of microorganisms to degrade harmful chemicals.

bioterrorism the use of disease-causing biological agents to kill or cause fear in the general population.

biotic interaction an ecological interaction between living organisms, such as competition or predation.

broad spectrum the general name given to antimicrobial drugs that are effective against a wide range of microorganisms.

budding the release of enveloped viruses from a host cell by a process in which the newly assembled viral particle protrudes through the plasma membrane and acquires a coating of plasma membrane—the viral envelope.

Calvin cycle a series of reactions used by photosynthetic eukaryotes, as well as some photosynthetic prokaryotes, including the cyanobacteria, to fix carbon dioxide into organic molecules.

capsid the protein coat of a virus particle.

capsomere one of the individual protein subunits making up the complete capsid of a virus particle.

capsule a polysaccharide-rich outer layer secreted by some bacteria, forming a thick, regular shell. The capsule may aid in adherence or immune evasion.

carbohydrate a class of biological molecule that always contains carbon, hydrogen, and oxygen. Carbohydrates are important in living things as sources of energy and as structural components. They are hydrophilic in nature.

carbon fixation the autotrophic process of incorporating carbon dioxide into organic carbon compounds.

carrier proteins transmembrane proteins that transport molecules across cell membranes.

case definition a list of common symptoms for a particular medical condition.

case-control study an epidemiological study used to determine the cause of a specific illness, in which a group of individuals suffering from the illness are compared with a similar population of healthy individuals.

cell the functional unit of life. A highly organized, typically microscopic structure, always surrounded by a cell membrane, and containing genetic material in the form of DNA.

cell lysis the rupturing of the cell membrane, resulting in cell death.

cell-mediated response the component of an adaptive immune response in which cytotoxic T cells detect and kill infected host cells.

cell membrane also called the plasma membrane. A membrane forming the outer boundary of all cells, composed of a bilayer of phospholipids in which proteins are embedded. The cell membrane regulates the transport of molecules and ions into and out of the cell.

cell nucleus the subcellular structure in eukaryotic cells that contains the genetic material. The nucleus is bounded by a double membrane.

cell respiration the gradual breakdown of "fuel" molecules, such as sugars, resulting in the production of ATP.

cell theory the theory stating that all living things are composed of one or more cells.

chemical bond The general term for any type of linkage between atoms.

chemically defined medium growth medium for microorganisms of a defined composition, containing specific amounts of particular nutrients required for growth.

chemoautotroph an autotrophic organism in which the energy source for the synthesis of organic molecules is chemical energy obtained from inorganic chemicals.

chemotaxis movement toward a favorable chemical stimulus (positive chemotaxis) or away from a harmful chemical stimulus (negative chemotaxis).

chemotherapeutic agent a compound that can be used to treat a medical condition.

chloroplast the organelle in photosynthetic eukaryotic cells in which photosynthesis takes place.

chromosome a structure consisting of DNA and protein, which carries genetic information. Prokaryotic cells usually contain a single chromosome, whereas eukaryotic cells have multiple chromosomes.

chronic disease a disease in which symptoms are present for a prolonged period.

-cidal a suffix indicating that a compound or process kills. For instance, "bactericidal" compounds kill bacteria and "fungicidal" compounds kill fungi.

Ciliophora a group of protozoa characterized by the presence of cilia.

cilium short motile cellular projection composed of protein, present on some eukaryotic cells (plural: cilia). Cilia can be involved in the locomotion of the cell or in the generation of a current of fluid over the cell surface.

class a taxonomic level between order and phylum.

clone a population of cells or organisms that are exact genetic copies of each other. The term cloning refers to the creation of exact copies of a gene, cell, or entire organism.

cloning host a cell into which a cloning vector is introduced, and in which it can replicate.

cloning vector a self-replicating DNA molecule into which foreign DNA can be inserted for the purpose of making many copies of that DNA.

coccus a bacterium with a spherical shape (plural: cocci).

coding strand the strand of DNA in a gene that is transcribed into RNA during transcription.

codon a three-base sequence in mRNA that specifies either a particular amino acid or a stop command in the translation of mRNA into protein.

coevolution hypothesis a hypothesis proposing that viruses evolved along with cellular life as obligate intracellular parasites.

commensalism a relationship between two organisms of different species living in close association, in which one of the organisms is benefited and the other is neither benefited nor harmed.

common-source epidemic an epidemic caused when a large number of individuals become infected from the same original source.

competition an interaction between two species in the same habitat that are trying to acquire the same resources.

competitive exclusion the phenomenon observed in some competitive interactions, in which the better-adapted organism drives the less well adapted organism to extinction.

complex capsid a type of viral capsid, neither completely icosahedral nor helical in shape, and often consisting of multiple protein layers.

complex medium a growth medium for microorganisms, to which a variety of nutrients has been added.

composting a technique in which organic solid waste is degraded though the activity of microorganisms.

condensation reaction a reaction in which the individual building blocks of biological molecules (amino acids for proteins, or monosaccharides for carbohydrates, for example) are linked together. An H atom is removed from one molecule and a hydroxyl (OH) group is removed from the other, allowing the two molecules to make a chemical bond, with the release of a molecule of water.

conjugation the transfer of DNA from one bacterial cell to another involving cell-to-cell contact.

constitutive gene a non-regulated gene that is continuously expressed, independently of any environmental conditions.

contamination the unwanted presence of microorganisms.

control group a group of treatments in an experiment in which the experimental variable is not manipulated. The control group is compared with the experimental group, in which the experimental variable is manipulated. This permits the scientist to assess the effect, if any, of the experimental variable on the treatments.

control variable a factor that is maintained the same in both experimental and control groups. Control variables are those variables not being tested in the testing of a specific hypothesis.

coupled reactions two chemical reactions that occur simultaneously, such that an energy-releasing reaction provides the energy needed for an energy-requiring reaction to proceed.

covalent bond a type of chemical bond in which two atoms are linked through the sharing of electrons.

cross-reactivity the ability of some antibodies generated in response to a specific antigen to also bind an entirely different antigen.

curd the solid portion of milk that has been separated from the liquid portion during the production of certain dairy products such as cheese.

Cyanobacteria a lineage of photosynthetic bacteria that release oxygen as a photosynthetic waste product.

cyst a metabolically inactive stage in many protozoan life cycles.

cytokine biologically active proteins released by cells that act on the same or other cells to regulate an immune response.

cytopathic effect any adverse effect that a virus has on its host cell.

cytoplasm the material forming the bulk of a cell inside the cell membrane. The cytoplasm is composed of a semifluid mix of proteins and other biological molecules (the cytosol) in which are located various subcellular organelles (in eukaryotic cells).

cytoskeleton a network of protein fibers in eukaryotic cells that provide structure to the cell and enable movement of and within cells.

cytotoxic T cell (Tc cell) a type of T cell that detects and kills infected host cells that are displaying foreign antigen on their surfaces.

dark reaction in photosynthesis, those reactions that use the chemical energy carriers generated in the light reaction to provide energy for the incorporation of CO_2 into organic molecules. This latter reaction does not require light.

Darwinian medicine the idea that principles of natural selection can be used to artificially select for reduced pathogen virulence.

daughter cell either one of the two newly formed cells arising from the division of a "parent" cell.

daughter DNA a newly synthesized DNA molecule produced by replication of a "parent" DNA molecule.

decline phase the phase in microbial growth during which the overall number of cells is declining.

decomposer an organism that absorbs organic matter from dead organisms and degrades it into inorganic molecules, which are returned to the environment.

definitive host the host in a parasite's life cycle in which sexual reproduction of the parasite occurs.

denatured protein a protein that has unfolded, losing its specific three-dimensional shape. Because they are not folded into their proper conformation, denatured proteins show decreased or no activity.

dendritic cell the main immune-system cell that presents antigen to T cells and activates them.

deoxyribonucleic acid (DNA) the principal genetic material of all living things, with the exception of some viruses. It is composed of chains of covalently linked nucleotides, and carries genetic information encoded in the order, or sequence, of the nucleotides. This information specifies the proteins that a given organism can make. In cells DNA is always found as a double-stranded double helix.

direct contact transmission transmission via physical contact such as kissing, shaking hands, or sexual contact.

disaccharide two monosaccharides linked together.

disease the disruption in some way of normal host function.

disinfectant a chemical agent appropriate for the elimination of microorganisms from non-living material.

DNA library a collection of bacterial clones, each carrying a different DNA fragment from the same organism.

DNA ligase an enzyme able to repair breaks or gaps in the sugar-phosphate backbone of DNA.

DNA polymerase the enzyme that adds nucleotides to the newly synthesized growing DNA strand in DNA replication, using a strand of the parental DNA molecule as a sequence template.

DNA proofreading the ability of DNA polymerase to recognize and remove a mismatched nucleotide in DNA, and replace it with the correct nucleotide.

DNA sequencing any technique by which the order of nucleotides in a DNA molecule can be determined.

DNA virus virus in which the genetic information is encoded in DNA.

domain the most inclusive taxonomic category. All living things belong to one of three domains—the Bacteria, the Archaea, and the Eukarya.

double helix double-stranded DNA, in which two helical DNA strands are wound around each other, forming a double-helix structure.

ecological community the various populations of different species within a defined geographic area.

ecology the study of the relationship of organisms to each other and to their environment.

ecosystem a ecological community and the physical environment with which the community interacts.

effector T cell a T cell that has been activated to perform its immunological function.

electron a fundamental subatomic particle that carries a negative charge. Electrons are distributed around the nucleus of the atom in a series of concentric "electron shells."

electron affinity a measure of the strength of attraction that an atom has for electrons.

electron shell a specified distance from the nucleus of an atom, where that atom's electrons are found. Each shell holds a specific number of electrons. Larger atoms have more electron shells, increasingly distant from the nucleus, to accommodate the larger number of electrons.

electron transport a process in which electrons are serially transferred from one compound to another, generating the conditions in which ATP can be synthesized.

element a substance that cannot be broken down to simpler components by chemical processes.

emergent disease a new or changing disease that is increasing in importance.

endemic disease a disease that is always present in a particular geographic area, usually occurring at low level.

endergonic reaction a chemical reaction that requires an input of energy.

endocytosis a mechanism used by eukaryotic cells to import molecules into the cell by engulfing them to form internal membrane-enclosed endosomal vesicles.

endomembrane system a group of membrane-enclosed organelles in eukaryotic cells that are involved in both endocytosis and exocytosis.

endoplasmic reticulum a network of membrane-enclosed tubules in eukaryotic cells that is involved in a variety of cellular processes including intracellular transport and exocytosis. It is part of the endomembrane system.

endoplasmic reticulum lumen the space within the endoplasmic reticulum.

endospore a structure formed by some bacteria that allows the cell to persist in a metabolically inactive state when the environment is not conducive to growth.

endosymbiosis the theory that some eukaryotic cell organelles, specifically the chloroplast and the mitochondrion, were originally prokaryotic cells that entered into a symbiotic relationship with larger cells.

endotoxin the lipid portion of the lipopolysaccharide of the outer membrane of Gram-negative bacteria, which has toxic properties when released upon the death of the bacterial cell.

enology the science of wine making.

enzyme a biological catalyst; usually a protein, but some enzymes are RNAs. Enzymes speed up the rate of biological reactions so that they can occur at a useful rate in the conditions inside living cells.

enzyme inhibitor any molecule that can bind to an enzyme and inhibit its activity.

epidemic a sudden increase in the number of cases of a specific disease, beyond what is considered to be a normal number of cases.

epidemic threshold the minimum proportion of a population that must be immune to a specific pathogen to prevent an epidemic.

epidemiology the study of diseases in populations.

ergot A toxin produced by the fungus *Claviceps purpurea*.

escaped-gene hypothesis a hypothesis proposing that viruses were originally fragments of prokaryotic or eukaryotic nucleic acids that evolved the capacity to replicate independently.

Eukarya the taxonomic domain consisting of all organisms that have eukaryotic cells.

eukaryotic cell a cell in which the DNA is enclosed in a membrane-enclosed organelle known as the nucleus. Animals, plants, fungi, protozoa, and algae have eukaryotic cells and are often collectively referred to as eukaryotes.

eutroph a microorganism requiring relatively high nutrient levels.

eutrophication the depletion of oxygen in an aquatic environment, often caused by microorganisms in response to an input of excess nutrients.

evolution the genetically based development of new properties in organisms over successive generations.

exergonic reaction a chemical reaction that releases energy.

exocytosis a mechanism in eukaryotic cells by which material can be released from the cell. It involves the fusion of a transport vesicle with the plasma membrane and the release of the vesicle's content to the extracellular environment.

exon a block of nucleotide sequence within a eukaryotic gene that codes for an amino acid sequence. Genes generally have more than one exon. Exons are separated by introns, nucleotide sequences that do not code for amino acids.

exotoxin a general term for the protein toxins produced by many pathogenic bacteria.

experimental group a group of treatments in an experiment in which the experimental variable is manipulated. The experimental group is compared with the control group, in which the experimental variable is not manipulated. This permits the scientist to assess the effect, if any, of the experimental variable on the treatments.

experimental variable the factor being tested in the testing of a specific hypothesis. The experimental variable will be manipulated in the experimental group, and any effect of this variable will be detected by comparison with a control group, in which the experimental variable was not manipulated.

extracellular environment the environment outside a cell.

extremozyme an enzyme that retains its ability to function under certain extreme physical conditions.

F⁻ cell a bacterial cell able to act as a recipient during conjugation.

F⁺ cell a bacterial cell able to act as a donor during conjugation.

facultative anaerobe an organism that can grow with or without oxygen.

family a taxonomic level between genus and order.

fastidious describes organisms that make relatively few of the biological compounds that they require for growth.

fermentation the anaerobic breakdown of organic molecules such as carbohydrates for the purpose of ATP synthesis, in which the final electron acceptor is an inorganic molecule.

filamentous fungus a fungus made up of an extensively branched system of tube-like filaments.

fimbria also called a pilus. A short protein fiber extending beyond the cell wall of many bacteria (plural: fimbriae). Fimbriae aid in adherence to specific surfaces.

final electron acceptor a molecule that accepts electrons at the end of the electron transport chain.

flagellum a long motile protein fiber, extending beyond the cell wall of many bacteria (plural: flagella). Flagella-bearing bacteria move by rotating their flagella. Some eukaryotic cells use flagella for movement, but the structure and mechanism of action of eukaryotic flagella are quite different from those of bacterial flagella.

fleshy fungi a group of multicellular fungi characterized by the production of reproductive structures composed of tightly packed hyphae.

fluid mosaic model of membranes the accepted model of membrane structure, in which membrane proteins are embedded in a fluid phospholipid bilayer.

fomite an inanimate object serving as the vehicle of transmission during indirect contact transmission of a pathogen.

food chain the path of nutrient flow between different species in an ecological community, for example the route linking primary producers and predators.

food or waterborne transmission transmission of a pathogen via contaminated food or water.

food web a network diagram that aims to show all the routes of nutrient flow between the species, or a group of species, in an ecosystem. A network, rather than a food chain, results because some species can feed at different trophic levels.

frameshift mutation a mutation caused by either the deletion of a nucleotide or the insertion of an extra nucleotide into a DNA sequence, which upsets the correct amino acid coding.

free radical a highly reactive, electrically charged compound that can bind to and damage other molecules.

fusion a mechanism used by some enveloped viruses to enter host cells, in which the viral envelope fuses with the host cell's plasma membrane, releasing the capsid into the host cell's cytoplasm.

gamete a specialized sex cell produced by eukaryotic organisms for sexual reproduction.

gel electrophoresis a technique used to separate a mixture of molecules on the basis of their movement in a gel-like material. How far it moves depends on the size and electrical charge of the molecule.

gene a stretch of DNA carrying the encoded genetic information necessary to specify the structure of one protein or portion of a protein.

gene expression the overall process of converting the encoded information in a gene's nucleotide sequence into a specific protein.

gene probe a short, single-stranded length of DNA with a sequence that is complementary to part of a gene being sought, and so will hybridize with it and identify it.

gene therapy the treatment of genetic disease by introducing normal copies of a gene that is defective in the individual affected.

generation time the time required for a cell or individual organism to divide or replicate.

genetic engineering The purposeful direct manipulation of DNA to create living things with novel genetic characteristics.

genetic recombination a process in which an exchange of genetic information occurs between two chromosomes or DNA molecules, resulting in the production of new genotypes.

genome the entire sum of an organism's genetic material.

genomic library a DNA library containing clones that together contain all the DNA from a particular organism.

genomics the field in biology that attempts to sequence and study the complete genomes of organisms.

genotype the specific genetic makeup of an individual for any particular characteristic.

genus a group of closely related species. The level of classification between species and family.

germ theory of disease the concept developed in the 19th century that microorganisms could cause disease.

glycocalyx t gel-like layer composed of polysaccharides, secreted by many bacteria to form a layer exterior to the cell wall.

glycolysis the first stage of cell respiration in which carbohydrate substrates are oxidized to pyruvate without a requirement for oxygen.

glycoprotein a protein that is covalently linked to carbohydrates.

Golgi apparatus an organelle found in many eukaryotic cells. Part of the endomembrane system that is involved in protein modification and transport.

Gram-negative bacteria bacteria that have both a thin peptidoglycan cell wall and an outer membrane, and which do not retain Gram stain.

Gram-negative outer membrane a second membrane present in Gram-negative bacteria, located beyond the plasma membrane and the thin cell wall.

Gram-positive bacteria bacteria with a cell wall composed of multiple peptidoglycan layers but lacking the outer membrane of Gram-negative bacteria, and which retain Gram stain.

Gram staining a staining technique allowing differentiation between Gram-positive and Gram-negative bacteria.

green nonsulfur bacteria a lineage of anaerobic, photosynthetic bacteria.

green sulfur bacteria a lineage of anaerobic, photosynthetic bacteria that use sulfur compounds during photosynthesis.

halophile an organism that grows best at relatively high concentrations of salt.

helical capsid a common type of viral capsid in which the capsomeres form a spiral around the nucleic acid.

helminth a general term for parasitic worms.

helper T cell (Th cell) a type of T cell responsible for coordinating an adaptive immune response.

herbivore a member of an ecological community that eats producers (e.g. plants or photosynthetic algae).

heterotroph an organism that obtains the materials for biosynthetic reactions by consuming organic molecules rather than constructing them from inorganic precursors.

host the organism in or on which a parasite lives.

host range the spectrum of hosts that a pathogen can infect.

host-to-host epidemic an epidemic in which the disease is transmitted from infected to non-infected individuals.

humoral response that component of an adaptive immune response in which antigen-specific antibodies are produced by plasma cells.

humus the top layer of soil, rich in organic material.

hydrocarbon a compound composed of carbon and hydrogen only.

hydrogen bond a type of non-covalent bond common in or between proteins. It can be formed between a hydrogen with a partial positive charge and certain atoms with negative charge in another chemical group.

hydrolysis reaction a reaction in which a covalent bond is broken by the involvement of a molecule of water, adding H to one reaction product and OH to the other. For example, hydrolysis breaks down polymeric biological molecules such as proteins and polysaccharides into their component building blocks.

hydrophilic property of a molecule indicating that it will interact with water. Hydrophilic molecules have partial or complete electrical charges, which interact with the partial charges on water molecules.

hydrophobic property of a molecule indicating that it will not interact with water, because it has no charges to interact with the partial charges on water molecules.

hypha a tube-like filament that is the typical structural unit of most fungi (plural: hyphae).

hypothesis a tentative explanation for a particular question about the natural world. A proper hypothesis must be used to generate predictions about future events, which can then be tested by experiment or observation, allowing the hypothesis to be either accepted or rejected.

icosahedral capsid a common type of viral capsid in which each of its 20 faces is an equilateral triangle.

immunological memory the ability of an adaptive immune response to grow stronger with repeated exposure to the same pathogen.

immunology the scientific study of host defenses against pathogens.

inactivated virus a virus that is no longer able to infect host cells and replicate.

incidence of disease the number of new disease cases over a specified length of time.

inclusion body a darkly staining body seen in the cytoplasm of cells infected with certain viruses. It is composed of viral proteins that have failed to assemble correctly.

incubation period the time between initial exposure to a pathogen and the onset of signs and symptoms.

indirect contact transmission transmission of a pathogen via an inanimate object .

inducer a molecule that can stimulate expression of an operon by binding a repressor protein and removing it from the operator.

inductive period the time between infection and the onset of an adaptive immune response, during which lymphocytes are dividing and developing into effector cells.

industrial fermentor a large vat in which the microbial production of a desired metabolite, such as an antibiotic, takes place on a commercial scale.

inert gas an element, existing as a gas at normal temperature and pressure, that is nonreactive because its atoms have full outer electron shells and so do not form bonds with other atoms.

infection a situation in which microorganisms are reproducing in or on a host.

infectious dose-50 (ID$_{50}$) the number of microorganisms that must enter a host to establish an infection in 50% of exposed hosts.

inflammation a host response to infection and tissue damage that is characterized by vasodilation and increased permeability of small blood vessels at the site of damage.

innate immunity inborn host defenses that are not specific for a particular type of pathogen.

intercalating agent a chemical that produces mutations by inserting itself into a replication fork during DNA replication.

interferon an antiviral cytokine released by cells in response to viral infection.

intermediate host the host in a parasite's life cycle in which sexual reproduction of the parasite does not occur.

intertidal zone an ecological zone in a marine environment located between the low and high tide marks.

intron a block of nucleotide sequence within a eukaryotic gene that does not code for an amino acid sequence. An intron separates two exons, which are the regions within a gene that encode amino acids.

ion a charged atom, produced by the gain or loss of one or more electrons.

ionic bond a chemical bond formed by the mutual attraction of negatively charged and positively charged ions.

in vitro a term denoting that a process was carried out under laboratory conditions.

in vivo a term denoting that a process was carried out in a living organism.

isomers molecules with the same chemical formula but different structures.

isotope a form of an atom defined by the number of neutrons in its nucleus. Different isotopes of the same element have different numbers of neutrons but the same number of protons.

kingdom a taxonomic level between phylum and domain.

Koch's postulates criteria that must be demonstrated to show that a particular microorganism causes a particular disease.

Krebs cycle a pathway that can occur in cell respiration, in which two-carbon molecules are oxidized using NAD^+ and FAD to produce NADH, $FADH_2$, and CO_2. It is also called the citric acid cycle.

laboratory fermentor a relatively small vessel in which combinations of physical and nutritional factors can be manipulated in the laboratory to determine the optimum conditions for microbial metabolite production.

lag phase the phase in microbial growth, before the log phase, during which there is no apparent population growth.

latent infection a type of viral infection, during which the virus does not replicate.

lethal dose-50 (LD$_{50}$) the number of pathogens that must be used to infect a host in order to kill 50% of the infected hosts.

leukocyte general name for a white blood cell. There are several different types of leukocytes, which are involved in defense against infection and immune responses.

light reaction in photosynthesis, the processes that capture the energy of sunlight and convert it into chemical energy carriers, and which can only occur in the light. Compare dark reaction.

lipase any enzyme that breaks down lipids by hydrolyzing ester bonds.

lipid a class of biological molecules that primarily in energy storage and as structural components of cells. Fats, oils, and sterols are examples of lipids. They are largely hydrophobic in nature.

lipid A the lipid portion of the lipopolysaccharide of Gram-negative bacteria. It anchors the lipopolysaccharide to the outer membrane.

lipopolysaccharide a complex molecule composed of lipid and polysaccharides, found in the outer membrane of Gram-negative bacteria.

liposome a small, hollow spherical structure, formed by phospholipids when placed in water. Liposome-like structures may have been the precursors of the first cells.

littoral zone the shallowest part of the photic zone, nearest to the shore, through which light penetrates right to the bottom.

log phase the phase in microbial growth during which population growth is exponential.

lymph the fluid present within the lymphatic system.

lymph node a small organ rich in lymphocytes, in which adaptive immune responses are generated. Lymph nodes are connected to each other by the lymphatic system.

lymphatic system a network of vessels that returns extracellular fluid to the blood and is involved in carrying pathogens from the tissues to lymph nodes.

lymphocyte a type of white blood cell that generates adaptive immune responses.

lysis the bursting of a cell membrane, resulting in cell death.

lysogeny infection of a bacterium by a bacteriophage in which the bacteriophage does not replicate.

lysosome an organelle of eukaryotic cells that is involved in the digestion of worn-out cellular components, and of foreign material imported into the cell.

lytic cycle the replicative cycle of a bacteriophage inside an infected bacterium, which results in lysis and death of the infected bacterium and release of newly formed bacteriophages.

macrophage the main type of phagocytic cell in vertebrates. It scavenges dead cells; it recognizes, ingests, and kills microorganisms; and it can also act as an antigen-presenting cell.

major histocompatibility complex (MHC) a cluster of genes encoding proteins known as MHC molecules, which are required for the presentation of antigen to T cells.

malt germinated barley seeds in which the starch is digested to simpler sugars.

Mastigophora a group of protozoa characterized by one or more flagella. Also known as flagellates.

matrix in a mitochondrion, the compartment enclosed by the outer membranes and in which the reactions of the Krebs cycle take place.

maximum growth temperature the highest temperature at which an organism can grow.

mechanical vector transmission a mode of pathogen transmission by an invertebrate vector, in which the pathogen adheres to the invertebrate's body without replicating or developing in any way.

memory B cell a long-lived B cell that responds to a second infection of the host by a given pathogen by generating a secondary humoral immune response.

memory T cell a long-lived T cell that responds to a second infection of the host by a pathogen by generating a secondary immune response.

mesophile an organism that grows best at intermediate temperatures.

messenger RNA (mRNA) an RNA molecule transcribed from a protein-coding gene.

metabolic pathway a series of enzyme-catalyzed chemical reactions in which a starting compound is ultimately converted into a final product.

metabolism the biochemical reactions carried out by a cell or organism.

metabolite an intermediate compound or final product produced in a metabolic pathway.

methanogen a member of a lineage of anaerobic archaea that produce methane gas as a metabolic waste product.

microaerophile an organism that grows best at low oxygen concentrations.

microbe *see* microorganism.

microenvironment the environment immediately surrounding a microorganism.

microorganism a living thing too small to be seen with the unaided eye. Microorganisms include bacteria, archaea, fungi, protozoa, algae, and viruses.

minimal medium a growth medium for nonfastidious microorganisms, containing a carbon source and inorganic compounds.

minimum growth temperature the lowest temperature at which an organism can grow.

minus-strand RNA virus a virus in which the genetic information is an RNA molecule with a sequence complementary to that of the RNA that encodes the viral proteins.

mismatch repair the process in which DNA repair enzymes recognize, remove, and replace mispaired or damaged DNA nucleotides in a molecule of DNA.

missense mutation a mutation in which a change in a single nucleotide results in a change in the amino acid encoded by the affected codon.

mitochondrion an energy-generating organelle found in most eukaryotic cells in which most of the cell's ATP is produced.

mode of transmission the manner in which a pathogen travels from an infected host to new, uninfected, and susceptible hosts.

model organism a species that has been intensively studied in the laboratory. Biological principles revealed through the study of model organisms may then be extrapolated to other living things.

mold a type of fungus characterized by a loosely organized mycelium.

molecule a chemical entity composed of two or more atoms held together by covalent chemical bonds.

monoculture an agricultural area dominated by a single crop.

monosaccharide any of the sugars that serve as the basic carbohydrate building blocks. Monosaccharides can be linked together to form larger molecules, such as disaccharides (two monosaccharides), trisaccharides (three monosaccharides) or polysaccharides (many saccharides).

motility the capacity for active movement.

multicellular consisting of many cells.

must the juice extracted from crushed grapes at the first step in wine making.

mutagen a chemical or a physical agent (such as radiation) that increases the likelihood of mutations in DNA.

mutation a change in the nucleotide sequence of DNA.

mutualism a biological interaction between two organisms that provides benefits for both.

mycelium the entire network of hyphae in a filamentous fungus.

mycorrhizal association a mutually beneficial relationship between a fungus and the roots of a plant.

mycosis a fungal infection in animals.

narrow spectrum the general name given to antimicrobial drugs that are effective only against very specific microorganisms.

native protein conformation the active form of a protein when it is folded up into its specific three-dimensional shape.

natural selection the mechanism through which biological evolution is thought to occur.

negative taxis movement away from an unfavorable stimulus.

neutral mutation a mutation in which the change in a particular DNA sequence does not change the sequence of amino acids specified by the DNA.

neutralization the inactivation of pathogens or toxins by the binding of a specific antibody.

neutron a fundamental subatomic particle present with protons in an atom's nucleus. Neutrons do not carry an electrical charge.

neutrophil a type of phagocytic cell that engulfs and kills microorganisms in an immune response.

nicotinamide adenine dinucleotide (NAD⁺) a molecule that acts as a hydrogen or electron acceptor in many biochemical reactions.

nicotinamide dinucleotide phosphate (NADP⁺) a molecule similar to NAD⁺ that is used as a final electron acceptor in the electron transport chain of the photosynthetic light reaction.

nitrification the microbial conversion of ammonium to nitrite and nitrate.

nitrogen fixation the conversion of atmospheric nitrogen into a form that plants can absorb. In nature, only a relatively small number of bacterial species can fix nitrogen.

nonenveloped virus a virus lacking the outer phospholipid bilayer present in enveloped viruses.

nonfastidious describes organisms that synthesize many or most of the biological compounds that they require for growth.

nonpolar covalent bond a type of covalent bond in which the two bonded atoms have an equal affinity for electrons. The electrons are consequently shared equally, resulting in a lack of any partial charges on the bonded atoms.

nonsense mutation a change in a single nucleotide that results in the conversion of a readable codon into a stop codon.

normal flora those microorganisms that colonize a host without causing disease.

nosocomial infection an infection acquired in a hospital.

nuclear envelope the double-layered membrane surrounding the nucleus in eukaryotic cells.

nucleic acid DNA or RNA, which are together involved in the storage and readout of genetic information. Nucleic acids are formed of chains of covalently linked nucleotides. In all cellular forms of life and in many viruses, DNA serves as the store of genetic information, while RNA assists in converting the information encoded in DNA into proteins.

nucleocapsid the capsid and associated nucleic acid of a virus particle.

nucleoid area the area within the cytoplasm of a prokaryotic cell, in which the DNA is found.

nucleotide the building-block of a nucleic acid, consisting of a sugar molecule, covalently bound to a phosphate group and to a nitrogen-containing base (either a purine or a pyrimidine).

nucleus of a eukaryotic cell, see cell nucleus. Of an atom, the central core of an atom, composed of protons and neutrons. The nucleus makes up almost all of an atom's mass.

O polysaccharide a molecule forming the polysaccharide portion of the lipopolysaccharide found in the outer membrane of Gram-negative bacteria.

obligate aerobe an organism that requires oxygen for survival and growth.

obligate anaerobe an organism for which oxygen is toxic.

oligotroph a microorganism adapted to low nutrient levels.

omnivore an animal that eats both plants and other animals.

oncogene a cancer-causing gene, of either viral or host origin.

oocyst the infective, cyst-like structure in the life cycle of some protozoa.

operator a regulatory nucleotide sequence present in an operon. When the operator is bound by a repressor protein, the operon is not expressed.

operon a type of gene organization found mainly in prokaryotes in which a group of adjacent genes is induced or suppressed as a unit. When induced, the genes are transcribed together as a single mRNA from which the individual proteins are translated.

opportunistic infection disease caused by a microbe that does not usually cause disease. It occurs in unusual circumstances, such as when the host's immune system is weakened.

opsonin a molecule that forms a physical link between a phagocytic cell and the material to be engulfed, facilitating phagocytosis.

optimum growth temperature the temperature at which an organism's growth rate is greatest.

order a taxonomic level between family and class.

organ in multicellular eukaryotes, a discrete structure composed of more than one tissue working together to carry out a particular function.

organelle general term for a membrane-enclosed structure within a eukaryotic cell that carries out a distinct function.

organic molecule any molecule built on a skeleton of two or more linked carbon atoms.

origin of replication the point on a DNA molecule where the two parental strands are separated in preparation for DNA replication.

oxidation the addition of oxygen to a molecule, or the loss of electrons or hydrogen.

pandemic technically, a world-wide epidemic. Commonly used to describe any widely distributed, multi-continent epidemic.

parasite an organism that lives in or on another organism, from which it obtains a place to live and reproduce, and a source of nutrients.

parent cell the original cell, before cell replication, which gives rise to two daughter cells.

parental DNA refers to a double-helical DNA molecule before DNA replication, each strand of which gives rise to a new molecule of double-helical daughter DNA.

pasteurization a process that eliminates most pathogens or organisms causing spoilage from liquids through brief exposure to high temperature.

pathogen a disease-causing microorganism.

pathogenesis the capacity of a microorganism to cause disease.

pattern recognition the ability of cells of the innate immune system to recognize certain general types of molecules that are unique to microorganisms.

pelagic zone the deeper portion of the photic zone in an aquatic environment, below which lies the profundal zone into which light does not penetrate.

pentaglycine bridge a short peptide consisting of five glycine residues, which links one layer of peptidoglycan to another in bacterial cell walls.

peptide a short chain of linked amino acids, either occurring independently or forming a portion of a protein.

peptide bond the covalent bond linking two adjacent amino acids in a peptide or protein.

peptidoglycan a complex molecule composed of carbohydrate and peptides, found in the cell walls of many bacteria.

peripheral protein a membrane protein that is attached to the membrane surface or is embedded within one layer of the membrane without passing through to the opposite side.

personal characteristics a description of the personal attributes of those individuals who become ill with a particular medical condition.

pH scale a numerical scale used to represent the concentration of protons (H^+) in solution. The pH scale thus provides a relative measure of the acidity of a solution.

phagocytosis a process by which relatively large particles (such as bacterial cells) are engulfed by a eukaryotic cell, with the plasma membrane extending around the particle to form a transport vesicle that is brought into the cell.

phenotype the outward appearance or other detectable property of an organism in regard to a specific trait. It is the result of the interaction of the organism's genotype for that trait and environmental influences.

phosphate group a functional group with the chemical formula PO_4^{3-}.

phospholipid a type of lipid consisting of glycerol covalently linked to two fatty acids and one phosphate group.

phospholipid bilayer a structure consisting of two adjacent layers of phospholipids, both with the hydrophilic phosphate groups of the phospholipids projecting outward into an aqueous environment, and hydrophobic fatty acid tails projecting inward, away from water. Biological membranes are composed of phospholipid bilayers.

phosphorylation the process of adding a phosphate group to a molecule.

photic zone the vertical zone in an aquatic environment through which light can penetrate, permitting photosynthesis.

photoautotroph an autotrophic organism in which light is used as the original energy source for the synthesis of organic molecules.

photosynthesis the process by which certain prokaryotes and eukaryotes capture energy from sunlight and use it for the synthesis of organic molecules from inorganic precursors.

phylogenetics a technique used to classify living things based on their similarity in DNA sequence, which reflects evolutionary relationships.

phylum a taxonomic level between class and kingdom.

phytoplankton Photosynthetic, planktonic microorganisms.

pilot plant fermentor a fermentor in which microbial metabolite production is tested, before the use of a large-scale industrial fermentor.

pilus also called a fimbria. A short fiber made of protein that extends beyond the cell wall of many bacteria (plural: pili). Pili aid in bacterial adherence to specific surfaces.

place characteristics a description of where individuals who become ill with a particular medical condition are found.

plankton small, aquatic organisms, either prokaryotic or eukaryotic, that drift through their environment.

plasma cell an antibody-producing cell that develops from a B cell in the adaptive immune response.

plasma membrane also called the cell membrane. A phospholipid bilayer membrane forming the outer boundary of cells. It also contains proteins and which regulates the transport of materials into and out of the cell.

plasmid a small self-replicating circle of DNA that is present in many bacterial cells in addition to the main chromosome.

polar covalent bond a type of covalent bond in which the two atoms involved have different affinities for the shared electrons. The electrons are consequently shared unequally, imparting a partial negative charge to the atom with higher affinity and a partial positive charge to the atom with lower affinity.

polymerase chain reaction (PCR) a technique used to make a very large number of copies of a particular DNA "target" sequence.

polysaccharide a large carbohydrate molecule composed of many monosaccharides covalently linked together into a branched or unbranched chain.

population all the members of a particular species that live in a defined geographic area.

porin a channel-like protein found in the outer membrane of Gram-negative bacteria, through which small, hydrophilic molecules can cross the outer membrane.

portal of entry the route by which a pathogen enters a host.

portal of exit the route by which a pathogen leaves a host.

positive-strand RNA virus a virus in which the genetic information is encoded in RNA, which also codes for viral proteins.

positive taxis movement toward a favorable stimulus.

predation the act of catching and consuming another organism.

primary consumer also called a herbivore. An organism that eats producers.

primary immune response the adaptive immune response that occurs upon first exposure to a pathogen.

primary metabolite a metabolite for which the rate of synthesis closely mirrors population growth.

primer a short single-stranded DNA sequence used in PCR that is complementary to one end of the target sequence and is used to start DNA replication at the beginning of the required region of the DNA.

prion an infectious protein, thought to lack all nucleic acid. Prions are thought to reproduce by forcing normal proteins to take on the shape of the aberrant prion protein.

probiotic a food product containing live microorganisms that are thought to promote good health.

producer an organism that converts inorganic carbon into biological molecules through the process of carbon fixation.

product the substance formed in a chemical reaction. In living organisms, the conversion of a substrate into a product is catalyzed by an enzyme.

profundal zone the vertical zone of water below the photic zone in an aquatic environment, in which photosynthesis cannot occur because light does not penetrate.

prokaryotic cell a cell in which the DNA is not enclosed within a membrane-enclosed nucleus. Bacteria and archaea have prokaryotic cells and are often collectively referred to as prokaryotes.

promoter a nucleotide sequence found at the beginning of a gene that indicates where RNA polymerase binds and transcription of the gene begins.

prophage bacteriophage DNA when it is part of the bacterial chromosome during lysogeny.

protease enzymes that digest proteins by hydrolyzing the peptide bonds that link the amino acids together.

protease inhibitor a type of drug that inhibits the protease of retroviruses, thus inhibiting replication and transmission of the virus.

protein a class of large biological molecules, each composed of a folded chain of linked amino acids. There are many different kinds of proteins, each of which performs distinct functions in the cell and organism, for example as components of cell structure, enzymes, transporters and gene regulators. Many bacterial toxins are proteins.

proteobacteria a lineage of bacteria characterized by a Gram-negative outer membrane.

proton a fundamental subatomic particle found in the nucleus of an atom. Protons carry a positive charge, and the number of protons determines what type of element an atom is.

proton gradient the buildup of protons (H^+ ions) on one side of a membrane, relative to the other side. Proton gradients across membranes can be used as an energy source to transport materials across the membrane, and also as a source of energy for ATP synthesis.

psychrophile an organism that grows best at relatively low temperatures.

psychrotolerant describes organisms that can continue to grow at relatively low temperatures.

pure culture technique the process of growing microorganisms in the laboratory in a prescribed manner such that cultures contain only a single, desired type of organism.

purine a class of double-ringed, nitrogen-containing organic bases that form part of some nucleotides found in DNA and RNA. The two purines present in both DNA and RNA are adenine and guanine.

pyrimidine a class of single-ringed, nitrogen-containing bases that form part of some nucleotides found in DNA and RNA. Thymine and cytosine are the two pyrimidines present in DNA; uracil and cytosine are the two pyrimidines present in RNA.

pyruvate the three-carbon molecule that is the end product of glycolysis.

reactivation of viruses, the return of a latent virus to active replication.

reactivity a measure of the tendency of an atom or molecule to bond with other atoms or molecules. The greater the reactivity, the greater the likelihood of such bonding.

recombinant DNA a DNA molecule formed by combining DNA from two different sources.

recombinant vaccine a vaccine produced through the use of recombinant DNA technology.

Red Queen hypothesis the hypothesis that in host–parasite interactions, an evolutionary adjustment on the part of one species requires a subsequent evolutionary adjustment by the other species. Consequently, continuing evolution is required by both species to ensure survival.

redox reaction a reaction in which one molecule is oxidized and another molecule is reduced.

reduction the addition of hydrogen or electrons to a molecule, or the loss of oxygen.

re-emerging disease a disease that was considered to be under control but is now causing increasingly serious problems.

regressive hypothesis a hypothesis proposing that viruses evolved from intracellular prokaryotic organisms.

regulated gene a gene whose expression can be controlled, as environmental conditions warrant.

regulatory T cell (T_R cell) a type of T cell that may be involved in inhibiting adaptive immune responses.

release in the replicative cycle of a virus, the stage at which newly assembled virus particles are released from the host cell.

replication fork the exact site on a replicating DNA molecule where nucleotides are being added to the growing, newly synthesized DNA strands.

repressor any protein that inhibits expression of an operon in prokaryotes and of individual genes in eukaryotic cells.

reservoir a place where a pathogen can survive indefinitely without causing disease, and from which it can infect new hosts when they become available.

restriction enzyme an enzyme that cuts DNA at specific nucleotide sequences.

retrovirus a type of RNA virus that makes a DNA intermediate that is integrated into a host cell chromosome. The virus is replicated by assembly of viral proteins and genomic RNA expressed by the viral DNA.

reverse transcriptase a retroviral enzyme that produces DNA by transcription of an RNA template, a process called reverse transcription.

reverse transcriptase inhibitor a type of drug that inhibits the activity of retroviral reverse transcriptase, thus preventing viral replication.

reverse transcription the process of transcribing RNA into DNA using the enzyme reverse transcriptase.

rhizosphere the soil environment immediately surrounding plant roots.

ribonucleic acid (RNA) along with DNA, one of the two types of nucleic acid found in cells. RNA functions mainly to convert the genetic information stored in DNA into proteins and RNAs that carry out the functions of the cell.

ribosomal RNA (rRNA) a type of RNA that, along with specific proteins (ribosomal proteins), forms ribosomes.

ribosome the site of protein synthesis in a cell. It is a multimolecular structure composed of proteins and ribosomal RNAs.

ribozyme enzyme made of RNA rather than protein.

RNA-dependent RNA polymerase an enzyme used by both plus- and minus-strand RNA viruses to synthesize RNA, using RNA as a template.

RNA polymerase the enzyme that unwinds the DNA double helix during transcription and assembles a chain of RNA nucleotides complementary to the exposed DNA sequence.

RNA splicing the removal of introns from a molecule of mRNA, after transcription and before translation.

root nodule a specialized structure found on the roots of some plants, which is colonized by nitrogen-fixing bacteria.

rough endoplasmic reticulum endoplasmic reticulum with ribosomes on its cytoplasmic surface.

salt a compound held together by ionic bonds between the atoms.

saprophyte an organism that obtains nutrients from dead organic material.

Sarcodina a group of protozoa that move by extending portions of their cytoplasm.

saturated fat a type of fat in which all the carbon atoms in the fatty acids are linked together by single bonds.

scale-up a sequence of stages in which production of a metabolite by a microorganism is transferred from small-scale to large-scale production.

science a process of learning about nature through observation and experiment.

scientific method a defined series of steps in which the scientist learns about nature. The scientific method requires the construction and testing of a proper hypothesis.

secondary immune response an adaptive immune response that occurs upon a second exposure to a particular pathogen, and which is usually stronger and more rapidly effective than the primary response.

secondary metabolite a metabolite that is only synthesized at maximum rate once the microorganism producing it has entered the stationary phase of growth.

selective toxicity a property of some antimicrobial drugs, describing the ability to kill microorganisms without harming the host.

sepsis a bacterial infection of a normally sterile area of the body such as the blood.

sex pilus a long hollow fiber made of protein that is produced by a bacterium to connect it to another bacterium (plural: sex pili). Sex pili bring two bacterial cells of the same species together for conjugation and the exchange of genetic material.

sexual reproduction a type of reproduction in which genes from two parent organisms are combined in the offspring.

single-cell protein bulk microorganisms that are used as a source of protein for humans or animals.

slime layer a glycocalyx with an irregular and diffuse structure, produced by bacterial cells. It aids in their adherence and/or protection.

smooth endoplasmic reticulum endoplasmic reticulum lacking ribosomes.

solute the material dissolved in a solvent to form a solution. In biological systems, the solute is usually hydrophilic molecules dissolved in water (the solvent).

solution a mixture of two or more substances in the same phase (solid, liquid, or gas). In biological systems most solutions consist of water (the solvent) and hydrophilic material dissolved in the water (the solute).

solvent the substance in which a solute is dissolved to form a solution.

Southern blotting a technique that uses short, single-stranded DNA "probes" to detect a particular DNA sequence among a large number of DNA fragments that have been separated by electrophoresis.

species the most fundamental level of classification within a hierarchical taxonomy.

spirillum a type of bacterium characterized by a spiral shape (plural: spirilla).

spirochete a type of bacterium characterized by a helical shape.

spleen an organ in mammals in which adaptive immune responses to blood-borne pathogens are generated; it is also the organ in which worn-out red blood cells are destroyed.

spontaneous mutation any change in DNA sequence that arises in the absence of a mutagen, such as a random error in DNA replication.

staphylococcus common name for any member of the bacterial genus *Staphylococcus*, which are cocci that typically grow as grape-like clusters (plural: staphylococci).

-static a suffix indicating that a process or compound inhibits microbial population growth. For instance, "bacteriostatic" processes prevent bacterial replication.

stationary phase the phase in microbial growth after the lag phase, during which the population is neither increasing nor decreasing.

sterilization the process of completely eliminating all living things from an area or an object.

sterols a class of lipid molecules, examples of which are cholesterol and certain hormones.

sticky ends short, single-stranded, overhanging ends on double-stranded DNA molecules that are formed when DNA is cut with certain restriction enzymes. As the single-stranded portions are complementary in sequence to each other, two pieces of DNA cut by the same enzyme can be joined via pairing of these ends.

storage granule a structure composed of carbohydrate or lipid and used as a reservoir of nutrients in prokaryotic cells.

strain improvement a process of increasing the production of a metabolite in a strain of microorganisms beyond what a naturally occurring strain would synthesize.

streptobacillus a general description of any rod-shaped bacterium that forms long chains of cells (plural: streptobacilli).

streptococcus common name for any member of the bacterial genus *Streptococcus*, which are cocci that typically grow linked into long chains (plural: streptococci).

stroma in chloroplasts, the compartment enclosed by the outer membranes and in which the reactions of carbon fixation take place.

stromatolite a layered structure occasionally formed by mats of cyanobacteria and other microorganisms in warm shallow waters. Fossil stromatolites contain what are believed to be the earliest microbial fossils.

subclinical infection an infection in which no symptoms are apparent.

substrate the starting substance in an enzyme-catalyzed reaction, and which is converted into product in the reaction.

subunit vaccine a non-living vaccine consisting only of specific antigens from a pathogen.

synchronous growth a growth pattern in which all the cells in a population divide at the same time.

taxonomy the science of biological classification.

T cell the class of lymphocytes that carry out the cell-mediated part of an adaptive immune response. T cells develop in the thymus.

T-cell receptor (TCR) receptor protein present on the surface of T cells, which recognizes a specific antigen bound to an MHC molecule on the surface of an antigen-presenting cell.

termination sequence a nucleotide sequence found at the end of a gene that indicates where transcription will end.

thermocline a sharp demarcation in temperature between the top layer of warmer water and the colder, deeper water in deep lakes in temperate regions during the spring and summer.

thermophile a microorganism that grows best at high temperatures.

thermotaxis movement toward a favorable temperature (positive thermotaxis) or away from an unfavorable temperature (negative thermotaxis).

thylakoid membrane a folded membrane found in some autotrophs, containing the electron transport chain that carries out the light reactions of photosynthesis.

thymine dimer a covalent linkage of two adjacent thymine bases in a molecule of DNA.

thymus a lymphoid organ in mammals in which T cells mature.

time characteristics a description of when individuals are becoming ill with a particular medical condition.

tissue an assemblage of a characteristic set of cell types that carries out a particular physiological function in multicellular eukaryotes.

Toll-like receptor any one of a class of receptors found mainly on cells of innate immunity such as macrophages and dendritic cells. Toll-like receptors recognize molecular features of microbes that are not present in animal cells.

transcription the synthesis of RNA by using a strand of DNA as a sequence template. RNA is complementary in sequence to its template DNA.

transcription factor a protein that binds to promoter sequences in eukaryotic genes. After binding of the promoter by transcription factors, RNA polymerase then binds the transcription factors to begin transcription.

transduction the transfer of bacterial DNA from one bacterium to another by a bacteriophage.

transfer RNA (tRNA) a class of small RNA molecule that transports amino acids to the ribosome and matches them to the correct codon on mRNA during translation.

transformation the process in which DNA released by a dead bacterium is absorbed by a living bacterium and incorporated into its chromosome.

transgenic organism an organism into whose genome DNA from another species has been deliberately introduced.

transient flora those microorganisms that are present on a host temporarily, without causing disease.

translation the process of assembling amino acids into a specific protein, guided by the genetic information encoded by a molecule of the corresponding mRNA.

transmembrane protein a membrane protein that extends across the membrane, projecting into both the cytoplasmic and extracellular environments.

transport vesicle a small sac of membrane in which material is transported between the components of the endomembrane system within a eukaryotic cell.

transposon a short DNA sequence that can move from one site to another within the genome.

trophic level the position of an organism in a food chain.

trophozoite a general term describing the stage in a protozoan life cycle during which feeding and reproduction take place.

uncoating the release of the nucleic acid of a virus into the cytoplasm of the host cell.

unicellular consisting of a single cell.

unsaturated fat a type of fat in which there is at least one double bond between carbon atoms in the fatty acids. Because of the double bonds, unsaturated fats have fewer carbon–hydrogen bonds than saturated fats.

variolation an early means of vaccination against smallpox, in which material from an infected individual was used to inoculate potentially susceptible individuals.

vasodilation increase in the diameter of a blood vessel.

vector (1) An invertebrate that carries a pathogen from one host to the next. (2) An entity such as a virus or a plasmid used in genetic engineering to transfer genes into a cell.

vector-borne disease a disease transmitted by an arthropod or other invertebrate.

vertical transmission pathogen transmission from mother to child, either across the placenta or in breast milk.

viral envelope a phospholipid bilayer of host origin that is acquired by some types of viruses as they leave a host cell.

viral glycoprotein a glycoprotein embedded in the envelope of enveloped viruses, which has an important role in host cell recognition and entry.

viral specificity describes the particular set of cell types that a given virus can infect.

virion an individual complete viral particle.

virulence the capacity of an organism to cause disease.

virulence factor a characteristic that increases the disease-causing capacity of a microorganism.

virus an acellular life form, which is an obligate parasite and can only replicate itself inside a living cell. Viruses are the cause of many diseases.

whey the liquid portion of milk remaining after the separation of milk solids during the production of dairy products such as cheese.

wort the liquid portion containing sugars, recovered after malting of barley.

zooplankton consumer organisms in the plankton.

Figure Acknowledgments

FIGURE	ACKNOWLEDGMENT
Ch 1 Opener, 11.10	David Scharf/Photo Researchers, Inc.
1.1a	Dr. Tony Brain & David Parker/Photo Researchers, Inc.
1.1b, 1.10a, 12.6	Courtesy of Christine L. Case, Skyline College
1.1c	National Park Service
1.2, 1.5c, 3.6cfg, 8.27a, 9.26b, 10.8a, 10.15, 13.19	CDC
1.3a, 7.32, Ch 15 Opener	Scimat/Photo Researchers, Inc.
1.3b	© University of Newcastle upon Tyne
1.3c	Courtesy of Dr. Marc Brodkin
1.5a	CDC/Dr. Stuart Brown
1.5b	CDC/Dr. Francis W. Chandler
1.6	NIAID
1.7a	*PNAS* 95 (23):13363–13383. © 1998 National Academy of Sciences, U.S.A.
1.7b	USDA-APHIS/Photo by Michelle Crocheck
1.8a	CDC/Maryam I. Daneshvar, Ph.D.
1.8b	USDA-ARS/Photo by Peggy Greb
Ch 2 Opener	Wolfgang Baumeister/Photo Researchers, Inc.
2.4	Courtesy of William Herring MD, www.LearningRadiology.com
2.10	Courtesy of Dr. Carl S. Kirby
2.17	© Flagstaffotos
2.18	© James G. Howes
2.20, 3.19b	CDC/Courtesy of Larry Stauffer, Oregon State Public Health Laboratory
Ch 3 Opener	CDC/Peggy S. Hayes
3.4	Courtesy of Esther R. Angert, Cornell University
3.6a	CDC/W.A. Clark
3.6b	CDC/Joe Miller
3.6d, 10.20a	CDC/Dr. Richard Facklam
3.6e	CDC/Dr. V.R. Dowell, Jr.
3.6h, 3.8, 9.26a	Courtesy of Gary E. Kaiser, The Community College of Baltimore County
3.7	Courtesy of John Ruby, University of Alabama
3.9	Reprinted by permission from Macmillan Publishers Ltd: *Kidney International* 72:19–25 © 2007

FIGURE	ACKNOWLEDGMENT
3.10abc	Courtesy of Dr. Aizawa
3.14a	CDC/Dr. Mike Miller
3.14b, 8.22a	CDC/Dr. George P. Kubica
3.17	CNRI/Photo Researchers, Inc.
3.18	© 2009 Public Affairs Office, Joint Forces Headquarters-NY National Guard
3.20	FBI
3.21a	Courtesy of Caroline C. Philpott, NIH
3.21b	CDC/Dr. Mae Melvin
3.21c	Courtesy of Dr. Gene Gushansky
3.24	Reprinted by permission from Macmillan Publishers Ltd: *Nat Chem Biol* 3:117–125 © 2007
3.25a	Courtesy of Leilo Orci, University of Geneva, Switzerland
3.25b	Courtesy of D.L. Schmucker, Department of Anatomy, UCSF
3.28, 4.9a	CDC/Dr. Edwin P. Ewing, Jr.
Ch 4 Opener	Lee D. Simon/Photo Researchers, Inc.
4.3a	CDC/Dr. Shirley Maddison
4.3b	WHO/TDR/Haaland
4.3c	Lewis FA, Liang Y-s, Raghavan N, Knight M (2008) The NIH-NIAID Schistosomiasis Resource Center. *PLoS Negl Trop Dis* 2(7): e267
4.5	CDC/Dr. Stan Erlandsen
4.8	Reprinted by permission from Macmillan Publishers Ltd: *Nat Rev Microbiol* 7:237–245 © 2009
4.9b	CDC/Susan Lindsley
4.10, 7.20a, 8.9a	NASA
4.13abd	Courtesy of Dr. Steve Upton
4.13c	Courtesy of Yvonne R. Vaillancourt, University of Massachusetts at Boston
4.17b	CDC/Robert Simmons; Janice Haney Carr
4.20	US Fishery and Wildlife Service/Charles H. Smith, vergrößert von Aglarech
4.21	Courtesy of Henry Mühlpfordt
4.22	Courtesy of Jay W. Pscheidt, Oregon State University
4.23	USDA Forest Service/Photo by Robert L. Anderson
4.34a	Reprinted by permission from Macmillan Publishers Ltd: *Nature* 411:848–853 © 2001
4.34b	Courtesy of Ben Schumin, The Schumin Web
4.35b	Courtesy of Dr. Larry Goodridge
4.38	Jean-Loup Charmet/Photo Researchers, Inc.
4.39	WHO
Ch 5 Opener	Eye of Science/Photo Researchers, Inc.
5.3	Courtesy of Malka Raisz, Raisz Landform Maps
5.5, 5.18, 5.19, 5.21, 5.22, 5.25ab, 8.20, 12.9	Courtesy of the National Library of Medicine
5.7	Courtesy of the Oriental Institute of the University of Chicago

FIGURE	ACKNOWLEDGMENT
5.8	Courtesy of the U.S. National Archives and Records Administration
5.9	CDC/Frank Collins, Ph.D.
5.10	USDA/Photo by Scott Bauer
Ch 6 Opener	Dr. Kari Lounatmaa/Photo Researchers, Inc.
6.5a	G. Murti/Photo Researchers, Inc.
6.6a	Courtesy of Brian Wells
6.6b	© Smith College and the Center for Microscopy and Imaging (CMI)
6.15b	Reprinted by permission from Macmillan Publishers Ltd: *Nature* 17:967–970, © 2002
6.23	From *Science* (1970) 169:392–395. Reprinted with permission from AAAS.
6.31b	Courtesy of Dr. Charles Brinton
Ch 7 Opener	Michael Abbey/Photo Researchers, Inc.
7.2	Courtesy of Toksave
7.3	Courtesy of Nick Hobgood
7.16	Courtesy of Sandie Smith
7.20b	Courtesy of Marshall D. Sundberg, in collection of Botanical Society of America 1978
7.26	Courtesy of Lourdes Norman-McKay, Ph.D.
7.29	Reprinted from *Trends Plant Sci,* 13:610–617, © 2008, Cantu D, Vicente AR, Labavitch JM, Bennett AB, Powell ALT, "Strangers in the matrix: plant cell walls and pathogen susceptibility", with permission from Elsevier
7.30	Archives and Special Collections, Queen Elizabeth II Library, Memorial University of Newfoundland
Ch 8 Opener, Ch 10 Opener, Ch 14 Opener	SPL/Photo Researchers, Inc.
8.2	National Park Service/Photo by J. Schmidt
8.4a	OAR/National Undersea Research Program (NURP); NOAA
8.4b	Courtesy of J. Kuyken
8.9b	Courtesy of J. W. Schopf, UCLA
8.10b	Courtesy of Wayne Lanier, Ph.D.
8.11	Courtesy of Stuart Edwards
8.16	Courtesy of Raymond E. Goldstein, University of Cambridge
8.17	CDC/Dr. Ed Ewing
8.18	NASA/JPL/DLR
8.19	NASA/JSC/Stanford University
8.22b	Courtesy of Dr. Tim Berger, UCSF School of Medicine
8.25	Courtesy of M.W. Mules
8.26b	Courtesy of N.C. Hinkle
8.27b	LifeStraw® Family, Vestergaard Frandsen, 2008
8.29	In chapter titled: Cytoplasmic incomatibility in insects Page 45 Figure 2.1 from "Influential Passengers: Inherited Microorganisms and Arthropod Reproduction" edited by O'Neill S.L., Hoffmann A.A., Werren J.H. (1997). By permission of Oxford University Press.

FIGURE	ACKNOWLEDGMENT
Ch 9 Opener	Garry DeLong/Photo Researchers, Inc.
9.1	Courtesy of Monika Bright, University of Vienna
9.2	Courtesy of Nick Evans, Webmaster—Bishops Offley, UK village website, www.bishopsoffley.co.uk
9.6a	U.S. Geological Survey
9.6b	Courtesy of B. Pitts, Montana State University Center for Biofilm Engineering
9.6c	© Water Pik, Inc., Fort Collins, CO. Reprinted with permission.
9.9	*Appl Environ Microbiol* (1984) 48:1140–1150, reproduced with permission from the American Society for Microbiology
9.10a	Courtesy of Peter West, National Science Foundation
9.10b	National Science Foundation/Russ Kinne
9.13	NOAA
9.14b	NASA/GSFC, MODIS Rapid Response
9.16a	Courtesy of Peter J.S. Franks
9.16b	Woods Hole Oceanographic Institution
9.18	Ferreira, J.F.S. 2007. *J Agric Food Chem* 55(5):1686–1694
9.20	USDA-ARS/Photo by Markus Dubach
9.22	Courtesy of Babinet Damien
9.25	Courtesy of Jack Randall
9.27	SEM by Daniel Kadouri & George O'Toole, and colorized by Russell Monds, Dartmouth Medical School
9.28a	Courtesy of Bruce Jaffee
9.28b	Courtesy of Joey Spatafora
10.1	Reprinted by permission from Macmillan Publishers Ltd: *Nat Med* 5:492–493, © 1999
10.2a	USDA-ARS/Photo by Keith Weller
10.2b	Gregory G. Dimijian/Photo Researchers, Inc.
10.4	CDC/John Molinari, Ph.D., University of Detroit, Detroit, Michigan; Sol Silverman, Jr., D.D.S.
10.8b	CDC/WHO
10.9a	CDC/Brian Judd
10.10a	USDA-ARS/Photo by Scott Bauer
10.10b	© Dorling Kindersley Images/Photo by Frank Greenaway
10.16	Courtesy of Robert Colebunders
10.18	Photos courtesy of Allergan Pharmaceuticals. The BOTOX® Cosmetic is a prescription medication that should be used only under the supervision of a physician.
10.20b	Courtesy of Preston Maxim MD, Department of Emergency Medicine, San Francisco General Hospital
10.21	Courtesy of F.A. Murphy, University of Texas Medical Branch
10.24	WHO/TDR/Wellcome
Ch 11 Opener	Juergen Berger/Photo Researchers, Inc.
11.6c	*PNAS* 103(13):4930–4934 (2006). © 2006 National Academy of Sciences, U.S.A.
11.14b	Courtesy of W. van Ewijk

FIGURE	ACKNOWLEDGMENT
Ch 12 Opener	Biophoto Associates/Photo Researchers, Inc.
12.1	Courtesy of Kai Troester
12.3	Courtesy of St. Vrain Sanitation District
12.4	U.S. Fish and Wildlife Service
12.10	CDC/Cheryl Tryon
12.17	Courtesy of Dr. Stephanie Dellis, College of Charleston
12.20	Courtesy of Farm Sanctuary
Ch 13 Opener	CDC/Cynthia Goldsmith
13.1	WhoZoo, www.whozoo.org
13.4	Courtesy of Colin F. Sargent
13.12	Food and Agriculture Organization of the United Nations/Photo by P. Johnson
13.13a	Courtesy of Kevin Boudreaux
13.13b	CDC/Dr. Fred Murphy; Sylvia Whitfield
13.13c	CDC/P.E. Rollin
13.20	CDC/Ethleen Lloyd
13.21	Max Planck Institute for Infection Biology
14.1b	Courtesy of Heather Angel
14.4	Courtesy of Pacific Northwest National Laboratory
14.6	Courtesy of Colegota
14.13a	Courtesy of GloFish® Fluorescent Fish, www.glofish.com
14.13b	Will & Deni McIntyre/Photo Researchers, Inc.
14.16	USDA-ARS/Photo by Bruce Fritz
14.19a	Division of Plant Industry Archive, Florida Department of Agriculture and Consumer Services, www.bugwood.org
14.20	Courtesy of Dr. Charles Fisher
14.21	Courtesy of FoEI, www.foei.org
15.1	Courtesy of St. Sixtus Abbey, Westvleteren Belgium
15.4a	© 2006 by the California Integrated Waste Management Board. All rights reserved.
15.4b	Courtesy of Chris Schnepf, University of Idaho, www.bugwood.org
15.5, 15.17	© Kraft Foods Limited
15.6b	© UTEX -- The Culture Collection of Algae at The University of Texas at Austin
15.10	Andrew Brookes, National Physical Laboratory/Photo Researchers, Inc.
15.11b	Courtesy of Scott Crawford, Kipahulu Ohana
15.13	Courtesy of Alan K. Outram
15.14a	Courtesy of Alina Zienowicz
15.14b	Reprinted from *Applied and Environmental Microbiology*, 42:111–1118 © 1981
Ch 16 Opener	Volker Steger/Christian Bardele/Photo Researchers, Inc.
16.1a	© 2004, University Corporation for Atmospheric Research
16.5	Reprinted from *TIBTECH*, 11:334–352, Armin Heitzer and Gary S. Sayler, Monitoring the efficacy of bioremediation, (1993), with permission from Elsevier

FIGURE	ACKNOWLEDGMENT
16.7a	Courtesy of David B. Fankhauser
16.7bc	Bioengineering AG, Switzerland
16.7d	Courtesy of Joe Mabel
16.9	Reproduced with permission from *J Exp Biol* 206:3877–3885 © 2003
16.10	V. Berry, R.F. Saraf, "Self-Assembly of Nanoparticles on Live Bacterium: An Avenue to Fabricate Electronic Device". *Agnew Chem Int Ed* (2005), 44:6668–6673 © Wiley-VCH Verlag GmbH & Co. KGaA. Reproduced with permission.
16.11ab	Courtesy of Marc Deshusses, Duke University
16.12	Jerry Mason/Photo Researchers, Inc.
16.14	Epping Forest District Council
16.15	Courtesy of the City of Modesto Compost Facility
16.16ab	Courtesy of K. Sudesh, Ecobiomaterial Research Laboratory, USM
16.17	Courtesy of the City of Austin Solid Waste Services
16.18	Courtesy of Black-Line International

Index

A

Abiotic conditions, 230
   biotic interactions, 230, **Fig. 9.2**
Abiotic synthesis, 205–206
   alternative explanations, 205–206, **Fig. 8.4**
   energy sources, 204–205, **Fig. 8.2**
   Miller's experiment, 205, **Fig. 8.3**
Acetic acid, 124, 406
Acetobacter, 406, **Fig. 1.3, Fig. 15.12**
Acetone
   as fermentation waste product, 184
   industrial production, 117–118, **Fig. 5.12**
Acid-fast bacteria, 56
Acid-fast cell wall, 56
Acid-fast stain, 57, **Fig. 3.14**
Acidithiobacillus, 252
Acidithiobacillus ferrooxidans, 27, 28, 195
Acidophiles, 80, 195
Acid rain, 417, **Fig. 16.1**
Acids, 25–28, **Fig. 2.11**
Acquired immune deficiency syndrome *see* AIDS
Activation energy, 39, **Fig. 2.25**
Active site, enzyme, 39, **Fig. 2.26**
Acute diseases, 113–114, **Table 5.1**
Acute viral infections, 97, 277, **Fig. 4.33, Fig. 10.22**
Acyclovir, 332–333, **Fig. 12.15, Fig. 12.16**
Adaptive immune response, 286, 296–307, **Fig. 11.2**
   activation, 293–296, **Fig. 11.9**
   effects of HIV, 309, **Fig. 11.26**
   subsequent infections, 304–306, **Fig. 11.22**
   successful, 303–304
   unsuccessful, 304
Adenine, 40, 135, **Fig. 2.27**
Adenosine diphosphate (ADP), 171, **Fig. 7.5**
Adenosine triphosphate *see* ATP
Adherence, 270–271, **Fig. 10.13**
Aedes aegypti, 360, 361
Aerobes
   decomposition of organic wastes, 432

evolution, 209–210
growth rates, 188–189, 198, **Fig. 7.33**
   obligate, 192, **Fig. 7.27**
Aerobic respiration, 180, 182–183, **Fig. 7.13**
   compared to photosynthesis, 186, **Fig. 7.18**
Aerosol transmission, 268, **Fig. 10.9**
Aerotaxis, 52, **Fig. 3.11**
Aflatoxin, 163
Africa, European colonization, 114–115
Agar, 126
Agglutination, 300, **Fig. 11.19**
Agricultural microbiology, 9
Agriculture
   antibiotic use, 337–338, **Fig. 12.20**
   biotechnology applications, 389–391
   historical aspects, 116
   water pollution, 241, 242
Agrobacterium tumefaciens, 389–391, **Fig. 14.19**
AIDS
   see also Human immunodeficiency virus
   early investigation, 358
   first reports, 308–309, **Fig. 11.24**
   global epidemic, 308, **Fig. 11.23**
   Kaposi's sarcoma, 363
   progression to, 310–311, **Fig. 11.27**
Airag, 407, **Fig. 15.13**
Airborne transmission, 268, **Fig. 10.9**
Air pollution, 288, 417, **Fig. 16.1**
Alcohol(s), **Fig. 2.14**
   see also Ethanol
   disinfection, 321
Alcoholic beverages, 396, 402–405
   see also Beer; Wine
   in ancient history, 397–398, **Fig. 15.3**
Alcoholism, 262
Algae, 6, **Table 1.1**
Algal blooms, 241, **Fig. 9.13**
Alkaliphiles, 195
Alleles, 155, 165
Amantadine, 333, **Fig. 12.15**
American Civil War, 107
Americas, diseases imported to, 112–115

Amino acids, 35–36, **Fig. 2.21**
   assembly into proteins, 147–151, **Fig. 6.19**
   codons, 148–149, **Fig. 6.20**
   dietary supplements, 409
   as energy sources, 184, **Fig. 7.16**
   peptide bonding, 36, **Fig. 2.22**
   R groups, 35, **Fig. 2.21**
Ammonia (NH_3), 23–24, **Fig. 2.6, Fig. 2.7**
   as energy source, 188
   interaction with water, **Fig. 2.9**
Ammonification, 249, **Fig. 9.19**
Ammonium ion (NH_4^+), 249
Amoebic dysentery, 82, **Fig. 4.13**
Amoxicillin, 147–148, 149
Amphotericin B, 331
Ampicillin, 329, **Fig. 12.12**
Anaerobes
   aerotolerant, 193, **Fig. 7.27**
   decomposition of organic wastes, 432, **Fig. 16.14**
   evolution, 209
   facultative, 193, **Fig. 7.27**
   final electron acceptors, 181–182
   growth rates, 189, 198, **Fig. 7.33**
   obligate, 192–193, **Fig. 7.27**
   strict, 78
Anaerobic respiration, 183–184
Animal cells, **Fig. 1.3, Fig. 3.1**
   lack of cell wall, 63
   plasma membrane, 63, **Fig. 3.22**
   specialization, 62
   viral replication, 92–95, **Fig. 4.28**
Animal reservoirs, 264–265, **Fig. 10.6, Table 10.1**
   control measures, 267
Animals
   classification, 72, 75–76, **Fig. 4.4**
   domestic, antibiotic use, 337–338, **Fig. 12.20**
   history of infectious disease, 114, **Fig. 5.7**
   natural antimicrobial compounds, 338–339, **Fig. 12.21**
   transgenic *see* Transgenic animals
   viral infections, 97–99
Antarctica, 237–238, 240, **Fig. 9.10**
Anthrax, 61
   see also Bacillus anthracis

bioterrorism, 60, 61, 364, **Fig. 3.18, Fig. 3.20**
discovery of causative agent, 126
Antibiotic resistance, 335–339
  antibiotic selection and, 328
  development during treatment, 336–337, **Fig. 12.19**
  emergence, 335–336, **Fig. 12.18**
  evolution, 218–220, 336, **Fig. 8.24**
  genes, recombinant DNA technology, 380, 382, **Fig. 14.10**
  genetic basis, 159, 160, 162, 163
  strategies for circumventing, 338–339
Antibiotics
  acid-fast bacteria, 56
  bactericidal, 56, 325–326
  bacteriostatic, 56, 325–326
  broad-spectrum, 326–327, **Table 12.2**
  commercial production, 423–424
  definition, 324
  development of new, 338–339
  discovery of, 10, 129–130, 324
  effects on normal flora, 263, 326–327
  factors influencing selection, 325–328
  Gram-negative sepsis, 55–56
  human survival and, 323, **Fig. 12.7**
  identifying new, 418, 419, **Fig. 16.2**
  mechanisms of action, 328–331, **Fig. 12.11**
    antiribosomal, 59, 149, 330–331, **Fig. 12.13**
    cell wall inhibitors, 54, 328, 329–330, **Fig. 3.12, Fig. 12.12**
    DNA targeting, 331
  misuse, 335–338
  narrow-spectrum, 326–327, 424, **Table 12.2**
  restrictions on use, 338
  selection of appropriate, 147–148, 149
  selective toxicity, 323, 328
  side effects, 328
  speed of action, 325
Antibodies, 286, 299–300
  cells producing and secreting, 65, **Fig. 3.25**
  classes, 302–303, **Fig. 11.20**
  cross-reactive, 119
  functions, 300–302, **Fig. 11.19**
  membrane-bound, 299–300, **Fig. 11.17**
  primary and secondary response, 305–306, **Fig. 11.22**
  structure, **Fig. 11.17**
Anticodons, 149, **Fig. 6.21**
Antifungal drugs, 331
Antigenic drift, 353–354
Antigenic shift, 354, **Fig. 13.11**
Antigen-presenting cells, 294–295, **Fig. 11.10, Fig. 11.11**
  activation of helper T cells, 296–298, **Fig. 11.15**

migration to lymphatic organs, 295–296
Antigens, 294
Antimalarial drugs, 332, **Fig. 12.14**
Antimicrobial activity, testing molds for, 10, 11–12, **Fig. 1.11, Fig. 1.12**
Antimicrobial chemotherapy, 322–339
Antimicrobial compounds, natural, 338–339, **Fig. 12.21**
Antimicrobial drugs, 320, 322–339
  see also Antibiotics
  development of new, 334, 338–339
  history of development, 129–130, 323–324
  resistance *see* Antibiotic resistance
  selective toxicity, 323, 328, 331–332
Antiseptics, 320–322
Antiviral drugs, 97, 332–335, **Fig. 12.15**
Apicomplexa, 82–84
Appendix, ruptured, 275, 276
Applied microbiology, 8, 9, 417, 426–436, **Fig. 1.8**
Aquifex/Hydrogenobacter lineage, 77–78, **Fig. 4.6**
Arabidopsis thaliana, **Table 14.1**
Archaea, 4–6, **Table 1.1**
  classification, 75–76, **Fig. 4.4**
  diversity, 76
  marine environments, 244
  phylogeny, 79–80, **Fig. 4.11**
  ribosomes, 151
  transcription and translation, 145, 151
Arthrobacter simplex, 425
Arthrobotrys, 256, **Fig. 9.28**
Artificial selection, 217
Artwork, microbial damage to, 7, 9, **Fig. 1.9**
Ascomycota, 88, **Fig. 4.19**
Aseptic technique, 125
Asexual reproduction, 154, **Fig. 6.26**
Ashbya gossypii, 409, 420
Aspartic acid, **Fig. 2.21**
Aspergillus, 406
Aspergillus flavus, 163
Aspergillus niger, 407, 419
A subunit, 274, **Fig. 10.17**
AT3 anticlotting protein, 379
Athlete's foot, **Table 10.1**
Atmosphere
  oxidizing, development, 208–209, **Fig. 8.11**
  reducing, primitive Earth, 204–205
Atomic mass, 18, **Fig. 2.1, Table 2.1**
Atomic number, 18, **Fig. 2.1, Table 2.1**
Atoms, 17–24
  composition of, 18–19, **Fig. 2.1**
  reactivity, 19–20, **Fig. 2.2**
  shell configurations, **Fig. 2.2**
  stability, 19–20, **Fig. 2.3**
ATP (adenosine triphosphate), 171–172, **Fig. 7.5**
  synthesis
    aerobic respiration, 182–183

cell respiration, 176, 182–183, **Fig. 7.9, Fig. 7.13**
chemoautotrophs, 188
electron transport chain, 179–180, **Fig. 7.12**
fermentation, 184
glycolysis, 177–178, **Fig. 7.10**
Krebs cycle, 178–179, **Fig. 7.11**
photosynthesis, 186–188, **Fig. 7.21**
proton gradient, 182, **Fig. 7.12**
utilization, Calvin cycle, 188, **Fig. 7.22**
ATP synthase, 182, **Fig. 7.12**
Attenuated virus/vaccine, 306, 307
Attenuation, 128–129
Autoclave, 317, **Fig. 12.1**
Autotrophs, 86, 186–188
Azole antifungals, 331
Azospirillum, 249
Azotobacter, 249
AZT (zidovudine), 334, **Fig. 12.15**
Aztecs, 107–108, 112, **Fig. 5.2, Fig. 5.6**

B

Bacilli, 49, **Fig. 3.6**
Bacillus anthracis
  see also Anthrax
  as bioterrorism agent, 60, 61, 364, **Fig. 3.18, Fig. 3.20, Fig. 13.21**
  discovery of, 126
  endospores, 60, 61
Bacillus cereus, **Fig. 16.10**
Bacillus species, 61
Bacillus thuringiensis, 391, 426, **Fig. 16.9**
Bacteria, 4–6, 76–79, **Fig. 1.1, Table 1.1**
  asexual reproduction, 154, **Fig. 6.26**
  cells, **Fig. 1.3, Fig. 3.1**
  cell wall, 52–57
  classification, 75–76, **Fig. 4.4**
  cytoplasm and interior structures, 59–61
  cytoskeletal proteins, 64
  directly consumed as food, 398–401
  diversity, 76
  electron transport chain, 180
  extracellular structures, 50–52
  gene structure, 145–147, **Fig. 6.18**
  lacking a cell wall, 57
  metabolic efficiency, 49
  origins of, 77–78
  pathogenesis of disease, 272–276
  phylogeny, 76–79, **Fig. 4.6**
  plasma membrane, 58–59, **Fig. 3.15**
  recombinant, 380–383, **Fig. 14.10**
  ribosomes, 59, 151, **Fig. 3.17**
  shapes, 49, **Fig. 3.6**
  transcription and translation, 145, 151, **Fig. 6.23**
  unusually large, 48, **Fig. 3.2**
Bactericidal control measures, 317, **Fig. 12.2**

Bactericidal drugs, 56, 325–326
Bacteriophages, 99–102
    adsorption, 99, **Fig. 4.36**
    cholera epidemics and, 348
    as cloning vectors, 383, **Fig. 14.11**
    as first viruses, 213
    importance to humans, 101–102
    penetration, 100, **Fig. 4.36**
    replication, 99–101, **Fig. 4.36**
    structure, 99, **Fig. 4.35**
    therapeutic use, 339, **Fig. 12.21**
    transduction, 157, **Fig. 6.30**
Bacteriostatic control measures, 318, **Fig. 12.2**
Bacteriostatic drugs, 56, 325–326
Bacteroides gingivalis, 78
Badgers, 267
Bagels, 411–412, **Fig. 15.15**
Baking *see* Bread-making
Barley
    malting, 403, **Fig. 15.9**
    scotch production, 405
Barotolerant organisms, 244
Barriers to entry, 288–290, **Fig. 11.3, Fig. 11.5**
    anatomical, 289, **Fig. 11.5**
    genetic, 290
    microbial, 289–290
Bartonella henselae, 264–265
Base pairing, complementary
    DNA molecule, 135–136, **Fig. 6.4**
    transcription, 143–145
Bases
    chemical, 25–28, **Fig. 2.11**
    nitrogenous, 40, 135, **Fig. 2.27, Fig. 6.2**
Basic microbiology, 8–9, **Fig. 1.8**
Basidiocarp, 87, **Fig. 4.18**
Basidiomycota, 87–88, **Fig. 4.19, Fig. 4.22**
Batrachochytrium dendrobatidis, 85, 87, **Fig. 4.20**
BAV1, 430
B cells, 296, 298–300
    activation, 300, **Fig. 11.18**
    membrane-bound antibodies, 299–300, **Fig. 11.17**
    memory, 300, 304, 306–307, **Fig. 11.18**
Bdellovibrio, 255–256, **Fig. 9.27**
Beer, 403–405, **Fig. 7.2**
    in ancient history, 397–398, **Fig. 15.3**
    bottom and top fermentation, 403, **Fig. 15.10**
    brewing, 403–405, **Fig. 15.9**
    cold-filtered, 318, 319
    Trappist ales, 395–396, **Fig. 15.1**
Bell, Alexander Graham, 120, **Fig. 5.14**
Benebac®, 30, 33
Benthic zone, 239, 242, 244–245, **Fig. 9.11**
Bilayer, lipid, 58, 59, **Fig. 3.15**
Binary fission, 189–190, **Fig. 7.23**
Binomial nomenclature, 72
Bioconversion, of steroids, 424–425, **Fig. 16.8**
Biodegradable plastics, 432–433, **Fig. 16.16**

Biofilms, 234–235, **Fig. 9.6, Fig. 9.7**
    bioremediation using, 235, 428, 429–430
    wastewater treatment plants, 427, 429–430, **Fig. 16.11**
Biofuels, 433–435
Biogeochemical cycles, 245–253
Biological molecules, 29–41
    abiotic synthesis, 204–206, **Fig. 8.3**
    energy release from, 170–174
    respiration, 175, 185, **Fig. 7.17**
    synthesis by autotrophs, 186–188
Bioluminescent bacteria, 253–254, **Fig. 9.25**
Biomass, 232
Biomphalaria snails, 74–75, **Fig. 4.3**
Biopesticides, 426
Bioprospecting, 419
Bioremediation, 428–431
    approaches to, 429–430, **Fig. 16.12**
    in different settings, 430–431
    microbial gene exchange, 160
    recombinant organisms, 391–392, 429
    using biofilms, 235, 428, 429–430
    wastewater treatment plants, 427, **Fig. 16.11**
Biotechnology, 367–393
    see also Recombinant DNA technology
Bioterrorism, 60, 364, **Fig. 3.18**
Biotic interactions, 230
    abiotic factors affecting, 230, **Fig. 9.2**
Bird flu, 267, 354–355, **Fig. 13.12**
Birds
    see also Chickens
    botulism in aquatic, 238, 239–240
    Campylobacter jejuni
        transmission, 347, 348–349
    malaria, 82–83
    psittacosis, 66, 67
    riboflavin deficiency, 408–409
    West Nile virus, 356, **Fig. 13.13**
Black Death, 111–112, **Fig. 5.1, Fig. 5.5**
Bleach, 320, 321
Body odor, 254
Boiling, 317
Bonds, chemical, 20–24
    covalent, 22–24, **Fig. 2.6**
    ionic, 20–22, **Fig. 2.5**
Booster vaccinations, 306
Bordetella pertussis, **Table 10.1**
Borrelia burgdorferi, 270, 358, 378
BOTOX® cosmetic, 275, **Fig. 10.18**
Botryococcus braunii, 434
Botulism, 35, 37, **Fig. 2.20**
    see also Clostridium botulinum
    fermented meat products, 408
    toxin, 238, 275, 364, **Fig. 10.18**
    water birds, 238, 239–240
Bourbon, 405
Bovine spongiform encephalopathy (BSE), 102, 103–104
BRCA1/BRCA2 genes, 371–373, **Fig. 14.5**
Bread-making, 401–402, **Fig. 15.7**
    ancient history, 398, **Fig. 15.3**
    recipe, 411–412, **Fig. 15.15**

Breast cancer, familial, 371, 373
Breeding, selective, 367, **Fig. 14.1**
Brevibacterium, 409
Brewing *see under* Beer
Broad-spectrum antibiotics, 326–327, **Table 12.2**
Brueghel, Pieter, **Fig. 5.1**
BSE *see* Bovine spongiform encephalopathy
B subunit, 274, **Fig. 10.17**
Bt, 426, **Fig. 16.9**
Bt corn, 389, 391, **Fig. 14.18**
Budding, viral release by, 94–95, **Fig. 4.30**
Buildings
    microbial damage, 9, **Fig. 1.9**
    restoration of historic, 417, **Fig. 16.1**
    stabilization against earthquakes, 436
Bursa, 299
Butyric acid, 34

C

Calcium, 20
Calcium carbonate ($CaCO_3$)
    building materials, 417
    in carbon cycle, 246, 247, **Fig. 9.17**
Calories, 33
Calvin cycle, 187, 188, **Fig. 7.22**
Campylobacter jejuni, 347, 348–349
Cancer
    chemotherapy, 262
    vaccines, 2, **Fig. 1.2**
    viruses causing, 277, **Fig. 10.22**
Candida albicans, 88, **Fig. 1.5**
    opportunistic infections, 263, **Fig. 10.4**
Cannibalism, 102, 223–224
Capsid, 90, **Fig. 4.26**
Capsomeres, 90, **Fig. 4.26**
Capsule, bacterial, 50–51, **Fig. 3.8**
Carbohydrates, 30–33
    as source of energy, 32, 175–176
Carbon
    atoms, 18, **Fig. 2.1, Table 2.1**
    chains and ring structures, 29–30, **Fig. 2.13**
    covalent bonding, 22, **Fig. 2.6**
Carbon cycle, 246–248, **Fig. 9.17**
Carbon dioxide (CO_2), 22
    in bread-making, 401, **Fig. 15.7**
    carbonation of beer, 403–405, **Fig. 15.9**
    in carbon cycle, 246, 247–248, **Fig. 9.17**
    as carbon source, 188, **Fig. 7.22**
    as fermentation waste product, 184, **Fig. 7.15**
    as final electron acceptor, 181–182
    production
        Krebs cycle, 178–179, **Fig. 7.11**
        oxidation of carbon fuels, **Fig. 7.7**
    release during combustion, 173, 248, **Fig. 7.7**
Carbon fixation, 188, 246, **Fig. 7.22**

Carbon fuels, combustion of, 173, 248, **Fig. 7.7**
Carbon–hydrogen bonds (C–H bonds), 33
    cell respiration, 176, 178
    electron carriers, 179, **Fig. 7.8**
    redox reactions, 173, **Fig. 7.7**
Carnivores, 230, 232
Carrier proteins, 58–59
Carroll, James, 127
Case–control studies, 358–359
Case definition, 356–357
Catalase, 192–193, **Fig. 7.26**
Catalysts, 39
Cat scratch disease, 264–265, 279
Caulobacter crescentus, 1
CD4, recognition by HIV, 309, **Fig. 11.25**
CDC *see* Centers for Disease Control and Prevention
Cell(s), 3–4, 45–69
    basic concepts, 46–49
    first, 207–208, **Fig. 4.7, Fig. 8.6**
    sizes, 48–49, **Fig. 3.2**
    staining, 46–48
    surface-to-volume ratio, 49, **Fig. 3.5**
    types, 4, 46, **Fig. 1.3, Fig. 3.1**
    virus entry, 93
Cell-mediated response, 298, **Fig. 11.16**
    see also T cells
    primary and secondary response, 305–306
Cell membrane *see* Plasma membrane
Cell theory, 3, 45
Cellulase, 391, **Fig. 14.20**
Cellulose, 32, 63
    digestion by herbivores, 32–33, **Fig. 2.17**
Cellulosic ethanol, 434
Cell wall
    acid fast, 56
    antibiotics targeting, 54, 328, 329–330, **Fig. 3.12, Fig. 12.12**
    bacterial, 52–57
    bacteria lacking, 57
    Gram-negative, 55–56, **Fig. 3.13**
    Gram-positive, 53–54, **Fig. 3.12**
    plants and fungi, 63
    staining techniques, 56–57, **Fig. 3.14**
Centers for Disease Control and Prevention (CDC)
    investigation of outbreaks, 345–346, 356
    Morbidity and Mortality Weekly Report (MMWR), 308–309, **Fig. 11.24**
Central dogma of genetics, **Fig. 6.13**
Cephalosporins, mechanism of action, 329, **Fig. 12.11, Fig. 12.12**
Cervical cancer vaccine, 2, **Fig. 1.2**
Cheese, 184, 407–408
    in ancient history, 397

Greek feta, recipe, 412–413, **Fig. 15.16**
    varieties, **Table 15.1**
Chemical bonds, 20–24
Chemicals
    control of microbial growth, 320–322, **Table 12.1**
    mutagens, 162–163, **Fig. 6.33**
Chemistry of life, 17–42
Chemoautotrophs, 186, 188
Chemotaxis, 52
Chemotherapeutic agents, 320
    see also Antimicrobial drugs
Chestnut blight, 85, 88, **Fig. 4.23**
Chickenpox, 278, **Fig. 13.7**
Chickens
    normal flora, 260, 263, **Fig. 10.2**
    as reservoirs of bird flu, 267
Childbed fever, 125
Chinese Communists, 120
Chitin, 63, 85–86
Chlamydiae, 78–79
Chlamydia psittaci, 66–67
Chlamydia trachomatis, 78–79
    eye damage, 260, **Fig. 10.1**
Chloramphenicol
    appropriate use, 327
    mechanism of action, 330, **Fig. 12.11, Fig. 12.13**
Chlorine
    germicidal activity, 320–321
    ions, 20, 21, **Fig. 2.5**
Chlorophyll, 187, **Fig. 7.20, Fig. 7.21**
Chloroplasts, 67–68, 187, **Fig. 3.29, Fig. 7.20**
    endosymbiotic origin, 68, 211–212, 384–385, **Fig. 8.13, Fig. 8.14**
    stroma, 67
Chloroquine, 332
Cholera, 101, 273–274, **Fig. 4.38**
    see also Vibrio cholerae
    bed, **Fig. 10.16**
    epidemics, 348
    evolution of virulence, 222, **Fig. 8.27**
    reasons for symptoms, 280, 281, **Fig. 10.25**
    Snow's investigations, 344, **Fig. 13.2**
    vaccine, 388
Cholera toxin, 273–274
    B subunit, 274, **Fig. 10.17**
    A subunit, 274, **Fig. 10.17**
Cholesterol, 63, **Fig. 3.22**
"Christmas trees," transcription, 143, 145, **Fig. 6.15**
Chromatograph, **Fig. 14.8**
Chromosomes, 137
    bacterial (prokaryotic), 59, 137
        DNA organization, 137, **Fig. 6.5, Fig. 6.6**
        replication, 189–190, **Fig. 7.23**
    eukaryotic, 137, **Fig. 6.6**
Chronic diseases, 113, 272, **Table 5.1**
Chronic viral infections, 98, 277, **Fig. 4.33, Fig. 10.22**
Chytridiomycota, 87, **Fig. 4.19**

Cidal agents, 325–326
Cidal control measures, 317, **Fig. 12.2**
Cilia
    eukaryotic cells, 64, **Fig. 3.23**
    respiratory tract, 288, **Fig. 11.4**
Ciliophora, 82, **Fig. 4.13**
Cinchona tree, 332, **Fig. 12.14**
Citric acid, 406, 407
    industrial production, 419
Citrobacter freundii, identification, 184–185
Class, 72, **Fig. 4.1**
Classification, biological, 71–76
Claviceps purpurea, 279, **Table 16.1**
Cloning, 378, 379–383
Cloning host, 382
Cloning vectors, 380–383, **Fig. 14.10**
Clostridium, 249
Clostridium acetobutylicum, 118
Clostridium bifermentans, 431
Clostridium botulinum, 35, 37, **Fig. 2.20**
    see also Botulism
    endospores, 61
    freshwater lakes, 239–240
    toxin, 238, 275, **Fig. 10.18**
Clostridium perfringens, endospores, 61
Clostridium tetani
    endospores, 61
    portal of exit, 279
    reservoirs, 266
    toxin, 275
Clouds, 245
CO_2 *see* Carbon dioxide
Coagulase, 276
    natural selection trade-offs, 217, **Fig. 8.22**
Coastal ecosystems, 242–244, **Fig. 9.14**
Cocci, 49, **Fig. 3.6**
    radioresistant, 78, **Fig. 4.8**
Coccidioides immitis, 84, 88–89, **Fig. 4.15**
    portal of exit, 279
    reservoirs, 266, **Table 10.1**
Codons, 148–149, **Fig. 6.20**
    start, **Fig. 6.22**
    stop, 149, **Fig. 6.20**
Coevolved hypothesis, origin of viruses, 214
Coffee beans, 408
Cold sores, 98
Cold temperatures, 318
Collagenase, 276
Colonies, cell, 212, **Fig. 8.16**
Colonization, normal microbial flora, 260–263
Colony, **Fig. 1.1**
Commensalism, 254, **Fig. 9.24**
Commercial applications, 418–426
Common cold, **Table 10.1**
Community, ecological, 230
Competition, 254–255, **Fig. 9.24**
Competitive exclusion, 255
Composting, 431–432, **Fig. 16.13**
    backyard, 431, 432, **Fig. 16.13**
    municipal, 432, **Fig. 16.15**

Condensation reaction, 31, **Fig. 2.15**
Conjugation, bacterial, 51, 157–159, **Fig. 6.31**
    significance for humans, 159
Conservation science, 9
Consumers, 232
    primary, 230, **Fig. 9.3**
    secondary, 230, **Fig. 9.3**
Contact transmission, 268
    direct, 268
    indirect, 268
Contamination, 260
Control, 358
Control group, 11, **Fig. 1.12**
Controlled variables, 11–12, **Fig. 1.12**
Copper mining, microbial, 435, **Fig. 16.18**
Cordite, 117–118
Corn
    liquor production, 405
    transgenic, 389, 391
Cortés, Hernán, 107–108, 112, **Fig. 5.2**
Cortisone, **Fig. 16.8**
Corynebacterium, 254
Corynebacterium diphtheriae, 196, 274
Cotton, transgenic, 391
Coupled reactions, 171, **Fig. 7.4**
Course of infection, 303–304, **Fig. 11.21**
Covalent bonds, 22–24, **Fig. 2.6**
    nonpolar, 23–24, **Fig. 2.7**
    polar, 22–24, **Fig. 2.7**
Cow manure, 230, **Fig. 9.2**
Cowpox, 128, **Fig. 5.25**
Crenarchaeota, 80, **Fig. 4.11**
Creutzfeldt–Jakob disease (CJD), 102–103
    variant (vCJD), 104
Crick, Francis, 130, 133, 137, 138–139
Crowd diseases, 113–114
Cryphonectria parasitica, 88
Cultures, pure, 126
Curd, 407
Cyanobacteria, 79, **Fig. 4.10**
    evolution, 208–209, **Fig. 8.10**
    as food, 400, **Fig. 15.6**
    photosynthesis, 186, 187, **Fig. 7.20, Fig. 7.21**
Cyclosporin, **Table 16.1**
Cyprian, Great Plague of, 110–111
Cystic fibrosis, 20, 21–22, 388, **Fig. 2.4**
Cysts, protozoan, 81, 83–84, **Fig. 4.14**
Cytochrome oxidase, 180
Cytokines, 291, 292, 298, **Fig. 11.7**
Cytopathic effects (CPE), 98–99, 277
Cytoplasm, 46
    eukaryotic cells, 64–68
    prokaryotic cells, 59–61
Cytoplasmic (side), 58–59, **Fig. 3.15**
Cytosine, 40, 135, **Fig. 2.27**
Cytoskeleton, 64, **Fig. 3.24**
Cytotoxic T cells (Tc cells), 298, **Fig. 11.16**

D

Dairy products, fermented, 397–398
Darwin, Charles, 216, **Fig. 8.20**

Darwinian medicine, 222
Daughter cells, 154
Dead zones, 243, **Fig. 9.15**
Decline phase, 197, **Fig. 7.31**
Decomposers, 85, **Fig. 4.16**
    role in ecosystems, 231, **Fig. 9.3**
Decomposition, organic wastes, 431–432, **Fig. 16.14**
Deep sea, 244–245
    hydrothermal vents *see* Hydrothermal vents
Deer mice
    changes in numbers, 348
    as reservoir for hantavirus, 265, 326, **Fig. 10.6, Fig. 12.10**
Definitive host, 83, **Fig. 4.14**
Dehalococcoides, 430
Deinococcus radiodurans, 78, 385, 430, **Fig. 4.8**
Denatured proteins, 36–37, **Fig. 2.24**
Dendritic cells, 290–291, 294, **Fig. 11.10**
Dengue, 360–361, **Fig. 13.17**
Denitrification, **Fig. 9.19**
Dental plaque, 50, 51, 234, **Fig. 3.7**
Dental procedures, 262
Deoxyribonucleic acid *see* DNA
Deoxyribose, 40, 135, **Fig. 6.2**
Designer babies, 392
Designer drugs, 334
Desowitz, Robert, 259
Detergents
    control of microbial growth, 321–322
    microbially-produced enzymes, 425
    water pollution, 241, 242
d'Hérelle, Felix, 339
Diarrhea
    bacterial, 30, 33
    outbreaks in spring, 347, 348–349
    reasons for, 280, 281, **Fig. 10.25**
Dietary supplements, 408–410
Differential stains, **Fig. 3.14**
Digestive system
    barriers to entry, 289, **Fig. 11.5**
    beneficial bacteria, 30, 32–33
    normal flora, 262
    as portal of entry, 270, **Fig. 10.11**
Dinoflagellates, 243–244, **Fig. 9.14**
Diphtheria, 196, 274–275, 363
Diphtheria toxin, 274
Diplococcus, **Fig. 3.6**
Disaccharides, 31, **Fig. 2.15**
Disease, 2, 260
    see also Infectious disease
    germ theory of, 124–125, 127
Disinfectants, 320–322
Distilled alcoholic beverages, 405
Diversity of life, 215–225
DNA, 4, 40–41, 133–154
    3′ end, 135, **Fig. 6.3**
    5′ end, 135, **Fig. 6.3**
    amplification by PCR, 374–375, **Fig. 14.7**
    analysis of relatedness, 73–75
    analysis techniques, 368–379
    antibiotics targeting, 331

antiparallel strand orientation, 136–137, **Fig. 6.4**
cleavage by restriction enzymes, 368–369, **Fig. 14.2**
coding strand, 145, **Fig. 6.17**
daughter strand, 139, **Fig. 6.7, Fig. 6.8, Fig. 6.10**
direction of protein synthesis, 143–154
discovery of, 130
double helix, 40, 137, **Fig. 2.28, Fig. 6.4**
double strand, 135–137
encoding of information, 134–135, 137, 141–143, **Fig. 6.1**
fragments, gel electrophoresis, 369–370, **Fig. 14.4**
function, 138–154
mitochondria and chloroplasts, 67–68, 211, 384–385
mutations, 73–74, 160–165
organization into genes, 134–135, **Fig. 6.1**
packaging in chromosomes, 137, **Fig. 6.5, Fig. 6.6**
parental strand, 139, **Fig. 6.7, Fig. 6.8, Fig. 6.10**
recombinant *see* Recombinant DNA
repair, 163–164, **Fig. 6.35**
replication, 138–140, **Fig. 6.7, Fig. 6.8, Fig. 6.10**
    accuracy, 139
    origins, 140, **Fig. 6.11, Fig. 6.12**
replication forks, 139, **Fig. 6.9**
sequencing, 373–376
    amplification by PCR, 374–375, **Fig. 14.7**
    applications, 376
    Sanger method, 375–376, **Fig. 14.8**
strand, 135, **Fig. 6.3**
structure, 135–137, **Fig. 6.2, Fig. 6.3**
    in relation to function, 135–137, 138–139
technology *see* Recombinant DNA technology
transcription *see* Transcription
world, transition to, 207–208, **Fig. 8.7**
DNA library, 383
DNA ligase, 382
DNA mismatch repair enzymes, 164, **Fig. 6.35**
DNA polymerase, 138, 139, **Fig. 6.7, Fig. 6.8**
    errors, 162, **Fig. 6.32**
    proofreading ability, 163–164, **Fig. 6.35**
    Taq, 375, 426
DNase, 276
DNA viruses, 89
    replication, 95, **Fig. 4.29, Fig. 4.31**
Dobzhansky, Theodosius, 215
Domagk, Gerhard, 129, 323
Domains of life, 75–76, **Fig. 4.4**

Double helix, 40, 137, **Fig. 2.28,**
    Fig. 6.4
Drain cleaners, 425
Drosophila melanogaster, **Table 14.1**
Drug resistance *see* Antibiotic
    resistance
Drying, 319–320, **Fig. 12.4**
Duyser, Diana, 191

E

Earth
    origin, 77, 203–204, **Fig. 4.7**
    primitive, environmental
        conditions, 204–205,
        Fig. 8.3
Earthquakes, 84, 436
Eastern equine encephalitis (EEE)
        virus, 356, **Fig. 13.13**
Ebola hemorrhagic fever, 362–363,
        Fig. 13.20
Ecological interactions, 253–256,
        Fig. 9.24
Ecology, 229–256
    basic principles, 230–232
    defined, 230
    microbial, 232–235
*Eco*RI, 368, **Fig. 14.2**
Ecosystems, 230–232
    energy availability, 231, **Fig. 9.4**
    trophic interactions, 230–231,
        Fig. 9.3
Effector T cells, 298
Egg, 154, 155, **Fig. 6.27**
Egypt, ancient, 397–398, **Fig. 15.3**
Ehrlich, Paul, 129, 323
Electron(s), 18–19, **Fig. 2.1, Table 2.1**
    affinity, 21
    carriers, 174, 178
    shells, 18–19, **Fig. 2.1**
    transfer during redox reactions,
        172–173, **Fig. 7.6**
Electron acceptors, final, 180,
        181–182, **Fig. 7.12**
Electron microscopy, 128
Electron transport, 176, 179–180,
        Fig. 7.9
Electron transport chain, 180, 182,
        Fig. 7.12
    evolution, 209–210
    photosynthesis, 187, **Fig. 7.21**
Elements, 17–18, **Table 2.1**
Elephantiasis, 127
El Niño, 348
Emergent diseases, 359–363,
        Fig. 13.15
    categories, 362–363
    causes, 360–361, **Fig. 13.16**
Encephalitis of unknown origin,
        356–358, **Fig. 13.13**
Endemic, 352
Endergonic reactions, 170–171
Endocytosis, 66, **Fig. 3.27**
    viral entry via, 93, **Fig. 4.28**
Endomembrane system, 64–65
Endoplasmic reticulum (ER), 65–66,
        Fig. 3.25
    evolution, 211, **Fig. 8.13**
    lumen, 65, **Fig. 3.26**

protein transport, 65–66, **Fig. 3.26**
    rough, 65, **Fig. 3.25**
    smooth, 65
Endospores, bacterial, 60–61,
        Fig. 3.19
Endosymbiosis, 68, 210, 211–212,
        384–385
Endotoxin, 55–56, 273
    pathogenesis of disease, 275–276,
        Fig. 10.19
Energy
    activation, 39, **Fig. 2.25**
    availability in ecosystems, 232,
        Fig. 9.4
    flow through ecosystems,
        230–231, **Fig. 9.3**
    release
        from food molecules, 170–171
        oxidation reactions, 172–174,
            Fig. 7.6
    sources
        abiotic synthesis, 204–205,
            Fig. 8.2
        biological molecules, 32, 33
        cell respiration, 175–176,
            185–186
        chemoautotrophs, 188
        photoautotrophs, 186–188
    states of atoms, 19–20
    storage in ATP, 171–172, **Fig. 7.5**
Entamoeba histolytica, 82, **Fig. 4.13**
    pathogenesis of disease, 278–279
Enterobacter agglomerans, 253
Enterococcus, vancomycin resistant,
        219, 220
Entomopathogenic fungi, 256,
        Fig. 9.28
Enveloped viruses, 91, **Fig. 4.26**
    adherence to host, **Fig. 10.13**
    host-cell specificity, 92, **Fig. 4.27**
    release by budding, 94–95,
        Fig. 4.30
    replication, 93, 94, **Fig. 4.28,**
        Fig. 4.29
Environmental conditions
    industrial production, 420, 421,
        Fig. 16.4
    microbial growth rate effects,
        197–199, 233, **Fig. 7.33**
    microbial growth requirements,
        189, 191–196
    origin of life and, 204
    phenotypic effects, 153–154
Environmental reservoirs, 266,
        Table 10.1
Enzymes, 30, 31, 38–40
    action, **Fig. 2.26**
    active site, 39, **Fig. 2.26**
    commercial production, 425–426
    digestive, 31, 32, 34
    effects on activation energy, 39,
        Fig. 2.25
    inhibitors, 39–40
    tissue-damaging, 276, 279,
        Fig. 10.20
Epidemics, 345–355
    see also Outbreaks
    causes, 347–349

common source, 345–347,
        Fig. 13.3
    cycles, 350–351, **Fig. 13.7, Fig. 13.8**
    host-to-host, 347, **Fig. 13.3**
    influenza, 350–351, 352–355
    investigation, 355–359
    population immunity and,
        349–351
Epidemic threshold, 350–351,
        Fig. 13.8
Epidemiology, 343–365
    origins of, 344–345, **Fig. 13.2**
Epithelial tissue, **Fig. 3.21**
Epulopiscium fishelsoni, 48, 49,
        Fig. 3.2
Ergosterol, 63, 331
Ergot alkaloids, **Table 16.1**
Ergot poisoning, 279
Erie, Lake, 238, 241–242
Erwinia dissolvens, 408
Erythromycin
    appropriate use, 148, 149
    mechanism of action, 149, 151,
        330, **Fig. 12.11, Fig. 12.13**
    resistance, 336
    spectrum of activity, 327,
        Table 12.2
    speed of action, 325
Escaped-gene hypothesis, origin of
        viruses, 214
Escherichia coli, 8
    chromosome, 137
    conjugation, 157–159, **Fig. 6.31**
    genetic engineering, 379, 382, 383,
        392, **Fig. 14.9, Fig. 14.10,**
        Fig. 14.11
    genome, **Table 14.1**
    growth, 190–191, **Table 7.1**
    infections, 262
    O157:H7, 159, 376–377, 378
    origin of virulence factor genes,
        377, 378
    phage lambda, 101, **Fig. 4.37**
    phage T4, 99–101, **Fig. 4.35,**
        Fig. 4.36
    pili, 51
Estrogen, 425
Ethane, 29–30, **Fig. 2.14**
Ethanol (ethyl alcohol), 29–30,
        Fig. 2.14
    bread-making, 401, **Fig. 15.7**
    cellulosic, 434
    as fermentation waste product,
        184, **Fig. 7.2, Fig. 7.15**
    use as fuel, 433–434
    vinegar production, 406,
        Fig. 15.12
    winemaking, 402, **Fig. 15.8**
Ethics, genetic engineering, 392–393
Ethyl alcohol *see* Ethanol
Eukarya, 75–76, 80–81, **Fig. 4.4,**
        Fig. 4.12
Eukaryotes
    DNA replication, 140, **Fig. 6.12**
    electron transport chain, 180,
        Fig. 7.12
    evolution, 210–212, **Fig. 8.12,**
        Fig. 8.13, Fig. 8.14

gene structure, 145–147, **Fig. 6.18**
multicellular, 62–63, **Fig. 3.21**
regulation of gene expression, 153
transcription, 145
translation, 151
unicellular, 62, **Fig. 3.21**
Eukaryotic cells, 4, 46, 62–68, **Fig. 1.3**
chromosomes, 137, **Fig. 6.6**
cytoplasmic contents, 64–68
different types, 62, **Fig. 3.21**
plasma membrane, 63–64,
Fig. 3.22
ribosomes, 65, 151, **Fig. 3.25**
sizes, 48–49, **Fig. 3.2**
structure, **Fig. 3.1**
Eukaryotic microorganisms, 6,
Fig. 1.5, Table 1.1
diversity, 76
phylogeny, 80–89, **Fig. 4.12**
Eukaryotic pathogens
antimicrobial chemotherapy,
331–332
pathogenesis of disease, 278–279
Europa, **Fig. 8.18**
European explorers, 112–115
Euryarchaeota, 80, **Fig. 4.11**
Eutrophication
coastal ecosystems, 243, **Fig. 9.15**
lakes, 242, **Fig. 9.13**
Eutrophs, 240
Evolution, 4, 203–226
see also Natural selection
Darwin's theory, 216–217
drug resistance, 218–220, 336,
Fig. 8.24
explaining life's diversity, 215–225
first cells, 207–208, **Fig. 8.6**
first eukaryotes, 210–212,
Fig. 8.12, Fig. 8.13
first metabolic pathways, 208–210,
Fig. 8.10
first multicellular organisms,
212–213, **Fig. 8.15**
first prokaryotes, 208, **Fig. 8.8,
Fig. 8.9**
first viruses, 213–214
genome analysis for studying,
384–385
influence of microorganisms on
host's, 222–225
prions, 214
role of mutations, 165
virulence of pathogens, 220–222,
Fig. 8.27
Evolutionary relationships, 72–73
evidence from DNA analysis,
73–75
taxonomy reflecting, 72–73,
Fig. 4.2
Exergonic reactions, 170–171
Exobiology, 214–215
Exocytosis, 66, **Fig. 3.26**
Exons, 145–147, **Fig. 6.18**
Exotoxins, 273–275
Experiment, controlled, 11–13,
Fig. 1.12
Experimental group, 11, **Fig. 1.12**
Experimental variable, 11–12,
Fig. 1.12

Explosives, 430–431
Exponential growth, 190–191,
Fig. 7.24, Table 7.1
Extracellular, 58–59, **Fig. 3.15**
Extraterrestrial life
search for, 214–215, **Fig. 8.18,
Fig. 8.19**
sightings, 169, **Fig. 7.1**
Extremophiles, 79, 80
Extremozymes, 425–426

F

Facultative anaerobes, 193, **Fig. 7.27**
FAD/FADH$_2$
cell respiration, **Fig. 7.9**
electron transport chain, 179–180,
Fig. 7.12
Krebs cycle, 178–179, **Fig. 7.11**
Family, 72, **Fig. 4.1**
Fastidious organisms, 196
Fats
as energy sources, 184, **Fig. 7.16**
hydrophobicity, 33
saturated, 34–35, **Fig. 2.19**
structure, 33–34, **Fig. 2.19**
unsaturated, 34–35, **Fig. 2.19**
Fatty acids, 33–34, **Fig. 2.19**
cell respiration pathway, 184,
Fig. 7.16
saturated, 34–35, **Fig. 2.19**
unsaturated, 34–35, **Fig. 2.19**
F$^+$ cells, 158–159, **Fig. 6.31**
F$^-$ cells, 158–159, **Fig. 6.31**
Feline leukemia virus (FLV), 91
Female genital tract, normal flora, 262
Fermentation, 183–184, **Fig. 7.14**
discovery of, 124
food production, 395–396,
401–408
waste products, 184, 401, **Fig. 7.15**
Fermented foods, 184, 401–408
ancient history, 397–398, **Fig. 15.3**
animal-based, 407–408
plant-based, 401–407
recipes, 410–413
Fermentors
industrial, 422–423, **Fig. 16.6**
laboratory, 423, **Fig. 16.7**
pilot plant, 423, **Fig. 16.7**
Ferric chloride, 184, 185, **Fig. 7.16**
Fertilizers, water pollution, 241, 242
Feta cheese, Greek, 412–413,
Fig. 15.16
Fever, 292
Fiber, dietary, 32
Fibrin, 38
Filamentous fungi, 86, **Fig. 4.17**
Filtration, 318–319
Fimbriae, 51, **Fig. 3.9**
adherence by, **Fig. 10.13**
Final electron acceptors, 180,
181–182, **Fig. 7.12**
Finlay, Carlos, 117, 127
Fish
absence of, due to mining, 25–27,
Fig. 2.10
bad odor, 180–181, 182
drying, **Fig. 12.4**

Flagella
eukaryotes, 64
prokaryotic, 51–52, **Fig. 3.10**
protozoan, 82
Flagellates, 82
Flagyl *see* Metronidazole
Flashlight fish, 253–254, **Fig. 9.25**
Flavin adenine dinucleotide
see FAD/FADH$_2$
Flavobacteria, 78
Flavobacterium, 392
Flavobacterium johnsoniae, 436
Flavoring agents, 406–407
Fleas
myxoma virus transmission, 219,
Fig. 8.26
transmission of plague, 111
Fleming, Alexander, 10, 129–130, 324
Flesh-eating bacteria, 276, **Fig. 10.20**
Flora
normal *see* Normal microbial flora
transient, 262
Flu *see* Influenza
Fluid mosaic model, 59, **Fig. 3.16**
Fluorine, 20, 21, **Fig. 2.3**
Folic acid, 324, 325, **Fig. 12.8**
Fomites, 268
Food
digestion, 1–2
energy release from, 170–171
fermented *see* Fermented foods
fungi and bacteria eaten directly
as, 398–401
genetically modified (GM), 368,
389–391, 392, **Fig. 14.21**
safety, 127
Foodborne diseases
morning sickness and, 223,
Fig. 8.28
outbreaks, 345–347
virulence, 222, **Fig. 8.27**
Foodborne transmission, 269,
Fig. 10.9
Food chains, 230–231, **Fig. 9.3**
Food microbiology, 9, 395–413
Food poisoning
origin of symptoms, 273
outbreaks, 343, 355, **Fig. 13.1**
Food preservation
drying, 319–320, **Fig. 12.4**
gamma radiation, 319
pasteurization, 317–318
refrigeration, 318
salting, 195, 319–320, **Fig. 7.30**
spices, 396–397, **Fig. 15.2**
Food production, 396, 398–413
ancient history, 396, 397–398
fermentation *see* Fermented foods
recipes, 410–413
Food spoilage, 124, 396–397
acidophilic fungi, 195, **Fig. 7.29**
fungal, 87, **Fig. 4.21**
prevention *see* Food preservation
psychrotolerant organisms, 194
Food webs, 231, **Fig. 9.3**
Fore people, New Guinea, 102,
223–224, **Fig. 4.39**
Forsch, Paul, 128

Fossil fuels
  see also Carbon fuels
  alternatives to, 433–435
  combustion of, 248
Fossils
  eukaryotic cells, 210, **Fig. 8.12**
  "Martian bacteria," 215, **Fig. 8.19**
  microbial, 208, **Fig. 8.9**
Fowl cholera, 128
Fowlpox vaccine, 387
Free radicals, 163
Freeze-drying, 320
French territories, Americas, 115,
  Fig. 5.8
Freshwater ecosystems, 238–242
  mining-related damage, 25–27,
    Fig. 2.10
  pollution, 241–242, **Fig. 9.13**
Fructose, 31, **Fig. 2.15**
Fuel cells, microbial, 435
Fuels
  see also Fossil fuels
  alternative, 433–435
  carbon, combustion, 173, 248,
    Fig. 7.7
Fungal infections
  drug treatment, 331
  opportunistic, 262–263
  pathogenesis, 279
Fungi, 6, 63, 84–89, **Fig. 1.5, Table 1.1**
  abiotic factors influencing, 230,
    Fig. 9.2
  acidophilic, 195, **Fig. 7.29**
  body plans, 86, **Fig. 4.17**
  cell wall, 63
  classification, 75, **Fig. 4.4**
  directly consumed as food,
    398–401
  filamentous, 86, **Fig. 4.17**
  fleshy, 86, **Fig. 4.18**
  phylogeny, 87–89, **Fig. 4.19**
  predatory, 256, **Fig. 9.28**
  single-celled, 86, **Fig. 4.17**
  soil, 236–237
Fungicidal agents, 317
Fusion, viral entry via, 93, **Fig. 4.28**

G

Galactose, 31
Galen, 124
Galls, 390, **Fig. 14.19**
Gametes, 154, 155, **Fig. 6.27**
Gamma radiation
  DNA damage, 163
  sterilization using, 319
Garfield, President James A., 120–121,
  125, **Fig. 5.14**
Gastric ulcers, 195, 363
Gel electrophoresis, DNA fragments,
  369–370, **Fig. 14.4**
Gene(s), 41
  see also DNA
  alleles, 155, 165
  amplification by PCR, 374–375,
    Fig. 14.7
  cloning, 378, 379–383
  coding for proteins, 141–143
  constitutive, 151

function, knowledge about, 384,
  Fig. 14.12
identification of specific, 371–373,
  383, **Fig. 14.5**
introns and exons, 145–147,
  Fig. 6.18
organization of DNA into,
  134–135, **Fig. 6.1**
regulated, 151
structural, 152, **Fig. 6.25**
transcription, 143–147
transfer between bacteria, 157–
  159
Gene expression, 143, **Fig. 2.29,**
  Fig. 6.13
  regulation, 151–153
Gene gun, 391
Gene probe, 371–373, **Fig. 14.5**
Generation times, 190, **Table 7.1**
Gene therapy, **Fig. 14.17,** 388
Genetically modified (GM) foods, 368,
  389–391
  safety concerns, 392, **Fig. 14.21**
Genetically modified microorganisms
  industrial applications, 391–392,
    420–421, 429
  medical applications, 386
  techniques for creating, 379–383
Genetic barriers, 290
Genetic code, 148–149, **Fig. 6.20**
Genetic diseases, 388, 392
Genetic engineering, 9, 367, 379–384
  see also Recombinant DNA
    technology
  applications, 385–392, **Fig. 14.13**
Genetic information
  first cells, 207–208, **Fig. 8.6**
  flow of, 41, **Fig. 2.29**
  RNA world hypothesis, 206
Genetics
  central dogma of, **Fig. 6.13**
  microbial, 133–166
Genetic variation
  mechanisms in prokaryotes,
    155–160
  mutations as original source,
    160–165
  natural selection, 216, **Fig. 8.21**
  role in evolution, 160
  sexually reproducing organisms,
    154–155, **Fig. 6.27**
  sources of, 154–165
Genome, 141
  analysis, value of, 384–385
  sequencing, 376–379, **Table 14.1**
Genomic library, 383–384
Genomics, 376–379
Genotype, 153–154
Gentamicin, mechanism of action,
  331, **Fig. 12.11**
Genus, 72, **Fig. 4.1**
Germicidal agents, 317
"Germs," 1
Germ theory of disease, 124–125, 127
Giardia intestinalis, 76, 82, **Fig. 4.5**
  adherence to host, **Fig. 10.13**
  pathogenesis of disease, 278
  reservoir, **Table 10.1**

Gingivitis, 78
Global warming, 181–182, 248
Glucose, 31, 32, **Fig. 2.15**
  cell respiration, 175–176, **Fig. 7.13**
  energy generated from, 182–183
  oxidation, 177, **Fig. 7.10**
    incomplete, anaerobic
      conditions, 183–184
  synthesis, photosynthesis, 186,
    188, **Fig. 7.18**
Glutamic acid, 409
Glycerol, 33–34, **Fig. 2.19**
  cell respiration pathway, 184,
    Fig. 7.16
Glycocalyx, 50–51, **Fig. 10.13**
Glycogen, 32, **Fig. 2.16**
Glycolysis, 176, 177–178, **Fig. 7.9,**
  Fig. 7.10
  anaerobic conditions, 183–184
  evolution, 208
Glycoproteins
  eukaryotic cell membrane, 63,
    Fig. 3.22
  viral envelope, 91, 92, **Fig. 4.26**
Goats, transgenic, 379, 386, **Fig. 14.14,**
  Fig. 14.15
Golgi apparatus, 66, **Fig. 3.26**
  evolution, 211
Gonorrhea, 272
Gorgas, William, 117
Grains, fermented, 397–398, 403–405
Gram-negative bacteria, 55–56
  cell wall, 55, **Fig. 3.13**
  identification, 56–57, **Fig. 3.14**
  isolation, 254, 255, **Fig. 9.26**
  phylogeny, 79
  selection of antibiotics, 325–326
Gram-negative sepsis, 55–56
Gram-positive bacteria, 53–54
  cell wall, 53–54, **Fig. 3.12**
  identification, 56–57, **Fig. 3.14**
  isolation, 254, 255, **Fig. 9.26**
  phylogeny, 79
Gram stain, 53, 56–57, **Fig. 3.14**
Granules, storage, 59–60
Grape juice (must), 402, **Fig. 15.8**
Greece, ancient, 108–110, 124, **Fig. 5.3**
Greenhouse gases, 181–182, 248
Greenland, 112
Green nonsulfur bacteria, 77–78
Green sulfur bacteria, 78
Griffith, Frederick, 155–156, 157,
  Fig. 6.28
Growth, microbial, 188–199
  control of, 315–339
    bactericidal and bacteriostatic
      methods, **Fig. 12.2**
    chemical methods, 320–322,
      Table 12.1
    physical methods, 316–320,
      Table 12.1
  environmental influences,
    191–196, 233
  exponential, 190–191, **Fig. 7.24,**
    Table 7.1
  in industrial processes, 421
  logarithmic, 191, **Fig. 7.25**
  nonsynchronous, 191, **Fig. 7.25**

phases, 196–197, **Fig. 7.31**
population, 189–191
synchronous, 191, **Fig. 7.25**
Growth curves, 196–197, **Fig. 7.31**
environmental influences,
197–199, **Fig. 7.33**
Growth hormone, human, 382, 386
Guanine, 40, 135, **Fig. 2.27**
Guano, 252
Gunpowder, 117–118

H

Habitats, microbial, 235–245
Haemophilus influenzae, 377, **Table 14.1**
Halobacterium, 80, 195
Halophiles, 80, 195, 242
Ham, cured, 408
Hand washing, 269, **Fig. 10.9**
control of bacterial numbers, 321,
322, **Fig. 12.5**
introduction of, 125
Hantavirus
epidemics, 348
investigation of epidemic, 326,
358, **Fig. 12.10**
reservoir, 265, **Fig. 10.6, Table 10.1**
Heart attacks, 38
Heat
see also Pasteurization
sterilization, 317, **Fig. 12.1**
Heavy chains, **Fig. 11.17**
Heavy metals, control of microbial
growth, 322, **Fig. 12.6**
Helical viruses, 90, **Fig. 4.26**
Helicobacter pylori, 195, 363
Helium, 19, 20
Helminths, 120
Helper T cells (Th cells), 296–298,
Fig. 11.15
B-cell activation, 300, **Fig. 11.18**
cytotoxic T-cell activation, 298,
Fig. 11.16
effector, **Fig. 11.15**
HIV infection, 309, 311, **Fig. 11.25,
Fig. 11.27**
Hemagglutinin (H), 353–354,
Fig. 13.10
Hemolysins, 276, **Fig. 10.20**
Hepatitis A
outbreaks, 345–346
transmission, 269
Hepatitis B virus, 92, 98
vaccine, 306, 388
Hepatitis C, epidemic, 348
Herbivores
digestion of cellulose, 32–33,
Fig. 2.17
role in ecosystems, 230, 232
Herpesviruses
drug treatment, 332–333,
Fig. 12.16
reactivation, 98
Hesse, Fannie, 126
Heterotrophs, 86, 186
Hiker's disease, 76
*Hin*dIII, **Fig. 14.2**
History

of microbiology, 120–130,
Table 5.2
microbiology of, 107–120
HIV *see* Human immunodeficiency
virus
Homeostasis, 4, **Fig. 1.4**
Homosexual men, 308–309
Hoof-and-mouth disease, 128
Hooke, Robert, 45
Hops, 403, **Fig. 15.9**
Horses, beneficial bacteria for, 30,
32–33
Hospital sanitation, 125
Host, 220
definitive, 83, **Fig. 4.14**
evolution, influence of
microorganisms, 222–225
intermediate, 83, **Fig. 4.14**
normal microbial flora, 260–263
pathogenesis of disease, 272–276
range, 91
Host defense, 285–311
see also Immune response
causing symptoms of disease,
272–273
evasion of, 272
strategy, 286, **Fig. 11.2**
Host–pathogen relationship,
evolution, 220–222
Hot springs, **Fig. 1.1**
Houseflies, 270, **Fig. 10.10**
House mouse, **Table 14.1**
Human genome, 377, 384, **Table 14.1**
Human herpesvirus type 1 (HHV-1),
98
Human immunodeficiency virus
(HIV), 307–311, **Fig. 1.6**
see also AIDS
attachment to cells, 93
CD4 binding, 309, **Fig. 11.25**
infection, 266
drug treatment, 334–335,
Fig. 12.15
global epidemic, 308,
Fig. 11.23
opportunistic infections, 262–
263, 307, 311, **Table 11.1**
progression, 310–311,
Fig. 11.27
interference with immune
response, 309, **Fig. 11.26**
replication, 97
Human papillomavirus, vaccine, 2,
Fig. 1.2
Human reservoirs, 265–266, **Table
10.1**
Humidity sensors, 426, **Fig. 16.10**
Humoral immune response, 298–303
see also Antibodies
primary and secondary, 306–307,
Fig. 11.22
Humus, 236
Hyaluronidase, 276
Hybridization, DNA, 371
Hydrocarbons, 22, 29, **Fig. 2.13**
see also Petroleum (oil)-digesting
bacteria
long-chain, microbially produced,
434

modified, 29–30, **Fig. 2.14**
oxidation, 173, **Fig. 7.7**
Hydrochloric acid (HCl), 27, **Fig. 2.11**
Hydrogenated oils, 35
Hydrogen atoms (H), 18, 19, **Fig. 2.1,
Table 2.1**
covalent bonding, 22, 23–24,
Fig. 2.6, Fig. 2.7
Hydrogen bonds, 24, **Fig. 2.8**
Hydrogen gas (H$_2$), 22, 23, **Fig. 2.6**
atmosphere of early Earth, 204,
205, **Fig. 8.3**
as electron source, 187–188
Hydrogen ions (H$^+$) *see* Protons
Hydrogen peroxide (H$_2$O$_2$), 192–193,
Fig. 7.26
Hydrogen sulfide (H$_2$S), 188
digestion, sewage treatment, 427,
429–430, **Fig. 16.11**
as energy source, 188
environmental cycling, 251,
Fig. 9.21, Fig. 9.22
microbial synthesis, 181
Hydrolysis reaction, 31
Hydrophilic compounds, 25, **Fig. 2.9**
Hydrophobic compounds, 25, **Fig. 2.9**
Hydrothermal vents, 230
abiotic synthesis, 205–206, **Fig. 8.4**
siboglinids, 229, **Fig. 9.1**
Hydroxide (OH$^-$) ions, 27, **Fig. 2.12**
Hydroxyl (OH) group, 30, **Fig. 2.14**
Hygiene
see also Hand washing; Sanitation
prevention of typhus, 345
Hynek, J. Allen, 169
Hyphae, 86, **Fig. 4.17, Fig. 4.18**
Hypochlorite, sodium, 321
Hypothesis, 10
development into theory, 13
testing, 10–13, **Fig. 1.12**

I

Ice, 240
Ichthyophthirius multifiliis, 82
Icosahedral viruses, 90, **Fig. 4.26**
ID$_{50}$ (infectious dose-50), 270
IgA, 302–303, **Fig. 11.20**
IgD, 302
IgG, 302, **Fig. 11.20**
IgM, 302, **Fig. 11.20**
Immune response, 272, 286, 290–311
see also Host defense; Immunity
effects of HIV, 309, **Fig. 11.26**
likelihood of disease and, 311,
Fig. 11.28
primary, 305, **Fig. 11.22**
secondary, 306, **Fig. 11.22**
Immunity
see also Immune response;
Immunological memory
adaptive *see* Adaptive immune
response
innate, 286, 290–293, **Fig. 11.2**
pathogen persistence and, 264
population, 349–351, **Fig. 13.5**
epidemic cycles and, 350,
Fig. 13.7

epidemic threshold, 350–351,
   Fig. 13.8
   rate of transmission and,
      349–350, **Fig. 13.6**
Immunocompromised patients
   see also Human
         immunodeficiency virus
   opportunistic infections, 262–263
Immunoglobulins *see* Antibodies
Immunological memory, 286, 304
   induction by vaccines, 306–307
   subsequent infections, 304–306,
      Fig. 11.22
Immunology, 285, **Fig. 11.1**
Inactivated virus/vaccines, 306, 307
Incas, 112
Incidence, 344
Inclusion bodies, 59–60, 277,
   Fig. 10.21
Incubation period, 271, **Fig. 10.14**
Inducer, *lac* operon, 153, **Fig. 6.25**
Inducible operons, 153
Inductive period, 303, 305, **Fig. 11.21**
Industrial fermentors, 422–423,
   Fig. 16.6
Industrial microbiology, 9, 417–426
   genetic engineering, 391–392,
      420–421
   in history, 117–118
   large-scale production, 422–423,
      Fig. 16.6, Fig. 16.7
   medical products, 423–425,
      Table 16.1
   nonmedical products, 425–426
   safety aspects, 421, **Fig. 16.5**
   strain improvement, 419–421
   suitability of organisms for, 421,
      Fig. 16.4
Inert gases, 20, **Fig. 2.3**
Infection, 260
   course of, 303–304, **Fig. 11.21**
   site, antibiotic selection and,
      327–328
Infection control, 315–339
   lack of, in history, 121
Infectious disease
   basic principles, 260–263
   control
      advances, 127–128
      since mid-1800s, 322–323,
         Fig. 12.7
      targeting reservoirs, 266–267
   defense against *see* Host defense
   emergent, 359–363, **Fig. 13.15**
   Koch's postulates, 126–127,
      Fig. 5.23
   likelihood of, 285, 304, 311,
      Fig. 11.1, Fig. 11.28
   pathogen's perspective, 259–281
   process, 263–281
   requirements for causation,
      263–264, **Fig. 10.5**
   symptoms aiding transmission,
      280–281, **Fig. 10.24,
      Fig. 10.25**
Inflammation, 272, 291
Influenza (flu), 2, 290
   adaptive immune response, 293,
      294, 296, 297, 298

avian (bird flu) (H5N1), 267,
      354–355, **Fig. 13.12**
   barriers to entry, 288–289
   control measures, 267
   course of infection, 303–304,
      Fig. 11.21
   drug treatment, 333–334,
      Fig. 12.15
   endemic, 352
   epidemics, 350–351, 352–355
   fatal, 304
   HIV-infected patients, 308, 309
   host defenses, 286
   humoral response, 300, 302, 303
   innate immune response, 291, 292
   pandemics, 352, 354, **Fig. 13.9**
   Spanish flu H1N1 pandemic
      (1918), 352, 354, **Fig. 13.9**
   subsequent infections, 304–305
   swine flu H1N1 pandemic (2009),
      352, 355
   vaccination, 304, 306, 309, 352,
      354
   vaccines, 388
Influenza A virus, 353–355
Influenza virus, 91, 353–354
   antigenic drift, 353–354
   antigenic shift, 354, **Fig. 13.11**
   structure, 353, **Fig. 13.10**
Inhibitors, enzyme, 39–40
Injections, disease transmission via,
      348
Innate immune system, 286, 290–293,
      Fig. 11.2
Insects
   antimicrobial compounds,
      338–339
   biopesticides, 426
   predatory fungi, 256, **Fig. 9.28**
   species formation, 224, **Fig. 8.29**
   transmission of disease, 269–270
Insulin, recombinant human, 386,
      Fig. 14.13
Intercalating agents, 163
Interferons, 292–293, **Fig. 11.8**
Interleukin-1 (IL-1), 292
Interleukin-2 (IL-2), 298
Intermediate host, 83, **Fig. 4.14**
Intertidal zone, 242
Introns, 145–147, **Fig. 6.18**
In vitro, 323, 423–424
In vivo, 323, 423
Iodine, 320–321
Ionic bonds, 20–22, **Fig. 2.5**
Ionized state, 27
Ions, 20
Irish potato famine, 116, **Fig. 5.10**
Iron
   fertilization of seawater, 246–247,
      248
   role in diphtheria, 196
Iron oxide, 209, **Fig. 8.11**
Isomers, 31, **Fig. 2.15**
Isoniazid
   mechanism of action, 330,
      Fig. 12.11
   spectrum of activity, **Table 12.2**
Isotopes, 18

Ivanowski, Dmitri, 128
Ixodes scapularis, **Fig. 10.10**

J

Jenner, Edward, 128, **Fig. 5.25**
Justinian, Emperor, 111

K

Kaposi's sarcoma, 363
Kilbourne, Frederick, 127
Kingdom, 72, 75, **Fig. 4.1**
Klebsiella, 262
Koala, 339, **Fig. 12.21**
Koch, Robert, 126, **Fig. 5.22**
Koch's postulates, 126–127, **Fig. 5.23**
Komodo dragon, 343, **Fig. 13.1**
Korarchaeota, 80, **Fig. 4.11**
Koumiss, 407
Krebs cycle, 176, 178–179, **Fig. 7.9,
      Fig. 7.11**
Kriek beer, 396
Kuru, 102, 223–224, **Fig. 4.39**

L

Laboratory fermentors, 423, **Fig. 16.7**
Lac operon, 151–153, **Fig. 6.25**
Lactic acid/lactate
   fermented plant products, 401,
      405
   production during fermentation,
      184, **Fig. 7.15**
   in yogurt, 199, 407, 412, **Fig. 7.32**
Lactobacillus acidophilus, 197–198,
      Fig. 7.32
   growth under different conditions,
      199, **Fig. 7.34**
Lactobacillus brevis, 405
Lactobacillus bulgaricus, 401, 407, 412
Lactobacillus plantarum, 401, 405,
      Fig. 15.14
Lactobacillus sanfrancisco, 401
Lactococcus lactis, 413
Lactose, 31, **Fig. 6.25**
Lag phase, 197, **Fig. 7.31**
Lakes, 238–240
   Antarctic, 240
   ecological zones, 238–239,
      Fig. 9.11
   eutrophication, 242, **Fig. 9.13**
   seasonal changes in temperate,
      239–240, **Fig. 9.12**
   water pollution, 241–242
Lambda phage, 101, 383, **Fig. 4.37,
      Fig. 14.11**
Lambic ales, 395–396
Landfill sites, methane utilization,
      434, **Fig. 16.17**
Latent viral infections, 98, 277–278,
      Fig. 4.33, Fig. 10.22
Lawns, bacterial, 10, **Fig. 1.10**
Lazear, Jesse, 127
LD$_{50}$ (lethal dose-50), 270, **Fig. 10.12**
Leeuwenhoek, Anton van, 121,
      Fig. 5.15
Legionella pneumophila, **Table 10.1**
Legumes, 248, 249, **Fig. 9.20**
Leishmania, 82
Leishmania donovani, **Fig. 1.5**

Leishmaniasis, 82
Leprosy, 56
Lethal genes, recombinant DNA technology, 380, 382, **Fig. 14.10**
Leuconostoc mesenteroides, 405
Leukocytes, 290–291
Lewis, Sinclair, **Fig. 12.21**
Lice, body, 115, 345, **Fig. 5.9**
Lichens, 237–238, **Fig. 9.10**
Life
    diversity of, 215–225
    extraterrestrial *see* Extraterrestrial life
    origin of, 77–78, 203–215, **Fig. 4.7, Fig. 8.1**
Light chains, **Fig. 11.17**
Light-emitting bacteria, **Fig. 7.3**
Lightning, origin of life, 205, **Fig. 8.2**
Light reactions, photosynthesis, 186–188, **Fig. 7.20, Fig. 7.21**
Linnaeus, Carolus, 72
Linoleic acid, **Fig. 2.19**
Lipases, 34, 184
Lipid A, 55, **Fig. 3.13**
Lipids, 30, 33–35
    hydrophobicity, 33, **Fig. 2.18**
    structure, 33–34, **Fig. 2.19**
Lipopolysaccharide (LPS), bacterial, 55, **Fig. 3.13**
    pathogenesis of disease, 275–276, **Fig. 10.19**
Liposomes, 207, **Fig. 8.6**
Lipoteichoic acid, **Fig. 3.12**
Liquor, 405
Lister, Joseph, 125, 321
Listeria monocytogenes, 318
Listerine®, 321
Lithium, 19
Littoral zone, 239, 242–243, **Fig. 9.11, Fig. 9.14**
Livestock, antibiotic use in, 337–338, **Fig. 12.20**
Living things
    cells as basic components, 3–4
    observable characteristics, 4, **Fig. 1.4**
Lloyd George, David, 118, **Fig. 5.12**
Lockjaw, 61
Loeffler, Friedrich, 128
Logarithmic microbial growth, 191, **Fig. 7.25**
Log phase, 197, **Fig. 7.31**
Lophotrichous bacteria, **Fig. 3.10**
Louisiana Territory, 115, **Fig. 5.8**
Louse, body *see* Lice, body
Lyme disease, 270, 358, 362, 378
Lymph, 295
Lymphatic system, 295–296, **Fig. 11.12**
Lymph nodes, 295–296, **Fig. 11.12**
Lymphocytes, 296
Lysine, 409, **Fig. 2.21**
Lysis
    prokaryotic cells, 52
    virus-infected cells, 94, 101, **Fig. 4.36**
Lysogenic cycle, 101, **Fig. 4.37**

Lysol®, 321
Lysosomes, 66–67, **Fig. 3.27**
Lytic cycle, 101, **Fig. 4.36**

M
Macrophages
    antigen presentation, 294
    innate immune response, 290–291, **Fig. 11.6**
    lipopolysaccharide receptors, 275–276, **Fig. 10.19**
    phagocytosis by, 67, **Fig. 3.28**
Mad cow disease, 102, 103–104
Magnesium, 21
Magnesium fluoride, 21, **Fig. 2.5**
Major histocompatibility complex (MHC), 294, 298, **Fig. 11.11, Fig. 11.16**
Malaria, 63, 259
    discovery of vector, 127
    drug treatment, 332, **Fig. 12.14**
    epidemics, 348
    impact on history, 110, 117
    parasite, 82–83, **Fig. 4.13**
    reemergence, 363
Mallon, Mary, 266, **Fig. 10.7**
Malting, 403, **Fig. 15.9**
Maltose, 31, **Fig. 2.15**
Manure, cow, 230, **Fig. 9.2**
Mare's milk, fermented, 407, **Fig. 15.13**
Marine environments, 242–245
Mars, "fossil bacteria," 215, **Fig. 8.19**
Mary, Typhoid, 266, **Fig. 10.7**
Mashing, 403, **Fig. 15.9**
Mastigophora, 82, **Fig. 4.13**
Measles, 114, **Table 5.1, Table 10.1**
Meat
    drying, 319–320, **Fig. 12.4**
    fermented products, 408
    tenderizer, 425
Medical applications
    industrially produced metabolites, 423–425, **Table 16.1**
    recombinant DNA technology, 386–388
Medical implants, biofilms on, 235
Medical microbiology, 9
Medium
    chemically defined, 196
    complex, 196
    minimal, 196
Membrane proteins, 58–59, 63, **Fig. 3.22**
Membranes
    cell *see* Plasma membrane
    first cells, 207, **Fig. 8.6**
    first eukaryotic cells, 211, **Fig. 8.13**
Memory B cells, 300, 304, 306–307, **Fig. 11.18**
Memory T cells, 298, 304, 306–307
Mendel, Gregor, 133
Mercuric chloride, 322
Mercury, 430
Mesophiles, 193, 194, **Fig. 7.28**
Mesophilic-A, 413
Messenger RNA (mRNA), 148, **Fig. 6.21**

formation during transcription, 143–147, **Fig. 6.17**
    translation, 147–151, **Fig. 6.19, Fig. 6.22**
Metabolism, 4, 169–201
    cell size and efficiency **Fig. 3.5**, 49
    evolution, 208–210, **Fig. 8.10**
    microbial growth and, 188–199
    similarities between all living things, 170
Metabolites, 419
    with commercial potential, 419–421
    environmental influences on production, 420, 421, **Fig. 16.4**
    large-scale production, 422–423, **Fig. 16.6, Fig. 16.7**
    medical applications, 423–425
    primary, 420, **Fig. 16.3**
    recovery methods, 423
    secondary, 420, **Fig. 16.3**
    strain improvements, 419–421
Meteorites, 206, **Fig. 8.4**
Methane (CH_4), 22, 80, **Fig. 2.6**
    atmosphere of early Earth, 204, 205, **Fig. 8.3**
    as energy source, 188
    oxidation, 173, **Fig. 7.7**
    oxidizing organisms, 245
    released from sea floor, 245
    use as fuel, 434, **Fig. 16.17**
    as waste product, 181–182
Methanogens, 80, 181–182, 245
Metronidazole, 327, 332
Mexico, Spanish conquest of, 107–108, 112–113, **Fig. 5.2**
MHC *see* Major histocompatibility complex
Microaerophiles, 193, **Fig. 7.27**
Microbes, 1
Microbiology, 3–9
    applied, 8, 9, **Fig. 1.8**
    basic, 8–9, **Fig. 1.8**
    of history, 107–120
    history of, 120–130, **Table 5.2**
    in the news, 2, **Fig. 1.2**
    specialized subdisciplines, 7–9
    study of nonmicroorganisms, 7
Microenvironments, 233, **Fig. 9.5**
Microorganisms, 1, **Fig. 1.1**
    classification, 76–104
    discovery of, 121, **Fig. 5.15**
    major types, 4–6, **Table 1.1**
    model, 8, 130
    study of, 8–9
MicroRNAs, 153
Middle East, in ancient times, 397, 398
Military bases, bioremediation, 430–431
Milk
    Campylobacter jejuni contamination, 347, 348–349
    fermentation, 407–408, **Fig. 15.13**
    pasteurization, 318
    spoilage, 31, 37
Miller, Stanley, 205, **Fig. 8.3**

Mining
  damage to freshwater ecosystems, 25–27, **Fig. 2.10**
  microbial, 435, **Fig. 16.18**
Mismatch repair, 164, **Fig. 6.35**
Mitochondria, 67–68, **Fig. 3.29**
  electron transport chain, 180, **Fig. 7.12**
  endosymbiotic origin, 68, 211, 384–385, **Fig. 8.13, Fig. 8.14**
  matrix, 67
Mitomycin, **Table 16.1**
Model organisms, 8, 130
Modes of transmission, 268–270, **Fig. 10.9, Fig. 10.10, Table 10.1**
Molasses, 405, 407
Molds, 86
Molecular clock, 74–75
Monarch butterfly, 389, 391, **Fig. 14.18**
Monera, 75
Monkeypox, 265, 362, **Fig. 13.19**
Monoculture, 116
Monosaccharides, 31, **Fig. 2.15**
Monosodium glutamate (MSG), 409
Monotrichous bacteria, **Fig. 3.10**
Monounsaturated fatty acids, **Fig. 2.19**
Montague, Lady, 128
Monuments
  microbial damage, 9, **Fig. 1.9**
  restoration of historic, 417, **Fig. 16.1**
Morbidity and Mortality Weekly Report (MMWR), 308–309, **Fig. 11.24**
Morning sickness, 223, **Fig. 8.28**
Mosquito-borne diseases, 269–270
  discovery of, 117, 127
  impact on history, 110, 117
  weather-related epidemics, 348
Mosquitoes
  biopesticides, 426
  dengue transmission, 360, 361
  myxoma virus transmission, 219, **Fig. 8.26**
  West Nile virus, 356, 358, 361
Motile bacteria, 52, **Fig. 3.11**
mRNA *see* Messenger RNA
Mucous membranes, 288–289, **Fig. 11.5**
Mucus, 288–289
Mudflats, 251, **Fig. 9.22**
Multicellular fungi, 86
Multicellular organisms, 4, 62–63, **Fig. 1.3, Fig. 3.21**
  evolution, 212–213, **Fig. 8.15**
Mushrooms, 86, 87, **Fig. 4.18**
  as food, 399–400, **Fig. 15.4**
Mus musculus (house mouse), **Table 14.1**
Must, 402, **Fig. 15.8**
Mutagens, 162–163, **Fig. 6.33**
Mutations, 73–74, 160–165
  beneficial, 161
  causes, 162–163
  frameshift, 162, **Fig. 6.32**
  harmful, 161–162

induced, for strain improvement, 420
  missense, 161–162, **Fig. 6.32**
  natural selection for/against, 216, **Fig. 8.21**
  neutral, 161, **Fig. 6.32**
  nonsense, 162, **Fig. 6.32**
  repair mechanisms, 163–164, **Fig. 6.35**
  role in evolution, 165
  spontaneous, 162
mutT gene, 385
Mutualism, 253–254, **Fig. 9.24, Fig. 9.25**
Mutualistic fungi, 85, 86, 87, 88
Mycelium, 86, **Fig. 4.17**
Mycobacterium, steroid production, 425
Mycobacterium leprae, 56, 194
Mycobacterium tuberculosis, 56
  see also Tuberculosis
  discovery of, 126
  DNA identification, 374, 375
  drug-resistant strains, 154, 363
  genome analysis, 378
  natural selection trade-offs, 217, **Fig. 8.22**
  pathogenic nature, 260
  reservoirs, **Table 10.1**
Mycobacterium vaccae, 2
Mycolic acid, 330
Mycoplasma, 57
Mycoplasma pneumoniae, 57
  antibiotics acting on, 148, 149, 330
  genome, 384
Mycorrhizal associations, 87, 88, 237
Mycoses, 88
Myxococcus xanthus, 417
Myxoma virus
  introduction to Australia, 220, **Fig. 8.25**
  transmission, 219, **Fig. 8.26**
  virulence over time, 220, 221–222

N

N-acetylglucosamine (NAG), 53, **Fig. 3.12**
N-acetylmuramic acid (NAM), 53, **Fig. 3.12**
NAD⁺/NADH *see* Nicotinamide adenine dinucleotide
NADP⁺/NADPH *see* Nicotinamide adenine dinucleotide phosphate
Nanotechnology, 426, **Fig. 16.10**
Napoleon Bonaparte, 115
Narrow-spectrum antibiotics, 326–327, 424, **Table 12.2**
Nasonia wasps, 224, **Fig. 8.29**
Natural selection, 216–217, **Fig. 8.21, Fig. 8.22**
  vancomycin resistance, 218–220, **Fig. 8.24**
  virulence of pathogens, 220–222
Nauru, 251–252
Necrotizing fasciitis, 276, **Fig. 10.20**
Negri bodies, 99, **Fig. 10.21**

Neisseria gonorrhoeae
  adherence, 271, **Fig. 10.13**
  pathogenesis of disease, 272
Neisseria meningitidis, 262, 265
Neon, 20, **Fig. 2.3**
Neuraminidase (N), 353–354, **Fig. 13.10**
Neutralization, 300, **Fig. 11.19**
Neutrons, 18, **Fig. 2.1, Table 2.1**
Neutrophils, 290, **Fig. 11.6**
Newcastle disease vaccine, 387
New World, introduced diseases, 112–115
Nicotinamide adenine dinucleotide (NAD⁺/NADH), 174, **Fig. 7.8**
  cell respiration, **Fig. 7.9**
  electron transport chain, 179–180, **Fig. 7.12**
  fermentation, 183–184, **Fig. 7.14**
  glycolysis, 177, 178, **Fig. 7.10**
  Krebs cycle, 178–179, **Fig. 7.11**
Nicotinamide adenine dinucleotide phosphate (NADP⁺/NADPH)
  photosynthesis, 186–188, **Fig. 7.21**
  utilization in Calvin cycle, 188, **Fig. 7.22**
Nightingale, Florence, 125, 345, **Fig. 5.20**
19th century, 115–117, 121–129
Nitrate (NO₃⁻)
  environmental cycling, 249, **Fig. 9.19**
  as final electron acceptor, 181
Nitrification, 249, **Fig. 9.19**
Nitrite, 249, **Fig. 9.19**
Nitrobacter, 249, 254
Nitrofurantoin, 327–328
Nitrogen, 18, **Table 2.1**
  covalent bonding, 23–24, **Fig. 2.6**
  effects on plant growth, 248, **Fig. 9.18**
  gas (N₂), 249
Nitrogen cycle, 248–250, **Fig. 9.19**
Nitrogen fixation, 79, 249, **Fig. 9.19**
Nitrogen-fixing bacteria, 249–250, **Fig. 9.20**
Nitrosomonas, 249, 254
Nitrous acid, 162, **Fig. 6.33**
Nonenveloped viruses, 91, **Fig. 4.26**
  host-cell specificity, 92
  replication, 93, 94, **Fig. 4.28**
Nonfastidious organisms, 196, 429
Nonpolar covalent bonds, 23–24, **Fig. 2.7**
Normal microbial flora, 260–263, **Fig. 10.3**
  acquisition, **Fig. 10.2**
  effects of antibiotics, 263, 326–327
  opportunistic infections, 262–263
Norwalk virus
  incubation period, 271
  transmission, 268, 270
Nosocomial infections, 315
Nuclear envelope, 64
Nucleic acids, 30, 40–41
Nucleocapsid, 90, **Fig. 4.26**

Nucleoid area, bacterial, 59, **Fig. 3.17**
Nucleotides, 40, **Fig. 2.27**
    DNA strand formation, 135, **Fig. 6.3**
    encoding of information, 134–135, **Fig. 6.1**
    one-letter code, 135
    RNA molecule, 143–144
    structure, 135, **Fig. 6.2**
Nucleus
    atomic, 18
    cell, 46, 64, **Fig. 3.1**
Number of pathogens
    increase during incubation period, 271, **Fig. 10.14**
    LD_{50}, 270, **Fig. 10.12**
    likelihood of infection and, 270, 311, **Fig. 11.28**
Nutrasweet®, 409
Nutrients
    environmental recycling, 232
    flow through ecosystems, 230–231, **Fig. 9.3**
    freshwater ecosystems, 240, 241
    marine environments, 242, 243, **Fig. 9.14**
    promoting bioremediation, 429
Nutritional factors, affecting microbial growth, 195–196

O

Obligate intracellular parasites, 78–79, 90
Obligate parasites, **Fig. 1.6**
Ocean, 242–245
Octane, 173–174
Oenology, 403
Oil-digesting bacteria *see* Petroleum-digesting bacteria
Oils
    hydrogenated, 35
    hydrophobicity, 33, **Fig. 2.18**
Oil spills, cleaning up, 429
Oleic acid, **Fig. 2.19**
Oligotrophs, 240
Olives, 405, 406
Omnivores, 231
Oncogenes, 277
Oocysts, 83, **Fig. 4.14**
Operator, *lac* operon, 152, **Fig. 6.25**
Operons, 151–153, **Fig. 6.24**
    inducible, 153
    repressible, 153
O polysaccharides, 55, **Fig. 3.13**
Opportunistic infections, 262–263, **Fig. 10.4**
    HIV infection, 262–263, 307, 311, **Table 11.1**
Opsonins, 300–302, **Fig. 11.19**
Order, 72, **Fig. 4.1**
Organelles, 4, 46, 62, 64–68, **Fig. 3.1**
    evolution, 210, 211–212, **Fig. 8.13**
Organic molecules, 29–41
    see also Biological molecules
Organophosphates, 392
Organs, 62
Origin of life, 77–78, 203–215, **Fig. 4.7, Fig. 8.1**

Origins of replication, 140, **Fig. 6.11, Fig. 6.12**
Outbreaks
    see also Epidemics
    case definition, 356–357
    investigation of, 355–359
    time, place, and personal characteristics, 357–358
Outer membrane, 55, **Fig. 3.13**
Oxacillin, 336
    appropriate use, 53, 54, 55–56
    mechanism of action, 329, **Fig. 12.12**
Oxidation, 172–174, **Fig. 7.6**
    carbon fuels, 173, **Fig. 7.7**
Oxidizing agents, 192–193
Oxygen atoms (O), 18, **Table 2.1**
    covalent bonding, 22, 24, **Fig. 2.6, Fig. 2.7**
Oxygen gas (O_2), 22, 175
    atmospheric, 204, 208–209
    combustion of carbon fuels, 173, **Fig. 7.7**
    composting and, 432, **Fig. 16.13**
    depletion, polluted lakes, 241–242, **Fig. 9.13**
    dissolved, freshwater ecosystems, 238–239, **Fig. 9.12**
    as final electron acceptor, 180, 181
    microbial growth rates and, 198, **Fig. 7.33**
    release, photosynthesis, 187–188
    requirements of prokaryotes, 192–193, **Fig. 7.27**
Oysters, raw, 345–347
Ozone layer, 209

P

PABA (*para*-aminobenzoic acid), 324, **Fig. 12.8**
Paintings, microbial damage to, 7, 9, **Fig. 1.9**
Panama Canal, 116–117, **Fig. 5.11**
Pandemics, 112, 352
Para-aminobenzoic acid (PABA), 324, **Fig. 12.8**
Paramecium, 82, **Fig. 4.13**
Parasites
    nonmicroscopic, 120
    obligate, **Fig. 1.6**
    obligate intracellular, 78–79, 90
    pathogenesis of disease, 278–279
Parent cells, 154
Parrot fever, 66–67
Pasteur, Louis, 121–124, 125, 128–129, **Fig. 5.17, Fig. 5.18**
Pasteurization, 124, 317–318
    batch, 318
    beer, 319
    flash, 318
Pathogenesis, 272–276
Pathogens, 56, 220, 260
    adherence, 270–271, **Fig. 10.13**
    elimination by host, 303–304
    evasion of host defenses, 272
    mechanisms of host damage, 272–276
    multiplication, and disease, 271, **Fig. 10.14**

    number of *see* Number of pathogens
    portals of entry, 270, **Fig. 10.11**
    portals of exit, 279
    requirements to cause disease, 263–264, **Fig. 10.5**
    reservoirs, 264–267
    transmission, 268–270
Pattern recognition, 291, **Fig. 11.7**
PCR *see* Polymerase chain reaction
Pediculus humanus humanus (body lice), 115, 345, **Fig. 5.9**
Pelagic zone, 242–243, **Fig. 9.14**
Peloponnesian War, 108–110, **Fig. 5.3**
Penicillin
    discovery of, 10, 129–130, 324, **Fig. 12.9**
    industrial production, 419, 420
    mechanism of action, 54, 328, 329, **Fig. 3.12, Fig. 12.11, Fig. 12.12**
    resistance, 329, 336, 338, **Fig. 12.18**
    spectrum of activity, 326, **Table 12.2**
Penicillinase, 329, 336, **Fig. 12.12**
Penicillin G, 329, **Fig. 12.12**
Penicillium camemberti, 407
Penicillium citrinum, **Table 16.1**
Penicillium notatum, 10, 129–130, 419, **Fig. 1.10**
Penicillium roqueforti, 10–11, 408, **Fig. 1.10, Fig. 1.12**
Pentaglycine bridges, 53, **Fig. 3.12**
Pentamidine, 308
Peptide bonding, **Fig. 2.22**
Peptides, 36, **Fig. 2.22**
Peptidoglycan
    alterations in vancomycin resistance, 218, **Fig. 8.23**
    antibiotics targeting, 54, 328, 329–330, **Fig. 3.12**
    bacterial endospores, 60, **Fig. 3.19**
    Gram-negative cell wall, 55, **Fig. 3.13**
    Gram-positive cell wall, 53–54, **Fig. 3.12**
Peripheral proteins, 59, **Fig. 3.16**
Peritonitis, 276
Peritrichous bacteria, **Fig. 3.10**
Permeability, increased, 291
Personal characteristics, 357–358
Petroleum (oil)-digesting bacteria
    bioremediation, 429, **Fig. 16.12**
    genetically engineered, 379, 382, 392
Petroleum-like substances, microbial, 434
pH
    effects on microbial growth, 195, **Fig. 7.29**
    scale, 28, **Fig. 2.12**
Phages *see* Bacteriophages
Phage therapy, 339, **Fig. 12.21**
Phagocytic cells, 290–291, **Fig. 11.6**
Phagocytosis, 66–67, **Fig. 3.28**
    microorganisms resistant to, 66–67

role of antibodies, 300–302, **Fig. 11.19**
"Pharming," 386, **Fig. 14.15**
Phenol, 125, 321
Phenotype, 153–154
Phenylalanine deaminase test, 184–185, **Fig. 7.16**
Phosphate
    ion (PO_4^{3-}), environmental cycling, 252–253, **Fig. 9.23**
    organic, environmental cycling, 252–253, **Fig. 9.23**
    water pollution, 241, 253
Phosphate groups, 58, **Fig. 3.15**
    ATP, 171, **Fig. 7.5**
    DNA molecule, 135, **Fig. 6.2**
Phospholipids, 58, 59, **Fig. 3.15**
    formation of first cells, 207, **Fig. 8.6**
Phosphorus cycle, 251–253, **Fig. 9.23**
Phosphorylation, 332
Photic zone, 238–239, 242, **Fig. 9.11**
Photoautotrophs, 186
Photosynthesis, 67, 186–188, **Fig. 7.19**
    carbon cycle, 246, **Fig. 9.17**
    compared to aerobic respiration, 186, **Fig. 7.18**
    evolution, 208–209, **Fig. 8.10**
    light reactions, 186–188, **Fig. 7.20**, **Fig. 7.21**
    synthesis of organic molecules, 188, **Fig. 7.22**
Photosynthetic bacteria, 78, 79, **F ig. 4.10**
Phylogenetics, 73
Phylogenetic tree, 73, **Fig. 4.2**
Phylum, 72, **Fig. 4.1**
Physical methods, control of microbial growth, 316–320, **Table 12.1**
Phytophthora infestans, 116, **Fig. 5.10**
Phytoplankton, 238, 242, 243
Pickles, 184, 405–406
    recipe, 410–411
Pickling, 405–406
Pili, 51, **Fig. 3.9**
    sex, 51, 158, **Fig. 6.31**
Pilot plant fermentors, 423, **Fig. 16.7**
Pizarro, Francisco, 112
Place characteristic, 357
Placenta, as portal of entry, 270, **Fig. 10.11**
Plague
    of ancient Athens, 109–110
    bubonic, 111–112, 126–127, **Fig. 5.1**, **Fig. 5.5**
    of Cyprian, 110–111
Plant(s)
    based fermented foods, 401–407
    classification, 75–76, **Fig. 4.4**
    effects of nitrogen on, 248, **Fig. 9.18**
    evolution, 211–212
    habitats, 237
    photosynthesis, 186–188, **Fig. 7.20, Fig. 7.21**
    transgenic, 388, 389–391, **Fig. 14.19**

viral infections, 99, **Fig. 4.34**
Plant cells, **Fig. 1.3, Fig. 3.1**
    cell wall, 63
Plant vaccines, 388
Plasma cells, 65, 300, **Fig. 3.25, Fig. 11.18**
Plasma membrane, 46
    eukaryotic, 63–64, **Fig. 3.22**
    fluid mosaic model, 59, **Fig. 3.16**
    prokaryotic, 58–59, **Fig. 3.15**
        electron transport chain, 180, **Fig. 7.12**
        photosynthesis, 187, **Fig. 7.20**
        psychrophiles and thermophiles, 194
Plasmids, 59, **Fig. 6.6**
    cloning vectors, 380–382, **Fig. 14.10**
    transfer between cells, 157–159, **Fig. 6.31**
    transfer of antibiotic resistance, 159
Plasmodium, 82–83, **Fig. 4.13**
    pathogenesis of disease, 278
    reservoirs, **Table 10.1**
Plasmodium falciparum, genome, **Table 14.1**
Plastics, biodegradable, 432–433, **Fig. 16.16**
Pneumocystis carinii pneumonia, 262–263, 307, 308–309
Pneumonia, atypical, 147–148, 149
Poe, Edgar Allan, 89, 99, **Fig. 4.24**
Poland, German occupation of, 118–119
Polar covalent bonds, 22–24, **Fig. 2.7**
Polarity, 24
Polio (poliomyelitis)
    emergence in 20th century, 360, **Fig. 13.16**
    epidemic paralytic, 360
    epidemics, 351
    vaccines, 307
Poliovirus, 92
Pollution, 426–436
    see also Air pollution; Water pollution
    effects on historic buildings, 417, **Fig. 16.1**
    need for new approaches to, 427–428
    remediation *see* Bioremediation
Poly-β-hydroxyalkane (PHA), 433, **Fig. 16.16**
Polychlorinated biphenyls (PCBs), 430
Polyenes, 331
Polymerase chain reaction (PCR), 374–375, 383, 426, **Fig. 14.7**
Polysaccharides, 32, **Fig. 2.16**
Polyunsaturated fatty acids, **Fig. 2.19**
Population, 230
    immunity, and epidemics, 349–351
Porins, 55, **Fig. 3.13**
Portals of entry, 270, **Fig. 10.11, Table 10.1**
Portals of exit, 279

Potato blight, 116, **Fig. 5.10**
Potatoes, 116
Prairie dogs, 362, **Fig. 13.19**
Pravastatin, **Table 16.1**
Predation, 255–256, **Fig. 9.24, Fig. 9.27, Fig. 9.28**
Preempt®, 260, 262, 263, 410
Pregnancy
    morning sickness, 223, **Fig. 8.28**
    vertical transmission, 270
Primary consumers, 230, **Fig. 9.3**
Primary immune response, 305, **Fig. 11.22**
Primers, 375, **Fig. 14.7**
Prion protein
    aberrant form (PrP^{sc}), 102, **Fig. 4.40**
    normal form (PrP^{c}), 102–103, **Fig. 4.40**
Prions, 6, 102–104, **Fig. 1.7, Table 1.1**
    brain evolution and, 223–224
    origins of, 214
Probe, gene, 371–373, **Fig. 14.5**
Probiotics, 410
Producers, 230, 231, 232, **Fig. 9.3**
Profundal zone, 239, 242, 244, **Fig. 9.11**
Progesterone, **Fig. 16.8**
Prokaryotes
    see also Archaea; Bacteria
    classification, 75, 76–80
    DNA replication, 140, **Fig. 6.12**
    electron transport chain, 180, **Fig. 7.12**
    endosymbiosis, 211–212, **Fig. 8.13**
    origin of first, 208, **Fig. 8.8, Fig. 8.9**
    oxygen requirements, 192–193, **Fig. 7.27**
    regulation of gene expression, 151–153, **Fig. 6.25**
    reproduction, 189–190, **Fig. 7.23**
    sources of genetic variation, 155–159
    transcription and translation, 145, 151, **Fig. 6.23**
Prokaryotic cells, 4–6, 46, 49–61, **Fig. 1.3, Table 1.1**
    cell wall, 52–57
    chromosomes *see* Chromosomes, bacterial
    cytoplasm and interior structures, 59–61
    extracellular structures, 50–52
    lacking a cell wall, 57
    shapes, 49, **Fig. 3.6**
    sizes, 48–49, **Fig. 3.2**
    structure, **Fig. 3.1**
Promoter, 145, **Fig. 6.17**
    lac operon, 152, **Fig. 6.25**
Prontosil, 129, 323–324
Proofreading, by DNA polymerase, 163–164, **Fig. 6.35**
Propane, **Fig. 2.9**
Prophage, 101, **Fig. 4.37**
Propionibacterium, 408, 409
Propionic acid, **Fig. 7.15**
Protease inhibitors, 335, **Fig. 12.15**
Proteases, 38, 184

commercial production, 425
Proteins, 30, 35–37
    3-dimensional structure, 36, **Fig. 2.23**
    control of assembly, 40–41, **Fig. 2.29**
    denatured, 36–37, **Fig. 2.24**
    determination of cell characteristics, 141–143
    encoding in DNA, 134–135, 141–143, **Fig. 6.1**
    as energy sources, 184, **Fig. 7.16**
    export from cells, 65–66, **Fig. 3.26**
    membrane, 58–59, 63, **Fig. 3.22**
    native conformation, 36
    peripheral, 59, **Fig. 3.16**
    single-cell consumable, 400
    synthesis, 143–154
        see also Transcription; Translation
    transmembrane, 59, **Fig. 3.16**
    transport within cells, 65–66, **Fig. 3.26**
Proteobacteria, 79
Proteus, 249
Proteus vulgaris
    identification, 184–185
    OX19, 119, **Fig. 5.13**
Protista, 81
Protons (H⁺)
    atomic structure, 18, **Fig. 2.1, Table 2.1**
    concentration in water, 27–28, **Fig. 2.11, Fig. 2.12**
    electron transport chain, 180, 182, **Fig. 7.12**
    gradient, 182, **Fig. 7.12**
    redox reactions, 173
Protozoa, 6, 81–84, **Fig. 1.5, Fig. 4.13, Table 1.1**
    classification, 75
    disease-causing, 63, 82–84
    as eukaryotes, 62, **Fig. 3.21**
    phylogeny, 81–82, **Fig. 4.12**
    predatory, 256
Protozoan infections
    drug treatment, 331, 332
    pathogenesis, 278–279
Pseudomonas
    bioremediation, 429
    cytochrome oxidase, 180
    infections, 346
    vitamin B12 synthesis, 409
Psittacosis, 66–67
Psychrophiles, 193, 194, 244, **Fig. 7.28**
Psychrotolerant organisms, 194
Public health microbiology, 9
Pure culture technique, 126
Purines, 40, 135, **Fig. 2.27, Fig. 6.3**
Pus, 272
Pyrimidines, 40, 135, **Fig. 2.27, Fig. 6.3**
Pyrosome, **Fig. 7.3**
Pyruvate
    fermentation, 183–184, **Fig. 7.14**
    oxidation in Krebs cycle, 178–179, **Fig. 7.11**
    production in glycolysis, 177–178, **Fig. 7.10**

Q
Quinine, 332, **Fig. 12.14**

R
Rabbits, European, in Australia, 220, **Fig. 8.25**
Rabies, 89
    Negri bodies, 99, **Fig. 10.21**
    reservoir, **Table 10.1**
Rabies virus, 89, 91–92, 99
    replication, 97
Radiation
    control of microbial growth, 319, **Fig. 12.3**
    DNA damage, 163, **Fig. 6.34**
Radioactive wastes, 430
Radioresistant cocci, 78, **Fig. 4.8**
Rainfall, 245
Reaction product, 39
Reactive oxygen compounds, 192
Recipes, fermented foods, 410–413
Recombinant DNA, 369, **Fig. 14.3**
Recombinant DNA technology, 9, 130, 379–393, **Fig. 14.9**
    ethics and safety, 392–393, **Fig. 14.21**
    gene cloning, 379–383
    genome sequencing, 377–378
    insights derived from, 384–385
    practical applications, 385–392, **Fig. 14.13**
Recombinant microorganisms *see* Genetically modified microorganisms
Recombinant vaccines, 387–388, **Fig. 14.16**
Recombination, genetic
    prokaryotes
        mechanisms, 155–159
        significance for humans, 159–160
    sexually reproducing organisms, 154–155, **Fig. 6.27**
Redi, Francesco, 121, **Fig. 5.16**
Redox reactions, 172–174, **Fig. 7.6**
Red Queen hypothesis, 224, 225, **Fig. 8.30**
Red tides, 243–244, **Fig. 9.16**
Reducing atmosphere, primitive Earth, 204–205, **Fig. 8.3**
Reduction, 172–174, **Fig. 7.6**
Reed, Walter, 127
Reemerging diseases, 363
Refrigeration, 318
Regressive hypothesis, origin of viruses, 213
Regulatory T cells (T_R cells), 304
Rennin, 407, 413
Replication forks, 139, **Fig. 6.9**
Repression, 153
Repressor, 152–153, **Fig. 6.25**
Reproduction
    asexual, 154, **Fig. 6.26**
    binary fission, 189–190, **Fig. 7.23**
    sexual, 154–155, **Fig. 6.27**
Reproductive system, as portal of entry, 270, **Fig. 10.11**

Reservoirs, 264–266
    animal, 264–265, **Fig. 10.6, Table 10.1**
    disease control implications, 266–267
    environmental, 266, **Table 10.1**
    human, 265–266, **Table 10.1**
Resources, nonrenewable, 433–435
Respiration, cell, 175–186, **Fig. 7.9, Fig. 7.13**
    aerobic, 180, 182–183, **Fig. 7.13**
        compared to photosynthesis, 186, **Fig. 7.18**
    anaerobic conditions, 183–184
    different starting molecules, 184–185, **Fig. 7.16**
Respiratory system
    barriers to entry, 288, **Fig. 11.4, Fig. 11.5**
    as portal of entry, 270, **Fig. 10.11**
    upper, normal flora, 262
Restriction enzymes, 368–369, **Fig. 14.2, Fig. 14.3**
    gene cloning, 380–382, **Fig. 14.10**
    Southern blotting, 371, **Fig. 14.5**
Retroviruses, 97, **Fig. 4.32**
    see also Human immunodeficiency virus
Reverse transcriptase, 97, **Fig. 4.32**
Reverse transcriptase inhibitors, 334–335, **Fig. 12.15**
Reverse transcription, 97
R groups, 35, **Fig. 2.21**
Rheumatic fever, 260
Rhizopus nigricans, **Fig. 16.8**
Rhizopus stolonifer, 87, **Fig. 4.21**
Rhizosphere, 237, **Fig. 9.9**
Riboflavin
    deficiency, 408–409
    production, 409, 420
Ribonucleic acid *see* RNA
Ribose, 41, 143
Ribosomal RNA (rRNA), 149, **Fig. 6.21**
Ribosomes, 46
    antibiotics targeting, 59, 149, 330–331, **Fig. 12.13**
    bacterial, 59, 151, **Fig. 3.17**
    eukaryotic, 65, 151, **Fig. 3.25**
    protein synthesis, 147, 149, **Fig. 3.26, Fig. 6.19, Fig. 6.22**
Ribozymes, 206, **Fig. 8.5**
Rickettsia, evolutionary status, 213, **Fig. 8.17**
Rickettsia prowazekii, 115, 385
    cross-reactive antibodies, 119, **Fig. 5.13**
Rickettsia typhi, 345
Rifampin, 331, **Fig. 12.11**
Rivers, 240–241
RNA (ribonucleic acid), 41
    first cells, 207, **Fig. 8.6**
    gene regulation role, 153
    messenger *see* Messenger RNA
    production from DNA, 143–147, **Fig. 6.14**
    ribosomal (rRNA), 149, **Fig. 6.21**
    self-replicating ability, 206, **Fig. 8.5**
    structure, 143–144, **Fig. 6.16**

transfer (tRNA), 149, **Fig. 6.21,**
  Fig. 6.22
RNA-dependent RNA polymerase, 97
RNA polymerase, 143, 145, **Fig. 6.14,**
  Fig. 6.17
  regulation in operons, 152, 153,
    Fig. 6.25
RNA splicing, 147
RNA viruses, 89
  minus (–) strand, 96, 97, **Fig. 4.32**
  plus (+) strand, 96, 97, **Fig. 4.32**
  replication, 96–97, **Fig. 4.32**
RNA world hypothesis, 206, 207,
  Fig. 8.6
  transition to DNA world, 207–208,
    Fig. 8.7
Rocks
  beneath sea floor, 245
  biofilms on, 234, **Fig. 9.6**
  microorganisms living within,
    237–238, **Fig. 9.10**
Rodriguez-Navarro, Carlos, 417
Roman Empire, fall of, 110–111,
  Fig. 5.4
Rome, ancient, 397–398
Roosevelt, Theodore, 117, **Fig. 5.11**
Root nodules, 250, **Fig. 9.20**
Ross, Ronald, 117, 127
Royal Chapel, Granada, 417, **Fig. 16.1**
rRNA genes, 74–75, 76, **Fig. 4.6**
Rum, 405
Russia, Napoleonic invasion, 115
Rust infections, 88, **Fig. 4.22**
Rye bread, 401
Rye whiskey, 405

S

Sabin vaccine, 307
Saccharomyces, beer brewing,
  395–396, 403
Saccharomyces carlsbergensis, 76, 403
Saccharomyces cerevisiae, 76, 88,
    Fig. 3.21
  beer brewing, 403
  bread-making, 401, 411, **Fig. 15.7**
  genome, 377, **Table 14.1**
  Vegemite production, 400
  winemaking, 402, **Fig. 15.8**
Safety
  genetic engineering, 392–393,
    Fig. 14.21
  industrial microorganisms, 421,
    Fig. 16.5
Salami, 408
Salem witch trials, 279, **Fig. 10.23**
Salk vaccine, 307
Salmonella
  food poisoning, 273, 343, 355
  protection of chickens against,
    260, 263
  reservoirs, 265, 343, **Fig. 13.1**
Salmonella enteritidis, 343, 355
Salmonella typhi
  see also Typhoid
  reservoirs, 266, **Fig. 10.7,**
    Table 10.1
Salmonella typhimurium, LD$_{50}$,
  Fig. 10.12

Salt, table *see* Sodium chloride
Salt preservation, 195, 319–320,
  Fig. 7.30
Salts, 20, 21, **Fig. 2.5**
Salvarsan, 129, 323
Sanger, Frederick, 375, 377
Sanger method of DNA sequencing,
  375–376, **Fig. 14.8**
Sanitation
  effects on human survival, 323,
    Fig. 12.7
  hospital, 125
  polio emergence and, 360
  prevention of transmission, 269
  virulence of pathogens and, 222,
    Fig. 8.27
Saprophytes, 86
Sarcodina, 82, **Fig. 4.13**
Saturated fatty acids, 34–35, **Fig. 2.19**
Sauerkraut, 184, 405
  recipe, 410, **Fig. 15.14**
Scaffolding proteins, 137, 153, **Fig. 6.5**
Scale-up, 423, **Fig. 16.7**
Schistosomiasis, 74–75, **Fig. 4.3**
  in history, 120
  injected drug therapy, in Egypt,
    348
Science, 9
Scientific method, 10–13
Scotch whiskey, 405
Scrapie, 102, **Fig. 1.7**
Scrubbers, chemical, 427, 428
Seasonal factors, epidemics, 347
Seawater, 242
Secondary consumers, 230, **Fig. 9.3**
Secondary immune response, 306,
  Fig. 11.22
Selection
  artificial, 217
  natural *see* Natural selection
Selective media, 254, **Fig. 9.26**
Selective toxicity, 323, 328, 331–332
Semmelweis, Ignaz, 125, **Fig. 5.19**
Sepsis, bacterial, 55–56
Septum, 60
Serine, **Fig. 2.21**
17th century, 121
Severe combined immunodeficiency
  (SCID), 388
Sewage treatment plants, 427, 428,
  429–430, **Fig. 16.11**
Sex pili, 51, 158, **Fig. 6.31**
Sexual reproduction, 154–155,
  Fig. 6.27
  evolution of, 224–225
Sexual transmission, 268
Shamokin Creek Basin, Pennsylvania,
  25–27
Shellfish, 345–347
Shewanella, 434–435
Shigella, 378
Shigella dysenteriae, 280, 281,
  Fig. 10.25
Shingles, 278
Shubat, 407
Siboglinids, 229, **Fig. 9.1**
Silicon, 29
Silver salts, 322, **Fig. 12.6**

Single-cell protein, 400
Skin
  barrier to infection, 289, **Fig. 11.5**
  normal flora, 53, 262
  as portal of entry, 270, **Fig. 10.11**
Sleeping sickness, 114–115, 278
Slime layer, 51
Smallpox
  acute disease, 113, **Table 5.1**
  bioterrorism, 364
  eradication, 266–267, **Fig. 10.8**
  in history, 108, 112, **Fig. 5.6**
  vaccination, 128, **Fig. 5.25**
  variolation, 128
Smallpox virus, 89, 90
Smith, Theobald, 127
Smokers, 288
Smuts, 88
Snails, *Biomphalaria*, 74–75, **Fig. 4.3**
Snow, John, 344, **Fig. 13.2**
Soaps, 321–322
Sodium atoms, 19–20, **Fig. 2.3**
Sodium chloride (NaCl) (table salt),
    20, 21, **Fig. 2.5**
  see also Salt preservation
  effect on microbial growth, 195,
    Fig. 7.30
  solubility in water, 25, **Fig. 2.9**
Sodium hydroxide (NaOH), 27–28,
    Fig. 2.11
Sodium hypochlorite, 321
Soft drinks, 406–407
Soil, 235–238
  bacteria, gene exchange, 159–160
  environment, 235–237, **Fig. 9.8**
  microenvironments, 233, **Fig. 9.5**
  microorganisms, antibiotic
    production, 418, 419,
    Fig. 16.2
  stabilization, 436
Solute, 27
Solution, 27
Solvent, 27
Sourdough bread, 401
Southern blotting, 371–373, 383,
  Fig. 14.5
Soybeans, transgenic, 389, 391
Soy sauce, 406
Spanish conquest, 107–108, 112–113,
  Fig. 5.2
Species, 72, 224, **Fig. 4.1**
Sperm, 154, 155, **Fig. 6.27**
Spices, 396–397, **Fig. 15.2**
Spirilla, 49, **Fig. 3.6**
Spirochetes, 78, **Fig. 3.6**
Spirulina, 400, **Fig. 15.6**
Spleen, 296, **Fig. 11.12, Fig. 11.13**
Spontaneous generation, theory of,
    121–123, **Fig. 5.16, Fig. 5.17**
St. Louis encephalitis (SLE) virus, 356,
    358, 359, **Fig. 13.13**
Staining, simple, 46–48
Stanley, Wendell, 128
Staphylococci, 49, **Fig. 3.6**
Staphylococcus aureus
  coagulase-secreting, 217, 276,
    Fig. 8.22
  drug resistance, 160, 337, 363,
    Fig. 12.18

infection, 53
skin flora, 53
Staphylococcus epidermidis, 260, 262
Starch, 32, **Fig. 2.16**
based plastics, 433
Static agents, 325–326
Static control measures, 318, **Fig. 12.2**
Stationary phase, 197, **Fig. 7.31**
Stearic acid, 34, **Fig. 2.19**
Stem rust, 88
Sterilization
gamma radiation, 319
heat, 317, **Fig. 12.1**
Steroid hormones, commercial
production, 424–425,
Fig. 16.8
Sterols, 57, 63
Sticky ends, 368, **Fig. 14.2**
Stomach ulcers, 195, 363
"Stonewashed" jeans, 391, **Fig. 14.20**
Stop codons, 149, **Fig. 6.20**
Storage granules, 59–60
Strain improvement, 420–421, 429
Streams, 234, 240–241, **Fig. 9.6**
Strep throat, 335, 336–337
Streptobacilli, 49
Streptococci, 49, **Fig. 3.6**
Streptococcus, 38
penicillin resistance, 338
tissue-damaging enzymes, 276,
Fig. 10.20
Streptococcus mutans, 51
Streptococcus pneumoniae
capsule, 50
Griffith's transformation
experiment, 155–156, 157,
Fig. 6.28
pathogenesis of disease, 272–273
Streptococcus pyogenes
antibiotic therapy, 335, 336–337
childbed fever, 125
hemolysins, 276, **Fig. 10.20**
LD$_{50}$, **Fig. 10.12**
reservoirs, 265
rheumatic fever, 260
Streptococcus thermophilus, 401, 407,
412
Streptokinase, 38, 39, 276
Streptomyces, 423
Streptomyces caespitosus, **Table 16.1**
Streptomycin, 324
factors affecting selection, 327,
328, **Table 12.2**
mechanism of action, 331,
Fig. 12.11, Fig. 12.13
Stromatolites, 208, **Fig. 8.9**
Subatomic particles, 18–19, **Fig. 2.1**
Subclinical infections, 306, **Fig. 11.22**
Subsoil, 236, **Fig. 9.8**
Substrate, enzyme, 39
Subunit vaccines, 306, 388
Sucrase, 31, **Fig. 2.26**
Sucrose (table sugar), 31, **Fig. 2.15**
digestion, **Fig. 2.26**
role in dental plaque, 51
solubility in water, 25, 30
Sugars, 30–31
see also Glucose; Sucrose

preservation of food, 320
Sulfa drugs, 129, 324, 325
mechanism of action, 328,
Fig. 12.11
spectrum of activity, **Table 12.2**
Sulfanilimide, 129, 323–324, **Fig. 12.8**
Sulfate (SO$_4$$^{2-}$)
environmental cycling, 251,
Fig. 9.21
as final electron acceptor, 181
Sulfur cycle, 251, **Fig. 9.21, Fig. 9.22**
Sulfuric acid (H$_2$SO$_4$), 28
Sumerians, ancient, 397–398,
Fig. 15.3
Sunlight
DNA damage, 163
as energy source, 186–188
Superglue, biological, 1
Superoxide, 192
Surface-to-volume ratio, 49, **Fig. 3.5**
Sverdlovsk anthrax incident (1979),
364
Swamp gas, 169, 182
Sweeteners, low-calorie, 409
Swine flu, 2
Symbiosis, 211
Syphilis, 78, 360, **Fig. 4.9**
see also Treponema pallidum
discovery of salvarsan, 129, 323

T

T4 phage
replication, 99–101, **Fig. 4.36**
structure, 99, **Fig. 4.35**
Taiwan, 120
Taq polymerase, 375, 426
Taxis
negative, 52, **Fig. 3.11**
positive, 52, **Fig. 3.11**
Taxonomy, 71–76
evolutionary relationships and,
72–73
hierarchy, 72, **Fig. 4.1**
T-cell receptors, 297, 298, **Fig. 11.15**
T cells, 296
cytotoxic (Tc cells), 298, **Fig. 11.16**
effector, 298
helper *see* Helper T cells
memory, 298, 304, 306–307
regulatory (T$_R$ cells), 304
Teeth
see also Dental plaque
biofilms, 234, **Fig. 9.6**
Temperature
control of microbial growth,
316–317
maximum growth, 193–194
microbial growth rate effects, 198,
Fig. 7.33
microbial growth requirements,
193–194, **Fig. 7.28**
minimum growth, 193–194
optimum growth, 194, 316–317
Tenderizer, meat, 425
Tequila, 405
Termites, 253
Tetanus, 61
see also Clostridium tetani

Tetanus toxin, 275
Tetracycline
mechanism of action, 330,
Fig. 12.11, Fig. 12.13
resistance, 336
spectrum of activity, 326, 327,
Table 12.2
Texas red water fever, 127
Theory, scientific, 13
Thermoactinomyces vulgaris, 61
Thermocline, 239–240, **Fig. 9.12**
Thermocycler, 375
Thermophiles, 77–78, 80, 193, 194,
Fig. 7.28
extreme, **Fig. 7.28**
industrial applications, 425–426
Thermotaxis, 52
Thermus aquaticus, 375, 426
Thiobacillus ferrooxidans, 435
Threshold of disease, 271, **Fig. 10.14**
adaptive immune response,
303–304, **Fig. 11.21**
factors affecting, 311, **Fig. 11.28**
primary and secondary immune
response, 305, 306,
Fig. 11.22
Thrush, 88, 263
Thucydides, 109
Thylakoids, 67, 187, **Fig. 7.20**
Thymidine kinase, 332–333, **Fig. 12.16**
Thymine, 40, 135, **Fig. 2.27**
Thymine dimers, 163, **Fig. 6.34**
Thymus, 296, **Fig. 11.14**
Ticks, 270, 358, **Fig. 10.10**
Time characteristic, 357, **Fig. 13.14**
Tissue-damaging enzymes, 276, 279,
Fig. 10.20
Tissues, 62, **Fig. 3.21**
Toad, golden, 84–85, 87, **Fig. 4.20**
Tobacco mosaic virus, 128, **Fig. 4.34**
Toilet bowls, biofilms in, 234, **Fig. 9.6**
Toll-like receptors (TLRs), 276, 291,
Fig. 11.7
Tolypocladium inflatum, **Table 16.1**
Topsoil, 235–236, **Fig. 9.8**
Toxins
bacterial, 273–276
fungal, 279
Toxoplasma gondii, 81, 83–84,
Fig. 4.14
effects on fetus, 223
Trachoma, **Fig. 10.1**
Transcription, 143–147, **Fig. 6.14,
Fig. 6.17**
"Christmas trees," 143, 145,
Fig. 6.15
initiation at promoters, 145,
Fig. 6.17
reverse, 97
termination sequence/site, 143,
145, **Fig. 6.17**
Transcription factors, 145
Transduction, 157, **Fig. 6.30**
Transfer RNA (tRNA), 149, **Fig. 6.21,
Fig. 6.22**
Transformation, 157, **Fig. 6.29**
artificial induction, 157
discovery of, 155–157, **Fig. 6.28**

Transgenic animals, 368, 379
    creation, 386, **Fig. 14.14**
    ethical concerns, 392
    medical applications, 386–387,
        Fig. 14.15
Transgenic bacteria, 382, 391–392
Transgenic plants, 389–391
    creation, 389–391, **Fig. 14.19**
    edible vaccines, 388
Transient flora, 262
Translation, 143, 147–151, **Fig. 6.19,
    Fig. 6.22**
Transmembrane proteins, 59,
    Fig. 3.16
Transmission
    factors promoting, in epidemic
        causation, 347–349
    modes of, 268–270, **Fig. 10.9,
        Fig. 10.10, Table 10.1**
    rate, and population immunity,
        349–350, **Fig. 13.6**
    symptoms of disease aiding, 280–
        281, **Fig. 10.24, Fig. 10.25**
Transport, intracellular, 64–65,
    Fig. 3.26
Transport vesicles, 66, **Fig. 3.26,
    Fig. 3.27**
Transposons, 214
Trappist ales, 395–396, **Fig. 15.1**
Travel, international, 361, 362
Treponema pallidum, 78, **Fig. 4.9**
    see also Syphilis
    optimum growth temperature,
        194
    reservoirs, **Table 10.1**
Trichoderma reesei, 434
Trichomonas vaginalis, **Fig. 4.13**
Trimethylamine oxide, 182
Trinitrotoluene (TNT), 431
Trophic interactions, 230–231, **Fig. 9.3**
Trophic level, 230
Trophozoites, 81
trp operon, 153
Trypanosoma brucei, **Fig. 3.21**
Trypanosoma cruzi, **Table 10.1**
Trypanosomes, 278
Tryptophan, 153
Tsetse fly, 114–115
Tsuda, Atsushi, 246–247
Tuberculosis, 56
    see also Mycobacterium
            tuberculosis
    antibiotic therapy, 331
    as chronic disease, **Table 5.1**
    control measures, 267
    discovery of causative agent, 126
    in New World, 374, **Fig. 14.6**
    pathogenesis, 273, **Fig. 10.15**
    reemergence, 363
    untreatable, 154
Tube worms, 229, **Fig. 9.1**
Tulips, variegated, 99, **Fig. 4.34**
Typhoid
    evolution of virulence, 222,
        Fig. 8.27
    human reservoirs, 266, **Fig. 10.7**
    influence on history, 107
    vaccine, 388

Typhus
    in history, 115, 119, 345
    vector, **Fig. 5.9**
    Weil–Felix test, **Fig. 5.13**

U

UFO sightings, 169, **Fig. 7.1**
Ultrapasteurization, 318
Ultraviolet (UV) radiation
    control of microbial growth, 319,
        Fig. 12.3
    DNA damage, 163, **Fig. 6.34**
    origin of life, 205
Uncoating, 94, **Fig. 4.29**
Unicellular eukaryotes, colonial, 212,
    Fig. 8.16
Unicellular fungi, 86, **Fig. 4.17**
Unicellular organisms, 4, 62, **Fig. 1.3,
    Fig. 3.21**
    evolutionary advantages, 213
United States, 115, **Fig. 5.8**
Unsaturated fatty acids, 34–35,
    Fig. 2.19
Uracil, 41, 144, **Fig. 2.27, Fig. 6.16**
Uranium, 430
Urbanization, 360–361
Urinary tract, normal flora, 262

V

Vaccination
    discovery of, 128–129, **Fig. 5.25**
    population immunity and, 349,
        351
Vaccines, 306–307
    attenuated virus, 306, 307
    cancer, 2, **Fig. 1.2**
    development of first, 128–129
    inactivated virus, 306, 307
    plant (edible), 388
    recombinant, 387–388, **Fig. 14.16**
    subunit, 306, 388
Valine, **Fig. 2.21**
Valley fever, 84, 88–89, **Fig. 4.15**
Vampirococcus, 256
Vancomycin
    mechanism of action, 218, 329–
        330, **Fig. 8.23, Fig. 12.11**
    resistance, 218–220, **Fig. 8.23**
        genetic basis, 160, 162, 163
Van Leeuwenhoek, Anton, 121,
    Fig. 5.15
Varicella zoster virus, 277–278
Variolation, 128
Vasodilation, 291
Vector-borne diseases, 110
    discovery of, 127
    reasons for symptoms, 280,
        Fig. 10.24
    weather-related epidemics, 348
Vectors
    cloning, 380–383, **Fig. 14.10**
    disease, 81, 110
Vector transmission, 269–270
    biological, 269–270, **Fig. 10.10**
    mechanical, 270, **Fig. 10.10**
Vegemite, 399, 400, **Fig. 15.5**
    sandwich, 413, **Fig. 15.17**

Vegetables, fermented, 405–406,
    Fig. 15.11
Vertical transmission, 270
Vibrio, **Fig. 3.6**
Vibrio cholerae, 101
    see also Cholera
    genome, **Table 14.1**
    pathogenesis of disease, 273–274,
        Fig. 10.17
    reservoirs, 266
    seasonal variations in numbers,
        348
    symptoms of disease, 280, 281,
        Fig. 10.25
Vinegar, 124, 406, **Fig. 15.12**
Vinyl chloride, 430
Viral infections, 97–99
    acute, 97, 277, **Fig. 4.33, Fig. 10.22**
    antibiotic treatment, 337
    cancer-inducing, 277, **Fig. 10.22**
    chronic, 98, 277, **Fig. 4.33,
        Fig. 10.22**
    drug treatment, 332–335,
        Fig. 12.15
    interferons, 292–293, **Fig. 11.8**
    latent, 98, 277–278, **Fig. 4.33,
        Fig. 10.22**
    pathogenesis, 277–278
    plants, 99, **Fig. 4.34**
    types, 97–98, **Fig. 4.33**
Virions, 90
    assembly, 94, **Fig. 4.29**
    release, 94–95, **Fig. 4.29, Fig. 4.30**
Virulence, 220
    evolution, 220–222, **Fig. 8.27**
    likelihood of infection and, 270,
        Fig. 10.12
    threshold of disease and, 311,
        Fig. 11.28
Virulence factors, 51
    deducing origins, 377, 378
    gene transfer, 159
Viruses, 6, 89–102, **Table 1.1**
    see also Viral infections
    adherence to host, 271, **Fig. 10.13**
    attachment, 93, **Fig. 4.29**
    damage to host cells, 98–99, 277
    discovery of, 127–128
    DNA *see* DNA viruses
    evolutionary origins, 213–214
    host-cell specificity, 91–92,
        Fig. 4.27
    host range, 91
    as living things, 7, 90
    penetration, 93, **Fig. 4.29**
    reactivation, 98
    replication, 92–97
        drug targets, 332–335,
            Fig. 12.15
    RNA *see* RNA viruses
    size, 89, **Fig. 4.25**
    structure, 90–91, **Fig. 4.26**
    synthesis, 94, **Fig. 4.29**
    uncoating, 94, **Fig. 4.29**
Vitamin B12, 409
Vitamins, 409
Vodka, 405
Volvox, **Fig. 8.16**

W

Washtenaw County, Michigan, UFO
    sightings, 169, **Fig. 7.1**
Wasps, *Nasonia*, 224, **Fig. 8.29**
Wastewater treatment plants, 427,
    428, 429–430, **Fig. 16.11**
Water (H_2O), 25
    atmosphere of early Earth, 204,
        205, **Fig. 8.3**
    content of soil, 236
    drinking, safety of, 127, **Fig. 5.24**
    hydrogen bonding, 24, **Fig. 2.8**
    ionized state, 27
    molecules, 22, 24, **Fig. 2.6, Fig. 2.7**
    solubility in, 25, **Fig. 2.9**
    splitting, photosynthesis, 187
    treatment, 319, **Fig. 12.3**
    waste product of cell respiration,
        180
Water birds, botulism outbreaks, 238,
    239–240
Waterborne diseases
    epidemics, 346, 348
    evolution of virulence, 222,
        Fig. 8.27
    Snow's investigations, 344,
        Fig. 13.2
Waterborne transmission, 269,
    Fig. 10.9
Water pollution, 253, 417
    coastal ecosystems, 243–244,
        Fig. 9.15, Fig. 9.16
    freshwater ecosystems, 241–242,
        Fig. 9.13

need for new approaches, 428
Watson, James, 130, 133, 137, 138–139
Weather, epidemics related to, 348
Weil–Felix test for typhus, **Fig. 5.13**
Weizmann, Chaim, 118, **Fig. 5.12**
West Nile virus (WNV)
    New York epidemic (1999), 356,
        359, **Fig. 13.13**
    spread in United States, 361,
        Fig. 13.18
    time, place and personal
        characteristics, 357–358,
        Fig. 13.14
Whey, 407
Whiskey, 405
Wine, 402–403
    in ancient history, 397–398
    production, 402–403, **Fig. 15.8**
    sparkling, 403
    spoilage, 123–124, 406
    sweetness and alcohol content,
        402–403
Wolbachia bacteria, 224, **Fig. 8.29**
Wood, burning, 173, **Fig. 7.7**
World War I, 117–118
World War II, 118–119
Worms, 120
Wort, 403, **Fig. 15.9**

X

Xenopus laevis, 379, **Fig. 14.9**
X rays, DNA damage, 163

Y

Yeasts, 76, 86, 88, **Fig. 4.17**
    beer brewing, 395–396, 403,
        Fig. 7.2, Fig. 15.9
    bread-making, 401, **Fig. 15.7**
    brewer's *see Saccharomyces*
        cerevisiae
    consumable forms, 400
    fermentation waste products, 184,
        Fig. 7.15
    winemaking, 402, **Fig. 15.8**
Yellow fever, 360
    control, 267
    discovery of transmission by
        mosquitoes, 117, 127,
        Fig. 5.11
    impact on history, 114, 115, 117
    reservoir, **Table 10.1**
Yersin, Alexander, 126–127
Yersinia pestis, 126–127, **Table 10.1**
    see also Plague, bubonic
Yogurt, 407
    in ancient history, 397
    production, 197–198, 199,
        Fig. 7.32, Fig. 7.34
    recipe, 412

Z

Zanamivir, 333–334, **Fig. 12.15**
Zebra fish, fluorescent, **Fig. 14.13**
Zidovudine (AZT), 334, **Fig. 12.15**
Zoonosis, 265
Zooplankton, 238, 242
Zygomycota, 87, **Fig. 4.21**